Evaluation of human work

A practical ergonomics methodology

Second Edition

Edited by

John R. Wilson and E. Nigel Corlett

University of Nottingham

Taylor & Francis
Publishers since 1798

UK	Taylor & Francis Ltd, 4 John St., London WC1N 2ET
USA	Taylor & Francis Inc., 1900 Frost Road, Suite 101, Bristol, PA 19007

British Library Cataloguing in Publication Data

Evaluation of human work.
  1. Ergonomics
  I. Wilson, John, *1951–*
620.8′2

  ISBN–07484-0084-2 Pbk
     07484-0083-4 Hbk

Library of Congress Cataloging-in-Publication Data available

Cover design by Jordan & Jordan, Fareham, Hants

Typeset by Photo·graphics, Honiton, Devon
Printed in Great Britain by Burgess Science Press,
Basingstoke, Hants.

Contents

Preface to the Second Edition

Since the First Edition of *Evaluation of Human Work* was published, much has happened to change the way we view ergonomics methods and techniques. This has lead to the inclusion of several new chapters in this second edition, and the considerable revision of many others.

Technical, social, political, and legal changes have required continual development and improvement of ergonomics methods. For instance, the ever-increasing power and prevalence of computer systems and the diversity of their user interfaces necessitate parallel improvements in methods of analysis, design and evaluation. We can see this, for instance, in human–computer interfaces generally (revised chapter 12) and in specialized applications such as control rooms (new chapter 13). Social and political changes, mirrored by changes in the way industrial work and jobs are organized, have increased recognition of the gains possible from greater involvement of people in what they do and from providing employees with a greater degree of control over their own activities. One manifestation of this is participation, and participative approaches (new chapter 37) have a long and honourable tradition in ergonomics. Legal developments have had a profound influence on ergonomics in recent years, especially the health and safety regulations governing use of display screen equipment, manual handling work, work equipment and workplaces, which have come into force in Europe, Australia and to an extent in North America. Coupled with costs of compensation claims, such regulations have required structured ergonomics assessments at work (new chapter 30) within an ergonomics management programme (revised Chapter 1 and new chapter 35). In addition to the above, this edition includes a new chapter on measurement of physiological functions (chapter 29), and substantial revision of chapters on task analysis (6), verbal protocol analysis (7), product assessment and user trials (10), knowledge elicitation (14), computer aided methods (20), mental workload (25), and work stress (26).

There are increasing moves towards greater professionalisation in ergonomics, for example the Board of Certification in Professional Ergonomics (BCPE) and the Centre for Registration of Ergonomists in Europe (CREE). Such moves require a recognition that the methods we choose will influence what we find from any investigation, and that methods must produce findings that are valid, reliable and generalizable, meet the objectives of the investigation and be safe and ethical to apply. There can be little excuse for administering questionnaires that make no attempt to use previously validated scales, carrying out experiments without careful piloting, or rigid ill-informed use of assessment checklists.

Any experimentalist should have a good knowledge of statistics. We decided, after long discussion, not to include statistics in this volume – it is heavy enough already! Statistics are necessary not only for experimental design, data compression or testing of results, but for the understanding they bring to the nature of variability and the importance of interactions. Although methods are reported in a 'stand alone' manner, it is rare in ergonomics to find an influence or a cause which has an exclusive relationship with an effect. Hence it is important to retain the ergonomics approach of viewing the whole person within the total environment, an approach which will make it necessary to match a selected group of methods to the requirements perceived. We hope this book will assist ergonomists to do this.

John Wilson and Nigel Corlett
May 1995

Acknowledgements to the Second Edition

For this second edition we would like to thank all our authors for their considerable efforts in updating their existing chapters or writing new ones; thanks are due also to reviewers of the new chapters. Many readers of the first edition have sent in comments and criticism to authors and editors, and these have been accounted for wherever this improved the book.

Again we must acknowledge the support of Taylor & Francis, and especially Richard Steele and Robert Chaundy, and thanks are due to Chris Stapleton for her professional service on proof reading and the index. The first editor (JW) gratefully acknowledges the Department of Safety Sciences, University of New South Wales for the space and time to carry out most of the editing and writing of new chapters.

Finally, we are very grateful to all our colleagues in the University of Nottingham's Department of Manufacturing Engineering and Operations Management, and especially in the Institute for Occupational Ergonomics, for their collaboration, support and a positive environment. Of these, Lynne Mills has contributed the lion's share in terms of typing, editing, and organization – of the book and of us!

John Wilson and Nigel Corlett
May 1995

Preface to the First Edition

For a long time there existed few books on ergonomics or human factors methodology; Chapanis' *Research Techniques in Human Engineering*, published in 1959, was probably the earliest, as well as the best-known. Lately there has been a slow increase in what is available. For instance one of the contributors to this volume, David Meister, has produced two books dealing with methods (Meister, 1985, 1986) and the present editors have also been involved in two collections of conference proceedings concerned with new methods and techniques (Laboratory of Industrial and Human Automatics, 1987; Wilson *et al.*, 1987).

The books by Meister, excellent in many respects, concentrate upon investigations of large-scale (military) systems design, simulation and evaluation. The two sets of conference proceedings, whilst containing a range of methodological developments and applications, represent what was selected from the papers submitted for presentation, and cannot pretend fully to represent the field. Also produced recently is the authoritative *Handbook of Human Factors,* edited by Salvendy (1987). This does have much to say about methods and techniques, both as separate chapters or as parts of other chapters; nonetheless its intention is to be a comprehensive, general text, with explanation of theories, principles, data and application, as well as of methods.

Our aim with this volume on ergonomics methodology is to produce a text on methods and techniques that is both broad and deep. We intend it to be a companion to the major general textbooks on ergonomics and human factors, particularly and most recently those of Bailey (1982), Grandjean (1988), Kantowitz and Sorkin (1983), Oborne (1987), Salvendy (1987) and Sanders and McCormick (1987). All of these are well known to students, teachers and practitioners of ergonomics, as well as to many of those from other disciplines who take a personal or professional interest in ergonomics. There is, though, little opportunity in such texts to emphasize and make explicit the major part of methodology.

Therefore we have set out to produce a general text on ergonomics methodology. As the book's title implies we are primarily concerned with people at work and with applied rather than basic research. However, the former concern has not ruled out contributions relevant to people's activities at home, leisure or on the road; nor does the latter concern invalidate descriptions of laboratory-based methods—these can have outcomes that are as practically applicable as are those from field investigations.

The contents of the book are intended to be interesting and useful for a wide range of people, including: *students*, to give them a feel for ergonomics investigation and to complement their learning of theory and principles; *industrial and business personnel at all levels*, to allow them to understand better what ergonomics can do for them, why, and how; and *ergonomics practitioners, researchers and teachers*, to give them a compendium of methods and techniques available. For all these groups the contributions here will also point to further sources for more detail on specific topics.

Our text on evaluating human work has brought together experts from many branches of ergonomics theory and practice, and has allowed them the space to introduce and give detail on those methods and techniques of value to them. Since ergonomics is both a science and a technology, these methods can of course be concerned with collecting data or with applying their own or others' data. The primary thrust of each contribution may be general method (e.g. direct observation or protocol analysis), or particular fields of application for several types of method (e.g. mental workload or the climatic environment). Whilst there will no doubt be omissions—of branches of methodology or of techniques within one area—regretted by some readers, we trust that most will find the book to be a comprehensive, readable and useful source of ergonomics knowledge and practice. Certainly we believe that for those students or readers from industry who are relatively new to ergonomics, one of the most interesting and valuable ways to learn about it is through its rich and varied methodology.

July 1989

John Wilson and Nigel Corlett
University of Nottingham

References

Bailey, R.W. (1982). *Human Performance Engineering: A Guide for System Designers*. (London: Prentice Hall), pp. 656 + xxviii.

Chapanis, A. (1959). *Research Techniques in Human Engineering*. (Baltimore: John Hopkins Press), pp. 316 + xii.

Grandjean, E. (1988). *Fitting the Task to the Man: A Textbook of Occupational Ergonomics*, 4th Edition. (London: Taylor and Francis), pp. 363 + ix.

Kantowitz, B.H. and Sorkin, R.D. (1983). *Human Factors: Understanding People-System Relationships*. (New York: John Wiley and Sons), pp. 699 + xii.

Laboratory of Industrial and Human Automatics (1987). *New techniques and ergonomics*: Proceedings of an International Research Symposium. (Paris: Hermes).

Meister, D. (1985). *Behavioural Analysis and Measurement Methods*. (Chichester: John Wiley and Sons), pp. 509 + ix.

Meister, D. (1986). *Human Factors Testing and Evaluation.* (Amsterdam: Elsevier Science), pp. 424 + xi.

Oborne, D.J. (1987). *Ergonomics at Work,* 2nd Edition. (Chichester: John Wiley and Sons), pp. 386 + xvii.

Salvendy, G. (editor) (1987). *Handbook of Human Factors.* (New York: John Wiley and Sons), pp. 1874 + xxiv.

Sanders, M.S. and McCormick, E.J. (1987). *Human Factors in Engineering and Design,* 6th Edition. (New York: McGraw-Hill), pp. 664 + viii.

Wilson, J.R., Corlett, E.N. and Manenica, I. (1987). *New Methods in Applied Ergonomics.* (London: Taylor and Francis), pp. 283 + x.

Acknowledgements

Our first debt with this book is to our contributing authors, all of whom have responded to our various requests with great patience, and have produced chapters of high quality within, in some cases, a very limited time. Amongst these authors we must mention those who were with us in the initial discussions about the book at the 2nd International Occupational Ergonomics Symposium at Zadar, Yugoslavia; they were Lisanne Bainbridge, Colin Drury, Ted Megaw, Ken Parsons, Pat Shipley and Rob Stammers. Colin Drury in particular has contributed much in terms of individual chapters and the overall content and style of the book.

We would like to thank our colleagues at Nottingham University for contributing to a working environment in which we feel able to embark on and complete this and other publishing ventures. One of us (JW) must also thank the Department of Industrial Engineering and Operations Research, University of California, Berkeley, for allowing him time and facilities to work on this book during periods there as a visitor in 1987 and 1988.

Our editors at Taylor and Francis—David Grist, Sarah Waddell and, for most of the time, Robin Mellors—have been exceedingly supportive, even in the face of a project which seemed to grow exponentially! The style of the book has been enhanced tremendously by the artwork of Tony Aston and cartoons of Moira Tracy. Despite both editors being away from Nottingham for substantial periods of time the production of the book has rolled on relatively smoothly; our colleagues would say this was because we left this and much else in the hands of our excellent secretaries, Lynne Mills and Ilse Browne, to whom we are immensely grateful.

Chapter 1

A framework and a context for ergonomics methodology

John R. Wilson

"Methods? Pah! Its all common sense"

Introduction

In April 1993, in the United Kingdom, an incident took place which captured the attention of the world's media, injured no-one, yet cost the government about £6 million. Inquiries into the event identified a number of interacting causative factors, all of which had human factors components. These
included: poor design and maintenance of equipment; earlier rejection, on
grounds of physical environment conditions, of an improved design for this
equipment; an inadequate testing programme; blocked lines of sight; impossible auditory communications; confusion caused by an unusual but foresee-

able occurrence in the social environment; lack of any thought given to, or rehearsal of, a 'worst case scenario'; errors made by personnel in both omitting to do things (errors of omission) or doing them wrongly (errors of commission); and nobody taking charge of a deteriorating situation by assessing alternatives and using available back-up systems to restrict damage. The incident was the Grand National fiasco in which one of the UK's premier horse races was turned into a debacle when the starting gate stuck. This 70 metre strand of cotton tape, 3 cm wide, failed to rise quickly enough and wrapped itself around some of the jockeys. The officials failed to prevent a number of the horses running $4\frac{1}{2}$ miles in two laps of the track, whilst others never started and yet others ran part way before realizing something was amiss. At the investigation, attempts were made to place all blame on one low paid employee who was to act as a back-up in cases like this (The Guardian, 1993).

Phrases such as 'global laughing stock' and 'world's greatest pantomime' may hurt of course, and the UK government may not have been pleased at the lost tax revenue, but most people laughed. However, it is no laughing matter when events such as those described above occur in situations of potential injury or death, or when the financial consequences of poor design result in closure of a small enterprise, or when operators are blamed for what is really failure in design or management. Yet many examples of inappropriate design of equipment, workplaces, systems, jobs and organizations can be found in large and small companies, in offices and factories, in physical and mental work. The common denominator is that the abilities, needs and limitations of the people working in the system or with the equipment have not been understood and accounted for. On the other hand, successful products or work systems will usually show evidence that the needs of their users have been accounted for during design, implementation and operation.

Taking such account of people is the province of *ergonomics*. This book is about the *methods* that ergonomists use, particularly in evaluating work and organizations. Ergonomics has little value unless it is applied and so its practical methodology is of great importance.

The consequences of not applying ergonomics, or of wrongly applying ergonomics through inappropriate methodology, can increase risks of ill-health and injury, dissatisfaction, and discomfort for the workforce. For a company the consequences at the least can be a loss of competitiveness, in terms of productivity, quality, flexibility, timeliness and so on. However, this book is more concerned with the positive side of applying ergonomics methodology, with the improvements in well-being that can result for employers, workforces, producers, users, engineers, designers and, indeed, people in general. This application will be in both of what Kragt (1992) has distinguished as **product ergonomics** and **production ergonomics**—the usability of things people use and the viability of their jobs in making these things.

Definitions of ergonomics

Ergonomics is now the accepted term worldwide for the practice of learning about human characteristics and then using that understanding to improve people's interaction with the things they use and with the environments in which they do so. In North America, the equivalent term has been *human factors*, but ergonomists elsewhere have generally regarded this as being synonymous with ergonomics and even in the USA the relevant professional society is now known as the Human Factors and Ergonomics Society.

Different definitions of ergonomics exist but the differences are more to do with where to draw the boundary of ergonomics than with fundamental disagreements on approach. A wide view is that ergonomics is the 'study of human abilities and characteristics which affect the design of equipment, systems and jobs . . . and its aims are to improve efficiency, safety and . . . well-being' (Clark and Corlett, 1984, p. 2); a narrower view is represented by Wickens (1984, p. 3), that human factors is to do with designing machines that accommodate the limits of the user. Pithy definitions of ergonomics are that it concerns designing for human use, or that it is the approach which fits systems (or machines or jobs or processes) *to* people and not *vice versa* (Figure 1.1). More detailed definitions exist; a relatively comprehensive one is that ergonomics is 'that branch of science and technology that includes what is known and theorized about human behavioural and biological characteristics that can be validly applied to the specification, design, evaluation, operation, and maintenance of products and systems to enhance safe, effective, and satisfying use by individuals, groups, and organizations' (Christensen *et al.,* 1988). Such a definition emphasizes data collection or derivation (science) and application (technology), the input of ergonomics into all aspects of system life cycles, and the multiplicity of aims that we have.

Rather than providing a single precise definition of our discipline though it is more important that ergonomics should be seen as an approach (or as a philosophy) of taking account of people in the way we design and organize; in other words, as 'designing for people'. In this view, ergonomics itself is primarily a process, to an extent a meta-method, which makes the clear understanding and correct utilization of individual methods and techniques even more important.

Ergonomics inputs and the importance of methods

In its formative years, the ergonomics profession was prone to considerable heart-searching about what ergonomics was, where it was going, and how it was perceived by the outside world; such concerns are found in papers such as 'Quo Vadis Ergonomia?' (Chapanis, 1979) or 'Ergonomics: Where Have We Been and Where Are We Going' (Welford, 1976–Part I, and Christensen, 1976–Part II). At the other extreme, students of ergonomics

THE FAR SIDE   By GARY LARSON

"Well, there it goes again . . . And we just
sit here without opposable thumbs."

Figure 1.1. An excellent design from a technical viewpoint but which has not accounted for the needs and limitations of the potential users! © 1984 The Far Side cartoon by Gary Larson is reprinted by permission of Chronicle Features, San Francisco, CA. All rights reserved.

complained that they could never explain to friends or employers what their subject was about. More recently we can find contributions questioning how far we should extend the impact of ergonomics to complex societal issues (e.g., Moray, 1993; Sheridan 1991a, 1991b), and how well our data fulfill needs for prediction and evaluation—not well at all according to David Meister (e.g., Meister and Enderwich, 1992).

Nonetheless, ergonomics has matured and moved away from debates about defining its remit or boundaries, even its *raison d'être,* and onto considerations of operation. In the words of Alphonse Chapanis, 'whether human factors is or is not a "science" is an issue not worth further consideration' (Chapanis, 1992, p. 6). Methods are now the focus of attention, and nowhere more so than in the evaluation of work and work systems. Back in 1986, the NRC Committee on Human Factors reported the development of applied methods to be one of the major research needs for human factors (National Research Council, 1986), and nothing has occurred in the years since to change this. Ergonomics is both a science and a technology and thus has need of techniques for both data collection (basic or functional data) and application. The debts we owe to other disciplines are obvious: as Singleton (1982, p. 9) says, the two integral parts of ergonomics are an interdisciplinary research activity based upon anatomy, physiology and psychology and an operational activity which 'usually finds expression through one of the two established technologies of medicine and engineering'. However, as we gain experience and confidence and as our armoury of knowledge and methodology grows, so the debt to other disciplines is being repaid. Methods developed or adapted within ergonomics will be employed by psychologists or engineers or health care professionals in turn, just as we are constantly enlarging our own human performance database through results of human-machine systems evaluation.

Bearing in mind the applied nature of ergonomics, if we try to draw parallels with the basic processes of design—*analysis, synthesis and evaluation*—which iterate throughout the design process (e.g., Markus, 1969), and extend these, then we might divide ergonomics methodology into methods for five types of input: data on people, systems development, evaluation of system performance, assessment of effects on people, and the organization of ergonomics programmes (Wilson, 1994).

Methods for the collection of data about people

Our first methodological need is for data about people and this can cover all characteristics—their physical size and strength, endurance and physiological capacity, sensory characteristics, mental capacities, psychological responses and so on. Just as important as the collection and reporting of data is the generation of design and evaluation criteria from these. For example, given data on the population range of arm reaches in different directions, what advice can be given on placement of frequently used rotary controls? Or again, given data on working memory limitations, can these be adapted to form design guidance on numbers of different codes that should be used in a coding system? Methods used to produce data about people comprise the scientific base of ergonomics and, as pointed out above, methods have often been borrowed directly or adapted from other disciplines.

Methods used in systems development

The second input of ergonomics is its contribution to the design and development process. This overlaps with both basic data collection and with evaluation (see below) but here is meant methods to assist in the analysis and development stages of design or redesign of equipment, workplaces, software, jobs or buildings. In essence we need methods to *analyze* current or proposed systems (analysis strictly meaning to resolve the system into its constituent elements and critically examine these) and then to *synthesize* data (that is, build up a coherent whole by putting elements back together) into ergonomically sound concepts, prototypes and final designs. Specifications produced out of this process must also have reasoned justifications, in order that ergonomists can work sensibly with engineers and designers.

Methods to evaluate human-machine system performance

In part, analysis at the start of development may involve evaluations of an existing system's performance. Certainly we must evaluate system performance during and at the end of development. Many measures can be defined for this and one challenge facing many ergonomists today is the search for measures of system performance other than the ubiquitous but often sterile 'time and error' based ones. Manufacturing system performance, for instance, may be assessed by means of production output rates and product quality levels, but we could also use machine utilization rates, minimization of finished stocks or work-in-progress, raw material wastage, speed of response to changed schedules, accident rates, sickness or other absence, or job attitudes and job satisfaction measures. Similarly, although a computer interface can be assessed in terms of time taken and errors made in performing a sequence of tasks, more interesting measures might be 'extent of system explored', 'willingness to change direction', 'quality of finished work' etc. These are much more difficult measures to take, undoubtedly, but perhaps they provide a more valid measure of actual system performance.

Any evaluation of subsequent system performance, in cases of an ergonomics input into system development, is also, in an interesting closing of the loop, an evaluation of just how well ergonomics was applied to the design.

Methods to assess demands and effects on people

Ergonomics has twin aims in its contribution to design and development—improvements for the job-holder or user *and* improvements for the producer or employer (see later). As a result, any system assessment should be carried out in terms of the demands made on people and of effects on their well-being, as well as accounting for system performance. There is an argument that the demands made on people should in fact be viewed as an implicit part of system performance and as such their assessment should not be seen as separate; however, on balance it is more persuasive that such a distinction will emphasize the twin, yet interdependent, aims of ergonomics. Many methods can be applied to assess the effects that different environments, jobs or equipment have on people. Such impacts might be medical, physical or psychological in nature and methods will vary from direct recording of observable phenomena (e.g., heart rate) to indirect observation of people's affective states (e.g., boredom). In almost all circumstances however, the data collected are not useful by themselves but must be interpreted and any effects inferred, which is again a large part of the ergonomist's input. Moreover, if assessment methodology is developed appropriately then data obtained can be generalized to become part of our first input, the basic data on people.

Methods to develop ergonomics management programmes

Our fifth input, of especial importance in the particular context of this book, concerns the management of ergonomics programmes. Methods are required here for two situations, although there is not a clear distinction between them. Firstly, there are ergonomists who are working within companies—whether in product ergonomics (the ergonomics of the goods or services the company produces) or in work systems/production ergonomics (the ergonomics of the processes used to do so)—and they are often doing this in very small groups or even on their own. Secondly, we have what is termed 'devolving' (Wilson, 1994) or 'giving away' (Corlett, 1991) ergonomics expertise. It is unrealistic to expect all enterprises to employ only ergonomists (either as employees or outside consultants) to handle their ergonomics; it is inappropriate also in most circumstances. In many areas of ergonomics application—for instance, health and safety, job redesign, workplace layout—the ergonomics profession must provide methods which allow the development of appropriate strategies and which support the management of programmes for ergonomics which can be run as a part of normal company activities. Nobody is advocating that untrained staff handle all human factors in a company. However, design engineering, production engineering, health and safety, line management and production workers can all make considerable contributions to an ergonomics effort; the methods and support that ergonomists give to them must include enabling them to recognize when specialists must be brought in.

This fifth input is certainly the 'messiest' area of ergonomics methodology, and it embraces aspects of all other inputs as well as participative ergonomics, systems implementation and so on (see chapters 35, 36 and 37). Nonetheless, it is an increasingly important area as ergonomics moves more and more away from research laboratories and universities and into real companies with real problems.

Approaches and context for ergonomics application

Ergonomics may be defined or interpreted by teachers, researchers or practitioners in many different ways, and so we can place ergonomics methodology in a variety of contexts. Individual ergonomists may work with a different focus at different times; they may concentrate upon how to apply their
work, or they may focus upon the aims of such application or implications
for non-application. Other ergonomists base their work around models of
people and performance, and still others would place their activities within
some defined design process. We can term these different contexts or
approaches as being *application-oriented, objective-oriented, human performance-oriented* and *design process-oriented,* although there is much overlap between them.

Taking account here of these different approaches to ergonomics enquiry
has the advantage for this book of providing some context for an overall
appreciation of ergonomics methodology. It also allows an introduction to
ergonomics which takes a broad view of its concerns and coverage. By understanding something of these contexts (or perhaps even models) we can get
a feel for the range of issues, processes, applications and conditions in which
methods must work. Moreover, working in each of the approaches will present different requirements for methodology and will therefore give different
constraints and limitations.

*It must be emphasized that there is no suggestion that ergonomics must be seen in
only one of these four contexts or as taking only one of the approaches. The truth of
course is that ergonomics may be viewed in any or all of these ways in different
circumstances for good reasons, and that our methodology should be broad enough to
work within any of them.*

Application-oriented approaches to ergonomics

A traditional view of ergonomics is that, like the epidemiological model
'host-agent-environment' in disease control or accident prevention, it is concerned with interactions between people and the things they use and the
environments in which they use them. Most ergonomics and human factors
texts open with a simple illustration of the interface between people and the
processes with which they interact, whether this 'process' be a toothbrush,
training manual, motor car, or power plant control room. Leamon (1980)
expanded upon this and Figure 1.2a is an amended version of his model;
Figure 1.2b shows this in a more abstract form.

The person and the process form a closed loop system (but not a closed
system); the output characteristics of each (such as the person's hands, feet or
speech, and the process displays) must match the input characteristics (process
controls and human sensory mechanisms) of the other. If such a match is
achieved we talk of a user-system fit or of a successful human-machine interface; this is the subject of many ergonomics studies and the focus of many
methods. Generally the displays and controls themselves are regarded as com-

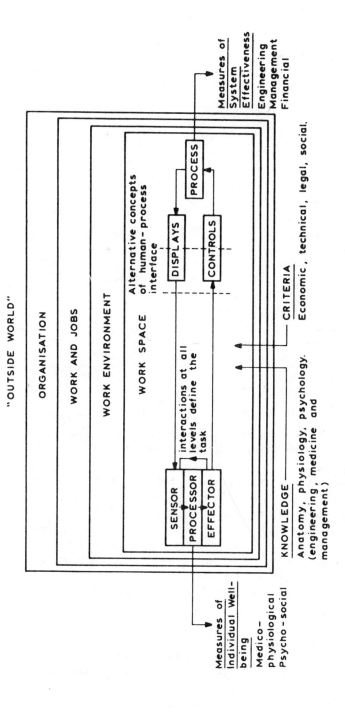

Figure 1.2. (a) Application-oriented model of ergonomics, from the viewpoint of application issues. Primarily this model stresses: the person–process interaction; that this interaction defines, and is defined by, tasks; the different concepts of the interface; the interactions with 'layers' of contextual factors; and the interaction with the outside world in terms of knowledge required, criteria to be accounted for, and measurements (adapted from Leamon, 1980)

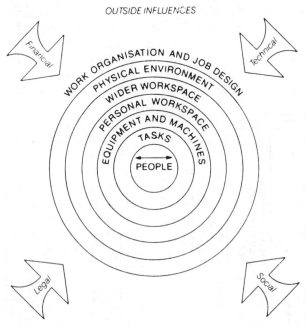

Figure 1.2. (b) Interactions of factors relevant to the application of ergonomics in work design. At the centre is the consideration of people, and of interactions between people. In turn we need to consider the tasks that they do—and physical and mental load as a consequence, the equipment and machines they work with (including software), and the various elements of the environment in which they do so. (Source: Grey *et al.*, 1987)

prising the interface; however in systems which are highly automated, where the operator acts as a monitor, the interface may be seen as lying between the person and the displays and controls with the latter considered to be part of the process.

Human–machine interaction does not take place in a vacuum; it will be affected by the physical workplace, the physical work environment and the social environment or organization of jobs and work, and also by factors from the world outside. Within such a model we can see ergonomics methodology as comprising the techniques needed to predict, investigate or develop each of the possible interactions, between person-task, person-process (hardware or software), person-environment, person-job, person-person, person-organization, and person-outside world.

Objective-oriented approaches to ergonomics

In most extensive definitions of ergonomics we will find a list of objectives or criteria that drive its application, for instance aiming for jobs, systems or products that are comfortable, safe, effective and satisfying. Aims of ergonomics are often divided into those of gains for the individual (employee or user), and those for the organization (employer or producer). *These aims, however, are neither mutually exclusive nor independent.* It is not a case of having either a more comfortable workstation or a more productive one, nor are the ways of achieving the former necessarily very different from those for the latter. For instance, and at a very simple level, the light intensity, position and colour rendering needed to give best possible performance at a product inspection task are likely to be very similar to those giving least potential risk of visual fatigue for the inspector: the preferred position, size and angle of pedals at an industrial sewing machine to improve work output and quality will be similar to settings to give operator comfort and convenience. Put more strongly, work or equipment designed to meet the needs of an employee or user will certainly not detract from performance effectiveness and in general will enhance it. To do this, and to prove it, is one of the main tasks facing ergonomists. Figures 1.3a and 1.3b illustrate these twin aims of ergonomics in the context of both work systems (or production) ergonomics and product ergonomics. For both we see that there is a direct connection

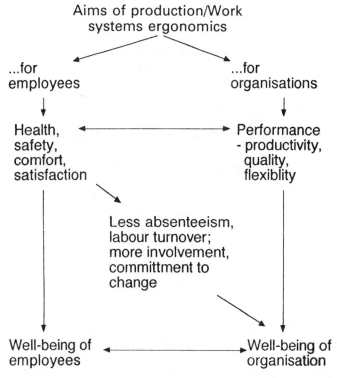

Figure 1.3. (a) and (b) Aims of ergonomics. Ergonomics can be seen in the context of its objectives, which are the well-being of people and of organizations. These might be seen as twin aims which are neither independent nor mutually exclusive and which have direct and systemic connections. This is illustrated for work systems or production ergonomics (1.3a) and product ergonomics (1.3b)

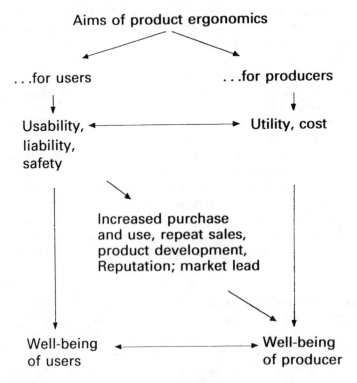

Aims of product ergonomics

...for users   ...for producers

Usability, ⟷ Utility, cost
liability,
safety

Increased purchase
and use, repeat sales,
product development,
Reputation; market lead

Well-being ⟷ Well-being
of users   of producer

Figure 1.3. (b)

between design and development criteria for people and organizations, and an indirect or systemic one also.

Generally, if we concentrate upon objectives of ergonomics then we require methods that go some way to helping us 'prove' that we have met certain aims or have achieved a certain level of improvement. Cost benefit analyses are of particular importance here (see chapter 34) as is the setting of usability metrics (chapter 12) or the setting of a criterion reliability value (see chapter 31) for instance.

Human performance-oriented approaches to ergonomics

A third way to approach ergonomics is to examine what people do, how they behave, in whatever domain. Then we could look at methods in the light of how they provide, improve, adapt and apply information gained from such a performance-oriented model. We will restrict consideration here to just one such model, the somewhat ubiquitous 'human information processor' (Figure 1.4). This is widely represented as a basis for explanation of how we behave in our environment, allowing us to test hypotheses about human performance.

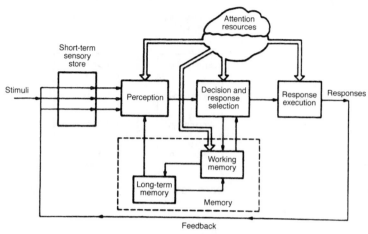

Figure 1.4. A model of human information processing. From Wickens (1984), reprinted by permission of the publisher

Basic human data (the input of ergonomics defined above) may be relevant to sensory, perceptual/cognitive and motor components of behaviour, and to associated attentional and memory processes. Application of ergonomics in design or evaluation can take into account the relevant stages of information processing. Any classification of methods here could be according to the information processing stage implied. Methods of relevance to a perform-ance-oriented view of ergonomics will allow us to interpret for instance how operators are representing a process—in terms of their mental (or conceptual) model (chapter 13) or the decisions or interpretations they make (see chapter 7 on verbal protocol analysis). At the other extreme, we will wish to know the physical and mental effects of work on their work capacity or affective responses (for example, posture, chapter 23, mental workload, chapter 25, or stress, chapter 26).

Design process-oriented approaches to ergonomics

There have been sufficient studies in recent years of 'how designers actually design' to make anyone defining a systems design process very cautious. Nonetheless, the fourth way of placing ergonomics in context and of dis-cussing its methodology is to use a design process-oriented model of ergo-nomics. In this view methods must be developed to support each stage of design, whether these be defined in general terms or in terms specific to ergonomics input. Many examples of a design process, and their complemen-tary methods, have been identified or proposed (see Jones, 1980 or Cross, 1984, for example). Already mentioned in this chapter is Markus' (1969) notion of the two dimensions of design, one progressive and one iterative

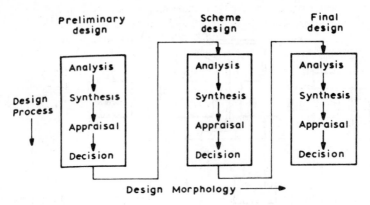

Figure 1.5. A two-dimensional model of the design process. Adapted from Markus (1969)

(Figure 1.5) wherein a sequence of analysis, synthesis and evaluation is found at all stages of the development process. At first sight we could, with profit, consider associating our methods with these generic stage descriptions, and perhaps also with particular phases of development. However, we would soon see that, particularly for analysis (which often also includes an evaluation of the existing situation) and evaluation, we could utilize almost all methods. It is interesting that within the domain of human-computer interaction Williges' model of iterative design (Williges *et al.,* 1987, see Figure 1.6) has a

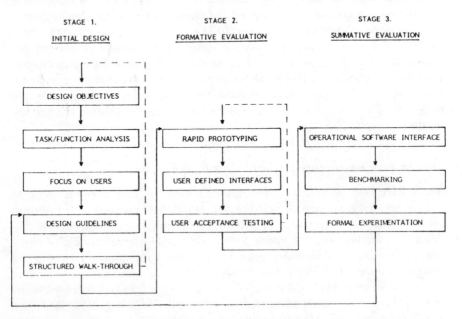

Figure 1.6. Flow diagram of the three stages in the design of human-computer software interfaces. (Source: Williges *et al.,* 1987, reprinted by permission of the publisher John Wiley and Sons)

similar two dimensional structure, with the notions of formative and summative evaluation to differentiate between assessment which is intended to enhance the design process and assessment which is to enable decisions to be made about the quality of prototypes or final versions.

Since ergonomics methodology must be seen as integral to the design and development process, and this book's central concern is the evaluation of work in practice, then human–machine systems design and development is enlarged upon below.

Human-machine systems development process

At the outset it must be said that all formalized descriptions of a human–machine systems development process are idealizations. It would be widely accepted by anyone who has been involved in such a process that it will never follow exactly a predefined path, and that any development will actually include the adaptation of planned stages, iterations around loops in the process, redefined objectives or means, and ever-changing priorities. This is not to say though that we should make no effort to specify a *flexible* process; some framework is needed at the least.

Early proposed processes for man–machine (nowadays human–machine) systems design were similar to that of Singleton (1974) whereby human and hardware subsystems are developed in parallel, followed by subsequent integration both in terms of the interface and then the operational system itself (Figure 1.7). It is probably no accident that models of this kind were proposed at a time of great interest in socio-technical systems design (see Cherns, 1987 for a review), with its emphasis on joint development of social and technical systems, although it has to be said that practice did not always match theory (e.g., Clegg and Symon, 1989). More recently, representations of the systems design process have handled the parallel hardware and human design needs by placing ergonomics considerations within all stages of the total systems design process. Thus Bailey (1989) and Sanders and McCormick (1987, p. 522) talk of stages—which overlap and are iterative—as: Stage 1—Objectives and performance specifications; Stage 2—Definition of system; Stage 3—Basic design; Stage 4—Interface design; Stage 5—Facilitator design; Stage 6—Testing. In other views, Kragt (1992, p. 15) and Kirwan and Ainsworth (1992, p. 28) define multiple overlapping human factors activities, within a 'parent' system design process flow of 'concept—specification/definition—design—production/commissioning—operation'. Kragt (1992) still makes explicit the parallel development of machine, jobs and interfaces (Figure 1.8).

In two further representations of the design process we can see a contrast. For product development a process has been suggested that defines ergonomics interventions within the total product design process (Figure 1.9), whereas the systems design process shown in Figure 1.10 is itself an ergonomics pro-

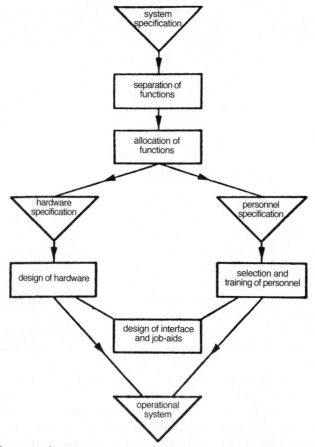

Figure 1.7. The systems design process. From Singleton (1974), reprinted by permission of the author

cess, developed as part of a programme to embed ergonomics within a company's system design activities (Aikin, Rollings and Wilson, 1994). It is constructed to parallel the existing system of design and approvals (for example, for safety testing, environmental impact, etc.). In addition to the above, chapters 10, 12 and 13 describe development processes for products, computer interfaces and control rooms.

Within representations of the human-machine systems design process is found a mixture of types of stage. 'Information collection' for instance can include any of a large number of methods, including those of direct and indirect observation, simulation and modelling and also examining other sources of data in the literature and elsewhere. 'Task analysis' might be regarded by some as a method in its own right, but also employs many other methods of data collection within it; in fact making decisions on data collection for task analysis provides an excellent illustration of the problems of method choice in ergonomics generally (see chapter 6).

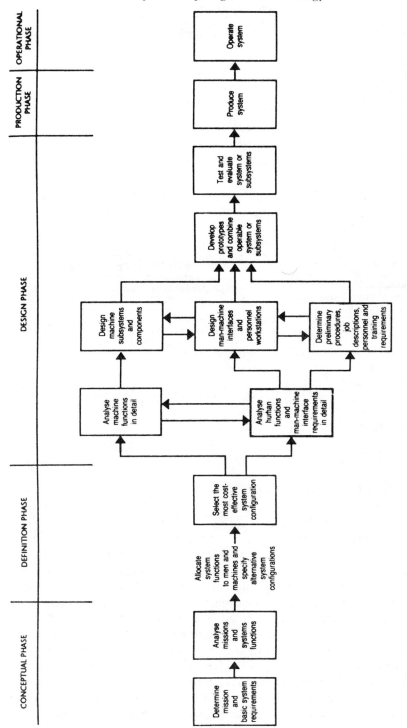

Figure 1.8. Design activities occurring during phases of the system development process (Kragt, 1992)

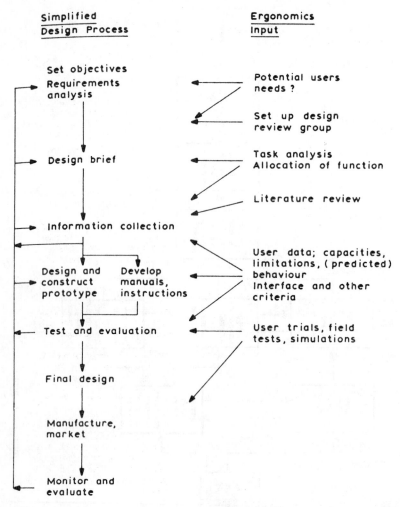

Figure 1.9. One possible simplified product design process showing types of ergonomics input (files of the author)

Allocation of functions is sometimes seen from an engineering perspective as comprising only a checklist method of selecting whether to plan functions to be carried out by people or machines. Engineers may use variations of the 'Fitts List' of comparative abilities, despite warnings against rigid application of such an approach (e.g., Jordan, 1963). Fuld (1993) questions the whole concept of function allocation in preliminary design, suggesting its use may be of greater value at a test stage. Recent and interesting developments, made possible by the potential of computer technology, are in flexible or dynamic allocation of function, whereby the extent of the role taken by the person can change according to personnel and operational conditions (Clegg,

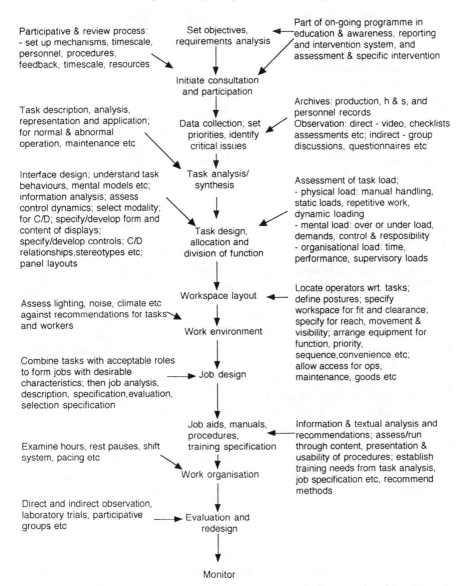

Figure 1.10. Ergonomics Design Process. Each of the stages of ergonomics input has itself got particular input requirements, which imply use of a large variety of methods in analysis, design and evaluation (files of the author)

Ravden, Corbett and Johnson, 1989). This gives a good example of how technology, approach, method and design are often intimately related or intertwined, whereby technical developments support a new design approach to account for human factors, which then allows formal integration into the system of a method for ergonomics design hitherto seen as being applied 'off-line'.

Classification of methods

It should be apparent from what is said above that attempts to classify ergonomics methodology in terms of the parts or stages of any models of ergonomics will be difficult, lead to much ambiguity, and probably not be a very useful or fruitful exercise anyway. Meister (1985) attempts a gross distinction in his book, *Behavioural Analysis and Measurement Methods,* and, as his title suggests, he divides his behavioural methods into the analytic techniques employed during the development of systems and the measurement methods employed to evaluate functioning systems. He does, though, recognize overlaps, particularly many 'measurement methods' also being used during system development.

It would seem, however, that there is utility in classifying ergonomics methods. It may make it easier for us to communicate the breadth, depth and detail of our activities to engineers, designers and managers. Comparative evaluations and situation-specific recommendations could also be made more easily, leading to better guidance on appropriate methodology for different needs. Also, we may be able to derive some form of human factors performance metric for particular products, an issue of particular concern with respect to human–computer interaction.

Edwards (1973) concludes that distinctions between psychological and physiological methods are of no value as a basis for a taxonomy of ergonomics assessment techniques and therefore classifies these partly by method of measurement but largely by the nature of the measure, although only within human performance evaluation (i.e., not explicitly including analysis or synthesis). His broad categories are of direct achievement measurement, operator loading measurement and correlated function measurement, and particular measures (or as he calls them techniques) within these are listed in Table 1.1.

Once all possible measures are considered, we begin to see some of the problems of classification and selection. Kantowitz (1992) lists 46 possible indicators of nuclear power plant safety, split into seven categories (e.g., operations, quality programs, management/administration) and reports that 'no single indicator was by itself an adequate measure . . . [nor were any] optimal for predicting plant safety' (p. 391).

Colleagues at the Ergonomics Information Analysis Centre (EIAC) at Birmingham University have developed an ergonomics classification to keep up with the changing nature of the discipline. The part of this currently dealing with methods is shown in Table 1.2.

Starting from Edwards' taxonomy, then reviewing the EIAC classification, and faced with a need to structure teaching in performance measurement and evaluation a few years ago, the author has attempted an initial taxonomy

Table 1.1. Taxonomy of assessment techniques (Edwards, 1973)

Assessment techniques (measures)	Examples
Direct achievement measurement	
Absolute scores	Distance travelled, numbers produced
Speed scores	Time, time per . . ., reaction time
Precision scores (continuous)	Tracking performance
Error scores (discrete)	Faults, missed signals
Information scores	Rate of information handling, redundancy
Multiple scores	Combinations of the above
Operator loading measurement	
Extracted outputs	ECG, EMG, EEG, GSR, O_2 uptake
Secondary tasks	Task loading through secondary activity
Alternative tasks	Decrement assessed on new task
Time function changes	Fatigue, learning
Operator reports	Interviews on, say, fatigue
Correlated function measurement	
Variability	Consistency as measure of skill
Operator reports	Critical incidents
Observer reports	Pilot assessment

of ergonomics methodology. This taxonomy recognizes differences in approach, method (or method group), technique, and measure. It may also be possible in future to add the criteria underlying the required measures. The present state of such a general classification scheme or taxonomy is shown in Table 1.3. Despite some redundancy, omissions and inconsistencies this taxonomy provides a useful framework.

Use and usefulness of methods

As far as methods are concerned, according to an old British expression, the proof of the pudding is in the eating. A method which to one researcher or practitioner is an invaluable aid to all their work may to another be vague or insubstantial in concept, difficult to use and variable in its outcomes. More than this, the validity, reliability and sensitivity of methods may well be application specific. Perhaps it is only by examination of utility and generalizability that we can select or prioritize between methods. (See also chapter 5 on choosing techniques and variables to measure.)

Meister (1986) reports a survey he conducted with members of the Human Factors Society interested in testing and evaluation (pp. 387–406). (Cursory examination of the list of 21 respondents reveals none from academia and 16 from military-based groups.) From an admittedly limited survey, it appears that those methods which were rated consistently highly on all three criteria of frequency, usefulness and ease of use, are those of indirect observation— questionnaires and rating scales for instance. The exception is attitude measurement, rated low for frequency and ease of use, but this could reflect

Table 1.2. Classification (part) from Ergonomics Abstracts

	Methods and techniques		
63	Approaches and methods	64	Techniques

63.1	Modelling and simulation 63.1.1 modelling human characteristics 63.1.2 modelling system characteristics 63.1.3 modelling environmental charac- teristics	64.1	Observation techniques 64.1.1 participative observation and group decision making 64.1.2 visible observation 64.1.3 unobtrusive observation
63.2	Use of simulators 63.2.1 use of test rigs	64.2	Checklists
63.3	Mock-ups, prototypes and prototyping	64.3	Classification systems and taxonomies
63.4	Manikins and fitting trials	64.4	Interviews
63.5	Systems analysis 63.5.1 task analysis 63.5.2 job analysis and skills analysis	64.5	Questionnaires and surveys
		64.6	Rating and ranking
		64.7	Application of test batteries
63.6	Human reliability and system reliability	64.8	Experimental equipment design 64.8.1 hardware design for experimen- tation
63.7	Physiological and psychophysiological recording		64.8.2 software design for experimen- tation
63.8	Work study 63.8.1 method study 63.8.2 work measurement	64.9	Critical incident technique
63.9	Data collection and recording methods	65	Measures
	63.9.1 human recording	65.1	Comparison of measures
	63.9.2 self recording	65.2	Time and speed
	63.9.3 instrument recording	65.3	Error, accuracy, reliability and frequency
	63.9.4 experimental design	65.4	Event frequency
	63.9.5 laboratory versus field	65.5	Response operating characteristics
63.10	Data analysis and processing methods 63.10.1 statistical analysis and psycho- metrics		65.5.1 sensitivity 65.5.2 response bias
	63.10.2 signal processing and spectral analysis	65.6	Output and productivity
	63.10.3 image processing	65.7	Combined measures and indices
	63.10.4 textual analysis and parsing	65.8	Subjective measures
63.11	Psychophysics and psychological scaling		65.8.1 ratings and preferences
63.12	Use of expert opinion		65.8.2 opinions
63.13	Protocol analysis	65.9	Usage
63.14	Approaches to equipment testing		
63.15	Cost benefit analysis		
63.16	Job appraisal		

the military bias in the sample! Interestingly, the relative difficulty in use reported for observation and for test plan do not seem to preclude their actual widespread use. Dynamic mock-ups, static mock-ups, automated data recording, activity analysis and computerized methods are all seen as useful but difficult to use, which leads to their relatively infrequent use.

Multiple methods

Given all that has been said about methods it is not surprising that we will often look to use more than one, and often several, in any one study. This

Table 1.3. A classification of methods, techniques and measures used with ergonomics. This classification has been split into categories of: 1. general methods; 2. collection of information about people; 3. analysis and design; 4. evaluation of human-machine system performance; 5. evaluation of demands on people; 6. management and implementation of ergonomics. It should be noted that the general methods at the start of the table can generally be used within any of the other categories

Method		Technique	Measure/outcome	Chapters in this text
Group	Subgroup			
1. General methods				
Direct observation (laboratory or field)	Unobtrusive, participative, or visible	Human recording; checklists, rating, ranking, critical incident technique, charts (time, spatial, sequence, link)	Event frequency, sequence Times, errors, accuracy Overload, underload Descriptive, evaluative, diagnostic measures of performance	2, 6, 10, 12, 21, 23
		Hardware recording; video, film, tape, event recorder, position/movement recording, computer real time recording		
Indirect observation (laboratory or field)	Psychometrics/scaling	Surveys, questionnaires, rating, ranking, scaling, diaries, critical incidents, checklists, group discussions, interviews	Attitudes, feelings, perceived effort, difficulties, advantages, disadvantages, preferences	3, 6, 10, 12, 13, 14, 15, 16, 17, 21, 23, 25, 26, 27
Archives and records	Production, medical, personnel, quality	Before/after comparison, trend data, use in cost-benefit calculations	Output, times, quality, etc. Absenteeism, labour turnover, injuries, sickness	4, 30, 32, 34
'Automated' methods				7, 9, 14
Experiments				5
Literature and data interpretation		Description, statistical analysis, model building		1, 3, 5, 8, 9
Standards and recommendations	Voluntary or enforced			4, 10, 12, 13, 16, 17, 18, 30
Multiple methods		Triangulation, validation		1, 21, 25, 26

Table 1.3. *Contd*

Group	Method		Technique	Measure/outcome	Chapters in this text
	Subgroup				
2. Collection of information about people					
Physical measurement	Anthropometry (static or dynamic)		Anthropometer, wall charts, video, photography, CODA, fitting trials, computer modelling	Dimensions, percentiles, other descriptive statistics	19, 20, 21, 23
	Biomechanics		Dynamometer, strength gauges, goniometer, stadiometer	Values, trends, descriptive statistics	21, 22, 23, 24
	Performance		Eye/body movements, acuity, CFF, TAF/CMF	Scores, counts, qualities	21, 27, 28, 29
Physiological measurement			ECG, EEG, EMG, ERP, HRV O_2 uptake, GSR, pupil diameter		9, 16, 22, 25, 26, 28, 29
Perceptual/cognitive assessment	Ability testing		Mental or cognitive tests (e.g. general aptitude test battery); perceptual tests	Prediction of performance	25, 26, 27
	Psychophysics		Method of limits, method of average error, method of constant stimuli, e.g. aesthesiometer, hearing loss audiometry, CFF, fitting trials	Thresholds and levels of perception, sensitivity	15, 17, 18, 19, 21, 27, 29
	Visual tests		Acuity, colour, stereopsis, heterophoria		27, 28
Knowledge acquisition	Knowledge elicitation (from expert)		Written, audio, video records; interviews (structured, unstructured) protocol analysis, conceptual mapping, goal decomposition, automatic techniques	'Rules', reasoning, explanations	7, 14, 16
	Other		Interpretation of records, standards, guidelines, criteria		
Models	Computer, mechanical, conceptual, mathematical				8, 16, 17, 18, 24, 30, 32

Table 1.3. *Contd*

Group	Method — Subgroup	Technique	Measure/outcome	Chapters in this text
3. Analysis and design				
Task analysis	Data collection methods — see all general methods; Description analysis and representation methods	Hierarchical TA, tabular TA, ability requirements analysis, TA for knowledge description, link analysis, cognitive TA, formal mappings (e.g. TAG), job analysis charts, operation sequence diagrams, flow charts, time charts	Requirements for people. Consequences of tasks; task sequences, times, probable error rates, criticalities, loading etc.	2, 6, 10, 12, 13, 21, 31
Expert analysis		Checklist, walkthrough, Delphi technique, method study techniques, expert systems, likelihood matrix, scored assessments	Qualitative reports, critical issues, weightings and priorities, approvals	2, 12, 13, 16, 21, 30, 31
Introspection/protocol analysis	Concurrent Retrospective	Written, audio, video records; diagrammatic, computational, debriefing, diary, critical incidents, shadowing	Explicit content, implicit content, behaviour transitions, rules, knowledge, models	7, 12, 14
User models		GOMS, CLG, TAG, etc. mental models		6, 8, 12, 13
Statistical analysis		Signal Detection Theory, information theory, reliability assessment	Performance measure, likelihoods	31
Models		Task network (SAINT, Siegel–Wolf etc.); control theory, deterministic (HOS, etc.); cognitive/information processing, (GOMS, etc.); behaviour (S-R-K)	Performance predictions	5, 6, 8, 12, 16, 17, 24
Simulation	Computational	Mathematical; computer, including CAD (e.g. man models); virtual reality		1, 8, 9, 10, 11, 20
	Physical	Physical mock-up, walkthrough		1, 10, 19, 37
Method study		Graphical analysis; process, flow, time charts; filming, micromotion	Movements, times, actions, frequencies	2, 6
Work measurement		Time study, activity analysis, synthetic analysis, electronic monitoring	Times, standards, task sequence, simultaneity, frequency	2, 34

Table 1.3. *Contd*

Method				
Group	Subgroup	Technique	Measure/outcome	Chapters in this text
Prototyping		Rapid prototyping, storyboarding, mock-ups		9, 12, 13, 37
Creative techniques		Brainstorming, decision groups, focus groups		10, 12, 35, 37
Participative methods		Design and follow-up groups, user involvement; training and awareness in ergonomics		10, 13, 30, 33, 35, 36, 37, 38

4. Evaluation of human-machine system performance

Group	Subgroup	Technique	Measure/outcome	Chapters in this text
Work systems analysis		Checklists, walkthrough, expert assessment		
Usability evaluation —hci, products, texts etc.	User trials	Techniques of direct and indirect observation, and physical performance measurement —individuals or groups	Time, reaction time / Accuracy, errors / Opinions, attitudes, responses / Physical fit / Workload, stress	10, 11, 12, 19, 21, 37
	Evaluation environment	CAFE OF EVE		12
	Introspection, expert analysis	Protocol analysis, checklists, group decision making		7, 10, 12, 13, 14, 31
Measurement by instrumentation		Light (illumination, glare, etc.), climate (temperature, humidity, air space, etc.), noise (sound intensity-weighted, frequency), vibration, workplace dimensions	Measurements *vs.* norms, Comparisons. 'Fit' of system to people's requirements. Acceptability	15, 16, 17, 18, 19, 21
Subjective assessment		Psychophysical techniques, scaling, rating, surveys, questionnaires, etc.	Comfort, annoyance, acceptability	3, 15, 16, 17, 18, 19, 21

Method	Description	Output	References
Performance measures	Speech intelligibility index, work-rate, standard psychomotor and mental tests, etc.	'Scores' against norms	5, 17, 27
	Speed/accuracy trade off		4, 5, 12, 27
Modelling and simulation	Computer, mechanical (mannikins), mathematical		8, 16, 20
Electronic monitoring	On-line record	Performance measures	
Self recording	Gripe button, diary, event recorder	Problems, incidents	12
Text analysis	Readability formulae (Gunning Fog, Flesch, etc.); cloze procedures; judgements (rate, rank, etc.); protocol analysis; scan/read tests; checklists	Normative scores, ratings	3, 7, 11
Human reliability analysis	SHERPA, GEMS, PHECA, etc. Fault tree, action tree, etc.	Errors, types, causes	31
Reliability estimation	THERP, HEART, SLIM, etc.	Descriptive and predictive charts Human error probabilities	
Accident reporting and analyses	Archive records, reporting system, in-depth follow up interviews, site analysis, statistical analyses	Incidence, severity, epidemiology and aetiology	4, 30, 32

Table 1.3. *Contd*

Group	Method		Technique	Measure/outcome	Chapters in this text
	Subgroup				
5. Evaluation of demands on people					
Physical workload			Subjective assessment, perceived exertion (e.g. Borg scale) Performance records, secondary or alternative tasks of psychomotor performance; physical changes (e.g. stadiometer)	Subjective ratings, performance decrement, physical measures	5, 9, 21, 22, 23, 30
Mental workload			Primary, secondary, alternative task; subjective assessment (e.g. SWAT, Cooper-Harper, TLX); physiological response (e.g. HRV); protocol analysis; CFF, TAF	'Performance' decrement Load (subjective or objective)	5, 7, 9, 25, 29
Posture analysis			Biomechanical (mathematical) models: optical methods (CODA, Selspot); notation—paper and pencil (posture target, body part discomfort); video; RULA, OWAS Psychophysical methods	'Postures' to compare with criteria, posture scores; discomfort ratings	19, 21, 23, 24, 30 23, 30
Physiological methods			HR, HR variability, O_2 uptake, air analysis, GSR, ECG, EMG, EEG, ERP	Objective data, to be interpreted against norms, criteria	12, 22, 23, 25, 26, 29
'Fatigue' measurement	Visual fatigue		Direct measurement of occular function, task performance, asthenopia (reported discomfort)	Eye movements, accommodation, vergence, blink rate, pupil size, times and accuracy	27, 28
Environmental response measures	Physiological, perceptual		Sweat rate, body temperature, heart rate; hearing loss; visual acuity, contrast sensitivity; sensation loss; body change	Measurements vs. norms Comparisons Acceptability	15, 16, 17, 18, 19, 21
Stress assessment	Subjective, physiological		GSR, etc.; indirect observation techniques (SACL, GWBQ, etc.)		26

Job and work attitude measurement	Techniques of indirect observation, especially rating scales, e.g. JDS, informal group or individual interviews	Satisfaction, needs, important job characteristics	26
Guidelines			30
6. Management and implementation of ergonomics			
Ergonomics management, organization analysis			1, 33, 35
Project management	Checklists, walkthroughs, project groups, consultation and participation procedures		35, 37
Implementation			36, 37
Cost-benefit analysis	Investment returns; productivity—life cost, revenue calculation; health and safety valuations	Financial measures	34, 38
Participative methods	Ergonomics working groups, design decision groups, user representatives, ergonomics programmes		30, 33, 35, 36, 37
Ethics in ergonomics			38

is particularly so when we are carrying out evaluations in the field. Technically this is known as triangulation (see Denzin, 1970 or Webb *et al.*, 1972) and, although the term can encompass data or investigators, triangulation is most often used to denote methodological triangulation. This is the use of two or more methods to improve the effectiveness of a study or our confidence in any findings; weaknesses in one method can be balanced by strengths in another. To take one simple example, only by questioning and observing operators in complex systems *and* also recording their concurrent verbal reports for subsequent protocol analysis can we begin to understand something about their decision making activities.

A multiple-methods study may utilize a mixture of qualitative and quantitative techniques, in field and laboratory settings. A typical battery of methods which could be employed in the evaluation of work and the production of recommendations for its redesign is:

Questionnaires

Attitude surveys and rating scales

In-depth, informal discussions with individuals

Group decision meetings

Written record of activities

Diaries kept by job holders

Video or audio recording

Photographic recording

Concurrent or post-hoc verbal protocol analysis

Physical measurement of workplace dimensions

Physical measurement of environmental variables

Physiological and psychophysiological recording

Computer workspace modelling

Simulation and test trials of systems or workplaces.

Only by use of several of these methods may a full evaluation be possible in any one situation, and thereby effective suggestions be made for redesigning job content, tasks, workstations and environments. However, a word of caution is in order. It is all too easy to fall into the trap, once an investigation has started, of measuring everything possible 'just in case'! This can lead to results that are difficult to interpret or, worse still, an analysis phase that uses large amounts of study resources to no clear or useful purpose.

Structure of the book

The problems of providing a complete, non-redundant and unambiguous classification of ergonomics methods have been alluded to already. One manifestation of this difficulty is knowing how to structure chapters in a book such as this. In the end a compromise has been reached; the structure comprises a mixture of classification by basic methods and techniques, general methods of analysis and evaluation, and context-specific groups of methods.

We open with a number of chapters on approaches and methods which are used in fields other than ergonomics, as well as being employed generally across all aspects of ergonomics endeavour. Therefore Part I includes contributions on direct observation techniques, indirect observation or subjective assessments, the use of archival data previously collected for another purpose, and an overview of how to design ergonomics studies. In Part II another group of methods and techniques, again general in that they can be applied in many situations of analysis or evaluation, but which—as they are reported here—are specifically relevant to ergonomics studies. Thus task analysis and protocol analysis, expecially the former, are useful across a wide spectrum of ergonomics study. The chapter on simulation and modelling, which obviously are available as methods in many disciplines, presents those techniques useful in ergonomics; computerized data collection also has a wide relevance but is looked at here specifically from an ergonomics viewpoint.

The next four parts cover four main areas of ergonomics activity: product or system design and evaluation (Part III); assessment and design of the physical workplace and environment (Part IV); the analysis of work activities, both physical and mental (Part V); and the analysis and evaluation of work systems (Part VI). Techniques relevant to all five inputs of ergonomics defined earlier, but especially to analysis, design and evaluation, have been considered together both in these parts and also in their constituent chapters, since methods are rarely applicable solely to one type of input.

Finally, we conclude the book with chapters concerned with project management and systems implementation, and a look to the future of ergonomics methodology (Part VII). Many of the methods, techniques and measures referred to in the body of the book have been placed in some sort of framework in Table 1.3. This stands as an, admittedly imperfect, frame of reference for methods used in the evaluation of human work. Other sources of information on methods are included in an appendix to this chapter.

Conclusions

The thin line which we must tread when we become involved in discussion of methodology is summed up in two contrasting views: '[psychology] . . . should not allow itself to be driven by obsession with method to the exclusion of the human problems that are its province' (Barber, 1988, p. 7, reporting

Maslow, 1946), but 'anyone who wishes to reflect on how they practise their particular art or science, and anyone who wishes to teach others to practise, must draw on methodology' (Cross, 1984, p. vii). Certainly ergonomists, if anyone, must be driven by human problems (or driven to improve the quality of people's lives), but also we are always concerned to educate others (designers, engineers, politicians, accountants, managers, public, media) in our approach and the necessity for it. If we truly believe that we are, above all, promoters of an approach and of a process, then it behoves us to pay great attention to the roots and current and future state of our methodology. The remainder of this book is but one step in doing this.

Acknowledgements

Thanks are due to Nigel Corlett, Christine Haslegrave and Ted Megaw for reading an earlier draft of this chapter and providing many helpful comments.

Further information

There are few texts in ergonomics and human factors specifically concerned with (particularly applied) methods. However, many of the best texts in the area do have varying amounts to say about method; also use of such sources is highly recommended for anyone wishing to apply the methods discussed in the present book. The brief descriptions below comprise one selection only; some good sources will have been omitted through ignorance. Also, there are many human factors texts now which deal with only a part of this rapidly expanding (in size and importance) field. Other than in human-computer interaction, these are not included.

Bailey, R.W. (1989). *Human Performance Engineering: A Guide for System Designers,* 2nd edition (London: Prentice Hall).

Very readable and a good juxtaposition of basic principles and findings with human factors criteria, this book is weakest in the areas of physical ergonomics (such as environment, work physiology and biomechanics). Over half the book discusses the application of human factors to design of various sorts and there are substantial chapters on evaluating preferences and performance through studies and tests and on conducting comparison studies. This is a good book for the areas (largely psychologically-related) it covers.

Chapanis, A. (1959). *Research Techniques in Human Engineeering* (Baltimore: John Hopkins Press).

Even 35 years on this text is still relevant and valuable. Coverage is general and largely domain-independent, thus making it usable in the context of modern systems. Specific chapters examine direct observation, the study of accidents and near accidents, psychophysical, statistical and experimental methods, and articulation testing. A good deal of information on the difficulties of using different methods, and their value is included. The book is highly recommended for anyone interested in methodology.

Clark, T.S. and Corlett, E.N. (1984). *The Ergonomics of Workspaces and Machines: A Design Manual* (London: Taylor and Francis).

A manual for students, professional designers and production engineers, covering the 'traditional' areas of ergonomics work—workspace, environment, displays and controls. Flow charts summarize design and selection criteria, and design principles are clearly explained.

Grandjean, E. (1988). *Fitting the Task to the Man: A Textbook of Occupational Ergonomics,* 4th edition (London: Taylor and Francis).

One of the staple teaching texts of ergonomics, it explains clearly with many diagrams much of the general area of traditional ergonomics. Emphasis is rather on consequences for people of work of certain types rather than upon methodology or re-design of work. A recommended basic introductory text for students or others interested in ergonomics/human factors.

Kantowitz, B.H. and Sorkin, R.D. (1983). *Human Factors: Understanding People-System Relationships* (New York: John Wiley).

Similar in some ways to the book by Bailey, this also makes little attempt to cover human physical activities in terms of energy expenditure, loading, etc. There is good coverage of other aspects of human factors; application information is in this case used throughout the text, as is information on methods.

Meister, D. (1985). *Behavioural Analysis and Measurement Methods* (Chichester: John Wiley).

This book was the first since Chapanis' in 1959 to present a useful and wide-ranging review of technique and methodology. Largely the book is split into techniques to aid in systems design, evaluations of systems effectiveness during and after development, and 'generic' methods. Intended as an introduction to behavioural measurement, the book has very full descriptions of many analysis and evaluation tools and approaches, and is—as is the author's wont—written from a pragmatic, applications-oriented standpoint. Its weakness is that the author's perspective is largely that of US armed forces needs and experiences.

Meister, D. (1986). *Human Factors Testing and Evaluation* (Amsterdam: Elsevier Science).

Not as well-written or produced as the same author's 1985 text, this book does however widen the range of methods and applications to include environmental evaluation.

Norman, D.A. and Draper, S.W. (1986). *User-Centred Systems Design: New Perspectives on Human-Computer Interaction* (Hillsdale, NJ: Lawrence Erlbaum).

This is a collection of contributions from workers and their colleagues at the University of California, San Diego. The focus is human-computer interaction; many chapters are concerned with what we should know about people to design interfaces and how we could obtain such knowledge. In particular there is a concentration upon indicating new directions for design and design-related activities, rather than upon empirical methods, quantitative rules or design processes.

Oborne, D.J. (1995). *Ergonomics at Work,* 3rd edition (Chichester: John Wiley).

Taking what its author describes as an evangelical stance to the communication of the ergonomics discipline, this book seems to an extent—and despite certain omissions and weaknesses—to bridge the gap between the human (especially physical) activity orientation of Grandjean and the systems (especially engineering) orientation of Sanders and McCormick. There is only a short chapter specifically devoted to method and ergonomics investigations, and some methodology is implied in earlier chapters. Readable.

Oppenheim, A.N. (1992). *Questionnaire Design, Interviewing and Attitude Measurement,* 2nd edition (London: Pinter Publishers).

Still the first port of call for many when wishing to access or develop an indirect observation technique, this well-written book covers survey and questionnaire design, checklists, rating scales, attitude assessment, and analysis.

Pheasant, S. (1986). *Bodyspace: Anthropometry, Biomechanics and Design* (London: Taylor and Francis).

Coverage here is similar to that in Clark and Corlett and in Grandjean—physical ergonomics—with one chapter only devoted to displays and controls. The measurement of human physical form and functions and use of such data in design are, however, the author's speciality and are very well covered. As far as method is concerned the use of anthropometric and biomechanical data in design is well covered; there are short sections upon collection of basic and performance data. The author's style is to write very clearly and simply in a somewhat idiosyncratic manner.

Ronan, W.W. and Prien, E.P. (1971). *Perspectives on the Measurement of Human Performance* (New York: Appleton-Century-Crofts).

The performance referred to in the title is that of interest within occupational psychology, referring largely to mental and psycho-motor tasks. The emphasis is upon shifting research attention from individual differences (as in psychological tests) to understanding human performance in the 'real' world. In addition to long sections devoted to performance reliability, dimensions of performance and organization performance and influence upon individual performance, there is also much on the reliability of performance observation and on criteria, with explanation of several methods and techniques.

Rubinstein, R. and Hersh, H. (1984). *The Human Factor: Designing Computer Systems for People* (Burlington, MA: Digital Press).

One of the best of the books which aim to increase the awareness of systems designers of human factors issues and to provide general guidance on how to improve user interfaces. The style is extremely readable and it is a well organized text. Chapters specifically concerning methods are on task analysis and models and on testing systems. Recommended.

Salvendy, G. (Ed.) (1987). *Handbook of Human Factors* (New York: John Wiley).

A mammoth tome, this is probably the most complete collection of general human factors approaches and knowledge publicly available. Specific sections devoted to methodology are ones on functional analysis (five chapters from surveys to task analysis to physical workload measurement), performance modelling (six chapters, but only some useful in terms of ergonomics methodology), and system evaluation (three chapters). Many other chapters discuss methods to varying extents. Admirable as this publishing effort has been to pull together as much as possible of current ergonomics knowledge, and despite much of the volume being surprisingly usable and readable for one with 103 authors, it is not geared primarily as a book covering the range of methods and techniques used in human factors and thus is not appropriate as such. A new edition will appear in 1996.

Sanders, M.S. and McCormick, E.J. (1992). *Human Factors in Engineering and Design,* 7th edition (New York: McGraw-Hill).

One of the most widely recommended teaching texts in ergonomics/human factors (previous editions were McCormick and Sanders), this takes an application-oriented view, much of its content being related to practical needs in engineering and design. Very thorough for its size, there are, though, omissions and/or weaknesses—in the areas of job design, computer dialogue design, and models, for instance. There is one chapter specifically devoted to methodologies—looking at field and experimental research, and criteria.

Definitely on any shortlist of useful texts it is nonetheless not written from the perspective of methodology.

Singleton, W.T., Fox, J.G. and Whitfield, D. (Eds) (1973). *Measurement of Man at Work* (London: Taylor and Francis).

Emanating from a conference as it does, the book suffers from the weaknesses of all collections of conference papers. Coverage is usually too full of gaps and at the same time has too much overlap. However, this collection suffers less than most and many of the contributions stand up well even two decades later. Many of the then authorities in the area have chapters in the book. Coverage is split into: '*Man*'—overviews of 'descriptors' of people useful in relation to human-machine interaction; '*Techniques*'—data acquisition procedures about people at work; and '*Applications*'. (The care given to explanation of the grouping (pp. xi–xii) shows this to be as difficult then as it has been for this book!—see earlier in this chapter.)

Singleton, W.T. (1974). *Man-Machine Systems* (Harmondsworth: Penguin).

Although short by today's standards, this is still an important book for anyone interested in the approach and stages of human-machine systems design. The whole book defines one methodological approach, with different methods or techniques comprising the stages.

Singleton, W.T. (Ed.) (1982). *The Body at Work: Biological Ergonomics* (Cambridge: Cambridge University Press).

Its title explaining well its content, this collection of contributions from various authorities takes each aspect of physical ergonomics (energy, biomechanics, vibration, climate, illumination and noise) in detail. In general each topic is discussed in terms of effects on people, measurement and application. It is a good review of methodology within the narrow field of ergonomics specified.

Webb, E.J., Campbell, D.T., Schwarz, R.D. and Sechrest, L. (1966). *Unobtrusive Measures: Non-Reactive Research in the Social Sciences* (Chicago: Rand McNally).

An excellent overview of indirect, direct, archival and physical measures, this still widely read book contains methods and techniques from the ordinary to the weird, solid to exciting. The title the book almost had, *The Bullfighter's Beard*—referring to the tendency for this to grow faster on the day of a bull fight, for which phenomenon there are alternative hypotheses—sums up the catholic approach of the authors.

Wickens, C.D. (1992). *Engineering Psychology and Human Performance*, 2nd edition (New York: Harper Collins).

Not really a book on applied methodology, this does however explain very clearly and comprehensively the type of cognitive experimental psychology research and results which could be applied to systems design problems. As a basic text it fills in well the gaps left by, say, Grandjean or Sanders and McCormick. The book is structured around the model of the human information processor; some design principles are given but no real guidance on their application. Nonetheless, an excellent book within its coverage.

Wilson, J.R., Corlett, E.N. and Manenica, I. (1987). *New Methods in Applied Ergonomics* (London: Taylor and Francis).

This is a record of the proceedings of the conference at which the idea for the present text was born. Many attendees are contributors to both. The earlier volume, however, apart from a few overview papers, contains contributions on 'new' methods or techniques, adapted methods, or explanations of 'technology transfer' of methodology. Particular coverage is given to subjective assessment, computerized techniques, HCI evaluation, thermal environment, task and safety analysis, and physical and mental workload assessment.

References

Aikin, C., Rollings, M. and Wilson, J.R. (1994). Providing a foundation for ergonomics: Systematic Ergonomics in Engineeering Design (SEED). Proceedings of the 12th Congress of the International Ergonomics Association. Toronto, August, **5**, 276–278.

Bailey, R.W. (1989). *Human Performance Engineering: Using Human Factors/Ergonomics to Achieve Computer System Usability,* 2nd edition (Englewood Cliffs, NJ: Prentice Hall).

Barber, P.J. (1988). *Applied Cognitive Psychology* (London: Methuen).

Chapanis, A. (1979). Quo vadis, ergonomia. *Ergonomics,* **22**, 595–605.

Chapanis, A. (1992). To communicate the human factors message you have to know what the message is and how to communicate it: Part 2. *Human Factors Society Bulletin,* **35**/1, 3–6.

Cherns, A. (1987). Principles of sociotechnical design revisited. *Human Relations,* **40**, 153–162.

Christensen, J.M. (1976). Ergonomics: where have we been and where are we going: II. *Ergonomics,* **19**, 287–300.

Christensen, J.M., Topmiller, D.A. and Gill, R.T. (1988). Human factors definitions revisited. *Human Factors Society Bulletin,* **31**, 7–8.

Clark, T.S. and Corlett, E.N. (1984). *The Ergonomics of Workspaces and Machines: A Design Manual* (London: Taylor and Francis).

Clegg, C., Ravden, S., Corbett, M. and Johnson, G. (1989). Allocating functions in computer integrated manufacturing: a review and a new method. *Behaviour and Information Technology,* **8**/3, 175–190.

Clegg, C. and Symon, G. (1989). A review of human-centred manufacturing technology and a framework for its design and evaluation. *International Review of Ergonomics*, **2**, 15–47.

Corlett, E.N. (1991). Some future directions for ergonomics. In *Towards Human Work: Solutions to Problems in Occupational Health and Safety*, edited by M. Kumashiro and E.D. Megaw (London: Taylor and Francis).

Cross, N. (1984). *Developments in Design Methodology* (Chichester: John Wiley).

Denzin, N. (1970). *Sociological Methods* (New York: McGraw-Hill).

Edwards, E. (1973). Techniques for the evaluation of human performance. In *Measurement of Man at Work*, edited by W.T. Singleton, J.G. Fox and D. Whitfield (London: Taylor and Francis), pp. 129–133.

Fuld, R.B. (1993). The fiction of function allocation. *Ergonomics in Design*, January, 20–24.

Grey, S.M., Norris, B.J. and Wilson, J.R. (1987). *Ergonomics in the Electronic Retail Environment* (Slough, UK: ICL (UK) Ltd).

Jones, J.C. (1980). *Design Methods*, 2nd edition (New York: John Wiley).

Jordan, N. (1963). Allocation of functions between man and machines in automated systems. *Journal of Applied Psychology*, **47**, 161–165.

Kantowitz, B.H. (1992). Selecting measures for human factors research. *Human Factors*, **34**, 387–398.

Kirwan, B. and Ainsworth, L. (1992). *A Guide to Task Analysis* (London: Taylor and Francis).

Kragt, H. (1992). *Enhancing Industrial Performance* (London: Taylor and Francis).

Leamon, T.B. (1980). The organisation of industrial ergonomics—a human machine model. *Applied Ergonomics*, **11**, 223–226.

Markus, T.A. (1969). The role of building performance measurement and appraisal in design method. In *Design Methods in Architecture*, edited by G. Broadbent and A. Ward (London: Lund Humphries).

Maslow, A.H. (1946). Problem-centring vs. means-centring in science. *Philosophy of Science*, **13**, 326–331.

Meister, D. (1985). *Behavioural Analysis and Measurement Methods* (Chichester: John Wiley).

Meister, D. (1986). *Human Factors Testing and Evaluation* (Amsterdam: Elsevier Science).

Meister, D. and Enderwick, T.P. (1992). Preface to special issue 'Measurement in Human Factors'. *Human Factors*, **34**, 383–385.

Moray, N. (1993). Technosophy and humane factors. *Ergonomics in Design*, October, 33–37, 39.

National Research Council. (1986). Research Needs for Human Factors. Report of the Committee on Human Factors. Washington, D.C.: National Academy Press.

Sanders, M.S. and McCormick, E.J. (1987). *Human Factors in Engineering and Design*, 6th edition (New York: McGraw-Hill).

Sheridan, T.B. (1991a). New realities of human factors. *Human Factors Society Bulletin*, **34**/2, 1–3.

Sheridan, T.B. (1991b). Methodology for tackling the bigger problems. *Human Factors Society Bulletin*, **34**/12, 2–4.

Singleton, W.T. (1974). *Man-Machine Systems* (Harmondsworth: Penguin).

Singleton, W.T. (Ed.) (1982). *The Body at Work: Biological Ergonomics* (Cambridge: Cambridge University Press).

The Guardian. (1993). Criticised tape start could stay; flagman blamed in panto horse race. June 14th.

Webb, E.J., Campbell, D.T., Schwartz, R.D. and Sechrest, L. (1972). *Unobtrusive Measures: Non-reactive Research in the Social Sciences* (Chicago: Rand McNally).

Welford, A.T. (1976). Ergonomics: where have we been and where are we going: I. *Ergonomics,* **19,** 275–286.

Wickens, C.D. (1984). *Engineering Psychology and Human Performance,* 1st edition (Columbus, OH: Charles E. Merrill).

Williges, R.C., Williges, B.H. and Elkerton, J. (1987). Software interface design. In *Handbook of Human Factors* edited by G. Salvendy (New York: John Wiley).

Wilson, J.R. (1994). Devolving ergonomics: the key to ergonomics management programmes. *Ergonomics,* **37,** 579–594

Part I

General approaches and methods

There are certain general groups of methods, which perhaps we can see as global methods, with which we can obtain insight and information in any of the human sciences. Particularly we can use many techniques of *direct observation* to measure 'directly' and assess behavioural phenomena, and techniques of *indirect observation* to collect data from subjects about their interpretations of what they are doing, feeling or thinking. In more detail, direct observation is the collection of information on subject performance either directly by the observers themselves or from objective recordings of subject behaviour; indirect observation involves the provision of reports of behaviour, attitudes and opinions by the subjects themselves, or by other associated individuals who are not the observer.

The former group of methods is sometimes seen as 'objective' methods, and the latter as 'subjective' methods. In this view it is assumed that direct observation produces 'true' records, with no biased interpretation by the subject whereas subjective assessment (as in the title of Sinclair's chapter 3) implies that the subject or other informant provide abstracted and interpreted information. However, this ignores the fact that data collected by direct observation will be re-analyzed, summarized and abstracted by the observers, from their memory, notes, tape or video recordings. Moreover, properly constructed and carried out, subjective assessment or indirect observation should be run so as to maximize as far as possible the data reliability and validity. In any case, much of psychological measurement is possible only through indirect observation; the phenomena involved cannot be directly observed or even inferred from direct observations.

Both groups of techniques can be used in laboratory or field studies, although only direct observation in the field can be naturalistic and non-interfering with behaviour and even then only if carried out carefully. The degree of interference will in part be determined by whether the observations are made visibly, participatively or unobtrusively. The last of these should provide a record of behaviour which is unaltered through reaction to an observer's presence, but any attempts to carry out such 'hidden' studies must beware a certain inflexibility of coverage which may result and, most importantly, must first entail careful consideration of the ethics of the particular circumstances (addressed in chapters 2 and 38). Of course, ethical issues will depend in part on the use to which data might be put but will also be situation specific; justification for unobtrusive observation may be easier for a study of pedestrian behaviour for instance, than for following an individual industrial worker.

Participatory observation, sometimes used in sociological studies, can allow insight into, and non-interference with, subjects' behaviour but sometimes leads to a 'worm's eye' view being taken. Also, the observer may change over time, identifying (or not) with those observed, and their presence may well be a catalyst for unusual behaviour. Generally then, in laboratory or field, the ergonomist relies upon being a visible observer, yet tries not to interfere with, or contaminate, behaviour by word, action or even mere presence.

The decision of whether to collect observable data directly in real-time, using storage in memory, on paper or event recorder and so on, or by using recording instrumentation such as video or audio tape, will depend upon the circumstances. Task type, behaviour to be observed, site conditions, study resources in terms of people, equipment and time, and the type of measurements or records needed will all be taken into account. Generally hardware recording will give unlimited data capture rates but be more limited in terms of geographical space covered. It will provide a permanent record to be reconsulted to check accuracy but requires considerable time to analyze down to a written record (generally over ten times the 'real' time for video for instance). Also it involves no human problems of data reinterpretation or misinterpretation, attention, learning and so on in the field, but these may be problems during analysis in the laboratory. (The trade-offs implied in these decisions are examined further in the context of protocol analysis in chapter 7.)

In chapter 2, Drury concentrates upon performance observation and largely upon techniques of observation and recording some of which are derived from work study. As ergonomists we have borrowed (and improved upon) much in work measurement and method study but our philosophy or approach is different—fitting tasks to people instead of vice versa which is the case in work measurement at least. Also, we have expanded the focus and methods of observation to embrace all kinds of human behaviour and system response, not just those concerned with effective performance.

In order to truly understand much of human behaviour we must not only observe people but must question them or obtain verbal and written reports from them also. This permits greater insight into what is being done and why. It is also the method by which we assess affective responses—feelings and attitudes. Indirect observation is reported as techniques of subjective assessment by Sinclair in chapter 3; this includes rating, ranking, questionnaires, interviews and checklists. Psychophysical techniques are also available for such measurement.

Ostensibly, indirect observation appears to be a simple approach to data collection. Need to ensure that subjective reports do indeed reflect accurately and completely the target behaviour, thoughts or knowledge though, means that development, administration and analysis must be treated very carefully. As well as minimizing any distortion, omissions or commissions on the part of the subjects, we must ensure as far as possible that the investigator and the method and techniques used do not introduce bias themselves and do not unnecessarily constrain the focus of enquiry.

Before embarking upon any of the extensive and expensive investigations often needed for direct or indirect observation, full use should be made of data that exist already. In chapter 4, Drury identifies several sources of archival data usually available within organizations, 'fixed' records (plans, annual reports, etc), production, industrial engineering, quality control, personnel, medical, and costing records. Other sources may be meetings minutes (e.g., worker-management meetings) or maintenance records.

The beauty of all these archives of course is that someone else has already taken the time and trouble to produce the data; all we have to do is extract and interpret what we need. Therein lies the downside, the dangers that the records themselves will be incomplete, variable over time or situation, selective or biased, or that we might misinterpret or selectively interpret them. Nonetheless, they are a very useful resource for any field study.

We might also include under the heading of archives our store of ergonomics principles, methods, data and criteria contained in textbooks, handbooks, journals and reports. Such literature archives are a necessary resource in any study.

Archival records from the organization will be used as part of field studies. However, the other two method groups can be applied in field or laboratory situations, in the case of the latter in the context of an experimental design. The ways of doing this, the factors to be taken into account and procedures to be followed, are laid out by Drury in chapter 5. We might prefer field studies for their greater realism and thus potentially their validity and generalizability, but the production of reliable and usable data will often necessitate use of the controlled laboratory environment. So that such studies and experiments do not end up as costly, sterile exercises, great care must be taken in their design and conduct.

A final word for this introduction, and a note for the whole book, is provided in Sinclair's conclusion to chapter 3 where he points out that although skilful use of methods requires practice we should beware of sticking rigidly to standard methods. Imagination, as well as scientific rigour, is an ingredient of successful ergonomics.

Chapter 2

Methods for direct observation of performance

Colin G. Drury

Introduction

When we collect data from the operational system by observing the system without changing it, we have an observational study. Such studies are classified in chapter 5 as having a high degree of face validity but a low degree of experimental control. For example, Cohen and Jensen (1984) had observers in two warehouses noting the behaviour of fork-lift truck drivers. In a different context, Schiro and Drury (1981) had emergency physicians observe the behaviour and performance of ambulance attendants in bringing patients into an Emergency Department. In both cases, the system (warehouse, hospital) was functioning normally except for the presence of an observer. Hence the face validity of the result was high because actual live events were recorded. But experimental control was very low as the experimenters did not change the system under study, or at least not at the time of the first observations. Causality is impossible to establish in such circumstances, but if the goal of an observational study is not to infer causality, then observation of behaviour is an appropriate technique. Issues of data contamination and inadvertent editing by choice of observed unit will be discussed later in this chapter when common concerns of all the observation methods are raised.

Initial observation of a system

Merely walking around an operating human/machine system to 'see what goes on' is hardly an observational method worthy of the name, yet it is an indispensable first step. The ergonomist needs to combine archival data (see chapter 4), such as plant layouts and job descriptions, with broad on-site observation to ensure that the objectives and interrelationships of the system are understood before any more formal investigation takes place. People's jobs are never simple, despite first appearances. There are subtleties of part placement, incoming quality variations and model changes which change the operation of even the most repetitive manual task. Skills developed by operators over months and years make assembly tasks look simple when they are not. The ergonomist should talk to operators, trainers and trainees to obtain a feel for what is important. If necessary, and safe, the job should be tried out. It is only by obtaining a thorough understanding of the complexities of a job or department that the ergonomist can make the main study reveal anything beyond the obvious. Remember that ergonomic insight comes from the interpretation of the system in terms of models and concepts of human functioning, so that a rapid 'walkthrough' will only yield the most simplistic insights.

Observation methods are useful in collecting not only quantitative data (e.g., parts per hour, errors per part), but also in collecting qualitative data on product routings, task sequences, causes of delays and error taxonomies (see chapter 1 also). In this latter context, an observational study is very close to a task analysis of an existing system, so that many of the concepts and methods of task analysis apply equally well to observation methods (see chapter 6) and vice versa.

Data collection methods classification

The individual methods will be described briefly in this chapter, with references to more complete sources, as each method can easily fill a chapter (Chapanis, 1953) or a book (Hansen, 1960). To help the ergonomist choose appropriately, the methods are classified by the type of data they produce. Observation of a system is necessarily an abstraction as there is too much detail available to record everything which happens. What is abstracted depends upon how the data are to be utilized. The typical charting methods which represent one outcome of an observational study can be classified by what they abstract from the raw data set, but first the raw data set needs to be defined.

Raw observation data are either of events or states. An event is an observable occurrence at a particular point in time and is usually recorded with its associated time. Thus in an observational study of queue formation in a bank, an event would be a customer arriving, a customer leaving the queue to start service, or leaving the teller after service. Each event would usually have an associated time (see Table 2.1).

Event/time records can be analysed further to provide information on sequence of events, duration of events, spatial movements or frequency of events.

The other method of data collection is to record system states at pre-specified times. A system state is a specified set of system conditions, typically

Table 2.1. Events and associated times in an observational study of queue formation at a bank

Event	Time (p.m.)
Customer arrival to queue	1:24:03
Customer leaves teller No. 3	1:24:15
Customer from queue to teller No. 3	1:25:20
Customer arrival to queue	1:25:38
Customer arrival to queue	1:26:01

constant between events. Thus to continue the previous example, system states will specify which teller positions are occupied, which occupied positions are serving customers, and how many customers are in the queue. If observations are taken at two-minute intervals, we may have the record shown in Table 2.2.

What is shown is that positions 1 and 3 are serving customers (S = serving) while 2 and 4 have no tellers (E = empty). As the queue reduces to zero, tellers 3 and then 1 are still there (O = occupied) but are not serving customers. The example shown only has one event happening between each pair of observations, but this is not typically the case.

Note that in the event/time record, data collection is event driven so that each event is recorded when it takes place. For the time/state record, data collection is time driven, with states being recorded on a predefined schedule (which need not necessarily be an equal time interval schedule). Note also that if a starting state is defined, then an event/time record allows the deduction of system states at any time during the observation period. The reverse is not true, as between state observation times many changes could have taken place. If it is wished to deduce the system state progression from the event/time record, it would be as well to allow for multiple recordings of system state as a check on the accuracy of recorded data.

Having collected the raw data, reduction and analysis are required to meet

Table 2.2. A record of system states in queue formation at a bank

		State			
		Teller positions			
Time (p.m.)	Queue length	1	2	3	4
3:00:00	2	S	E	S	E
3:02:00	3	S	E	S	E
3:04:00	2	S	E	S	E
3:06:00	1	S	E	S	E
3:08:00	0	S	E	O	E
3:10:00	0	O	E	O	E

the study objectives. Thus a time study would require means and standard deviations of element times to be calculated, while in an occurrence sampling study the relative frequencies of system states would need to be calculated. Some abstraction takes place on raw data collection, but more comes from data reduction and analysis. Each of the techniques preserves certain information while ignoring other aspects. Five types of information can be recognized.

1. *Sequence of activities:* which particular activity follows another activity, by the operator or by the item processed. For a fixed-sequence operation, there will only be one answer, but in more varied situations we need ways to summarize the sequences obtained.
2. *Duration of activities:* the length of time an activity takes, with statistical summaries as appropriate if the time is indeed a random variable.
3. *Frequency of activities:* how often an event occurs, e.g., how many breakdowns per day on a wave-solder machine.
4. *Fraction of time spent in states:* the fraction of time (of a person, machine, or work unit) spent in a particular state or activity.
5. *Spatial movement:* where a person, machine or work unit moves to and from during the daily activity.

These five types of information are either preserved (P), can be calculated with outside information (C), or are lost by each of the observational techniques described, as shown in Table 2.3. Note that frequency of occurrence of an *event* is only preserved in the raw event/time records. Frequency of occurrence of a *state* in occurrence sampling should not be confused with event frequency.

Each of the methods will be presented in turn, followed by some general considerations of observational studies. Note that more extensive coverages of the methods of use to engineers are available elsewhere (e.g., Kadota, 1982).

Table 2.3. Information preserved (P) for each observational method; C indicates that additional data can be incorporated to calculate the information

Information type	Observational method						
	Time study	Process chart	Process flow chart	Gantt chart	Multiple activity chart	Link chart	Occurrence sampling
Sequence		P	P	P	P	P	
Duration	P	C_1	C_1	P	P		C_2
Frequency							P
Fraction							P
Spatial		P				P	

C_1, durations on annotated process chart.
C_2, durations calculable from production records and occurrence sampling data.

Descriptions of the methods

Raw event/time records

Even the raw data themselves comprise, at times, a useful final form. Detailed accident and critical incident investigations for aircraft (e.g., by the FAA) are based directly on the time history of a single sequence of events. The time history is obtained from recordings of cockpit/air traffic control communications, radar plots, the flight data recorder and eye-witness accounts. These reports are often models of careful investigation and analysis, well repaying detailed study by any ergonomists studying errors or accidents (Figure 2.1). (See also the protocols in chapter 7.)

If time is eliminated from the event/time record, it is possible to count event frequencies. Thus we can record use/non-use of seat belts in automobiles (Matthews, 1982), non-work related behaviours (Salvendy *et al.*, 1984) or law-observance behaviour of bicyclists (Drury, 1978).

Time	IAS(kts)	Cockpit Recorder	Event
19:57:35	0	To ATC: Ok, is 63 clear to go?	
:42	30	ATC: Ok, cleared for takeoff,	
:47	30	good day	
:55	40	We all set?	
19:58:10	95	Eighty	
:20	110		
:22	95	Hang on guys	
:25	100	Lost all our airspeed	
:30	125	V-1, Rotate	
:31	130		Nose gear unloads
:37	140	Gonna get the damn power. . .	
:38	130		Power wire impact
:41	150	Keep it going, keep it going Tommy	
:42	145	Ok, you're clear out there just keep the airspeed	Full power on
:47	165	Ok, declare an emergency	
:48	165	To ATC: Ok, 63, we just got the wires and we're gonna be airborne	
:55		To ATC: We're gonna make it	

Figure 2.1. Extract from report of aircraft N32725, Tucson, AR, June 3, 1977, showing the effect of severe windshear from a 'dry microburst'. Adapted from NTSB report. IAS is indicated air speed

Time study

Any abstraction of a raw event/time record which produces statistical data on activity times could be called a time study, for example, a frame-by-frame film analysis of walking or manual lifting. Traditionally, though, time study has concerned itself most with finding representative times for repetitive activities. The standard reference books (e.g., Barnes, 1980; Mundel, 1978) provide overwhelming levels of detail on a method which has a history going back to the turn of the century. More evaluative presentations are also available, both in an ergonomics context (e.g., Konz, 1983: Ch 21–25) and from an engineering viewpoint (e.g., Salvendy, 1982: Ch 4.1–4.9).

For repetitive tasks, with multiple elements following each other regularly from cycle to cycle, the observation record of events and times is usually not a simple list like the earlier bank example but a matrix with rows representing cycles and columns representing elements. Figure 2.2 shows a typical form. Activities or elements are defined to occur between events. Thus the element 'get washer' might be defined to start at the event 'touch washer' and end at the event 'washer touches bolt'. Event definitions are not recorded on the observation sheet, only activity names. Any activity which occurs out of sequence (e.g., dropping a washer or answering a supervisor's questions) is dealt with by annotation as a 'foreign' element.

Analysis is performed to obtain mean values of element times and, less often, their standard deviations. If a simple event/time record is kept, statistics for element times can be recovered by calculating means and standard deviations for every event transition possibility. A repetitive task, even with some foreign elements, will only create a small number of different event transitions. If a computer program is used to record the events and times, then a simple program can be used to find the sample statistics for each transition. Figure 2.3 shows an event recording menu, the raw event/time record, and event-transitions output for a repetitive task using the programs in Drury (1987).

As noted in chapter 5, industrial time studies use two modifications to the representative time (usually the mean time but neglecting 'outliers') to arrive at a *standard time* for a task. Both modifications, rating and allowances, have no value for ergonomists, although we all need to be aware of them when using archival time study data.

Process charts

A *process chart* is nothing more than a plant (or office, etc.) layout with the materials movement for one or more processes marked on it. The materials movement can be a single line (as in a simple assembly line onto a base unit), multiple converging lines (as in a progressive assembly line with sub-assemblies) or even a set of quite disparate flows (as in a job shop). Spatial and sequence information are both preserved, while timing and workload

Time Study Observation Sheet

Department   Final Assembly

Operation   Mount Norwegian Blue

Workplace   Bench 23A

Operator   A. Notlob   **Analyst**   Bolton, A.

Part Nos   221657, 273318

Start Time   11:08 a.m.   **Start Date**   4/1/88

Element	1	2	3	4	5	6	7	8	9	10	Rem
1. NB from box to rail	.17	.16	.16	.17	.15	.16	.17	.15	.16	.17	
2. P U Nails	.23	.24	.23	.23	.22	.24	.24	.24	.24	.23	
3. Nail L	.14	.14	.14	.13	.14	.13	*.78	.17	.14	.13	*Dropped
4. Nail R	.13	.14	.14	.13	.13	.14	.55	.13	.14	.28	
5. Adjust to spec	.35	.36	.41	.30	.40	.51	.58	.30	.27	.31	
6. Rail + NB to cage	.41	.38	.42	.43	.34	.41	.50	.41	.39	.40	
7.											
8.											
9.											
10.											
11.											
12.											
TOTAL	1.43	1.42	1.50	1.40	1.43	1.59	2.82	1.40	1.34	1.52	

Selected Time   1.448   **Rating**   90

Normal Time   1.303   **Allowances**   10%

Standard Time   1.537 min

Figure 2.2. Time study observation sheet example

information is lost. At times, activity duration information can be superimposed to form an *annotated process chart*.

The main use of process charts is to provide a graphical indication of movement. They are a useful aid to visualizing the effects of changes in plant layout as they dramatize the often inefficient pattern of movements required to process a part from raw material to finished product. Figure 2.4 shows an

General approaches and methods

(a) Computer on-screen menu

1 Pick up wrench   6 Nut 1 to bolt 2

2 Wrench to nut 1   7 Start nut 1

3 Tighten nut 1

4 Wrench to bench

5 Pick up nut 1

TO STOP PRESS 0

PRESS ANY KEY TO GO

(b) Data set

598	1	959	2	1333	3	1583	4	1956	5
2282	6	2632	7	3041	1	3261	2	3491	3
3723	4	4268	5	4400	6	4699	7	5137	1
5564	2	5650	3	6109	4	6497	5	6740	6
7034	7	7563	1	7849	2	8076	6	8381	3
8705	4	8984	5	9399	6	9727	7	9825	1
10165	2	10440	3	10728	4	11114	5	11459	6
11724	7	11812	0						

(c) Data summarization

TIMES BETWEEN EVENTS IN 1/100 SEC UNITS

	N	SUM	SUM SQ	MEAN	STD.DEV
FROM 0 TO 1	1	598	357604	598	0
FROM 1 TO 2	5	1644	567086	328	81
FROM 2 TO 3	4	955	274177	238	124
FROM 2 TO 6	1	227	51529	227	0
FROM 3 TO 4	5	1553	514925	310	90
FROM 4 TO 5	5	1971	813535	394	95
FROM 5 TO 6	5	1461	473999	292	108
FROM 6 TO 3	1	305	93025	305	0
FROM 6 TO 7	5	1536	476146	307	32
FROM 7 TO 1	4	1474	648570	368	187

Figure 2.3. Computer-based data collection and summarization in time study (times in 0·01 sec. units)

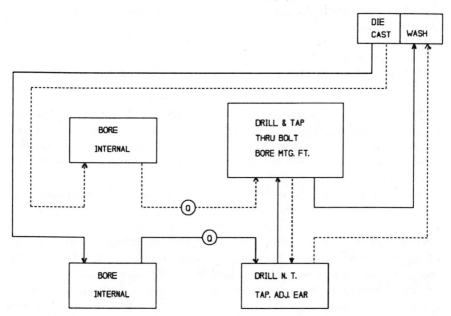

Figure 2.4. Process flow in a manufacturing cell, showing routes of two components

example of such a product flow change associated with the introduction of a 'just-in-time' manufacturing cell. If the process chart is only produced as a visual aid in presentations, an even more dramatic alternative is to videotape a part as it flows through the system. Before and after comparisons using this technique can be striking, even if the same information could be presented by simple numerical comparisons of time or distance moved.

A process chart is an ideal first technique in many studies as it allows the analyst to observe the system, ask questions, and talk to supervisors and operators in a non-threatening manner. It also serves a technical purpose in that it brings to light rework, backward flows, work in process storage areas, and multiple handling.

Note also that a process chart can be used to follow the progress of other entities apart from products. Continuous material flows (liquids, gases), movement of patients and staff in hospitals, and even information flows within a computer network are all legitimate process charting examples.

Flow process chart

A *flow chart*, or in its more restricted use, a *flow process chart*, removes the spatial information from a process chart and in its place uses conventional symbols to indicate particular aspects of the flow. The symbols are as follows:

○ *Operation,* where an object is intentionally changed (or □ can be used).

⇐ *Transport,* where an object is moved and the movement is not part of an operation

◇ *Inspection,* where an object is examined or tested for quality, quantity or identification.

D *Delay,* where an object waits for the next planned action.

△ *Storage,* where an object is stored for later use.

There are variations. As with a process chart, the flow process chart can refer to humans or information as easily as to objects being processed. Process symbols can be annotated with numbers to represent comments or the geographical location of activities, or with times to produce an *annotated flow process chart.*

Because they remove spatial location information, flow process charts can emphasize the logical structure of operations. They are useful for highlighting *delay and storage* elements which are expensive and in many ways the anathema of modern manufacturing. One aspect they do enhance is the structure of multi-component and branching product flows. Two branches flow together when a sub-assembly is added to a main assembly; two branches diverge whenever a choice is made at an inspection activity. Because of this ability to abstract the logic from a situation, the flow process chart has been adapted to many other areas such as computer programming (Phillips *et al.,* 1988), task analysis of complex operations (Drury *et al.,* 1987) and even signal flow graphs used in process control task analyses (Edwards and Lees, 1974). The same principles are used in charting precedence diagrams into networks. An example is the familiar PERT chart for project planning, a chart which is now available on even microcomputer project planning and control software.

In addition to their use as direct aids to visualizing a system, flow process charts for multiple product flows form the basic input data to link charts and from/to matrices (see later in this chapter). Figure 2.5 shows a flow process chart for an operation in assembly of automotive thin-film ignition components.

Gantt charts

Named after its inventor in the 1920s, the *Gantt chart* is a graphical depiction of the time relationships among several activities. Figure 2.6 gives an example from a project plan. The time axis may run vertically down or horizontally to the right. There are as many bars on the graph as there are activities to be depicted, with bars starting and ending at the appropriate times. Although the chart in Figure 2.6 has a time scale covering weeks, a Gantt chart can be used for any operation with multiple activities such as machinery maintenance or even a surgical procedure in the operating room.

The primary use of the Gantt chart is in visualizing several related activities. It is used extensively in scheduling (vehicles, orders in factories) and project planning. Microcomputer-based project planning packages use Gantt charts

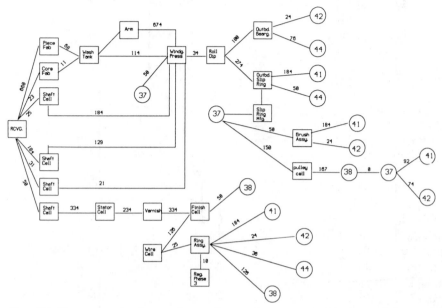

Figure 2.5. Flow process chart for manufacture of automotive components. Numbers on links represent flow volumes

as a principal output. Even if the ergonomist never has occasion to plot a Gantt chart for a system under study, it is an essential tool (with a PERT network) for ensuring that ergonomics projects are completed on time.

Multiple activity charts

The concept of the Gantt chart is modified somewhat for the *multiple activity chart*. Instead of each bar representing a single activity, the activities are grouped into continuous bars. Thus a single bar may represent the activities of a machine operator, while a second bar represents the status of the machine (e.g., off, operating, standby). Different shadings are used on the bar to represent the different activities or states. Figure 2.7 shows a multiple activity chart for the operation of a just-in-time cell. Front and rear housings are machined on six machines. One machine (column 1) operates on pairs (two fronts and two rears), two other machines operate on two rears while three deal with two fronts. The columns OP1 and OP2 show the two operators, with black representing times at which the operator is loading/unloading a machine (L) or gauging parts produced (G). Machine cycles either require the operator (O) or are unattended (CYCLE). Figure 2.7 was one of a series to determine throughput and operator workloads under different configurations of machines and operators in the cell.

A multiple activity chart is a general purpose tool. It can be annotated

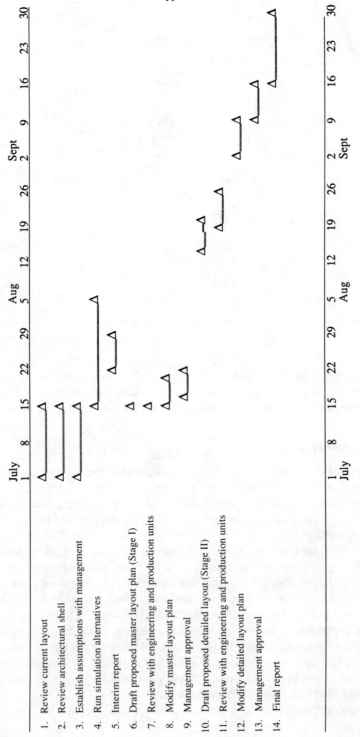

Figure 2.6. Gantt chart for design of a new facility

REARS AND DOUBLE   FRONTS

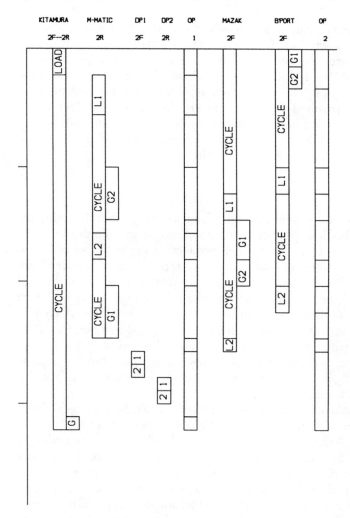

Figure 2.7. Multiple activity chart for a manufacturing cell. Time goes down the vertical axis

with exact times, comments and notes. The standard flow-charting symbols can be used in place of shading. The specific form shown in Figure 2.7 is known as the *man/machine chart* in industrial engineering texts (e.g., Salvendy, 1982). A common variant on a smaller scale is the *left hand/right hand chart* (e.g., Konz, 1983). Here the two bars represent the two hands of an operator in an assembly task, showing which activities are being performed and the co-ordination required. In a more obviously ergonomics context, similar charts have been devised which have separate bars for cognitive and information processing activities. Crossman (1956) proposed the *sensori-motor process chart,* which has seen some use in the field of manual skills training (e.g., Seymour, 1967).

A multiple activity chart is a uniquely useful way to visualize the co-ordination of several activities. It shows conflicts (where a resource is needed in two places at once) and idle time (where a resource is waiting for a prior operation to finish). We should note that 'idle time' does not mean that the operator is doing nothing. Most tasks have planning and other cognitive acts which are no less required for not being visually obvious. In Figure 2.7, blocks of time are needed for preventive maintenance, machine monitoring, and communications with other parts of the plant, even if these are not activities built into every cycle of operations.

Link charts

When multiple products flow through a system using multiple paths, the resulting process or flow process charts can become so complex as to be unreadable. In these cases, the separate product identities are sacrificed to preserve only the information on how many products flow between each pair of machines. The data can be recorded as a table with identical row and column labels, one for each machine, called a *from–to chart.* In such a chart, the number in each cell represents either the number of products which flow between the machines of row and column, or a weighted number, for example, total trays of parts or total kilograms of parts. For any strictly repetitive task, only one entry will appear in each row and column, showing a single flow path. A totally random flow would have approximately equal weights or numbers in each cell.

Any from–to matrix can be equally well represented by a network, with the row and column labels as nodes and the cell entries as links between the nodes. This is a *link chart.* In most applications, the flows between nodes are represented either by numbers appended to each link or by a code of heavier link lines for larger flows. Figure 2.8 shows a classic example, from Chapanis (1953) of human and machine communications in a battle cruiser during air attack. The matrix contains exactly the same information as the first diagram, but it is much easier to visualize the relationships in the link chart. The second link chart demonstrates the reduced flow lengths and reduced path interference when the layout was revised.

As with other charting methods, there are variations. Spatial information can be removed to create a multi-product analogue of the flow process chart, even to using the standard symbols. Annotation can indicate unusual circumstances for each machine or link, e.g., different link usage under start-up and operating conditions. The same analysis and charting can be applied to sequences other than product flows between machines. Thus control use sequences on a control panel can be made into diagrams in this way. Figure

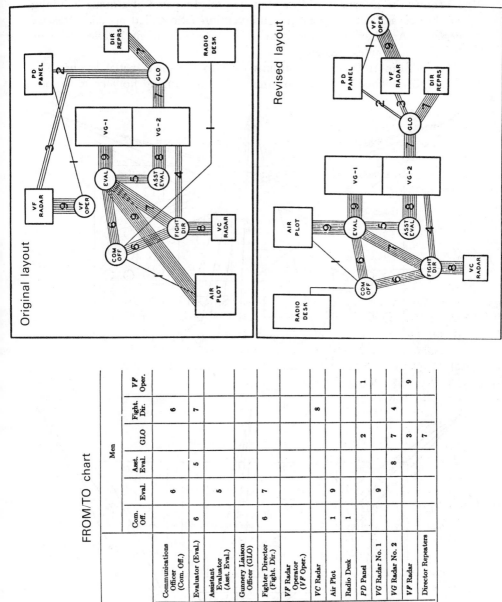

Figure 2.8. Link charts for a battle cruiser under air attack, from Chapanis (1953). The first chart represents the original layout, the FROM/TO chart extracts the movement information, and the second chart shows a revised layout with reduced congestion

2.9 shows the information flow sequence on the panel of a metal stamping line before and after ergonomic intervention (also called *operation sequence diagrams*). Another classic use is to record eye movements between instruments in a complex display.

For a from–to matrix, algorithms exist to optimize spatial layout. The science of *facilities layout* is concerned with minimizing flow costs (among

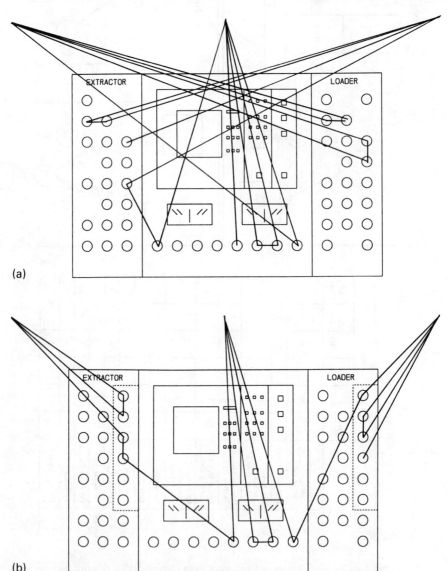

Figure 2.9. (a) Hand/eye movement chart of a control panel for a stamping operation. (b) Chart for revised panel layout where outer panels have been laterally reversed and sub-panels installed for major sequential task

other costs) by locating facilities optimally on a one–, two– or three–dimensional space. Simple computer programs (e.g., CRAFT, CORLAP) exist to improve plant layouts (Tomkins, 1982). Any of these can be applied to the control panel layout problem to minimize movement between controls.

Occurrence sampling

The idea of sampling system state at predetermined intervals rather than timing the instant of occurrence of events goes back to Tippett (1934), who wished to estimate the fraction of time an operator or a machine was working. Known earlier as *ratio delay* or *work sampling,* the method of *occurrence sampling* is thus qualitatively different from the others presented so far.

In occurrence sampling, the analyst observes the system at predetermined times. To perform the study we need to determine both the observation times and the system states or categories which we will record. Choosing the times means choosing the total number of observations (see later in this chapter), the time intervals between observations, and the statistical distribution of observations over time.

System states must be observable, unambiguous, mutually exclusive, and exhaustive. The analyst needs to understand the system well to develop a rational and useful set of categories. Observable states means that the analyst needs only to record what is seen, not what is inferred. Thus 'idle', 'thinking', 'planning', and 'daydreaming' all look very similar to the observer. A safer state to record would be 'no task seen' as it is both observable and unambiguous. Mutually exclusive means that on any single observation only a single response is possible so that states must be defined such that two cannot happen simultaneously. Thus for a secretary, it is entirely possible to be on the computer while answering the telephone. The analyst must record only what is being done at the instant of recording—a key being pressed (computer use) or a word being spoken (telephone). Exhaustive means that ANY system state can be recorded. The better you know the system, the better you can anticipate all possible system states. An 'other' category is always a safe solution, although if it captures more than an isolated reading or two, more named states should be added to the study.

The choice of number of readings will be considered later but the time interval between readings is an important variable. If observations are too close to each other in time, an activity may continue from one reading to the next. This introduces sequential dependencies into the data, destroying the assumption of independent observations. Thus readings should be far enough apart in time to ensure independence, a stricture which requires at least some foreknowledge of the expected durations of activities and states. On the other hand, if observations are too widely separated, the total study duration can become so long that changes (e.g., in product mix or workload) take place during the study, making the derived probabilities meaningless.

The other choice to be made is of the observation schedule—should it be random or fixed-interval? A fixed-interval schedule is attractive in that it allows the analyst opportunity to interleave another activity with the occurrence sampling study. However, if there is any natural periodicity in the system being observed, then there is a risk that each observation will coincide with a certain phase of the cycle, leading to highly biased data. The alternative is randomization and is always safer. If in doubt, randomize.

Having decided what to observe and when, the recording sheet needs to be designed to reduce recording errors. Such sheets should have a list of all system states at the top and leave a place for recording the code for each state against a predetermined time printed on the sheet. The alternative way of recording is to have columns representing the states and rows representing successive observation times. Here a check mark is placed in the appropriate column of each row in turn. The use of tally marks against system states is not recommended as it does not prompt the analyst with the time for each reading, leading to missing data.

A simple computer program is available for portable computers to conduct simple occurrence sampling studies (Drury, 1987). Figure 2.10 shows an example of the on-screen menu and the data file for a secretarial job. In setting up the study, the user is prompted for the total number of observations and the total study duration. The program then generates random times to give the correct number of observations in the given duration, providing an audible prompt when each observation is due.

It should be noted that inexpensive video recorders now allow the analyst to film the system throughout the study period unattended and then to perform the occurrence sampling study on the recording. This means that the

(a) Computer on-screen menu

1 Writing   7 Reference use
2 Typing   8 No task seen
3 Computer use   9 Absent
4 Telephone
5 File cabinet
6 Walk-in queries
A "beep" will tell you when to enter
each observation. PRESS ANY KEY TO START

(b) Data file

    1 Writing 1
    2 Typing 3
    3 Computer use 6
    4 Telephone 5
    5 File cabinet 2
    6 Walk-in queries 3
    7 Reference use 1
    8 No task seen 3
    9 Absent 1
total = 25

Figure 2.10. Example of computer-based occurrence sampling study for a secretary

study can take up much less of the analyst's time but only where analysis is relatively simple—see chapter 9 for discussion of analysis time to recording time ratios.

An occurrence sampling study only provides information on the fraction of time the system (human or machine) spends in the various states. Time and sequence information is lost. It is thus uniquely useful in systems with no repetitive cycles (e.g., maintenance), or with too many different repetitive cycles to conveniently keep track of (e.g., an attendant serving several looms). There have been uses of occurrence sampling to determine times for repetitive jobs by adding more information. Thus if 74% of the time is spent operating a machine, which produced 1532 parts in a 40-h week, the operating time per part is

$$\text{operating time/part} = \frac{40 \times 0\cdot74}{1532} = 0\cdot019 \text{ h}$$

or 1·16 min/part. There are easier ways to perform time studies, but this way has high face validity.

General considerations

Germane to all of the methods described above are issues of who or what to observe, how often to observe and how to define the observation events.

Choice of subjects and conditions

Choosing representative subjects and time periods in an observation study is difficult. If we choose 'normal' subjects and periods (however defined), we miss many of the unusual conditions which are such a strength of methods involving direct observation of a system in its natural state. However, if we want to predict future performance, we may want to avoid New Year's Day or an untrained operator, or the time when the machine shop was flooded after a thunderstorm. The general considerations of sampling (such as random, stratified and clustered) which are usually discussed with reference to questionnaire surveys (e.g., Sinclair, 1975 and chapter 3 in this book) apply equally well here. Indeed, they are explicit considerations in modern treatments of occurrence sampling (Richardson and Pape, 1982).

It may be preferable to perform different separate studies under different system conditions for many reasons. Not only would the final result be applicable more widely, but even such details as choice of events and system states may be different between conditions. The events appropriate to running a power station are quite different from those encountered in start-up or shutdown. A data recording technique to cover all conditions would be difficult to devise without it becoming unwieldy.

Ethical observation

Ethical considerations need to be explicitly addressed (see also chapter 37). We are observing an operating system and, if it is possible that the data collected may harm the participants, proper safeguards are needed. It should not be possible to tie particular data to individual subjects. This may be reasonably simple where the data recording is done by the ergonomist, but is very difficult to ensure with direct computer recording or videotaping. The workforce is now alerted to the potential misuse of keystroke recording in office tasks and timing of manual operations with video, but they also need to be informed of the objectives of your study and how you intend to protect their rights and privacy. For example, the author used videotaping of traders at a major stock exchange (Drury, 1980) and then interviewed each trader while the tape was playing. Traders were assured that the tapes would be erased after use, and this assurance had to be honoured when management realized that the tapes would make a fine training tool; if a training tool was needed, new tapes should be produced with the specific permission of the subjects.

Sample size determination

The total number of observations required in a study is a trade-off between accuracy (or risk of drawing the wrong conclusions) and the cost of collecting and analyzing the data. Accuracy can be expressed as a confidence interval within which the true value of a measured statistic should lie with a given probability. Thus we could say that the duration of a task should be estimated so that there is a 95% chance of the true mean time lying within ± 10 s of the calculated sample mean. To calculate sample size, we need to know the statistical properties of the process generating the data, the width of the confidence interval, and the probability of the interval containing the true value.

All of the methods described above produce either a time or a fraction, and hence we need to understand distributions of times and proportions. For estimating times, we typically assume that they are normally distributed, although log-normal would often be a closer approximation for the skilled operator (Dudley, 1968). If we assume normality, then we can calculate the number of readings, N, from

$$N = \left(\frac{2\, z_{\alpha/2}\, \sigma}{A} \right)^2 \tag{1}$$

where: $z_{\alpha/2}$ is the normal deviate corresponding to a confidence interval including a fraction $(1 - \alpha)$; σ is the population standard deviation of performance times; and A is the absolute size of the confidence interval in time units.

Thus if we wish to find the time for an operation where the 95% confidence interval is 10 s and the population standard deviation is 20 s, we have

$$z_{\alpha/2} = z_{0.025} = 1.96$$

$$\sigma = 20 \text{ s}$$

$$A = 10 \text{ s}$$

$$\therefore N = \left(\frac{2 \times 1.96 \times 20}{10}\right)^2 = 61.5.$$

Thus 62 readings will be needed.

For occurrence sampling, a fraction or proportion is to be estimated, thus we use the binomial distribution. For large sample sizes and proportions not too close to 0.0 or 1.0, we can use a normal approximation to the binomial and thus use a similar equation to that used for times.

The standard deviation (σ) of a binomial proportion (p) is given by

$$\sigma^2 = \frac{p\,(1-p)}{N}$$

so we have $N = \dfrac{2^2 p(1-p)\, z_{\alpha/2}^2}{A^2}$ (2)

where z, σ and A retain their previous definitions.

Thus to find the number of observations required to estimate a proportion to an accuracy of 0.02 with a probability of 0.90, when the expected proportion is 0.75, we have

$$z_{\alpha/2} = z_{0.05} = 1.645$$

$$p = 0.75$$

$$A = 0.02$$

hence $N = (0.75)(0.25)\dfrac{(2 \times 1.645)^2}{0.02^2} = 5073.8$

or 5074 observations.

Most texts will give alternative formulae for relative intervals (where A is to be a percentage of a mean), and different definitions of α, but the ones presented here cause the minimum confusion. Confidence interval length, A, is always the total length measured in the correct units (seconds or proportion), with no confusion about plus-or-minus intervals or definition of percentages.

It remains to ask where A, α, σ and p come from. The statistical parameters, A and α, are supposed to be given by the sponsors of the study but I have never found this to be the case in industry! In epidemiological work for government, one may be dealing with sponsors sufficiently versed in statistics for this to be possible, but in most industrial situations the sponsor is more concerned with the cost, which is proportional to N, than with details of precision. The ergonomist must be prepared to advise on sensible cost/accuracy trade-offs. Even a short BASIC or FORTRAN program to find N given A, α, σ (or p) can be useful in demonstrating how small increases in precision can have a large effect on study time and cost.

Estimates of σ and/or p are more difficult to obtain. The typical advice is to run a pilot test, which is always a good idea, but it does not help in costing and time estimation for studies on a new system. For research questions involving laboratory data, one can obtain gross estimates from the published literature, but in a working system consensus guesswork may be the only practical solution.

Choice of method

The variety of what to observe and how to observe it can leave the ergonomist either bewildered or practising tunnel vision to keep the number of alternatives manageable. An obvious way out of this embarrassing position is to return to the study objectives for guidance after the initial observation of the system. What are the main issues of interest—performance or well-being? If performance, how can simultaneous sense be made of speed and accuracy? Which of the methods are truly non-reactive in this system? (See chapter 4.)

Perhaps the most telling are outcome measures, the final results of system performance. Was the mission completed? What was production/quality for the month of March? Did the plant manage to remain in business? Unfortunately, a system rarely stays still while ergonomics interventions and measurements take place, so that such global measures are difficult to tie logically to the ergonomist's work. More often we rely on process measures, i.e., those which can be logically related both to ergonomics interventions and to final outcomes. Examples are error rates and productivity on specific jobs, improved process flow or reduced time spent waiting for work. These are certainly more comfortable for the ergonomist but need to be related directly to study goals before the study starts. It is important that the client agrees to the measures and their interpretation in mission terms if later problems of meaning are to be avoided.

Meaning is the key. The purpose of any study is the interpretation rather than the data themselves. In a scientific journal, uninterpreted data are (rightly) difficult to publish. Equally, in an applied study, unless the data contribute directly to the *client*'s goals, the study is sterile. As with anything an ergonomist does, it is the ergonomist's responsibility to make this interpretation. Hence there is no substitute for the ergonomist's knowledge

and understanding of both the system under study and the ergonomics literature.

References

Barnes, R.M. (1980). *Motion and Time Study: Design and Measurement of Work,*
7th edition (New York: John Wiley).
Chapanis, A. (1953). *Research Techniques in Human Engineering* (Baltimore,
MD: The Johns Hopkins Press).
Cohen, H.H. and Jensen, R.C. (1984). Measuring the effectiveness of an
industrial lift truck safety training program. *Journal of Safety Research,* **15,**
125–135.
Crossman, E.R.F.W. (1956). Perceptual activity in manual work. *Research,*
9, 42–49.
Drury, C.G. (1978). The law and bicycle safety. *Traffic Quarterly,* **32,** 599–
620.
Drury, C.G. (1980). Task analysis methods in industry. *Applied Ergonomics,*
14, 19–28.
Drury, C.G. (1987). Hand-held computers for ergonomics data collection.
Applied Ergonomics, **18,** 90–94.
Drury, C.G., Paramore, B., Van Cott, H.P., Grey, S.M. and Corlett, E.N.
(1987). Task analysis. In *Handbook of Human Factors,* edited by G. Sal-
vendy (New York: John Wiley), pp. 371–401.
Dudley, N.A. (1968). *Work Measurement: Some Research Studies* (London:
Macmillan).
Edwards, E. and Lees, F.P. (1974). *The Human Operator in Process Control*
(London: Taylor and Francis).
Hansen, B.H. (1960). *Work Sampling* (Englewood Cliffs, NJ: Prentice-Hall).
Kadota, T. (1982). Charting Techniques. In *Handbook of Industrial Engineering*
edited by G. Salvendy (New York: John Wiley).
Konz, S. (1983). *Work Design: Industrial Ergonomics* (Columbus, OH: Grid).
Matthews, M. (1982). Seat belt use in Ontario four years after mandatory
legislation. *Accident Analysis and Prevention,* **14,** 431–438.
Mundel, M.E. (1978). *Motion and Time Study,* 5th edition (Englewood Cliffs,
NJ: Prentice Hall).
Phillips, M.D., Bashinski, H.S., Ammerman, H.L. and Fligg, C.M. (1988).
A task analytic approach to dialogue design. In *Handbook of Human
Computer Interaction,* edited by M. Helander (Amsterdam: North-
Holland).
Richardson, W.J. and Pape, E.S. (1982). Work sampling. In *Handbook of
Industrial Engineering,* edited by G. Salvendy (New York: John Wiley).
Salvendy, G. (1982). *Handbook of Industrial Engineering* (New York: John
Wiley).
Salvendy, G., McCabe, G.P., Souminen, S.and Basila, B. (1984). Non-work
related movements in machine-paced and self-paced work: an industrial
study. *Applied Ergonomics,* **15,** 21–24.
Schiro, S.G. and Drury, C.G. (1981). Emergency medical communications:

an ergonomic evaluation. In *Case Studies in Ergonomics Practice,* Volume 2, edited by H. Maule (London: Taylor and Francis), pp. 65–81.

Seymour, W.D. (1967). *Industrial Skills* (London: Pitman).

Sinclair, M.A. (1975). Questionnaire design. *Applied Ergonomics,* **6**, 73–80.

Tippett, L.C.H. (1934). Statistical methods in textile research. *Shirley Institute Memoirs,* **13**, 35–93.

Tomkins, J.A. (1982). Plant layout. In *Handbook of Industrial Engineering,* edited by G. Salvendy (New York: John Wiley).

Chapter 3

Subjective assessment

Murray A. Sinclair

Characterizing the ergonomist's tasks

Assume you have to carry out some design project. Considering only the information and knowledge that you need to do this, we can picture it as in Figure 3.1, as an 'egg' of knowledge, divided as shown.

The important thing about the representation is that it emphasizes the need to gain knowledge as the project continues; almost never are you in a position to say, 'I know enough'. Of course, this applies not only to design projects, but to almost all projects.

The next general point to make is to characterize the activities of ergonomists. Let us take an example familiar to us all—the educational process. Imagine you have been asked to improve the efficiency with which statistics knowledge is transferred to students on a given course. A simple model of this might be Figure 3.2.

Given that this represents the process, you as ergonomist might have three roles to play:

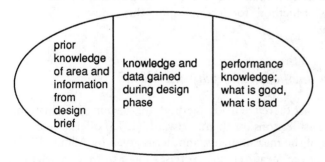

Figure 3.1. Representation of knowledge required for a design project

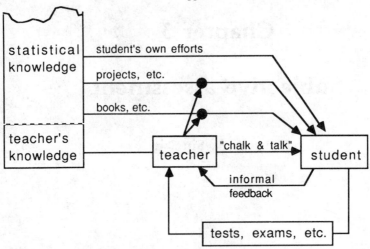

Figure 3.2. A simple model of system for transferring knowledge to student. The black blobs on the lines represent the teacher's ability to control these information flows

- *Explorer:* your first task is to show that this model does in fact represent what happens. One way in which you might wish to gain relevant information is by talking to people involved in the current system

- *Optimizer:* knowing the goals of the system, your task is to optimize the flow down the various channels to ensure both more efficient learning and control of the learning process. For evaluation purposes, you might wish to talk to those involved about their perceptions of the changes you have made

- *Innovator:* you introduce new channels of flow to improve even more the efficiency of the system—see Figure 3.3. Again you might wish to evaluate the changes by talking to people involved.

For each of these activities, then, you will have to gain knowledge. This chapter is about a class of methods for accomplishing this, known as 'Subjective Methods'.

What are subjective methods, and why use them?

The first thing to say is that there are many methods for gleaning information, and there are several classifications for them. Edwards (1973) offers one based on the intrinsic nature of the measurement, another is given by Alluisi (1975) based on the purpose of the measurement exercise, and a third is by Meister and Rabideau (1965) based on the source of the data. An edited version

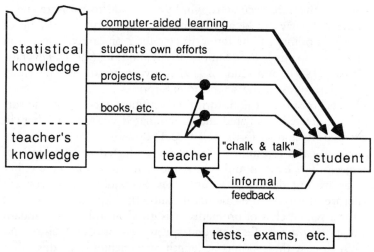

Figure 3.3. The learning model with innovation: computer-aided learning has been added

of the last is given in Figure 3.4 (see also the discussion in chapter 1, and Table 1.3).

The first class, 'Observational methods', includes most of the methods that are derived from those used in the so-called 'hard' sciences, such as physics. Laboratory experiments, activity sampling methods and so on are examples of these. The thing in common among these methods is that some degree of formal 'objective' measurement is involved; the idea is that you are reaching out to touch 'reality' as directly as possible. It should be noted however,

Observational methods

- Experimental methods
    - laboratory expts.
    - simulations & games
    - field experiments
- Observational studies

Database methods

- consult books, journals, etc.
- study system records
- obtain advice from experts

Subjective methods

- questionnaires & interviews
- rankings & ratings
- critical incident techniques, etc.

Figure 3.4. Classification of data-gathering methods, edited from Meister and Rabideau (1965). It should be noted that class boundaries in this scheme are fuzzy

that while you may measure accurately what you decide to measure, what you decide to measure may not be a good measure of reality. Discussions of this and other related points will be found in chapter 2 on direct observation in this book and in Chapanis (1967), Simon (1976) and Jung (1971). Feyerabend (1975), Ziman (1980), and Hudson (1972) give other interesting views on the role of experiments in the scientific enterprise.

The second class constitutes the historical class, where records of one sort or another or the collected, conflated and disseminated wisdom of others in books, journals, reports and so forth are used. The main problems here are firstly, that in many cases the information is not collected for the kinds of purpose that ergonomists have in mind and important types of data are either not collected, or are lost during data conflation. Secondly, where you are consulting expertise, it is often the case that because the expertise is intended to be generic for a whole class of problems, it is good on rules and standard procedures, but poor on specific facts and specific situations. Meister and Rabideau (1965) discuss these points at length and chapter 4 in this book discusses the use of archival data.

The third class constitutes the subjective methods. The common thread here is to use the people involved in the system that you wish to study as measuring instruments. In effect, you rely on people to come to some sort of conclusion about the system, then access that conclusion as a measurement of the system. This class contains any method that draws its data from the psychological contents of people's heads.

This chapter is concerned with the last of these classes. This particular class of subjective measurements is important for a number of reasons:

1. There are size limits to what can be measured 'objectively'. Without enormous resources, how would you measure, assemble and digest all the objective data about the activities occurring simultaneously, interdependently, in public and in private, on a ship such as the *Queen Elizabeth II* in order to assess the efficiency of the ship's crew?
2. There are type limits as well. Many types of information can be measured objectively, as we all know, but there are some that cannot be— at least, not easily. For example, how would one measure the skills involved in figure-skating without using human judges? Rather more germane to our problems, we often wish to assess human decision-making skills. For this, one needs to know what the decision-maker was thinking, not just what data were presented and what the resulting actions were. This kind of information is not accessible without subjective methods.
3. As a sage once remarked, 'Seeing is believing, but experience gives you an original truth'. The point of this is that people who are part of a system (for example, the 'teacher' and the 'student' in Figure 3.2) will have different views of the system, both from each other and from an external observer. Furthermore, their perceptions will almost certainly

be deeper, more complex and more subtle than those of the external observer as far as their local area of the system is concerned, and therefore will constitute a valuable source of information.

4. There are accuracy limits. In many cases, instruments can be made more sensitive, more reliable, more precise and more accurate than humans, but this is not always so, especially where qualitative judgement is concerned, or where a particular measurement is multivariate. Taste is a good example; it is not for nothing that most food and beverage companies allocate considerable resources to taste panels. Certainly, there are problems of human bias, but the subjective methods are intended to reduce, and it is hoped to nullify, such sources of error.

5. Finally, validity. Irrespective of the methods used, we would like to have valid data. The most common approach to try to ensure this is to use the notion of 'Convergent Validity'. The principle is that if you use two different, independent methods to get data about the same topic, and both produce the same (or nearly the same) results, then it is likely that the data are valid. Subjective methods comprise an independent group of measures.

The arguments in favour of subjective measurement thus lie in the independence of the measures and the ability to acquire data that cannot be obtained, or cannot be obtained easily, by other methods. The two main arguments against the use of such methods are firstly the inherent biases in human judgement, and secondly the resource requirements necessary to get reliable data. These two arguments will be considered later, after a discussion of some of the basics of subjective assessment.

Typical methods used in subjective measurement

The methods available are legion. Of these, only a few will be discussed, the ones used most frequently by practising ergonomists. These are:

Ranking methods

These methods are concerned with questions of the type:

'Given four typefaces, [Courier, **Helvetica**, Times, Palatino], rank them from first to last for ease of reading.' What you obtain are data that distinguish between the examples quite well, but do not tell you for example whether the best is actually easy to read.

Rating methods

These methods deal with questions of the type:

'Given the same four typefaces, rate each one for ease of reading on the following 5-point scale:

| | Very easy to read |
| Easy to read |
| Acceptable |
| Difficult to read |
| Very difficult to read |

These methods will provide information on the perceived ease of reading of the examples, but are less sensitive to differences between them.

Questionnaire methods

Questionnaire methods make use of the two kinds of question above, as well as several other classes of question. Typically, questionnaires presume a fixed series of questions, often with a fixed range of alternative answers. As far as the respondent is concerned, they are typically either paper-based or interviewer-based. Questionnaires are the most common method for collecting subjective data.

Interviews

Interviews are at the opposite end of a continuum from questionnaires. They are distinguished from questionnaires by the relative lack of rigidity in the questions and acceptable answers. In other words, the interviewer has more discretion and flexibility over the question and the course of the interview.

Checklists

Checklists are much as the name suggests; they are list of events that you expect might occur in a given situation, and you wish to record their occurrence. In use, the observer of the situation ticks off against the list whichever event is observed to occur.

The common characteristic among the methods is the environment in which they operate. This is shown in Figure 3.5. The generic problem faced by anyone in obtaining the relevant information is that the query must be transmitted to the respondent in such a manner that the respondent can understand what is required. Then the respondent must delve into long-term memory to retrieve the information required. This may not be a textual memory;

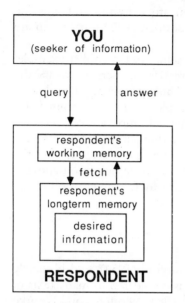

Figure 3.5. Representation of the information-seeker's basic problem. In most cases the information sought requires the respondent to access long-term memory, retrieve the information and then report it accurately to the information-seeker

it could be a sensation of comfort or of anger, or a visual image. The respondent must then turn this memory trace into a suitable form of words. In all of these steps it is possible for errors or bias to creep in.

Discussion of the scaling methods

Theoretical details regarding the various classes of method listed above are not discussed here; whole books are devoted to these methods, and for such details you might wish to consult some of the texts listed in the references, such as Coombs *et al.* (1970), Guilford (1954) and Torgerson (1958).

Instead, there is an outline of the practical principles of the method and a brief discussion of the characteristics of the techniques. In case there is some difficulty with the jargon in this section, some definitions follow:

entity   The objects that are going to be scaled. For instance, if you wished to assess cookers for ease of use, each cooker that is assessed would be an entity. Similarly, if you assess people for intelligence, each person would be an entity. If you are assessing statements typed on cards, each statement is an entity.

attribute   The property of the entity that you are scaling. For ease of use of cookers, you might be scaling the 'ease of controlling temperature'. This is the attribute that you are assessing, and it is the extent to which each cooker possesses that attribute that you are measuring with your scale.

subject The people that you use to do the scaling. In the cooker example, you might recruit people off the street as 'lay persons' to try the cookers and register their opinions. These people are the subjects. If these people are assessing other people (for instance politicians), the latter would be entities, from the definition above.

respondent Usually used in connection with questionnaires. Again, they are the people who give their opinions which are recorded on the questionnaire, and thus are subjects, from the definition above.

judge A subject used for special purposes, usually for the creation of the scales in the first place.

Ranking methods

Details of ranking methods will be found in Guilford (1954), and Nunnally (1970). The analysis of the rankings produced is considered in many texts of non-parametric statistics; examples are Meddis (1984), Mosteller and Rourke (1973), Siegel (1956) and Tukey (1977). Guilford is particularly appropriate.

A single question as in the example about typefaces earlier is rare; usually, there is a series of questions forming a 'battery', probing a number of different attributes of the situation in question. These batteries are usually used in experimental situations, where the subjects have a number of different alternative entities to rank. Entities are usually physical objects, but there is no reason why people, software products and suchlike should not be included. The essentials of the method involve presenting the entities to the respondent in such a way that the attribute in which you are interested can be assessed. This may involve user trials, etc. (see chapter 10, as well as Kirk and Ridgeway (1970, 1971) for examples of what this entails). The entities should be presented in a random order for each subject involved, for experience. The respondent should then be presented with all the entities together, and be asked to rank them. If there is a large number of entities to be ranked, it is sometimes better to ask the respondent to rank the best ones, then the worst, and then the ones in the middle.

Characteristics

- For each attribute to be ranked in the battery, each entity should possess the attribute in some degree. While this is an obvious statement, it can be difficult to establish in practice.
- It may well be the case that for several of the entities, a subject may not have any real preferences. The ranking process disguises this for an individual subject, and its occurrence can only be detected by using

Coombs' Unfolding Technique (or some derivative—see Coombs *et al.*, 1970) or by using many subjects and examining the results statistically.
- It is commonly accepted that ranking methods only record real preferences for the first two or three ranks, and the last two or three ranks. In between, the rankings are possibly unreliable, hence it is unwise to give too many objects to a subject for ranking; about nine is usually taken to be the upper limit.

Rating methods

Rating methods have been studied extensively since the early 1900s. The literature is replete with texts describing exhaustively the details of the various methods for generating scales. The most commonly-used methods are:

- Simple rating scales
- Thurstone's Paired Comparisons Technique
- Thurstone's Equal-appearing Intervals Method
- Likert's Summated Ratings Method

Good introductory texts to these techniques are Oppenheim (1966) and Edwards (1957). The latter is especially good. More technical texts are Guilford (1954) and Torgerson (1958).

Simple rating scales

Simple rating scales are as shown in Figure 3.6. The 'parking bays' representation is typical.

The method requires appropriate questions to be generated about the attributes of the entities involved, typically in the form of a battery of questions,

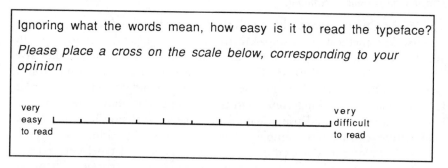

Figure 3.6. Example of a simple rating scale. It features the question to the respondent, instructions to the respondent, and the scale, as the standard 100 mm line, in this case with tick marks. Scale 'anchors' are also given at each end, but no intermediate labels

as for rankings. Scales are then created for these questions. The important points here are that the researcher must decide whether it is necessary to have a genuine 'neutral' region in the middle of the scale (the example does not have one), and how the two ends of the scale shall be 'anchored' (i.e., given unambiguous labels that will be interpreted the same way by the majority of users of the scale). The rating scale itself is usually shown as a 100 mm line; subdivisions are optional, as are labels for the subdivisions. As for the ranking methods, the respondents should be presented with the entities in a random order, and should be given an opportunity to experience the attribute of interest. The respondents should then place a mark on the scale to indicate their opinions.

CHARACTERISTICS OF RATING SCALES

- This is the most common technique used by ergonomists. This is because of its ease of use, particularly for respondents.
- You must take care about the meanings of the scale anchor points and the labels used along it. It is not easy to establish clear anchor points, nor is it easy to get good labels; if in doubt, leave them off. It appears that almost all respondents are quite good at dividing the line into equal subjective intervals, and in any case, if you use a 100 mm line, the usual way to get the rating converted to a numeric value is to measure to the nearest cm. This unit of measure will encompass most of the variability shown by respondents.
- There are a number of biases that can affect rating scales. A list will be found in Guilford (1954). The worst of these is the 'leniency' effect, where respondents are unwilling to be critical. A second problem is the 'halo' effect, where the respondent has already decided that one of the entities is better than the rest, and unconsciously 'adjusts' his or her ratings to demonstrate this clearly.

Paired Comparisons Technique

The Paired Comparisons technique is the 'standard' technique, against which other scaling methods are compared. In this technique, subjects are typically asked to compare two entities, A and B, and make a decision whether A is 'greater than' or 'less than' B. The entities are taken two at a time from a range of entities provided by the researcher, and the judgements are recorded for each pair. By making some statistical assumptions about the nature of human judgements, and using a pool of subjects, it is possible to derive a quantitative scale from the simple judgements, with the positions of the entities marked along it. For example, if we have compared typefaces two at a time for ease of reading, we would know where each typeface fell on an 'ease of reading' scale, and have a numerical value for that point. If instead of objects we use verbal statements, we can create a scale for measuring other objects, to be used by subjects in a field situation, where each statement has a numerical value. For example, we may have created a scale for sitting comfort. A subject would then sit at a workstation, for example, and assess how comfortable he or she feels, then select a statement from the scale set which best represents that assessment. The numerical value for this statement is then allocated to the workstation seat.

CHARACTERISTICS OF PAIRED COMPARISONS

- During creation of the scale, the judgement required of subjects is a relative judgement, and does not require the subject to assess 'by how much' one is greater than the other. This is a simple judgement, which most subjects find fairly easy to make about almost anything.
- Under normal circumstances, more data than strictly necessary are collected. This 'overdetermination' allows a number of internal checks for consistency to be applied to the resulting scale.
- The number of judgements required per subject rises by a factor of $[n(n-1)/2]$ as n, the number of entities, rises. This can require considerable subject time. There are penalties in this; a bored subject towards the end of the session is a very different person from the keen, perhaps slightly apprehensive, subject who started the session. There are ways of reducing the number of judgements required; see Torgerson (1958).
- The scale that results is usually taken to be unidimensional, but there is no guarantee of this (for example the comfort scale might be made up of two scales, one that measures ease and well-being, and other which measures localized pain in the rump). Small departures from unidimensionality may not be detected by the internal checks.
- Because of the resource problems it is commonly accepted that there should not be more than 9 entities to be compared.
- The success of the scale hinges on the selection of entities. If the attributes being assessed are not commensurable, or if the entities are too clearly dissimilar to each other, the scale will be invalid. In the case of scales of statements, the choice of statements is critical. One requires a set of statements that cover the range of the scale, each of which has a relatively fixed point on the scale. Most of the work should be allocated to this initial choice of statements.
- Care must be taken to ensure that the group of subjects used to create the scale is equivalent to the group of subjects who will use the scale subsequently, otherwise the scale values allocated may not be accurate.

Thurstone's Equal-appearing Interval Technique

The method of equal-appearing intervals produces the same scales as for Paired Comparisons, but is much quicker. However, there is less opportunity for internal checks. It was developed by Thurstone to provide a quick means of obtaining scales for use in the field, typically in questionnaires. The method emphasizes the selection of statements to comprise the final scale; it assumes that a pool of statements can be created by the researcher, and provides a means of reducing this pool to the final selection. Once created, each scale is represented as a randomized list of statements, with no scale values. The subject using the scale is asked to select that statement which best represents his or her judgement.

CHARACTERISTICS OF THURSTONE'S TECHNIQUE

- The scales that are produced by this technique are almost comparable with the Paired Comparisons approach for validity and reliability.
- The data produced by this technique are very easy to analyze.
- Subjects can experience problems if their feelings do not quite match the statements given. This can create the impression that words are being put in their mouths, which can have fatal effects on their motivation.

Likert's Summated Ratings Method

The Likert scaling technique is based on a very different approach, but is taken to be as powerful as the Thurstone techniques. Whereas in the Thurstone techniques the subject is asked to select a statement with which he or she most agrees, in the Likert technique the subject must respond to every statement, showing his or her disagreement with it. Typically, each statement will have a simple 5-point scale associated with it as shown in Figure 3.7.

Each point has a scale value (for example 1 to 5), and, using the point selected, the scale values are summated and the result represents the subject's opinion. Clearly, there is a premium in the selection of statements and in ensuring that the points allocation is matched to the statement. What one seeks is that if a subject's response to one statement obtains a score of 5, then a similar score will be obtained from a similar statement. For example, if a subject strongly agreed with the statement in Figure 3.7, scoring 5 points,

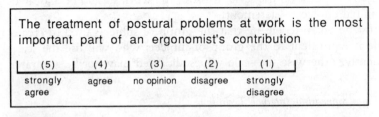

Figure 3.7. Example of a Likert scale item. It contains a statement with which the respondent can either agree or disagree, and indicate this opinion on the scale. Normally, the three scale labels in the centre are not included. The points values in brackets are never shown; they have been included here for illustration purposes only

one might expect the subject to disagree to some extent with the statement in Figure 3.8, again scoring high points. This enables individuals with differing viewpoints to be discriminated from each other by accumulation of these scores. Some 10 or 20 statements are used to create the scale.

CHARACTERISTICS OF LIKERT'S METHOD

- It is said that subjects prefer the Likert scaling technique, because it is 'more natural' to fill in and because it maintains the subjects' direct involvement.
- The Likert approach is said to require less effort to generate scales compared to the Thurstone techniques.
- There appears to be no great difference between the Likert and Thurstone techniques in validity and reliability.
- Thurstone techniques are said to be better at measuring 'snapshot' views, whereas Likert techniques are better at measuring changes over time.

More complex methods using ratings and rankings

The methods outlined above are used as described. However, it is more common to find several scales bundled together to make a 'battery' of scales, as mentioned earlier; it is seldom that you will wish to assess any group of entities just for one attribute alone.

Such batteries may be found in questionnaires, as discussed below. At this point, two other techniques will be mentioned.

Semantic Differential Technique

This technique is described in detail by its originators, Osgood *et al.* (1957) and shorter descriptions will be found in most textbooks on attitude measurement (e.g., Moser and Kalton, 1971). Their original intention was to discover how groups interpreted the meaning of words; since then the technique has been used for all sorts of purposes, including evaluation of household goods.

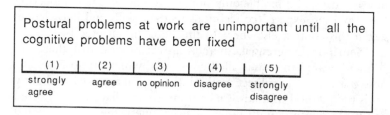

Figure 3.8. A second example of a Likert scale. Since the statement conflicts with the example in Figure 3.7, one would expect the (invisible) scoring for this item to be reversed, as shown

A series of rating scales is generated which describes the class of entities to be studied. By convention, the scales are usually seven categories long. The end points of the scales are given anchors which are single word adjectives, and are 'polar opposites' (e.g., good-bad; strong-weak; fast-slow). The subject then rates each entity according to these adjectival scales. Generally, the data resulting from a group of subjects carrying out such an exercise are subjected to a Factor Analysis, to determine what 'dimensons' are being used. The assumption in this is that the scales represent deeper underlying evaluation dimensions that can be ascertained from the adjectival scales. These dimensions may not be apparent beforehand, nor within the normal vocabulary of the subjects involved, whereas the adjectival scales are fairly commonplace. The differences between this technique and the Likert technique are twofold. First, superficially in the Likert technique there is a range of statements but only one version of the scale, whereas in the Semantic Differential there are different versions of the scales but only one entity to be evaluated at a time. Rather more deeply, in the Likert technique the statements are all representations of a single undelying issue, whereas in the Semantic Differential an individual scale may be examining a unique aspect of the entity unrelated to any of the other scales.

The strength of the Semantic Differential lies in its explanatory power, in elucidating the underlying dimensions and, once these are obtained, in showing the relationships between the entities in the n-dimensional space constructed from these dimensions. These are not necessarily apparent before the exercise, though you, the investigator, may have some shrewd suspicions (and wrong ones), and it is this power to reveal what you might otherwise miss that is the appeal. Of course, the power depends critically on the quality of the data collected; statistical techniques by themselves do not provide explanatory power.

Repertory grids

This technique was developed by Kelly (1955); one of the best technical and operational descriptions will be found in Fransella and Bannister (1977). Kelly's aim was to provide a psychiatric tool to help understand an individual's representation of his or her environment. The tool has been widened in its use since the early days, and has become popular in knowledge engineering as a means of exploring experts' conceptions of their skills and knowledge; it is discussed in this context elsewhere in this book (in chapter 14).

As with the Semantic Differential, there is an assumption that there are underlying dimensions, in this case called constructs, which are to be established. Because this is a single-subject method, the typical approach is slightly different; this is the method of triads. Once some entities have been selected (let us assume we are dealing with different telephones), they are presented three at a time to the respondent (e.g., telephones A, B and F). The respondent is asked to arrange them so that two are similar and the third is different, and to state what the difference is (e.g., 'these two are small (B and F), and that one is bulky (A)'). The subject repeats this with the same triad (e.g., 'these two are modern (A and B) and that one (F) is old-fashioned'), until no further differences are found. Another triad is then presented (e.g., A, B, and G), and the process is repeated, until no new differences are recorded. The more common, or most interesting, of these differences are then selected, and for each of these differences (e.g., 'Small-Bulky') the respondent ranks the entities along the implied continuum. A standard statistical analysis (usually a 'Factor Analysis', using correlations between the rankings) is then used to elucidate the underlying dimensions.

Questionnaires

The appeal of questionnaires lies in their ability to obtain large amounts of information from large numbers of people, at relatively low cost, and relatively quickly. Two good texts are Oppenheim (1966) and Moser and Kalton (1971). The former is easier to read, while the latter is more comprehensive and detailed.

However, there are problems with questionnaires. Unless the information required is strictly factual and fairly easily checked, their reliability and validity can be quite low, so considerable care must be taken. Furthermore, they do make a substantial demand on resources. Designing and administering a high quality questionnaire is a skilled task; a specialist in one of the behavioural sciences, a statistician, a computer expert and a graphic designer working as a team may be needed. Then there is the fieldwork; interviewing respondents is a time-consuming and skilled task, requiring fairly large numbers of people. There are problems of getting the data into a computer system for analysis and producing reliable analyses as well. All of this costs money; as a rule of thumb, if the project is carried out commercially, the cost per completed questionnaire would be about £22 (estimated 1992 costs).

In what follows, the assumption is made that the questionnaires will be postal questionnaires, and that they are intended for groups of more than one hundred. Many questionnaires are administered to respondents by interviewers, but this is omitted from the discussion here because the topic is discussed in the next chapter.

Know thy respondent

Eliciting accurate, reliable information from respondents is not an easy thing. Some observations, based on many people's experience, that indicate the necessity of careful planning of your work, are given below.

– Data-gathering presupposes a fairly high level of inter-personal under-

standing, a common culture, and a common language. Your questionnaire must stay within the boundaries imposed by this.
- Clumsy presentation may lead respondents to 'close down', for example
  by giving minimal answers, devious answers, or by outright refusal.
- The opportunities for misinterpretation are much greater than you
  might suppose. This is considered in more detail under the heading,
  'Questionnaire construction', and is illustrated in Figure 3.9.
- Your respondent's knowledge may not be organized usefully, it may be
  limited to a small range of circumstances, and it may be wrong. It is
  extremely difficult to guard against these problems.
- Respondents often have only a partial knowledge of the extent of their
  own knowledge; this is usually the case when opinion is passed off as
  truth. Careful question design can ameliorate this problem, typically by
  the use of closed questions (discussed later).
- The most common characteristic of respondents' knowledge is that they
  have a good grasp of generalities and rationales, but a poor grasp of
  particular, necessary facts. There are no solutions to this.
- Particularly for non-verbal knowledge (e.g., driving skills), a respondent
  may not know what the skill is, let alone be able to describe it.

Planning the questionnaire

Typically, when you administer a questionnaire to a respondent you are
implicitly asking him or her to reconstruct from memory the environment
or the problem area that you are interested in, and then to pluck from it the
information or opinions that you require. This is not necessarily an easy task
(how much detail can you recall of your journey to work today?) and it
requires a careful, methodical approach to make your efforts worthwhile.
The steps are discussed briefly below.

Definition of objectives and resources

This is the most essential and hardest step. First, you must define your objectives *in detail*. For example, it is not sufficient to have as your objective: 'Find
out about accidents on grinding machines.' You should be able to define
precisely what you mean by 'accident' (does it include damage to clothing?
Does it include near-accidents? What about minor accidents such as bruised
fingers, which are almost never reported?), and what range of machines is
covered by 'grinding machines' (all of them, or just the 'common' ones?
What does 'common' mean?). At this point, there are three important questions to be answered:

- What are the results supposed to show?
- What level of accuracy is required?

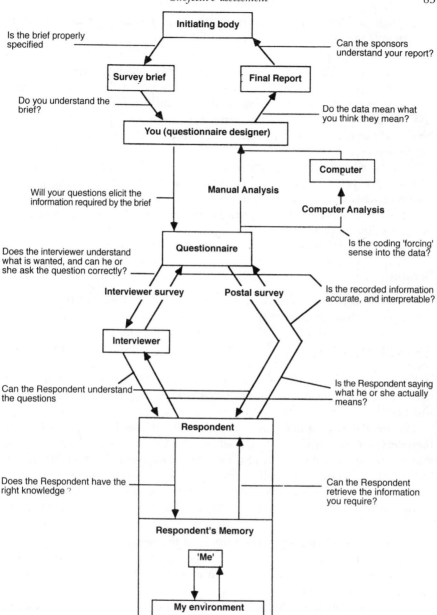

Figure 3.9. Illustration of the communication problems in questionnaire design. The arrows represent communication processes, and the questions associated with them indicate some of the problems that can occur. It should be noted that once you have designed your questionnaire, any errors are 'frozen in'. Redrawn from Sinclair (1975)

    – What additional data will be required to link this survey with other
      people's work?

This stage must be done thoroughly, down to deciding the form of the
questions you will be asking. As a rule of thumb, you will have spent enough
time and effort at this stage when the questions virtually write themselves,
and the format of the final report is clear. Another rule of thumb is that
between a third and one-half of the total time available should be spent on
this first stage.

    There is no substitute for this part of the design work. No matter how
sophisticated you are from here on, if you start with fuzzy thinking you will
continue with fuzzy questions, and however ingenious the analyses, at best
you will finish with sophisticated fuzzy answers, and a fairly clear perception
of the futility of your efforts.

Sampling

This step deals with whom you will survey. First, you must define the population you wish to investigate (for example, which organizations your respondents work for, and what skills and/or grades they must possess), and then
decide on the sample size (which could be 100%). This is determined by the
resources available and the accuracy required—see Guilford (1954). Sample
size is discussed relatively formally in chapter 5 on ergonomics studies and
experiments, and pragmatically in the context of user trials in chapter 10 of
this book.

    The next step is to generate a 'sampling frame', which is a list of all the
members of your population (e.g., constructed from personnel records), from
which you then draw your sample. You should be aware that sampling frames
are not always accurate, usually because the sources are not up to date.

    The aim is to eliminate any systematic bias (of course, if you have decided
upon a 100% sample, this is not a problem). Bias can arise from three sources:

    – Non-random sampling. For example, if the personnel records have been
      arranged in order of seniority, a decision to sample every tenth name
      in a small workforce will result in a sample with a bias towards lower
      levels of management.
    – Bias in the sampling frame. This is usually due to the sources not being
      up to date.
    – Non-response. In any population there are people who refuse to
      respond, as a matter of principle. There are also those who are never
      available, such as salesmen and drivers. Exclusion of these groups inevi-
      tably means introducing bias.

The chief means of overcoming bias is by random sampling. This allows you to alleviate the effects of the first two sources, but only perseverance on your part will deal with the third.

'Stratification' is often used with random sampling, to improve the representativeness of the sample. The population is arranged into strata on the basis of age groups, sex, income levels, etc. You then sample at random within the strata, but note that you need this information about each individual in your sampling frame right at the beginning, in order to sort them into strata.

Another approach used is quota sampling. This is particularly useful where you cannot generate a sampling frame. In this case each interviewer is given a quota of people to interview, subject to them falling into certain age groups, income levels, and so on. The interviewer is then left to find the people to fill the quota. This method is open to interviewer bias, but produces about as accurate results as the stratification method. There are also panels, usually groups of people to whom you make repeated reference over a period of time, to measure changes in opinions or judgements over time. The approach has its own problems of attrition or conditioning of panel members.

Having selected the sample, there remains the problem of ensuring maximum participation in the survey. This is important, because many experiments have shown that non-responders are not typical of the total population. The main causes of this together with suggested remedies are as follows:

- Units outside the sampling frame—someone who should be in the sampling frame but is not available (e.g., dead). In this case, select more people from the sampling frame.
- Units unsuitable for interview—too old, too deaf, no relevant knowledge, or language barrier. There are generally fewer instances of this than field workers claim.
- Movers. In this case the people who have moved in to replace those who moved out should be used.
- Refusals to answer. This is an important source of bias. A 'good' response rate would be between 60-80% for mail questionnaires, but to obtain this figure it is generally necessary to do follow-ups of one sort or another.
- People on holiday. In this case follow-up calls should be made.
- People who are away at the time of call. Accepted practice is to make two follow-up calls, at different times, to ensure that shiftworkers and the like are included. After this, the return from further follow-ups is usually not justified in terms of expense.

Questionnaire construction, and question wording

At this stage you should know specifically what questions are going to be asked, from stage 1. Question wording has to do with how you ask for the information. The problem is discussed below and in Wright and Barnard (1975), and is exemplified in Figure 3.9.

If each arrow in the diagram is regarded as a process of communication, it is obvious that there is ample scope for error. It is important to realize that your only chance to control most of these errors is in the wording of the questions. Once the questionnaire is printed your control over the information is almost nil. What complicates the matter further is that the wording of questions is an art, rather than a science, and consequently there are no rules, just a few guidelines. Quite apart from the problems in Figure 3.9, you have to ensure that at the same time:

- the respondent is motivated to respond;
- the respondent has the particular knowledge required;
- the questionnaire takes into account the respondent's limitations and personal frame of reference, so that he or she will understand easily the aim and meaning of the questions;
- the respondent has produced an adequate answer, from his or her own knowledge.

Questionnaire sequencing

Typically, a questionnaire is made up of four parts:

   (i) A prologue, which introduces the topic to the respondent, provides any information he or she will need, and tries to motivate the respondent to answer the questions. It may also include examples of difficult questions and instructions for answering them.
  (ii) An information section, in which you ask your questions (e.g., 'How long have you worked on this machine?' 'How many of your fingers has it chopped off?').
 (iii) A classification section, in which you obtain personal data and any other background information (e.g., 'How old are you?' How long have you been doing this job?').
  (iv) An epilogue, which thanks the respondent for his efforts, and includes any further instructions, if necessary.

Within the information section, the questions should be arranged in consistent groups which follow each other in a reasonably logical way. This should be from the respondent's point of view, not from yours; it is important that the respondent should feel comfortable.

You should also consider the use of 'filter' questions. These serve two purposes: firstly, they determine whether your respondent has relevant knowledge or not (e.g., 'Have you taken a training course for. . .?'), and secondly they can guide your respondent past sections that are not applicable in his or her case (e.g., 'If you have had back pain, answer this section; if not, go to section. . .').

Degree of structure in questions

After a while it becomes apparent that questions, irrespective of content, can be classified into certain structured classes. Two the classes have already been discussed, ratings and rankings, and some of the sub-classes within these have been outlined. Others are: factual questions, for example asking for birthdates, gender, etc. (Question 10 in Figure 3.10 is an illustration); questions with mutually exclusive answers (Question 5 in Figure 3.10); questions with multiple non-exclusive answers (Question 6 in Figure 3.10); matrix questions (Question 4 in Figure 3.10); open questions (Question 2 in Figure 3.10); and closed questions (Question 1 in Figure 3.10). There are many more. Your questionnaires will most likely be composed of a variety of these, as in the example. This structured classification is of use subsequently in organizing the analytical routines.

Irrespective of the class of question, it must be worded correctly. For ratings and rankings the methods discussed above will have ensured this. It is with the other classes of questions that we are now concerned. Belson (1968) and Kalton *et al.* (1978) have a number of important, interesting and practical comments to make about the problems. The first decision that must be taken is whether the questions should be open (i.e., subjects compose their own answers) or closed (subjects choose an answer from a given set).

The advantages of closed questions are:

- They clarify the alternatives for the respondent and avoid snap responses being given.
- They reduce keying errors in analysis, and eliminate the need for people to code the answers.
- They eliminate the useless answer (e.g., 'How long have you done this job?' 'Since I moved here').

The disadvantages are:

- It is difficult to make the alternatives mutually exclusive.
- They must cover the total response range (this presupposes that the researcher will have a good idea of what answers are likely to appear— hence the importance of pilot studies).
- They create a forced-choice situation which rules out marginal or unexpected answers.
- All the alternative answers must seem equally logical or attractive, for fear of biasing the results.
- In complex or difficult questions, subjects may dive for the safety and ease of the 'don't know' alternative.

PART 1 – General

1　What are the risks in your brigade area?
　Please show the distribution of risks by putting a
　percentage in the box. (e.g. if your brigade is a
　county with B-D risk, the B risk might represent
　60% of the area, C 25% and D 20%　　High risk ☐ %
　　　　　　　　　　　　　　　　　　A risk ☐ %
　　　　　　　　　　　　　　　　　　B risk ☐ %
　　　　　　　　　　　　　　　　　　C risk ☐ %
　　　　　　　　　　　　　　　　　　D risk ☐ %
　　　　　　　　　　　　　　　　　　E risk ☐ %
　　　　　　　　　　　　Remote rural risk ☐ %

2　Do you consider that there is a suitable production
　commercial vehicle chassis currently on the market
　that will meet the demands of the Fire Service? Yes ☐
　　　　　　　　　　　　　　　　　　　　　　No ☐
　If Yes, who manufactures it?
　_____

　If No, what features render them unacceptable for your use?
　_____
　_____
　_____
　_____

3　Do your drivers receive training similar to the
　programmes given to police drivers?　Yes ☐
　　　　　　　　　　　　　　　　　　No ☐
　If Yes, please describe the training procedure.
　_____
　_____
　_____
　_____

4　How frequently are your vehicles involved in accidents?

	Total Calls	Total Accidents		
		On the road	Off the road	At the fire ground
1970 ...				
1971 ...				

5　In fighting fires in recent years how many times have
　you used open water, with either a main or portable
　pump?　　　　　　　More than 100 times per year ☐
　　　　　　　　　　　　　50-100 times per year ☐
　　　　　　　　　　　　　11-50 times per year ☐
　　　　　　　　　　　　　0-10 times per year ☐

6　Have you had any vehicle failures in the last year?
　　　　　　　　　　　　　　　　　　　Yes ☐
　　　　　　　　　　　　　　　　　　　No ☐
　If Yes, please tick the relevant boxes
　　　　　　　　　　　　　　　Mechanical
　　　　　　　　　　　　　　　　Engine ☐
　　　　　　　　　　　　　　Transmission ☐
　　　　　　　　　　　　　　　　Axles ☐
　Other (please specify)
　_____

　　　　　　　　　　　　　　　Electrical
　　　　　　　　　　　　　　　Batteries ☐
　　　　　　　　　　　　　　Generators ☐
　　　　　　　　　　　　　　　Wiring ☐
　Other (please specify)
　_____

　　　　　　　　　　　　　　　Ancilliary
　　　　　　　　　　　　　　　　Pump ☐
　　　　　　　　　　　　　　Compressors ☐
　　　　　　　　　　　　　　　Tank ☐
　　　　　　　　　　　　　　Ladders ☐
　Other (please specify)
　_____

7　Do you have particular problems in obtaining spare
　or replacement parts for your fleet?　Yes ☐
　　　　　　　　　　　　　　　　　　No ☐
　If yes, are there any particular parts which
　are always difficult to obtain?
　_____
　_____
　_____
　_____

8　Are your vehicles out on the road performing fire
　prevention and other duties during the day?　Yes ☐
　　　　　　　　　　　　　　　　　　　　　No ☐
　If Yes, can you give an approximate percentage of
　those on the road against those giving cover at
　Fire Stations?　☐ %

9　What was the percentage of special calls against
　total calls in your brigade last year?　☐ %

10　As brigade procedure do crew members don BA
　sets on the way to a call?　　　　　Yes ☐
　　　　　　　　　　　　　　　　　No ☐

Figure 3.10. A page from a well-designed questionnaire, illustrating different question types, the use of different typefaces for different purposes, a layout to optimise both the use of space and the clarity of the questions, and the pleasing overall appearance of the document. (Source: Gray, 1975)

Points to be considered in question wording

In framing the questions the following important points must be considered:

- Question specificity. The requirement is that your questions should be precise and unambiguous. Fortunately, if you have carried out the planning stages thoroughly, this is not likely to be a problem. It implies that where you have a difficult topic, you may have to ask a number of questions to obtain the information you want, rather than try to get it in one omnibus question. As a general rule, avoid omnibus questions like the plague.
- Language. It is essential to use language relevant to the population, to make the questions easily understood by all. You should also be aware of localized interpretations. For example, consider the word 'tea'— which in Britain may mean a 'pot of tea', 'high tea', or 'the main evening meal', depending on where in Britian you are. It is important to use short words rather than long ones, and to be aware of the dangers of the unconscious use of scientific or professional jargon. In this context, look at some of the editorials in the so-called 'popular press'. As a method of communication, the style is superb, whatever might be thought of the content!
- Clarity. It is a cardinal rule that questions should be short. This rule has two useful consequences; firstly, it ensures that you clarify your thinking and remove unnecessary words; secondly, it reduces the chance of overloading the respondent with too much information to digest. Complex questions can lead to error, as can those that contain such vague phrases as 'on the whole', 'generally', 'normally' and 'frequently'. Double-barrelled questions should be avoided, such as 'Do you suffer from headaches or stomach pains?' (what would the answer 'yes' mean?).
- Leading quetions. Clearly these must be avoided. Obvious examples such as 'Do you agree that the policies of the present government are unfair?' (which invites the answer 'yes') are quite easy to detect. More insidious examples are questions that contain such loaded words or phrases as 'get involved in', 'student' and the like. You should be aware of questions that become leading questions because of the nature of the questionnaire. As an example of this, the question 'How many cigarettes do you smoke in a day?' may be innocuous in a questionnaire about household expenditure, but may produce different answers in a questionnaire concerned with medical matters.
- Prestige bias. This is a bias that can arise in questions that involve socially desirable behaviour. Thus, a question such as 'Which magazines have you looked at recently?' is likely to reveal that such journals as *The Economist* and the *Literary Review* have a considerably larger readership than is actually the case. Great care is required to overcome this sort of bias; filter questions and careful wording should be used, so that low-prestige answers appear equally as acceptable as high-prestige ones. In

the example above, you might introduce a filter question such as 'In the past seven days, have you had any time to read magazines?' to identify those who have not opened a magazine, but are not prepared to say so.

- Embarrassing questions. There is seldom an easy way to obtain information of a personal nature. If such questions must be asked, they should be placed some distance into the questionnaire, and the whole tone of the questionnaire should be personal, relaxed and permissive. This requires considerable skill in question wording: it is very easy to appeal to one segment of the population and totally offend another at the same time. Further, the use of euphemisms should be considered in the place of blunt questions.

- Hypothetical questions. These are the 'What would you do if. . .' type of question. They almost never yield reliable results, and should be avoided. There is usually a noticeable difference between people's self image in a particular set of circumstances and their actual behaviour.

- Impersonal questions. These questions tend to produce spurious answers because the respondent becomes disengaged from the subject-matter, and consequently can lose interest in the questionnaire, sometimes to the extent of refusing to answer any more questions.

Layout of the questionnaire

By this we mean the arrangement of words on the page. This aspect is almost invariably overlooked by questionnaire designers and yet it can make a difference of about 20 per cent in the response rates. Figure 3.10 illustrates a well laid-out questionnaire, and you should note the two-column layout, the use of different weights of type, the use of italics for instructions, the use of boxes to restrict the size of written answers, and the pleasing aesthetic appearance (for its date) of the page.

Piloting the questionnaire

This stage is vital. It is here that the last chance occurs to discover the fallacies and unnoticed assumptions in your thinking. If is here that the respondent's understanding of the questions and the problems of analysis are revealed. It is also the last occasion on which remedial action can be taken. It is necessary to test all aspects of the questionnaire at this stage: the introductory passage, the questions, the alternative answers (or coding frames for interpreting open questions), and the form of the analysis.

This is best done in three stages:

(i) Individual criticism: the questionnaire should be handed to a colleague or several colleagues who have experience of questionnaires (but not of this particular one) for comment.

(ii) Depth interviewing: once the criticisms generated above have been examined and any appropriate changes made, the questionnaire should be given to a small sample of respondents (about ten) for their reaction. On completion of the questionnaire, each respondent should be questioned in detail about the answers to the questions, to ascertain what the respondent understood the question to be asking, and the exact meaning of the responses given.

(iii) Finally, the questionnaire should be given to a larger sample of respondents to investigate the implications of the desired analysis, and to detect whether any invalid or meaningless patterns of answers are occurring. This also enables estimates to be made of the reliability and validity of the questionnaire, the reliability of the sampling frame, and the likely non-response rate. This stage should be repeated until the questionnaire appears to be error-free (but you will never get rid of them all).

Fieldwork

For many surveys, where the number of respondents required exceeds 100 and the survey is to be interviewer-administered, the most appropriate course of action may be to obtain the services of one of the commercial groups. In view of the expertise and service that is provided, they are good value for money. As was stated earlier, interviewing is a job requiring training and experience: co-opting students or other 'odd-job people' does not tend to produce reliable or accurate results. Hence, if the time, trouble and cost of obtaining a field force, training and maintaining it is taken into account, there is seldom a cheaper alternative to the commercial organizations. Where the numbers are below 100, it is feasible (and highly instructive) for the designer to do the interviewing personally.

Where you intend to use a mail questionnaire, the considerations are different; the problems become those of layout and design of the questionnaire, and its retrieval. If the layout and the introduction to the questionnaire are good, the problems of retrieval are usually reduced. Nevertheless, it is usually necessary to consider some means of improving the response rate. Firstly, there are encouraging letters, to act as reminders. It is advisable to send these within one week after the expected return date, together with a duplicate questionnaire, in case the first one was mislaid. Secondly, a shortened version of the questionnaire may be sent, to ask for the most important information. Thirdly, the use of raffles or prizes to encourage the return of questionnaires might be considered. However, the cost of these can offset the chief advantage of postal questionnaires, which is their cheapness. The final method of follow-up is to contact the non-respondents by telephone. This method has received increasing use in recent years.

You should note that for any of these follow-up methods to work, it is necessary to know who has responded and who has not, which necessitates removing the cloak of anonymity from respondents. It appears, however, that this is not generally a serious cause for non-response provided the questionnaire is well designed.

Analysis

Having completed the fieldwork and collected the completed questionnaires, there is still the problem of analysis. There are two major areas of interest here: the first is data editing, to identify any inconsistencies in the responses to questions and to take appropriate action, and the second is data tabulation, where the tables and statistics required for the report are generated.

A very brief overview of this follows. For simplicity, let us assume that the analysis will be computer based, and that the information will be transferred from the questionnaires to a computer file, for further analysis. The transfer of the data from the questionnaire into a file is normally accomplished via a human agent, and this stage is therefore prone to error. It is worth noting that computer files may also be lost, so a duplicate record of the data should always be available.

At this stage the data for a single survey unit (a data set) can be considered to be in 'raw' form; the data sets may now be edited to remove logical errors (for example a man in the household has seemingly given birth to a baby) and out-of-range errors (for example the man is 300 years old instead of 30). When these errors are checked and removed, and various new data produced (for example, volume, created from data on height, length and breadth), it may be said that one now has 'treated' data sets, which in turn should be filed.

It is the treated data sets on which the analysis is performed. If you consider the data to be in matrix form, with data items across the top and respondents (or survey units) down the side, one may then conveniently use the manipulative power of matrix algebra to simplify the analysis, and do most of the analytic sub-routines. There is a wide range of computer-based statistical packages available to accomplish this; a popular and widely-available one is SPSS (Nie *et al.*, 1975). Recently, it has appeared in a version suitable for personal computers. Whichever package is selected, it is essential to ensure that the arithmetical and statistical operations performed on the data are valid, as it is very easy to arrive at false conclusions by the use of inappropriate analyses.

Interviews

In a sense, interviews can be considered to be questionnaires carried out face to face by an interviewer, and the comments above regarding the need for planning, sampling, and so on apply here as well. However, because there is an interviewer present, the typically rigid structure of the questionnaire can be relaxed; this of course brings its own dangers unless the interviewer is skilled and experienced, and the interview is carefully planned beforehand. While interviews have always been part of the repertoire of ergonomists, the emergence of knowledge engineering has lent a new importance to this technique (see chapter 14). An excellent, practical discussion of interviewing can be found in Macfarlane Smith (1972), written from the perspective of market research.

The big danger in deciding to carry out interviews is to use the rationale that because the interviewer is an intelligent person, he or she can adapt to the needs of the situation, and do the right thing, thereby reducing the need for careful planning at the outset. Unless you have an extremely knowledgeable and skilled interviewer, this is unlikely to happen. All too often, this argument is used as an excuse to ease the mental pain at the outset. It almost never works well; the sporting cliché, 'No pain, no gain' still applies. You will have to plan your interviews with the same dedication as for questionnaires.

However, it is because there is an intelligence in the interviewer that certain benefits do accrue, provided care is taken. In some circumstances, interviews are the best way to capture information. It is in situations where matters are very personal to the respondent, or where complex information is involved, or where you think respondents might need to have some help in giving their answers, or where different people may have markedly different views about reality, or where you genuinely do not know what is involved (as in the initial stages of piloting a questionnaire), that interviews are most helpful.

There is a continuum of interviewing styles, ranging from directed interviews to non-directed interviews. Directed interviews are those where the questions to be asked, and the order of questions, are specified beforehand. Interviewer-administered questionnaires are an example. Non-directed interviews are those where essentially the respondent controls the interview, and the interviewer is there to help the respondent express himself or herself, hopefully eliciting the important information during the process. This latter style is close to that used in psychiatry. The questions are not formulated in any detail beforehand, and the main role of the interviewer is to help the conversation along, with interjections such as 'Really?' and so on. If the respondent stops, the interviewer may start the process again by using a non-directive question, for example, by repeating the respondent's last phrase with a questioning tone of voice.

The former technique has been criticized for its rigidity and its intolerance of unanticipated individual differences, whereas the latter has been criticized for the time required, and, to quote a memorable phrase, for 'leaving behind a posse of cured souls, but not necessarily producing much worthwhile information'. Most interview techniques fall between these two extremes, where some direction of the respondent occurs, if only to direct the respondent towards the areas that you want discussed. In some cases, the initial questions might be defined, and so on.

- The use of an interviewer can serve to direct and accelerate the information flow
- The interviewer can explore unexpected information, or unexpected occurrences.
- A well-trained interviewer will be sensitive to the individual needs of the respondents, and will adjust his or her behaviour accordingly, thereby improving the quality of the information flow.
- Interviewers can help to motivate respondents to give more information about the topic during the interview.
- For the advantages above to occur, the interviewer must be well-trained in interview technique, should have at least some knowledge of the topic areas (the more the better), and must be sensitive to people. Collectively, these criteria are not easy to meet.
- It can be difficult to find and schedule people for the interview session.
- Interviewer's bias may creep in; this might be due to the interviewer's own knowledge of the topics, interpersonal relationships between the interviewer and the respondent, or to more mundane things such as fatigue, and so forth. This constitutes an extra source of error.
- Systematic recording of data is difficult, and in some cases impossible. In certain instances (an example is knowledge elicitation from experts), it may take up to three times as long to sort and assimilate the data as it took to obtain it (Shadbolt and Burton also warn us of this in chapter 14 of this book).

'Critical incident' techniques

This approach was first used by Flanagan (1954), in a study of near-accidents in aircraft, hence its title. He could not study accidents themselves, because there weren't many respondents available. The technique is basically a semi-directed interview technique in which all parties know what areas are going to be covered, and by and large the interviewer will control and direct the interview and may ask certain specific questions. However, since the antecedents of near-accidents are often unique, there has to be some room for variance. One of the main characteristics of this method is that it is not intended to be carried out with individual respondents, but within small groups. The reasoning behind this is that individuals in such situations can, usually inadvertently, encourage and assist each other to recall more information than they might as individuals. This technique, suitably adapted, is used in market research at the product definition phase, and in many other circumstances.

THE MAIN CHARACTERISTICS OF THE CRITICAL INCIDENT TECHNIQUE

- It is possible to get data on rare events, or ephemeral events, that cannot be obtained easily in other ways.
- The methods can reveal much about abnormal or otherwise memorable situations, but it typically has little to say about normal conditions.
- There is a danger that because a group is involved, a single consensus view may be all that is obtained. There is a real danger of this if one of the group of respondents is a dominant character.
- For certain situations, the respondents are in effect asked to give evidence against themselves. This will only occur in the right group environment, and it requires the interviewer to have highly polished interpersonal and group skills.
- Good interviewers for this are hard to find.

Checklists

Checklists come in a variety of forms, for a variety of purposes. There are simple lists, which may serve as procedural reminders to make sure that the right sequence of activities has occurred, there are complex hierarchical lists, using 'if. . .then branch. . .' constructions, there are activity sampling checklists, for recording behaviour at particular times during a shift, and so on. This discussion deals with the latter kind of list, for recording activities; other discussions will be found in Guilford (1954), Konz (1979) and Oppenheim (1966). The text by Konz contains many practical hints.

The first requirement for a list to record activities is that you should know comprehensively what activities you wish to record. This is a non-trivial problem. Firstly, there may be a number of activities in which you are not interested, which may be disregarded. Secondly, there are the activities which are unusual, but have an important effect that you ought to record—accidents, and near-accidents are examples. These are very difficult to foresee, and therefore are difficult to record in a list a priori. While it is easy to have catch-all categories in your list such as 'accidents', the very broadness of such categories reduces considerably their usefulness when it comes to analysis. Thirdly, there are the activities in which you are interested, which you have on your list, and which you can record. The first difficulty here is to establish the correct level of detail for the activities. For example, is it enough to have a category of 'monitoring behaviour'? Since this can encompass anything between alert scanning to zombie-like behaviour, you may wish to discriminate between various types. The second difficulty here lies in ensuring that you can observe and record these activities. One reason why you might not be able to do so is because the activity concerned might not be visible—perhaps the person you are studying has inadvertently turned away from you. A second reason is that other simultaneous activities might distract you.

Another reason is that the preceding and following activities may be misleading.

Having generated a list of activities to record, it is essential that starting and ending cues for each activity are identified, and that these should be clearcut. As an example, consider 'monitoring behaviour'. What establishes when it has started? Standing in front of a display does not necessarily mean that monitoring is occurring. If the activity sampling is to be undertaken by other people, these points are critical to the quality of data that will be garnered.

CHARACTERISTICS OF CHECKLISTS, ESPECIALLY IN ACTIVITY SAMPLING

- There are problems in generating unequivocal lists of behaviours and/or activities that you wish to observe.
- The problem of incomplete lists is endemic.
- They record superficial aspects of activities, not necessarily the way in which the activities are modified into sequenced skills, nor their grouping into higher level, sometimes more obscure, activities.
- The lists themselves serve efficiently as aides-memoire to the observer, reminding the observer of what should be recorded.
- The method can produce quite high correlations between observers of the same set of activities. This implies that what is recorded has some degree of validity.
- The method is relatively easy to create and administer.

Conclusion

A number of subjective assessment methods have been explored briefly in this chapter. If you wish to make use of any of the methods, the texts given in each section should be sufficient to enable you not only to get started with the method, but also to achieve reasonable results. As in most areas, using these methods skilfully requires lots of practice, not just for polish, but to make the blunders that all the experts have made in their time (including the editors and authors of this book!), from which you will learn the real lessons about the methods. As Rasmussen (1985) has said rather more cogently, it is only by making errors that you learn skills.

You should be warned: a trap into which you might fall is to stick rigidly to the methods outlined in the texts, because there is a fairly strong scientific basis to them, and a wide acceptance of them. While these are laudable attributes of the methods, what counts in the end is the quality of the data that are gathered, and what is revealed about the subject of your interest. In this regard, it is suggested that you should read the text by Webb *et al.* (1966); in its own way it is a highly entertaining book for the originality of its thoughts and the subtlety of its methods.

Finally, it should be remembered that even though the methods may seem laborious, and the statistical complexities daunting, the data that result are usually worth all the trouble that was caused, and the warm feeling engendered by this will certainly outweigh the pain you experienced at the time.

References

Alluisi, E.A. (1975). Optimum uses of psychobiological, sensorimotor, and performance measurement strategies. *Human Factors,* **17**, 309–320.

Belson, W. (1968). Respondent's understanding of survey questions. *Polls,* **3**, 52–70.

Chapanis, A. (1967). The relevance of laboratory experiments to practical situations. *Ergonomics,* **10**, 557–577.

Coombs, C.H., Dawes, R.M. and Tversky, A. (1970). *Mathematical Psychology: an Elementary Introduction* (Englewood Cliffs, NJ: Prentice Hall).

Edwards, A.L. (1957). *Techniques of Attitude Scale Construction* (New York: Appleton-Century-Crofts).

Edwards, E. (1973). Techniques for the evaluation of human performance. *Measurement of Man at Work.* W. T. Singleton, J. Q. Fox and D. Whitfield (London: Taylor and Francis) pp. 129–134.

Feyerabend, P.K. (1975). *Against Method: Outline of an Anarchistic Theory of Knowledge* (London: New Left Books).

Flanagan, C. (1954). The critical incident technique. *Psychological Bulletin,* **51**, 327–386.

Fransella, F. and Bannister, D. (1977). *A Manual for Repertory Grid Technique* (London: Academic Press).

Gray, M. (1975). Questionnaire typography and production. *Applied Ergonomics,* **6**, 81–89.

Guilford, J.P. (1954). *Psychometric Methods,* 2nd edition (New York: McGraw-Hill).

Hudson, L. (1972). *The Cult of the Fact* (London: Cape).

Jung, J. (1971). *The Experimenter's Dilemma* (Toronto: Harper & Row).

Kalton, G., Collins, M. and Brook, L. (1978). Experiments in wording of opinion questions. *Applied Statistics,* **27**, 149–161.

Kelly, G.A. (1955). *The Psychology of Personal Constructs,* volumes I & II (New York: Norton Press).

Kirk, N.S. and Ridgeway, S. (1970). Ergonomics testing of consumer products 1; General considerations. *Applied Ergonomics,* **1**, 295–300.

Kirk, N.S. and Ridgeway, S. (1971). Ergonomics testing of consumer products 2: Techniques. *Applied Ergonomics,* **2**, 12–18.

Konz, S. (1979). *Work Design* (Columbus, OH: Grid Press).

Macfarlane Smith, J. (1972). *Interviewing in Market and Social Research* (London: Routledge & Kegan Paul).

Meddis, R. (1984). *Statistics Using Ranks* (London: Blackwell).

Meister, D. and Rabideau, G.F. (1965). *Human Factors Evaluation in System Development* (New York: John Wiley).

Moser, C. and Kalton, G. (1971). *Survey Methods in Social Investigation,* 2nd edition (London: Heinemann).

Mosteller, F. and Rourke, R.E.K. (1973). *Sturdy Statistics* (New York: Addison–Wesley).

Nie, N.H., Hull, C.H., Jenkins, J.G., Steinbrenner, K. and Bent, D.H. (1975). *SPSS—Statistical Package for the Social Sciences* (New York: McGraw-Hill).

Nunnally, J.C. (1970). *Introduction to Psychological Measurement* (New York: McGraw-Hill).

Oppenheim, A.N. (1966). *Questionnaire Design and Attitude Measurement* (London: Heinemman).

Osgood, C.E., Suci, G.J. and Tannenbaum, P.H. (1957). *The Measurement of Meaning* (Urbana, IL: University of Illinois Press).

Rasmussen, J. (1985). Trends in human reliability analysis. *Ergonomics*, **28**, 1185–1196.

Siegel, S. (1956). *Non-parametric Statistics for the Social Sciences* (New York: McGraw-Hill).

Simon, C.W., (1976). Analysis of Human Factors Engineering Experiments. NTIS Document. AD-A038-184/8GA.

Sinclair, M.A. (1975). Questionnaire design. *Applied Ergonomics*, **6**, 73–80.

Torgerson, W.S. (1958). *Theory and Methods of Scaling* (New York: John Wiley).

Tukey, J. (1977). *Exploratory Data Analysis*. (New York: Addison–Wesley).

Webb, E.J., Campbell, D.T., Schwartz, R.D. and Sechrest, L. (1966). *Unobtrusive Measures* (Skokie, IL: Rand McNally).

Wright, P., and Barnard, P. (1975). Just fill in this form—a review for designers. *Applied Ergonomics*, **6**, 213–220.

Ziman, J. (1980). *Reliable Knowledge* (Cambridge: Cambridge University Press).

Chapter 4

The use of archival data

Colin G. Drury

Introduction

This chapter covers the collection and the use of data which are pre-existing in the sense that the ergonomist had no part in their original collection. The ergonomist's role is to access, copy, edit, compile and interpret data collected for other purposes. Obvious examples are the use of medical records to estimate crash injury severity (e.g., Baker *et al.*, 1974), the use of company accident records to assess causes of accidents (e.g., Saari, 1976) or the use of quality control records to measure inspector performance (Drury and Addison, 1973).

Sources are available for most existing systems the ergonomist needs to study, such as factory, office, hospital and transportation systems. Archival data look attractive in terms of collection effort, but have their own unique pitfalls. Hence the chapter is organized by sources of data followed by general comments on suitability of consulting the archives.

Data sources

Fixed data

All organizations have sources of data which are relatively invariant over typical project time-scales and which are essential to provide proper background to an ergonomics investigation. A factory or office will usually have a layout drawing, given in more or less detail, showing the physical size of work areas involved, sizes of storage areas and locations of fixed facilities (e.g., toilets, cafeterias). More detailed layouts will include location of services (e.g., electricity, water, drains, compressed air), positions of lighting fixtures or permissible floor loadings. It is difficult to conceive of working in a plant (or airport or hotel) without such layouts, particularly if equipment is to be replaced or relocated. The accuracy of such fixed information is rarely in doubt, although locations and names of machines or other movable equipment may show differences between drawing and reality. It pays to check, and even have your own set of updated drawings, in order to plan lighting, noise or space surveys. As organizations convert to CAD (computer-aided design) systems, changes and updates should be easier for both the factory and the ergonomist.

Another type of fixed information is that provided by company annual reports or other legal documents. These show sales, profits, production volumes, inventories, fixed assets and total employment for a company, or equivalently numbers of beds and patients for a hospital, or numbers of aircraft and passenger miles for an airline. Such data are a great help in putting the ergonomics work into context before, during and after the study. At start-up the data show where to choose the various sites for ergonomics interventions to have the most generality. During the study, the fixed data show reasonable breakdowns for data collection, e.g., by department, line or product. After the study the data can be used to obtain multiplying factors to judge the overall impact on the company as implementation proceeds.

An obvious danger with the use of fixed data is that they can easily be outdated without the ergonomist realizing it. There is no substitute for checking 'on the ground' that the machine shop really does have three Heald ICF90 grinders, and so on.

Fixed records represent the input resources an organization has with which to achieve its goals. There is no guarantee that the appropriate quantities of machines, buildings, capital and personnel will actually achieve the goals, but it is often true that insufficient input resources will of themselves prevent goals being reached. The float process in flat glass manufacture is a good example. Until it was developed, all companies competed with different processes, but once one company had developed it, the cost and quality advantages were so overwhelming that soon it was licensed by all major competitors.

A final form of useful fixed information is found in planning documents. Strategic plans and operating plans based on where the enterprise and its products are headed are found in most organizations. They represent a useful source of future information about product types, changes in technology or marketing strategy and product volumes anticipated. Hence the ergonomist will be able to find out whether markets for products are growing or shrinking, whether individual items are going to need larger or smaller casting, whether new chemicals are to be introduced and how long production runs will be between product changes. All of these can impact production process design. More rarely there will be planning documents related to such ergonomics matters as payment policy or total employment expected, but talking to key people in the personnel department and union office will often make the ergonomist aware of any obvious trends.

Production records

If the fixed records represent the 'order of battle' for an organization, the day to day records represent the communications during the battle. A company must know how many items it is ordering, receiving, producing or shipping if it is to meet delivery targets, control inventory and schedule its internal processes. In manufacturing, the record site of production is often known as *order execution,* and is in itself the object of much study. We now frequently encounter computer-based *factory information systems* (FIS) and *management information systems* (MIS). Order execution proceeds from receipt of an order through a *materials requirements planning* (MRP) phase to ensure that the correct raw materials are available at the appropriate times. As jobs are released into the manufacturing shop either a push system or pull system is used to ensure that the correct machining and assembly operations are carried out. Finally, completed goods will be accumulated until the order is complete, checked and shipped.

Records are generated at each stage. For example the original order is entered onto a form as a database, purchase orders for raw materials are released and so on. On the manufacturing floor jobs are sequenced with routing slips or cards. Each line on the card is an operation, so that an operator signs for receiving a certain number of parts and for completing one operation on another number of parts. Incoming and outgoing numbers are different because of scrap, rework, missing parts and so on. Often a copy of the operator's part of the routing document will be retained by the operator to prove completion of a task, and hence get paid for that work. In continuous, as opposed to batch, production, each unit is covered by a separate record, rather than each batch. The record will specify, for example, a car's colour, engine type and options.

Routing slips show what work was performed at what times on what machines. Thus they give information on production levels, times, scrap rates and machine utilization. What they do not usually include is downtime, set-ups vs. running time, inventory levels or process logs, all of which are (or can be) recorded separately.

Production records in other situations can be very different, but the same principles apply. Thus a patient in a hospital has a time record by the bed to show tests and medications given and to record vital signs. Similarly, each airline flight has a manifest with passengers carried, route, weights and places to record safety-related task completions.

While records have a tremendous air of authenticity, they can conceal rather than reveal reality. On a production floor, a worker can 'bank' routing slips on a good day to have some excess production in hand for more difficult days. Recent press reports have shown that workers' time can be charged to different jobs to conceal cost overruns, and this has been found in jobs other than just the military ones.

Industrial engineering records

The industrial engineering (IE) or production engineering department of a plant or service enterprise is the source of many useful records, but with some caveats. Long associated with 'methods study and time study' (e.g., Barnes 1980; Konz, 1983), the IE department has traditionally been concerned with the details of how operators perform their jobs. Records are kept of task and sub-task breakdowns of jobs which form a remarkably good starting point for any task analysis (e.g., Drury, 1980 and chapter 6 of this book). However, the ergonomist should be aware of the fact that most task breakdowns were made for the purpose of estimating times for task completion. Hence, there is rarely any information on errors and variability or on processes difficult to time (e.g., cognitive tasks). Additional sources of data which could contribute to task analysis are in quality control and from training and safety, both of which are typically personnel functions (see later).

There are two major uses for task time completion data: planning and control. For planning purposes (bidding, estimating delivery dates, scheduling) we need accurate forecasts of time taken on each operation. For control purposes (e.g., production control, efficiency measurement, incentive payments) we need times which everyone concerned will accept as a reasonable yardstick against which to be measured. Thus planning and control lead to different potential biases in the times produced, and these biases can easily be the undoing of the naive ergonomist. Particularly if these times are for control purposes as they will directly affect people's measurable performance, so that pressures towards estimating larger times than actual will be beneficial to those whose performance is measured. As a 'standard time' for a job is composed of a measured time, a rating by the analyst of how effectively the operator is working and an allowance for non-productive time, there are ample opportunities for bias. Many time estimates are now made with predetermined motion time systems (PMTS), either by hand (e.g., MTM-2, WOFAC) or on a computer (e.g., MOST, ADAM), but the ergonomist should be aware that these same biases were built into the databases on which PMTSs are based. Just because a computer gives the answer to the nearest millisecond does not mean that the answer is necessarily either valid or reliable.

Having said that, time study data have a particularly valuable part to play in increasing the precision of ergonomic studies. Any job which is variable from minute to minute or day to day is particularly difficult to measure, for example, before and after comparisons of ergonomics change. Examples of such jobs are in batch production, where the product is not always the same, and in maintenance, where jobs recur only rarely. Having a standard time for each job allows the ergonomist to measure *relative* productivity rather than *absolute* production. Productivity is defined as a ratio of output to input, and thus is often expressed as a percentage of the expected production or inversely as allowed hours divided by actual hours for a given volume produced. Drury and Wick (1984) were able to demonstrate productivity improvements due to ergonomics changes in a shoe factory, despite frequent shoe style changes at each workplace, by measuring percentage productivity (rather than number of units produced) before and after the change.

Such normalization can also help in other ways, for example by allowing us to aggregate performance across different tasks such as set-up and production. Normalization is, of course, only as trustworthy as the standards used to normalize. Existing standards should never be accepted uncritically.

Quality control records

Quality is assessed, or ensured, at many different points in an organization. Often the work of a department labelled 'Quality Control' is only a fraction of the total quality effort. Hence in using quality measures for ergonomics studies, the ergonomist needs to be aware of the danger of inferring too much from departmental or job titles. Particularly with the current push toward ensuring quality production, rather than employing an army of inspectors to check a fault-prone production (Taguchi, 1978), quality control departments are tending to take on more of an advisory role while the primary quality responsibility rests with the operator.

Overall quality is measured by fulfilment of customer requirements and so logically includes design aspects as well as manufacturing aspects. Within an organization, however, a more restricted definition, such as 'quality of conformance', is accepted (e.g., Drury, 1982). This means that specifications exist against which quality can be measured to ensure conformance. With this definition, quality is measured in two ways: by attributes and by variables (e.g., Schilling, 1982). Attributes inspection refers to whether or not a particular attribute is present and typically leads to a dichotomous measurement scale of conforming/non-conforming, acceptable/rejectable, good/defective, etc. Measures include fraction defective, process yield (= $1 \cdot 0$ − fraction defective), defects per 100 units, as well as more detailed data on frequencies of particular types of defects. Variables data are amassed where a continuous variable is measured (e.g., length, temperature, viscosity, resistance) and sample statistics (mean, range, standard deviation) are used to infer population characteristics from sample measurements. Both attributes and variables can form the basis for two distinct control systems. Acceptance sampling answers the questions 'Should this batch be released to the next stage or customer?' while control charts answer 'Has the process producing these items changed sufficiently to require adjustment?'. The former is an after-the-fact judgement, whether applied to incoming parts, work in-process, or finished goods. Hence it represents a relatively slow-acting control loop. Control charting is a form of in-process quality control, which is aimed at ensuring that defective items are never produced. Such a fast-acting control loop is obviously preferable in a competitive manufacturing environment, although it requires both training and acceptance of responsibility by the operator.

Given the measures which exist, how can the ergonomist use them in studies? The first way requires no counting, just classification. In any process control study the first essential is to document all the variables involved. Defect lists from operators, quality control and customer returns make an excellent starting point. As has been noted before (e.g., Drury and Sinclair, 1983) the discipline of collecting, and agreeing upon, defect names and standards is an arduous process, but a necessary precursor to successful studies in process control inspection. The variables documented can be treated as head events in a fault tree analysis (Brown, 1976) or output variables in a signal flow graph of the process (Edwards and Lees, 1974). Again, do not accept data at their face value as defect names are often surprisingly local, and causes of defects can be shrouded in process mythology.

To use quality information in a quantitative way means using the fault rates, yield value and variable means described above. Most processes have multiple testing or inspecting functions, each of which acts as a data capture point for the ergonomist (Sinclair, 1984). Existing data should be usable in tracing error rates, and because re-inspection is not uncommon, revealing rates of inspection errors. Drury and Addison (1973) show how re-inspection data can be used to reconstruct initial yield and inspection error rates.

Having said that error applies to the inspection process itself, it should come as no surprise to be warned that inspection data are at least as imperfect as other existing records. A major caveat is that departments and individuals will do what they must to avoid ownership of specific faults. If there is doubt over the origin of an error, the blame may be freely passed from person to person before coming to rest, often on a person absent at the time. Less blatant biases occur when previous processes have to share the responsibility for the faults, and time-accumulation scales differ between departments. The author has seen examples where the defect rate goes negative every Friday, because this is when a previous process sends a representative to take responsibility for certain defects. All previous defects collected over a *week* are substracted from one *days* total, providing data which can easily confuse the unwary.

An example of an error-prone inspection task from archival data is shown in Figure 4.1. Prior to ergonomic intervention, data had been collected in a steel bar rolling mill to see how rapidly to pass bars through a 'Magnaglo' inspection system. Here, bars are dipped into a fluorescent liquid and then magnetized. Subsequent viewing under ultra-violet light allows the inspector to see cracks or other surface blemishes as glowing areas. The archival data consisted of a table of throughput rates (T bars/minute) on different weeks, with the numbers of bars inspected (N) each week and the number of faults (F) missed. As the task was expected to be one of visual search, the data were transformed to give a conventional speed–accuracy trade off (SATO) function plotting probability of fault detection $(N-F)/N$ against time in minutes per bar $(1/T)$. This enabled a search curve to be fitted (see Chapter 2) to make explicit the strategies for performance improvement.

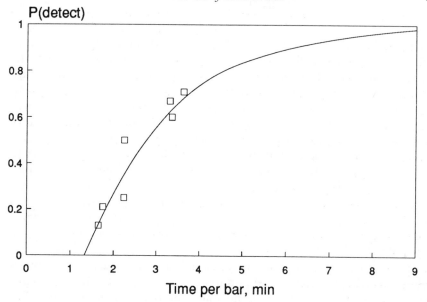

Figure 4.1. Speed Accuracy Trade off for Magnaglo inspection of steel bars

Personnel data records

Three types of data typically come from the personnel or human resources function: fixed data on jobs, data on individual employees and safety data. All have their corporate uses, and cross-referencing data between compartments can often yield useful insights.

Job data

In order to hire workers, place and promote them, a company needs descriptions of the jobs to be performed. Such job descriptions should (theoretically) exist for every job in the company; they should be detailed and regularly updated. Such Utopia is rare, but job descriptions do exist in sufficient numbers to make them a useful initial tool in systems and task analysis. Although typically lacking in ergonomic detail, some schemes for job descriptions can tell much about the job. Job evaluation schemes of the 1970s had operator and specialists rate jobs on many dimensions to determine parity of job difficulty and hence of pay scales. Even more detailed are the schemes based on one of the ergonomic systems developed in the USA (e.g., Position Analysis Questionnaire or PAQ; McCormick *et al.*, 1977) or in Germany (the Arbeitswissenschaftlichen Erhebungsuerfahoens zur Tatigkeitsanalyse of AET; Landau and Rohmert, 1981). Unfortunately such systems are not yet in widespread use so that the ergonomist may have to install one rather than find data in the archives.

Individual data

Personnel departments keep records of each individual employed, from which can come useful data. However these data on individuals are particularly sensitive and all of the ethical considerations of using such data must be fully realized before the study starts. Discriminations in hiring, placement and advancement are all too easy: the ergonomist should be seen to be sensitive to such issues. Data on age, gender, race, work history, medical history and company evaluations may provide useful co-variates or matching variables in ergonomic studies, but think ahead to the outcome. If you help to show that males or older workers or immigrants are unsuited to certain jobs, even as a by-product of your main hypothesis, you may put yourself and the company into an embarrassing position, as well as deny advancement to individuals. One must also be aware of subjectivity and bias in interview records, evaluation documents, citations for awards and even medical absence data.

In many companies and countries, research access to data as sensitive as medical records is restricted without appropriate safeguards. Even so, such data on pre-conditions, smoking and injury history, can be used to select homogeneous groups of subjects to reduce inter-subject variability and hence keep down sample sizes.

Accident and injury data

Human resources departments are often the home of safety and/or medical functions, both of which provide useful archival data. Medical histories have already been discussed under Individual Data, so that accident and injury data, whether in medical or safety departments, will be considered here.

In most endeavours, injuries rather than accidents are the events which precipitate a record in the archives. Despite pious articles on 'Total Loss Control', the typical enterprise will only record those events which it is legally required to report. In the USA, there are specific criteria which define what must be reported in industry (via OSHA), in aviation (via FAA), in mining (Bureau of Mines) and so on. As many writers (e.g., Hale and Hale, 1972) have pointed out, accidents are not synonymous with injuries, but merely a necessary condition (see chapter 32). Indeed, the injury is seen as one event in an accident sequence by Monteau (1977), by Haddon (1973) and by the US Consumer Product Safety Commission (CPSC, 1975).

Recorded injuries are thus the typical archival substitute for accident data, whether accessed by medical logbook, accident report forms or in-depth investigations. Because of this, there are many potential biases or artifacts in such data. First, non-injury producing accidents do not get reported, or go by another route such as scrap report. Second, not all injuries are reported. The '2000 Accidents' study (Powell *et al.*, 1971) clearly showed that low-severity injuries are remarkably under-reported. So are the less visible injuries, such as back strains and mental health problems. One suspects that *all* amputations are reported!

Given that an injury is reported, it may not be recorded, particularly if it is of low severity. OSHA in the USA requires all visits to a medical facility to be recorded in a medical log, but only injuries meeting certain criteria stimulate an accident or injury report. Beyond this only very serious injuries cause an in-depth investigation to be undertaken. Thus the ergonomist has a progressively filtered system, with broad but shallow coverage in the medical log, more detailed coverage of fewer events in the accident/injury report and deep coverage of a minimal number of accidents via in-depth investigations.

From a reasonable accident/injury report comes certain standard information on the victim (age, gender, height, weight, job category), the task (job at time of accident, accident precipitating event, injury event), the equipment used (agent of accident, agent of injury), and environment (time of day, unusual conditions). The care with which forms are devised and filled out are both highly variable. Most forms are strong on medical facts and individual differences and blame-pinning and weak in details of task, equipment and environment. The archetypal form reports that the victim sustained an (alleged) back sprain due to improper lifting technique and that the remedial action taken was to instruct the victim not to lift that way again. Here we have examples of blaming the victim (who was after all the only person involved), scepticism (strains are always alleged, amputations never are), circular arguments (for a back injury to occur, the lift *must* have used the wrong technique) and inappropriate intervention (which assumes that the employee did not have the knowledge or motivation to prevent the injury). One can go into any plant and find at least one such report within the thirty most recent reports.

If we are sceptical of the archival accident/injury data, why do we use it at all? First, it is of immediate economic importance. Each report represents a loss of revenues as well as large direct costs to the company and the employee. Second, accident/injury reports have a high level of face validity. After the events at Three-Mile Island everybody believed that there were dangers in current nuclear power stations. Third, archival reports can be used in a boot-strap procedure to devise better classifications (Drury and Wick, 1984), to see which departments are in most urgent need of ergonomic help (Drury and Wick, 1984), and to derive new and more detailed accident investigation procedures (Monteau, 1977, Drury and Wick, 1984). If the ergonomist looks at the accident/injury reporting system with detachment and foreknowledge, useful data will be found. (See also chapter 32.)

Costing data

It is often important to make a direct financial case for ergonomic change (see chapter 34), although we all recognize that the most visible cost elements are rarely the whole story. Organizations keep detailed financial records from the annual balance sheet down, but ergonomists are rarely equipped to understand the concepts and terminology of the cost accounting systems employed. Excellent standard texts are available (e.g., Horngren, 1967) and most introductions to business (e.g., McGarrah, 1963) and engineering management (e.g., Garrett and Silver, 1973) provide useful, if shallow, primers.

Unfortunately, the questions an ergonomist wishes to ask are rarely the ones the costing system is designed to answer. We typically want to know how much it costs now to perform a certain operation, and how much less it will cost if specified changes are made. Direct costs are relatively easy to measure. Labour costs per hour, scrap costs per unit and even energy costs per cycle, should present no difficulties. But the operation we are studying does not exist in isolation, indirect costs are involved. A lathe operator will use a certain amount of space, environmental energy (heat, light), inventory, quality control, maintenance, personnel, washrooms, pension plans, sales executives and even ergonomists. These indirect costs, or overhead burden, are notoriously hard to allocate to direct operations, but ultimately all must be paid for from product sales and all need to be accounted for. In one company, where workers were paid at $15 per hour, the time cost of an hour's work was pegged at $60 for any labour-saving calculations as a more reasonable value for the true cost of employing an operator.

Overheads may be assigned by labour hour, by square feet of space occupied, by product being produced or by any other formula the company desires. The intent is to stress the factor which is critical to operations, whether it be direct labour or space. With the advent of plentiful and powerful computers, it is now possible to allocate overheads in many different ways to more properly reflect the true costs of doing business. Some functions which were previously financed from overheads are now being asked to operate as cost centres. Thus the ergonomics department, personnel offices or medical centre must hire out its services to the operating departments, and at least break even each year. Such systems focus attention onto the business essentials, but do have the obvious side-effect of discouraging a long-term view.

How does all this help the ergonomist to find the cost savings potential of an intervention? In truth, it does not; but largely because the ergonomist is not skilled in formulating the questions which still need to be asked and answered. Perhaps the best advice is to read up on this subject, to understand at least the rudiments, and then develop a close relationship with the costing functions in any organizations you enter.

The joys and perils of archival data

It should be obvious by now that archival data represent a valuable resource for the ergonomist, but are full of hidden traps for the unwary.

Measurements from archival data complement the more usual ergonomic surveys and experiments in many ways. Primarily they represent a long-term view of the effects of change over months or years rather than a snapshot of one particular process over a few days. This long-term view enhances the face validity, because the real system is operating, with all of its day to day idiosyncrasies as opposed to a 'best behaviour' controlled experiment. Additionally, the raw data are already being collected, and so the additional data collection costs look minimal. Inexpensive, face-valid data which reach back into the past and will reach forward into the future are certainly worth considering, but this is not the whole story.

Existing data were *never* collected for ergonomics purposes and so will never answer the ergonomist's particular questions except by fortunate coincidence. We must accept a high level of aggregation, with, for example, accident rate being aggregated across many individuals whose jobs vary to a greater or lesser extent. Similarly, productivity or quality data are accumulated over weeks or months of highly variable conditions, about which the ergonomist is likely to remain forever ignorant. Thus cause and effect are difficult to establish. The author has participated in studies in which scrap rates and accident rates halved from the year before the intervention to the year after, but it is naive to think that the ergonomics intervention was the sole innovation by the organization in two years. All of the workers responsible for poor quality or high accident rates (assuming such people existed) could have been fired in the second year, but without a very close relationship on an on-going basis it would not be obvious to the ergonomist that such a change had taken place.

Additionally, there is much scope for bias and distortion in organizational records. Thus a particular customer (e.g., the Government) may require particular cost reporting procedures which differ from the regular ones. Alternatively, accident causation can be classified by a supervisor in a way different from that used by an ergonomist investigating the same accident. People play games with quality and productivity records, legal and ethical considerations prevent access to certain data sources. As the world moves towards computer integration in manufacturing and other enterprises, disparate data sources can now be integrated and correlated. While there is much potential for increased depth and breadth of understanding of as complex an organism as the human at work, there is also the potential for increased control and abuse. Ergonomists must know where they stand, and be prepared to back up their decisions with appropriate action.

References

Baker, S.P., O'Neill, B., Haddon, W. and Long, W.B. (1974). The injury severity score: a method for describing patients with multiple injuries and evaluating emergency care. *Journal of Trauma*, **14**, 187–196.

Barnes, R.M. (1980). *Motion and Time Study: Design and Measurement of Work*, 7th edition (New York: John Wiley).

Brown, D.B. (1976). *Systems Analysis and Design for Safety* (Englewood Cliffs, NJ: Prentice Hall).

CPSC (1975), *In-Depth-Investigations* (Washington, DC: Consumer Product Safety Commission).

Drury, C.G. (1980). Tak analysis methods in industry. *Applied Ergonomics*, **14**, 19–28.

Drury, C.G. (1982). Improving inspection performance. In *Handbook of Industrial Engineering*, edited by G. Salvendy (New York: John Wiley).

Drury, C.G. and Addison, J.L. (1973). An industrial study of the effects of feedback and fault density on inspection performance. *Ergonomics*, **16**, 159–169.

Drury, C.G. and Sinclair, M.A. (1983). Human and machine performance in an inspection task. *Human Factors*, **25**, 391–400.

Drury, C.G. and Wick, J. (1984). Ergonomic applications in the shoe industry. *Proceedings of 1984 International Conference on Occupational Ergonomics*, pp. 489–493.

Edwards, E. and Lees, F.P. (1974). *The Human Operator in Process Control* (London: Taylor and Francis).

Garrett, L.J. and Silver, M. (1973). *Production Management Analysis* (New York: Harcourt, Brace, Jovanovich).

Haddon, W. (1973). Energy damage and the ten countermeasure strategies. *Human Factors*, **15**, 355–366.

Hale, A.R. and Hale, M. (1972). *A Review of the Industrial Accident Research Literature* (London: HMSO).

Horngren, C.T. (1967). *Cost Accounting—A Managerial Emphasis* (Englewood Cliffs, NJ: Prentice Hall).

Konz, S. (1983). *Work Design: Industrial Ergonomics* (Columbus, OH: Grid).

Landau, K. and Rohmert, W. (1981). *Fallbeispiele zur Arbeitsanalysi* (Bern: Hans Huber).

McCormick, E.J., Mecham, R.C. and Jeanneret, P.R. (1977). *PAQ Technical Manual* (West Lafayette, IN: PAQ Services).

McGarrah, R.E. (1963). *Production and Logistics Management: Text and Cases* (New York: John Wiley).

Monteau, M. (1977). *A Practical Method for Investigating Accident Factors* (Luxembourg: Commission of the European Communities).

Powell, P.I., Hale, M., Martin, J. and Simon, M. (1971). *Two Thousand Accidents* (London: National Institute of Industrial Psychology).

Saari, J. (1976). Typical features of tasks in which accidents occur. In *Proceedings of 6th Congress of IEA, Human Factors Society*, Santa Monica, pp. 11–16.

Schilling, E.G. (1982). *Acceptance Sampling in Quality and Control* (New York: Marcel Dekker).

Sinclair, M.A. (1984). Ergonomics of quality control. *Workshop for the International Conference on Occupational Ergonomics*, Toronto.

Taguchi, G. (1978). Off-line and on-line quality control systems. *Proceedings of International Conference on Quality Control*, Tokyo, Japan.

Chapter 5

Designing ergonomics studies and experiments

Colin G. Drury

Introduction

This chapter will address the broad issues of designing studies so as to reduce the chances of the ergonomist leaping to obvious solutions without fully considering the alternatives. Thus the design and conduct of experiments *per se* are treated as a special case of ergonomic studies, to be used where a conscious decision has been made that they are appropriate.

There are two major choices to be made in study design: what to measure and how to measure it, known respectively as *measurements* and *methods*. Unfortunately, they are not entirely independent, so that the process of choosing alternatives is often iterative. Starting with the methods is most logical, as major decisions must first be made on the overall technique (observation of natural behaviour vs. designed experiment) before decisions on independent and dependent variables must be made.

As with any goal-oriented activity, designing studies is largely a matter of accurate goal specification, followed by logical (and/or economic) choice of steps to meet that goal. What is the goal of a study? Typically, this would be set by those employing the ergonomist, e.g., to test different computer keyboard features for user acceptance. In many circumstances the goals will be defined in a formal request for proposal (RFP), but in other cases the ergonomist will have some leeway to question the study goals. Perhaps the greatest freedom of goal-choice is in academia, where a student or faculty member can run studies with self-generated goals, subject only to their ability to find funding and satisfy human subjects' protection rules. Any good textbook must advise the ergonomist to question the goals of the study, but must also caution that much effort has been wasted on this activity over the years. We will assume that you have a set of goals and proceed as if these were now engraved on stone tablets. The object of this chapter is to provide an orderly way in which these goals can be turned into a designed study, in much the same way as a statistical textbook will demonstrate how to turn a research hypothesis into a statistical hypothesis.

If a goal is the input then the output is a study design. The specificity of the design is particularly important as minor details can have major consequences. One only has to look at critical evaluations of the research literature on vigilance (e.g., the special issue of *Human Factors,* **24.6**, 1988) to see that after 40 years of work, differences between major theories still hinge on the minutiae of study design. A good test of the completeness of the design is that it can be given to a competent technician to carry out, and it will always be done in the way the originator intended. In statistical design of studies, the hypotheses to be tested are defined in terms of the measurements to be taken and the tests to be performed on the data, *before* any data are collected. A well-designed ergonomics study should meet the same criteria—you must know what to do with the data you have collected if you are to avoid the statistical (and economic) error of collecting data *and then* looking for interesting interrelationships. This does not mean that all studies require a formal research hypothesis to justify every measurement. Data collection is expensive, so that it may be possible to augment the study by including some measurements which can be conveniently collected but for which no formal hypotheses are made. In that case, these augmented variables cannot be used to test hypotheses after the fact, but they can be used to formulate new hypotheses for testing in subsequent, independent experiments. While the major goals of any study must always be tested hypotheses, it would be foolish to overlook the opportunities presented by major studies to learn more about human interaction with the environment.

Choice of technique

Chapanis' (1953) original methodology text classified the techniques available to the ergonomist into observation and experiment, the distinction being whether the world was observed by the observer or manipulated by the experimenter. This distinction is still important, but we should now recognize more than just two levels. The major difference between the levels is in their reactivity—how much or how little they change the system under study.

At the extreme low end of reactivity are *task analytic methods*. At the earliest levels of system design, they do not even rely on a system to study. Tasks are described and analyzed for their ergonomic impact by logically deducing the task description from system goals and functions and by deriving task demands and human capabilities from the logic and the literature. Later stages of task analysis may well observe the real system, or even call for experiments (e.g., Drury *et al.*, 1987) but by then we are using the techniques which follow (see chapter 6 in this book).

At the next higher level of system manipulation are the methods of *direct observation.* Here a functioning system is studied by reading its records or observing its behaviour (see chapters 2 and 4). We often think that such techniques are non-reactive, but they do involve actively interfering with the system, i.e., no observations would be taken were the study not taking place. The reactivity may only be potential, as in studying ambulance effectiveness from patient records (Baum and Drury, 1976) or it may be very real, as in a stop-watch time study (Konz, 1983).

Questionnaires and *rating scales* (see chapter 3), the next higher level, obviously affect the system but are often used as if they did not. For example, a study of emergency telephone responses (Baum and Drury, 1976) probably caused many subjects to think through their emergency response behaviour for the first time as less than 15% had ever used the telephone to call for fire, police or ambulance.

Finally, any *direct manipulation* of the system must be classed as an experiment, whether it is a multifactorial design used in a research laboratory or a case-study (or other pseudo experiment; Kerlinger, 1986) run in a factory. The system is deliberately changed so that the results of the changes can be observed. This can never be done in a non-reactive manner.

Given that manipulating and changing systems must be costly and potentially dangerous, what does this increase in reactivity buy for the ergonomist? The two major advantages of a more highly reactive design are:

(1) The ability to be in the right place at the right time to observe (Chapanis, 1953). This is particularly important in ergonomics studies where the system behaviour observed is rare and unexpected, e.g., accidents or breakdowns.

(2) The ability to use more obviously invasive, but information-rich, measurement techniques. For example, in inspection research, the response to each individual item inspected can be of considerable detail in an experiment (e.g., Drury and Sinclair, 1983), whereas in the real situation only a simple accept/reject response is often given.

If the ergonomist gains in experimental control and measurement detail by using highly reactive designs, what else is lost? The major loss is in face validity. If we observe a system in its natural state, those associated with the system, and possibly those who commissioned the study, can be convinced that the study is realistic, however that is defined by those to whom it is important. An experiment, particularly one performed in a laboratory with artificial stimuli and non-representative subjects, requires much more persuasion on the part of the ergonomist to gain acceptance. The author was once involved in two studies of fork-lift truck control. One (Drury and Dawson, 1974) involved real drivers using real fork-lift trucks in a real warehouse to study lateral control behaviour. The other (Drury *et al.*, 1974) involved real drivers controlling a toy train in a laboratory to study longitudinal control behaviour similar to Fitts law tasks. It is obviously much easier to quote the former study to convince warehouse managers of its design implications.

There are many degrees of realism within the experimental paradigm. Experiments can be conducted using the operational system (e.g., real aircraft), a faithful simulation (e.g., flight simulator) or a task only logically related to the operational system (e.g., tracking or dichotic listening). The issues of fidelity in experimentation are essentially the same as those in simulator design (see chapter 8). Only a thorough knowledge of the ergonomics literature can guide the experimenter on what it is safe to leave out of an experiment. As an example, there are thousands of references on the component processes of an industrial inspection task (visual search, signal detection, vigilance) but only a few dozen which meet the stringent requirements of experimental techniques used by real inspectors with real products.

These issues of reactivity, control, measurement detail, and validity are summarized in Table 5.1.

Choice of independent variables

Independent variables are what you change, while dependent variables are how you measure the effects of the change. Choice of independent variables comes primarily from the hypotheses: if you hypothesize that control–display compatibility affects reaction time, you have chosen compatibility as your independent variable. If life were that simple, this section would not be needed. In order to measure reaction time under different values of the independent variable, you must get down to specifics on other matters: who will be the subjects? What will be the instructions? Will the experiment be run sitting or standing, morning or evening, and so on? Thus choice of independent variables is not merely a matter of making obvious deductions from the hypotheses.

It will be simpler to talk of independent variables as if they were manifested in an experiment, implying that the ergonomist actively implements the choices of values of each variable. However, the same principles apply if the technique is observation or questionnaire although choice is more a matter of selecting, from among already existing alternatives, those to be included

Table 5.1. Comparison of techniques by different criteria

Technique	Reactivity	Face validity	Control	Measurement detail
Task analysis	zero	high	—	—
Observation/records	low	high	zero	low
Questionnaires/ratings	medium	medium	low	medium
Experiments	high	low	high	high

in the study. In a similar manner, some of the language of experimental design can be applied to observation and questionnaires. Thus an independent variable is a factor and the value which the independent variable takes is the level of that factor. Both naming conventions will be used throughout this section.

The design of any study first consists of choosing levels of all of the independent variables of relevance. Because we must be specific in our study design, then we must specify not only what we will vary, i.e., what levels of which factors, but what we will not vary, i.e., what we will keep constant. One has only to look at the *post hoc* rationalizations of unexpected study results in our journals to see that ergonomists do not always recognize what must be kept constant. Taken with the frequent finding of insignificant effects of a major factor, it recalls the scientist's version of Murphy's law: 'variables won't and constants aren't'. The least that an uncontrolled variable can do is to contribute to random experimental errors: typically it does more and introduces real biases into a study.

Systematic techniques are required if we are to bring ergonomics knowledge to bear upon the design of effective and efficient studies. We need a technique for listing all possible factors which can affect the outcome of the experiment, and a technique for deciding what to do with each factor.

To generate a list of all possible factors which could affect the dependent variables, it is simplest to use the categories which ergonomists regularly use: human, machine and environment. To these should rightly be added task, to cover the instructions, restrictions and goals under which the system operates. Thus the four categories are *task, operator, machine* and *environment*. Under each of these is listed the factors likely to affect performance. For example, in a study of compatibility and reaction time obvious factors under operator are: age, experience of similar systems, training on system under test, and national stereotypes for control/display relationships. Less obvious factors, but ones which can certainly affect reaction time, are: intelligence, visual capabilities, risk acceptance/aversion, and motor co-ordination. Some factors *may* affect performance, but our insight tells us that the probability is small: body size, gender, and time since last meal.

This list could obviously continue well into the trivial, as could similar lists for task, machine and environment. Note that the lists will be different for different hypotheses; in manual materials handling systems, body size and gender would be of primary importance.

We now have four somewhat orderly lists of possible factors and face the question of what to do with each factor on each list. There are only five alternatives:

1. *Build the factor into the experiment at multiple levels.* This ensures that our hypotheses will be general across the whole range of this factor used in the experiment. Thus if we have five age groups, covering the decades of the working population (i.e., 15–25, 25–35, 35–45, 45–55, 55–65 years), we can

be sure that our compatibility conclusions are valid for all working ages, or we will know that they only apply to older workers, depending upon the specific outcome of the study. This is the preferred treatment of each factor, as it maximizes the impact of our studies. It is also by far the most expensive option, especially as each factor tends to multiply the size of the experiment by the number of its levels. Typically, only factors specifically named in our hypotheses will be built in at multiple levels.

2. *Treat the factor as a co-variate.* Often it is not feasible to choose levels of a major variable in advance, as we may need expensive measurements to determine what the level actually is. For example, although age and body size are simple enough to measure or even to ask for by telephone when scheduling subjects, more subtle factors such as perceptual style or visual reaction time will require the subject to undergo tests before being allowed into the experiment. In such cases, the factor of interest can be measured, typically before the main experiment, and used as a co-variate in an analysis of covariance design. This means that any correlation between the dependent variable and the co-variate is taken out as a specific term in the analysis, typically before the other factors are analyzed (Nie *et al.*, 1975). The co-variate option is usually used for operator variables, although it is not limited to them. In observational and questionnaire studies, many of the variables are treated as co-variates, such as number of miles ridden per year in studies of motorcycle accident frequency. At times co-variates are the major focus of a design, as for example the classic Fleishman and Rich (1963) study of factors affecting task performance at different stages of learning. At the very least, carefully chosen co-variates reduce the waste associated with any study which finds that individual differences account for a large portion of the overall study variance. To give an example, significant correlations between perceptual style and inspection performance (Gallwey, 1982) have enhanced our understanding of the inspection task. Obviously, if multiple subjects are to be a part of the design for other reasons, then treating a factor as a co-variate is much less expensive than building that factor in at multiple levels.

3. *Fix the factor at a single level.* Here a factor is treated as a constant. We could run our study with a single, narrow, age group; we could choose only third-generation Americans; we could run all experiments in a well-lighted room at constant temperature. By having only a single level of a variable, we do not increase the size of our study but the price we pay is that the results will not generalize beyond the constant conditions we specify. We may wish to extrapolate the study results, based on other data and other models, but false extrapolation has plagued many disciplines including ergonomics. Human performance has a way of constantly surprising us. Typically, the fixing of a factor at a single level is used for those factors we know are important, but about which we do not need to generalize.

4. *Randomize the effects of a factor.* If a factor may be important, but we cannot control it in one of the ways above, then randomization will prevent

the factor from biasing the study results. By randomly assigning subjects to factor level combinations or stimulus presentation order to subjects, any uncontrolled variability is given an equal chance of affecting all levels of the particular factor. Thus the random variability may be increased (resulting in a weaker study) but we will avoid reaching biased conclusions. A well-known example of the use of randomization are the random assignment of subjects to treatment groups, rather than putting the first volunteers in group A, the rest in group B, and so on. People who volunteer early may be different from those who volunteer late and randomization ensures that this volunteer bias does not result in a biased study.

5. *Ignore the factor.* This final strategy is only included for the sake of completeness. It is never safe to ignore a factor in studies involving entities as complex as human beings. The old adage, 'If in doubt, randomize,' applies here. Ignore factors at your peril.

At this point, we have an assignment problem—how to assign each factor to each alternative. In practice, the assignment is usually quite straightforward, because the real work was performed in listing the potential factors in an organized and ordered manner. What has been achieved is that a systematic technique has been used to reduce the chance of the study finding unexpected conclusions for the wrong reasons. The whole procedure may seem laborious, but it forces the ergonomist to think through the issues *before* the study starts and to use ergonomics insights to advantage. A useful exercise to practise these skills is to take a paper from the literature and to use these techniques to see how you would have designed the study. Here you have the advantage of hindsight—you know what the results were. In practice you need to develop foresight if you are ever to get a second chance at designing a study.

Choice of dependent variables

Traditionally, ergonomists have measured the effects of their factors on a single variable (e.g., reaction time, error percentage, heart rate), but advances in the ability to record and analyze multivariate data have to some extent stopped this trend. Now we tend to think in terms of sets of dependent variables, each illuminating a different aspect of human/machine fit. Despite this trend, the current section will concentrate on single measures, as the choice of sets of measures depends to some extent on the branch of ergonomics involved. For example, it is pointless to run a signal detection experiment without measuring (or controlling) both type 1 and type 2 errors (see later).

Before classifying and describing possible measures, it is necessary to have available some criteria by which to judge the adequacy of measures. Social science texts (e.g., Kerlinger, 1986) see three main criteria:

1. *Validity.* Does the measure have a direct relationship to the phenomena described in the hypothesis? Is heart rate a valid measure of metabolic cost? Only under particular circumstances such as lack of heat stress, static muscular tension and emotional stress. Is reaction time a valid predictor of the relative performance of three computer keyboards? Only when certain major variables such as information content per stimulus are kept constant and only when minimum response time has some relevance to the final criteria of the system evaluation. Technically, validity is the correlation between the variable measured and the phenomenon represented in the hypothesis. A measure may be valid a priori (face validity) because it patently measures the phenomenon. Thus accident frequency is a face-valid measure of plant safety. A measure may be valid because it can be logically related to the phenomenon (construct validity). Thus reaction time is a construct-valid measure of control-display compatibility because it is the reciprocal of information processing rate (for constant information per stimulus) and thus is logically related to compatibility. Finally, a measure may be proven valid experimentally. Thus Borg (1982) has validated two scales of rated perceived exertion against heart rate by experimentally correlating the two. Clearly, choosing an invalid measure will ensure that we answer the wrong question.

2. *Reliability.* Does the measure give consistent results when used repeatedly? Is heart rate too variable from pulse to pulse or minute to minute to be a reliable measure of metabolic load? Do we get the same reaction time to the three computer keyboards if we measure them today, tomorrow and next week? If validity asks how well a measure correlates with an external phenomenon, reliability asks how well the measure correlates with itself. While there are many reliability measures, the two most common are test/retest reliability and split-half reliability. Test/retest reliability correlates the measure obtained on a subject in the first (test) period, with that obtained during a subsequent (retest) period. Split-half reliability correlates two halves of the measure obtained during a single time period. Thus in an intelligence test, the score on even-numbered questions should correlate well with that on odd-numbered questions. Similarly, the reaction times measured in the first and second halves of a trial with computer keyboards should correlate highly. Reliability coefficients close to 1·0 are desirable, with values of 0·8 and upwards being used regularly in social science work. Reliability sets a limit to the internal consistency of our studies. Unreliable measures, even if they are valid, cannot prove a hypothesis because we cannot have faith in them to tell how the phenomenon of the hypothesis changes.

3. *Sensitivity.* Does the measure react sufficiently well to changes in the independent variable? It is quite possible that the measure chosen may be valid and reliable, but will not show a large enough effect to be measured easily. Thus, reaction time may not differ between different keyboards because it is not sufficiently sensitive to record the subtleties of keystroke force/distance characteristics. Similarly, heart rate may not react by more

than a few beats per minute to differences in the design of handles in manual lifting tasks (Deeb and Drury, 1986), requiring a relatively large design to obtain statistical significance.

Any measure must pass all three tests (validity, reliability and sensitivity) before it can be used, although not every measure will need to perform equally on all three criteria. Measures themselves fall into two broad classes:

(1) performance—measures of the effect of the human on the system; and
(2) stress—measures of the effect of the system on the human, or cost of the performance to the human (see discussion of evaluating effects on performance and on people in chapter 1).

In general, it is difficult to conceive of measuring one of performance or stress without measuring (or at least controlling) the other; at times this control is implicit. Thus, when we measure reaction times in a laboratory, it is usually implied that the subject performs at the maximum level, i.e., at a constant level of stress or cost. Similarly, in a manual lifting task, the task is often fixed (box size, lift distance, speed) so that the physiological cost can be estimated using oxygen consumption, heart rate or even subjective ratings. At other times, the joint measurement of performance and stress will be explicit, as in multivariate assessment of jobs where both performance and stress need to be measured separately in a realistic setting to determine how subjects choose to allocate their processing resources to tasks of variable demand.

A final consideration in choosing a dependent measure is the level of aggregation of the measure. Ergonomists use data from a variety of levels, from the minute (e.g., time taken for a single cycle of a single component of a task) to the gross (e.g., the annual injury rate of a plant). Measurements can be aggregated over organizational components and over time. Table 5.2 shows typical levels.

In choosing a level of aggregation, there are trade-offs to be considered. The higher levels of aggregation, such as the plant's annual quality costs, are more obviously meaningful to the organization, often referred to as 'system

Table 5.2. Classification of levels of aggregation in dependent measures

Level	Organizational component	Time period
Lowest	Individual task component	Single cycle
	Individual task	Batch or lot
	Individual job	Shift
	Work group	Day
	Plant	Week
	Organization	Month
Highest	Industry	Year

relevant' measures. However, they may be difficult to interpret unambiguously due to the many other factors which can influence these measures. Thus the annual quality costs for the plant may not only reflect ergonomic changes, but those due to industry-wide quality changes demanded by customers, changes in personnel, or even a new accounting system. The discussion of the dangers of archival measures (chapter 4) is relevant here. Whenever a high level of aggregation is used, special pseudo-experimental designs (Kerlinger, 1986) may be needed to make realistic comparisons against control groups which have not had the intervention, e.g., Cohen and Jensen (1984).

Performance measures

Performance itself can be subdivided into two aspects; speed and accuracy. Speed refers to the time for task (or sub-task) completion, to the amount produced in a given time period (e.g., output per shift) or to the speed of movement (e.g., driving). At times it can be a straightforward measure such as task completion time, or it can be normalized with respect to some external criterion, for example speed expressed as a percentage of maximum speed. Normalization can be a source of increased accuracy, as in measurement of accident rates rather than accident frequencies, or it can be a source of potential confusion, as in measuring industrial efficiency with respect to time standards. Productivity is another normalized speed measure, typically defined as output from a system for a given level of input resources.

Accuracy is the other aspect of performance. Dictionaries define accurate as correct, truthful, or conforming with standards, of which the last provides a useful working definition. If there are no standards we cannot measure accuracy. In discrete tasks, we can measure accuracy as freedom from discrete errors, and in continuous tasks as the ability to match output to input. Both will be considered in turn, but first the phenomenon of the speed-accuracy trade-off (SATO) must be covered. SATO occurs in certain tasks, known as resource-limited tasks. In these, the more time of effort which is expended, i.e., more resources devoted to the task, the more accurate is the response. In contrast, data-limited tasks are those where the input data limits the accuracy, no matter how many resources are brought to bear. An example of a resource-limited task is visual search, where the more fixations an inspector uses to view a product, the more likely any defect is to be detected. A typical data-limited task would be deciding whether to play red or black at roulette, where no amount of extra thought will help predict the next spin of the wheel. Figure 4.1 in chapter 4 gives an example of a visual search task exhibiting a SATO effect. From theoretical models of visual search, we expect that the probability of detection has an exponential relationship to the time allowed for searching (Morawski *et al.*, 1980). Fitting such a relationship to the data in Figure 4.1 gives:

$$p(detect) = 1 - \exp(-0{\cdot}69 \, (time - 1{\cdot}33))$$

where 'time' is the time per bar in minutes. This equation fits well ($r^2 = 0.91$) and tells us that there is a 1.33 minute fixed time per bar, and that the bar is searched at a rate of 0.69/min, i.e., that in a time of 1.45 min (= 1/0.69) a fraction (1/e) or 38.8% of the search will be completed. From such an equation we can improve matters by decreasing the fixed time (e.g., by improved bar handling and data recording) or by increasing the search rate (e.g., by improving the lighting or the inspector training). We could even decide upon an optimum time for inspection of each bar, knowing the costs of errors, the costs of the inspector's time, and the probability of a defect.

In a more theoretical vein, there are good discussions of SATO in Wickens (1992) and Sperling and Dosher (1986). From the point of view of measurement the implications are straightforward: if the task studied is likely to be resource-limited, then it will be necessary to consider the SATO function before deciding upon the measurement scheme. We have one of three alternatives, if we do not take the dangerous step of ignoring one or the other aspect of SATO:

1. Fix speed and measure accuracy.
2. Fix accuracy and measure speed.
3. Let speed and accuracy be chosen by the operator and sort out the effects during analysis.

The first alternative represents a paced task. An example from the movement control literature is that of Beggs and Howarth (1972) who had subjects aim at a target in time to a metronome, and measured the resulting error distribution of deviations from the target. In an unpaced task, by contrast, the subject is given all the time needed to achieve accurate performance. For accurate movements, the classic Fitts' tapping task is an example (Fitts, 1954), where the subject is given targets of a fixed size and must hit each target with perfect accuracy, with time per movement being the measured dependent variable. Alternative 3 has some attraction in that it gives the operator choices of balance between speed and accuracy, but interpretation is difficult because of the resulting variability. It is not recommended.

One aspect of the SATO phenomenon which is important to those who would measure performance is that the typical SATO curve shows diminishing returns for accuracy as time increases. Thus in Figure 4.1, the first minute of search (after the fixed time) can be expected to capture 49.8% of the faults, whereas the second minute only gives another 25.0%. During the seventh minute (assuming that the SATO curve can be safely extrapolated that far) only 0.8% of faults will be detected. Thus, if we demand 'perfect' accuracy of our operators in a resource limited task, the times taken will be very long, and highly variable depending upon how each operator interprets 'perfect'. It is more realistic to have a fixed, non-perfect, accuracy criterion. In Fitts' tapping studies, for example, one miss in ten trials may be a useful definition. One can even use a payoff-matrix, to give the operator a better idea of the relative weighting of speeds and errors, although this can be somewhat artificial in industrial tasks.

Dealing with the SATO curve has another important advantage in interpretation: any improvements may be taken as speed or accuracy changes, provided the operator is given appropriate instructions about the desired balance. Thus in a task where accuracy is of extreme importance, such as inspection of airliners for structural cracks, it is possible to measure task time in an experiment, and later interpret any performance improvement in terms of accuracy. For such an interpretation to be valid, a model of human performance in the system must be available (e.g., the visual search model in Figure 4.1), and be accepted by all concerned. In practice, models are often the ergonomist's key to understanding the task, so that they will be available anyway.

An error rate or error probability is defined as

$$p(\text{error}) = \frac{\text{number of errors}}{\text{number of opportunities}}.$$

The measurement of both numbers can be an error-prone activity itself. Thus errors have reporting biases and classification problems. Number of opportunities can be easy to calculate in some circumstances but difficult in others. For repetitive tasks, error rate is typically defined per cycle, so that we have sewing errors per 100 pairs of jeans, near misses per flight, or polarity errors per component mounted on a circuit board. For some repetitive tasks, such as inspection, there is more than one input and thus more than one error rate. An inspector receives both good components and faulty components and thus has two error rates with different denominators as well as numerators:

$$\text{Type 1 error} = p(\text{reject/good}) = \frac{\text{number of good items rejected}}{\text{number of good items}}.$$

$$\text{Type 2 error} = p(\text{accept/faulty}) = \frac{\text{number of faulty items accepted}}{\text{number of faulty items}}.$$

In such tasks, an overall error rate, obtained by dividing all errors by the total items inspected, is a meaningless quantity, although techniques do exist for combining both errors into other measures with more meaning (e.g., Drury, 1982).

For highly variable tasks, choosing the error rate denominator becomes more difficult. Thus in driving, process control, or logging we can define two denominators with different meanings:

Output based: errors per mile driven
errors per barrel of oil processed
errors per ton of logs harvested

Time based: errors per hour driven
errors per month of oil processing
errors per day of logging.

There are interesting moral questions in the choice of denominator; an output base has meaning for the company while a time base is more important to the individual at risk.

To some extent each task has its own unique error characteristics, as can be seen by examining defect reporting sheets in industrial processes. However, there are some useful, more general error taxonomies in use in ergonomics.

Classification of errors, particularly human errors, has received considerable attention in recent years, with books such as Reason (1990). If any attempt is made to do more than merely count errors, then reference should be made to such works, and also to summaries such as chapter 32 in this book. A useful attempt has been made to classify the classification schemes, by Senders and Moray (1991). They suggest classifying errors into four types of taxonomy:

1. Phenomenological   e.g., omissions, substitutions
2. Internal Processes   e.g., capture errors, overload, decision errors
3. Neuro-psychological   e.g., forgetting, stress, attention
   Mechanisms
4. External Processes   e.g., poor equipment design.

All may be used as appropriate error classification schemes, either singly or, if there is sufficient error data, in combination. One useful classification which incorporates many of these is that of Rasmussen (1982) also considered in chapter 32 (Figure 32.1). A convenient strategy where there are relatively large numbers of recorded errors, accidents or incidents is to develop Hazard Patterns (Drury and Brill, 1983) which group together incidents by their immediate, or proximal, cause. For example, Drury and Brill found that chain saw accidents could be classified into:

Loss of control of chain saw
Loss of body balance
Kickback
Loss of attention

with an appropriate description of each included in each classification. Other ways of classifying errors are discussed in chapters 31 and 32.

All of these taxonomies provide a framework for studying systems errors, but in most experiments and observational studies much finer classification is needed. For example, Kasprysk *et al.* (1979), in their study of calculator notation, broke errors into sequence errors, where the key being pressed was logically wrong, and keyboard errors, where a closely adjacent key was pressed. This enabled the authors to make a much more detailed comparison between algebraic and reverse Polish calculator operating systems.

Error measurement of a different kind is required for a continuous task such as tracking. Here, a person (aided or unaided) attempts to follow an input, time-varying, course, $x_i(t)$, with a vehicle or follower tracing an output track, $x_o(t)$. At any instant, the error is the difference between the output and the input:

$$x_e(t) = x_o(t) - x_i(t).$$

In order to quantify this error, it is necessary to combine values of $x_e(t)$ over some finite time period, O–T. The error function $x_e(t)$ has amplitude, frequency and phase information, although only rarely are all preserved in any omnibus measure. Amplitude and frequency information are preserved in the power spectral density (PSD) function which is a transform of the autocorrelation function of $x_e(t)$ into the frequency domain. The PSD will show which frequencies are present in the error function, and how rapidly error power falls with increasing frequency. Both are useful characteristics, diagnostic of servomechanism functioning, but are often too complex as a single error measure.

The area under the PSD curve represents the total error power, essentially preserving amplitude information while relegating frequency information to a weighting role.

Other less sophisticated continuous error measures are possible, although with inexpensive and rapid computers with Fast Fourier Transform programs, they are required less than in the past. Time–on–target measures, which define an arbitrary target width and count the percentage of time spent within that width, were discredited many years ago, although they retain some traditional uses such as in pursuit rotor tests. Any more direct measure of $x_e(t)$ needs to distinguish between the mean error value, or bias, and the amount of error variability around that mean value. Bias is typically defined as:

$$\text{Bias} = \frac{1}{T} \int_o^T x_e(t) \, dt$$

or, if x_e is measured at N discrete instants,

$$\text{Bias} = \frac{1}{N} \sum_{j=1}^{N} x_{e_j}$$

Obviously, positive and negative errors tend to cancel out so that the bias is small in most tracking tasks. Of more interest is a measure related to the variability about the mean. This is measured by squaring the error to remove negative signs, averaging the squared error, and then taking the square root to preserve the original measurement units. As such, it is known as the root mean square (RMS) error where:

$$\text{RMS error} = \left(\frac{1}{T} \int_{0}^{T} x_e^2(t)\ dt\right)^{1/2}$$

or for discrete instants

$$\text{RMS error} = \left(\frac{1}{N} \sum_{j=1}^{N} x_{e_j}^2\right)^{1/2}$$

defining both for zero bias.

In general RMS error is an easily understood and easily calculated measure of tracking performance, and as such it has gained widespread acceptance (e.g., Sheridan and Ferrell, 1974).

Stress measures

Measures of the cost to the operator of achieving the desired measure of performance are referred to here as stress measures. Technically, stress may be defined (Cox, 1978) as the difference between the perceived demands of the task and a person's perceived capacity to cope when coping is important. This is a narrower definition of stress than 'cost to the operator', and is covered in more detail elsewhere in chapter 26. Here only a briefer, broader survey is attempted.

Stress can be measured only by its effects. Thus the heart rate required to perform a continuous lifting activity and the muscle tension required to swing a cricket bat are both effects on the body of the stresses demanded by their respective tasks. The concept of validity is especially important in stress measures. If the measure reflects changes in the bodily sub-system stressed, then we have face or construct validity.

For example, heart rate measures cardiovascular demand (at least above a level at which stroke volume is constant) while EMG in the flexor muscles of the fingers reflects the force demands of a gripping task. Such direct physiological measures are easy to validate, but others are not. For example, both heart rate variability and eye pupil response have been correlated with information-processing load ('workload'). They have a certain degree of construct validity but rely in the main on experimental validations, many of which have been equivocal (e.g., Wierwille and Casali, 1983).

In addition to physiological measures of stress, there are a number of well-validated rating scales for different aspects of stress. Examples include *rated perceived exertion* (Borg, 1982)—see chapter 22, *body part discomfort* (Corlett and Bishop, 1976)—see chapter 23, and the modified Cooper-Harper scale of workload (Wierwille and Casali, 1983) shown in Figure 5.1. This last scale has also been used successfully by the author to evaluate equipment changes in manufacturing industry in the following simpler form:

How easy or difficult is this task?

1   Very Easy, Highly Desirable
2   Easy, Desirable
3   Fair, Mild Difficulty
4   Minor but Annoying Difficulty
5   Moderately Objectionable Difficulty
6   Very Objectionable, but Tolerable Difficulty
7
8   Major Difficulty
9
10   Impossible

Currently, there is considerable interest in the NASA Task Load Index, or TLX (Hart and Staveland, 1988), which is available in software form for use in computer-assisted data collection (see chapter 9). This Index uses weighted responses on six rating scales:

Mental Demand
Physical Demand
Temporal Demand
Performance
Effort
Frustration

Weightings of the scales are obtained from task experts and are used to give a weighted sum of the subject's scale responses so that an overall index of task load may be obtained. Of course, the individual scales can be used separately to determine the source of the workload more precisely.

Finally, there are outcome measures related to stress. In a factory or office, increases in stress can be logically (and often experimentally) related to sickness, absence, tardiness and accidents. Such highly aggregated measures are only available in large-scale studies but have great face validity and what the Americans call bottom-line impact. Further discussion of stress measures may be found in chapter 26 and, in the context of mental workload, in chapter 25.

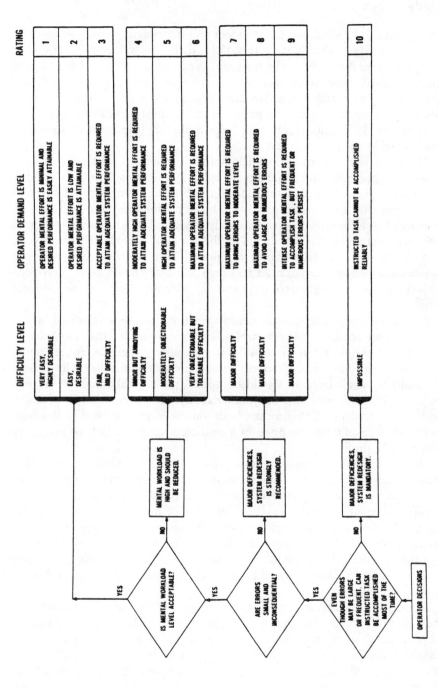

Figure 5.1. Modified Cooper Harper (MCH) scale of Task Difficulty

Study design

While choice of technique, independent and dependent measures have to a large extent fixed the design of the study, there are still more decisions to be taken before the study can be finalized. We must know how many subjects are to be used, how many data points per subject are to be collected, and how the experimental conditions and subjects are to be assigned to each other. These next steps are the province of statistical design of experiments (e.g., Winer, 1972), but this section helps bridge the gap between the ergonomics and statistical concepts.

The main issue in study design is human variability and how to obtain reliable results despite this variability. People differ from each other in anthropometry, in physiological performance, in information processing ability, and in their reaction to external stressors. This is obvious and any study design must take these facts into account. Equally obvious, but less often considered, is the fact that any individual differs on the same measures from year to year, from day to day, and from minute to minute. Thus we have two sources of variability which any design must take into account, known as: (1) between-subject, or inter-subject variability; and (2) within-subject, or intra-subject variability.

As discussed previously, both variability sources can be minimized by the correct choice of independent variables. Thus choosing all females for a backpacking task would reduce inter-subject variability somewhat in the energy costs measured. Similarly, choosing trained users of spreadsheets in a computer-aided calculation task would minimize trial-to-trial variability in performance and so reduce intra-subject variability. However, within any single, even narrowly-defined, population, both sources of variance will be too large to ignore. In addition, the price of a restricted sample is lack of generality in the study findings.

Human variability is explicitly recognized in how we interpret our results. Statistical tests are used to establish the statistical significance (or otherwise) of a result by estimating the probability of so extreme a result having been obtained through pure chance. The pure chance here is the inter- or intra-subject variability. For example, if use of backpack A gives a mean oxygen consumption 0·1 l/min higher than the use of backpack B, we would interpret the fact differently if we have four subjects with an inter-subject standard deviation of 0·15 than we would if 40 subjects with an 0·05 standard deviation had been tested.

A test statistic (such as t or Mann–Whitney's 'U') is typically calculated as the ratio of the size of an effect to the appropriate variability. The test statistic increases in absolute magnitude as the size of the effect increases and as the variability decreases. Size of an effect is represented by the difference between two means (or the variance between several means for an F test) and is thus a physical fact, although subject to sampling error. For example, given enough subjects and measurements, the true difference between backpacks A and B may be 0·12 l/min. We will only conclude that such a difference is significant, however, if the variability is low enough for our test statistic to be in the region of rejection.

The appropriate variability for a test statistic is the standard error of the difference between two means, which in simple cases is:

$$SE = \frac{\text{standard deviation}}{(\text{number of datapoints})^{1/2}} = \frac{SD}{N^{1/2}}$$

where SD is the standard deviation of a set of N data points. Thus the test statistic is:

$$\text{Test statistic} = \frac{M_1 - M_2}{SE} = \frac{(M_1 - M_2)\,N^{1/2}}{SD}$$

where M_1 and M_2 are the means of the conditions being compared.

It is now obvious that the three ways to increase the size of the test statistic are:

1. Increase $M_1 - M_2$.
2. Increase N.
3. Decrease SD.

Each represents a valid option in study design and will be considered in turn.

Increasing difference between means

A large effect will of course be easier to detect than a small effect. But ergonomics is a 'real-world' discipline and the size of effect may be fixed. However, there are a number of steps that can be taken to make the effects larger.

1. *Use a more sensitive measure,* as previously outlined. Oxygen consumption may be less sensitive than, for example, ratings of perceived body part discomfort in comparing two backpack designs. Part of the ergonomics insight comes from a careful reading of the literature to be able to predict which measures will be sensitive in a new study.

2. *Make the two conditions more extreme.* This implies that each mean is physically as different as possible from the other. One tried and tested method is to choose subjects who are extreme on a measure likely to be related to the task if the difference in means is between subject characteristics. Thus ageing research often compares 20–30 year olds with 60–70 year olds, rather than comparing the 30–40 with the 40–50 age group. Similarly, rather than split subjects at the median in perceptual style for study of an inspection task

(Schwabish and Drury, 1984), it would have been possible to choose only subjects scoring in the upper and lower quartiles on the impulsive/reflexive scale. With task variables, an example would be choosing two weights in a box handling task which would be different enough to show a difference in heart rates. Deeb and Drury (1986) used 7 kg and 13 kg to represent average and difficult tasks based on a large survey of box weights handled in industry. More extreme weights have often been used, e.g., by Fish (1978), who had subjects lift barbells weighing either 0·2 kg or 20 kg to measure disc compressive forces. The danger is that one or the other extreme is unrealistic or dangerous. Physicists often refer to this technique as increasing the magnitude of the forcing function.

3. *Increase task difficulty.* This is not just making M_1 more different from M_2 but increasing the difficulty of both tasks so that differences are more likely to show up. Thus if backpacks A and B differ in the amount of padding, then testing both at heavy pack loads will show up the differences more clearly. Another example is in road tests of automobiles where the test drivers negotiate a tight slalom course and measure the performance time. Pushing cars to their limits can reveal differences that may be unimportant in ordinary driving but which become suddenly critical in emergency situations.

Variations on this theme are legion. Instead of changing the task difficulty itself, environmental stressors can be added to push subjects nearer to the limit and hence observe subtle performance changes. High noise levels, poor lighting, external pacing, or even competition have all been used in this way.

The dangers, as with making the conditions more extreme, are that the task becomes unrealistic or dangerous. Both apply directly to ergonomics practice and may contravene human subjects' protection laws. A more insidious danger is that of hitting a 'ceiling effect', which occurs when both conditions are so difficult that performance is equally impossible in both. Backpacks weighing 100 kg would represent an obvious ceiling effect. It should be noted that the opposite of a ceiling effect is a 'floor effect' where both conditions are so easy that either 100% performance is achieved whatever the condition or that some other factor limits performance. In a series of studies of self-paced tracking reviewed by Drury (1985), a floor effect is seen whenever the width of the track is so great that speed and errors are limited by vehicle maximum speed or human willingness to go faster.

Increasing number of data points

It is a truism in statistical testing that any difference, no matter how small, could be found significant given a large enough sample size. Ergonomists pride themselves on practicality and hence the need to ask how large a difference must be in order to achieve practical significance. We should thus aim to make the difference which achieves practical significance the same as that which meets the criteria for statistical significance. Doing this we can solve the equation defining the test-statistic for the sample size N and hence have a rational way of limiting our sample. In order to solve for N, we need to know:

1. The practical difference we wish to detect.
2. The critical value of the test statistic for a specific level of type 1 (alpha) error.
3. The appropriate standard deviation.

Unfortunately, while 1 and 2 may be relatively straightforward to specify, no data may exist on the variability to be expected. One can at times obtain it from the literature, but rarely has the same set of conditions been used as you want to test. The recommended method is a pre-test to estimate the variability, but such a recommendation is cold comfort when a proposal is being costed and the prototype is as yet unbuilt. In such a case, many ergonomists guess, although they will rarely admit it in print. There are obviously some broad guidelines. Upper limits on N are provided by cost and lower limits by the face validity. If you showed by calculation that only two subjects are needed to give the appropriate level of significance, your sponsor may not feel that such a low number is representative. Such misuse of statistics causes despair among statisticians who rightly point out that if the calculations were correct and the two subjects were indeed chosen randomly, there is no need to run more subjects. Life is not always kinder to statisticians than it is to ergonomists. Journals, for example in the behavioural sciences, may need much convincing that small numbers of subjects are adequate, although in anatomical and physiological studies sample size seems to be less of a controversial issue.

Note that the test statistic formula includes N as a square root term. Thus to double the test statistic means to increase sample size, for instance, from 25 to 100, as $100 = 2^2 \times 25$. Increasing sample size as a strategy is likely to meet with rapidly diminishing returns. However, the cost of a study is usually composed of a fixed cost and a variable cost linear in the number of data points. If the fixed cost is relatively low, then four times as many subjects will mean almost four times the cost for the study. However, if the fixed cost is large, a very large increase in the number of data points will only have a small effect on total study costs. For example, the study already quoted which used 30 subjects in a manual materials handling task (Deeb and Drury, 1986) had most of its time (and hence cost) spent on the experimental set-up, pre-testing and analysis. Decreasing or increasing the number of subjects by 25% would have had less than a 10% effect on total time or total cost. In a more recent extension of that study, each subject took two half days to test and about two weeks to reduce and analyze the data fully. Total study time and cost was closely related to sample size in this case.

Decreasing appropriate standard deviation

In the previous discussion of sample size, the meaning of N was left rather ambiguous. Did it mean the number of subjects, the number of replications of a data point on one subject, or both? This ambiguity was deliberate as the arguments used apply to both meanings of sample size, as do arguments concerning the 'appropriate standard deviation'. Not only are there general ways of decreasing variability, which will be considered first, but also ways of altering which is the 'appropriate' standard deviation by varying the experimental design.

First we consider variance reduction techniques. Careful attention to detail is the only way to minimize variability. All of the arguments which apply to the choice of independent variables need to be reviewed to ensure that there are no unexpected sources of variability. All variables which can affect the dependent variable(s) under study should now have been fixed or otherwise controlled. There are, however, some obvious precautions which apply to all studies. All involve standardizing the experimental procedure so that it does not differ between or within subjects. This means consistent, and consistently presented, instructions to subjects. General admonitions to 'do your best' rarely produce the consistency that specific instructions on speeds and errors can provide. Subjects must choose some speed–accuracy trade-off, but if left to their own devices will choose trade-offs which are different from subject to subject and even from trial to trial for the same subject. Specifying a payoff-matrix would be ideal, but many experiments cannot be reduced to dollars or pounds or yen. As important as the specification of instructions is their consistency of presentation. Written or tape-recorded instructions ensure that each subject receives the same set, without unwanted variables such as the inflexions of the experimenter's voice. When the experimenter is conducting studies with speakers of another language, he or she must be particularly aware of consistent and clear instructions.

Following consistent instructions are written procedures for running trials on subjects. Experimenters learn during an experiment so that the first subject is unlikely to be treated in the same way as the last unless written procedures are adhered to. Obviously these procedures must be perfected on pilot subjects. Such pre-testing of subjects (whose results are not included in the final data) not only removes bugs from the experiment but gives the experimenter enough practice to prevent adding to the experimental variability. The same concept applies to analysis. When any data reduction must be performed, again the experimenter learns. Analysis of pilot subjects' data allows the experimenter to learn analysis without adding to final bias or variance.

Types of design

The choice of appropriate standard deviation is a major factor in experimental design. Between-subjects variability almost always exceeds within-subjects (or trial to trial) variability and thus any test in which the appropriate standard deviation includes only within-subject variability will be more likely to detect significant differences than one which includes between-subject variability. This section is not meant to be a treatise on experimental design, for comprehensive treatments have wide circulation (e.g., Winer, 1972), but the aim is to point out the major choices available to the ergonomist, with their advantages and pitfalls.

Broadly, the choice is between a within-subjects design in which each subject performs in a number of experimental conditions and a between-subjects design in which different subjects are chosen for each condition, although a third design route will also be mentioned.

Between-subjects designs

Here each subject is only tested in a single condition. For example, Laughery and Drury (1979) used a between-subjects design in a study of optimization skills because it was suspected that techniques learned during the solution of one type of optimization problem might transfer in an inconsistent manner to other problems, with an adverse effect on bias and variability. Thus five subjects were used in each condition, which meant that any comparison between conditions had to be made against between-subject variability. The groups were kept reasonably homogeneous (engineering students) but this in turn limits the generalizability of the results. Because between-subjects variability was large, only large effects could be found with the given sample size.

Within-subjects designs

Here each subject receives many experimental conditions or, to put it another way, each subject is their own control. Thus a subject with a slow reaction time is likely to be slow under all conditions when reaction time is measured. Techniques such as analysis of variance or regression can estimate the size of the between conditions effect by using only the deviations from the subject's own mean performance. Hence the slow subject will not add to the standard deviation used to test differences between conditions. As an example, a study of the biomechanics and physiology of handle positions on boxes used ten subjects, each performing a box holding task using ten handle positions. The within-subjects design allowed small differences to be detected despite the limited sample size.

An obvious question is why were the experimenters concerned about inconsistent transfer in optimization but not in box holding? The answer is equally obvious: no changes to the subject were expected during the box holding experiment, but changes were expected in optimization. Change occurs in humans in the short-term as they fatigue and in the long-term as they adapt or learn. With appropriate rest periods, no fatigue was expected (or found) in the box holding task and certainly an hour or two of experimentation on a well-practised task is unlikely to change either a subject's body strength (adaptation) or box holding technique (learning). Here a biomechanical and physiologically limited task is unlikely to exhibit what Poulton's famous (1974) paper called asymmetrical transfer effects.

The same cannot be said for most intellectual skills. What you learn in first solving one calculus or chess problem is quite likely to affect your performance in solving the next. The transfer can be positive, if the same solution techniques are useful in both problems, or negative if the solution to the first problem is inappropriate in solving the second. An optimization task is a priori likely to be closer to an intellectual task than to a biomechanical one, hence the choice of a between-subjects design.

Any human functions, even anatomical ones, will adapt or change given time, but the key question is not whether or not change will occur but whether enough will occur to bias or desensitize the experimental comparison. This depends upon both the length of the experiment and the resistance to change of a function. We can run short studies to minimize the change, but we are limited by other experimental constraints of how many data need be collected. We can increase resistance to change by either choosing to experiment on systems which are inherently resistant to change (e.g., bone length) or by deliberately ensuring that a performance plateau has been reached. Thus athletes make good subjects for physiological tests as their aerobic capacity, for example, is well trained and unlikely to change during even relatively long experiments. Similarly, experienced bus drivers are unlikely to develop new bus driving skills during a few hours of tests on buses they have already driven. Finally, for 'new' skills, we can give subjects sufficient practice in all conditions to ensure that plateaus have been reached in each condition. Often the learning/training process required to achieve this level can become of interest in itself (e.g., Bishu and Drury, 1985) so that the time spent achieving stable performance is not time wasted.

Much of the above discussion has centred around creating conditions under which a within-subjects design can be used. Such a design is clearly more efficient, but the fact that we need to take special precautions means that this efficiency is bought at the price of potential danger. Transfer between conditions is always possible; what we do is reduce its probability and magnitude. A between-subjects design is always safer, and sometimes no other is possible. For example, an experiment comparing training techniques in inspection (e.g., Czaja and Drury, 1981) must always use different subjects in different conditions because a fact or skill can only be learnt once. Conversely, some experiments must always use a within-subject's design. If it is desired to derive a functional relationship for each individual subject, then clearly multiple conditions must be given to each subject. Examples are the measurement of the speed–accuracy trade-off or the utility function for each individual subject.

Matched-subjects designs

There is a third option for experimental design. If we could somehow clone a subject we could present different conditions to each clone, preserving both the elimination of between-subject variance and the lack of unwanted transfer effects. Obviously this is currently impossible, as well as being morally and legally dubious. We can do the next best thing however, and carefully match subjects between conditions. Thus two conditions could be compared using pairs of identical twins, although it would be rather difficult to find sufficient subjects who were both, say, airline pilots. Comparing three conditions would lead us to triplets and so on.

A less perfect form of matching is to use unrelated (non-family) subjects who are likely to respond in the same way; the art lies in discovering who will be likely to react similarly. If the skill being tested is one of intellectual activity, then one would expect IQ or educational level to determine similarity of response rather than shoe size. On the other hand, shoe size is likely to be correlated with body size and hence strength, making it a likely (although unusual) matching variable for experiments where strength was the limiting human subsystem. Technically, the benefits of matching are determined by the correlation between the matching measure and the experimental dependent variable. If this is denoted by r, the reduction in between-subjects variance is directly proportional to r^2. Hence we need to choose matching variables correlating highly with the dependent variable. Again the only guide is the literature, often embodied in models of the human appropriate to the current experiment. Matched subjects are used extensively in medical and epidemiological studies. For example, Whitfield (1954) studied accidents in coal miners using subjects carefully matched in job type and experience, one of whom was involved in an accident and one who was not. Similarly Saari (1976) matched not subjects but situations to study task and environmental factors in accident causation.

It should be noted that instead of matching individual subjects, groups of subjects can be matched. Thus in many studies parameters such as the mean age, height and weight of each group are kept constant to reduce group differences. Group matching is not as powerful statistically as individual matching. Finally, a caution should be raised that the tests used to establish matching criteria should not in themselves produce inadvertent transfer to the experiment proper. Thus a pre-test of a tracking task on a flight simulator to determine RMS tracking error would almost certainly transfer to future simulator tasks for novices.

Conclusion on types of design

What then is the conclusion? Should we go with the costly between-subjects design or the more efficient within-subjects design, or match our subjects? There are no universal answers. If experiments are very costly and subjects are limited in number, we may have no alternative to the within-subject design. For example, Drury (1973) measured the speed–accuracy trade-off for inspection using four subjects inspecting four batches under four speed conditions with a Graeco-Latin square design. There were only four skilled inspectors in the plant so that this was the only design possible. The complex design used only four trials per subject to measure the effects of subjects, trials, batches and speeds. Clearly obtaining four main effects in 16 readings forced the author to make many untested assumptions which he would be unwilling to make in the less demanding environment of a research laboratory. But answers were needed and without the experiment there would be none, so that design decisions would be made without ergonomics input. In that case, it was better to make the assumptions than to abdicate responsibility.

Conversely, if transfer effects are known to exist in a situation, there will be no alternative to the between-subjects design; mention has already been made of training experiments in this context. For any between-groups experiment, matching is always good practice, i.e., it never hurts. Whether it is worth the time and effort depends upon the correlation, if known to the experimenter in advance, between the matching variable and the experimental measure. This in turn depends, like so many things in designing ergonomics studies, upon the breadth and depth of the ergonomist's knowledge.

References

Baum, S. and Drury, C.G. (1976). Modelling the human process controller. *International Journal of Man–Machine Studies*, **8**, 1–11.

Beggs, W.D.A. and Howarth, I.A. (1972). The accuracy of aiming at a target. *Acta Psychologica*, **36**, 171–177.

Bishu, R.R. and Drury, C.G. (1985). A study of a location task. *Proceedings of the Human Factors Society 19th Annual Meeting,* Santa Monica, CA.

Borg, G.A.V. (1982). Psychological bases of perceived exertion. *Medicine and Science in Sports and Exercise,* **4**, 377–381.

Caplan, R.D., Cobb, S., French, J.R.P., Van Harrison, R. and Pinneau, S.R. (1975). *Job Demands and Worker Health* (Washington, DC: US DHEW/NIOSH, Superintendent of Documents).

Chapanis, A. (1953). *Research Techniques in Human Engineering* (Baltimore: The Johns Hopkins University Press).

Cohen, H.H. and Jensen, R.C. (1984). Measuring the effectiveness of an industrial lift truck safety training program. *Journal of Safety Research,* **15**, 125–135.

Corlett, E.N. and Bishop, R.P. (1976). A technique for assessing postural discomfort. *Ergonomics,* **19**, 175–182.

Cox, T. (1978). *Stress* (London: Macmillan).

Czaja, S.J. and Drury, C.G. (1981). Training programs for inspection. *Human Factors,* **23**, 473–484.

Deeb, J.M. and Drury, C.G. (1986). Hand positions and angles in a dynamic lifting task. Part 2. Psychophysical measures and heart rate. *Ergonomics,* **29**, 769–778.

Drury, C.G. (1973). The effect of speed of working on industrial inspection accuracy. *Applied Ergonomics,* **4**, 2–7.

Drury, C.G. (1982). Improving inspection performance. In *Handbook of Industrial Engineering,* edited by G. Salvendy (New York: John Wiley).

Drury, C.G. (1985). The influence of restricted space on manual materials handling. *Ergonomics,* **28**, 167–175.

Drury, C.G. and Brill, M. (1983). Human factors in consumer product accident investigation. *Human Factors,* **25**, 329–342.

Drury, C.G. and Dawson, P. (1974). Human factors limitations in fork-lift truck performance. *Ergonomics,* **17**, 447–456.

Drury, C.G. and Sinclair, M.A. (1983). Human and machine performance in an inspection task. *Human Factors,* **25**, 391–400.

Drury, C.G., Cardwell, M.C. and Easterby, R.S. (1974). Effects of depth perception on performance of simulated materials handling task. *Ergonomics,* **17**, 677–690.

Drury, C.G., Paramore, B., VanCott, H.P., Grey, S.M. and Corlett, E.N. (1987). Task analysis. In *Handbook of Human Factors,* edited by G. Salvendy (New York: John Wiley), pp. 370–401.

Fish, D.R. (1978). Practical models of human postures and forces in lifting. In *Safety in Manual Materials Handling,* edited by C.G. Drury, DHEW (NIOSH) Publication No. 78–185, pp. 72–77.

Fitts, P.M. (1954). The information capacity of the human motor system in controlling amplitude of movement. *Journal of Experimental Psychology,* **47**, 381–391.

Fleishman, E.A. and Rich, S. (1963). Role of kinaesthetic and spatial visual abilities in perceptual-motor learning. *Journal of Experimental Psychology,* **66**, 6–11.

Gallwey, T.G. (1982). Selection tests for visual inspection on a multiple fault type task. *Ergonomics,* **25**, 1077–1092.

Hart, S.G. and Staveland, L.E. (1988). Development of a NASA-TLX (Task Load Index): results of empirical and theoretical research. In *Human Mental Workload,* edited by P.A. Hancock and N. Meshkati (Amsterdam: North-Holland).

Kasprysk, D.M., Drury, C.G. and Bialas, W.F. (1979). Human behavior and performance in calculator use. *Ergonomics,* **22**, 1004–1019.

Kerlinger, F.N. (1986). *Foundations of Behavioural Research,* 3rd edition (London: Holt, Rinehart and Winston).

Konz, S. (1983). *Work Design: Industrial Ergonomics* (Columbus, OH: Grid).

Laughery, K.R. and Drury, C.G. (1979). Human performance and strategy in a two-variable optimisation task. *Ergonomics,* **22**, 1325–1336.

Meister, D. (1971). *Human Factors: Theory and Practice* (New York: John Wiley).

Miller, R.B. (1963). Task description and analysis. In *Psychological Principles in System Development,* edited by R.M. Gagre (New York: Holt, Rinehart and Winston).

Morawski, T., Drury, C.G. and Karwan, M.H. (1980). Predicting search performance for multiple targets. *Human Factors*, **22**, 707–718.

Nie, N.H., Hull, C.H., Jenkins, J.G., Steinbrenner, K. and Bent, D.H. (1975). *Statistical Package for the Social Sciences* (New York: McGraw-Hill).

Poulton, E.C. (1974). *Tracking Skill and Manual Control* (New York: Academic Press).

Rasmussen, J. (1982). Human errors. A taxonomy for describing human malfunction in industrial installations. *Journal of Occupational Accidents*, **4**, 311–333.

Reason, J. (1990). *Human Error* (Cambridge: Cambridge University Press).

Saari, J. (1976). Typical features of tasks in which accidents occur. *Proceedings of 6th Congress of IEA, Human Factors Society*, Santa Monica, CA, pp. 11–16.

Schwabish, S.L. and Drury, C.G. (1984). The influence of the reflective-impulsive cognitive style on visual inspection. *Human Factors*, **26**, 641–647.

Senders, J.W. and Moray, N.P. (1991). *Human Error: Cause, Prediction and Reduction* (Hillsdale, NJ: Lawrence Erlbaum Associates).

Sheridan, T.B. and Ferrell, W.F. (1974). *Man–Machine Systems* (Cambridge, MA: MIT Press).

Sperling, G. and Dosher, B.A. (1986). Strategy and optimization in human information processing. In *Handbook of Perception and Human Performance*, edited by K.R. Boff, L. Kaufman and J.P. Thomas (New York: Wiley).

Swain, A.D. and Guttman, H.E. (1980). *Handbook of Human Reliability Analysis with Emphasis on Nuclear Power Plant Applications* (Washington, DC: US Nuclear Regulatory Commission).

VanCott, H.P. and Kincade, R.G. (1972). *Human Engineering Guide to Equipment Design* (Washington, DC: US Superintendent of Documents).

Whitfield, J.W. (1954). Individual differences in accident susceptibility among coalminers. *British Journal Industrial Medicine*, **1**, 126–130.

Wickens, C.D. (1992). *Engineering Psychology and Human Performance*, 2nd edition (Columbus, OH: Charles Merrill).

Wierwille, W.W. and Casali (1983). A validated rating scale for global mental workload measurement applications. *Proceedings of the Human Factors Society 27th Annual Meeting*, pp. 129–133.

Winer, B.J. (1972). *Statistical Principles in Experimental Design* (New York: McGraw-Hill).

Part II

Basic ergonomics methods and techniques

There are some groups of methods and associated techniques which are particularly a part of ergonomics methodology, or are widely used throughout
ergonomics investigation. As an early and vital part of ergonomics studies
we have *task analysis* (chapter 6, Stammers and Shepherd), the term often
denoting both a stage in such studies and also the set of recording and
reporting techniques available. General ergonomics opinion, followed in the
text here, is that task analysis comprises both the description and analysis of
tasks; within task analysis, data are collected, represented in an appropriate
description form, and are analyzed to assess task requirements for the person,
expected behaviour, and task and environmental demands on them.

Task analysis is widely used since it can be applied during the analysis
of existing systems, the design of new ones, and the evaluation carried out
subsequently. Information gained is useful both in development and as criteria
against which to assess what is developed. What is represented and analyzed
is what must be done in order to fulfil certain goals, within constraints from
the task environment and from individual or general human limitations. Task

analysis and its techniques were originally distinguished from method study both by underlying purpose but also by their concentration upon operator decisions as much as upon actions. Lately this has been extended and we have much interest in cognitive task analysis, which obviously demands different methods of data collection, reporting and interpretation. We must find out about the information the subjects attend to, how they interpret it and make decisions to act upon it; multiple methods of direct and indirect observation, expert analysis and so on are required. What then is particular to task analysis are the description formats or representations selected to best allow requirements analysis; many such formats are described by Stammers and Shepherd.

One methodological approach which can be used within task analysis, but which has a wider utility, is *verbal protocol analysis*. As described by Bainbridge and Sanderson in chapter 7, the concentration is upon reports made or people 'thinking aloud' whilst carrying out a task, explicitly stated as concurrent rather than retrospective reports. There is a relationship here with cognitive task analysis in that we are seeking insight into non-observable processes, the thinking which may or may not lead to subsequent action. As with task analysis, stages of data collection, representation and analysis are implied. Data are collected in the form of verbal—often audio taped—reports; reports will be transcribed and placed in the context of a simultaneous record of behaviour (often from video) to produce the actual protocols, divided into phrases or phrase groups; subsequent analysis will be made of explicit and implicit content and of inferred connecting material.

Bainbridge and Sanderson quite clearly point out many of the problems of verbal protocol analysis, both in its operation but also in the very concept. However, verbal protocols offer one of the few ways, if not the only way, presently practicable for making inferences about cognitive processes and for understanding complex behaviour in relation to similar past events, present circumstances or predictions of the future.

Both task and protocol analysis techniques will be applied when human performance can be observed or reported or at least, in the case of task synthesis, relatively easily predicted. When this is not the case, or when the range of potential behaviour is large, then we must turn to other means. Behaviours, or task performance, and also related equipment and events, must be modelled or simulated, differentiated by Meister in chapter 8 as involving physical and symbolic representation respectively. There must be some degree of overlap between the two methods though, in that models may be used in the simulation of systems and simulators may be built around models. For instance, computer workspace modelling, considered in chapter 20, is a modelling technique which can also produce simulations of activities in the workspace. As Meister puts it in his introduction, simulations lead to observations which may give models; models can be superimposed on simulations.

Simulation can involve game or role playing but is described by Meister as especially the physical representation of reality. Hardware plus software simulations with a good deal of fidelity to the actual system are used in system design and evaluation and in training. Expert systems are also seen by Meister as a form of simulator. At a lower level of sophistication simulations can involve expert analysis, including 'walkthrough', of two-dimensional (drawings) or three-dimensional 'mock-up' models of equipment or work places.

Models of various types will be employed usually when an actual system or even a physical simulation is not available, and thus models are used generally in new system development or in needs analysis. As seen by Meister, models of most use to ergonomists are of behaviour in human–machine systems, and are symbolic representations of human performance. The mathematical, usually computer-based, nature of the model allows manipulation of variables to determine system outcomes; they may be used within task analysis (or synthesis) in assessing task requirements or consequences.

The last chapter in this section, chapter 9 by Drury, also covers computer-based methodology. A wider look is taken at computerized data collection and analysis in ergonomics. Many of the techniques described earlier as a part of direct or indirect observation methodology entail the collation of information about a number of variables from a large number of data points or over a long time period. Automated or instrumented observation techniques can be controlled by computer in terms of what data are collected and when, and the measurements taken fed directly into the computer for subsequent reduction and analysis.

Just as developments in computer technology have enabled great improvements in model and simulator power and utility, so can general data collection and analysis be much more efficient and comprehensive. However, within Drury's chapter lie the seeds of a growing feeling about the need for caution. In contrast to the large multi-measure evaluations possible now, the more limited investigations previously possible did have the advantage of producing relatively simple statements on performance or stress which colleagues in design, implementation and management might appreciate and utilize. For all the methods described in this section there is a danger of methodology acquiring a momentum of its own, of analysis being performed way beyond the needs of the situation and of the ergonomist losing sight of the original problem. Methods and techniques are only tools, to be selected and applied with a clear understanding of the design or evaluation objectives.

Chapter 6

Task analysis*

Robert B. Stammers and Andrew Shepherd

Introduction

The observation of people engaged in work is a central feature of ergonomics. In order to optimize human activities in systems, to design interfaces, to produce training programmes and to design equipment, we must always have observed others at work. Prior to systematic studies of human work, very few exact notation or recording systems were developed; examples can be found in the arts, e.g., for recording music. Alternatively, depictions of human activities can be found in such things as military drill books. Systematic observation of tasks, with some specific purpose in mind, has more recent origins. It has become a central part of the human resource disciplines which have developed this century. The development of task analysis methods is inextricably tied up with the development of ergonomics.

Some of the earliest approaches to systematic recording of human activity are to be found, in the early part of the century, in the work-study approaches developed by Gilbreth and Taylor. These techniques were used for recording observable actions, and the approach was to codify them in a written form. Gilbreth's system used a set of symbols to represent different actions. These symbols were termed therbligs (the source for this term being 'Gilbreth' spelt backwards). These symbols, rather than a longhand description, were then used to represent actions. The symbols could also be placed in temporal order to show the sequence of actions in a task. Such techniques were mainly restricted to perceptual motor tasks, of a highly procedural and repetitive nature (see chapter 3).

The overall objective of the work-study approach was 'efficiency'. It was used in order to determine an optimum sequence of actions and to reduce inefficient or wasteful activity. This whole approach of 'scientific management' came in for criticism for the way it focused only on efficiency in using

---

*The writing of this chapter benefited from the earlier version in the 1st Edition, co-written with M.S. Carey and J.A. Rajan.

human labour, and did not address the ergonomics and social content of the work. For the purposes of this chapter the most important criticisms were that it represented human skill as being very static, focusing on manual activities with little reference to the central processing demands of real life speed skills. These aspects were recognized in the 1950s (e.g., Conrad, 1951) at a time when theories of skilled performance were emerging. The bringing together of elements of work study with a deeper understanding of the nature of perceptual-motor skills was achieved by Crossman (1956). He saw the need to record 'mental therbligs', which included planning and controlling activities. He produced the 'sensori-motor' process chart. This records skill cycles in a time-based way, showing the important links between planning, initiating and executing actions. Although this technique does not appear to have been widely used in industry, developments of it can be seen to have influenced the approach of Seymour (1966). Seymour's technique has become more commonly known as 'skills analysis'. It requires the analyst to record task elements and draws attention to such things as the senses and hands used in carrying out different aspects of the task. It was specifically directed at training, and in the 1960s and early 1970s was probably one of the more widely known techniques in manufacturing industry.

Alternative approaches came from 1950s developments in the military human factors field, particularly in the United States. This led to a recognition of the need for more systematic approaches, through task analysis, to training and to equipment design. The work of Miller (1962) is probably the best known in this area. This approach attempted to represent tasks in charts using various codifying techniques which also took into account temporal patterns of the tasks. Some techniques in this area used psychological categories as an underlying taxonomic approach to classifying performance. Such techniques have had a continued development and the taxonomic approach has more recently been revived (Fleishman and Quaintance, 1984). Two camps have emerged, one concerned with theoretical robustness and elegance of such taxonomic schemes, the other more concerned with the utility of such schemes for applied contexts (Miller, 1967). Developments of this kind in the last three decades have led to a multitude of techniques being developed in a variety of contexts.

Developments in the UK have been more closely allied to the development of techniques for training applications. One such approach, Hierarchical Task Analysis (HTA), which was first proposed by Annett and Duncan (1967), is described in more detail later. This technique has been applied to a number of ergonomics contexts as well as in training (Astley and Stammers, 1987; Piso, 1981; Shepherd, 1985). It uses a hierarchical representation of tasks, with more detailed task information being recorded in an accompanying table. It is a technique that has shown itself to be very adaptable to different contexts and has therefore been chosen for detailed coverage in this chapter. However, it is important to be aware of the broad range of approaches available to cover the range of activities that go under the heading of task analysis, some of which are summarized later (see also Kirwan and Ainsworth, 1992).

This overview has so far concentrated on techniques for *representing* information. Parallel with this have been developments in techniques for *collecting* task information. We have moved through a period of focusing mainly on simple observational techniques to using recording media such as videotape. In addition to this, questioning techniques have been refined. These range from those based on informal discussions through to structured interviews. In more recent years attempts to get at the cognitive processes of individuals, through the use of such techniques as verbal protocols, have extended the role of the observer's comments as sources of information (see chapter 7). These techniques have recently received increased interest with the concern to elicit job incumbents' knowledge so that it can be utilized in expert systems (see chapter 14). Approaches to gathering task information are discussed in a later section.

Task analysis, therefore, has a fairly short history, closely tied to the development of ergonomics disciplines and their concern with 'data' on which to base design decisions. This is marked by the wide range of techniques that have been produced and a lack of agreement on any single approach being more appropriate to one context than another. In the following sections attempts are made to clarify some of the underlying concepts in the field and later in the chapter a particular approach will be discussed in more detail.

The definition and process of task analysis

Despite the broad usage of task analysis techniques in ergonomics, there is limited agreement over what can be called a task analysis technique and what the component stages of the task analysis process are. Also, there is no agreed definition of what constitutes a task. To some extent the resolution of such issues can be viewed as theoretical and unrelated to the main concerns of practitioners, whose interests would be the applicability of a technique, its ease of use and its efficiency. However, in order to be able to make decisions on the appropriate use of techniques, a basic understanding is required of what is to be analyzed and how the analysis is to be performed.

Task analysis as a tool for system design and evaluation

Task analysis techniques set out to represent information that is to be used in either the design of a new human/machine system or in the evaluation of an existing system design. This is achieved through the systematic analysis of the tasks required of the user. In the design of a new system, the existing operator tasks need to be examined in order to provide a basis for design. However, there are likely to be changes in both the range and the nature of tasks that are to be performed. Task analyses needs to be applied during the design process to evaluate these future task demands. Using task analysis for evaluation will be more straightforward; information may be collected at a number of points during the operational life of the system.

In the design context, the way in which the task analysis is conducted depends on how closely the functions of the new system are envisaged to mirror those in the existing system, and on the extent to which they are accessible for examination. If there are changes in the user functions or in the interface for instance, then the existing task context will remain as the first stage, but will be of limited value if many changes are proposed.

Different ways in which task analysis can be used in the design context are illustrated in Figure 6.1. In case 1, a limited modification of a system is envisaged. Here can be found the most direct use of task analysis. If there are only limited changes, then most of the task information that can be collected will be relevant. The existing users can be involved in the process and there are likely to be similar features in the way that training and documentation are organized. Caution should still be exercised, as even quite small changes in the task can have broader effects on the task and its context.

The situation in Case 2 (Figure 6.1), where a new system design is proposed, presents more difficulty for any task analysis technique. It is possible to derive some relevant information from an existing system, but it is likely to be of limited value if major changes are planned. It is more likely to be the case that the task information that can be gathered will be at a generalized

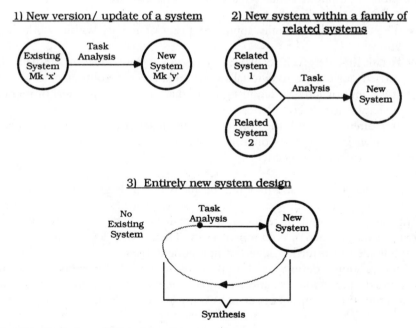

Figure 6.1. Applications of task analysis within system design

level of detail. The extra detail that is likely to be required for the design of the system will need to be generated from a process of task synthesis. This latter approach is detailed below.

Task synthesis is needed not only in the above situation where only part of an existing system is relevant, but also where a system is completely novel, or where it is impossible to gain access to a particular system (Case 3 in Figure 6.1). In these situations, it is necessary to refer to the proposed design of the system to synthesize a description of tasks to be performed. This assumes that it is possible to determine what the human task demands are going to be. However, the design process is likely to be an evolutionary one, with task design being a part of it. A range of human functions, and ways of achieving them, are likely to come under consideration as the overall system design proceeds.

The role of task analysis in system evaluation does not present such great difficulties as those found in design. With evaluation methods, it is a question of identifying problems with existing or proposed tasks. Task analysis can be used in assessing part or all of an existing system or for comparison of alternative system designs or configurations of tasks.

Task requirements and human behaviour

Despite the centrality of the concept of the 'task' to the process of task analysis, there is little consensus on the meaning and scope of the term. Common assumptions about what constitutes a task are that:

- The term 'task' generally applies to a unit of activity within work situations
- A task may be given to or imposed upon an individual or alternatively carried out on the individual's own initiative and volition
- It is a unit of activity, requiring more than one *simple* physical or mental operation for its completion
- It is often used with the connotation of an activity which is non-trivial, or even in some cases onerous in nature
- It has a defined objective.

The concept of a task is not limited in its use to large-scale activities or to simple roles. In a number of task analysis approaches, the overall activity of the user is defined in terms of an overall task and this is then broken down into a number of component tasks. These lower levels may, in turn, be subdivided into further levels of component tasks.

Whilst a simple definition of a task may be possible for everyday use, in the context of ergonomics a more complex view must be taken. In particular, three interacting components of tasks must be considered:

- task requirements

- task environment
- task behaviour.

Task requirements refer to the objectives or goals for the task performer. For example, a word processor user, having completed a document, may be required to save the document on to a permanent storage device. Similarly, the operator in a power station dealing with a sudden loss of power in a unit which they are controlling may be required to take actions that will minimize loss.

Task environment refers to factors in the work situation that limit and determine how an individual can perform. This may be through either restricting the types of action that can be taken and their sequencing, or by providing aids or assistance (e.g., checklists) that channel user/operator actions in a particular way.

Task behaviour refers to the actual actions that are carried out by the user within the constraints of the task environment in order to fulfil the task requirements. Behaviour of the user may be limited by inherent psychological or physiological factors, or a lack of appropriate knowledge or skill. The actions employed may also have been developed through experience to optimize efficiency and to minimize effort.

Distinction among these three aspects of tasks is important for the process of task analysis. The first two aspects are determined by the system context; this incorporates the organizational context, the operating requirements and the limitations of the technology involved, the prescriptive elements of training, the structure of the interface, the operating procedures, the environmental conditions and the influence of other externally connected events. Almost all these elements can be observed, recorded or predicted accurately. Thus the basic framework of task activity, in so far as it is determined by its context, can be accurately documented. Task behaviour, on the other hand, can vary greatly between individuals and with experience. A consensus may have to be reached then on what optimal performance will involve. This is particularly the case when a task is largely cognitive in nature.

The process of task analysis

There is uncertainty about terminology in the definition of the task analysis process. Miller (1962) distinguishes between two processes; task description, which contains just task requirements and a separate process of task analysis, which is an analysis of the behavioural implications of the tasks identified in the description. More recently, the terms have come to be confused, with the term task analysis used to cover the combined description and analysis processes.

Taking the latter view, the process of task analysis can be represented as three main stages of activity. The first stage is data collection, which involves the collection and documentation of various sources of information about the system under scrutiny. The task description document is then created, either from the task data or by the process of task synthesis. The task description provides the information for the final analysis stage generating the task specific data required by the analyst. This is represented in Figure 6.2 below.

Any particular analysis may vary from this idealized model in a number of ways. For example, in practice the process can involve iteration. Vital pieces of information may be found to be missing when generating the task description, and further data collection may be necessary. Whilst iterations of this kind can be inefficient they may be unavoidable. Another difference that often exists between the model and actual practice is that some stages may be omitted or brought together. It is possible, for example, to miss out the formal description phase, generating the analysis from task information using a subjective understanding of the task. This is possible with analyses of very simple tasks or where only partial task information is required (e.g., error data). However, it is always desirable to generate a task description document, to ensure completeness and to provide a means of retracing how decisions leading to recommendations and design were made. Similarly, the task description can be generated without any formal description of the information collected, such that the data collection and description processes are carried out at the same time. Figure 6.2 represents the formal stages of task analysis. Whilst a formal description of the task is the required output of the analysis, it is also important for the task analyst to develop a personal under-

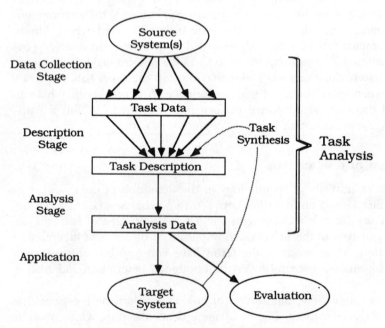

Figure 6.2. The processes employed in task analysis

standing of the task. Additionally, it is desirable that those involved in the design or redesign activity possess an accurate model of the future users and what their task needs are.

The final stage in Figure 6.2 consists of the application of the task analysis output, to a new or redesigned system or taking the form of a system evaluation. Typical of the range of uses for task information are:

- System design/evaluation—The collection of information concerning the allocation of functions between human and system components, assessments can also be made of the benefits of additional task aids to users (see chapter 1).
- Training design/evaluation—The generation of information on user tasks so that new training programmes can be designed or existing programmes improved.
- Interface design/evaluation—The collection of information on user tasks, task frequency, etc. This should assist in the design of a human-machine interface or in the evaluation of an existing design (see chapter 13).
- Job/team design—The provision of information on tasks within an existing system or on those that will be carried out in a future system. This information can be used to predict individual workloads and aid in the effective allocation of tasks within a team.
- Personnel selection—The collection of information about the mental and physical demands of tasks. This can then be used in the derivation of personnel selection criteria.
- System reliability analysis—The prediction of the future reliability of a system based on the application of error data to each identified task component. A reliability analysis may also be used to identify tasks which are critical to system reliability and safety (see chapter 31).

So far, the discussion has not addressed the type of information that may be collected in a task analysis. The different roles of the analysis as described above will require different types of data for their execution. The eventual application of the information is very important in determining the selection of an appropriate technique and in the way it is used. A 'complete' task analysis is unlikely to be necessary in every application.

The identification of what the task information requirements are likely to be is an important component in ensuring the efficiency of the task analysis process. Some of the types of information that might be required are as follows:

- Identification of component tasks—A listing of the activities involved with a task.
- Grouping of component tasks—An organized, often hierarchical listing of the activities involved in a task, showing how sub-tasks cluster according to functional or temporal considerations.

- Commonalities and interrelationships between component tasks—An indication of the extent to which component tasks have features in common and are linked to each other, or the way that sub-tasks may cluster in terms of the information they use or the methods that are used in their execution.
- Importance or priorities of component sub-tasks—Formal or informal procedures may be used to assess the criticality of sub-tasks.
- Frequency of component sub-tasks—Information can be gathered on the relative frequency of occurrence of sub-tasks under different conditions.
- Sequencing of component sub-tasks—The ordering of sub-tasks is usually recorded. Some components may be serially organized, but other forms of sequencing, e.g., branching, are possible.
- Decisions made in the execution of component sub-tasks—Part of the recording of sequences may reveal the need to record decisions that have to be made in order to choose one branch of activity rather than another.
- 'Trigger' conditions for sub-task execution—Many sub-tasks are initiated as a result of the completion of a previous task or following a decision; it is also possible for the execution of a task to depend upon a particular stimulus or command originating within the task environment and these may need to be recorded.
- Objectives or goals of each sub-task—A key feature of an analysis, for any application, is the recording of the objectives of each component.
- Performance criteria for each sub-task—The recording of objectives may include statements about the performance criteria.
- Information required by each sub-task—The items of information needed, and their sources in displays etc., may be recorded.
- Information generated by each sub-task—The information that the user inputs to the system may need to be collated.
- Knowledge employed in making decisions—The information the user utilizes in decision making, both from displayed sources and from memory, may be sought.
- Knowledge of system employed in performing sub-tasks—The understanding that the user has of how the system functions may be deemed important, and attempts may be made to assess this.

The type of task information required determines the nature of the information collection techniques that can be used. Observation or activity sampling (chapter 2), for example, could be used to obtain information about the identities of tasks, their frequencies, sequencing and success/error rates. Observation could be done directly or via recordings, both audio and video. Interview techniques (chapter 3) can provide a wide range of information, though such techniques are limited in the extent to which they can give accurate or complete information on the more cognitive aspects of tasks, i.e., those situations where direct observation and questioning cannot reveal much about the nature of the information processes of the task incumbents. In fact, the cognitive elements of task information are particularly difficult to obtain. The most comprehensive technique that can be employed is verbal protocol analysis, though it does require a high level of access to the task situation or an off-site simulation. In this technique, the user is asked to provide a running commentary on what he or she is doing and why. Through these descriptions, hypotheses about the user's cognitive processes can be derived (see chapter 7). A large amount of information regarding task requirements and the task environment may be available in operational manuals, training documents, policy documents, and operational procedures.

Once obtained, the task description must meet four criteria of usefulness:

- It is important that the description is complete, so the form of documentation chosen must be capable of easily revealing omissions to the analyst.
- As a basic form of validation, the description should be shown to current users for their comments on its accuracy and completeness. The form of representation must therefore be relatively easy to understand.
- The information must be presented in a clear and concise manner so that subsequent users of the document can find what they need easily.
- The task description should be designed to communicate the analyst's understanding of the task domain to the reader.

There is a wide variety of approaches that can be adopted for documenting task information. The most commonly used forms of task representations are diagrammatic and tabular methods, but there are other alternatives including computer-based methods. There is no reason why only one form of representation should be employed. In order to meet the multiple objectives outlined above for a task description, a set of interrelated representations may be optimal. The production problems involved in their generation are rapidly being reduced by the application of increasingly sophisticated computer packages.

The final step of extracting information for the purpose in hand may involve the drawing up of summary documents or lists of conclusions. In many cases it will involve further analysis such as the derivation of estimates of physical or mental workload (see chapters 22 to 30). It is not possible to give specific procedures for doing this as it will depend to a great extent upon the specific application and its context. In many cases, this final analysis process may be based upon analysts interpreting the task description, using their intuition and experience to draw conclusions.

Some analysis approaches

As mentioned earlier, there are many task analysis approaches to choose from; some are general in their approach, but some are directed towards specific applications such as training or interface design. Factors that can influence the choice of a method have been discussed above. Some methods are better documented than others, some have been widely applied and have been shown to be effective in use and application. Rather than attempt to review a range of approaches in detail, the tactic adopted here is briefly to mention some general classes of approach, and then to cover one in more detail. This is Hierarchical Task Analysis (HTA), and it has been chosen because it is an approach that has been shown to have general applicability and is the one with which the authors have most experience. It is briefly introduced below, other methods are also mentioned in outline and then HTA is dealt with in more detail.

Hierarchical Task Analysis (HTA)

A number of task analysis approaches use hierarchies for analyzing tasks. A task can be broken down into task components at a number of levels of description, with more detail of the task being revealed at each level. Hierarchies allow a simple and convenient form for representing tasks. Certain aspects of the task can be highlighted and studied in detail. Some techniques restrict the number of levels in the hierarchy, for example GOMS (Card *et al.*, 1983). HTA is not prescriptive in the number of levels allowed and utilizes 'stopping rules' to help the analyst decide when analysis is complete. The approach has been widely applied in the process and other industries, for both training and ergonomics. It has been developed to meet the needs of the changing roles of the user in line with increased automation and advances in technology (Shepherd, 1976, 1989).

Other approaches

An alternative approach which has been used for task analysis is that of the taxonomy. A wide variety of taxonomies has been developed to categorize task elements for different purposes (e.g., Fleishman and Quaintance, 1984). A task is analyzed by breaking it down into elements which are allocated to the categories of the taxonomy.

Many taxonomies are aimed at describing behaviour, others break the task down in a more analytical way. Familiar methods include Miller's (1967, 1974) taxonomies, and Gagné's (1977) taxonomy for learning tasks. The PAQ approach of McCormick (1976) involves the use of a questionnaire, where job elements are rated in terms of psychological and contextual factors, and it has been widely used in the personnel selection area (e.g., Sparrow *et al.*, 1982).

A number of approaches have been developed in the context of human–computer interaction. The focus here is on the cognitive rather than the physical aspects of task performance. One of these is Task Analysis for Knowledge Description (TAKD) (Johnson *et al.*, 1984; Diaper, 1989). This sets out to be more formal than many task analytical methods and provides both an abstract description of the task in question and a way of representing the knowledge required to perform the task as a 'grammar'. The aim is to produce a task analysis that is not task specific and which can therefore be more easily used in design.

Other approaches of this kind have utilized such things as 'language grammars', for example, Command Language Grammar (CLG), (Moran, 1981). The Task Action Grammar (TAG) approach, developed by Payne and Green (Payne, 1984; Payne and Green, 1989) provides a psychological notation for a rigorous approach to task definition. The analysis defines the mapping from tasks to actions. The primary application of TAG is to provide a model of user knowledge which helps to predict the learnability of command languages and helps to provide a description of an interface.

Other approaches have focused more on the way in which task information can be formally presented. Such methods can be used in isolation or can be incorporated in other approaches. One example would be Link Analysis (Chapanis, 1959). A record is made of all the possible links between items in a workplace, together with the frequency with which they occur. Additionally, each link can be given a weighting indicating its importance. The weightings and the frequency information can be combined to yield a value which can indicate the importance of the relationships between elements. This information can be used to decide upon the effectiveness of existing layout and/or communication links.

Full details of all approaches are not given here, but a comprehensive coverage is provided in Kirwan and Ainsworth (1992) and a review with a training emphasis is provided by Patrick (1992, chapters 5–7).

Hierarchical Task Analysis and its application

Hierarchical Task Analysis (HTA) emerged from ideas developed by Annett and Duncan (1967). Their original accounts discussed strategies for analyzing tasks, based on the notion of hierarchical task description, where operating goals are progressively re-described in terms of sub-ordinate goals and an organizing component, referred to as a 'plan'. An important feature, fully exploiting the hierarchical description, is that the analyst redescribes only as far as is necessary in order to establish useful hypotheses for enabling competent operation. Earlier examples of this approach focused on training design, but it is equally suited to handling a wide range of ergonomics issues.

Initially, HTA was justified in terms of a hierarchical model of human performance. However, it is not easy to justify the product of a hierarchical task analysis as a psychological model, nor is this necessary in order to take advantage of the benefits of the approach. An alternative explanation is that HTA should be seen as a model of a system's goals developed in sufficient detail to enable an analyst to identify each of the features pertinent to making ergonomics design decisions. In this fashion, the analyst redescribes a goal into increasing detail until each aspect of the task is understood sufficiently to offer an appropriate design hypothesis.

As redescription develops, the analyst is able to:

- express the system's goals more explicitly;
- identify appropriate features of the context with more precision;
- establish methods for accomplishing the overall goal.

The question of deciding on an appropriate level of detail is one of the difficult issues in HTA. The original approach of Annett and Duncan (1967) envisaged the consistent use of 'stopping rules'. These are to be applied at each level of description, and for each goal at that level. They are typically couched in terms of the complexity of the goal being examined and/or its criticality to the overall system goals. A rule needs to be decided on in a particular context which will enable the minimum amount of detail to be collected (Shepherd, 1985).

HTA is conducted using a variety of data collection/task analysis methods to examine the specific features of the task as they emerge. For example, to gather data the analyst might use observation or verbal protocol analysis; to examine sources of human error, the analyst might employ a cognitive modelling approach, such as GEMS (Reason, 1987)—see chapter 32 also. These different forms of analysis are more effective when conducted within a framework such as HTA, since the context of the task element under scrutiny is in better focus, and the combined insights provide a coherent 'whole' through a framework such as HTA.

Conducting an HTA entails the following stages.

- Consider the system's goals.
- Examine the tasks that have to be carried out to attain these goals.
- Judge whether current system performance is satisfactory.
- If current system performance is unsatisfactory, then seek to manipulate task components to ensure satisfactory performance—that is, look for interface design, job-aiding, training and other ergonomics hypotheses.
- If an ergonomics hypothesis cannot be proposed, redescribe the goal in terms of sub-goals and an organizing plan.
- If the goal cannot properly be redescribed, examine the relaxation of current constraints.

HTA has been used in a wide range of contexts, including managerial, supervisory and operating tasks in the process industries, office environments, warehousing, maintenance and para-medical professions. It has been applied to problems associated with issues of task design, training, documentation, and human reliability analysis.

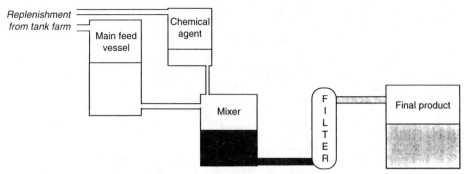

Figure 6.3. A simplified process plant

An illustration

HTA will be illustrated by reference to a simple (and fictitious) process plant, shown in Figure 6.3. In this plant a chemical agent is introduced to a main feed in suitable small quantities, before filtering it to the final product tank. Pumps, valves, indicators and controllers have been left off to simplify presentation. The choice of this example is simply to enable various aspects of HTA to be illustrated concisely.

1. The illustration in Figure 6.4 starts with a key goal: 'maintain target production to the correct formulation'. In practice it would probably be part of a larger goal. All goals in HTA are stated as instructions to the operating system to attain a goal. It is inappropriate to view them narrowly as elements of human behaviour.

If further examination of a goal is warranted, the HTA progresses by stating subordinate operations and a plan. The plan is not a psychological entity, but a summary statement of the conditions when each subordinate goal should be attained.

2. Plans deals with a number of different sorts of relationship between

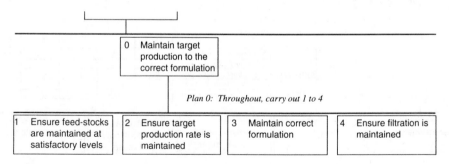

Figure 6.4. Part of an HTA, illustrating goal hierarchy

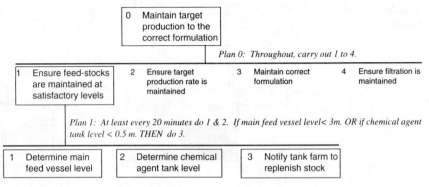

Figure 6.5. Part of an HTA, illustrating plans

subordinate goals. For example, plan 0 in Figure 6.5 requires that all sub-goals are maintained throughout; plan 1 entails time scheduling and different contingencies.

An important feature of HTA is that it records integration between plans. Thus, the subordinate goals of plan 1 are carried out in the context of each of the other sub-goals of plan 0.

3. Plans may be stated in different ways. This may entail a different redescription or simply a preferred form of expressing a plan. Figure 6.6 shows a fault–symptom matrix used to set out how different actions are chosen

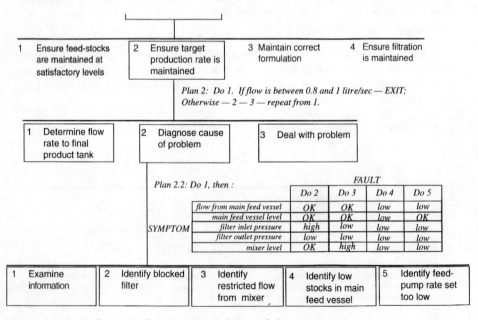

Figure 6.6. Part of an HTA, illustrating plans including a fault symptom matrix

against different conditions. Fault symptom matrices are a neutral way of setting out decision tasks and are helpful in designing job aids and simulation training.

Figure 6.7 shows a different approach where the information collection activities are explicitly included as sub-goals. Two different versions of plan 2.2 are shown. These are equivalent to one another—one in the form of a rule list and the other represented as a flow graph. Both of these represent a specific procedure. Such representations are based either on a procedure elicited from an expert at the task or on a logical analysis of the fault-symptom information.

Both Figures 6.6 and 6.7 represent the case where the decision set is finite. However, it is often the case in decision-making tasks that the choices that must be distinguished are not finite—e.g., the common case where the operator must diagnose from an unspecified fault set. The constituent decision process could not be represented using HTA and so some form of cognitive analysis could be beneficial, such as the TAKD approach referred to above. HTA is still useful in providing the framework in which such cognitive analysis should be made.

Hierarchical Task Analysis and ergonomics design

Hierarchical Task Analysis serves as a framework for examining tasks and handling ergonomics design decisions. Part of this facility rests with the decision to cease further redescription at any point. A major benefit, how-

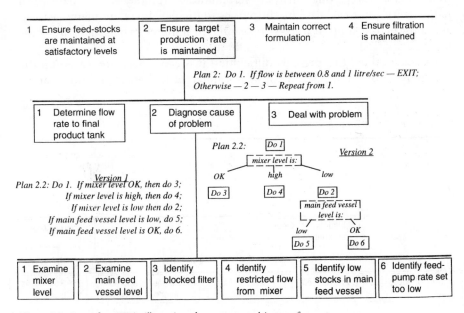

Figure 6.7. Part of an HTA, illustrating plans represented in two forms

ever, rests with using the hierarchy to organize different design decisions and to make sure they are consistent with each other. Additionally, consistent use of a stopping rule for redescription should lead to economy in the level of detail represented in the analysis.

Examining plant operability

HTA enables procedures to be anticipated and to satisfy managers that plant can be operated in accordance with design intentions. This can help clarify safety issues, productivity issues and job design issues. Plans with conditional statements, such as plan 4 in Figure 6.8 must be judged against likelihood of events occurring and plant capacity to cope.

In the analysis of one plant, stating this sort of plan showed that the filter the company planned to install was so small that it would probably need cleaning every 45 minutes with the attendant loss of production this entailed.

Allocation of function

As a goal is redescribed the analysis becomes constrained to a technology for attaining it. In the case in Figure 6.9 the first technology to be considered is a human operator taking and analyzing samples (the operator is already employed, so why not use him or her?). This redescription shows that the operator: must be conscientious and sample every 4 hours (when there are competing activities—see top plan); must open vessels to take samples (safety risk); must carry out analysis scrupulously (problems of perceptual judgement); must calculate modifications (difficult—there are many factors); and must be relied upon to re-sample after 10 minutes (when there are competing activities).

Generally then, this presents difficulties. It is easy, therefore, to go up the hierarchy to operation 3 and consider a different technology—and so auto-

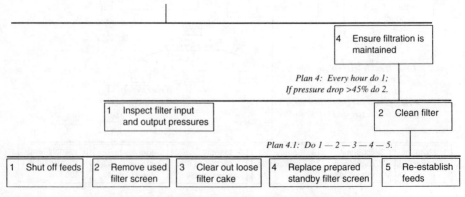

Figure 6.8. Part of an HTA, illustrating a plan with a conditional statement

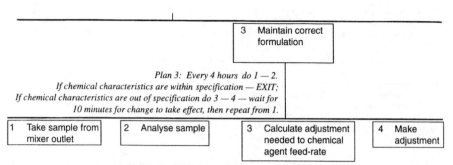

Figure 6.9. Part of an HTA, illustrating plan which suggests a re-allocation of function

matic sampling and adjustment can be justified, if they are technically possible.

Job design

In the illustration of 'Allocation of function', reviewing the technology may
have proven fruitless. There may be no automatic way of in-line analysis and
it may not be possible to adjust the flow of the chemical agent automatically.
This may warrant using a human operator to carry out the routine of 'maintaining correct formulation'. However, there are still problems of time stress,
distraction and expertise to address, especially if workloads across the whole
task are considered. The aggregation of plans throughout the hierarchy,
together with some sort of modelling of event likelihoods, enables these
judgements. This means that job-loading issues can be addressed. The overall
team function may be partitioned using the hierarchy. The result entails operators responsible for higher level jobs giving instructions to operators responsible for subordinate functions; the subordinate relies on instructions as cues
to act.

Developing information requirements

By developing the analysis to a point where information sources and controls
are specified, the analyst can record the information and control actions
necessary for operators to accomplish their tasks. For example, the redescription in Figure 6.10 indicates the requirement for:

– a level indicator on the main feed vessel
– a level indicator on the chemical agent tank
– a means of notifying the tank farm
– a clock
– a target level indication.

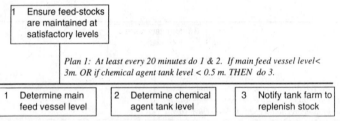

Figure 6.10. Part of an HTA, illustrating information needs requirement

For decision-making tasks, HTA alone cannot provide suitable information and some form of cognitive modelling, allied to suitable training conditions, will complement the information requirements specification (see Shepherd, 1993).

Interface design

Task description via HTA has a number of features that can be exploited in interface design:

1. information requirements specification;
2. information allocations to different job holders' workstations;
3. functional groupings of task elements;
4. consistency checking, to ensure that similar interfacing tasks are accomplished in a consistent way (similar to TAG, see above);
5. interrelationships between task elements (through plans—monitoring, time-sharing, sequential, etc.) to aid accessibility of related screens etc. (see Figure 6.11);
6. identification of key parameters which cue responses for safety displays.

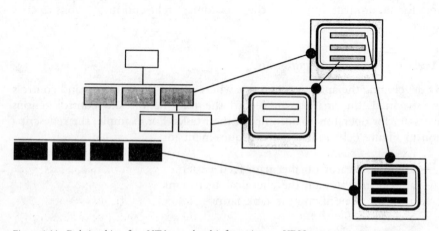

Figure 6.11. Relationship of an HTA to related information on VDU screens

Developing job aids

Task description via HTA has a number of features that can be exploited in job-aid design (see also, Shepherd, 1989):

1. instructions are based on systematic examination of task;
2. description is expressed in active voice and in order of mention;
3. task elements are organized through plans;
4. the hierarchical organization of activities translates to job-aid structure (see below);
5. the hierarchical organization of activities links training to job aids.

Figure 6.12 shows how an HTA may be translated into a form of written documentation.

Developing training

Task description via HTA has a number of features that can be exploited in training design (Shepherd and Duncan, 1980):

1. systematic examination of the task in detail supports training hypotheses;
2. systematic examination of knowledge requirements, e.g., through knowledge elicitation during examining plans or from formal approaches such as TAKD (Johnson *et al.*, 1984);
3. task structure assists in part-task training design; this is illustrated in Figure 6.13;
4. hierarchical structure provides a functional context for training;
5. information requirements specification helps generate features for part-task simulation design;
6. plan hierarchy generates scenarios for simulation training.

Assessment

Task description via HTA has a number of features that can be exploited in assessment of both designs and individuals:

1. hierarchical structure of plans assists in the generation of scenarios;
2. identification of common elements enables representative samples of scenarios to be tested;
3. functional hierarchy enables system performance to be assessed at higher levels, possibly prompting more detailed examination (see Figure 6.14);
4. the functional hierarchy provides a rationale for performance audits.

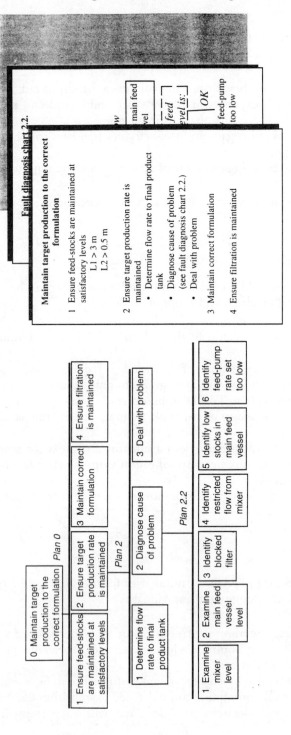

Fault diagnosis chart 2.2.

Maintain target production to the correct formulation

1 Ensure feed-stocks are maintained at
  satisfactory levels
      L1 > 3 m
      L2 > 0.5 m

2 Ensure target production rate is
  maintained
  • Determine flow rate to final product
    tank
  • Diagnose cause of problem
    (see fault diagnosis chart 2.2.)
  • Deal with problem

3 Maintain correct formulation

4 Ensure filtration is maintained

Figure 6.12. Illustration of how an HTA can be translated into written documentation

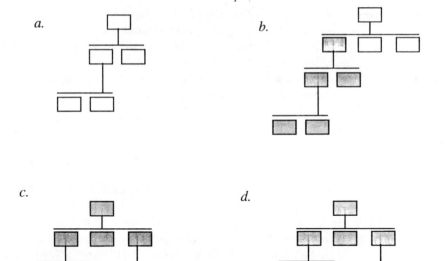

Figure 6.13. Illustrating how an HTA structure can indicate the sequence of part-task training (in (a), the trainee practises the unshaded area, this is then cumulatively added in (b) to new unshaded areas which become the practice units, and so on)

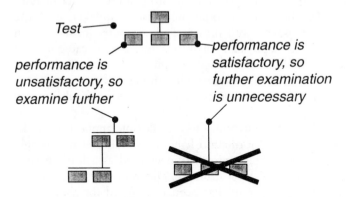

Figure 6.14. Illustration of how an HTA can provide the basis for an assessment procedure (failure in one component can allow for a more detailed examination of performance at lower levels)

Conclusions on HTA

- HTA is a framework for examining tasks. It is not a dogmatic procedure but a strategy for exploiting a hierarchical description.
- HTA is not a separate stage in ergonomics design which precedes stating the design hypotheses; generating design hypotheses is integral to HTA.

- The approach can be used throughout the system life-cycle to develop ergonomics solutions according to system constraints.
- Conducting HTA entails application of a wide range of ergonomics methods, including other techniques of task analysis.
- Hierarchical organization provides a flexible framework for examining tasks and provides a structure to be exploited in design.

Future developments

This review of task analysis has discussed the current state of the art, but it has also revealed the need for more consideration to be given to future activities. It is hoped that more researchers and practitioners will report their use of techniques and show their advantages and disadvantages in particular applications. It would be valuable if existing techniques were more widely tried out by practitioners, rather than their continually developing new techniques to tackle new problems. A forum also needs to be established in which such applications can be discussed.

It is to be hoped that new technology will make an impact on this area. In chapter 9 of this book, ways in which computer techniques are being used in ergonomics are discussed. Three particular roles emerge for computers in the task analysis area. One is in the recording of information. This could be for analyzing data or for the on-line recording of information. In data analysis, computer techniques can be used to codify information quickly; e.g., the coding of questionnaire responses into prearranged computer formats, or in assisting the coding of videotape material into a number of categories of behaviour. On-line recording has been tried in a number of situations where keyboard inputs can be used to log events. The timing of these events can then be automatic. Voice recognition techniques could have a role to play here.

In terms of information representation, the very flexible application packages that are now available for information handling should be increasingly utilized in the future. For example, information entered as a list can subsequently be presented as a hierarchical structure. Different ways of codifying human activity in computer programs has been recognized for some time. For example, Rigney and Towne (1969) outlined a system that enabled hierarchies of sub-tasks to be very quickly produced from task analysis material. Such a process was enhanced, as they envisaged, by alerting the analyst to information that had been missed and enabling additional information to be easily added. Current developments in object oriented programming offer great potential in this area and need to be exploited.

The final area to be considered is the extent to which computer-based modelling techniques can be used as an adjunct to task analysis activities (see chapter 8). This is clearly not an analysis technique in itself, but is one way in which task synthesis for future system design can be performed. Developments of these techniques, together with databases of information on times for task actions and error probabilities, are a way in which the whole process of task synthesis and modelling of behaviour in situations will become more feasible in the very near future.

References

Annett, J. and Duncan, K.D. (1967). Task analysis and training design. *Occupational Psychology*, **41**, 211–221.

Astley, J.A. and Stammers, R.B. (1987). Adapting hierarchical task analysis for user-system interface design. In *New Methods in Applied Ergonomics,* edited by J.R. Wilson, E.N. Corlett and I. Manenica (London: Taylor and Francis), pp. 175–184.

Card, S.K., Moran, T.P. and Newell, A. (1983). *The Psychology of Human-Computer Interaction* (Hillsdale, NJ: Lawrence Erlbaum).

Chapanis, A. (1959). *Research Techniques in Human Engineering* (Baltimore: Johns Hopkins University Press).

Conrad, R. (1951). Study of skill by motion and time study and by psychological experimentation. *Research*, **4**, 353–358.

Crossman, E.R.F.W. (1956). Perceptual activities in manual work. *Research*, **9**, 42–49.

Diaper, D. (1989). Task Analysis for Knowledge Descriptions (TAKD); the method and an example. In *Task Analysis for Human-Computer Interaction,* edited by D. Diaper (Chichester: Ellis Horwood) pp. 108–159.

Fleishman, E.A. and Quaintance, M.K. (1984). *Taxonomies of Human Performance* (New York: Academic Press).

Gagné, R.M. (1977). *The Conditions of Learning,* 3rd edition (New York: Holt Rinehart and Winston).

Johnson, P., Diaper, D. and Long, J. (1984). Tasks, skills and knowledge: task analysis for knowledge based descriptions. In *Interact '84: Proceedings of the First IFIP Conference on Human-Computer Interaction,* edited by B. Shackel (Amsterdam: North-Holland) pp. 23–27.

Kirwan, B. and Ainsworth, L.K. (Eds) (1992). *A Guide to Task Analysis* (London: Taylor and Francis).

McCormick, E.J. (1976). Job and task analysis. In *Handbook of Industrial and Organizational Psychology,* edited by M.D. Dunnette (Chicago: Rand McNally) pp. 651–696.

Miller, R.B. (1962). Task description and analysis. In *Psychological Principles in System Design,* edited by R.M. Gagné (New York: Holt, Rinehart and Winston) pp. 187–228.

Miller, R.B. (1967). Task taxonomy: science or technology? In *The Human Operator in Complex Systems,* edited by W.T. Singleton, R.S. Easterby and D.J. Whitfield (London: Taylor and Francis) pp. 67–76. (Also: *Ergonomics*, **10**, 167–176.)

Miller, R.B. (1974). A method for describing task strategies (Alexandria, VA: American Institute for Research) Rep. No. AFHRL-TR-74-26.

Moran, T.P. (1981). The command language grammar: a representation for the user interface of interactive computer systems. *International Journal of Man-Machine Studies*, **15**, 3–50.

Patrick, J. (1992). *Training: Research and Practice* (London: Academic).

Payne, S.J. (1984). Task-action grammars. In *Interact '84: Proceedings of the First IFIP Conference on Human-Computer Interaction*, edited by B. Shackel (Amsterdam: North-Holland) pp. 39–45.

Payne, S.J. and Green, T.R.G. (1989). Task-Action Grammar: the model and its development. In *Task Analysis for Human-Computer Interaction*, edited by D. Diaper (Chichester: Ellis Horwood), pp. 75–107.

Piso, E. (1981). Task Analysis for process control tasks: the method of Annett *et al.* applied. *Occupational Psychology*, **54**, 247–254.

Reason, J. (1987). Generic error-modelling system (GEMS): a cognitive framework for locating common human error forms. In *New Technology and Human Error*, edited by J. Rasmussen, K.D. Duncan and J. Leplat (Chichester: Wiley) pp. 63–83.

Rigney, J.W. and Towne, D.M. (1969). Computer techniques for analysing the microstructure of serial-action work in industry. *Human Factors*, **11**, 113–122.

Seymour, W.D. (1966). *Industrial Skills* (London: Pitman).

Shepherd, A. (1976). An improved tabular format for task analysis. *Occupational Psychology*, **49**, 93–104.

Shepherd, A. (1985). Hierarchical task analysis and training decisions. *Programmed Learning and Educational Technology*, **22**, 162–176.

Shepherd, A. (1989). Analysis and training in information technology tasks. In *Task Analysis for Human-Computer Interaction*, edited by D. Diaper (Chichester: Ellis Horwood) pp. 15–55.

Shepherd, A. (1993). An approach to information requirements specification for process control tasks. *Ergonomics*, **36**, 1425–1437.

Shepherd, A. and Duncan, K.D. (1980). Analysing a complex planning task. In *Changes in Working Life*, edited by K.D. Duncan, M.M. Gruneberg and D. Wallis (Chichester: Wiley) pp. 137–151.

Sparrow, J., Patrick, J., Spurgeon, P. and Barwell, F. (1982). The use of job component analysis and related aptitudes in personnel selection. *Journal of Occupational Psychology*, **55**, 157–164.

Chapter 7

Verbal protocol analysis

Lisanne Bainbridge and Penelope Sanderson

Introduction

There are many complex jobs in which the outcome of thinking does not emerge in observable action. For example, one can think out a plan of action, assess it, and decide it is inadequate for the purpose, or one can work out the implications of a situation, and memorize the decision for use later. If we want to train and support these types of work, then we need information about these mental processes. One apparently obvious way of getting this information is to ask people to 'think aloud' while they are doing the task. These verbal reports are called 'verbal protocols'. To distinguish them from other knowledge elicitation techniques (discussed by Shadbolt and Burton in chapter 14 of this book), such protocols essentially report mental processes used during a task, rather than being the answers to questions about the task which are given while the person is not actually doing it. This chapter will focus on the analysis of naturalistic, 'on-line' (concurrent) verbal protocols, rather than retrospective, walkthrough, or prompted verbal reports.

This review of verbal protocol methodology will be in four sections:

1. The status of verbal protocols as evidence; their validity as data, and what types of data can be obtained.
2. Technique; practical aspects of collecting verbal protocols, and devising task situations.
3. Analysis; how to extract valid and useful information from the verbal protocols.
4. Tools; different hardware and software media that can help researchers perform verbal protocol analysis.

From the outset it should be emphasized that there is as yet no simple, quick method of collecting and analyzing protocols, and the classic techniques of Newell and Simon (1972) and Ericsson and Simon (1984/1993) are likely to remain practical only for research. However verbal protocol analysis can be performed at different levels of detail, and there are signs that some simpler versions of the technique can be useful in well-chosen applied settings, especially if appropriate supporting software tools are used.

Verbal protocols as evidence

As there is no way of observing someone's mental behaviour directly, it is not possible to test whether there is a correlation between what someone actually thinks and what they say they think. Consequently, when verbal reports do not fit a theory of mental behaviour one does not have to reject that theory. Because the main test for the scientific value of data is that they can be used to reject theories, some academic psychologists feel strongly that verbal data are useless.

In more practical contexts, there are two main uses for verbal protocol data. First, they can be a source of hypotheses about cognitive processes, and so of predictions about non-verbal behaviour. The predictions can be tested without involving verbal reports. Suppose one notes from verbal reports that someone doing a task uses prediction and complex working memory. One can then compare performance using equipment designed on the assumption that prediction is used, and equipment designed assuming that prediction is not used, and test which is best.

Secondly, if someone says something, evidently they do have this knowledge somewhere in their heads. The question then becomes: Is this reported activity and knowledge actually what is being done and used at the time that it is reported, or is it epiphenomenal? We need to determine when verbal reports are likely to correlate sufficiently highly with cognitive processes for them to be useful for practical purposes, and when they are not. Several studies give indirect evidence of the circumstances under which people do give valid reports of their thought processes.

In a highly influential paper, Nisbett and Wilson (1977) reported studies which should be taken seriously by anyone involved with knowledge elicitation, and which have developed into a major research area in social psychology. Their studies are of people's 'self-perception' of the influences on their attitudes and judgements. Nisbett and Wilson carried out experiments in which people were observed acting in various circumstances. They asked the people, in questionnaires, what they would do if faced with these situations, and what they thought was influencing their behaviour. Nisbett and Wilson concluded that people can give the same verbal report for two situations in which they actually behave quite differently. People can also claim that factors influenced their behaviour which actually did not, and can deny that factors influenced their behaviour which actually did. They concede that an individual does know more than an observer about private facts, not only the results of their thinking but also their focus of attention at a given time, current sensations, emotions, evaluations, plans, intentions and personal history. Nisbett and Wilson concluded that people are conscious of the results of their thinking, such as their attitudes or judgements, but have no special access to the process of thinking, such as how they made those judgements. Overall, Nisbett and Wilson's research strongly suggests that people are not very good at reporting the factors which influence their decisions.

This suggests that knowledge elicitation techniques which concentrate on the content of thinking are more likely to be valid than reports which claim to be observations of the processes underlying thinking. (See Shadbolt and Burton, chapter 14, for further discussion of appropriate techniques under different circumstances.) Verbal protocols mainly contain factual statements, especially when people are under time pressure in a real task. The contents of verbal protocols and interviews differ in syntax and type. This is illustrated in these report fragments:

1. 'I shall have to cut (furnace) E off, it was the last to come on, what is it making by the way? E makes stainless, oh that's a bit dicey, I shall not have to interfere with E then.'
2. 'If a furnace is making stainless, it's in the reducing period, obviously it's silly, when the metal temperature and the furnace itself is at peak temperature, it's silly to cut that furnace off.'

These two examples come from the same furnace operator, the first while he was making decisions about the task, the second during a lull in activity a few minutes later. The verbal protocol contains mainly observed facts and results of decisions. These are the data from which an analyst infers the underlying cognitive processes and knowledge used, rather than getting direct information about them. Cuny (1979), who explicitly compared verbal protocols and interview data from the same process operators, came to the conclusion that interviews give more general information about strategies but omit the details of a particular working context which can mean that behaviour differs.

The research by Nisbett and Wilson (1977) did not go unchallenged. The most important replies to it were from Ericsson and Simon (1980, 1984/1993) who, while acknowledging that verbal reports could misrepresent mental processing, argued that in certain circumstances verbal reports should be expected to represent the products of mental processing rather accurately. They detailed different types of distortion, and suggested a simple human information processing model of how verbal reports are generated. They proposed that if the results of mental processing are normally in short-term memory (even if the person is not thinking aloud), if these short-term memory contents are in an easily verbalizable code (e.g., not spatial or motor code) and if the short-term memory contents are not changing so fast that they cannot be articulated fast enough, then verbal protocols have a much better chance of being valid (see Simon and Kaplan, 1989, for a recent summary). However, valid verbal protocols depend upon other factors as well. In the next section we will outline the types of distortion that can affect verbal protocols which a researcher should take into account before undertaking a verbal protocol study.

We might add—though this is a hypothesis based on experience rather than being based on careful experiments—that the information people report about their behaviour in familiar working situations may be more accurate than the information they report about their behaviour in general situations. More detailed and accurate information is available to them about the state of the working environment and the results of their behaviour. In addition, in many working environments people are explicitly—and usually verbally— trained to respond to particular aspects of the environment. Experienced workplace instructors, in particular, are very fluent at expressing their behaviour and thinking as they carry out their work.

Types of distortion, and types of information obtainable

In practice, we can consider the ways in which a verbal report is likely to be distorted, with the aim of minimizing these effects. Note that many types of distortion could occur in all verbal reporting situations, not just with verbal protocols.

First, having to give a verbal protocol changes the task and may change the way the task is done. When there are alternative methods of doing the primary task, the need to give a verbal protocol may influence the person to use a method which is more easily described. Someone could do a task in a mechanical rather than an abstract way, as the former is better fitted to the pace of speech, or do things in sequence rather than doing several things at the same time. They may use a version of the task which is more verbal in form, for example, a beginner's method which has been explained in words, or they may follow the official regulations more closely than usual because these are in verbal form.

Second, there are time constraints when giving a verbal protocol. In problem solving situations, many things may quickly pass through people's minds and be forgotten before there is time to report them. If people are working under time pressures, there may not be time to mention everything that is relevant. Moreover, they may not mention information which they collect while reporting other activities, which may lead to unexplained behaviour later.

Third, giving a verbal protocol is a social situation involving self-presentation, and so can be influenced by social biases. People can select what they think it is appropriate to say. People in experiments are often overly co-operative, and try to say what they think the experimenter wants to hear. If the person reporting feels that the listener has a superior status, and perhaps is powerful, then there may be pressures to present a non-habitual approach to the work, such as appearing rational, knowledgeable and correct or, conversely, being uncooperative and unforthcoming. An experienced worker will talk much more freely and fully if they think that the listener will understand and value what they are talking about. On the other hand, the worker may overestimate the listener's understanding and may not mention points which seem to them to be obvious.

Fourth, distortions arise when the person usually does the job in a nonverbal way, but is asked to give a verbal report. People are not consciously aware of how they carry out perceptual–motor skills, for instance when skating or swimming. Parts of cognitive skills are automatic, and there is evidence that words, pictures and the feel of movements are each dealt with by different parts of the brain. Thus, images and movements may not be accurately represented when they have to be reported in words. Moreover, if the person's vocabulary is limited, they may not be able to find an expression for all that they know. The knowledge expressed may therefore appear limited, whereas actually it is the language which is limited, not the skill.

Fifth, knowledge about the components, mechanisms, functions and causal relations in a machine, memories of specific past events, and helpful categories will be mentioned explicitly only if the task involves some problem solving that requires the person to review this sort of evidence. Otherwise the researchers will only be able to infer such knowledge indirectly from the fact that the person would not have been able to act in a certain way if they had not had particular knowledge. This underlines the general point that verbal protocol evidence may provide only a limited sample of the total knowledge available to the person being studied. Even if the task is well-controlled, it will not be possible to test a large enough number of situations to explore all of the person's knowledge. When collecting protocols in a real task situation, the researcher must be aware that the evidence collected will be limited to what happened during the recording.

In summary, there are many factors that can compromise the validity of verbal protocols. Rather than reject the technique entirely, researchers should critically consider each new situation in light of the above factors to see if valid data might be expected from verbal protocols. The same critical scrutiny should be applied to the results of the protocol analysis. If there is reason to suspect that verbal protocols may be subject to distortion, then this suggests that verbal protocols should be combined with more conventional behaviour observation techniques (see chapters 2 and 3) to check on these distortions and to provide potentially corroborating data. If protocols are likely to be distorted because of the task or social situation, this has implications for the technique used in collecting the protocols, which will be mentioned below.

Getting the most out of verbal protocol analysis

Clearly, verbal protocols are more suited to obtaining some types of information about cognitive processes than others. We have already seen that they may give data on the outcomes but not the processes, of skill. As the protocols are made while performing a task, they can give explicit information about the sequence of items considered. From this the strategy being used may be inferred, and also the working memory contents. As we will see, if there are enough data it is possible to identify decision-making choice points, and what determines how the decisions are made. Even leaving aside all the problems outlined in the previous section, verbal protocol analysis is anyway a very difficult and time-consuming technique. There is no one accepted way of performing verbal protocol analysis—no 'canon'—and researchers must adapt existing methods to new situations, and maybe even develop new methods of their own. First, when embarking on a new study the researcher should seriously examine whether verbal protocol analysis is needed at all. The answer is often 'yes', but there are often situations which are served equally well by more structured knowledge elicitation techniques. Second, the researcher needs to decide how to sample the data: how many subjects will be used, how much data will be collected, under what conditions, etc. Third, the researcher needs to have some a priori idea of the general kind of approach he or she will take to the analysis of the data. Some approaches are considerably more time-consuming than others. On the one hand, the researcher may wish to document, and loosely describe, certain types of problematic situations. This is less time-consuming because a detailed analysis of every phrase may not be needed to identify and analyze the situation, and only certain parts of the protocol need to be analyzed. On the other hand, if the researcher wishes to build a complete computational model of how the subject performs the task, a much more microscopic approach is needed. Sanderson and Fisher (1993) have discussed the 'analysis time:sequence time' (AT:ST) ratio, which indicates how much analysis time will be required for every unit of recorded time (sequence time). Estimates range from 10:1 to 1000:1, and transcription alone can account for 5:1 or more. Thus, researchers should keep approximate estimates of the AT:ST ratios they experience for different types of analysis, and set realistic goals for new research projects.

Verbal protocol analysis process

Figure 7.1 diagrams the typical verbal protocol analysis process, showing how protocols are used to move from research or design questions through to statements or conclusions that answer those questions. The Figure divides verbal protocol analysis into the three stages of (1) collecting and storing protocol data, (2) preparing protocol data for analysis, and (3) analyzing the explicit and implicit content of the protocol. We will use this structure to present protocol analysis in the sections that follow.

Techniques of collecting the protocols

In order to collect protocols, the person is asked to 'think aloud' whilst performing some task of interest, and their comments are recorded and usu-

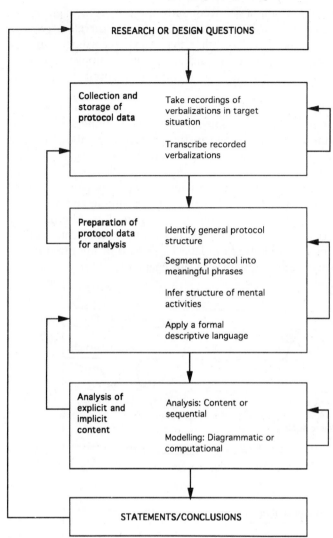

Figure 7.1. Flow chart of the verbal protocol analysis process

ally transcribed. Where possible it is useful to make short recordings which can then be transcribed and checked with the speaker during the same session; the speaker can read through the transcripts to correct mistakes and clarify vague passages (though this can involve distortions of the interview situation).

It is possible to make a much more detailed analysis of the data if a simultaneous recording is made of the information the person receives and the actions they make. The more context present, the easier it is for the researcher to interpret ambiguities. This is useful in at least two ways:

1. People talking while doing a job frequently use pronouns, e.g., '*it*'s at 55'. The objective record shows what is 'at 55', and therefore what is being referred to. Taking video rather than audio recordings can also help when people use general anaphoric references supplemented by pointing, such as '*that's* too high so I'll lower *this* until it's between *these*'.
2. It is useful to have information about the total task situation which may later be used to see if people's behaviour is influenced by the situation without being mentioned in the protocol (see above on the completeness of protocol data, and below on analysis).

This observation record can be made in several ways, but does require considerably more work to collect and analyze than a simple audio record. An observer can log the events (for example, see Figure 7.2). Alternatively, if the investigation is in a high-technology environment, the data can be captured and stored in a data file that can later be printed or imported into a software protocol analysis environment (see also chapter 9). Moreover, video

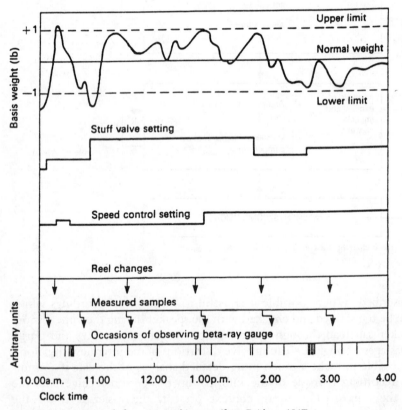

Figure 7.2. Activity graph for paper machineman (from Beishon, 1967)

recordings of the workplace and the operators' movements often substitute for audio recordings entirely. Video preserves the audio, but reveals aspects of the task that may not be verbalized. Video greatly increases the amount of data that a researcher can analyze if he or she so wishes, and the researcher must exercise discipline in choosing what to analyze.

The task and social situations

Asking people to 'think aloud' is asking them to do something unfamiliar and awkward. Some people find it a very uncomfortable requirement. It helps if the researcher first demonstrates what is meant by giving an on-the-spot example—for instance, a verbal protocol of doing complex multiplication. The subject can then try the example for themselves and this usually breaks the ice.

Another technique that helps promote more fluent verbalization is to ask two or more people to work together, particularly where there is a problem-solving situation, or to ask one person to guide another in how to do the task. Communication of thoughts and knowledge, or admissions of lack of knowledge, can then emerge naturally from the people's interaction. However the verbalizations will now be influenced by interpersonal communicative purposes, rather than being solely a report of mental contents. This will affect how the knowledge is expressed and how its contents might be inferred by the researcher (for discussions of social factors, see research on conversation analysis and interaction analysis: Suchman, 1987; Jordan and Henderson, in press).

There are two principal activities in protocol analysis once data have been collected, which are deeply interwoven. The first activity is the preparation and description of the protocols, which includes developing and applying a system of summarization to them. The second activity is the analysis of the explicit and implicit content and structure of the protocols, always with respect to the research or design question at hand.

Preparing the protocol for analysis

As many of its practitioners have commented, verbal protocol analysis is a highly iterative process. Although it is usually presented as a linear process, in practice researchers usually go through several reworkings of how the data are characterized and analyzed before they find the most cogent and useful descriptions and formalizations of the results. Throughout the verbal protocol analysis process, it is helpful if a researcher can remain as closely in contact as possible with the raw verbal and behavioural data. This serves as a 'reality check' for the higher-order transformations being imposed on the data.

We have discussed limitations to the validity of verbal protocols in their raw form as evidence of thought processes. There are also problems with the validity of the analyses that are performed on the protocols. There are no objective independent techniques for doing these analyses. This means that investigators have to use both their own natural language understanding processes, and their knowledge of the task, in order to make sense of what is going on, to infer missing passages, and to interpret the results of summary analyses. This worries many investigators, who wonder whether all they are doing is rediscovering their own knowledge. There are several replies to this. First, researchers do find activities, strategies and structure which had not been anticipated. Secondly, the analysis can give a description of the behaviour which is consistent with other data or theories, so providing converging evidence. Ideally, the analysis should be made by several investigators who work independently and then come together to agree on a final version.

Identifying general protocol structure

The first stage of protocol analysis is to identify the structure of the recorded material. This can be done at several levels:

1. finding the most general stages of activity over the period recorded, such as 'taxi out', 'take off', 'fly to cruising altitude', 'straight and level flight', 'diversion around bad weather', 'descent', and 'landing'.
2. finding units of activity within each stage of activity, such as 'request clearance', 'configure aircraft for landing', 'decide to make diversion'.
3. dividing the material in each unit of activity into a sequence of elemental phrases, each of which represents a separate thought or activity.

The structure can be noted with pencil marks, or if working with a word processor by the use of spaces, indenting, or use of outlining capabilities.

Segmenting material into phrases

The next stage is to identify meaningful units within the protocol that may suggest discrete mental processes. The following piece of report will be used in the examples below (the dots indicate pauses):

C is on oxidation now that's something you can make an estimate for it's a quality so I must leave it alone . . . oxidation average length is one hour 30 minutes for C and started at time zero no it didn't it started at time 33 minutes how confusing of it so it's got nearly one and a half hours to run . . . I'd better check that oxidation for C one hour 30 minutes started 50 minutes ago so it's got 37 minutes to go . . .

Segmented into phrases this becomes:

1. C is on oxidation

2. now that's something you can make an estimate for
3. it's a quality
4. so I must leave it alone
5. oxidation average length is one hour 30 minutes on C
6. and started at time zero
7. no it didn't
8. it started at time 33 minutes
9. how confusing of it
10. so it's got nearly one and a half hours to run
11. I'd better check that
12. oxidation for C one hour 30 minutes
13. started 50 minutes ago
14. so it's got 37 minutes to go.

In this section of protocol a steel-works operator is talking about a furnace called 'C'. Phrases 1–4 are concerned with general properties of the 'oxidation' stage through which the furnace is going, and phrases 5–14 make an estimate of the time at which C will finish oxidizing. The average stage length is given in a job aid booklet, and the time the stage started is displayed on the operator's panel. The segmentation of the text into phrases (which might loosely be described as minimum grammatical units, though the language in such reports is often not at all well formed) is done by natural language understanding. This segmentation can often be done by people who have no knowledge of the specialist content of the material.

Inferring structure of mental activities

The structure of mental activities can be inferred by combining phrases into groups. There are two methods for doing this, both of which make use of the semantic content. The first approach is to identify the pronominal referents, since these indicate cross-references between phrases. There are three ways of identifying the pronoun referents. The most reliable requires an independent record (e.g., video) of the states of the environment during the task, from which what 'started at time 33 minutes' can be identified. Another method involves going through the report immediately afterwards with the speaker to check on the meaning. The third method is to use judges' semantic knowledge of the task.

The second approach to combining phrases into groups is possible after the links between phrases made by the pronouns have been identified. At this point further groupings can be made on the basis of judges' knowledge of what items go together in the task. The result of doing this for the material above is shown in Figure 7.3. One can also take advantage of the necessity to use judges' knowledge, by making an explicit record of one's own knowledge which was used in the analysis, and using this as a record of the knowledge one is attributing to the speaker. For example, lines 1–4 list properties

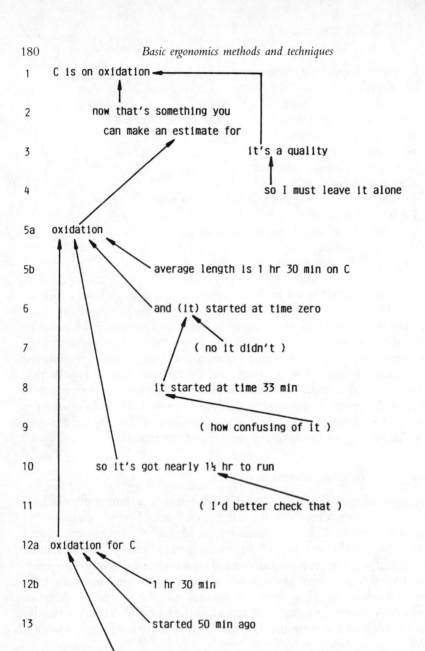

Figure 7.3. Cross-references between phrases in protocol analysis

of C furnace, and lines 5–14 recount a method of calculation. These points are also what one needs to mention to a person unfamiliar with the task before they can understand the report, as in the task explanation given earlier.

The semantics of an individual phrase may imply that the speaker is referring to wider background knowledge. Consider the phrases (from a process operator):

> 'I'll try to run the temperature down to about 400 degrees . . . no trouble, it's far away now.'

The first phrase implies the operator knows the required future process state (400 degrees). The second phrase appears to assess the ease of making the change, which implies that he knows how to make both this change and this assessment. A full protocol analysis could detail the types of more general knowledge which can be inferred in this way. Unfortunately, although one can infer from the protocol that such knowledge exists, there may be very little direct evidence in the protocol about its structure and use. Information about this might be obtained by following the protocol study with interviews on these points.

Applying a formal descriptive language

Having reached this point, many researchers wish to use a formal descriptive language to summarize and systematize, in the context of a chosen research question or framework, what they are seeing in the protocol. Examples of research frameworks are 'Knowledge about system relations', 'Interpersonal warmth', or 'Strategies for control'. The research framework points to the kinds of distinctions the researcher wants to make, such as 'friendly', 'cold', 'helpful', and 'authoritarian', or 'knows X' and 'knows Y', where X and Y represent specific classes of fact a subject has mentioned. These distinctions form a set of standardized categories that can be applied to elemental phrases or units of activity in the protocol. This process is known as 'encoding'.

Encoding is not a necessary step for extracting meaning from protocols, but it is frequently used to systematize the researcher's understanding of the protocol. Moreover, a formal encoding lends itself to symbolic and statistical manipulations which can bring out features of the situation recorded that would otherwise be missed. The categories developed, to be useful, must be ones that encourage further inferences about the material. For example, the frequencies with which certain categories appear may suggest emphases in the way the speaker thinks about the topic. Alternatively the categories can be used simply as a preliminary sort before further analyses. For example, the analyst could look at all instances of the phrases that have been categorized as 'comment on own behaviour' to see if the phrases have any common properties. If the categories are based on semantic aspects of the reports, one might want to count only the occurrence of overall concepts (e.g., birds) or to count the frequency of individual instances (e.g., robins vs. blackbirds) (content analysis will be discussed further below in the section on analysis). Different aspects of syntactic structure could be differentiated. For example, one might wish to distinguish between active and passive voice as reflecting different emphases by the speaker. Alternatively, one could differentiate between different types of conditional statement, perhaps comparing the incidence of 'A therefore B' with 'B because A'. Whether one does this depends not only on whether one is interested in this level of analysis but also on whether the concepts occur sufficiently often to make a frequency count something more informative than a simple listing. The categories may also imply inferences about the types of cognitive process underlying them, for example, 'statement of fact' and 'comment on own behaviour' may imply different types of underlying cognitive activity.

The categories are therefore always a function of particular empirical questions. If there is a set of categories that can be applied in many different circumstances, they are likely to be so general that the results of using them will not be very rich. Indeed, one of the main analytic problems is the development of categories which are both relevant to the investigation and reliably usable by the judges who assign the material to them. Rasmussen and Jensen (1974) describe the iterative method that must be used. First, several independent judges each develop a set of categories. Then they attempt to use eachother's categorization schemes. This both combines the inferences the judges have made about the important distinctions to be made, and also tests whether different people can make the same allocation of material to categories. If not, then analysis using these categories cannot give reliable data, and the categories must be revised, again with the judges working independently during development, and coming together for assessment. The categories are repeatedly defined and revised by trying to use them to analyze the protocol, until the category definitions and phrase assignments stabilize. This procedure is arduous and repetitive but must be done with precision. The degree of success can be measured with reliability statistics, such as Cohen's kappa (1960). Suen and Ary (1989) provide a comprehensive account of how reliability can be promoted and measured in behavioural encoding.

To take a further look at categorizing, examples of identifying characteristic structures within phrases at two levels will be examined: the types of referent word that it is useful to identify within the phrases, and the characteristic patterns in which these occur. For example, the phrase:

'average length is 1 hour 30 minutes'

can be interpreted as a statement of the general form:

'VARIABLE has VALUE'.

The phrase: 'steam pressure is 101' can then be categorized as a statement of the same type. Having identified all statements of this type, one could then look at the category members further, e.g., identifying how many different VARIABLEs occur, and with what frequency, or concentrating on the VALUEs and seeing how accurately they are specified. One would have to consider whether to categorize 'there's steam temperature—it's rising' as a statement of the same type, or whether this would lose some of the information in the report, so this phrase should be identified as 'VARIABLE has CATEGORIZED VALUE'. This also shows how categorizing can retain information considered important for one type of analysis, but can lose other aspects. This phrase is syntactically different from the previous instances, but it has been assumed that this is not relevant to the question of how VARIABLE VALUEs are thought about by the speaker. The syntactic change rather than the semantic one might, of course, be the emphasis of an analysis being made for other purposes. Simple phrases also occur as a component in more complex ones, for example:

'I'll try to run the temperature down to about 400 degrees'

can be interpreted as:

'ACTION causes VARIABLE has VALUE'.

Statements of this type show the speaker's knowledge about how changes in the outside world can be made. Speakers may also give information about their knowledge of the conditions in which certain effects occur, for instance,

'We have to have 50 degrees superheat before we can run it up'

could be described as:

'when VARIABLE has VALUE then ACTION causes VARIABLE has VALUE'.

Or more simply, if one is not interested in distinguishing between different ways of expressing conditional knowledge (e.g., to test whether different expressions occur in different task contexts) the same phrase can be described as:

'ACTION given VARIABLE VALUE'.

These examples have categorized material within phrases. Related methods can be used to study categories of phrase type, or of groups of phrases. As an example, the phrases in Figure 7.3 might be categorized as follows:

1. Fact
2. Strategy
3. Fact
4. Strategy
5. Fact
6. Fact
7. Comment
8. Fact
9. Comment
10. Prediction
11. Strategy
12. Fact
13. Fact
14. Prediction.

These categories are more general, and they use a simple one-word notation. In this case there may be an even more distant relation between the actual material in the report and the way in which it is described. Therefore more care may be needed to ensure that the categories are unambiguous to judges. Again, it may be useful or interesting to distinguish sub-categories, for example, between statements of fact about past and present, or between comments expressing general strategy compared with rules for behaviour at this particular time.

In some tasks, it is not very useful to categorize individual phrases, because similar task situations do not recur at this level of detail. In this case, analysis is done at the level of groups of phrases. As an example, the explanatory notes given for Figure 7.3 are equivalent to a categorization of types of activity in that piece of report. Rasmussen and Jensen (1974) and Umbers (1981) give examples of this type of analysis in maintenance and process control tasks, respectively.

The above examples have shown two different types of encoding notation: the first was a semantic net notation, and the second a single word notation. There are other variants, such as coding each phrase or phrase group with multiple unconnected words, or using the kind of relational notation advanced by Ericsson and Simon (1984/1993). Choosing an encoding notation can be as difficult as choosing the actual encoding categories. The choice of notation will depend upon the meanings that must be captured, and the types of analysis the researcher wishes to perform with the encodings. As encoding categories are developed in the process of analyzing the data, neither the data nor the categories can be justified in terms of each other, and some external tests of objectivity are necessary. One test is that the categories can be applied to the data, and with the same results, by judges who were not involved in developing the categories. Another test is that it should be possible to apply the categories meaningfully to other data. The latter test is difficult to carry out because there are too few case studies and they tend to be for purposes that are too different for comparisons to be useful. One solution to this problem is to develop the categories using one half of the data, and test their adequacy on the other half.

Inferring what is not spoken

As mentioned earlier, there are many reasons why verbal protocols may be incomplete reports. First, material may be glossed over because events are proceeding faster than they can be verbalized. Second, certain material may be encoded spatially, acoustically, or haptically, and thus not be in a form amenable to verbalization. Third, significant activity may be verbalizable, but it may be more effective, appropriate or conventional to communicate it through silence, pauses, tonality, or indirect speech acts (utterances whose communicative significance is different from their literal content). It can be possible, however, to reconstruct some types of intervening material, when the speaker says something that must be the result of some thinking that they have not mentioned. For example, in lines 10 and 14 of the steelmaking example, the controller states the result of a calculation, so he must have made this calculation in some way. The report does not necessarily indicate how the intervening processes were carried out, only that they must have been done. A question can arise about how deeply these inferences can be taken and still be useful. One way is to concentrate on describing behaviour at a level at which there are some explicit clues to constrain the range of possible underlying mechanisms suggested, but some people would not agree with this. In tasks in which the same situation recurs frequently, it may be possible to combine what is said on different occasions to obtain a fuller account of what is happening. In the example, lines 5–10 can be interpreted as:

5.   read STAGE LENGTH
6–8. read STAGE START TIME
10.   TIME TO GO = X.

(Lower case indicates operations that have been inferred, upper case items that are explicitly mentioned. This version in terms of basic operations has already involved considerable inference about underlying processes, which will be discussed more fully later.)

When lines 5–10 are combined with lines 12–14, a complete picture of the processes can be inferred:

(5)   12.   read STAGE LENGTH
(6–8)   read STAGE START TIME
              read time now
       13.   TIME SO FAR = time now − start time
(10)   14.   TIME TO GO = stage length − time so far.

Together with the results from the identification and grouping of phrases, such inferences provide the material that is used in further analyses. In other cases, inferring the unspoken material involves developing an interpretation of the communicative significance of what was said, rather than filling in its surface content.

Analysis of explicit and implicit content

Analysis and encoding are not the same, although they overlap considerably. In order to determine the categories to use in encoding, the researcher must work within some analytical framework, deciding what the most important distinctions are that should be represented if the protocol is to speak to the research issue at hand. Encoding is the application of categories to the raw protocol material, whereas analysis is the process of inferring the ultimate significance of the protocols for the research issue. Sometimes a description of the protocol data provides this, but very often the significance of the protocols is not fully appreciated until qualitative and quantitative manipulations have been performed on the encoding. We shall now look at analysis techniques, all of which can be used for non-verbal just as much as verbal protocol analysis.

Content analysis

There are many styles of analysis called content analysis. Content analysis typically involves counting frequencies of occurrence of chosen words or encoding categories. For example, in the steelmaking example a researcher could count the number of times certain VARIABLEs such as 'steam pressure' or 'average length' were mentioned, or the number of phrase groups encoded as 'Prediction'. Further examples have been given in the previous section. Many of the software tools listed in the final section of this chapter can help researchers to perform content analysis. Moreover, if the segments and phrases of the protocol (and encoding) have been associated with time-stamps, then duration analysis can be performed as well. For example, the researcher could calculate the total amount of time, or total duration, that the subject was performing activities categorized as 'Prediction'.

The researcher can analyze the frequency of occurrence of particular categories of phrase types, or types of phrase groups, or within phrases. These categories, their instances, and the frequency counts can be used as the basis for further analyses, such as comparing the protocols of subjects who have received different training, or with different information management systems.

Sequential analysis

One of the most important features of on-line verbal protocols is that—if elicited appropriately—they represent the normal sequential flow of working and thinking. The person, however, does not talk about this sequential flow while it is unfolding: instead, it emerges from his or her talk. Thus, we can characterize the analysis of sequential aspects of protocols as the analysis of structure which is implicitly rather than explicitly expressed by the subject. Looking for sequences in the material can be done either at the level of individual phrases or of groups of phrases: the choice depends on the material. For example, in the fairly constrained 'world' of controlling a simple industrial process, similar sequences of phrases may occur sufficiently frequently to allow the researcher to identify a standard sequence phrase by phrase. However, sequential structure does not always entail repetitions of the same literal sequence. For example, in Rasmussen and Jensen's (1974) verbal protocol study of fault diagnosis, each technician was working on a different piece of equipment with a different fault. Consequently, the fault finding behaviour was not repeated at the level of individual phrases. To search for the common properties of the behaviour in these disparate situations, Rasmussen and Jensen had to look at the data more globally and think about functioning at a more abstract level.

Sequences of phrases. The sequences in which individual items are mentioned in the report can indicate the standard 'routines' or 'programmes' with which the speaker thinks about a particular topic. This is the main approach of workers who use verbal protocols as the basis for simulations of cognitive processes. There is extensive literature on this type of theory development, of which Newell and Simon (1972) is a classic example. It is only possible to reach strong conclusions about these activities, however, if there are several examples of each type of behaviour. The reports are always incomplete at some level, and one may have several hypotheses about the processes intervening between two phrases. Unless one has other examples of the same behaviour, the analysis remains very speculative and unwieldy. For example, in Figure 7.3 it would be possible for phrases 8 and 10 to be linked by the following serial calculations:

$$\text{Stage end time} = \text{stage start time} + \text{stage length};$$

$$\text{Time to go} = \text{stage end time} - \text{time now}.$$

However, we have another instance of the same behaviour, in which phrase 13 shows which of the two possible linking calculations is being used.

Sequences of groups of phrases. Identifying the sequences of groups of phrases is more difficult because the researcher wants to infer what influences the speaker to move from one topic to another. One cannot take the speaker's word for it, because Nisbett and Wilson (1977) have shown that speakers do not necessarily have good access to this type of information about their behaviour. To identify sequences of groups of phrases, the researcher needs to take into account earlier behaviour by the speaker, because previous items can affect the choice of later behaviour. Also a record of the environment is almost essential in studying dynamic tasks (ones which change over time), because it is often these changes which influence what is the most appropriate thing for the speaker to think about next. A first step is to identify transitions from one type of behaviour to others. For example, behaviour A may be followed by behaviour B or by behaviour C. One then looks at the whole context of the speaker's behaviour and the environment to see whether any dimension of the task or previous thinking consistently has one value when A is followed by B and another when A is followed by C. If so, then one infers that the value of this dimension determines the choice of behaviour at this point.

An example comes from a process control task. The operator frequently

Table 7.1. Frequency of assessments at different total power values

Total Power	Above	Alright	Below
31–40		1	3
41–50	2	2	2
51–60	2	3	1
61–70	2		
71–80			1

Table 7.2. Frequency of assessments of discrepancy meter readings

Discrepancy	Above	Alright	Below
−41 to −50			1
−31 to −40			
−21 to −30			
−11 to −20			3
−6 to −10		3	2
−1 to −5		2	
1 to 5	1	1	
6 to 10	1		
11 to 20	2		
21 to 30			
31 to 40	1		
41 to 50	1		

made remarks such as 'it's above now', 'it's below', and the problem was to identify what dimension of the process he was using to make this judgement. There were two main candidates; the total power being used at the time, or the discrepancy between present power usage and target power usage. Table 7.1 shows the distribution of judgements at different readings of the total power display, and Table 7.2 shows the distribution of judgements at different readings of the discrepancy meter. It is clear that the judgements correlate with the discrepancy meter reading and not with the total power display, so the speaker is assumed to be using the discrepancy meter when making these judgements. (Note that this example is not drawn from an

Table 7.3. General classes of tools for protocol analysis

Pencil and paper

General software applications:
- Word processors
- Spreadsheets
- General multimedia software

Specialized protocol analysis software

analysis of phrase sequences, as no compact example is available. Here the technique has been used to identify pronominal referents. This illustrates that most of the techniques used in analyzing verbal reports can be used to study any level of organization in the material.)

Sequential statistics and exploratory techniques. There are certain statistical techniques that can sometimes help the researcher extract informative sequential structure in protocol data (see van Hooff, 1982; Bakeman and Gottman, 1986; Gottman and Roy, 1990 for details). These techniques need to be used with care in verbal protocol analysis. An important problem is that for full statistical validity a great deal of data usually needs to be available, and this is often not the case in protocol studies.

A commonly used procedure for analyzing sequences is to perform a Markov analysis, i.e., to find the probability of transition from one item to another, or from a strict sequence of items to the next (Kemeny and Snell, 1960; Howard, 1971). An example was given in the previous section. Markov analysis can be a useful preliminary technique to determine which transitions are most frequent, and therefore most likely to be rewarding to study further. However, this method gives a very limited description of the properties of a sequence, hiding some properties completely. First, the statistical structure of a person's sequential behaviour can change during a course of protocol, and this will not be picked up unless special techniques are used. Second, a person's behaviour often includes considerable amounts of interpolated material, or noise, that makes it difficult to see sequential structure with a technique like Markov analysis which requires exact instances of sequences, rather than approximate instances.

Other techniques make weaker assumptions about sequences, and can be used for exploratory analyses just as much as confirmatory ones. Lag sequential analysis is a technique for finding dependencies between events that are separated in time by one, two, or more steps (Bakeman and Gottman, 1986). Lag sequential analysis does not require events to fall into strict sequences, as Markov analysis does. For example, a researcher may find, more often than by chance alone, that subjects make a judgement three or four phases after reading a discrepancy meter, but not one or two phases after reading it. Such a pattern of results would suggest that some other activity is taking place before the judgement can be made. This other activity can be sought, and its meaning for the sequence inferred.

Further sequential techniques are even more exploratory, helping the researcher find potentially informative patterns in the data. Some researchers who perform protocol analysis in UNIX environments have successfully used regular expressions to define and seek loose patterns in protocol data. Regular expression notation allows the researcher to describe a pattern, such as 'find all sequences that start with "read STAGE LENGTH", followed by one or more statements of indeterminate nature, and conclude with "TIME TO GO = " '. The computer finds all such patterns, and prints them out or performs some other operation on them. The researcher can now see all the different ways people handle the calculation of 'time to go' in a stage, and start to see commonalities and differences.

Another exploratory computer-based technique involves finding 'maximal repeating patterns' in a protocol. The computer scans the protocol for all sequences of categories that repeat, such as 'ABC' in the sequence 'FDABCGABC' and tries to find whether these repeating patterns form parts of even larger repeating patterns. In this way both local and global structure can be induced, although ways have to be found for handling noise if this technique is to be completely useful.

Analysis through modelling

Many researchers use symbolic modelling techniques to analyze protocols and to convey the results of those analyses. There are two principal approaches: diagrammatic modelling and computational modelling.

Diagrammatic modelling is a qualitative technique used to capture the content, structure, and/or sequential flow of a protocol. Figure 7.4 represents a verbal protocol from Roth, Bennett and Woods (1988) in a study in which technicians tried to diagnose a faulty electromechanical device with the help of an expert system. The boxes represent groups of phrases, the lines represent connections between different parts of the protocol, and other visual enhancements have been used to emphasize the structure that the researchers have inferred. Figure 7.5 moves further away from the physical form of protocols themselves. It is a subjective 'influence diagram' that shows the results of a study by Sanderson, Verhage and Fuld (1989). Subjects were asked to think aloud while controlling a simulation of a thermal hydraulic system. Their protocols were combed for statements about the effect of variables on other variables in the system. The combined results were superimposed on a diagram of the correct cause–effect relations, so that the most common patterns of verbally-expressed understanding could readily be seen.

Computational modelling based on protocol analysis is a vast topic that cannot be addressed here. Briefly, however, the idea is to analyze protocols in sufficient detail that a computer program can be written of how the subject performed the task (Simon and Kaplan, 1989; Ritter, 1992). The computer program is then actually run, and it generates a 'protocol' of how it performs the task. This protocol can include actions and also an indication of what the current mental contents should be, which the researcher would expect to see reflected in a verbal protocol. The computer's 'protocol' can then be compared with the protocols of the original subjects or, even better, with the protocols of subjects whose data were not used to build the computer model, to see how well they fit. To date, the best examples of this type of modelling have been performed in the context of basic, comprehensive cognitive models such as ACT (Anderson, 1983) and SOAR (Newell, 1990),

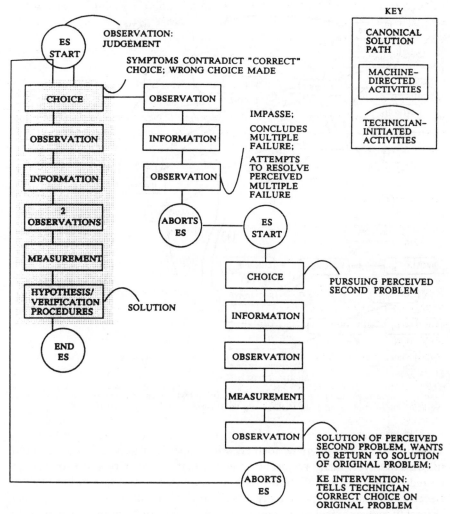

Figure 7.4. Summarized protocol of aided troubleshooting session (from Roth, Bennett and Woods, 1988)

but the principle holds for any model that might generate traces of behaviour and assumed mental activity.

Tools for aiding protocol analysis

As pointed out at the beginning, verbal protocol analysis is very time-consuming, with AT:ST ratios of anywhere from 10:1 to 1000:1, depending on the type of data, the research question, and the level of detail of the analysis.

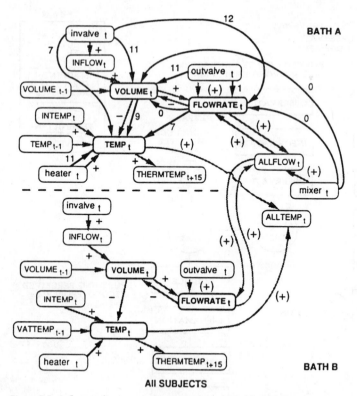

Figure 7.5. Influence diagram combining twelve subjects' knowledge about system relations in a thermal hydraulic process (from Sanderson, Verhage and Fuld, 1989)

Traditionally it has been performed with paper and pencil and very often still is. However, many researchers believe that AT:ST ratios can sometimes be reduced by the use of appropriate software (Sanderson and Fisher, 1993). Because software can potentially turn verbal protocol analysis—a typically unwieldy research technique—into a fully viable one, some space will be devoted to this topic. Of the software tools mentioned here, some have been developed for verbal and nonverbal protocol analysis and some for broader types of observational analysis. The remainder are general software applications which were not designed for any kind of behavioural data analysis, but which happen to have features that can be successfully exploited in protocol analysis. The majority of the tools mentioned below are for the Macintosh™. However, tools for many other machines and operating systems are available as well. The treatment below offers comments about the strengths and weaknesses of the different approaches. It cannot be emphasized too strongly that researchers should exercise critical judgement when contemplating the use of software to support protocol analysis. Each researcher should find a tool—or combination of tools—that supports how he or she wants to analyze protocols, rather than adopt a tool because someone else has found it useful, or because it purports to be designed for protocol analysis. The software designer's intentions and the tool's success elsewhere do not guarantee that it will be suitable for the needs of a given research programme. If a tool is not helpful this does not mean that the researcher has failed to follow some canon for performing protocol analysis, but instead simply that the tool was not helpful.

Pencil and paper

Traditionally, verbal protocols have been transcribed, printed out, and then analyzed with paper and pencil. There are distinct disadvantages but also equally distinct advantages of this approach. On the one hand, if the researcher wishes to change his or her categories during encoding, or to change the way they have been applied to protocol segments or units, then these changes are very tedious to make on paper. Moreover, if statistics are to be performed then the necessary data must be extracted manually, which is highly unreliable and extremely time-consuming. On the other hand, the paper and pencil medium provides a flexible, high-quality visualization of the transcript that is unsurpassed for basic annotation, such as chunking the phrases and units of the protocol with hand–drawn boxes, connecting spatially separate but thematically related parts of the protocol with lines and arrows, and making informal comments in the margins. Moreover, the whole annotated transcription can be spread out on a desktop so that its overall structure is apprehended in a way that cannot be achieved on even the largest computer screens.

It is widely felt that anyone contemplating using verbal protocol analysis should first train themselves with paper and pencil analysis so that they fully understand the types of manipulation and types of thinking that go into protocol analysis. Moreover, at least some paper and pencil analysis is highly recommended at the outset of a new research programme. Only then can the full data manipulation requirements be appreciated, and an appropriate software environment chosen from the many that are available. Paper and pencil coding may be all that is needed for an entire research programme. This is particularly true if a researcher has a stable set of categories, does not plan to do any statistical analysis or qualitative modelling, and if he or she will be using a style of analysis that relies heavily upon connections between widely disparate parts of the protocol.

General software applications

Researchers turn to various types of general software application for help with specific aspects of protocol analysis. These applications include word processors, spreadsheet programs, and multimedia applications. Their most general advantages are proven reliability, good documentation, and the availability of many expert users who can help researchers find the best way to use the application for protocol analysis.

Word processors

Word processors have been used by some researchers for certain aspects of protocol analysis. After transcription, some of the features of advanced word processors such as Microsoft's Word™ can be put to good use. Formatting features, such as indentations, line feeds, font and font size, allow researchers to give meaningful visual structure to a protocol. Annotations and categories can be added if tabular or columnar formats are used. Word processors designed for collaborative writing and multi-author editing add the ability to mark and link certain parts of the text visually and to make marginal annotations (e.g., PREP: Chandhok, 1993). Furthermore, categories can be stored as glossary items to save time typing them fully every time they are used and to promote standardization. Glossary changes or find and replace capabilities can be used to change the categories, if global changes are needed. Finally, document outlining capabilities can also be used to good effect. For example, encoding categories can be handled as headings at one or more levels, and inserted in the raw protocol. Then the document can be displayed using outlining to show just the top level headings, the top *n* level headings, or all headings plus raw protocol for all or some of the whole document. However, there are considerable disadvantages. Word processors do not allow statistical analysis, modelling or certain types of querying to be performed, and they do not offer any connection through software to the raw data in audio or video form. Moreover, their ability to conditionalize word counts and textual changes on other features of the headings or protocol is limited. Finally, they do not make it easy to juxtapose and align multiple streams of information, such as verbalizations, actions, environmental states, researchers' comments, and formal encoding categories. In summary, although word processors provide some of the benefits of paper and pencil and ease certain kinds of manipulations, they are useful principally for formatting and encoding.

Spreadsheets

Spreadsheet programs such as Microsoft's Excel™ overcome some of the limitations of word processors, but add other constraints. Spreadsheets provide a row/column array of cells into which various kinds of data can be entered. Thus spreadsheets make it easy to handle multiple streams of information and to keep them properly aligned, relieving the researcher of this concern. The researcher can define selected parts of the spreadsheet as containing data of certain types, such as timestamp, text, integer, or float data, so that different kinds of calculation can be performed on them. Moreover, because the cells of a spreadsheet can contain formulae that use information in other cells, the spreadsheet can store summary information about itself.

Spreadsheets often incorporate many of the features of word processors, such as find and replace, the use of font and font size, and outlining capabilities. However, they have added advantages because cells conforming to certain descriptions can be selected so that further operations such as counting, printing, and formatting can be performed on them alone. Finally, some spreadsheets allow the researcher to perform queries, such as 'every time the subject mentions neutron flux, print out the corresponding cell in the "activity" column'.

The limitations of spreadsheets arise from the fact that they were not designed to handle events occurring at different points in time. For example, because of the row/column grid, it is difficult to enter, display, and perform manipulations on events in different columns that occur out of synchrony with each other, as things will in the real world. Moreover, although some spreadsheets offer time series analysis, they do not offer the types of sequential data analysis that a researcher more usually wants to use for protocol analysis. Finally, like word processors they do not offer any connection through software to the raw data in audio or video form. In summary, though, spreadsheets can be very effectively used for the more straightforward kinds of protocol analysis (Hoeim and Sullivan, in press), and a competent user can quickly adapt them to this purpose.

As a final note, it is possible to annotate word processing or spreadsheet documents with digitized sound or video inserts. This tends to serve presentation purposes better than analysis purposes because the insert is placed at a fixed location rather than being associated continuously with a timeline, and because only small inserts can usually be handled. However it is a useful capability if analysis is being shared among investigators who wish to bring certain episodes to each other's attention during the analysis process.

General multimedia software: digitized audio and video

Many researchers are currently exploring the feasibility of digitizing their audio or video, and accessing and annotating it using audio and video editing programs. In principle this can cut down on the need for transcription if there is good support for the type of thinking required in protocol analysis. Editing tools for digitized audio such as SoundEdit™ or SoundDesigner™ often provide timeline visualizations of a sound signal, showing the 'on–off' pattern of verbalization in a way that makes it easy to navigate through the protocol. Some tools allow sections of the signal to be marked, annotated, and played in any order, and the annotation associated with any part of the timeline easily retrieved. Such fluent contact with the raw data can be of enormous help to the researcher doing protocol analysis. If relying upon audio, it is important to find software that stores in quickly accessible form the amount of sound recording required—some will only handle a few seconds or minutes at a time. Moreover, audio and video editing programs are seldom designed to support large amounts of transcription and annotation, and so it is often best to run them alongside a word processing or spreadsheet program. This means there are no logical links between points in the sound signal and points in the program document, and the researcher must handle the coordination. Finally, such programs do not include querying and statistical capabilities of the kinds usually needed for protocol analysis.

The growth of multimedia has brought many video digitizing and editing programs onto the market, building on products such as Apple Macintosh's Quicktime™ movies. Because many of these products are hypermedia based, or designed for end-user authoring, such as Hypercard™ and MacroMind Director™ they are easily moulded to a variety of uses. Many of these video editing tools have the same limitations as the audio editing and sound analysis programs discussed above, and the storage requirements for digitized video are far greater—usually quite prohibitive if anything more than a few minutes are to be recorded. CD storage circumvents the problem, but only the most well funded research institutions can afford one-time pressings of many hours of observational data.

An alternative to digitization that removes the storage problem at the cost of increasing access time, is to control external analogue devices such as audio cassettes and VCRs. Analogue video is easily timestamped, which goes part of the way towards handling coordination between document and medium. Many tools linking VCRs and computers have developed for the video production industry. At present, software tools for video production that might be adapted to protocol analysis are usually very costly and often require expensive supporting equipment. However, many of the specialized tools for protocol analysis now being developed use this approach, and will be discussed in the next section. Fewer software tools link computers to audio cassettes, although some have been developed in the context of office automation, as stenographic aids.

Specialized protocol analysis software

For over 20 years there have been efforts to automate protocol analysis, or at least to facilitate it through software (see Sanderson, James and Seidler, 1989; Ritter, 1992; Harrison and Baecker, 1992; Roschelle and Goldman, 1991 for details of these and related efforts). The ideal tool would seemingly combine the advantageous features of pencil and paper, word processors, spreadsheets, database programs with good querying and browsing features, multimedia editing tools, hypermedia, and statistical programs, but of course it is no simple matter to achieve all this. Nonetheless, promising specialized tools are in various stages of development, and there is space to mention only some of them here. Many are research tools rather than commercially available products, but some authors are willing to share their software or to distribute it for a small fee. Sources for further information are given at the end of the chapter.

The simplest tool is probably CVideo (Roschelle, 1992) which coordinates a customized text editing tool with an external VCR. The VCR is controlled from the computer, and tape locations can be found either by moving a marker up and down a scroll bar on the computer screen or by selecting some text in the editor and commanding the VCR to find the corresponding time. Timestamps can be captured from the videotape, inserted into the word processor, and any subsequent text will be associated with that timestamp. No support for sophisticated formatting, encoding, multiple streams of activity, or data analysis is provided. However, the sort of arrangement found in CVideo can help protocol analysis a great deal. The logical connection between the VCR and the document seen in CVideo is at the heart of many of the more sophisticated tools.

Another class of tool provides support for the encoding and analysis phase of protocol analysis, assuming that transcription has been done elsewhere and can be either viewed in another application, or imported as a text file. PAW (Fisher, 1988, 1991) and SHAPA (James and Sanderson, 1991; Sanderson *et al.*, 1989) are software environments for performing verbal protocol analysis very much in the style outlined in Ericsson and Simon (1984/1993). Both tools allow complex vocabularies of encoding categories to be established, altered, and applied to the data. To differing degrees, both allow the data— or more precisely the encoding—to be filtered, queried, and submitted to some of the sequential data analysis techniques listed in an earlier section of this chapter. However, neither offers connections with external recording devices, and neither naturally allows multiple streams of information to be represented.

A further class of tool combines the capabilities of the above two classes and adds further capabilities. Tools such as EVA (MacKay, 1989), VideoNoter (Roschelle and Goldman, 1991), MacSHAPA (Sanderson, 1993), VANNA/ Timelines (Harrison and Baecker, 1992), and the Observer/Tracker/ Reviewer suite (Hoeim and Sullivan, in press) support text entry and editing, connection with external video devices, multiple streams of information, and the establishment and use of vocabularies of encoding categories. These tools differ in other respects. EVA, VideoNoter and VANNA/Timelines offer particularly sophisticated video manipulation, MacSHAPA offers a broad set of sequential and non-sequential data analysis routines, EVA and VideoNoter offer powerful ways to link data in different forms, and both the Observer/Tracker/Reviewer suite and MacSHAPA offer particularly sophisticated querying and filtering capabilities.

Certain tools for protocol analysis have been developed to perform the kind of cognitive modelling that often accompanies protocol analysis. Very early, Waterman and Newell (1971, 1972) developed software tools (PAS I and II) to perform the types of verbal protocol analysis reported in Newell and Simon (1972). Bhaskar and Simon (1977) later developed a protocol analysis support tool called SAPA that adjusted its model of human cognitive processes as it learned more about how the human performed a task. Some current protocol analysis tools form part of a suite of programs centred around a strong model of cognition (such as SOAR, grammar-based, or production system models of human performance) which support model development and manipulation as well as the comparison of behavioural results with a model's predictions. For example, Ritter's (1992) SMT provides researchers with a suite of tools for aligning and measuring the degree of fit between two 'traces' (protocols): one generated by a human and the other by a computational model representing the researcher's idea of how the task might be done and what the human should be doing and saying at each point.

There is a considerable amount of related software that can also be used for certain types of protocol analysis. First, hypermedia-based tools have been developed for ethnographic studies (Fielding and Lee, 1992; also see special issue of *SIGCHI Bulletin,* 1989, on the use of video for HCI). AI-oriented tools have been developed to support knowledge elicitation in the context of knowledge engineering, and some handle continuous verbalization (see special issues of *International Journal of Man-Machine Studies,* 1987, on knowledge elicitation). For studying computer-supported cooperative work, a system named Conversation Editor has been developed that digitizes and separates the speech signal from multiple speakers, and represents the signal in timeline form so it can be edited, explored and analyzed (Sellen, 1992) and a related system has been developed by van der Velden (1992). Finally, an extensive and powerful set of tools called Childes-Clan (MacWhinney, 1991) has been developed for studying child language which, although it takes considerable training to learn because of its complexity, could be equally well used for verbal protocol analysis.

In conclusion, researchers interested in using software to aid protocol analysis should keep abreast of developments in such journals as *Behavior Research, Methods, Instruments and Computers, Behaviour and Information Technology, Human-Computer Interaction,* and particularly in the proceedings of annual conferences of the Association of Computing Machinery's Special Interest Group on Computer-Human Interaction (ACM SIGCHI) and the Human Factors and Ergonomics Society (HFES) in the United States.

Conclusion

This methodological review has taken an optimistic approach to the difficulties of collecting and analyzing verbal protocols. Analyzing verbal reports is not easy, and even with the best software support it can still be time consuming and difficult. Ironically, a thorough verbal protocol analysis study of a task may remove the need to use verbal protocol studies in future investigations with the same task. If the task becomes well known, and the range of strategies used for it identified, then in future studies the influence of different variables may be detected by more focused probes of subjects' knowledge, such as questionnaires, interviews, or walkthroughs of situations that are certain to be sensitive to the manipulation, rather than through the more tortuous route of verbal protocol analysis. Sometimes the process of verbal protocol analysis can be sufficient training for the researcher to understand behaviour with less dense data. However verbal protocol analysis may be needed again when different classes of question, or different systems, are being used.

A person's choice of complex behaviour is influenced not only by immediate circumstances but also by planning in relation to the predicted future or by reference to similar past events. It is difficult or impossible to get sufficient evidence from observed nonverbal behaviour to suggest or constrain hypotheses about such cognitive activities. As a consequence, we know very little about the processes underlying complex behaviour. Verbal protocol analysis is currently one of the richest ways of investigating the nature of behaviour that is a function of past, present, or future. The methods described here illustrate the flexible analytic techniques that can be used, and the software tools outlined provide a wide range of options as to how the researcher might proceed.

References

Anderson, J.R. (1983). *The Architecture of Cognition* (Cambridge, MA: Harvard University Press).

Baecker, R. (1993). Timelines software. Dynamic Graphics Project, University of Toronto, Toronto, Ontario, Canada.

Bainbridge, L. (1979). Verbal reports as evidence of the process operator's knowledge. *International Journal of Man-Machine Studies, 11*, 411–436.

Bainbridge, L. (1985). Inferring from verbal reports to cognitive processes. In *The Research Interview*, edited by M. Brenner, J. Brown and D. Canter (London: Academic Press), pp. 201–215.

Bainbridge, L. (1986). Asking questions and accessing knowledge. *Future Computing Systems, 1*, 143–149.

Bakeman, R. and Gottman, J. (1986). *Observing Interaction: An Introduction to Sequential Analysis* (Cambridge: Cambridge University Press).

Beishon, R.J. (1967). Problems of task description in process control. *Ergonomics, 10*, 177.

Bhaskar, R. and Simon, H.A. (1977). Problem-solving in semantically rich domains: an example from engineering thermodynamics. *Cognitive Science, 1*, 193–215.

Chandhok, R. (1993). The PREP editor. Department of Computer Science, Carnegie-Mellon University, Pittsburgh, PA.

Cohen, J. (1960). A coefficient of agreement for nominal scales. *Educational and Psychological Measurement, 20*, 37–46.

Cuny, X. (1979). Different levels of analysing process control tasks. *Ergonomics, 22*, 415–425.

Duncan, K.D. and Shepherd, A. (1975). A simulator and training technique for diagnosing plant failures from control panels. *Ergonomics, 18*, 627–641.

Ericsson, K.A. and Simon, H.A. (1980). Verbal reports as data. *Psychological Review, 87*, 215–251.

Ericsson, K.A. and Simon, H.A. (1984/1993). *Protocol Analysis: Verbal Reports as Data* (Cambridge, MA: MIT Press).

Fielding, N.G. and Lee, R.M. (1992). *Using Computers in Qualitative Research* (London: Sage Publications).

Fisher, C. (1988). Advancing the study of programming with computer-aided protocol analysis. In *Empirical Studies of Programmers, 1987 Workshop,* edited by G. Olson, E. Soloway and S. Sheppard (Norwood, NJ: Ablex Publishing Corporation).

Fisher, C. (1991). Protocol Analyst's Workbench: Design and evaluation of computer-aided protocol analysis. Unpublished Ph.D. thesis. Department of Psychology, Carnegie-Mellon University, Pittsburgh, PA.

Fisher, C. and Sanderson, P.M. (1993). Exploratory sequential data analysis: traditions, techniques and tools. Report of the CHI '92 workshop. *SIGCHI Bulletin,* **25**, 31–40.

Gottman, J.M. and Roy, A.K. (1990). *Sequential Analysis: A Guide for Behavioral Researchers* (Cambridge: Cambridge University Press).

Harrison, B.L. and Baecker, R.M. (1992). Designing video annotation and analysis systems. *Proceedings of the Graphics Interface '92 Conference.* Vancouver, BC., May 11–15.

Howard, R. (1971). *Dynamic Probabilistic Systems* (New York: Wiley).

Hoeim, D. and Sullivan, K. (in press). Data collection and analysis tools: Challenges and considerations. Manuscript accepted for publication in *Behaviour and Information Technology.*

James, J.M. and Sanderson, P.M. (1991). Heuristic and statistical support for protocol analysis with SHAPA version 2.01. *Behavior Research Methods, Instruments and Computers,* **23**, 449–460.

Jordan, B. and Henderson, A. (in press). Interaction analysis: foundations and practice. *Journal of the Learning Sciences.*

Kemeny, J.G. and Snell, J. (1960). *Finite Markov Chains* (New York: Van Nostrand).

Leplat, J. and Bisseret, A. (1965). Analyse des processus de traitement de l'information chez le controleur de la navigation aerienne. *Bulletin du CERP,* **1–2**, 51.

MacKay, W.E. (1989). EVA: An experimental video annotator for symbolic analysis of video data. *SIGCHI Bulletin,* **21**, 68–71.

MacWhinney, B. (1991). *The CHILDES Project: Tools for Analyzing Talk* (Hillsdale, NJ: LEA).

Newell, A. (1990). *Unified Theories of Cognition* (Cambridge, MA: Harvard University Press).

Newell, A. and Simon, H.A. (1972). *Human Problem Solving* (Englewood Cliffs, NJ: Prentice Hall).

Nisbett, R.E. and Wilson, T.D. (1977). Telling more than we can know: verbal reports on mental processes. *Psychological Review,* **84**, 231–259.

Rasmussen, J. and Jensen, A. (1974). Mental procedures in real-life tasks: a case study of electronic trouble shooting. *Ergonomics,* **17**, 293–307.

Ritter, F.E. (1992). A methodology and software environment for testing process models' sequential predictions with protocols. Unpublished Ph.D. thesis. Department of Psychology, Carnegie-Mellon University, Pittsburgh, PA.

Roschelle, J. (1992). *CVideo Manual* (Palo Alto, CA: Institute for Research on Learning).

Roschelle, J. and Goldman, S. (1991). VideoNoter: A productivity tool for video data analysis. *Behavior Research Methods, Instruments, & Computers,* **23**, 219–224.

Roth, E., Bennett, K.B. and Woods, D.D. (1988). Human interaction with an "intelligent" machine. In *Cognitive Engineering in Complex Dynamic Worlds,* edited by E. Hollnagel, G. Mancini and D.D. Woods (New York: Academic Press).

Sanderson, P.M. (1993). Designing for simplicity of inference in observational studies of process control: ESDA and MacSHAPA. *Proceedings of the Fourth European Conference on Cognitive Science Approaches to Process Control (CSAPC '93): Designing for Simplicity,* Frederiksborg, Denmark, August 25–27.

Sanderson, P.M. and Fisher, C. (1993). Exploratory sequential data analysis: Foundations. Manuscript accepted for publication in *Human-Computer Interaction.*

Sanderson, P.M., James, J.M. and Seidler, K.S. (1989). SHAPA: An Interactive Software Environment for Protocol Analysis. *Ergonomics,* **32**, 1271–1302.

Sanderson, P.M., Verhage, A.G. and Fuld, R.B. (1989). State space and verbal protocol methods for studying the human operator in process control. *Ergonomics,* **32**, 1343–1372.

Sellen, A. (1992). Speech patterns in video-mediated conversation. *Proceedings of the ACM Conference on Human Factors in Computing Systems.* New Orleans, 27 April–2 May. (New York: ACM Press).

Simon, H.A. and Kaplan, C. (1989). Foundations of cognitive science. In *Foundations of Cognitive Science,* edited by M. Posner (Cambridge, MA: MIT Press).

Suchman, L. (1987). *Plans and Situated Actions: The Problem of Human-Machine Communication* (New York: Cambridge University Press).

Suen, H.K. and Ary, D. (1989). *Analyzing Quantitative Behavioral Observation Data* (Hillsdale, NJ: LEA).

Umbers, I.G. (1981). A study of control skills in an industrial task, and in a simulation, using the verbal protocol technique. *Ergonomics,* **24**, 275–293.

van der Velden, J. (1992). *Delft-WITLab: Research issues and methods for behavioral analysis.* CSCW '92 Technical Video Program, Toronto: ACM SIGGRAPH Video Review.

van Hooff, J.A.R.A.M. (1982). Categories and sequences of behavior: methods of description and analysis. In *Handbook of Methods in Nonverbal Behavior Research,* edited by K.R. Scherer and P. Ekman (Cambridge: Cambridge University Press).

Waterman, D.A. and Newell, A. (1992). Protocol analysis as a task for artificial intelligence. *Artificial Intelligence,* **2**, 285–318.

Waterman, D.A. and Newell, A. (1973). *PAS-II: An Interactive Task-Free Version of an Automatic Protocol Analysis System* (Pittsburgh, PA: Department of Computer Science, Carnegie-Mellon University).

Chapter 8

Simulation and modelling

David Meister

Introduction

Both simulation and modelling are forms of representation. Simulation is physical, modelling is symbolic, representation of equipment, events and task performances. Both are most often implemented by computer, although some models are purely conceptual and some physical simulations—primarily static mock-ups—do not involve computers. Although not every simulation and model involves human performance, and most simulations in fact describe physical processes only (e.g., manufacturing, Mellichamp and Wahab, 1987; and marine operations, Park and Noh, 1987), the simulations and models ergonomists are interested in do involve human performance. Simulators are operated by and train students; and functions in human–machine system models are performed by symbolic operators and technicians.

Although most simulators and many models replicate already existent systems (e.g., a particular type of aircraft, a particular type of ship), research simulators and models can be programmed to represent systems in general or ones not yet designed.

There is a close relationship between simulation and modelling. One simulates in order to observe and from the observation to model (although there are other uses for simulation, e.g., training, measurement). One models to understand but the understanding may derive from what one simulates. A model can be superimposed on a simulation to try to understand what went on in terms of covert entities. A model can be used to explain what goes on in a black box whose inner activities are not observable.

The assumption underlying this chapter is that simulation and models are constructed to be *useful*. Usefulness can be defined in a number of ways. To a researcher who is interested primarily in understanding a phenomenon, usefulness may result from the generation of new ideas and ideational connections. To the training specialist usefulness resides in developing a skill. To the system developer usefulness is when answers to specific design-related questions are produced. The orientation of this chapter is more to the training specialist and the system developer, because human performance simulation and models are designed largely for pragmatic purposes.

Simulation

Simulation is something we do all the time. When children play games, they simulate adult activities. Actors simulate the persons they are supposed to represent in the written play. When ergonomists talk about simulation they usually refer to physical simulation of human–machine systems and this is what will be described in this chapter. If, however, one defines simulation in general as an attempt to represent reality in various forms, many simulation-like activities should also be mentioned. These include what are generally termed 'games', e.g., experimental games, in-basket simulations and role playing (Cunningham, 1984). Simulation is important to ergonomists because it enables them to secure answers that might not be available by other methods of data acquisition and analysis. Kirwan and Ainsworth (1992) point out that simulation techniques have been used in a variety of contexts, for example, comparing different procedures for executing a task, or determining the human reliability of nuclear power operators solving a diagnostic problem (Woods *et al.*, 1990).

Simulator history and uses

The very first simulators were the primitive devices used for flight training before and during World War I. The Link trainer was developed before World War II for instrument flight training. The first specific aircraft simulator was that of the Lockheed Hudson bomber developed in 1942. Electronic simulators were first used in the early 1950s by commercial airlines. The military services followed closely on.

Training simulators for systems other than aircraft have proliferated, including (the following is only a partial listing): automobiles, trucks, railroads, ship propulsion and collision avoidance systems, submarine and surface warfare systems, air traffic control, tanks, artillery, missiles, military command control, nuclear power, mining and fire fighting.

In addition to their use as trainers, simulators have been used extensively for systems research (see Parsons, 1972) and for research on workload, decision making, stress, and performance assessment (see Jones *et al.*, 1985, for a more complete list). Simulators are also used for equipment and system design, development, testing and evaluation (to 'try out' alternative design configurations), and for licensing of personnel. Training simulators have many advantages over operational equipment: *cost* (less in a simulator); *availability* (the operational equipment may not be available); reduced *hazard* (less dangerous to practise emergency procedures in a simulator); training *enhancement* (one can stop events in a simulated mission to emphasize a point, back up and repeat an activity, change conditions rapidly). Because simulators are so expensive, few are built solely for research and considerable research is performed in simulators when students are not being trained (Beare and Dorris, 1983).

The computer in simulation

Most simulation (other than that involving no equipment) incorporates one or more computers, so that human–machine simulation can also be called computer simulation. The advantages of the computer over other means of dynamically representing phenomena are too well known to require elaboration here. Whicker and Sigelman (1991) point out that the computer functions at the intersection of simulation and modelling. This section builds on their work. There are two types of computer simulation, that involving human–machine interaction and that involving only the machine or all-computer simulation. There are five elements in a computer simulation: (1) assumptions on which the simulation is constructed; (2) parameters, or fixed values; (3) inputs, or independent variables; (4) algorithms or process decision rules; and (5) outputs of the simulation, or dependent variables. Computer simulations are similar to standard experimental designs in that both include independent, dependent, and control variables. The difference is that in all-computer simulations the values of the dependent variables are determined by the algorithms driving the simulation rather than by observation and measurement of the humans working in the simulation. One can think of the computer simulation as an abstract experiment.

Although all-computer simulations do not involve the physical presence of operators, in ergonomics those operators are represented in the algorithms driving the computer software. For example, in a workload simulation human performance could be included in the form of a distribution of task times. The all-computer simulation is of value when the operators and/or the equipment are unavailable for study or when one wishes to study the effect of many human performance variables simultaneously; or one wishes to test a theory or model. For measurement of actual operator performance a human–machine simulator is of course necessary.

Simulation characteristics

Simulators vary along two continua; their *realism* or fidelity to the operational systems they represent, and their *comprehensiveness,* which is the extent to which operational functions and environmental characteristics, etc. are reproduced in a simulator.

Fidelity, which presents the greater problem, will be discussed later. If, however, simulators are properly designed for their training function, their comprehensiveness should vary simply as a function of what is to be learned. For example in the early stages of procedural training, when students learn sequences of discrete procedural steps, it is hardly necessary to provide full scale simulation; alternatively, a total mission can be practised only in a full scale simulator.

At one extreme the simulator may be quite abstract, for example a computerized display of a ship propulsion system (Hollan *et al.*, 1984). At the other extreme the simulator may consist of the actual operational equipment to which only special control hardware and software have been added. Most simulators fall between these extremes.

A special form of simulator is the 'expert' system (Hayes-Roth *et al.*, 1983). This attempts to replicate the thinking or judgemental processes that underlie an individual's expertise (a step up from merely replicating physical events and task performance). Most expert systems even today are still research tools; when fully developed they will be aids used in a great variety of activities in which judgement and decision making are required.

One need not think of a simulator as consisting solely of hardware and software. The mock-up commonly used in equipment design consists of plywood or styrofoam and may even make use of paper drawings of controls and displays (Buchaca, 1979). A form of simulation that makes use of engineering drawings or static mock-ups as simulators is the *walkthrough,* which is performed to verify the adequacy of procedures and human–machine interface design. The walkthrough is a quasi-symbolic rehearsal of the way in which an equipment under design will ultimately be operated; an engineer points to or touches a mocked up control or display and describes the actions to be taken with these.

A special form of simulation is the development of computer-assisted behavioural design tools, such as CAFES (*C*omputer-*A*ssisted *F*unction Allocation *E*valuation *S*ystem; Parks and Springer, 1975). These automate the behavioural analyses involved in predicting workload, determining anthropometric requirements, allocating functions, and laying out control panels and workstations.

More complex simulators generally have mission scenarios, a schedule or script that drives the sequence of events with which the simulator is exercised. Part of the simulator's software generates the scenario events; another part keeps track of these events and the subject's responses. These events may have a predetermined sequence unaffected by operator actions or they may vary in accordance with algorithms in response to what the operator does.

The models to be discussed in the next major section of this chapter are often used in human-in-the-loop simulations to represent exogenous and endogenous variables and to drive the simulator's processes. A simulator also includes sub-systems, for control, for monitoring progress and measuring subject performance.

Simulation fidelity

The evolution of simulation has been primarily a matter of technological advances to make simulators more accurate and hence more realistic representations of a specific system. It seems logical to developers that the more the simulator is like its real world prototype, the more confidence one can have that personnel performance, including the learning of skills, will be equivalent to personnel performance in the actual system.

Pragmatically, the more realistic the simulation, the fewer problems one should have in designing it, because one need only reproduce existing design. Complete simulation fidelity *apparently* eliminates the need to worry about the relationship between operator characteristics and simulation features; one trains or measures the same way one would if an actual system were being used. (The term 'apparently' suggests that this is only a hypothesis.) As soon as a simulation is less than completely faithful to its original, questions arise concerning which system characteristics to modify and in what way, and what the effect of these modifications will be on the student's skill acquisition and performance.

Complete duplication of a system is costly, much of the cost arising from the need to simulate the characteristics of the system's operational environment and system-related events, such as the interaction of other systems with one's own. This is why the visual sub-system and the presentation of other aircraft are so difficult and expensive in military aircraft simulations. Because of the difficulty and expense there is strong motivation to determine how far one can deviate from complete realism without unduly reducing simulator training value and performance.

Not least of the problems presented by this is the question of how one defines *simulation fidelity*. Kinkade and Wheaton (1972) suggest that there are three types of fidelity: (1) *equipment fidelity*, or the degree to which the simulator duplicates the appearance and 'feel' of equipment; (2) *environmental fidelity*, or the degree to which the simulator duplicates the sensory stimulation from the task situation; and (3) *psychological fidelity*, or the degree to which the simulation task is perceived as being a duplicate of the operational task. Major attention has been paid to the first two, it being assumed implicitly that if the first two are provided, the third will follow automatically.

One might expect from the extensive research on simulation fidelity (see Table 8.1 for a list of the major studies performed in relation to the factors affecting fidelity) that guidelines would now be available relating simulation characteristics to performance outcomes. Whatever guidelines are available are impossibly general, as can be seen later. Perhaps the relationships we are seeking are highly specific, meaning that one must investigate the effect of each simulator's characteristics on performance in advance of simulator design or opt for as much fidelity as one can afford.

Since the purpose of the training simulator is to provide the conditions necessary for learning or for eliciting human performance very similar to what would be elicited operationally, two principles can be derived: (1) characteristics and methods of using simulators should be based on behavioural objectives; (2) physical realism is not the only, or even the optimal, means of achieving these objectives.

Table 8.1. Summary of research on fundamental problems in human–simulator interaction (taken from Jones *et al.*, 1985)

1.	*Fundamental Behavioural Processes*	
	Structure and acquisition of skills	Gagné (1954)
		Muckler *et al.* (1959)
	Cognitive skills	Prophet *et al.* (1981)
	Motivation and learning	Gagné (1954)
		Muckler *et al.* (1959)
	Behavioural mechanisms of transfer	Gagné (1954)
	Perceptual learning	Hennessy *et al.* (1980)
	Visual perception	National Research Council (1982)
2.	*Fidelity of Simulation*	
	Important and unimportant simulator characteristics	Miller (1954)
	Effects of fidelity on transfer	Muckler *et al.* (1959)
	Interaction of fidelity with	
	(a) Instructional variables	
	(b) Experience level	Smode *et al.* (1966)
	Fidelity Requirements	McCluskey (1972)
		Huff and Nagel (1975)
		Hays (1981)
		Gaffney (1981)
	Departures from fidelity of dynamics to compensate for simulator deficiencies	NATO-AGARD (1980)
		Adams (1973)
3.	*Visual Simulation*	
	Visual display characteristics	Muckler *et al.* (1959)
		Smode *et al.* (1966)
		Huff and Nagel (1975)
		National Research Council (1975)
		Hennessey *et al.* (1980)
		NATO-AGARD (1980, 1981)
		Kraft *et al.* (1980)
	Scene content and visual cues	Matheny (1975)
		National Research Council (1975, 1982)
		Thorpe *et al.* (1978)
		Hennessy *et al.* (1980)
		NATO-AGARD (1980, 1981)
		Prophet *et al.* (1981)
4.	*Vehicle Motion*	
	Motion cues	Muckler *et al.* (1959)
		Huff and Nagel (1975)
		NATO-AGARD (1980)
		Prophet *et al.* (1981)
	Interaction of motion and vision	Smode *et al.* (1966)
		Matheny (1975)
		National Research Council (1975)
	Interaction of motion skill level	Smode *et al.* (1966)
	Effects of motion on transfer	Smode *et al.* (1966)
		Matheny (1975)

Table 8.1. Contd

5.	*Performance Assessment*	
	Criteria for performance	Gagné (1954)
		Center for Nuclear Studies (1980)
	Performance measurement	Gagné (1954)
		Muckler *et al.* (1959)
		National Research Council (1975)
		United States Air Force (1978)
		Center for Nuclear Studies (1980)
		Gaffney (1981)
		Prophet *et al.* (1981)
	Automated performance monitoring	United States Air Force (1978)
		Center for Nuclear Studies (1980)
	Measurement of team performance	Parsons (1972)
		Gaffney (1981)
		Prophet *et al.* (1981)
6.	*Modelling*	
	Models of visual and motion simulation	Waag (1981)
	Sensory system modelling	United States Air Force (1978)
	Model of multisensory spatial orientation	NATO-AGARD (1980)
	Models of visual environment to identify variables relevant to training	NATO-AGARD (1980)
	Models to predict training effectiveness	Prophet *et al.* (1981)
7.	*Training*	
	Critical characteristics for transfer	Miller (1954)
		Caro (1977)
	Effects of change in task characteristics on transfer measurement of transfer effectiveness	Muckler *et al.* (1959)
		Smode *et al.* (1966)
		Williges *et al.* (1973)
		NATO-AGARD (1980)
	Systematic method for developing training requirements that provide guidance for simulator design	NATO-AGARD (1980)
8.	*Training Methods*	
		Gagné (1954)
		Human Factors Operations Research Laboratories (1953)
		United States Air Force (1978)
		Center for Nuclear Studies (1980)
	Sequence of training	McCluskey (1972)
	Use of feedback and guidance	Prophet *et al.* (1981)
	Instructional features	Prophet *et al.* (1981)
	Instructional training	Prophet *et al.* (1981)
	Simple (part task) versus complex (whole task) simulators	Muckler *et al.* (1959)
		Smode *et al.* (1966)
		National Research Council (1975)
	Generic versus specific simulation	Center for Nuclear Studies (1980)
9.	*Other*	
	Documentation and use of lessons learned from past simulators	Smode *et al.* (1966)
		Caro (1977)
	Experimental design	Muckler *et al.* (1959)
		Williges *et al.* (1973)

Adams (1979) discusses psychological principles which should underlie the design and use of any simulator. The usefulness of these principles can be illustrated by the following examples: since human learning depends on knowledge of results (KOR), every simulator should incorporate KOR; if a response is to be made to a stimulus, then the stimulus and the control for the response to it must be in the simulator; transfer is greatest when the similarity between the simulator and the actual equipment is high, although transfer is possible with low similarity.

Micheli (1972) makes the point that it may be more important how a device is used than how it is designed. The training effectiveness of a simulator is a function of the total training environment and not merely the characteristics of a particular piece of equipment. Wheaton *et al.* (1976) concluded after an extensive review of the fidelity literature that fidelity *per se* is not sufficient to predict device effectiveness and that the effect of fidelity varies as a function of the type of task to be trained.

These principles have limited usefulness. A possible reason for this is that no one has managed to define objectively in physical terms and to quantify what simulation fidelity actually consists of in a simulator, for example, whether it is number of identical components, number of displays in common with the operational equipment, or common units of information. What is needed are quantitative equations relating amount of fidelity (however defined) to personnel performance. Ideally, one would develop a number of simulators differing in amount of fidelity and then test personnel training and performance. However, this is impractical because of the costs involved, but if one could operationalize the amount of fidelity in the form of a metric, one could differentiate the available simulators in terms of that metric and correlate the metric with the resultant personnel performance.

Simulator development and validation

The development of a simulator, particularly a training simulator, is not or should not, be simply a matter of requiring engineers to reproduce an operational system. The proper development of a simulator begins with an analysis of what one wishes the simulator to do. This includes an analysis of the system mission, its tasks, and most particularly, what is supposed to be trained. If, for example, the simulator is designed to teach nomenclature and basic principles of equipment operation, it need not replicate the complete operational equipment. Unfortunately, many trainers are not used for the purposes for which they were ostensibly built, and many simulators satisfy their training objectives only in part.

The logical endpoint of simulator development is test and evaluation. Since the simulator is only a tool, its validation is merely verification that it accomplishes the purpose for which it was developed. Adams (1979) has pointed out that one cannot utilize high physical fidelity as a measure of simulator effectiveness. Validation is especially important for trainers because what goes on in the simulator, e.g., the number of hours spent in practice, is not necessarily the same as skill acquisition. Validation of a research simulator requires that operational personnel perform in the simulator as they do in the operational system.

The most meaningful method of evaluating the effectiveness of a trainer is by means of transfer of training studies, because the objective of training is the transfer from training to the operational job of the skills required to perform the latter. The transfer of training paradigm requires two groups of trainees: one (experimental) that receives training on the device before proceeding to perform on the operational equipment, and another (control) that receives an equivalent amount of training but only on the operational equipment. Both groups are tested on the operational system and the effectiveness criterion is performance on that system. If the experimental group performs as well as, or preferably, better than the control, the simulator is presumed to be effective. Equal performance for the two groups may represent success for the simulator, since the device is to be preferred because the cost of training with it is usually much less than that involved in using the operational equipment. The two groups must be comparable in terms of relevant prior training, experience and skill, and neither group must be permitted to engage in activities likely to influence their performance unequally on criterion tasks.

In order to employ the transfer of training paradigm effectively, measurements should be made to determine the extent to which training objectives have been met. With measures such as training time or number of training trials, it is necessary to determine (1) the training effort (TE) required to learn the job on the operational equipment without the aid of the simulator (TE − SIM), and (2) the training effort needed to learn the job when some of the training is undertaken using a simulator (TE + SIM). The difference between (1) and (2) is a measure of the training resources saved by the use of the simulator. However, the value of any savings must be considered in relation to (3) the amount of training effort required to learn the task in the simulator (TE in SIM). Roscoe (1971) has derived a transfer effectiveness ratio (TER) with the following equation:

$$\text{TER} = \frac{(\text{TE} - \text{SIM}) - (\text{TE} + \text{SIM})}{(\text{TE in SIM})}$$

The transfer design is particularly advantageous because it is sensitive to both positive and negative transfer effects. Several variations of the transfer design are possible for these (see Rolfe and Caro, 1982; Meister, 1985).

Modelling

A behavioural model of a human–machine system is a symbolic representation of the actions performed by personnel in the operation and maintenance of that system. The representation must allow manipulation of extrinsic and intrinsic variables to permit the determination of the effect of model variables. The model is almost always mathematical and is exercised with a computer. One attempts to model two different kinds of activity: (1) that which is objectively observable; and (2) that which is objectively non-observable. An example of the first is command/control decision making aboard a carrier. An example of the second is neural function. In the first the attempt is to understand by simplifying, making the major parameters specific; in the second the attempt is to understand what goes between a set of stimuli and a set of responses when the interim activity is not observable. The purpose in both cases is a better understanding of what has occurred.

A model is not a theory. Although models often make use of theories, they are not *per se* theories of behaviour. The purpose of a theory is to describe functional relationships. A model obviously incorporates such relationships, but the model goal is to predict and control the effects resulting from the *manipulation* of model variables. Nor is the model any of the varieties of computer-assisted tools, such as computer aided design (CAD) or computer aided manufacturing (CAM). These latter are not true models because they do not describe systems, nor perform the missions these systems perform. Although CAD, for example, utilizes a computer and a software program, CAD is involved with only a small part of a total system, like a control panel, nor does it exercise that part in performing a system task.

The criteria that an effective model must satisfy are listed in Table 8.2.

Many different kinds of human performance model have been developed. They are differentiated along a number of important dimensions: output ver-

Table 8.2. Criteria of model effectiveness

Criterion	Definition
Validity	Agreement of model outputs with actual system performance
Utility	The model's ability to accomplish the objectives for which it was developed
Reliability	The ability of various users to apply the model with reasonable consistency and to achieve comparable results when applied to similar systems
Comprehensiveness	Applicability to various types of systems, to various kinds of system devices and to various stages of design
Objectivity	Requires as few subjective judgements as possible
Structure	Explicitly defined and described in detail
Ease of use	Ease with which analyst can readily prepare data, apply and extract understandable results
Cost of development/use	Includes both time and money
Richness of output	Number and type of output variables and forms of presentation

sus process orientation, predictive versus descriptive, prescriptive (normative) versus descriptive, top-down versus bottom-up, and single task versus multi-task.

Output versus process

This dimension refers to the degree to which a model focuses on system output versus the processes by which the output is generated. An output model predicts the systems outputs resulting from a set of inputs. There is little concern for the internal mechanisms of the model. The model merely has to produce useful outputs for particular inputs. The process model describes the processes by which an output is generated rather than merely predicting the results of human actions; hence they are more complete descriptions. Typically models combine output prediction with some degree of process description. How much process description is needed depends on what the model is applied to.

Prediction versus description

There are two distinct ways of employing models: (1) predicting human-system performance prior to data collection, and (2) fitting the model to human-system performance by adjusting free parameters of the model to conform to existing data. All models have some parameters that must be estimated from empirical data, but predictive models are of greater value than those that merely describe data.

Prescriptive versus descriptive

Models for human performance can either describe how a human is likely to perform or predict ideal behaviour. The former (descriptive) is obviously of greater value, because humans do not behave in ideal ways.

Top-down versus bottom-up

The top-down/bottom-up distinction refers to the extent that a model is determined either by system goals or by human performance capabilities. The top-down approach begins with system goals, then progressively decomposes these into subgoals and functions until the modeller reaches a level at which functions are considered primitives (no longer decomposable). The bottom-up approach begins by defining a set of primitive elements at both the human performance and engineering levels, after which the model is developed. Top-down models focus on system performance (outputs) and bottom-up models focus on the processes leading to outputs.

Single versus multiple tasks

Obviously the model may describe a single task with sub-tasks; alternatively it may be much more comprehensive, describing many tasks. One would prefer to have multitask models, but these are much more difficult to develop, harder to validate, and present more problems in adapting them to a specific system application.

Model uses

Neelamkavil (1987) points out that 'mental modelling is a basic human activity that simplifies planning and decision making'; we use mental models in our daily life for very common activities like recognizing each other. The reasons why models are developed are much the same as those for developing a simulator. Models become necessary when the actual system or a physical simulation are not available, which is usually the case when a new system is being developed and there is need to explore system parameters whose values are unknown. For example, the parameters of a new system under development can be exercised on a computer to suggest how well the future system will perform under specified conditions. In this respect the model is used the same way in which a wind tunnel is used for examining wind stresses on aircraft.

One also uses a model when the phenomena and conditions being studied are too complex to be studied operationally, e.g., predictions of available manpower in the future. One of the major characteristics of the model is that it is a *simplified* representation of a system, process, behaviour or theory. Modelling has certain advantages over the real world; for example, it can reduce the latter's complexity by holding certain variables constant and varying others systematically, in much the same way in which an experiment is designed. In fact, most uses of a model are experimental. Although their functions overlap, most physical simulations are for training and most models are for research, although the research is usually highly applied, supporting concrete systems development and operations. However, models can also be used to train personnel, as in management games to train managers. Because the model simplifies its relationship to the real world phenomena it represents, it cannot be perfect. How much the model outputs differ from those of the actual system must be considered when the model is, as it must be, validated.

An effective model is developed for a specific purpose and has certain goals that it is intended to achieve. The assumptions that underlie model operations, for instance, that workload is some specified function of time required and time available to perform a job (see Siegel and Wolf, 1969), must be clearly stated so that they can be examined critically. The model also has to have clear-cut rules for, among other things, specifying the kind of data it requires, as well as for exercising system functions and for synthesizing the effects of system operations.

When used for support of new systems under development or system modification, behavioural models enable one to answer the following questions: (1) Will personnel be able to complete all required tasks within the time allotted? (2) Where during system operations will personnel be most over- or underloaded, and where are they most likely to fail? (3) How will task restructuring or reallocation of functions affect system functioning? (4) How much will performance degrade when personnel are fatigued or stressed? (5) How will the system's extrinsic and intrinsic variables affect operator and system performance?

Type of model

Models are of various types, the most important of which are task-network or event-driven models, control-theoretic or manual control models, and microprocess or deterministic models. Cognitive models are presently largely experimental. There are also queuing and composite models.

The task network approach emerged from operations research and is oriented toward the sequencing of a large number of discrete tasks arranged in a network to achieve a particular goal; such models focus on the time required to complete tasks and task error probabilities. Copas *et al.* (1985) further subdivide the task-network model into those utilizing analytic methods and those using simulation methods. The primary example of the analytic model is the human reliability predictive process represented by THERP (Swain and Guttmann, 1980) which uses combined database statistics to aggregate task data, while the simulation network model exercises task sequences repeatedly in order to obtain performance outputs (see chapter 31 for more detail). The Siegel-Wolf and SAINT models described in greater detail below are primary examples.

Control theory models come from engineering and are oriented toward continuous time descriptions, optimization of closed-loop human-system performance, and state estimation or manual control. Event deterministic models are based on psychological theories of human information processing and the assumption that human performance can be modelled by the aggregation of the internal (micro) processes required to perform the procedures that define the task. Knowledge-based or cognitive theoretical models are rooted in cognitive psychology and in computer science/artificial intelligence. Cognitive science develops representations of human cognitive processes (decision making, problem solving, planning) in the form of computer algorithms or in simulations of the processes undertaken by the human.

Queuing models require an analysis of system dynamics in terms of the traffic characteristics of multiple tasks. Depending on the formulation given to these, the number of tasks, their arrival rate, and their distribution in time are important independent variables. Task priorities must be specified in order that the cost of negating tasks can be represented mathematically. Human limitations must be defined in terms of the operator's performance as a 'server'. Service rate is an important variable in these models. Because one cannot describe human performance independent of the system context, the generality of this type of model is reduced. In terms of outputs it is possible to predict the number of tasks completed and the average waiting time. If the system permits the operator to have idle time, it is possible to predict operator workload as a function of type of task. One serious limitation of queuing models is that the system must satisfy rigorous mathematical conditions. Difficulties are encountered (Rouse, 1980) if tasks do not arrive independently, because an analytic solution of the model equations may not be possible. Lack of a theoretical database also means that many parameters are left free to vary. These considerations imply that the fundamental requirements for the validity and generality of these models are rarely satisfied.

Finally, the use of composite models has frequently been advocated as a means of expanding system and human performance prediction (Sheridan and Ferrell, 1974; Rouse, 1980). Control and queuing models might, for example, be combined. Or two or more models may be combined, counteracting the deficiencies of one type of model with the strengths of another. The motivation for composite modelling may stem from the inability of control theoretical modelling to incorporate monitoring and supervisory behaviour. Copas *et al.* (1985) have attempted to assess the utility of the various models for system development; their estimate, slightly abridged, is shown in Table 8.3.

Table 8.3. Comparative summary of human performance models

Early design	Relevant for application	Simulation human–computer interaction	Forecast of design solutions	Accommodates personnel requirements	Team behavior
a. Simple analytic models	H	L	L	L	L
b. Siegel & Wolf Naval models	M	L	M	M	M
c. SAINT	M	L	H	?	?
d. HOS	L	P	H	L	L
e. Optimal control model	?	L	H	L	L
f. Queuing models	?	P	H	P	?
g. Composite models	?	L	H	L	L

H = High utility
M = Medium utility
L = Low utility
P = Potentially useful
? = Unknown

Task-network models

Task-network or event-driven simulations sequentially simulate the performance of the sub-tasks of a given task or the events to be performed during
a specified segment of activity. The goal of network techniques is to go from
a statement of the functional relationships among individual task elements to
an estimate of the performance of some defined aggregate of those elements
(the task, the job or the system as a whole). The performance parameters of
interest are typically one or more of the following: (1) the time required to
complete the task aggregate; (2) the probability that the task aggregate will
be completed in a given time; (3) the probability that the aggregate will be
completed correctly.

All network techniques require a thorough description (frequently in diagrammatic form) of actual or intended tasks and of the interrelationships
among those tasks. This means that one must decompose the total system
into its component tasks and sub-tasks (the so-called 'top-down' approach).
All models are data intensive, requiring measures of performance of each task
included in the aggregate, either in the form of measures of central tendency
or in the form of various distributions, e.g., normal, Poisson. The lack of data
creates serious difficulties for the model developer. There is also a problem of
how one combines point estimates or data distributions of individual tasks
to derive data describing total system performance.

The network approach has not been suitable for modelling of continuous
activities (this has been left to the control-theoretical model). There appears
however to be no inherent limitation to continuous task modelling via network formulations, since SAINT (Pritsker *et al.*, 1974) has been successfully
extended to include certain families of continuous variables. The networks
are not models of human performance, though they describe system structures
in which performance is embedded.

Two major network models are Siegel–Wolf and SAINT. The Siegel–
Wolf (Siegel and Wolf, 1969) technique, which was the first systematically
to exploit computer simulation of human performance as a systems analysis
tool, drew on PERT (*Program Evaluation Review Technique*) concepts, in
which a project is conceived to be made up of a network of tasks and sub-
tasks, each of which has estimated completion times—or time distributions—
and probabilities of successful completion. To this Siegel–Wolf added more
sophisticated, psychologically-oriented concepts to examine the impact of
such factors as task-induced stress. This technology proved sufficiently attractive that the US Air Force sponsored the development of a special purpose
simulation language, SAINT, specifically for the purpose of implementing
network-based human performance models. SAINT has had wide acceptance
and usage.

The basic approach to modelling represented in SAINT and other event-
oriented modelling languages is to analyze the system flow chart into a series
of processes related to each other through a contingently branching structure.

A specified distribution of completion times and a probability of successful completion is assigned to each process. These values may be modified while the model is being run. Each time the model is exercised, a sample of each random variable is chosen and the contingent effects are assessed so that a unique trace through the multi-path flow chart is completed.

After a task has been completed, a decision is made as to which task should next be selected. Five decision rules are implemented in SAINT: (1) *deterministic*, in which all *n* branches are selected; (2) *probabilistic*, in which one branch is selected on the basis of a random number drawn from a uniform distribution; (3) *conditional, take first*, in which the first branch satisfying a specified condition is selected; (4) *conditional, take all*, similar to the preceding rule except that all branches satisfying a stated condition are taken; (5) *modified probabilistic*, similar to (2) above, except that branch probabilities are qualified by the number of previous completions of the task from which the branches emanate. Aggregated values for completion time and probability of completion are then derived from repeated runs of the simulation.

Control-theoretic models

The use of control-theoretic or manual control models is restricted to systems in which tracking plays a major role. In control theory the interactions between the operator and the system are represented by servo-control models. Unlike the task-network models, control theory models have a limited concept of human performance—an operator behaves in such a way that errors are minimized within fixed performance constraints. In this concept the operator is an information processing and control-decision element who relates to the system in closed-loop fashion. Feedback is central, comparing actual response with predicted or desired response. Control-theoretic models are more quantitative than other model types. Because of the explicit nature of their assumptions, inputs and outputs, they have been more thoroughly and carefully validated. The models do not, however, attempt to deal with discrete operator inputs, with monitoring or decision making, nor with the procedural aspects of tasks, all of which makes them somewhat unsuitable for describing total job performance.

Microprocess models

The outstanding example of a microprocess or deterministic model is the *H*uman *O*perator *S*imulator (HOS; Streib and Wherry, 1979). Microprocess models are very detailed representations of the operator in terms of the physical and psychological processes that are involved in carrying out a task. These are 'bottom-up' models, which is to say, they synthesize larger segments of human performance from a sequence of molecular fundamental activities such as bodily movements, reaction times, recall of events, and so on. Consequently they encounter the difficulty that the molecular phenomena they model must be combined in some manner to represent the higher order task.

In contrast to task-network and control-theoretical models microprocess models assume an operator's behaviour is explainable and not random and that an operator's actions and the times those actions will take are determined fully by the state of the system and the operator's goal at any particular point in time. Microprocess models are basically deterministic (although individual microprocesses may contain random components). In another striking contrast to the other models, HOS assumes that trained operators rarely forget procedures or make procedural errors. This means that HOS avoids the problems of dealing with error processes. HOS internally constructs an activity by using micromodels of behaviour describing information absorption, recall of information, mental calculation, decision making, anatomical movements, control manipulation and relaxation.

Cognitive models

Cognitive models are, properly speaking, problem solving or diagnostic models in which the behaviours modelled are those involved in, for instance, determining the cause of an equipment malfunction or diagnosing the cause of an illness. What is modelled is: (1) the task environment—that is, the objective problem to be solved—the rules to be applied in the solution, the information representing the status of the system; and (2) the program developed to solve the problem. Material for development of the model is often derived from verbal protocols of personnel performing diagnostic tasks.

The systems to which many cognitive models have been applied are highly sophisticated, computerized, 'intelligent' systems, which means that they possess or mimic human qualities and characteristics. Such systems are designed to solve problems, either on their own or in cooperation with a human operator. Because of the attempt to replicate human processes in intelligent systems, there has been considerable interest among computer-oriented researchers in models of human cognition and communication. Cognitive models have certain common themes: (1) Since the model for intelligent system design is the human, the cognitive models attempt to describe how humans perceive, store and retrieve information from short- and long-term memory, manipulate information, make decisions and solve problems, and implement responses. The human processes hypothesized to underlie cognition and communication are then analogized to equivalent or parallel computer processes. (2) Another theme in these models is 'what you see is not really what is there'; in other words, there is no one-to-one correspondence between the state of the world and the human's perception or understanding of that state. (3) Internal processing of information occurs on several levels. Bobrow and Norman (1975), for example, propose an information processing system based on memory structures called schemata; they suggest that these structures function hierarchically. The lower levels deal with 'event driven' processing which is bottom up and seeks structures to which sensory inputs can be transmitted. The higher levels involve 'concept driven' top-down processing and are stimulated by goals. (4) 'Advanced' thinkers in this field model not only the human but also the world or the system which represents that world. The world is an indeterminate (uncertain) one, so that to represent it correctly the system must examine the state of the world to determine what the nature of the evidence before it actually is.

Cognitive models may be physiological or performance-based. It is unclear how useful these models are, since their application is to highly sophisticated, intelligent systems, and such systems, if they exist, are only in an experimental state.

Stages of model development

These include the following steps, with the understanding of course that the process of model development, like system development itself, is iterative, possessing multiple feedback loops. The following section is taken largely from Chubb *et al.* (1987).

(1) *Problem formulation:* the developer must ask, what questions must the model answer? (2) *Model building:* system operations must be decomposed into tasks, sub-tasks, events, etc. An appropriate simulation language must be selected. (3) *Data acquisition:* data to apply to algorithms, e.g., estimates of task duration and probabilities of task success completion, must be gathered. (4) *Model translation:* the model must be prepared for computer processing in accordance with whatever simulation language is selected. (5) *Verification:* the developer must demonstrate that model operations and outputs correspond to actual system operations and outputs. (6) *Exercise planning:* the developer must establish the experimental (design) conditions for running the model. (7) *Exercise implementation:* the developer must run the model on a computer. (8) *Analysis* of run results. (9) *Utilization:* the developer or someone else implements the decisions resulting from the model exercise. (10) *Documentation:* the developer describes the model and the results of the exercise in written form.

Model validation

Since the model is only a representation of reality (not the reality itself), the effectiveness of that representation must be demonstrated. Tests of model validity may range from informal demonstrations that it produces reasonable results to formal tests of how well model predictions fit independently derived data.

The problem of model validation is a difficult one. In one sense a model is valid if it can be used to arrive at reasonable decisions; accuracy may not be very important. More often, one looks for experimental, quantitative validation or tests of model accuracy. In this case it is necessary to compare experimental results with model predictions and to apply both engineering and formal statistical tests of the null hypothesis to determine whether or not the model should be considered valid.

If the model is to be useful as a design tool, it must be validated prior to, or at least concurrently with, the development of a system simulation. Under these conditions, system performance data against which to validate the model may not be available, in which case one must seek less rigorous means of examining model validity. If, in the course of building the model, the developer generates hypotheses about how behaviour will change with changes in critical system parameters, at a minimum he or she should be able to predict the *direction* of changes in system performance measures, if not their magnitude (Pew *et al.*, 1977).

Simulation approaches

To select a model one must consider whether the activities being modelled are discrete, continuous or combined.

(1) *Discrete* event simulation occurs when the dependent system variables change by fixed amounts at specified points in simulated time, referred to as 'event' times. The system is modified and time updated only by the occurrence of an event (normally the start or completion of an operator task in human operator modelling). To build an event model one defines the variables that portray the system's status and then identifies those events that can change system status; this is done by defining system status. The state of a system is defined by the values assigned to the attributes of system entities; these last can be, for example, physical objectives, personnel, and so on.

(2) In a *continuous* simulation model system state is represented by continuously changing dependent variables (commonly referred to as 'state variables'). This model describes only tasks involving continuous monitoring or control (e.g., tracking or maintaining process variables with narrow tolerances over substantial time periods). However, many task–network models involve continuous simulation components which lead to combined models.

(3) In *combined* (discrete–continuous) models the variables in the model may change both discretely and continuously. The behaviour of the system model is simulated by recomputing the values of the continuous variables into small time steps and recomputing the values of discrete event variables.

Selection of model variables

Since there are usually more variables than it is feasible to include in any individual model, a choice must be made. Criteria of variable selection (modified from Siegel and Wolf, 1981) suggest that a preferred variable is one which (a) is backed by available empirical data; (b) has high data reliability; (c) is sensitive to changes in system dynamics; (d) is capable of empirical (objective) measurement; (e) is free of unwarranted assumptions, excessive processing time or requirements for large amounts of memory storage; (f) is generalizable to a range of modelled situations; (g) is easily understood by model users; and (h) is most useful for answering the questions the user of the model wishes to ask. Of course, few variables ever satisfy all of the above criteria to the extent desired.

Advantages and disadvantages

Network models have a number of disadvantages. (1) They are data limited. Values for all input parameters must be derived by estimation or empirical sources. (2) Task interactions are a problem. Although simulation methods do not incorporate combinatorial statistics, the performance of a single task may differ from that which it presents when it is embedded in other tasks. Practice and fatigue effects must be presumed. SAINT, however, allows task performance to be moderated as a function of other variables. (3) Cognitive processes are difficult to model. On the other hand, the network approach is generally applicable, as long as the system can be analyzed into discrete tasks.

If one uses HOS as a primary exemplar of information processing models (and most writers on the subject do), there are some significant difficulties with such models. For example, HOS assumes that the operator makes no errors, follows procedures properly, and that microprocess times are additive. The more operators deviate from an optimum sequence of actions, the less accurate HOS predictions are likely to be. In addition, details of validation studies are difficult to find. The outputs of the model are deterministic, so that it is difficult to describe human variability. On the other hand, the use of human performance sub-models can be applied to any situation with little modification required. However, these models require a procedural system description and it is not clear how that description is achieved.

As for control theory, alternative system designs may be readily evaluated and compared because control theory regards human and system performance as inextricable. Detailed pre-modelling task analysis of operator activities is not required. Estimation of human performance parameters is not required because tasks are not individualized. On the other hand, one of the most serious deficiencies of control modelling is the neglect of individual differences and team performance.

Simulation languages

This section also leans heavily on Chubb *et al.* (1987). The selection of a simulation language is frequently based on convenience (i.e., familiarity with the language and its availability). However, the important factors to consider in comparing simulation languages are the training required to use the language, its ease of coding and debugging, its portability from one model type to another, its flexibility, statistical capability and report production capabilities, its reliability of software, and speed of execution.

Special languages have been developed to fit the various types of models. Nevertheless, a simulation can sometimes be implemented by more than one language.

Discrete languages

The general purpose simulation system (GPSS) exists in a number of variations, GPSS/360 and GPSS/H being most widely used. The principal appeal of GPSS is its modelling simplicity. A GPSS model is constructed by combining sets of standard blocks into a block diagram that defines the logical structure of the system. Entities are represented in GPSS as transactions that move sequentially from block to block as the simulation proceeds. GPSS provides almost all basic simulation functions and has extensive data collection and summarization capabilities. On the other hand, GPSS works more slowly (and is consequently more expensive to run) and has a limited capability for generating random variates.

Q-GERT, developed by Pritsker (1979) has a network consisting of nodes and branches. A branch represents an activity that models a processing time or delay. Nodes are used to separate branches and to model decision points milestones and queues. Flowing through the network are entities called 'translations'. The procedures for constructing a model in Q-GERT are much the same as those for GPSS.

SIMSCRIPT developed by Kiviat *et al.* (1969) has five levels, ranging from a simple teaching language to introduce programming concepts, to statement types that are comparable in power to FORTRAN, ALGOL or PL/1, to levels that utilize entity, attribute and set concepts and provide, for example, for time advance and event processing. SIMSCRIPT is primarily event oriented, system state being defined by entities, their associated attributes and groupings of entities known as 'sets'. System structure is described by defining the changes that occur at event times. SIMSCRIPT is particularly attractive because of its free-form and English-like syntax. Like GPSS, SIMSCRIPT requires its own compiler and is therefore not always available at every computer facility.

Continuous languages

A wide variety of continuous system simulation languages (CSSL) has been developed. Most recent CSSLs are equation-oriented and have FORTRAN-like syntax (see Graybeal and Pooch, 1980).

Combined languages

Only the general activity simulation program (GASP IV) is widely used. The more recently developed Simulation Language for Alternative Modelling

(SLAM) is based on GASP IV. GASP IV allows system descriptions to be written either as discrete event models, continuous models, or a combination of the two (Pritsker, 1974). It specifies procedures for writing differential or difference equations as well as methods for defining the logical conditions that affect system status variables. SLAM II (Pritsker, 1984) incorporates the process-oriented features of Q-GERT and the combined discrete-continuous features of GASP IV. SLAM employs a network structure composed of nodes and branches, which model elements in process, such as queues, servers and decision points. Their symbols are combined into a network model that pictorially represents the system with system entities flowing through the network model.

In concluding this section on simulation languages, special attention must be drawn to Micro-SAINT, a microcomputer version of SAINT (Laughery, 1985), which captures many of the features of SAINT but is designed to represent operator task networks in a more straightforward and simple manner.

Summary

The questions raised about physical human–machine simulation, such as the amount of fidelity required, are essentially research questions. Notwithstanding any problems with finance, improved computers and their graphics will permit one to simulate any system to the point that it reproduces the operational system and the system environment almost perfectly. The questions raised at the start of this chapter are important only to reduce cost and to focus simulator development more precisely.

It is otherwise with models. Human performance modelling is still plagued by most of the same weaknesses it has had for many years. The range of human behaviours which models address adequately is still quite limited. There has been little effort to determine the validity and generality of those models that are available. A considerable amount of technical knowledge is still needed to make use of most models. And most models are normative and do not adequately describe individual differences or the sources of error in operator performance. It appears that there has been far less design use of these than one would wish.

As applied to system design, users should not expect models to provide detailed design solutions or the final selection of the best design option. Actually, they are tools to help reduce a large number of design options to a more manageable subset for evaluation in mock-ups or experiments. The qualitative pay-off from models may in fact be more valuable than any quantitative results. Even where the model makes specific quantitative predictions, its primary value is the identification of those design parameters that are important as against those that are not. Of course, quantitative results are important when the ergonomist is defending human factors requirements to the design team, because unless the concepts and benefits can be made quite specific, i.e., defended with numbers, they will tend to be ignored.

System designers would dearly love to have general simulation models which include the human, but validated quantitative data on a sufficiently wide range of human behaviour are not available, even if one limits oneself to operator control (Chubb *et al.*, 1987). This should not be considered a criticism of simulation models of human performance. If validated, they are useful both as a theoretical tool in the study of performance and as a means of predicting human behaviour in a system. If model descriptions deviate from actual behaviour, they may still be acceptable as long as they do not lead to gross errors, or when they are somewhat conservative. It is different when the model promises more than can be expected from the human and if there are important factors about which the model can say nothing. Sanders (1988) suggests that, given the complexity of human performance, it is better to content oneself at present with small scale simulations of fairly specific tasks rather than aiming at general purpose simulation. Examples of the actual use of human performance models and simulations are perhaps the best instructors; a paper by Hulme and Hamilton (1988) is very useful in this regard.

Further reading

It is impossible within one short chapter to describe fully all the variables that enter into two such complex technologies as simulation and modelling. To assist readers in further explorations of these topics it is suggested that they consult the following: Naylor (1969), Emshoff and Sisson (1970), Van Horn (1971), Reitman (1971), Shannon (1975), Wilson and Pritsker (1982), and Card *et al.* (1983). Three more recent sources are Copas *et al.* (1985), McMillan *et al.* (1988) and Baron *et al.* (1990). See also Salvendy (1987).

References

Adams, J.A. (1973). Preface. Special issue on flight simulation. *Human Factors,* **15**, 501.

Adams, J.A. (1979). On the evaluation of training devices. *Human Factors,* **21**, 711–720.

Baron, S., Kruser, D.S. and Huey, B.M. (Eds) (1990). *Quantitative Modeling of Human Performance in Complex Dynamic Systems* (Washington, DC: National Academy Press).

Beare, A.N. and Dorris, R.E. (1983). A simulator-based study of human errors in nuclear power plant control room tasks. *Proceedings of the Human Factors Society Annual Meeting,* pp. 170–174.

Bobrow, D.G. and Norman, D.A. (1975). Some principles of memory schemata. In *Representation and Understanding,* edited by D.G. Bobrow and A. Collins (New York: Academic Press).

Buchaca, N.J. (1979). *Models and Mockups as Design Aids.* Technical Document 266/Revision A (San Diego, CA: Naval Ocean Systems Center).

Card, S.K., Moran, T.P. and Newell, A. (1983). *The Psychology of Human–Computer Interaction* (Hillsdale, NJ: Lawrence Erlbaum).

Caro, P.W. (1977). *Some Factors Influencing Transfer of Simulator Training.* HUMRRO Technical Report TR-77-2 (Alexandria, VA: Human Resources Research Organization).

Center for Nuclear Studies (1980). *Nuclear Power Simulators: Their Use in Operator Training and Requalification.* Report B0421-8 (Oak Ridge, TN: US Department of Energy).

Chubb, G.P., Laughery, Jr., K.R. and Pritsker, A.A.B. (1987). Simulating manned systems. In *Handbook of Human Factors,* edited by G.S. Salvendy (New York: John Wiley), pp. 1298–1327.

Copas, C.V., Triggs, T.S. and Manton, J.G. (1985). *Human Factors in Command-and-Control System Procurement* Report HFR-15 (Melbourne: Monash University).

Cunningham, J.B. (1984). Assumptions underlying the use of different types of simulation. *Simulation and Games,* **15**, 213–234.

Emshoff, J.P. and Sisson, R.L. (1970). *Design and Use of Computer Simulation Models* (London: MacMillan).

Gaffney, M.E. (1981). Bridge simulation: trends and comparisons. In *Proceedings of the U.S. Institute of Navigation Annual Meeting,* Washington, DC.

Gagné, R.M. (1954). Training devices and simulators: some research issues. *American Psychologist,* **9**, 95–107.

Graybeal, W. and Pooch, U.W. (1980). *Simulation: Principles and Methods* (Cambridge, Massachusetts: Winthrop).

Hayes-Roth, F., Waterman, D.A. and Lenat, D.B. (1983). *Building Expert Systems* (Reading, MA: Addison-Wesley).

Hays, R.T. (Ed.) (1981). *Research Issues in the Determination of Simulator Fidelity.* Technical Report 547 (Alexandria, VA: US Army Research Institute).

Hennessey, R.T., Sullivan, D.J. and Cooles, H.D. (1980). *Critical Research Issues and Visual System Requirements for a V/STOL Training Research Simulator.* Report NAVTRAEQUIPCEN 78-C-0076-1 (Orlando, FL: Naval Training Equipment Center).

Hollan, J.D., Hutchins, E.L. and Weitzman, L. (1984). STEAMER: an interactive inspectable simulation-based training system. *The AI Magazine,* **5**, 15–27.

Huff, E.M. and Nagel, D.C. (1975). Psychological aspects of aeronautical flight simulation. *American Psychologist,* **30**, 426–439.

Hulme, A.J. and Hamilton, W.I. (1988). Human engineering models: A user's perspective. In *Applications of Human Performance Models to System Design,* edited by G.R. McMillan, D. Beevis, E. Salas, M.H. Strub, R. Sutton and L. Van Breda (New York: Plenum Press) pp. 487–500.

Human Factors Operations Research Laboratories (1953). *Flight Simulation Utilization Handbook.* Report 43 (Bolling Air Force Base, Washington, DC: Human Factors Operations Research Laboratories).

Jones, E.R., Hennessey, R.T. and Deutsch, S. (Eds) (1985). Committee on Human Factors, National Research Council. *Human Factors Aspects of Simulation* (Washington, DC: National Academy Press).

Kinkade, R.G. and Wheaton, G.R. (1972). Training device design. In *Human Engineering Guide to Equipment Design,* edited by H.P. Van Cott and R.G. Kinkade (Washington, DC: US Government Printing Office), pp. 667–699.

Kirwain, B. and Ainsworth, L.K. (Eds.) (1992). *A Guide to Task Analysis* (London: Taylor and Francis).

Kiviat, P.J., Villaneuva, R. and Markowitz, H. (1969). *The SIMSCRIPT II Programming Language* (Englewood Cliffs, NJ: Prentice Hall).

Kraft, C.L., Anderson, C.D. and Elworth, C.L. (1980). *Psychophysical Criteria for Visual Simulation Systems.* Report AFHRL-TR-79-30 (Williams Air Force Base, AZ: Air Force Human Resources Laboratory).

Laughery, K.R. (1985). Network modeling on microcomputers. *Simulation,* **38**, 10–16.

Matheny, W.G. (1975). Investigation of the performance equivalence method for determining training simulator and training methods requirements, AIAA paper 75-108. *AIAA 13th Aerospace Science Meeting,* Pasadena, California, American Institute of Aeronautics and Astronautics.

McCluskey, M.R. (1972). *Perspectives on Simulation and Miniaturization.* CONARC Training Workshop, Fort Gordon, Georgia.

McMillan, G.R., Beevis, D., Salas, E., Strub, M.H., Sutton, R. and Van Breda, L. (Eds.) (1988). *Applications of Human Performance Models to System Design* (New York: Plenum Press).

Meister, D. (1985). *Behavioral Analysis and Measurement Methods* (New York: John Wiley).

Mellichamp, J.M. and Wahab, A.F.A. (1987). Process planning simulation: an FMS modelling tool for engineers. *Simulation,* **48**, 186–192.

Micheli, G.S. (1972). *Analysis of the Transfer of Training, Substitution and Fidelity of Simulation of Transfer Equipment.* Report 2 (Orlando, FL: Training Analysis and Evaluation Group, Naval Training Equipment Center).

Miller, R.B. (1954). *Psychological Considerations in the Design of Training Equipment.* Technical Report 54-563 (Wright-Patterson Air Force Base, OH: Wright Air Development Center).

Muckler, F.A., Nygaard, J.E., O'Kelly, L.L. and Williams, A.C. (1959). *Psychological Variables in the Design of Flight Simulators for Training.* Technical Report 56-369 (Wright Patterson Air Force Base, OH: Wright Air Development Center).

National Research Council (1975). *Visual Elements in Flight Simulation.* Assembly of Behavioral and Social Sciences (Washington, DC: National Academy of Sciences).

National Research Council (1982). *Automation in Combat Aircraft.* Air Force Studies Board, Commission on Engineering and Technical Systems (Washington, DC: National Academy Press).

NATO-AGARD (1980). *Fidelity of Simulation for Pilot Training.* Advisory Group for Aerospace Research and Development, Advisory Report 159 (Neuilly sur Seine, France).

NATO-AGARD (1981). *Characteristics of Flight Simulator Visual Systems.* Advisory Group for Aerospace Research and Development, Advisory Report 164 (Neuilly sur Seine, France).

Naylor, T.H. (1969). *The Design of Computer Simulation Experiments* (Durham, NC: Duke University Press).

Neelamkavil, F. (1987). *Computer Simulation and Modelling* (Chichester: John Wiley).

Park, C.S. and Noh, Y.D. (1987). A port simulation model for bulk cargo operations. *Simulation,* **48**, 236–246.

Parks, D.L. and Springer, W.E. (1975). *Human Factors Engineering Analytic Process Definition and Criterion Development for CAFES.* Report D-180-18750-1 (Seattle, WA: Boeing Aerospace Company).

Parsons, H.M. (1972). *Man Machine System Experiments* (Baltimore, MD: The Johns Hopkins University Press).

Pew, R.W., Baron, S., Feerher, C.E. and Miller, D.C. (1977). *Critical Review and Analysis of Performance Models Applicable to Man–machine Systems Evaluation.* BBN Report 3446 (Cambridge, MA: Bolt, Beranek and Newman).

Pritsker, A.A.B. (1974). *The GASP-IV Simulation Language* (New York: John Wiley).

Pritsker, A.A.B. (1979). *Modeling and Analysis of Q-GERT Networks* (New York: Halstead).

Pritsker, A.A.B. (1984). *Introduction to Simulation and SLAM II* (New York: John Wiley).

Pritsker, A.A.B., Wortman, D.B., Seum, C.S., Chubb, G.P. and Seifert, D.J. (1974). *Systems Analysis of Integrated Networks of Tasks, SAINT,* Volume I. Report AMRL-TR-78-126 (Wright-Patterson Air Force Base, OH: Aerospace Medical Research Laboratory).

Prophet, W.W., Shelnutt, J.B. and Spears, W.D. (1981). *Simulator Training Requirements and Effectiveness Study (STRES): Future Research Needs.* Report AFHRL-TR-80-37 (Brooks Air Force Base, TX: Air Force Human Resources Laboratory).

Reitman, J. (1971). *Computer Simulation Applications* (New York: John Wiley).

Rolfe, J.M. and Caro, P.W. (1982). Determining the training effectiveness of flight simulators: some basic issues and practical developments. *Applied Ergonomics,* **13**, 243–250.

Roscoe, S.N. (1971). Incremental transfer effectiveness. *Human Factors,* **13**, 561–567.

Rouse, W.B. (1980). *Systems Engineering Models of Human–Machine Interaction* (Amsterdam: Elsevier).

Salvendy, G. (Ed.) (1987). *Handbook of Human Factors* (New York: John Wiley).

Sanders, A.F. (1988). Human performance models and system design. In *Applications of Human Performance Models to System Design* edited by G.R. McMillan, D. Beevis, E. Salas, M.H. Strub, R. Sutton and L. Van Breda (New York: Plenum Press) pp. 475–486.

Shannon, R.E. (1975). *System Simulation: The Art and Science* (Englewood Cliffs, NJ: Prentice Hall).

Sheridan, T.B. and Ferrell, N.R. (1974). *Man–Machine Systems* (Boston: MIT Press).

Siegel, A.I. and Wolf, J.J. (1969). *Man–Machine Simulation Models: Psychosocial and Performance Interactions* (New York: John Wiley).

Siegel, A.I. and Wolf, J.J. (1981). *Digital Behavioral Simulation—State-of-the-Art and Implications* (Wayne, PA: Applied Psychological Services).

Smode, A.F., Hall, E.R. and Meyer, D.E. (1966). *An Assessment of Research Relevant to Pilot Training.* Report AMRL-TR-66-196 (Wright–Patterson Air Force Base, OH: Aeromedical Research Laboratories).

Streib, M.I. and Wherry, R.J. (1979). *An Introduction to the Human Operator Simulator.* Technical Report 1400.02-D (Willow Grove, PA: Analytics).

Swain, A.D. and Guttmann, H.E. (1980). *Handbook of Human Reliability with Emphasis on Nuclear Power Plant Applications.* Report NUREG/CR-1278. (Albuquerque: Sandia National Laboratories).

Thorpe, J.A., Varney, N.C., McFadden, R.W., LeMaster, W.D. and Short, L.H. (1978). *Training Effectiveness of Three Types of Visual Systems for KC-135 Flight Simulators.* Report AFHRL-TR-78-16 (Williams Air Force Base, AZ: Human Resources Laboratory).

United States Air Force (1978). Scientific Advisory Board. *Ad Hoc Committee on Simulation Technology* (Washington, DC: US Department of Defense).

Van Horn, R.L. (1971). Validation of simulation results. *Management Science,* **17**, 247–258.

Waag, W.L. (1981). *Training Effectiveness of Visual and Motion Simulation.* Report AFHRL-TR-79-7 (Brooks Air Force Base, TX: Human Resources Laboratory).

Wheaton, G.R., Rose, A.M., Fingerman, P.W., Korotkin, A.L. and Holding, D.H. (1976). *Evaluation of the Effectiveness of Training Devices.* Research Memorandum 76-6 (Alexandria, VA: US Army Research Institute).

Whicker, M.L. and Sigelman, L. (1991). *Computer Simulation Applications: An Introduction* (Newbury Park: Sage Publications).

Williges, B.H., Roscoe, S.N. and Williges, R.C. (1973). Synthetic flight training revisited. *Human Factors,* **15**, 543–560.

Wilson, J.R. and Pritsker, A.A.B. (1982). Computer simulation. In *Handbook of Industrial Engineering,* edited by G.S. Salvendy (New York: John Wiley).

Woods, D.D., Pople, H.E. and Roth, E.M. (1990). *The Cognitive Environment Simulation as a Tool for Modelling Human Performance and Reliability.* Report NUREG/CR-5213 (Washington, DC: US Nuclear Regulatory Commission).

Chapter 9

Computerized data collection in ergonomics

Colin G. Drury

Introduction

We live in an era where serious data collection *without* a computer sounds slightly perverse, yet even 15 years ago the computer was considered novel in ergonomics data collection, and 25 years ago only well-equipped laboratories and military test and evaluation sites had on-line systems. Three legitimate questions can be asked about *data collection* to help us understand this phenomenon, which has the proportions of a revolution:

1. How did we collect data in the past?
2. Where is computer-aided data collection now?
3. What new techniques and capabilities await the future ergonomist?

Data analysis of any detail or size is almost never performed without a computer and many of the most complex and precise analysis methods available would be impossible, or at least costly and time-consuming, by any other method. There is no need to review past data analysis methods, but the second and third questions above apply equally to data analysis as they do to data collection.

Obviously, to provide comprehensive answers to these questions would require a distinguished panel of science historians, computer scientists and futurists. The aim of this chapter is somewhat more modest—to provide a review from the viewpoint of the working ergonomist who is a consumer of computer hardware and software rather than a computer specialist. Data collection will be considered first, followed by data analysis.

Data collection

Historical data collection techniques

Ergonomics has its roots in psychology, human physiology and industrial engineering, so that our primary instruments have been (like ergonomics itself) imported from other disciplines. All data collection in experiments involves providing the subject with an input (stimulus, challenge) and observing the results (response, output). Our primary measuring tools were thus time, force and distance measuring devices and chemical analysis systems, for example the stop-watch, metre rule, dynamometer and oxygen analyzer. Unfortunately, many phenomena of interest such as reaction times, tracking errors and force transients happen too quickly or over too small a physical range to be measured conveniently in this way. Two solutions are possible: measure many examples of the phenomenon (repeated trials) or find more precise measuring equipment.

A good example of timing multiple trials is the use of reciprocal tapping by Fitts (and many others since) to study the speed/accuracy trade-off in target aiming tasks. Another example is the use of the pursuit rotor which counts the number of times (or total time) a small target is lost rather than measuring the more complex position/time history of target and follower. The 'multiple trials' strategy is still used in class demonstration experiments and in measuring industrial performance from production records.

More precise measuring equipment for small times and distances typically involved mechanical or electrical amplification and a time-record imprint of the measure onto a continuously moving chart. Times in the millisecond range could be obtained from paper chart recorders operating sufficiently rapidly, as could transients in physiological phenomena. Strip-chart recorders were produced in increasing complexity, with less intrusive instrument dynamics and multiple channels. However, the data transcription process can be lengthy, tedious and error-prone. The advent of high quality tape recorders (including multi-channel FM recorders) did not solve the problem and only allowed the user to extend the time scale for recording very rapid events (e.g., fine-structure of EMGs) or very slow events (e.g., diurnal variations in physiological output and performance).

As will be discussed later, the purpose of data collection is ultimately data interpretation, so that the analysis stage cannot be ignored. The rise of statistical design and analysis methods for experimentation during the 1930s to 1950s meant an increased demand for data collection in sufficient volume to provide statistical significance. Typically, this required larger sample sizes and analysis of individual data points rather than mean values. Happily, digital computers made their debut during this time, allowing the experimenter to attempt complex analyses of variance and regression in a reasonable time frame. However, the burden of transcription between the data collection device and the computer input device still fell onto the experimenter or onto technicians. The dedication of an expensive computer to control an experiment and collect the data was only possible with considerable money and technical help, usually only available to answer military questions.

For the laboratory experimenter in general, the rise of transistors enabled logical elements to be used in relatively inexpensive, general purpose control systems. Input devices, such as response keys, photocells and voice-sensing units, could be interfaced with logical elements such as steppers, relays and AND/OR/NOT gates and with recording devices such as timers and counters. These same logic systems could control more complex stimulus presentation devices such as coloured light displays, numerals and even slide projectors, but the outputs were still not readable by computers.

One exception was the series of event timers and recorders which produced punched paper tape output. SETAR, and later METRA, recorded input events and times at which they occured, with a speed of up to 10–30 event/s. Although of low reliability when these data rates were temporarily exceeded, these machines enabled experimenters to collect complex event timing data over long periods and removed the chore of data transcription. At this point the experimenter needed computer programming skills to interpret the paper tape, particularly when dealing with the inevitable errors. Experimental data collection and data interpretation were at last coming together.

Current data collection techniques

Perhaps the real revolutions in data collection over the last 25 years have been caused by dedication of the computer itself to laboratory experimentation and, more recently, by the ability of different computers to exchange data.

To earlier experimenters 'computer dedication' meant receiving a dispensation from the operating staff of a mainframe computer to take over sole running of the machine during particularly inconvenient hours. With the advent of minicomputers it became feasible for a department, or even a laboratory, to have its own computer and hence to dedicate it to real-time data acquisition and experimental control. All that has changed in the past few years in that the price of computing power has decreased by about two orders of magnitude, and it has been recognized that the software for computer use has to be used by people other than computer enthusiasts. Microcomputers are now available in large numbers at prices even universities can consider. High-level languages have evolved which can access real-world data as well as provide display functions and calculation abilities. At one end of the scale languages such as 'BASIC' make relatively crude programmers out of experimenters, while at the other end of the scale highly structured languages such as 'C', 'PASCAL' and 'MODULA2' enable complex programs to be written by teams rather than individuals. Newer languages, which are object-oriented and event-driven, (e.g., VISUAL BASIC) allow the experimenter to interface easily and graphically with the subject, often in a consistent Graphical User Interface (GUI) with which the subject is familiar. Even older languages such as 'FORTRAN' now have the ability to display graphics (e.g., DI-3000) and communicate with external devices. Programming environments now include editors, debuggers and compilers with relatively 'friendly' user interfaces. Full scale data collection environments are now available for microcomputers (e.g., 'LABVIEW') which generate working display panels for complex systems. These can be used in studies where a degree of realism is required (such as rapid prototyping), as well as in more general studies where complex stimuli and responses are required. Much of the programming for these environments can be performed by manipulation of graphic icons, as in connecting logical elements with 'wire' icon, rather than by writing lines of text-based code.

Networking of computers is another possibility opened up by cheap computing power and the advent of various character- and file-transmission standards. Most major computing centres have data networks which allow the small, dedicated laboratory computers to exchange files with large multi-user mainframes. Hence data collected on a machine of appropriate size for collection (mini- or microcomputer) can be analyzed using a machine of appropriate size to handle massive data handling packages (mainframe or super-computer).

We now can run an experiment in a totally computer-controlled environment. Instructions, stimulus timing and feedback can all be handled by the same machine as that which records the data. In this way, complex tasks can be undertaken, bringing a sense of realism into the laboratory where appropriate. Recent examples include complex process control tasks and microscope-based inspection tasks, e.g. Synfelt and Brunskill (1986). Certain data collection methods are still relatively expensive, however. Data rates appropriate to multiple-channel physiological recording can be in the tens of thousands of samples per second. These can be difficult to handle on microcomputers, both at data input and at data storage, if collection goes beyond a few seconds. Another expensive example is the recording of three-dimensional position data over a reasonable volume of space at reasonable precision and input rates. Typical systems still cost an order of magnitude more than a microcomputer workstation in an ergonomics laboratory.

The prospect of generating data at rates of thousands of samples per second is both daunting and exciting for the ergonomist. Large, multi-measure experiments are enjoying something of a vogue (Drury and Deeb, 1986a, b) but they give complex and often messy results. In the past, a choice of a single performance or stress measure meant that system evaluation and ergonomics research had a certain happy simplicity. Thus system A performed better than system B at high speeds but not at low speeds. Ergonomists could make the relatively simple statement often demanded of them by others (usually hardware engineers) on the design team. Now with multiple measures, the picture is rarely so clear, even though techniques such as factor analysis can be used to organize data interpretation. A recent experiment on human and algorithm interaction in job-shop scheduling showed overall that humans were a necessary part of any effective system, but the various measures of effectiveness often gave directly conflicting results (Kondakci, 1985). Ergonomists and their clients are having to adapt to this new reality.

The future of data collection

While we continue to bask in our good fortune in working at a time when technological marvels aid us in the laboratory, it is necessary to examine how our tools are likely to evolve and how they will influence the data we collect and the tasks chosen for study. Outside the laboratory, computers are revolutionizing our home and professional lives while word processors have largely replaced typewriters, and spreadsheets have at least complemented calculators. Communications programs allow us to obtain on-line weather, news and shopping services as well as access to vast data bases. Even *Ergonomics Abstracts*, the bibliographic journal published by Taylor and Francis, has computer access. Today we can sort our electronic mail and fill our friends' mailboxes with electronic memos. For the future, some clear trends can be seen in the short-term.

Portability

As power requirements for integrated circuit chips decrease and the energy densities of batteries increase, portable computing becomes possible. We have used simple portable computers based on 8-bit microprocessors to enhance data collection in the field (Drury, 1986). It is now possible to design a questionnaire on a portable computer, run subjects and edit the questionnaires based on an analysis of the responses. Data collected can be passed to a microcomputer or a mainframe computer for analysis using standard packages. Similarly, a portable computer the size of a small telephone directory and weighing less than 2 kg can provide sophisticated event timing functions and even run an occurrence sampling scheme. We need no longer talk about ergonomists as if they were shackled to a laboratory. The range and complexity of portable computers is increasing. Discounting the 10 kg machines, which need mains power, the portable 16-bit and 32-bit microprocessors are highly compatible with desk-top microcomputers. Should you ever need to recalculate a large spreadsheet on the beach, this is now possible on a machine weighing less than 3 kg. New ranges of powerbook and hand held computers—personal digital assistants (PDAs)—provide even greater physical convenience (see Wilson, 1995).

A major advantage of the simpler portable computers is their ease of programming and interfacing. Analogue and digital interfaces, similarly battery powered, are inexpensive; data transfer to a desk-top or mainframe computer is largely menu-driven. At this level of simplicity, even occasional computer users can write programs (in high-level languages) carefully tailored to the user's needs. And ergonomists can now take to the road without abandoning modern data collection techniques.

Better displays

As microprocessor power increases into the millions of floating-point oper-ations-per-second range, so more computing power is available for the com-putation-intensive needs of display generation. Coupled with decreasing memory costs, for both RAM and CD-ROM, ergonomists can look forward to being able to use more complex graphics displays. Already the 1000 × 1000 pixel display is an inexpensive reality. With further advances in com-puter architecture (e.g., parallel processing) and software (e.g., graphics algorithms) colour and animation techniques, presently only available to Hol-lywood and the military, will be affordable in the ergonomist's laboratory. Instead of the computer controlling the presentation of complex images to the subject by manipulating slide projectors and videotape, the regular display screens will provide the same resolution with vastly increased flexibility. Inci-dentally, increased computer power and graphics knowledge are already merging different media. We can capture and digitize actual scenes as well as produce computer-animated scenes on videotape. Computers are already a standard method of display prototyping in the military; using computers and displays only twice to three times the price of desk-top microcomputers, rapid changes can be made in display panels and their performance assessed before the hardware is built.

For the future high resolution colour and animation will allow us to move back from what Crossman (1960) called 'symbolic' displays. Perhaps we can create a new reality for controllers of complex processes, mixing real scenes with colour-coded analogue representation of important parameters, allowing rapid changes to be displayed and even producing predictor displays. All of these remarks apply equally well to the auditory sense. Modern musicians frequently have computers between their instrument and the speakers; per-haps this technology will be found useful in studying human auditory responses and producing complex auditory displays.

With the advent of Virtual Reality (VR), the prospect of interacting directly with objects in a three-dimensional computer environment is with us. Using an instrumented data glove, the subject can manipulate objects within the display directly, without the need for more cumbersome input devices (see next section). Although it is most obviously applicable to realistic scenes, such as items superimposed onto a helicopter pilot's outside visual field, it can also be used at higher levels of abstraction. For example, direct manipulation of control devices in a complex but virtual chemical plant is now possible, or even manipulation of pieces of code in a computer program-ming task. Beware, however, that actions in a VR environment may not always transfer to a realistic task.

Better input devices

The much-loved and much-hated typewriter keyboard is merely the tra-ditional computer input device for human commands and can hardly be hailed as a triumph of ergonomic design. Just as the computer display is moving away from alpha–numerics into the more analogue representations of animated graphics and complex sounds, so inexpensive computing power is allowing us to utilize alternative human input devices. No longer is the required analogue-to-digital conversion and interfacing the province of computer hackers. Currently available are graphics tablets, joysticks, light-pens, trackballs, mice (for foot as well as hand use), data gloves, handwriting recognition, optical character reading and voice entry. The data capture and collection possibilities are seemingly endless. Subjects can make responses directly onto a display screen (light-pen), perform tracking tasks (e.g., joysticks) and give verbal responses. With more specialist input devices such as frame-grabbers and bar-code readers, we can collect data on location much more easily. Keyboards may be retained for many years, but ergonomists are no longer limited by their capabilities.

Intelligent data collection

Just because we *can* collect data does not mean that it *should* be collected. As more complex and realistic ergonomics models are developed, data collection can be guided using these models. In a strictly scientific sense data collection without an underlying model is unproductive, but in a rapidly growing field such as ergonomics how can an experimentalist keep up with the latest models from the underlying disciplines? One solution is the artificial-intelligence (AI) inspired creation of expert systems to advise the practitioner, crude forms of which already exist. Mir (1980) devised an ergonomics audit program to evaluate the level of ergonomics found in the workplaces at a plant (Drury, 1990). Many of the mathematical models and logical constructs used regularly (e.g., biomechanics) are now having their computer programs repackaged as 'Expert Systems'. Even the humble checklist appears to be headed in this direction.

A better focusing of data collection efforts can be expected as AI concepts become increasingly common. Error taxonomies can be devised with the help of an expert system. Models of human response (e.g., to thermal loads) can be used to predict what measures to take and when to take them. Standard symptoms of industrial malaise (accidents, quality, labour turnover) can be used as input to indicate probable causes and hence guide data collection efforts. Perhaps all ergonomists will carry an annually reprogrammable 'automated aide'.

Intelligent tools

As microcomputers invade more and more products (e.g., telephones, cameras, TV sets), they can easily provide the hardware for automated data capture by the ergonomist. If your automobile has an interface to the repair shop's diagnostic computer, this same interface can be used selectively to monitor the same functions during actual use. There are obvious moral questions concerning privacy, but if the subjects' informed consent is obtained then there is no reason why these interfaces should not become a relatively standard data collection tool. Already a running shoe is available whose on-board chip interfaces with your desk-top computer to analyze your daily run. Office chairs have chips to remember adjustment preferences for several individuals. How long will it be before your hammer indicates whether you are striking the nails correctly, or your pen organizes your work/rest schedule to avoid writer's cramp? More seriously, we are seeing the start of a product design revolution at least equal to the revolution caused by computers in research and industry. The possibilities for ergonomics data collection can only increase.

Data reduction and analysis

We have already alluded to the revolution in data handling caused by computers. If one re-reads experiments in the literature of the 1930s and 1940s, the absence of what we now regard as commonplace statistical analyses is striking. Where massive data gathering, reduction, and analysis efforts are reported, for example Blackwell's (1946) studies of vision, one can only marvel at the patience and dedication of the experimenters.

The trend, also noted earlier, is for the integration of data collection, data reduction and data analysis. We have now reached the point where this cycle can be performed rapidly enough to provide meaningful feedback to the experimental subject and even control of the subsequent stimulus presentation. Even in the 1970s, minicomputers made possible the study of adaptive training systems where task difficulty is continuously adjusted to current performance level. The studies by Wiener (1973) in vigilance are good examples.

More traditionally though, the three functions are separated in time so that it is still useful to consider data reduction and data analysis separately.

Data reduction

There are three aspects of data reduction which need to be considered. For data which are not collected directly by computer there must be a data entry stage, followed by data verification and finally, post-processing.

Data entry ease and accuracy should be considered at the experimental design stage. An easily read source document for the subject may not be an easily read source document for the data entry person. For observational and questionnaire data it is important to minimize transcription errors, so a well-designed data entry program is essential. Three alternatives are available (all are assumed to be used on a microcomputer dedicated to the data entry task; mini and mainframe machines can be used, but connect time is usually chargeable; simple file transfer programs are available to allow use of mainframes for later analysis). First, a word processor, spreadsheet, or text editor can be used to create a data file, typing in the numerical and text entries in order with suitable delimiters. Although error correction will be easy, there is a considerable chance of forgetting the location of each item in the data file. A second, more compatible method, is to use a database program where each unit of data (a completed questionnaire, a set of workplace dimensions or observation records) is entered onto a custom-designed data input screen. Figure 9.1 shows such a screen for data entry in a study of postal operations. Some databases also allow the user to enter data in a tabular format. Figure 9.2 shows further data from a postal study where body angles were entered for later calculation of biomechanical body stresses. Both methods constantly remind the entry person of the layout of the data. If it makes sense, hard copies of the format shown in Figure 9.1 can be used as field data collection forms, giving a one-to-one position correspondence between hard copy and entry screen. Third and finally, data entry programs can be written in high-level computer language (e.g., FORTRAN, C, BASIC) to accept data at particular prompts. Most sophisticated database programs now have input screen manipulations exceeding the abilities of non-expert programmers. For example, writing a program in FORTRAN to allow full-screen editing is no simple task unless a good library of sub-routines is available.

If this represents the present situation for data entry, what of the future? The aim should be to eliminate the data entry process entirely by computer collection of data as noted previously. However, not all data are as easy to capture as switch closure, analogue signal or key press. For example, although direct data collection in a questionnaire format is possible using a keyboard or menu-and-pointer system, there will always be some subjects who find that the computer interface is unsuitable. Non-readers and physically handicapped individuals will have problems unless special solutions are available. The computer can be unnecessarily obtrusive in the work situation; even recording with a lap-top portable would be difficult in a trading crowd at a stock exchange, or in a military aircraft. Pencil and paper or voice transcription are likely to remain with us despite our best efforts.

Given that data entry is a necessary evil, what aids can we expect? Computers will certainly have better interfaces to minimize transcription errors and will be faster and better interconnected to speed data on their way to the analysis programs, but these are only small evolutionary changes. The larger changes will come from improved optical character recognition and speech recognition algorithms which allow direct entry of data which has been collected by writing on forms or recording on tape. With such devices, we can become at last independent of one of the least rewarding chores associated with research.

The second aspect of data reduction is verification. In the days of card input, this often referred to having two clerks enter data independently and then comparing cards for errors. Transcription errors were caught unless the two clerks happened to make identical errors. With full-screen editing and

WORKPLACE EVALUATION. PHASE II
SUBJECT: 1   HEIGHT: 61
SEX: 0   WEIGHT: 130
AGE: 39   MONTHS: 168
EYES: 1   SITE: 1
                              INT. MOD: 0

ENVIRONMENT
1. SUFF SPACE: 0
2. CROWD: 0
3. WK ITEMS: 1
4. STORE: 1
5. COUNTER: 1
6. ACCESS: 0
7. LIGHT: 1
8. GLARE: 0
9. NOISE: 1
10. DISTRACT: 0
11. TEMP SUM: 9
12. TEMP WINT: 1
13. APP WORKPLACE: 1
14. APP STATION: 0
15. EQUIPMENT: 0

WORKPLACE/EQUIPMENT
1. REACH KEYS: 0
2. READ LABELS: 0
3. READ SCREEN: 0
3A. READ SCALE: 0
4. READ ZIP: 0
5. CHARACTERS: 0
6. COUNTER HT: 0
7. DIFF DRAWERS: 0
8. WP ADEQUATE: 1
9. DIFF PACKAGES: 0
10. REACH SCALE: 0
11. USE SCALE: 1
12. EXCESS WALK: 0
13. EQUIP PROBS: 0
14. SPACE FORMS: 1
15. SPACE PARCELS: 0
16. STORE PARCELS: 2
17. SLOT LAYOUT: 0
18. ACCESS FORMS: 0
19. EXCESS BENDING: 0
20. EXCESS REACH: 0

Figure 9.1. Example of customized input screen for database entry

instant print-outs, this is now an easier problem to tackle, but it is at times neglected, resulting in unreliable data and hence unreliable conclusions.

A broader problem in data verification is the logically impossible or implausible data point. In a student sample, any age having other than two digits is suspect. Similarly, sibling birth dates closer together than the gestation period raise questions of data validity. Heart rates above 200 beats/min and negative blood pressures are other obvious examples, but there are less glaring

SITE	BEFAFT	SUBJECT	TASK	LR	BACK	UPPER	LOWER
4	1	15	1	1	22	159	69
4	1	15	1	2	-8	156	99
4	1	15	2	1	9	165	105
4	1	15	2	2	-11	137	100
4	1	15	3	1	15	127	105
4	1	15	3	2	-10	152	112
4	1	15	4	1	2	127	99
4	1	15	4	2	3	151	88
4	1	15	5	1	10	140	77
4	1	15	5	2	-18	174	87
4	1	15	6	1	5	162	124
4	1	15	6	2	-9	168	126
4	1	15	7	1	2	166	122
4	1	15	7	2	-6	169	127
4	1	15	8	1	-4	165	107
4	1	15	8	2	-18	165	112
4	1	15	10	1	40	188	150
4	1	15	10	2	-2	159	103

Figure 9.2. Example of tabular database entry format

data errors which must be detected. There is always some question of how far the ergonomist should go in editing data. Unusual values will occur legitimately from time to time, indeed there is a whole branch of statistics concerned with extreme values sampled from populations (Gumbel, 1958). Some students are over 60; premature births do occur; very rapid typists exist to cause outlying data points in word processing studies.

The ideal place for data verification is at the time of recording, when the subject is still present and the data point can be repeated if necessary. All subsequent data verification which involves data editing can be called into question. Only logically impossible data can be removed from the database and still withstand serious questioning. Reversed dates for starting and ending employment, mechanical impossibilities, or off-scale responses are all 'safe' to remove from the database, but how can they be replaced? If the error is a transcription error, we can re-enter the data, but if it is a recording error, generally it will not be possible to replace the reading at a later time. A subject's response cannot always be re-collected, as the subject may now have changed, leaving the problem of missing data, which is discussed later.

In the future, data verification has the potential to become more consistent as well as less time-consuming. Current database programs already have built-in error checking routines, so that data can be tested as they are input to determine whether they meet certain criteria. In this way logical impossibilities are automatically flagged so that a decision can be taken on their inclusion. The same filtering of data can also be applied on-line to computer-recorded data. For example, heart rates based on timing the EKG 'R' wave occasionally count a spurious beat or miss a beat, leading to heart rates double or half the correct value. Data filtering routines can be built into the recording software to detect such anomalies and remove them from the final record. Ease and consistency in such processes can be expected to increase with the more widespread adoption of artificial intelligence techniques for data verification. It should eventually be possible to mimic the decisions of an intelligent and well-informed experimenter to ensure data veracity.

The final step in data reduction is post-processing—the stage between data entry and statistical data analysis. Although many hypotheses, and hence statistical analyses, are based directly on the measured data, others require transformed data. This may be as simple as subtracting the starting date from the leaving date to determine a respondent's period of employment or it may be as complex as calculating the error power spectral density function in a tracking task. Other examples are calculating signal detection theory parameters (A', d', x_c, etc.) from hit and false alarm rates, finding disc compressive forces from a biomechanical model of a manual lifting task, finding envelope measures of EMG, or even calculating heart rates from R–R interval times.

In all cases, data from various data points are combined to give a measure of greater meaning. The main post-processing decision is whether to perform this post-processing on-line as data are being collected or off-line when data are retrieved from a storage medium. On-line processing obviously requires powerful computational resources, but, as the processed data are usually less extensive than the raw data, storage requirements are minimized. As an example, if four muscles are recorded using EMG techniques at 200 Hz, we need to take $4 \times 200 = 800$ data points each second. If, however, we wish to obtain only fatigue data, then the ratio of high-to-low frequency components (e.g., Kroemer, 1984) can be calculated and recorded much less frequently, perhaps twice per second for each muscle. Our data storage requirements are reduced 100-fold, but high-capacity on-line processing is needed to filter the data by frequency and calculate ratios.

The main advantage of on-line post-processing is that it can provide useful feedback to either subject, experimenter or both. The main danger is that the data cannot be used to answer any questions which require different post-processing. To continue our example, level of effort, represented by average or envelope EMG, would not be available if only frequency ratios were recorded. For this reason, recording of the original raw data is always good practice: only record the subject's actual response, particularly in observation or questionnaire data; do not attempt to convert weeks to days just because the questionnaire requires an answer in days and the subject gives the answer in weeks; record original data long-hand and process them later without time pressure.

The future can only bring advantages for post-processing. Higher computing power, data handling rates and storage capacities in computers mean that both original data storage and on-line processing will be available simultaneously.

Data analysis

At this stage, our data are collected, entered into the computer, verified, and the calculations performed so that we are ready for analysis. A number of choices are now open for data analysis and all of the following discussion is based on computer analysis, although it is rare that no hand-work is used. Specific analyses may well be beyond the capabilities of the software package: graphs with meaningful but unusual scales may need to be drawn; a rapid calculation of chi-square for a contingency table may be preferable to re-running a batch job on a mainframe. Most ergonomists still own a pocket calculator and a box or two of graph paper.

The typical course of analysis is statistical analysis of data followed by comparison between some non-statistical model and the summarized data. Thus Chapman and Sinclair (1975) ran an experiment on inspection of food products, varying both speed of the conveyor and inspector's time on task. An ANOVA established significant differences between speeds and times on task, so that graphs were plotted to show both the speed/accuracy trade-off and the vigilance decrement. Any computer package for data analysis should support both of the activities and most do.

Additionally, the handling of missing data is an important topic in its own right and may be different for different packages. In many packages, missing data within one cell of an ANOVA (e.g., four readings instead of five) can be handled easily, while the simpler packages insist on equal numbers in each cell. In other cases there may be different types of missing data. People may have no answer to a question on their age because they do not remember their age, they refuse to provide this information, the experimenter fails to record it, or an obviously wrong value is entered. A good package will allow the analyst to treat these differently, while still denoting all of them as 'missing'. Some missing data may cause other relevant data to be eliminated; e.g., a missing age in a correlation between job satisfaction score and age would cause that subject's job satisfaction score to be omitted from the correlation calculation. If there are more than two attributes and only one is missing the experimenter must choose between omitting *all* data from the subject with the attribute missing or just omitting data from calculations impossible to perform without that data.

Before considering the statistical packages, it is worth noting that some spreadsheets (e.g., Lotus 1-2-3, QUATTRO-PRO, etc.) and many database programs (e.g., REFLEX) for microcomputers perform rapid data summarization (means, standard deviations), some statistical functions (regression), and simple graphical plots (line, bar, scatter) which may be all that are required in simple experiments. Given means and standard deviations of two samples, a *t*-test can be performed on a hand calculator without recourse to a statistical package.

More detailed statistical analyses require specialist statistical packages. Even a few years ago, these were larger mainframe-based, but now microcomputers offer acceptable analysis speed for even moderately large applications. One advantage with microcomputer packages is that the data are always under the analyst's control, so that errors in data handling, translation, and storage are reduced. Following microcomputer–led trends, these packages now offer graphical user interfaces, or at least menu choices, instead of somewhat arcane command lines. Presumably, these changes will further reduce the costs of learning a package, or relearning it after some absence. A good example is MINITAB, which is usable in both microcomputer and mainframe environments, but others such as SYS-STAT, SPSS-PC, and STAT-PAC are available. Statistical packages try to balance generality against ease of use. The more statistical procedures possible, and the more choices which must be specified before analysis, the more complex the ergonomist's task in performing the analysis.

Major packages in general use include MINITAB, SPSS-X, SAS and BMDP. All will perform *t*-tests, ANOVAs, contingency table analysis and regression. All will provide summary statistics, tabular output, histograms and graphs. Their detailed capabilities will not be compared as new versions are brought out frequently, rendering such comparisons instantly dated. However, each package has its own flavour and uses. MINITAB is a rapid interactive system which is easy to learn and remember between uses. It is ideal, for example, for running regressions on means and plotting the results instantly. However, it is limited in the types of ANOVA which it can tackle and is unsuited for large databases. SPSS-X was originally developed for social science research so that its forte is in handling multivariate data, for example by cross-tabulation or factor analysis. The days when its ANOVA capabilities were limited are long past. For graphical output, the conventional wisdom points to SAS as the package of choice. Finally, almost any model of ANOVA can be analyzed by BMDP, originally a biomedical package.

There is still no universal package which meets all needs, but the ergonomist should not need to learn more than two to handle almost all needs. On-line help facilities have expanded, and interactive processing has supplemented batch processing, but getting the most out of a statistical package is still a process with a lengthy learning period. One final caution is that not all analyses are performed by a package. If the package does not meet your needs and assumptions, analyze by hand. A few hours with a calculator can produce even complex ANOVAs on a few hundred data points, whereas locating the correct package, reading the manuals, and consulting a resident expert can take much longer. Never let the package constraints dictate your experimental design or data analysis.

References

Blackwell, H.R. (1946). Contrast thresholds of the human eye. *Journal of the Optical Society of America*, **36**, 624–643.

Chapman, D.E. and Sinclair, M.A. (1975). Ergonomics of inspection tasks in the food industry. In *Human Reliability in Quality Control*, edited by C.G. Drury, and J.G. Fox (London: Taylor and Francis).

Crossman, E.R.F.W. (1960). *Automation and Skill* (London: HMSO).

Drury, C.G. (1986). Hand held computers to collect field data. *Engineering Progress*, **6**, 10–12.

Drury, C. G. (1990). The ergonomics audit. In *Contemporary Ergonomics*, edited by E.J. Lovesey (London, Taylor and Francis).

Drury, C.G. and Deeb, J.M. (1986a). Hand positions and angles in a dynamic lifting task. Part 1, Biomechanical considerations. *Ergonomics*, **29**, 743–768.

Drury, C.G. and Deeb, J.M. (1986b). Hand positions and angles in a dynamic lifting task. Part 2, Psychophysical measures and heart rate. *Ergonomics*, **29**, 769–778.

Fitts, P.M. (1954). The information capacity of the human motor system in controlling the amplitude of movement. *Journal of Experimental Psychology*, **47**, 381–391.

Gumbel, E.J. (1958). *Statistics of Extremes* (New York: Columbia University Press).

Kondakci, S. (1985). *An interactive approach to scheduling of job shops with dual constraints*. Unpublished Ph.D. Dissertation, State University of New York at Buffalo.

Kroemer, K.H.E. (1984). Engineering anthropometry. In *Handbook of Human Factors*, edited by G. Salvendy (New York: John Wiley).

Mir, A. (1980). *An ergonomics audit program*. Unpublished M.Sc. Thesis, State University of New York at Buffalo.

Synfelt, D.L. and Brunskill, C.T. (1986). Multi-purpose inspection task simulator. *Computers and Industrial Engineering*, **10**, 273–282.

Wilson, J.R. (1995). Is there a role for Personal Digital Assistants on the factory shopfloor? In *Mobile Personal Communication and Cooperative Working*, edited by P. Thomas. London: Alfred Waller.

Wiener, E.L. (1973). Adaptive measurement of vigilance performance. *Ergonomics*, **16**, 353–363.

Part III

Techniques in product or system design and evaluation

The first two parts of this book contained chapters describing fundamental groups of methods and techniques, and also implicitly proposed certain approaches to ergonomics study. This third part has within it descriptions of how the basic methods and approaches may be modified and combined in five particular examples of application in design and evaluation: products, text, human–computer interfaces (HCI), interfaces in human–machine systems, and expert systems. In some cases the techniques normally available are not sufficient and so special ones have been developed.

It will be apparent that there is overlap between the five chapters in terms of the type and application of appropriate method, and that much of the methodology delineated has potential value in many other domains. However these five have been selected so as to provide an interrelated set of application domains, and as likely areas of importance into the foreseeable future. Within this whole part we will find all the methods of chapters 2-9, direct and indirect observation, archives, experiments, task analysis, protocol analysis, modelling, simulation and computerized data collection—with explanation of their strengths and weaknesses in the particular circumstances.

We start in chapter 10 with McClelland on an application area that has been of interest to ergonomists for 20 or 30 years or more—user trials. Such an approach has wide applicability but has been particularly valuable in the development and assessment of products and equipment. Linkage with HCI evaluation is obvious and current concern with usability engineering in the whole area of information technology implies amongst other things that user trials must be an integral part of an iterative design process. User trials are then seen as a way of bringing basic measurement principles into the product development process, and that their use provides both quantitative and qualitative information on product effectiveness.

In a product development process described earlier in chapter 1 the parallel development of product and manuals and instructions was proposed. This might seem a little optimistic or excessive but the usability and even market success of a product can be impaired by inadequate instructions for use and maintenance. Nowhere is this more so than in the world of modern consumer products—videos, cameras, dishwashers etc.—where manuals have historically been the subject of mirth or frustration or both. Hartley in chapter 11, summarizes the evaluation of text, describing the sometimes specialized, sometimes general, methods to do this.

At this time much ergonomics effort is being put into the development of human–computer interfaces that match the needs and limitations of users and tasks. After a period of producing guidelines and criteria, the HCI community is now much taxed by the problems of evaluation (related of course to criteria, etc.) and of what should be evaluated. In doing so they are confronting problems, compromises and gaps in knowledge that have faced ergonomists in other domains for years. What is interesting is that the enormous concentration of ergonomics resources in HCI has allowed exploration of the use of virtually any method or technique—from psychological through to physiological measurement—and the combination of some of these into broad-based approaches to HCI evaluation. Out of much of this work are emanating results of interest and significance for all ergonomists. Christie *et al.* in chapter 12 give a comprehensive account of the area.

For 50 years, ergonomics text books have included chapters on 'controls and displays'. Of late there has often been distinction between human–computer interfaces and so-called 'traditional' interfaces, with the latter including dials and scaled displays, status and warning information, switches, levers, knobs and valves, for instance. Such distinction is not really valid any longer since increasingly all interfaces are computer–based and the computer interfaces are not just VDU–based. Nowhere is the blurring of display and control technology more apparent than in control rooms, and chapter 13 looks generally at human–machine interfaces—analysis, design and evaluation—but with specific reference to systems control applications.

Within the sphere of computer technology one particular advance of great interest to ergonomists is that of expert systems. In particular we have a role to play in attacking what Shadbolt and Burton in chapter 14 call the bottleneck in expert systems—acquiring the knowledge necessary to build any knowledge-based system. A major, and generally *the* major source of such knowledge, will be the experts themselves, and knowledge elicitation is the name for gathering the relevant information. In some ways parallels can be drawn between this and the four preceding chapters in that all the techniques described are to elicit information, opinion and insight from people in the first three cases from users or user 'representatives', and in the last from those whose skills the system is intended to duplicate, enhance or even replace.

Common to all methodological approaches in these five chapters is the need to understand and allow for who the users will be and what type of tasks they will carry out. McClelland, Christie *et al.* and Wilson and Rajan address explicity the question of 'what users?'. Between them they define attributes that need to be identified to specify and characterize the target user group as: personal characteristics; physical, cognitive and attitudinal characteristics; specific skills, training, previous experience; the characteristics of their jobs, and roles; and personal preferences. Perhaps most critical is to know whether the users will be beginners, novices, regular users or experts, whether use will be frequent or not, and the degree of discretion they have in using the product or system. Shadbolt and Burton coin three descriptors for classes of experts that might be useful in the wider context of user characterization: 'academics', 'practitioners', and 'samurai', the last being pure performance experts with little generalized theory and few heuristics.

A second main underlying need within analysis, design and evaluation is to understand what it is the users will have to do, the tasks they must undertake. In most cases task analysis will have allowed such identification, and the user characterization in terms of experience, frequency and discretion of use will also be relevant. The number, complexity, sequence, simultaneity, loading and criticality of the tasks will, *inter alia*, be important to the making of design and development decisions, in determining how each of the products or systems covered in this section must be evaluated, and in selecting the appropriate methods and techniques.

Chapter 10

Product assessment and user trials

Ian McClelland

Introduction

This chapter opens with an introduction that provides an overview of the main components of user trials. In the second section user trials are discussed in more detail. The third and last section considers some of the main issues that will influence future directions for this type of evaluation.

What is a user trial?

The fact that it is possible to measure the interaction between people and the products, equipment, environments and services they use is the fundamental principle that underpins all ergonomics. A close second is the principle that these interactions can be measured in ways which can guide the specification, design and evaluation of the artifacts we all use. These principles have given rise to both a considerable body of data and to a range of investigative techniques which are central to the effective implementation of ergonomics. This expertise and knowledge continues to develop, and not least because many recognize the need for more effective methods to support the application of ergonomics to system and product development (Meister and Enderwick, 1992). The term user trial started to become common currency in the 1970s as part of a vocabulary more suited to the environment of industrial product development. 'User trials' is regarded here as synonymous with the terms 'user testing' and 'usability testing' insofar as these terms also refer to the evaluation of some artifact under controlled conditions involving users. The challenge was, and still is, to market the idea that 'experiments' with people could be helpful in the design and development of products. But the term 'user trials' is not just a new label for 'experiments'. The continuing use of the term also reflects real changes in the practice of ergonomics; changes aimed at accommodating the pressures and priorities that come from working within a product development environment.

The most significant development in recent years has been the emergence of the concept of 'usability engineering'. This concept has been developed primarily as a vehicle for the application of human factors to products within the 'information technology' area (Gould, 1988a,b; Nielsen, 1993, 1994; Preece *et al.*, 1990; Shackel, 1986, 1987; Whiteside *et al.*, 1988). The concept is now incorporated into ISO 9241 Part 11 (currently in Committee Draft form). Although ISO 9241 is primarily concerned with the design of equipment incorporating Visual Display Units for office use, Part 11 deals with principles which are applicable to a wide range of products and systems. The central purpose behind 'usability engineering' is to make the process of designing user requirements into products more systematic. The concept involves firstly, incorporating 'usability requirements' into product specifications. The 'usability requirements' should be specified in terms of how the user must perform in order to satisfy the goal (see ISO 9241 Part 11, and Chapanis and Budurka, 1990, for examples of how usability requirements can be specified). Secondly, user trials should be an integral part of an iterative design process so that a product can be checked against the usability goals and the performance requirements as it is developed. Bewley *et al.* (1983) and Boies *et al.* (1987) provide good examples of how different forms of user trial contributed to the development of specific products in this way. In general though, products are not developed on this basis at present. However, the incorporation of user trials in human computer interface development started to become common practice in the mid 1970s (Grudin, 1990). The development of 'usability engineering' principles provides clear guidance on how the concept of user testing can be incorporated into the development of products in a wide range of industries as well as the information technology area.

So we can say that user trials have become a way of bringing the essential principles of measurement and systematic evaluation into the product development process. The central idea is that the application of these principles is product focused; i.e., that we can measure the effectiveness of products both from quantitative and qualitative points of view. User trials can provide information which is at least useful and, at times, essential if products are successfully to accommodate users' requirements. This involves using methods often associated with 'experiments', and indeed, the same basic principles of experimental design, measurement techniques and data analysis that apply to formal experimental work also apply to user trials. Chapanis (1959) describes the experimental method as follows:

'It is a series of controlled observations undertaken in an artificial situation with the deliberate manipulation of some variables in order to answer one or more specific hypotheses.'

In summary, user trials may be described in the same way with the one modification; ...to answer one or more specific questions about the effectiveness of the product.

Many of the principles involved will be discussed later in this chapter, but for a more extensive discussion of the principles and practice of test and measurement in human factors work the reader is referred to Meister (1986). Bailey (1982), Chapanis (1959), Cushman and Rosenberg (1991), Gould (1988b), Holleran (1991), Kirk and Ridgeway (1970, 1971), Macleod (1992), Rennie (1981), Rubinstein and Hersh (1984) and Whiteside *et al.* (1988) also provide discussions on the application of ergonomics to product testing.

Setting up a user trial is primarily about creating an environment that enables the interaction between product and user to be systematically examined and measured. In creating this environment it is important to recognize that a product is usually a 'tool' of some description which helps the user to achieve some goal or objective. The context in which the product is used (actual or intended) must also be carefully considered in relation to the aspects which critically influence the way in which the product is used. Using the product involves the user in carrying out some tasks. It is the combination of these tasks and the related product attributes that together make up the process of interaction between product and user. The tasks involved can generally be described as a cycle of user action, product response, user decision, user action and so on. In design terms this process of interaction is central to what is now often referred to as the 'user-interface' aspect of the product. So evaluating the effectiveness of a product from a user's point of view means evaluating the design of the user-interface.

The principal components in a user trial are (Figure 10.1):

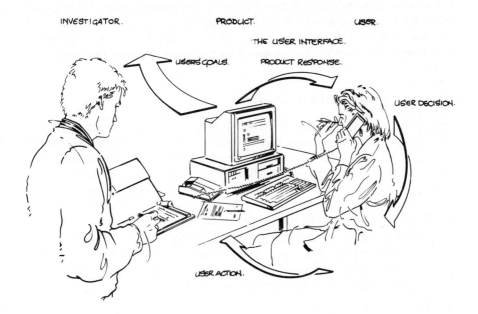

Figure 10.1. The general scenario of a user trial

- A group of users;
  members of the current of intended user population.
- A product;
  the device(s) to be used in the trial.
- A set of tasks;
  what the users must do to achieve their goals that involves the use of the product in some way.
- User performance criteria;
  the criteria which are used to judge the effectiveness of the product.
- Measurement techniques;
  the techniques that are used to measure the effectiveness of the user interface in enabling the user to complete their tasks successfully in terms of the criteria selected.
- The site for the user trial;
  the location where a product is to be evaluated.
- An investigator;
  the person who is responsible for the design and conduct of the user trial.

How simple or complex a user trial becomes will be determined by the circumstances surrounding the demand for the investigation. Many user trials can be successfully carried out based on simple pragmatic approaches, but even when running the simplest of trials all these basic aspects need to be considered.

Circumstances will also dictate which aspects present most problems and the order of priority in which the problems must be overcome. From both a technical and practical point of view all these aspects must be regarded as interrelated and must be incorporated into a comprehensive test plan.

Questions a user trial can answer

User trials can answer many different types of question about products. The character and form a particular user trial takes will depend on many considerations, not least the technical ones; the product to be tested, the measurement techniques and so on. However, what is of critical importance is to be clear about the purpose of the user trial and the types of question that need to be answered. The purpose of a user trial may be diagnostic; what faults exist in the design, and how might they be rectified? The trial may be exploratory; what usability issues are relevant to a particular group of users? User trials can be used to explore the expectations users have of early design concepts. The purpose may be to measure performance; does a particular design meet specific performance requirements? For example:

Who asked for a user trial to be carried out?

- Manager of a development department;
  a choice has to be made between two technologies; does it matter to the user which we choose?
- Marketing;
  particular product comparison data may be required; market research cannot identify reasons for users' complaints about a product; is the design at fault?
- Design engineers;
  what are the critical dimensions in the layout of a particular workstation?
- Software engineers;
  a particular user group is accustomed to a particular style of interaction; will a new style be acceptable and are there any negative transfer effects?
- Training specialists;
  how long should customer training programmes take and where might particular difficulties with the product occur?
- Maintenance engineers;
  a specific time limit has to be met for the exchange of particular machine components; will the design meet this requirement?

Does a performance standard have to be met?

- Is there an international standard, national standard, industry standard, or company standard that should be met?
- Does user experience with existing products imply certain standards of performance which new products should meet?
- Should a performance standard be developed if one does not exist?

How does one product compare with another?

- How does the product of one manufacturer compare with the products from the competition?
- How does an existing range of products compare with a new range of prototypes?

Is there a need to improve the design of a product?

- Have accidents or critical incidents occurred with users that may be attributable to the design?
- Have users reported particular problems with a product?

User trials in more detail

Although the components of a user trial are discussed separately it is worth emphasizing that all these aspects are closely interrelated and should be regarded as the basic ingredients that go to make the complete recipe.

Users

Selection of users

The fact that users vary in their characteristics and requirements will come as no great surprise, but what is important to understand is that a user trial requires a sample of users which in some way reflects the product user population as a whole. This means selecting a group of users who do not just have the same characteristics as the user population but who reflect the extent to which these characteristics vary. Here it is important to remember that many of the characteristics that need to be considered are not just basic physical or psychological aspects but also related to the aspect of interaction to be examined in the trial, as well as the basic questions the trial is attempting to answer.

There are two major aspects to obtaining a sample. The first is to develop a profile of the type of people who are in your user population. The profile should identify as clearly as possible what user characteristics are actually relevant to the product under investigation. Where possible data should also be identified which describe how the population varies with respect to these characteristics. Obtaining relevant data will often not be easy, and sometimes will be impossible. In the first case the users of the product may be difficult to identify. Secondly, even when the user population can be identified with reasonable precision, it is often difficult to say which user characteristics have the most impact on the successful use of the product. In some cases data are available. For example data describing the physical characteristics of people such as anthropometric data. However, if experience in the use of the product is an important consideration then some data might be available that can be used, but there are many instances where data simply will not be available. Descriptions of user populations in terms of statistically reliable data is a persistent problem. The problem is particularly acute when the product is new and the user group is largely unknown. There is no easy solution. If reliable data are limited or cannot be found then the investigator must make a best estimate based on the data that are available. Many sources of useful information can be considered. Apart from the professional ergonomics literature, information sources could include marketing, service, sales, application specialists, training departments, occupational health specialists, etc. All may be able to offer useful guidance. Inevitably a number of assumptions about the user population will have to be made. These assumptions should be discussed in some detail, and agreed as reasonable, with those who asked for the user trial. In general terms user characteristics are usually grouped under the following headings:

- General descriptive characteristics;
  age, gender and other demographic data.
- Basic physical characteristics;
  sensory processes such as visual and auditory capabilities, tactile and ves-

tibular sensitivity.

body dimensions such as height and weight.

movement control such as balance, coordination and reaction time.

- Cognitive characteristics;

  general intellectual capabilities such as problem solving and decision making.

- Attitudinal characteristics;

  motivation, personality and initiative.

- Specific skills and training;

  education level.

  experience in specific types of work and responsibilities undertaken.

- Experience with a particular product or product type;

  novice, intermediate or experienced users.

The second main aspect concerns the actual process of selection. From the user profile a set of descriptors must be identified which can be used to describe the type of participants required for the user trial. The descriptions used to recruit users need to take account of the recruitment method. If users select themselves, for example by responding to an advertisement, then the descriptors must be understandable to the user population and presented in terms which make it easy for users to know whether or not they comply. If recruitment of users is by interview, for example by telephone, then more precise descriptions can be used. Although even with such a characteristic as height one might expect that people would generally know how tall they are, in practice people are quite inaccurate about their height. Consequently selection criteria tend to be described in relatively loose and general terms. If necessary users can be screened on a post-hoc basis to check for consistency with the user profile. In contrast, if users can be recruited from a 'captive population', such as a particular group of workers, and the investigator has the possibility of screening prior to conducting a user trial then, correspondingly, the selection criteria can be quite precise.

Comparing users with the user population

Whenever possible or practicable it is always good advice to collect descriptive data from the sample of users which enable comparisons to be made with the user population as a whole, but comparisons can only be made if the corresponding data which describe the user population are available in an appropriate form.

Screening users

If the validity of the investigation relies on users complying with specific requirements, then participants in a trial should be appropriately screened. Just because users say they comply it would be unwise to assume this to be the case if it is possible to check. In many cases the user will simply not know. For example if there is a requirement that an auditory display be tested with users who have normal hearing ability, then the hearing of users should be examined using standard test procedures.

User numbers

Providing general advice on the number of users required is always a difficult issue. To make accurate estimates requires specific statistical information and the use of statistical procedures, associated with formal experimental design, for making estimates of sample sizes (Winer, 1971; Collins, 1986; see also Drury, chapter 5 of this book). Making reliable estimates depends on the answers to the following interrelated questions:

- How many variables need to be considered in relation to user character-istics, product characteristics, and environmental characteristics?
- Can the variation of the population within each characteristic be statisti-cally described?
- What level of accuracy or precision is required of the results?

What we usually find is that the number of variables that could be included in a user trial is large. It is also often the case that accurate statistical infor-mation about how the characteristics of users within a population vary is not available and estimates based on experience have to be made. In short, this usually results in estimates of user numbers being very large, sometimes alarm-ingly so, even at moderate levels of accuracy.

Against this we have to balance the practical questions of costs and time-scales. A product development programme can seldom tolerate elaborate and extensive user trials unless the need is critical, for instance if the consequences of making incorrect design choices could be very expensive in terms of potential accidents, lost sales etc. As a result ergonomics practitioners often have to resort to simplifying the design of user trials in order to work within practical constraints. Usually this means examining products with relatively small numbers of users and a restricted set of variables. Results from such trials have to be seen at the level of providing estimates, directions for devel-opment, detecting major design weaknesses etc.

In answering the question 'how many users do I need?' one must be guided by practical experience and the constraints imposed by local circumstances. The basic objective of a user trial is to go one better than estimates based on opinions and to obtain evidence of how well users can perform, and whether the level of performance achieved is acceptable. On that basis experi-ence does show that dependable results can be obtained from as few as five users in a single trial (Rubinstein and Hersh, 1984). More than five users will generally always be beneficial but how many more is an open question. Virzi (1992) reports a study relating the proportion of usability problems identified to the number of subjects used in evaluation trials. His main conclusions are that 80% of usability problems can be detected by 4–5 subjects, more subjects tend to reveal less and less new insights as the numbers increase, and the first few subjects detect the most severe problems. This study was concerned with a limited range of telecommunications applications. Whether or not the conclusions drawn have wider validity requires investigation. However the results reflect a widely held view that small numbers are often sufficient to identify the most severe problems. There are many examples where individual trials have involved no more than 30 users yet in some cases samples in excess of 100 are quoted. Drury, in his chapter on designing ergonomics studies and experiments, considers sample selection and size elsewhere in this text.

The product

The products that people use come in many shapes and sizes all of which in principle could be tested with users. In this chapter the term 'product' is meant to embrace (Figure 10.2):

– The physical attributes associated with traditional three dimensional products such as household appliances, vehicles, office equipment, medical apparatus etc.;

Figure 10.2. A user's interaction with a product is made up of many attributes; the physical activities, the perceptual and cognitive processes involved in using the product, and the manuals, instructions etc., which support its use

- The control/display systems from simple 'knobs and dials' concepts to full software-based control systems found in information technology products;
- The material required to support the use of a product such as user instructions or help facilities.

We tend to think of products as they are sold, a complete working design as we might see it in the High Street shop window, and indeed many products are tested in this form. However, products can be successfully tested at much earlier stages in their development and the form in which a product is presented to a user can vary considerably. The form depends upon the development stage, the product type and the aspect which requires investigation. (Christie *et al.* in chapter 12 of this book discuss test methods relevant to stages in development of human–computer interfaces.) The forms commonly tested include:

- 'Paper-based' descriptions of product concepts;
  These can enable initial explorations of ideas on product functionality to be made, important usability characteristics to be identified, or 'walkthrough' studies of protocols for product control systems to be conducted.
- Part prototypes or simulations;
  Part prototypes are used to simulate specific functional attributes of a design. The prototype may look nothing like the final product but it is essential that those aspects which are investigated are accurately represented. Part prototyping of software for product control systems through the use of rapid prototyping tools is one area where this type of testing is becoming more common.
- Full prototypes;
  Full prototypes perform as the final product is intended to perform and incorporate the complete functionality and appearance of the product.
- Complete products;
  Complete products enable the complete user-interface to be examined. This opens up the possibility of field investigations, comparative studies with other products, in-service studies etc. to be carried out.

User trials can be employed, in principle, at all stages in the development process of a product. The type of user trial carried out, its specific objectives and the criteria used, all have to be discussed as project specific issues. The product development process illustrated in Figure 10.3 is simplified and schematic, but it serves to emphasize the evolutionary nature of product development work. The product development process should be seen as a series of phases which often involve ill defined borders, and which involve many iterative cycles. Product testing can directly contribute to the cycle of development by focusing attention on user issues which are critical to the effective-

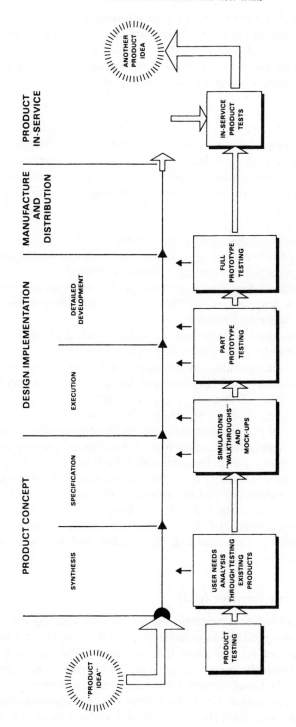

Figure 10.3. The contribution of product testing to the product development process

ness of the product. Each test produces fresh information and the experience from one test will help to refine the design and also define the needs for the next development stage. Product testing can also be used to evaluate in–service products. This can generate valuable information which can contribute directly to subsequent generations of product designs.

Product variables

The simplest form of trial is where users test a product as it stands without the investigator varying the conditions under which it is examined in any systematic way. It is more usual to control systematically the way in which the product is presented to the user. In general terms variables can be grouped into four categories:

- Evaluate possible design options for a particular design concept. The issue may be a basic technology choice, refining design details, or specifying the range of adjustment for a particular control.
- Comparing independent design solutions such as comparing a complete prototype with competitors' products, or radically different design solutions for the same end product;
- Procedures associated with the use of a product such as evaluating different methods for the presentation of product instructions;
- Circumstances surrounding the product itself such as the effects of changes in environmental conditions; lighting, acoustic or mechancial conditions.

The ergonomist can play an important role in identifying which aspect of the product would benefit from a user trial. A preliminary assessment of the product or the 'problem' can help to identify which aspects of the product are likely to affect the user's ability to perform satisfactorily. If the ergonomist is familiar with the capabilities and limitations of the product, as well as the way it is used, the ergonomist will then be better equipped to make the assessment. The ergonomist can then collaborate with a design team in a more productive way in understanding the options available for change or modification.

Tasks and protocols for user trials

The tasks carried out by users represent the assumptions that the investigator makes about how the product, in practice, is, or will be, used. The tasks are the 'script' that the 'actors' (the users) have to follow. The tasks will determine the way in which the users interact with the product, what 'view' the users get of the product and therefore how they respond when the investigator measures the interaction. Correct selection of tasks is consequently vital to the success of the user trial; they must involve the use of all the product attributes that are of interest to the investigator, they must accurately simulate the pattern of product use that occurs in practice or is predicted, they must be relevant to the user, and the patterns of behaviour associated with the task must be measurable in some way. For example Fulton and Feeney (1983) evaluated the design of powered lawnmowers in relation to reported accidents. They examined typical patterns of cleaning procedures to ensure that the tasks to be carried out in a user trial reflected actual patterns of use.

Task sequence

The tasks performed in a user trial generally should be carried out in a sequence which is logical and relevant to the way in which a user would expect to use the product. This does not necessarily mean that all users carry out all tasks in the same sequence. The sequence would normally be varied in some systematic way in order to control for 'order effects' such as transfer of learning from task to task. This is particularly important where a user would normally expect to choose between tasks in which case a randomized or systematically varied order of task sequence is necessary.

It is important that the procedure for the trial should be consistent for each user. This includes all user instructions, measurement procedures, any response forms used, and any interviews carried out during the trial. Maintaining consistency is often easier if the investigator works from detailed notes or even reads instructions.

User trials are usually organized in the following way:

- An introduction to the trial; the purpose of the trial, the product to be used including a demonstration if necessary, any associated equipment that may be involved, the tasks to be undertaken, the measurement procedures to be followed. Carry out screening procedures. Sign a 'confidentiality agreement' and an 'information disclosure agreement'.
- A walkthrough of the essential aspects of the trial with the user; this usually clarifies whether or not the user has understood what is required of him or her.
- Data collection; the user carries out the sequence of tasks including a pilot run if necessary, and the investigator takes measurements as appropriate;
- Debriefing the user; it is always useful to enquire about the users' views of the trial even if this is not part of the data that are being collected; users also appreciate some feedback on 'how well they did' and how the results are likely to influence the product they have just used.

The set-up for a user trial should also take account of the physical or mental effort involved. If the effort is significant, rest periods should be incorporated into the procedure, unless of course fatigue is what the trial is set up to examine. Another issue is that of boredom. In long trials of 90 minutes or more users can find it difficult to maintain concentration. It is therefore worthwhile considering building some variety into the procedure in order to maintain the interest of users.

User performance criteria

Selecting user performance criteria

Measuring how users respond when they use a product is the essence of a user trial. The response is a consequence of the tasks the user has carried out and the specific set of circumstances under which the trial was undertaken. It is, of course, essential that the criteria chosen provide a realistic indication of the effectiveness of the product. Deciding on which criteria to use to evaluate a product can be one of the most difficult aspects of setting up a user trial. Kantowitz (1992) summarizes the role of measurement in Human Factors research as; 'Measures are the gears and cogs of that make empirical research efforts run. No empirical study can be better than its selected measures'. In some cases the appropriate criteria may be self-evident but often the choice of the best criteria to use is an issue which in itself must be evaluated. Indeed studies are sometimes set up to evaluate which criteria are relevant, often when the technologies are new and usage patterns are still evolving; see, for example, Egan *et al.*, 1989, and Francik and Akagi, 1989.

As noted earlier, the reasons for running a user trial, the questions to be answered, and the priorities to be addressed need to be discussed with the client. This should enable the investigator to judge which criteria are most appropriate and realistic. Which criteria are appropriate will also depend on the particular aspect of the user-interface which is of interest. For example Dillon (1992) discusses, in a review of studies on reading from paper versus screens, a number of criteria used to evaluate reading performance including; speed, accuracy, fatigue, comprehension, preference, eye movements, manipulation (of material), and navigation. The reason for the trial may be to check whether the basic physical, perceptual and cognitive characteristics of users are satisfactorily accommodated. There may be conflicts between what the user can satisfactorily cope with and what the product demands, that make the product unacceptable. Another reason may be to examine the use of the product in relation to the users' goals; can users achieve their goals easily or with difficulty? Is a product enabling the users to be effective in their jobs? Bailey *et al.* (1988), for example, evaluated a new range of Tektronix oscilloscopes in terms of productivity. It may be that weaknesses in the design of the product have been identified already and the purpose of the trial is to bring about improved user performance. In addition constraints derived from the context of use may make demands on the design of the product which influence the behaviour and perception of users. A user trial could be set up to check that the design satisfactorily takes such constraints into account.

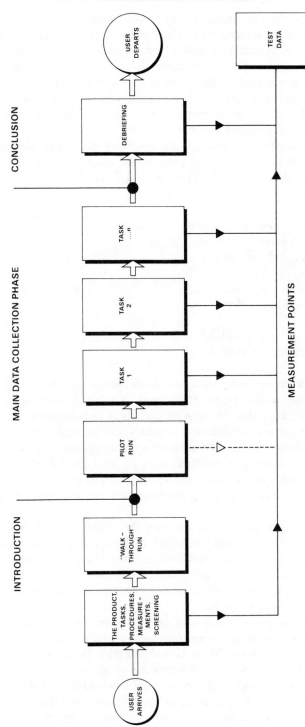

Figure 10.4. The main phases in a typical user trial. Data are usually collected during or on completion of each task, but may also be collected from a pilot run (although a pilot run is generally only used as a practice phase for the user)

If it is not clear which criteria should be used then the possibilities should be considered in general terms first. The options will then require interpretation to obtain a set of criteria which are appropriate to the particular investigation. The aim is to express each criterion as a specific question that can be answered in a user trial. Each question should incorporate, implicitly or explicitly, the dimension on which the performance of the user is to be measured. In the design of formal experiments this dimension would be referred to as the 'dependent variable'.

There are a number of general criteria which are frequently used to evaluate user-interfaces. Below are listed some of the more common ones which are frequently used in user trials. They are not listed in any particular order of priority:

- time; speed of response, activity rate, etc.
- accuracy and errors
- convenience and ease of use
- comfort
- satisfaction
- physical health and safety
- physical fit
- physical effort and workload
- stress and mental workload.

How such criteria are translated into more specific questions that can be answered in a user trial must be considered. It is usual practice to include several measures rather than to rely on one. It is also common practice to include both objective and subjective measures. For example objective measures may be used for criteria such as time, accuracy, physical fit and physical workload whereas subjective measures may be used for other criteria such as convenience and ease of use, comfort and satisfaction; see Muckler and Seven (1992) for a more extensive discussion on the selection of performance measures.

Sometimes it is not possible to measure the criteria directly and several measures have to be combined in order to arrive at an assessment. The measurement of stress or mental workload (see chapters 25 and 26) are examples where this approach is often taken; objective measures or work rate and error rates may be combined with subjective reports to assess whether the level of stress is acceptable or not.

The identification of criteria can be considered in very general terms. For example, in the area of information technology products and the design of computing systems, Bailey (1982), Gould (1988b), Rubinstein and Hersh (1984), Shackel (1986, 1987) and Whiteside *et al.* (1988) discuss a variety of criteria and related methods that can be used in evaluation. ISO 9241 Part 11, the measurement and specification of usability, proposes the criteria effectiveness, efficiency, and satisfaction, as the principal categories for the evaluation of screen based human–computer interfaces. The standard also includes examples of how these general statements can be interpreted for a telecommunications device. Ravden and Johnson (1989) propose the following general criteria for the evaluation of computer interfaces; visual clarity, consistency, informative feedback, explicitness, appropriate functionality, flexibility and control, and user guidance and support (see also Johnson *et al.*, 1989).

Where a product or product family is more clearly defined, more specific criteria and their related measures can be considered. For example, Spradlin (1987) discusses a range of general and specific criteria that went into the development of flight deck instrumentation for Boeing's 757/767 aircraft. This article also shows the value of considering data on accidents, critical incidents and perceived problems in determining the appropriate criteria. Fulton and Feeney (1983) provide an additional example where background data were used in this way; here accident data and interviews with accident victims were used to help develop evaluation criteria in an investigation into powered domestic lawnmowers.

For the evaluation of a specific product the criteria and the corresponding measures have to be quite specific. Here are two examples of criteria used in relation to the evaluation of specific products.

The first was the development of the Digital Equipment Corporation (DEC) mouse. The programme of work involved comparative testing of proprietary products and prototypes. The final design emerged after several cycles of user testing with successive generations of prototypes (Hodes and Akagi, 1986). Subjective criteria used were ease of grasp, degree of hand comfort and degree of hand, wrist, and forearm strain. Objective criteria were fit in relation to hand size including hand length, palm length, palm width, and angle of palm at rest, rate of movement of the cursor over the screen (pixels/sec) and number of button pushes.

In the second case four designs of magnetic stripe readers were evaluated; two were 'insertion' designs and two were 'slot' designs (Lewis, 1987). The results showed that the 'slot' designs resulted in superior throughput times but that there was little difference in other respects. The subjective criterion was user preference; the objective criteria were reading rate (reads/min) and percentage of good reads.

Evaluation criteria can also be developed as part of a consistent programme of product testing that extends over several generations of a product, enabling criteria to be standardized and refined to high levels of sensitivity. Repeated use of standardized criteria in controlled tests and the data that result allow benchmark performance standards to be set up as part of product usability specification procedures. This also enables improvements in the products to be monitored over time. For example, Xerox Corporation carry out routine 'operability testing' on their photocopier products (Hartman, 1987). These tests incorporate standardized procedures for machine families which test prototypes in terms of operator preferences, error rates and task times. Similarly Philips also incorporate regular user testing into the development of

Philishave electric razors; criteria used have been refined over several years and very specific measuring techniques have been developed particular to the product family.

Collecting data

The methods used to collect data from a user trial fall into three basic categories:

- interview and questionnaire methods to collect data on attitudes and subjective assessments from users;
- direct observation of user actions;
- collection of objective data.

The extent to which a user trial makes use of these data collection methods will vary although in practice a blend of all three categories is often used. The blend that is most appropriate has to be judged in relation to the particular investigation.

Useful data can be collected effectively without the need for developing complicated interview schedules, questionnaires, etc. Many user trials are successfully undertaken on the basis of quite pragmatic and simple approaches. For example, questionnaires may be no more than a few questions answered using basic rating scales. Observations of user actions may consist only of counting the number of times a particular event occurs. On the other hand an investigation might require a much more elaborate approach where a full questionnaire must be developed, observation metrics compiled and interviewers trained. In this form a user trial can become a major exercise. Even within each category there are many choices to be made and some general remarks will follow about the wide variety of techniques which can be employed.

User interviews

Even at a basic level most user trials involve some form of interview, such as asking users their general opinions of the trial and the product they used. In practice the emphasis given to an interview may be high, it might be the focal point of the whole investigation, whilst on other occasions it may only play an incidental part, but it is usually advisable to incorporate some form of interview. Approached in the right way an interview can be an extremely productive way of collecting information.

The relationship between user and product is a dynamic process which has many facets, concerning the perceptual and cognitive aspects of product use; how information is absorbed from the product, how it is interpreted, and what user actions follow as a result. Such information can usually only be obtained directly from users themselves and is most effectively recorded through some form of structured interview. A full interview usually involves users in:

- evaluating the product in general terms;
- explaining difficulties they may have experienced during the course of the trial;
- explaining why they took certain actions;
- assessing their own physical or psychological condition;
- commenting on how the product might be improved.

An important consideration when conducting interviews is to ensure that the user is put at ease. A user trial is an unfamiliar environment for many and, however well they are briefed, they seldom know quite what to expect. The way the interview is handled is important if the co-operation and interest of the user is to be obtained and maintained. To achieve this is not difficult providing some basic guidelines are borne in mind:

- receive the user with courtesy;
- keep the user informed of what is to happen;
- take a positive lead in directing events;
- ensure that the user feels that his contribution is valuable;
- aim at giving the interview the flavour of a conversation rather than an inquisition.

An interview often requires users to make judgements about product characteristics which they seldom, if ever, have considered. In everyday life people use one product or another more or less continuously without giving them much thought, unless the person has significant negative or positive experiences with using the product. The product may delight the user in some way, it may simply breakdown, or it may interfere with the ability of the user to achieve their tasks. Often users tend to notice negative inter- ference with task completion rather than positive support. Users are seldom able to alter or modify the design to their liking. Consequently a challenge for the investigator is to stimulate the user to become critically aware of the product, and to think creatively about the quality of interaction experienced in using the product. Achieving this is largely a question of professional expertise and experience in the use of interview techniques. An important aspect is to carry out the interview in the immediate vicinity of the product under investigation. This is particularly important where the product itself is discussed and the user is required to explain actions taken or make compari- sons between products or design options and so on. The presence of the product will be an invaluable aid to explaining points and prompting remarks which might otherwise be overlooked. Stimulating critical awareness applies to what the user does as well as the type of questions asked. Here are some approaches that can be taken to achieve this:

- ensure that judgements are based on the experience of actually using the product;
- the interview should have clear structure and direction and not be simply open ended;
- be prepared to prompt and question the user in depth;
- encourage thoughts to be put into words as the user carries out the tasks;
- ask the user to make choices through rating scales, ranking scales, etc;
- use contrasting design solutions to highlight differences where choices have to be made;
- require users to write down their views as a trial progresses.

Interview formats

The traditional format for an interview is one to one but group interviews are being used more frequently in product evaluation work. One of the main stimulants for the evolution of these approaches has been the desire to explore more productive formats for obtaining information from users.

The format of one investigator and two users has been employed, sometimes referred to as 'co-discovery'. In one example it was used to investigate the problems of users installing a small computer (Comstock, 1983). More generally this format has been advocated as an aid to protocol analysis (see chapter 7) where users are required to verbalize their interpretation of a product as a trial progresses (Rubinstein and Hersh, 1984). Lewis (1982), Olson *et al.* (1984) and Ericsson and Simon (1980) also discuss the use of this approach as part of using verbal reports in user interface evaluation.

One or two investigators and several users is an approach that is also commonly used. For instance in a study by Spicer (1987) on solid fuel space heaters for domestic use, users individually carried out a set of tasks with the products under investigation and gave their own assessments. Subsequently they were brought together in small groups to compare their experiences. Conventional methods of subjective assessment techniques were used to record users' attitudes.

The concept of group interviews has been extended to include 'user participative' approaches to design. O'Brien (1982) has used this approach in the identification of the main user requirements and subsequent evaluations using product simulations for the design of control rooms; the work reported here has since been extended (O'Brien, 1987). Similar approaches to the design of workstations involving groups have been adopted by Davies and Phillips (1986), Murphy *et al.* (1986), Pikaar *et al.* (1985), Stubler and Bernard (1986) and Wilson (1991). Although these have involved well defined and smaller user groups the principles could be equally well applied to user groups representing larger populations. (See also Eason chapter 36 of this book.)

Unfortunately there has been little discussion in the literature on the strengths and weaknesses of interview techniques in product development work, and more particularly any systematic evaluation of the approaches

Figure 10.5. More than one user can be interviewed at a time. Bringing users together in this way can be a great aid to stimulating discussion

(McClelland, 1984; Meister 1986). However, most experience with interview techniques shows them to be extremely important in the evaluation of user interface design, but that there is still considerable scope for improvement.

Questionnaires

Questionnaires are often an integral part of a structured interview and the basis for documenting the results of a user trial. Typically they are paper-based, incorporating a series of questions designed to be answered in a predefined order; in their simplest form they may be only a short list of questions requiring Yes/No responses. Questionnaires can be completed in two basic ways, by the user or by the investigator; often a combination is used. The types of question used are either open ended or closed. Open ended questions are usually used to obtain free responses from users without pre-defining the answer; they may receive a wide variety of responses which then require post-hoc coding during analysis. For large surveys this type of question can be very time-consuming to analyze. In closed questions the user responds to a set of predefined categories. A variety of these may be used, for example:

- Factual statements;
  where simple yes/no responses or specific items of information are required.
- Multiple category questions;
  where categories are specified and the user chooses which of the categories apply. They can be answered on a yes/no basis or using some form of rating scale.
- Rating scales;
  used to assess the attitudes of users to specified product attributes. Commonly used to assess comfort, degrees of convenience, ease of use, perceived degrees of difficulty and so on.
- Ranking scales;
  used to indicate the relative order of a set of conditions or products according to a specified attribute.

There is a considerable variety of methods for the presentation, wording and analysis of questionnaires. The classic texts of Moser and Kalton (1971) and Oppenheim (1966 new edition 1992) still provide valuable advice on the design and use of questionnaires. More recent overviews are provided by Morton-Williams (1986) and Sinclair in chapter 3 of this book.

Observations of user actions

The effective use of observation techniques (see chapter 2) relies on a clear understanding of the purpose of the observation. Three types can be identified (adapted from Meister (1986)):

- descriptive techniques, where the observer simply records events as they take place (e.g., time based, frequency based, event sequence, postures adopted, controls used, etc);
- evaluative techniques, where the observer evaluates the outcome or consequence of events that have taken place (e.g., degrees of difficulty, incidence of hazardous events, errors of judgement, etc.); and
- diagnostic techniques where the observer identifies the causes that give rise to the observed events (e.g., positioning of controls, inadequacy of displays, poor user instructions, etc.).

Descriptive observations are generally the easiest to set up and conduct. The degree of difficulty increases with evaluative observations and is greatest with diagnostic observations. This is primarily because of the type of on-the-spot judgements the investigator is required to make during the observation. Ability to conduct evaluative or diagnostic observations depends greatly on the background knowledge of the investigator. Detailed knowledge of the tasks in hand and the objectives behind using the product, is usually required before evaluative or diagnostic observations can be successfully carried out. It is essential that the events to be observed, what data must be recorded and how, are clearly identified beforehand. There is usually a great deal of data that could be collected but much will be irrelevant. The greatest problems come from the speed of events and the number which occur in parallel.

The simplest form of observation to make is where events occur in strict sequence and at a rate which allows them to be recorded accurately. People have limited capacity to record events that occur in parallel or where the intervals between events are very short. It often surprises people who are new to observing user behaviour just how quickly the accuracy and reliability of an investigator can break down when there are too many items to monitor. Even moderately experienced investigators, when planning investigations, often find themselves being far too ambitious in the variety of events they wish to record. Consequently it is most important that the method of observation is carefully piloted whenever possible.

Still the most common method for recording data is paper based, the effectiveness of which can be greatly influenced by the precise way in which the data are to be recorded. Wherever possible graphic methods should be used; ticking boxes, marking diagrams, use of graphic codes and so on. Spicer *et al.* (1984) provide an example of this approach where observations were made of disabled and elderly people entering and leaving cars.

Apart from paper based methods video recording is used frequently and is now regarded as standard equipment in any product usability laboratory (e.g., Benel and Pain, 1985; Cantwell and Stajano, 1985; Neal and Simons, 1985; Nielsen, 1994). Video is easy to set up and to use; it can provide a comprehensive record of events and of course it enables one to replay events. The great advantage of video is that a particular sequence of events can be analyzed several times from different viewpoints, enabling an investigator to overcome many of the problems associated with analyzing events which occur in parallel or very quickly. An example of this as a major consideration was in a study into warning systems for railway workers working alongside the track (McClelland *et al.*, 1983). Strommen *et al.* (1992) used this approach in a study of three year old children using a computer games controller. In this case, because of the age of the children, the investigators had to rely heavily on observations drawn from video recordings to evaluate the behaviour of the children. However, a few cautionary words are in order. Firstly, detailed analysis of video recordings can be extremely time-consuming. Ratios of between 1:5 and 1:10, recorded time:analysis time, are commonly quoted (MacKay *et al.*, 1988), but see Bainbridge and Sanderson in chapter 9 who report analysis time:sequence time ratios of 10:1 to 1000:1. Secondly, the need for detailed preparatory work is not reduced because a comprehensive record of events is available. The use of video requires at least as much care as paper based approaches in deciding beforehand what events are to be observed and the form in which the data are to be analyzed.

Objective measures

It is always wise to obtain objective data to support data from other sources whenever it is possible and practicable, notably the subjective data from interviews and questionnaires. Sources of objective data that are frequently used can be divided into three categories. Firstly, direct objective measurements of the user; these would include body dimensions, physiological measurements such as heart rate, oxygen intake, body temperature, and so on. Secondly, directly recorded data resulting from user actions, registered by the investigator or by some remote means such as video or automatic event recording. Data generated by user actions can come in several forms. They can be time based (e.g., task duration, event duration, response time, reaction time) or based on scores of errors/accuracy (e.g., mistakes in procedures, incorrect responses to stimuli, error rates in relation to time or events). They may also be based on the frequency of events such as the rate of specific responses or task completion rates. The third type of objective data is that measured directly from the product on completion of or during the trial; for example the position of seats, shelves, or controls, the accuracy of cursor control devices in an interactive computer system, numbers of keystrokes made, lighting levels, colour quality, brightness, contrast settings on visual displays or sound levels.

Location of a user trial

Investigators must take account of many factors when selecting the appropriate location for a user trial. Perhaps the first and most obvious consideration is that the trial is carried out in a place where the data required can be collected in a cost effective way. Other important considerations are discussed below.

The product

The form in which the product is to be presented to users can significantly influence choice of location. For example, if the trial is of an early simulation of the product control system, backup equipment to run the simulation may be required. If a special hardware interface has to be built it may have to be built in the location where the trial is to be carried out. This could limit the simulation to a particular usability laboratory or specially prepared test area. On the other hand if the simulation can be run on a standard PC or a portable PC the investigator may have an unrestricted choice. The need for technical support to be on hand may also be an important consideration. Trials involving prototypes often involve purpose built equipment which has an unnerving habit of breaking down at the most inappropriate moments. Effective technical support may be critical.

The users

How are users to be contacted and recruited? Many basic practical considerations can influence the choice of location. An investigator may wish to recruit members of the general public. Having contacted them, many questions arise; how do users get to the location? do they know where it is? how are users to be scheduled? etc. On the other hand the study might require users with very specific characteristics. For example they may be part of a particular work force or some other 'captive population', in which case it may be more appropriate for the investigator to take the trial to the user group. However it is possible that the organization hosting the trial was never prepared for such an activity. Significant practical problems may arise, from the trivial such as arranging schedules for users, to the more significant such as repeated unavailability due to higher priority activities. In short sourcing appropriate users may be eased or complicated by the correct choice of location.

Complementary tasks

The tasks that are carried out in a trial may be interdependent with the use of other artifacts and their associated tasks. For example, evaluating the user interface for an office receptionists' database which is used to support the answering of incoming telephone calls. The validity of the trial may depend on integrating the use of these other artifacts into the test procedures. An investigator may have to bring the other artifacts to the test location, or, alternatively, go to the location where the other artifacts can be found. This may not be the same place as the users' normal environment. For example, Johnson *et al.* (1992) report a study on the evaluation of basic car radio controls involving users driving a car over a specially selected route.

Context of use; simulation or realism?

Investigators often carry out user trials in a usability laboratory, or some similar location, where circumstances can be carefully controlled by the investigator. This means simulating the context in which the product is to be used in practice. Ideally the investigator must be sure that any contextual characteristics that may critically influence the behaviour of the user, and consequently their performance, are correctly simulated. The physical environment (visual, acoustic, thermal, and mechanical aspects etc.) is an obvious example where the investigator may need, and indeed is able, accurately to simulate circumstances easily. It is often less easy to simulate patterns of product use or social considerations. Often an investigator does not have hard evidence available which describes the critical contextual variables and their relative importance. As a result investigators frequently have to work on the basis of making reasonable assumptions based on their own professional experience and the best information available. It is important that any assumptions made are understood by, and agreed with, those who have requested the trial to be carried out. If incorrect assumptions are made the validity of the trial may be undermined. Whiteside *et al.* (1988) provide an interesting discussion on this topic in relation to human–computer interface development. There is a growing conviction amongst practitioners within the field of human–computer interaction development that studies conducted in real circumstances provide opportunities to gain insights that would not be revealed in the laboratory (see Wolf *et al.*, 1989).

Data collection

The quality, quantity and type of data required from a trial can be a significant consideration in determining where the trial should be located. Investigators now enjoy the advantages of highly portable means of collecting data in a variety of forms. The most obvious development in recent years has been the evolution of video, and especially the capability to take video and audio recording apparatus into the field that can be carried in a shoulder bag. The evolution of portable PCs also opens up possibilities for on-site data collection and analysis that only a few years ago would not have been considered realistic. See also the discussion by Drury in chapter 9 of this book.

Confidentiality

Can the prototype be seen by 'outsiders'? Confidentiality may be another and very important consideration in determining where to locate a user trial.

The investigator

User trials should be set up and designed by personnel qualified in the design of experiments and who have knowledge of the ergonomics issues appropriate to the product under investigation. Traditionally such personnel have included people qualified in applied physiology and applied psychology as well as ergonomics or human factors. However it is clear from the wide variety of work reported that many other professional groups can successfully carry out user trials, including, for example, anthropologists, social psychologists, industrial engineers and marketing personnel. Many of the tasks involved in the day-to-day conduct of user trials can, on occasions, be under-taken by other personnel but specific training will often be required before-hand to ensure results of a satisfactory quality. Where an organization does not have the necessary in-house expertise, appropriate staff should be recruited, or alternatively, an outside agency contracted which has experience in this type of work.

Ideally the investigator should retain a degree of independence from the product under test. Often, through a lack of resources, the investigator has a stake in the product and, as a result, the outcome of the trial. It is sometimes difficult for the investigator to remain detached from the data collected, especially where the data are mainly subjective or based on the observations of the investigator. The investigator may seek to confirm previously held expectations of the result. This is not so much an ethical question, although sometimes ethical considerations will arise, but more a question of bias induced by a close association with the product under test. The primary objective of the investigator should be to execute a reliable user trial, not to confirm their own prejudices or those of their colleagues.

Investigators must also play the role of communicator. At some stage the results of a trial need to be absorbed and understood by the audience. The audience will have to make use of the results in some way, and this often means translating the results into specific design decisions. It is worthwhile for the investigator to spend some time considering how the trial, and the subsequent reporting, might be set up to ease the knowledge transfer prob-lem. The value of the trial will depend on ensuring that the implications of the trial are correctly understood and interpreted by the audience. There are many ways of achieving this some of which are:

- active involvement of the audience in contributing to the design of the trial; promotes an understanding of the goals and the reasons for the methods chosen;
- audience observes the trials (all or some) first hand; the audience sees at first hand the difficulties and experiences users have of the product. The observers also can hear the comments of users. It is now widely accepted that direct observation of users has a powerful impact on the attitudes of designers to the use and value of user trials;
- present preliminary results through participatory discussion within a few working days after the trial is complete;
- set up a debriefing workshop to disseminate the results and discuss the implications of the trial;
- do not rely only on written reports, they tend not to be read, and even when read, easily misinterpreted.

The investigator also has another important communication role to play in relation to the users who participate in a user trial. As noted earlier in this chapter, when users first arrive at the location of the trial they will probably have little idea of what will be expected of them. The first goal will be to ensure that the user is well briefed and understands what will happen, and what he or she will be expected to do, during the trial. This demands that the investigator is skilled in the clear presentation of instructions, both verbal and written. The investigator will also have little time to 'put the user at ease' and establish a reasonable rapport with the user. It is therefore important that the investigator has the necessary social skills to achieve this rapport quickly. The user must feel comfortable, unintimidated, and willing to cooperate in undertaking whatever tasks are presented to him or her.

Future directions

Interest in the incorporation of user testing into product development has been increasing in recent years. As products become more complex, and commercial pressures become more severe, the need to examine the usability aspects of products is growing. There is increasing interest in methods which enable products to be assessed systematically and judgements made about their qualities which are based more on 'facts' and less on 'opinions'. This is precisely where the application of user trials can play a vital part in product development, but their relevance to mainstream product development will depend in large measure on the development of testing and evaluation methods that suit such an environment.

Ensuring the relevance of user trials

From a technical standpoint the most notable contribution in recent years has been from Meister (1986). As already noted he provides an extensive discussion of the methods available. Meister also notes the lack of attention given to evaluating the methods used and calls for further efforts from the ergonomics research community to improve our understanding of the methods that are used (see also Anderson and Olson, 1985). However many of the technical issues discussed are rather long term and not within the scope of most practitioners within industry to resolve. Practitioners within a product development environment must, of necessity, take a more pragmatic view over the short to medium term. In this context the successful implementation of user trials is likely to depend more on demonstrating their effectiveness in terms of:

- the extent to which the results from user trials with prototypes can predict how a product will perform in practice;
- how the performance of products in practice is improved as a result of user trials being carried out;

- the cost-benefit to product development of improving the quality of products and, in particular, ensuring that significant design deficiencies do not get to market;
- evaluating products in relation to the goals the user wants to achieve by using the product as well as the specific design details of the user-interface itself;
- demonstrating the importance of giving proper consideration to end user requirements during product development.

Organisational and technical barriers

But user testing is not an end in itself. User trials provide only one of several means of implementing a systematic approach to designing satisfactory levels of usability into products. In the longer term the value of user trials lies in the acceptance by the managers of product development that achieving satisfactory levels of usability is a necessary part of their product development strategy. In other words usability and, in the widest sense, user requirements should be incorporated into the specific design objectives which product developers need to meet. On such a basis it is then possible to maximize the benefits of user trials by accurately focusing on the design issues in question, and delivering timely and useful results. The principles that should be adopted to achieve such a systematic approach are easily stated but, in practice, are very difficult to implement. Difficulties are often due to organizational and technical aspects rather than the ability of human factors and usability special-ists to deal with the design problems that face development teams. Gould (1988a) presents a valuable discussion of this topic (see also Grudin, 1990). But the organizational and technical barriers need to be overcome. There is considerable scope for research work aimed at assessing the most effective methods for ensuring usability issues are taken into account within pro-duct development.

Reduction of response times

An important aspect to overcoming the barriers is, in part, the speed of response of human factors groups to requests for user tests to be carried out. Tools and methods are required which reduce the time taken by investigators to execute the various aspects of user testing including the set up of user trials, the collection, collation, analysis and reporting of the data gathered. The data gathered will often include audio and video recordings as well as more traditional forms of data. Audio and video recording technologies offer the investigator many opportunities to gather a wealth of data, but there is also the potential handicap of gathering many more data than are necessary.

Too many data also bring with them added problems of collation and analysis. Improvements in the tools and methods now available need to be developed which help to reduce response times.

What is the best product evaluation method?

Are user trials the best form of evaluation method at our disposal? Several research studies have been published recently which have explored the relative value of user trials in comparison with other methods of evaluating user interfaces. These studies suggest that other options are open to the ergonomist which may sometimes be at least as effective. For example Jeffries and Desurive (1992) review recent literature that compares heuristic methods of evaluation (specialists inspecting user interfaces and evaluating the usability aspects based on their expertise) with usability testing. These authors conclude that complementary evaluation techniques are needed by the practitioner, that heuristic methods require several experienced usability specialists to provide comparable results with usability testing, and that the usability issues raised using heuristic methods are different from those found in usability tests. In an earlier study Jeffries *et al.* (1991) reported that usability testing revealed more severe problems, more recurring problems and more global problems than heuristic evaluation methods. This study also compared guidelines and cognitive walkthroughs but found that these methods gave less favourable results. This view is confirmed by Karat *et al.* (1992) who report that, in comparison with walkthroughs, user testing revealed the largest number of usability problems and identified a number of severe problems that were missed by using walkthrough methods. When the authors examined the cost benefit data they found that user testing required less time to identify each problem when compared with walkthroughs. Karat (1989) also reports on a case study which illustrates the cost benefits of usability testing.

On the subject of heuristic methods alone Nielsen and Molich (1990) suggest that the chances of detecting usability problems are greatly improved if several evaluators are used. In a later paper Nielsen (1992) estimates that if the evaluators are specialists in the application area as well as in usability issues then 5 evaluators may detect ±90% of the usability problems. Whitfield *et al.* (1991), advocate that assessment of user interfaces is valuable in all phases of product development but to be effective investigators must be prepared to adapt their techniques to suit local circumstances. The basic techniques these authors suggest include observation methods, specialist inspections, cognitive walkthroughs, as well as user trials.

The studies cited above were primarily aimed at evaluating methods used in the field of human–computer interaction. Whether the conclusions are equally valid for other types of user interface needs investigation. However the data suggest that the investigator in other product areas might gain benefits from considering the use of other evaluation methods as well as user trials.

International and national standards

A number of international and national standards have emerged in recent years which require product manufacturers to pay more attention to user issues. Increasingly user performance standards, and the corresponding need formally to evaluate design solutions for compliance, are being integrated into standards as the basis for approving products. One early example where this approach is used is for reclosable pharmaceutical containers in the UK (BS 5321; 1975). ISO 9241 Part 11 (currently a Committee Draft) proposes to take the principle of incorporating user performance requirements some-what further. ISO 9241 is primarily concerned with visual display units in office environments. Part 11 proposes principles for the specification and measurement of usability which have much wider applicability to the usability of products than only the use of VDTs. Part 11 outlines an approach for defining the user, their tasks, the context of use, usability criteria, and actions which will be necessary during the course of development to ensure that the necessary levels of usability are achieved. A particularly important part of this approach is that the description of usability should be used in a proactive way to guide and lead product development, rather than be used only to evaluate designs following development. The clear implication is also that application of the standard will necessarily involve user trials as a method of judging compliance. Other sections of ISO 9241, Part 3 (Visual display requirements) and Part 4 (Keyboard requirements - currently in Draft form), also incorporate procedures for user testing. The principle of user testing is also reflected in the draft ISO/IEC DIS 9126 standard 'Information Tech-nology - Software product evaluation - Quality characteristics and guidelines for their use'. The European Telecommunications Standards Institute have also recently drafted a 'Guide for Usability Evaluations' (ETSI 1991). It is clear from the introductory sections to this guide that usability evaluations are expected to play an increasingly important role in the development of telecommunications products.

References

Anderson, N.S. and Olson, J.R. (1985). Methods for designing software to fit human needs and capabilities. In *Proceedings of the workshop on software human factors* (Washington, DC: National Academic Press).

Bailey, R.W. (1982). *Human Performance Engineering* (Englewood Cliffs, NJ: Prentice Hall Inc.).

Bailey, W.A., Knox, S.T. and Lynch, E.F. (1988). Effects of Interface design upon user productivity, *CHI 88, Conference Proceedings*, Association for Computing Machinery Press (New York: Addison-Wesley), pp. 207–212.

Benel, D.C.R. and Pain, R.F. (1985). The human factors usability laboratory in product evaluation, *Proceedings of Human Factors Society, 29th Annual Meeting* (Santa Monica, CA: Human Factors Society).

Bewley, W.L., Roberts, T.L., Schroit, D. and Verplank, W.L. (1983). Human factors testing in the design of Xerox's 8010 "Star" office workstation, *CHI '83 Proceedings*. Association of Computer Manufacturers.

Boies, S.J., Gould, J.D., Levy, S., Richards, J. and Schoonard, J. (1987). The 1984 Olympic messaging system, *Communications of the ACM*, September 1987.

BS 5321:1975 Reclosable Pharmaceutical Containers Resistant to opening by children (London: British Standards Institution).

Cantwell, D. and Stajano, A. (1985). Certification of software usability in IBM Europe, *Ergonomics International 85 Proceedings of the Ninth Congress of the IEA*, edited by I.D. Brown, R. Goldsmith, K. Coombes and M.A. Sinclair (London: Taylor and Francis).

Chapanis, A. (1959). *Research Techniques in Human Engineering* (Baltimore, MD: The John Hopkins Press)

Chapanis, A. and Budurka, W.J. (1990). Specifying human-computer interface requirements, *Behaviour & Information Technology*, **9**, 479–492.

Collins, M., (1986). Sampling, In *Consumer Market Research Handbook*, edited by R.M. Worcester and J. Downham (Amsterdam: North-Holland).

Comstock, E. (1983). Customer installability of computer systems, *Proceedings of the Human Factors Society, 27th Annual Meeting* (Santa Monica, CA: Human Factors Society).

Cushman, W.H. and Rosenberg, D.J. (1991). *Human Factors in Product Design.* (Amsterdam: Elsevier).

Davies, D.K. and Phillips, M.D. (1986). Assessing user acceptance of next generation air traffic controller workstations, *Proceedings of the Human Factors Society, 30th Annual Meeting* (Santa Monica, CA: Human Factors Society).

Dillon, A. (1992). Reading from paper versus screens: a critical review of the empirical literature, *Ergonomics*, **35**, 1297–1326.

Egan, D.E., Remde, J.R., Landauer, T.K., Lochbaum, C.C. and Gomex, J.M. (1989). *CHI '89 Conference Proceedings*, Association for Computing Machinery (New York: Addison-Wesley), pp. 205–210.

Ericsson, K.A. and Simon, H.A. (1980). Verbal reports as data, *Psychological Review*, **3**.

ETSI (1991). *Guide for usability evaluations*, European Telecommunications Standards Institute, ETSI/TC-HF(91)4.

Francik, E. and Akagi, K. (1989). Designing a computer pencil and tablet for handwriting, *Proceedings of the 33rd Annual Meeting of the Human Factors Society*, pp. 445–449.

Fulton, E.J. and Feeney, R.J., (1983). Powered domestic lawnmowers: design for safety, *Applied Ergonomics*, **14**, 91–95.

Gould, J.D. (1988a). Designing for usability: the next iteration is to reduce organizational barriers, *Proceedings of the 32nd Annual meeting of the Human Factors Society*, pp. 1–9.

Gould, J.D. (1988b). How to design usable systems, In *Handbook of Human–*

Computer Interaction, edited by M. Helander (Amsterdam: North-Holland).

Grudin, J. (1990). The computer reaches out: the historical continuity of interface design. *CHI 90 Conference Proceedings*, Association for Computing Machinery Press (New York: Addison Wesley), pp. 261–268.

Hartman, W. (1987). Head Industrial Design/Operability Department, Xerox Corporation, Personal Communication.

Hodes, D. and Akagi, K. (1986). Study, development and design of a mouse, *Proceedings of the Human Factors Society, 30th Annual Meeting* (Santa Monica, CA: Human Factors Society).

Holleran, P.A. (1991). A methodological note on pitfalls in usability testing. *Behaviour & Information Technology*, **10**, 345–357.

ISO/IEC DIS 9126, *Information Technology–Software product evaluation–Quality characteristics and guidelines for their use*. International Organization for Standardization and the International Electrotechnical Commission.

ISO 9241, *Ergonomic requirements for office work with visual display terminals*. International Organization for Standardization.

Jeffries, R. and Desurive, H. (1992). Usability testing vs heuristic evaluation: was there a contest?, *SIGCHI Bulletin*, **24**(4), 39–41.

Jeffries, R., Miller, J.R., Wharton, C. and Uyeda, K.M. (1991). User interface evaluation in the real world: a comparison of four techniques. *CHI 91 Conference Proceedings*, Association for Computing Machinery Press (New York: Addison Wesley) pp. 119–124.

Johnson, G.I., Clegg, C.W. and Ravden, S.J. (1989). Towards a practical method of user interface evaluation, *Applied Ergonomics*, **20**, 255–260.

Johnson, G.I. and Vianen, E.P.G. (1992). Comparative evaluation of basic car radio controls. In *Contemporary Ergonomics*, edited by E.J. Lovesey, (1992) (456–462) (London: Taylor and Francis).

Kantowitz, B.H. (1992). Selecting measures for Human Factors research, *Human Factors*, **34**, 387–398.

Karat, C.M. (1989). Iterative usability testing of a security application, *Proceedings of the 33rd Annual Meeting of the Human Factors Society*, pp. 273–277.

Karat, C.M., Campbell, R. and Fiegel, T. (1992). Comparison of empirical testing and walkthrough methods in user interface evaluation, *CHI 92 Conference Proceedings*, Association for Computing Machinery Press (New York: Addison Wesley), pp. 397–404.

Kirk, N.S. and Ridgeway, S. (1970). Ergonomics testing of consumer products, 1: General considerations, *Applied Ergonomics*, **1**, 295–300.

Kirk, N.S. and Ridgeway, S. (1971). Ergonomics testing of consumer products, 2: Techniques, *Applied Ergonomics*, **2**, 12–18.

Lewis, C. (1982). Using the 'thinking allowed' method in cognitive interface design. IBM Research Report RC 9265.

Lewis, J.R. (1987). Slot versus insertion magnetic stripe readers; user performance and preference, *Human Factors*, **29**, 465–476.

Mackay, W.E., Guindon, R., Mantel, M.M., Suchman, L. and Tatar, D.G. (1988). Panel session: Video: Data for studying human-computer interaction, *CHI '88 Conference proceedings* (Reading, MA: Addison-Wesley).

Macleod, M. (1992). *An introduction to Usability Evaluation*, National Physical Laboratory, Department of Trade and Industry.

McClelland, I.L. (1984). Evaluation trials and the use of subjects, In *Contemporary Ergonomics 1984 Proceedings of the Ergonomics Society Conference*, edited by E.D. Megaw (London: Taylor and Francis).

McClelland, I.L., Simpson, C.T. and Starbuck, A. (1983). An audible train warning for track maintenance personnel, *Applied Ergonomics*, **14**, 2–10.

Meister, D. (1986). *Human Factors Testing and Evaluation* (Amsterdam: North-Holland).

Meister, D. and Enderwick, T.P. (1992). Measurement in Human Factors; Special issue preface, *Human Factors*, **34**, 383–385.

Morton-Williams, J. (1986). Questionnaire design, In *Consumer Market Research Handbook*, edited by R.M. Worcester and J. Downham (Amsterdam: North-Holland).

Moser, C.A. and Kalton, G. (1971). *Survey methods in social investigation*, (London: Heinemann Educational Books).

Mountford, J., Vertelney, L., Bauersfeld, P., Gomoll, K. and Tognazzini, B. (1990). Designers: meet your users. Panel discussion, *CHI '90 Conference Proceedings*, Association for Computing Machinery Press (New York: Addison Wesley), pp. 439–442.

Muckler, F.A. and Seven, S.A. (1992). Selecting performance measures: 'objective' versus 'subjective' measurement, *Human Factors*, **34**, 441–455.

Murphy, E.D., Coleman, W.D., Stewart, L.J. and Sheppard, S.B. (1986). Case study: Developing an operations concept for future air traffic control, *Proceedings of the Human Factors Society, 30th Annual Meeting* (Santa Monica, CA: Human Factors Society).

Neal, A.S. and Simons, R.M. (1985). Evaluating software and documentation usability, In *Ergonomics International 85 Proceedings of the Ninth Congress of the IEA*, edited by I.D. Brown, R. Goldsmith, K. Coombes and M.A. Sinclair (London: Taylor and Francis).

Nielsen, J. (1992). Finding usability problems through heuristic evaluation, *CHI 92 Conference Proceedings*, Association for Computing Machinery Press (New York: Addison-Wesley), pp. 373–380.

Nielsen, J. and Molich, R. (1990). Heuristic evaluation of user interfaces, *CHI 90 Conference Proceedings*, Association for Computing Machinery Press (New York: Addison-Wesley), pp. 249–256.

Nielsen, J. (1993). Usability Engineering (Boston, MA: Academic Press).

Nielsen, J. (1994). Usability Laboratories, Special Issue, *Behaviour and Information Technology*, **13**, 1&2.

O'Brien, D.D. (Ed.) (1982). Design Methods; Seminar on Control Room Design, Lancashire Constabulary HQ.

O'Brien, D.D. (1987). Personal communication from the UK Government Home Office, London.

Olson, G.M., Duffy, S.A. and Mack, R.L. (1984). Thinking-out-loud as a method of studying real-time comprehension processes, In *New Methods in Reading Comprehension*, edited by D.E. Keiras and M.A. Just (Hillsdale, NJ: Lawrence Erlbaum).

Oppenheim, A.N. (1966). *Questionnaire Design and Attitude Measurement* (London: Heinemann Educational Books).

Pikaar, R.N., Lenior, T.M.J. and Rijnsdorp, J.E. (1985). Control room design from situational analysis to final layout; operator contributions and the role of ergonomists, *2nd IFAC/IFIP/IFORS/IEA Conference; Analysis, Design and Evaluation of Man-Machine Systems*, edited by G. Mancini, G. Johannsen and L. Martensson (Oxford: Pergamon Press).

Preece, J., Davis, G. and Keller, L. (Eds) (1990). *A Guide to Usability*, Milton Keynes: The Open University in association with the Department of Trade and Industry (UK).

Ravden, S.J. and Johnson, G.I. (1989). Evaluating Usability of Human Computer Interfaces: A Practical Method (Chichester: Ellis Horwood).

Rennie, A.M. (1981). The application of ergonomics to consumer product evaluation, *Applied Ergonomics*, **12**, 163–168.

Ross, E.H. (1983). Human factors in system development: project team approaches and recommendations, *Proceedings of the Human Factors Society, 27th Annual Meeting*. (Santa Monica, CA: Human Factors Society).

Rubinstein, R. and Hersh, H.M. (1984). *The Human Factor* (Digital Equipment Corporation: Digital Press).

Shackel, B. (1986). Usability–context, framework, definition, design and evaluation, *Proceedings of the SERC CREST Course; Human factors for informatics usability*, HUSAT Research Centre, University of Technology, Loughborough.

Shackel, B. (1987). Human Factors for Usability Engineering, ESPRIT '87; Achievements and impact, *Proceedings of the 4th annual ESPRIT conference*, Brussels, September 1987, edited by Commission of the European Communities (Amsterdam: North-Holland).

Spicer, J. (1987). Personal communication from the Institute for Consumer Ergonomics, Loughborough.

Spicer, J., Wilkinson, S. and McClelland, I.L. (1984). Access to cars by disabled and elderly people; Report No. 3, Car trials. Working Paper WP/VED/84/10, Transport and Road Research Laboratory, Department of Transport.

Spradlin, R.E. (1987). Modern air transport flight deck design, *Displays*, October 1987, 171–182.

Strommen, E.F., Razavi, S. and Medoff, L.M. (1992). This button makes you go up: three-year-olds and the Nintendo controller, *Applied Ergonomics*, **23**, 409–413.

Stubler, W.F. and Bernard, T.E. (1986). Office ergonomics: design methodology and evaluation, *Proceedings of the Human Factors Society, 30th Annual Meeting* (Santa Monica, CA: Human Factors Society).

Virzi, R.A. (1992). Refining the test Phase of Usability evaluation: how many subjects is enough? *Human Factors*, **34**, 457–468.

Whiteside, J., Bennett, J. and Holtzblatt, K. (1988). Usability engineering: our experience and evolution, In *Handbook of Human-Computer Interaction*, edited by M. Helander (Amsterdam: North-Holland).

Whitfield, A., Wilson, F. and Dowell, J. (1991). A framework for human factors evaluation, *Behaviour & Information Technology*, **10**, 65–79.

Wilson, J.R. (1991). Design decision groups - a participative process for developing workplaces. In *Participatory Ergonomics*, edited by K. Noro and A. Imada (London: Taylor and Francis), pp. 81–96.

Winer, B.J. (1971). *Statistical Principles in Experimental Design* (New York: McGraw-Hill).

Wolf, C.G., Carroll, J.M., Landauer, T.J., John, B.E. and Whiteside J. (1989). The role of laboratory experiments in HCI: Help, hindrance, or ho-hum?, Panel discussion *CHI 89 Conference Proceedings*, Association for Computing Machinery Press, (New York: Addison-Wesley), pp. 265–268.

Woodson, W.E. (1981). *Human Factors Design Handbook: Information and guidelines for the design of systems, facilities, equipment and products for human use* (New York: McGraw-Hill).

Chapter 11

Is this chapter any use? Methods for evaluating text

James Hartley

Introduction

How can we evaluate a piece of text? What questions must we ask, and how can we answer them? The literature in this area reveals a concern with at least four issues. We can ask questions about a text's content; about the way it is presented (in terms of its typography and layout); about the way it uses devices (such as headings) and illustrative materials (such as diagrams); and about its suitability for its intended audience. These four issues form the basis of this chapter.

Evaluating content

We can approach the task of assessing the content of textual materials from many different points of view. When the material is a procedural document, a form to be completed, or an explanatory note to accompany some other material, then the procedures used to evaluate its content will be rather different from those used to evaluate a chapter in a textbook. Our main concern will be with whether the text is fit for the job. One common way of assessing this (in addition to making one's own judgement) is to ask other people, experts and users, for their opinions. If, however, the text we are assessing is, say, a textbook or a chapter in a textbook such as this one, then our main concern will be to decide whether or not the content is sufficient for our purposes. To assess this we typically skim through the material, study the headings and subheadings, examine the illustrative materials, and perhaps read the introduction and/or concluding summary. We then decide whether the coverage is sufficiently detailed and sufficiently interesting for us to pursue it in more detail. And, since we may be evaluating a text for others to use, rather than ourselves, we will look to see if there are outdated materials, errors of facts or principles, unjustified inferences, important omissions, and biases of any kind—academic, national, racial and sexual (Zimet, 1976).

This may seem obvious, but what is not obvious is how we can be sure that the content meets such qualifications. Not only are the above descriptions hard to specify but also at times they may be misleading. Users of technical documents sometimes complain, for instance, that such documents contain *too much* information—that is, that they contain more than is needed for the task in hand. The same kind of problem can also arise with textbooks. How much background information, of interest historically but now out of date or even inaccurate by today's standards, should go into, say, an ergonomics textbook? And how can we be assured that selective bias is not present, except through our own rather subjective attempts to interpret today's shifting nationalistic, racial and sexual standards?

Evaluating content is at best a difficult activity, and usually a subjective one. One way to increase its objectivity is to increase the number of judges, and to provide some sort of checklist to help ensure that the judges all evaluate the same issues. Such an approach is commonly used for evaluating textbooks in countries which have state-controlled school systems. An unpublished report to Unesco from the Educational Products Information Exchange Institute entitled *Selecting Among Textbooks* contains, and critically comments on, a dozen such checklists which teachers and subject matter experts have used when they have been assessing or making comparisons between texts. Figure 11.1 provides an illustration. Although such checklists are useful in making the judges' ratings more systematic and consistent, there are no *standard* scales that have been widely adopted. Farr and Tulley (1985) reported, for instance, that the average number of items on the checklists that they studied was 73; the longest had 180 items and the shortest 42.

Such checklists are usually completed *before* recommending a particular textbook for use, but there is, of course, no reason why such information could not be collected *after* the texts have been used by readers and by teachers. Information gained in this way would be useful in deciding whether or not to use the book again, and it would be helpful to authors who are planning subsequent editions. Such feedback sheets can occasionally be found in scholastic textbooks and academic journals.

The layout of text

Figure 11.1 shows that readers are often asked to judge the technical quality of text as well as the content. Generally speaking it is difficult without expert knowledge to evaluate the minutiae of typographic practice. However, it is possible to look out for some problems. Questions might be asked, for example, concerning the density of the type (are the lines too close?) and the excessive use of typographical cueing (does it look too messy?). Sless (1984) points out that the cover page shown in Figure 11.2 uses five different unre-

A. Format of book
1. General appearance
2. Practicality of size and colour for classroom use
3. Readability of type
4. Durability and flexibility of binding
5. Appeal of page layouts
6. Appropriateness of the illustrations
7. Usefulness of chapter headings
8. Usability of index
9. Quality of the paper

B. Organization and content
10. Consistency of the organization and emphases with the teaching and learning standards of the school
11. Consistency of the point of view of the book with the basic principles of the subject area for which the book is being considered
12. Usefulness in stimulating critical thinking
13. Aid in stimulating students forming their own goals and towards self-evaluation
14. Usefulness in providing situations for problem solving
15. Usefulness in furthering the systematic and sequential program of the course of study
16. Clarity and succinctness of the explanations
17. Interest appeal
18. Provision for measuring student achievement
19. Adequacy of the chapter organization
20. Adaptability of content to classroom situations and to varying abilities of individual students
21. Degree of challenge for the reasonably well-prepared students
22. Usefulness for the more able students
23. Usefulness for the slow learners
24. Adequacy of the quality and quantity of skills assignments
25. Provision for review and maintenance of skills previously taught

Figure 11.1. A checklist for assessing textbooks. (In the original version each item is rated on a five-point scale from very good to very bad.)

lated typefaces in the space of seven lines. Other examples of multiple cueing (and its adverse effects upon the reader) can be seen in publications by Shebilske and Rotondo (1981) and Tukey (1977).

When considering the density of text it is appropriate to bear in mind that readers like text to be spacious and well structured. How can typography help one to achieve these aims? I have written frequently and at length elsewhere on how the layout of complex text can be so configured that it con-the visual literacy center

presents the

PROVOCATIVE PAPER SERIES
#1

Visual Literacy, Languaging, and Learning

by

JOHN L. DEBES and CLARENCE M. WILLIAMS

Figure 11.2. An example of multiple typographic cueing (Reproduced with permission)

veys at a glance its underlying structure. Here I shall only repeat the basic arguments. (Readers are referred to Hartley, 1987a and 1994a for more detailed discussions.) These arguments are as follows:

1. Both the horizontal and the vertical spacing of the text needs to be regular and consistent;
2. This means that interword spacing should be regular (as in typescript); and
3. That the interline, inter-paragraph and inter-section spacing should each be consistent throughout the text;
4. To achieve this the text should be set with a ragged right hand edge technically called unjustified composition), and
5. With a variable baseline; i.e. the text should not necessarily have the same number of lines on each page;
6. The stopping point for text should be determined (horizontally) by syntactic considerations and (vertically) by sense. (For example, one should not start a page with the last line of a paragraph.)

Figures 11.3 and 11.4 provide before and after illustrations of this approach in practice.

Readers will notice at this point that I have emphasized the consistent spatial display of the text before commenting on the niceties of typefaces and typesizes. These latter issues are important, and useful discussions of them can be found in Hartley (1994a) and Misanchuk (1992). However, to my mind, choice of typefaces and typesizes is secondary to deciding on the spatial

INSULATING GLOVES

1. GENERAL

1.01 This section covers the description, care and maintenance of insulating gloves provided for the protection of workmen against electric shock, and the precautions to be followed in their use.

1.02 This section has been reissued to include the D and E Insulating Gloves.

2. TYPES OF INSULATING GLOVES

2.01 All types of insulating gloves are of the gauntlet type and are made in four sizes: 9-1/2, 10, 11 and 12. The size indicates the approximate number of inches around the glove, measured midway between the thumb and finger crotches. The length of each glove, measured from the tip of the second finger to the outer edge of the gauntlet, is approximately 14 inches.

2.02 There are various kinds of insulating gloves. The original ones were just called Insulating Gloves. After that B, C, D and E Insulating Gloves were developed. As described below, the D Glove replaced the original Insulating Gloves and the E glove replaced the B and C Gloves.

2.03 **Insulating Gloves** are thick enough to eliminate the need for protector gloves and are intended for use without them. These gloves have been superseded by the D Insulating Gloves.

Figure 11.3. The first page of a piece of technical text in its original format (Reproduced with permission)

arrangement of the text. After all, the first thing that a designer does is to choose the size of the page on which the text is to be printed. This choice of page size constrains virtually all the other remaining decisions about page design (see Hartley, 1994a).

There are several ways of assessing the effects of changes to text. In fact a whole battery of procedures may be used as appropriate. I will discuss the measures used later but, for the present, I shall note that they cover search and retrieval tasks, and measures of reading speed, comprehension, performance and reader preferences.

Structural devices and illustrative materials

Access structures

Readers come to text with many different purposes: they need to be able to skim, to search, to look back, to look ahead, and to read in detail. They

INSULATING GLOVES

1.0   General

1.1   This section covers the description,
care and maintenance of insulating gloves
provided for the protection of workmen
against electric shock, and the precautions
to be followed in their use.

1.2   This section has been reissued to include
the D and E Insulating Gloves.

2.0   Types of Insulating Gloves

2.1   All types of insulating gloves are of the
gauntlet type and are made in four sizes:
9-1/2, 10, 11 and 12.
The size indicates the approximate number of
inches around the glove, measured midway
between the thumb and finger crotches.
The length of each glove, measured from
the tip of the second finger to the outer edge
of the gauntlet, is approximately 14 inches.

2.2   There are various kinds of insulating gloves.
The original ones are just called Insulating
Gloves.
After that B, C, D and E Insulating Gloves
were developed.
As described below, the D glove replaced
the original Insulating Gloves and
the E glove replaced the B and C gloves.

2.3   Insulating Gloves are thick enough to eliminate
the need for protector gloves and are intended
for use without them.
These Gloves have been superseded by
the D Insulating Gloves.

Figure 11.4. The same page with a revised typographical layout

may want to read the text themselves or to choose it for another person. Such readers need to be able to find their way around a text easily. This 'textual navigation' can be aided by good typographic practice and by the effective use of what Waller (1979) called *access structures*—devices which facilitate the readers' access to the text. Such access structures may be found at the beginning and at the end of texts (e.g., contents pages, glossaries, indexes, summaries) and embedded in the texts themselves (e.g., chapter indicators and titles at the tops of pages, page numbers, text headings, figure and table numbers and captions).

In evaluating a piece of text, one looks for the presence of devices such as these. It is assumed that it is easier to use a piece of text that has access structures than it is to use one that has not. Indeed, the little research that has been done does suggest that access structures are useful aids in text. It appears, for instance, that reading glossaries before a piece of technical text can help people's understanding of it (Black, 1987); that devices such as advance organizers and summaries can aid people's comprehension (Hartley and Trueman, 1982; Jonassen, 1982a); and that headings aid search and retrieval and, to a lesser extent, recall (Hartley and Jonassen, 1985; Hartley and Trueman, 1985). Of course, judges evaluating text have to bear in mind that there is no guarantee that because a device is provided it will always prove effective. Poor headings can mislead the reader (Swarts *et al.*, 1980), prior knowledge may be important (Wilhite, 1989; Lambiotte and Dansereau, 1992) some devices (such as boxed asides) can be distracting (Hartley, 1987b), and some texts may use these devices excessively (e.g., by providing a heading for every paragraph).

Illustrative materials

In this chapter I shall for convenience refer to tables, graphs, diagrams, examples and illustrations as 'illustrative materials'. I have listed in Table 11.1 the authors of more detailed summaries of research on each of these separate textual features. The points I wish to make here, however, are basically the same, whatever the particular feature being discussed. These points are as follows:

1. To judge the effectiveness of illustrative materials one needs to know about good practice.
2. Two main features to look for in all illustrative materials are *simplicity* and *clarity*.
3. The design and presentation of repeated illustrative materials should be *consistent*.
4. The positioning of illustrative materials is important: such materials should appear as soon as possible after their first textual reference and they should not get misplaced, or out of sequence.
5. Captions can be used to *instruct* as well as inform: captions can tell the readers what they are supposed to see rather than just label a table or a figure.

Figure 11.5 illustrates a contravention of point 3. Here, within the space of five pages, the authors used four different ways of presenting a graph. Such an approach can only be confusing for the reader. A good example of point 4 occurs in Cleveland's otherwise excellent book on graphs (Cleveland, 1985). At the beginning of Chapter One, each figure occurs not in its own text section but in the following one, thus all of the figures are out of step. Admittedly it is not always possible to follow my counsel of perfection— especially when there are several illustrations and little text, or when the

Table 11.1. Useful research reviews on how to present illustrative materials in printed and electronic text

	Books	Articles
Graphs	Tufte (1983) Cleveland (1985) Kosslyn (1994)	Macdonald-Ross (1977a) Macdonald-Ross (1977b) Bryant and Somerville (1986) Spence and Lewandowsky (1991) Wainer (1992)
Tables	Tufte (1983)	Ehrenberg (1977) Wright (1982) Norrish (1984) Hartley (1991) Wainer (1992)
Diagrams	Lowe (1993) Dwyer (1987)	Larkin and Simon (1987) Winn (1987) Winn (1990)
Illustrations	Fleming and Levie (1978) Dwyer (1985, 1987) Willows and Houghton (1987) Houghton and Willows (1987)	Levie and Lentz (1982) Alesandrini (1984) Anglin (1987)
Cartoons	Harrison (1980)	Sewell and Moore (1980) Bryant *et al.* (1981)
Examples		Mandl *et al.* (1984) Walczyk and Hall (1989)
Flow Charts	Wheatley and Unwin (1972)	Wright (1982) Guri-Rozenblit (1989) Michael and Hartley (1991)
Captions		Hertzog *et al.* (1989) Bernard (1990)

Useful general texts which cover many of these features are: Briscoe (1990), Easterby and Zwaga (1984), Hartley (1994a), Helander (1988), Horton (1991), Jonassen (1982b, 1985) and Misanchuk (1992).

illustrations are exceptionally large—but designers could try harder to minimize mismatches between text and illustration.

Again there are several ways of assessing the effectiveness of both access structures and illustrative materials and perhaps the most common of these is to carry out actual experiments. In my own research I have found this to be the most useful approach, although it is not without its limitations. Usually one cannot test a sufficient number of variants of the device being tested and, although it is relatively easy to show that revised materials may be easier to use than their original versions, it is harder to compare effectively subtle variations in the revisions.

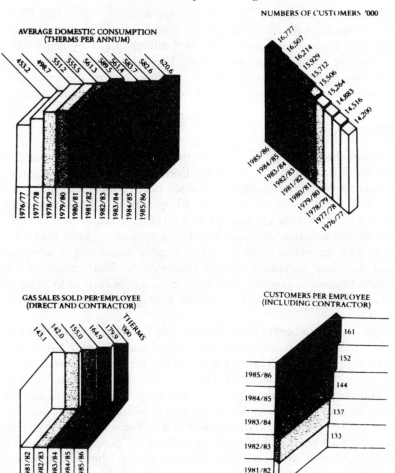

Figure 11.5. A series of bar charts from a promotional pamphlet. Note how the orientation of each chart changes. (The original charts were colour coded in shades of blue) (Figure reproduced with permission)

The suitability of the text

Perhaps the most common question asked about a piece of text concerns how suitable or appropriate it is for its intended readers. Sometimes, when the readership is well known, it is easy to arrive at an answer. However, if the text is to be used by multiple users for a variety of different reasons, then the task becomes more difficult.

Several methods can be used to evaluate the suitability of a text for its intended audience. Most measures can be used by *authors* when they are producing their own text (to ensure that it is effective) and by *judges* when they are assessing the suitability of published text for others. It is perhaps more common for most of these measures to be used in the latter case, especially when the judges want to revise or improve the published text.

Readability formulae

Most measures of reading difficulty require the author or the judge to employ readers of one kind or another, preferably from the text's target population. There is one approach, however, which does not require readers, and this is to apply one of the many measures of reading difficulty. These readability formulae aim to predict the reading age that a reader needs to have in order to understand the text that the formula is being applied to. Most readability formulae in fact are not as accurate in this respect as one might wish, but the figures that they provide do give a rough guide. Furthermore, if one uses the same formula to compare two different pieces of text, or to compare an original and a revised version, then one does get a good idea of relative difficulty.

There are several readability formulae available (see Harrison, 1980; Klare, 1974–5) but each one typically combines, with a constant, two main measures: the average sentence length of samples of the text and the average word length in these samples (usually considered in terms of the number of syllables). The basic underlying idea is that the longer the sentences and the more complex the vocabulary, then the more difficult the text will be. Clearly this notion, whilst generally sensible, has its limitations. Some technical abbreviations are short (e.g., DNA) but difficult for people who have not heard of them. Some words are long but, because of their frequent use, become quite familiar (e.g., ergonomics in this context). Word, sentence and paragraph order is not taken into account by the formulae and nor are the effects of other devices used to aid comprehension such as typographical layout, access structures and illustrative materials. Also, most importantly, the readers' motivation and prior knowledge are ignored. Davison and Green (1987) provide other, more technical criticisms.

A readability measure then provides a *rough* guide to text difficulty. If the score goes off the scale (as is often the case with government documents) then it is clear that the text is too difficult for most readers. Several studies have been carried out which demonstrate the advantages of producing more readable text (see Hartley, 1981a, 1994b; US Department of Commerce, 1984; Cutts and Maher, 1986); but it is true to say that in the majority of these studies more than just the readability of the text has been manipulated. (Revised layouts, new illustrative materials and wholesale deletions are fairly common in such reports.)

Some readability formulae are quite complex to calculate by hand (e.g., the Flesch Reading Ease Index) but there are tables available to ease the calculations (see Hartley, 1994a, p. 52). Other formulae are simpler to calculate. One of the simplest, the Gunning Fog Index, is as follows:

(a) take a sample of 100 words;
(b) calculate the average number of words per sentence in the sample;
(c) count the number of words with three or more syllables in the sample;
(d) add the average number of words per sentence to the total number of words with three or more syllables; and
(e) multiply the result by 0·4.

The answer that one gets is the reading grade level as used in American schools. (Grade 1 = 6 years old, Grade 2 = 7 years old, etc.) One can add 5 to this answer to obtain an equivalent British reading age (if you think that British and American children's schooling is similar).

Today, with the advent of word processing systems, it is easier to apply more complex readability tests. For example, the Style and Diction programs of the IBM Xenith text formatting system can be applied to text to provide sets of readability data, and various other measures such as the number of sentences used and the average sentence and word lengths. Table 11.2 shows the output obtained when this program was run on the first 20 sentences of this chapter. It can be seen that the four readability formulae are roughly in agreement, and that the general picture suggests that the text is suitable for 16–19 year olds. Other kinds of computer program are probably more useful for writers and revisers. Table 11.3 lists a sample of the measures available in *Grammatik 5,* a programme which is available in both British and US versions. Studies of writers, using such programs have indicated that users like them, but their value seems less certain (Hartley, 1993).

Table 11.2. Output from the Style program of the IBM Xenith Text Formatting System applied to the first 20 sentences of this chapter

Number of words	471
Average sentence length	23·5
Average word length	4·75
Percentage of short sentences (<19 words)	40%
Percentage of long sentences (>34 words)	20%
Longest sentences (No. 4)	44 words
Sentence types:	
simple	40%
complex	30%
compound	15%
compound–complex	15%
Readability formulae (US grade levels):	
Coleman–Liau	10·9
Kincaid	12·5
Auto	12·7
Flesch	13·4

Table 11.3. A sample of computer programs available in Grammatik 5 from Apple Macintosh

Programs which indicate grammatical errors:

- Adjective Errors
- Adverb Errors
- Article Errors
- Clause Errors
- Comparative/Superlative Use
- Double Negatives
- Incomplete Sentences
- Noun Phrase Errors

- Object of Verb Errors
- Possessive Misuse
- Preposition Errors
- Pronoun Errors
- Sequence of Tense Errors
- Subject-Verb Errors
- Tense Changes
- etc.

Programs which indicate mechanical errors:

- Spelling Errors
- Capitalization Errors
- Double Word
- Ellipsis Misuse
- End of Sentence Punctuation
  Errors

- Incorrect Punctuation
- Number Style Errors
- Question Mark Errors
- Quotation Mark Misuse
- Similar Words
- Split Words
- etc.

Programs which indicate stylistic errors:

- Long Sentences
- Wordy Sentences
- Passive Tenses
- End of Sentence Prepositions
- Split Infinitives
- Clichéd Words/Phrases
- Colloquial Language

- Americanisms
- Archaic Language
- Gender-Specific Words
- Jargon
- Abbreviation Errors
- Paragraph Problems
- Questionable Word Usage
- etc.

Cloze procedures

Another measure, similar in some respects to a readability formula, but this time requiring readers, is the cloze procedure (Taylor, 1953; Rye, 1982). With this technique samples of passages are presented with every nth word missing, and readers are required to fill in the missing gaps. Technically speaking, if say every 6th word is deleted, then six versions should be prepared with the gaps each starting from a different point. However, it is more common ------ prepare one version, and perhaps ------ to focus the gaps on ------ words. Whatever the procedure, the ------ are scored either by marking ------ correct those responses which directly ------ what the original author actually ------, or by accepting these and ------ synonyms. Since the two scoring methods correlate highly, it is more objective to use the tougher measure of matching words. (In the case above these are: to, even, important, passages, as, match, said, acceptable.) Scores can be increased slightly by varying the length of the lines to match the lengths of the missing words, providing dashes to match the number of letters missing in each word, or by providing the first of the missing letters (Hartley and Trueman, 1986). These minor variations, however, do not affect the main purpose of this measure which is to assess the reader's comprehension of the text, and, by inference, its difficulty. The cloze procedure has a strong advantage over readability formulae in that it can be used to assess the effects of the presence of other features (such as illustrations or underlining) on the comprehension of text (e.g., see Hartley *et al.*, 1980a; Newton, 1983; Reid *et al.*, 1983).

Readers' judgements

A rather different, but useful measure of text difficulty is to ask readers to judge it for themselves. One simple procedure is to ask readers to circle on the text those areas, sentences, or words that they think *readers less able than themselves* will find difficult. In my experience if you ask readers to point out difficulties for *others*, respondents are much more forthcoming than if you ask them to point out their own difficulties. An elaboration of this technique, of course, is to use *protocol analysis* (see chapter 7 by Bainbridge and Sanderson in this book). Here readers are asked to verbalize what they are thinking about as they are reading or using text. This technique has proved extremely useful in evaluating instructional manuals (see e.g., Swaney *et al.*, 1981; Sullivan and Chapanis, 1983; Komoski and Woodward, 1985). Some critics have suggested that talking about the task whilst trying to do it can cause difficulties: such problems can be partly overcome by getting readers to work in pairs and discussing together the problems they are facing (Miyake, 1986) or by videotaping the procedure and then getting respondents to talk through the resulting tape (Schumacher *et al.*, 1984).

In this section of the chapter we may also consider *reader preferences*. When I discussed preference measures in the first edition of *Designing Instructional Text,* I dismissed their value because I considered preferences to be 'untutored judgements'. I argued that people's preferences can be based on inappropriate considerations: for example, a person might prefer a car on the basis of its colour rather than its technical quality. Today I am not so dismissive of preference measures. My own studies have shown that such measures are sensitive to differences between novices and experts, and that they can be used to assess the effects of instruction (Hartley, 1981b; Hartley and Trueman, 1981). Furthermore, people have clear views about what they like in texts, and how they expect texts to perform (Wright and Threlfall, 1980). So, first impressions might colour one's attitudes to texts. A text which looks dull, dense and turgid is not going to encourage readers, no matter how important the content. So I now use preference judgements as one of a battery of measures. Preferences can provide additional information—they can tell you whether a revised text is preferred to the original, whether people see no difference, or whether people prefer the original even though (in your eyes) it is not as effective. Such information needs to be considered, but along with other measures. (See also chapter 10 by McClelland in this book for a view on incorporating preference measures into a series of evaluations.)

One useful tool to use here if you require preference rankings for a number of say typographic designs which vary in different ways is the method of *paired comparisons*. Suppose, for example, you have 15 designs. You could ask participants to judge them (overall, or on a particular aspect) and to make paired comparisons. Essentially, this involves each judge comparing design 1 with 2 and recording the preference, then 1 with 3, 1 with 4, 1 with 5 and so on, until 1 with 15 is reached. Then the judge starts again, this time comparing 2 with 3, 2 with 4, 2 with 5 and so on until 2 with 15. This procedure is replicated in terms of 3 with 4, 5, 6, etc., 4 with 5, 6, 7, etc., until all the designs have been systematically compared. Finally, the number of recorded preferences for each design is totalled to see which one has been preferred the most often. Reliability can be assessed by asking people to do the task again backwards. Examples illustrating this approach can be found in Hartley *et al.* (1979) and Hartley (1979–80). (See chapter 3 for discussion of rating and ranking generally.)

Writing and revising text

Before moving on to consider experimental methods of evaluating text, it is perhaps appropriate at this stage to say something about techniques for producing readable writing. Hayes and Flower (1986) argue that writing can be considered as a complex activity or skill, and that like all skills, writing is made up of subcomponents which have to be put together in such a way to guarantee smooth performance. Many of these subcomponents are organized hierarchically and some have priority over others, but even during the writing of a simple sentence, the writer shifts attention from one subcomponent to another. Thus, for example, the following issues competed for my attention whilst I was writing this particular paragraph.

Who is this chapter for? Will they be involved in writing or revising?

Where should I put this bit? Should it come before or after the section on readability?

Will readers understand what I mean by subcomponents and such components being hierarchically organized? Should I explain that one has *long-term* goals—deciding what to put in the chapter; *medium-term* goals—deciding where to put it; *ongoing* goals—deciding how to say it; *specific* problems—deciding how to spell hierarchically; and *immediate* goals—writing clearly so my secretary can read what I have written.

When writing is considered in this way it is possible to suggest that in order to improve writing, writers need (a) to practise, and (b) to focus on different goals at different times. Thus, for example, one can first concentrate on getting the material down in a rough and ready fashion and then one can work on polishing it. As Wason (1983) put it, 'First say it, and *then* try to say it well'.

There are numerous guidelines on how to write clear text (see e.g., Klare, 1979; Hartley *et al.*, 1980b; Armbruster and Anderson, 1985) and also on how to revise published text. In my own work with secondary school children I use the guidelines for revision listed in Figure 11.6. These guidelines are based upon current work on the revising process (see Hayes and Flower, 1986). Noticeable changes occur when procedures such as these are applied to published text. Figure 11.7 provides an illustration. Here the text presented in Figures 11.3 and 11.4 has been revised using guidelines similar to those provided in Figure 11.6.

The advent of word processing has, of course, helped in this respect. It is now much easier to revise text than it was. Studies indeed suggest that writers using word processors spend less time planning and more time revising than they did before (see Hartley, 1993).

1. Read the text through quickly.
2. Read the text through again but this time ask yourself:
- *Who is the text for?*
- *What is the writer trying to do?*
3. Read the text through again, but this time ask yourself:
- *What changes do I need to make to help the reader? How can I make the text easier to follow?*
4. To make these changes you may need:
- *to make big or* global *changes (e.g., re-write sections yourself); or*
- *to make small or minor* text *changes (e.g., change slightly the original text).*
You will need to decide whether you are going to focus first on global changes or first on text changes.
5. Global *changes you might like to consider in turn are:*
- *re-sequencing parts of the text;*
- *re-writing sections in your own words;*
- *adding in examples;*
- *changing the writer's examples for better ones;*
- *deleting parts that seem confusing.*
6. Text *changes you might like to consider in turn are:*
- *using simpler wording;*
- *using shorter sentences;*
- *using shorter paragraphs;*
- *using active rather than passive tenses;*
- *substituting positives for negatives;*
- *writing sequences in order;*
- *spacing numbered sequences or lists down the page (as here).*
7. Keep reading to see if you want to make any more global or text changes.
8. Finally repeat this whole procedure some time after making your initial revisions (say 24 hours) and do it without looking back at the original text.

Figure 11.6. My guidelines for revising text

INSULATING GLOVES

1.0   General

1.1   This section describes how to care for
and maintain the insulating gloves
that will protect you from electric shocks.

1.2   The section has been revised to include
the D and E Insulating Gloves.

2.0   Types of Insulating Gloves

2.1   All insulating gloves are made
in the gauntlet style.
There are four sizes: 9½, 10, 11, 12.
The size indicates the approximate number
of inches around the glove across the palm.
Each glove is about 14 inches long
from the bottom of the gauntlet to the top
of the second finger.

2.2   There are various kinds of insulating gloves.
The first kind were originally just called
Insulating Gloves.
After that the B, C, D and E Insulating Gloves
were developed.
As described below, the D glove replaced the
original insulating gloves, and the E glove
replaced the B and C gloves.

2.3   So **Insulating Gloves** have now been replaced
by D Insulating Gloves.
(Insulating Gloves could be worn without
protector gloves.)

Figure 11.7. The same material as that shown in Figure 11.3, only this time both the typography *and* the text have been revised

Experimental comparisons

The methods described above can be used relatively informally: some researchers, however, may wish to carry out more formal experimental comparisons, and to use more precise measuring instruments. Some people, for instance, might be interested in measuring reading speed, ease of location and retrieval, ease of use, or ease of comprehension and recall. Some methods might be quite precise involving, for example, the use of eye-movement recording devices, or videotapes of readers using instructional texts.

Over the last ten years or so my colleagues and I have used a variety of measures (and combinations of them) to evaluate instructional text. Table 11.4 tries to encapsulate some of the strengths and some of the limitations of these measures which I have divided into five main groups. Opposite these five blocks are my estimates of the reliability of these measures and some

Table 11.4. Methods of evaluation in assessing text, and estimates of their reliability

Method	Estimated reliability	Test-retest correlations obtained in our studies	Comments
Reading aloud	High	0·78 to 0·95	Insensitive to differences in layouts
Reading aloud (text upside down)	Very high	0·87 to 0·99	Slows reading right down; larger spreads of scores with males than females
Scanning technical material*	High	0·75 to 0·91	Good for technical materials: sensitive
Scanning prose* (short intervals)	Moderate	0·49 to 0·68	Moderately useful and sensitive
Scanning prose* (wide intervals)	Low	0·36 to 0·83	Poor: searchers 'get lost'
Silent reading speed (without test)	Fairly high	0·53 to 0·96	The researcher does not know what has been read
Silent reading speed (with test to follow)	Fairly high	0·70 to 0·82	Knowledge of forthcoming test slows readers down markedly
Comprehension† (cloze test)	Fairly high	0·55 to 0·95	Useful for assessing relative difficulty
Comprehension (recall questions)	Low	0·46 to 0·73	Too specific to make comparisons with, if different materials are used
Preferences	Fairly high	0·72	Useful as an additional measure
Readability formulae	Very high	(not applicable)	Useful as an additional measure

*Scanning involves providing the reader with a list of phrases drawn from the text, each with a word missing. The reader has to scan (or skim) the text, find the phrase, and write in the missing word.
†The cloze test involves providing the reader with a text with, say, every sixth word missing. The reader has to supply the missing words.

comments about them. Table 11.4 shows that some measures are more reliable than others, and common sense indicates that some measures are more suitable than others for different purposes. Thus oral reading measures give detailed information about specific reading difficulties; search and retrieval tasks are appropriate for evaluating the layout of highly structured text; comprehension measures are more appropriate for evaluating the effectiveness of continuous prose; and readability and preference measures are useful as additional sources of information.

Experimental evaluations of text require one to start off with a specific problem, to prepare a variety of solutions, to select the ones that seem most plausible, and then to evaluate their effectiveness. The evaluation will always be limited because (1) it is not possible to evaluate and compare every possible solution to a problem, (2) the methods which we have at our disposal for evaluation in this area tend to be somewhat limited, and (3) different measures have their own in-built assumptions. (For example, most of the techniques listed in Table 11.4 seem to assume that the reader starts at the beginning of a text and reads through it steadily to the end.) None the less, these experimental comparisons do provide evidence which is of value. And, this is especially the case when several measures are combined (Hartley, 1994b; Wright, 1987).

The following case history demonstrates that combined measures are more informative than single ones. In this study comparisons were made between full length (three page) versions of Figure 11.3 and Figure 11.7. The results were as follows:

1. In terms of *readability* the first 100 words of the original document (excluding headings) had a reading age level of 19·5 years, whereas the first 100 words of the revised version had a reading age level of 15 years (Gunning FOG index).
2. In terms of *reading speed* there was no significant difference between the average times taken by two groups of ten university students to read the three pages set in either version.
3. In terms of *factual recall* however, these same groups of students recalled an average of 5·4 out of 10 for the original and 7·9 out of 10 for the revised version. (This difference was statistically significant.)
4. In terms of *preferences*, seven out of ten university colleagues chose the revised version in preference to the original when asked to judge which figure they found 'the clearer'. (This difference, whilst pleasing, was not statistically significant.)
5. When the reading speed and factual recall measures were used again in a *replication* study with a further 20 students, the results were repeated almost exactly.

This composite picture allows me to suggest that the revised version was easier to read, that the students extracted more information per unit time, and that judges were more likely to prefer the revised version.

We may finally note in this section that a different kind of measure, not listed in Table 11.4, is that of the *cost* of production. In our experiments we have sometimes shown reductions in the cost of production without any loss in comprehension and sometimes great improvements in comprehension for a slight increase in cost (see Hartley, 1994a). Results such as these point to the hidden costs of badly designed materials and to the fact that *cost effectiveness* is an important consideration. (These issues are expanded on in chapter 34 of this book.)

Cyclical testing and revising

Finally I want to point out that as far as an ergonomics approach is concerned, we may not be interested so much in making *comparisons* between texts, as in using the measures available to help us to *improve* a text. One of the most useful approaches here is that of cyclical testing and revising. This approach requires us to test the text with appropriate readers, revise it on the basis of the results obtained, test it again with another set of readers, revise it again and so on, until the text achieves its objectives. This iterative approach was used extensively in a study reported by Waller (1984) that described how he and his colleagues at the Open University set about improving a form that was to be used by unemployed people claiming supplementary benefit in the UK.

A prototype form was developed and piloted by the Department of Health and Social Security. The form was small in format (165 × 204 mm) with eight pages organized as a folded concertina (see Waller, 1984). Although the respondents found the form attractive to look at, they found it difficult to use. About 75% of the forms were completed unsatisfactorily in one way or another. This meant that the forms had to be returned and/or respondents followed up in some way before an assessment of benefit could be made and that this was a very expensive procedure.

The main sources of error in the prototype form appeared to be:

1. Problems of relevance and contextual interpretation: the form did not elicit enough information for an assessment to be made, and appeared irrelevant to a large number of claimants.
2. Problems of reading sequence: many sections did not apply to many claimants, but the form gave inadequate directions concerning which parts were to be completed.
3. Problems in graphic design: poor design practice also contributed to the problems of sequencing.

The problems of the form ranged from the fairly obvious to the subtle and the debatable. In order to redesign the form the prototype was first tested with small groups of appropriate respondents. (Interestingly, part of this assessment included the use of an eye movement recorder to assess which pieces of text were read and in what order.) The aim of this first assessment was to isolate the main causes of difficulty, and to collect data against which the redesigned forms could be compared. It appeared from this first testing that the form was not asking the right questions to gather the information that the civil servants needed. In addition, many questions were ambiguous.

Thus a redesigned version was prepared with emphasis on revising the language and the sequencing of the form: thus the content was well spaced and simply designed so that the confusions brought about by poor design practice (noted earlier) could be avoided at this stage. The revisions focused on making the branching instructions more explicit, and on giving users clearer instructions when they first encountered such a branching instruction.

This redesigned version was tested (again with small groups of appropriate respondents). It was clear that improvements had been made, but that more could be done. So a third version was then prepared. Headings for the different sections were added, and the routing instructions were further improved. The testing of this third version showed that this had solved most of the problems. Thus a fourth and final version was prepared. This version used colour coding for the main headings (earlier versions had been in black and white), a larger page-size (200 × 330 mm) and yet another re-sequenced order. Once the logical and linguistic problems were sorted out, the typography could be enhanced.

This final version was tested with larger groups of appropriate respondents. The results now indicated that about 75% of the forms were completed satisfactorily (as opposed to the original 25%). Today, after further revisions, the successful completion rate is estimated to be over 80%. These impressive results have led to massive cost benefits for the Department of Health and Social Security.

Summary

This chapter has discussed the evaluation of texts from four different viewpoints:

1. Evaluating content.
2. Evaluating typographical layout.
3. Evaluating structural devices and illustrative materials.
4. Evaluating the suitability of the text.

In each section a variety of methods for assessing text has been mentioned. Content can best be evaluated by what might be called survey techniques. Typographical layout, structural devices and illustrative materials can be evaluated both by survey techniques and by laboratory and field experiments. The suitability of text can be assessed by applying readability formulae, but such an approach would seem unduly restrictive when there are other more useful survey, field and experimental approaches. In many cases the aim will be to improve the text. Thus those approaches that emphasize combining methods and those approaches that use a test-revise-retest model are likely to be the most helpful for ergonomists.

Acknowledgements

I am grateful to colleagues who commented on earlier drafts of this chapter, and to Margaret Woodward and Dorothy Masters for their repeated processing of the text.

References

Alesandrini, K.L. (1984). Pictures and adult learning. *Instructional Science*, **13**, 63–77.

Anglin, G.J. (1987). Effects of pictures on recall of written prose: how durable are picture effects? *Educational Communication and Technology Journal*, **35**, 25–30.

Armbruster, B.B. and Anderson, T.H. (1985). Producing 'considerate' expository text: or easy reading is damned hard writing. *Journal of Curriculum Studies*, **17**, 247–274.

Bernard, R.M. (1990). Using extended captions to improve learning from instructional illustrations. *British Journal of Educational Technology*, **21**, 215–225.

Black, A. (1987). Lexical support in discourse comprehension. Unpublished Ph.D. Thesis, University of Cambridge.

Briscoe, M.H. (1990). *A Researcher's Guide to Scientific and Medical Illustrations* (New York: Springer-Verlag).

Bryant, J., Brown, D., Silberg, A.R. and Elliot, S.C. (1981). Effects of humorous illustrations in college textbooks. *Human Communication Research*, **8**, 43–47.

Bryant, P.E. and Somerville, S.C. (1986). The spatial demands of graphs. *British Journal of Psychology*, **77**, 187–197.

Cleveland, W.S. (1985). *The Elements of Graphing Data* (Monterey, CA: Wadsworth).

Cutts, M. and Maher, C. (1986). *The Plain English Story* (Whaley Bridge, Stockport: Plain English Campaign).

Davison, A. and Green, G. (Eds) (1987). *Linguistic Complexity and Text Comprehension: A re-examination of Readability with Alternative Views* (Hillsdale, NJ: Erlbaum).

Dwyer, F.M. (1985). *Strategies for Improving Visual Learning* (Learning Services: Box 784, State College, Pennsylvania 16804).

Dwyer, F.M. (1987). *Enhancing Visual Information: Recommendations for Practitioners* (Learning Services: Box 784, State College, Pennsylvania 16804).

Easterby, R. and Zwaga, H. (Eds) (1984). *Information Design* (Chichester: John Wiley).

Ehrenberg, A.S.C. (1977). Rudiments of numeracy. *Journal of the Royal Statistical Society A*, **140**, 227–297.

Farr, R. and Tulley, M.A. (1985). Do adoption committees perpetuate mediocre textbooks? *Phi Delta Kappan*, **66**, 467–471.

Fleming, M. and Levie, W.H. (1978). *Instructional Message Design* (Englewood Cliffs, NJ: Educational Technology Publications).

Gunning, R. (1968). *The Technique of Clear Writing* (New York: McGraw Hill).

Guri-Rozenblit, S. (1989). Effects of a tree diagram on students' comprehension of main ideas in an expository text with multiple themes. *Reading Research Quarterly*, **24**, 236–247.

Harrison, C. (1980). *Readability in the Classroom* (Cambridge: Cambridge University Press).

Harrison, R.P. (1981). *The Cartoon: Communication to the Quick* (Beverly Hills: Sage).

Hartley, J. (1979–80). Designing journal content pages: the role of spatial and typographic cues. *Journal of Research Communication Studies*, **2**, 83–98.

Hartley, J. (1981a). Eighty ways of improving instructional text. *IEEE Transactions on Professional Communication*, **PC-24**, 17–27.

Hartley, J. (1981b). Sequencing the elements in references. *Applied Ergonomics*, **12**, 7–12.

Hartley, J. (1987a). Designing electronic text: the role of print based research. *Educational Communication and Technology Journal*, **35**, 3–17.

Hartley, J. (1987b). Typography and executive control processes in reading. In *Executive Control Processes in Reading,* edited by B.K. Britton and S.M. Glynn (Hillsdale, NJ: Erlbaum).

Hartley, J. (1991). Tabling information. *American Psychologist*, **46**, 655–656.

Hartley, J. (1993). Writing, thinking and computers. *British Journal of Educational Technology*, **19**, 4–16.

Hartley, J. (1994a). *Designing Instructional Text* (3rd edition) (London: Kogan Page).

Hartley, J. (1994b). Three ways to improve the clarity of journal abstracts. *British Journal of Educational Psychology*, **64**, 331–343.

Hartley, J. and Jonassen, D.H. (1985). The role of headings in printed and electronic text. In *The Technology of Text,* Volume 2, edited by D.H. Jonassen (Englewood Cliffs, NJ: Educational Technology Publications).

Hartley, J. and Trueman, M. (1981). The effects of changes in layout and changes in wording on preferences for instructional text. *Visible Language*, **XV**, 13–31.

Hartley, J. and Trueman, M. (1982). The effects of summaries on the recall of information from prose: five experimental studies. *Human Learning*, **1**, 63–82.

Hartley, J. and Trueman, M. (1985). A research strategy for text designers: the role of headings. *Instructional Science*, **14**, 99–155.

Hartley, J. and Trueman, M. (1986). The effects of the typographic layout of cloze-type tests on reading comprehension scores. *Journal of Research in Reading*, **9**, 116–124.

Hartley, J., Trueman, M. and Burnhill, P. (1979). The role of spatial and typographic cues in the layout of journal references. *Applied Ergonomics*, **10**, 165–169.

Hartley, J., Bartlett, S. and Branthwaite, J.A. (1980a). Underlining can make a difference—sometimes. *Journal of Educational Research*, **73**, 218–224.

Hartley, J., Trueman, M. and Burnhill, P. (1980b). Some observations on producing and measuring readable writing. *Programmed Learning and Educational Technology*, **17**, 164–174.

Hayes, J.R. and Flower, L.S. (1986). Writing research and the writer. *American Psychologist*, **41**, 1106–1113.

Helander, M. (Ed.) (1988). *Handbook of Human–Computer Interaction* (Amsterdam: North Holland).

Hertzog, M., Stinson, M.S. and Keiffer, R. (1989). Effects of caption modification and instructor intervention on comprehension of a technical film. *Educational Technology Research & Development*, **37**, 2, 59–68.

Horton, W. (1991). *Illustrating Computer Documentation: The Art of Presenting Information Graphically on Paper and Online* (New York. Wiley).

Houghton, H.A. and Willows, D.M. (Eds) (1987). *The Psychology of Illustration,* Volume 2. *Instructional Issues* (New York: Springer).

Jonassen, D.H. (1982a). Advance organisers in text. In *The Technology of Text* edited by D.H. Jonassen (Englewood Cliffs, NJ: Educational Technology Publications).

Jonassen, D.H. (Ed.) (1982b). *The Technology of Text* (Englewood Cliffs, NJ: Educational Technology Publications).

Jonassen, D.H. (1985). *The Technology of Text Volume 2* (Englewood Cliffs, NJ: Educational Technology Publications).

Klare, G.R. (1974–5). Assessing readability. *Reading Research Quarterly*, **X**, 62–102.

Klare, G.R. (1979). Writing to inform: making it readable. *Information Design Journal*, **1**, 98–105.

Komoski, P.K. and Woodward, A. (1985). The continuing need for learner verification and revision of textual material. In *The Technology of Text, Volume 2*, edited by D.H. Jonassen (Englewood Cliffs, NJ: Educational Technology Publications).

Kosslyn, S.M. (1994) *Elements of Graph Design* (New York: Freeman).

Lambiotte, J.G. and Dansereau, D.F. (1992). Effects of knowledge maps and prior knowledge on recall of science lecture content. *Journal of Experimental Education*, **60**, 189–201.

Larkin, J.H. and Simon, H.A. (1987). Why a diagram is (sometimes) worth ten thousand words. *Cognitive Science*, **11**, 65–99.

Levie, W.H. and Lentz, R. (1982). Effects of text illustrations: a review of research. *Educational Communication and Technology Journal*, **30**, 195–232.

Lowe, R. (1993). *Successful Instructional Diagrams* (London: Kogan Page).

Macdonald, N.H. (1983). The Unix writer's workbench software: rationale and design. *Bell System Technical Journal*, **62**, 1891–1908.

Macdonald-Ross, M. (1977a). How numbers are shown: a review of research on the presentation of quantitative data in texts. *Audiovisual Communication Review*, **25**, 359–409.

Macdonald-Ross, M. (1977b). Graphics in text. In *Review of Research in Education,* Volume 5, edited by L.S. Shulman (Itasca, IL: Peacock).

Mandl, H., Schnotz, W. and Tergan, S.O. (1984). On the function of examples in unstructured texts. Paper available from the authors at Deutches Institut fur Fernstudien an der Universitat Tubingen, Hauptbereich Forschung, Bei der Fruchtschranne 6, 7400 Tubingen 1.

Michael, D.E. and Hartley, J. (1991). Extracting information from flow charts and contingency statements: the effects of age and practice. *British Journal of Educational Technology*, **22**, 84–98.

Misanchuk, E.R. (1992). *Preparing Instructional Text: Document Design Using Desktop Publishing* (Englewood Cliffs, NJ: Educational Technology Publications).

Miyake, N. (1986). Constructive interaction and the iterative process of understanding. *Cognitive Science*, **10**, 151–177.

Newton, L.D. (1983). The effect of illustrations on the readability of some junior school textbooks. *Reading*, **17**, 43–54.

Norrish, P. (1984). Moving tables from paper to screen. *Visible Language*, **XVIII**, 154–170.

Reid, D.J., Briggs, N. and Beveridge, M. (1983). The effect of picture upon the readability of a school science topic. *British Journal of Educational Psychology*, **53**, 327–335.

Rye, J. (1982). *Cloze Procedure and the Teaching of Reading* (London: Heinemann Educational).

Schumacher, G.M., Klare, G.R., Cronin, E.C. and Moses, J.D. (1984). Cognitive activities of beginning and advanced college writers: a pausal analysis. *Research in the Teaching of English*, **18**, 169–187.

Sewell, E.H. and Moore, R.L. (1980). Cartoon embellishments in informative presentations. *Educational Communication and Technology Journal*, **28**, 39–46.

Shebilske, W.L. and Rotondo, J.A. (1981). Typographical and spatial cues that facilitate learning from textbooks. *Visible Language*, **15**, 45–54.

Sless, D. (1984). Visual literacy: a failed opportunity. *Educational Communication and Technology*, **32**, 224–228.

Spence, I. and Lewandowsky, S. (1991). Displaying proportions and percentages. *Applied Cognitive Psychology*, **5**, 61–77.

Sullivan, M.A. and Chapanis, A. (1983). Human factoring, a text editor manual. *Behaviour and Information Technology*, **2**, 113–125.

Swaney, J.H., Janik, C.J., Bond, S.J. and Hayes, R. (1981). *Editing for Comprehension: Improving the Process Through Reading Protocols.* Technical Report No. 14. Document Design Center, Carnegie-Mellon University, Pittsburgh, PA.

Swarts, H., Flower, L.S. and Hayes, J.R. (1980). How headings in documents can mislead readers. Paper available from the authors, Department of Psychology, Carnegie-Mellon University, Pittsburgh, PA, 15213, U.S.A.

Taylor, W.I. (1953). Cloze procedure: a new tool for measuring readability. *Journalism Quarterly*, **30**, 415–433.

Tufte, E.R. (1983). *The Visual Display of Quantitative Information* (Cheshire, CT: Graphics Press).

Tukey, J. (1977). *Exploratory Data Analysis* (New York: Addison Wesley).

U.S. Department of Commerce (1984). *How Plain English Works for Business: 12 Case Studies* (Washington: Office of Consumer Affairs, US Department of Commerce).

Wainer, H. (1992). Understanding tables and graphs. *Educational Researcher*, **21**, 12–23.

Walczyk, J.J. and Hall, V.C. (1989). Effects of examples and embedded questions on the accuracy of comprehension self-assessments. *Journal of Educational Psychology*, **81**, 435–437.

Waller, R.H.W. (1979). Typographic access structures for instructional text. In *Processing of Visible Language,* edited by P.A. Kolers, M.E. Wrolstad, and H. Bouma (New York: Plenum).

Waller, R. (1984). Designing a government form: a case study. *Information Design Journal*, **4**, 36–57.

Wason, P.C. (1983). Trust in writing. Paper available from the author. Department of Phonetics and Linguistics, University College, London.

Wheatley, D.M. and Unwin, A.W. (1972). *The Algorithm Writer's Guide* (London: Longman).

Wilhite, S.C. (1989). Headings as memory facilitators: the importance of prior knowledge. *Journal of Educational Psychology*, **81**, 115–117.

Willows, D.M. and Houghton, H.A. (Eds) (1987). *The Psychology of Illustration*, Volume 1, *Basic Research* (New York: Springer).

Winn, W. (1987). Using charts, graphs and diagrams in educational materials. In *The Psychology of Illustration*, Volume 1, *Basic Research*, edited by D.M. Willows and H.A. Houghton (New York: Springer).

Winn, W.D. (1990). A theoretical framework for research in learning from graphics. *The International Journal of Educational Research*, **14**, 553–564.

Wright, P. (1982). A user-oriented approach to the design of tables and flow charts. In *The Technology of Text*, edited by D.H. Jonassen (Englewood Cliffs, NJ: Educational Technology Publications).

Wright, P. (1987). Issues of content and presentation in document design. In *Handbook of Human–Computer Interaction*, edited by M. Helander (Amsterdam: North-Holland).

Wright, P. and Threlfall, M.S. (1980). Readers' expectation about format influence on the usability of an index. *Journal of Research Communication Studies*, **2**, 99–106.

Zimet, S.G. (1976). *Print and Prejudice* (London: Hodder and Stoughton).

Chapter 12

Evaluation of human–computer interaction at the user interface to advanced IT systems

Bruce Christie, Robert Scane and Jenny Collyer

Introduction

This chapter is concerned with evaluation of the interaction between humans and IT (Information Technology) systems used in work and work-related contexts. By 'IT systems' is meant systems based on computers and telecommunications. The chapter follows on from a chapter by Christie and Gardiner (1990), in an earlier edition of this book, being now updated and modified in scope to take account of developments which have taken place in the field since then. It focuses in particular on the use of personal computers and workstations in the context of office and business systems, but the points made are relevant to a much broader range of domains in which IT systems are used. Office and business systems are taken as a focus because this is the largest sector in which IT is used, in terms of the number of people employed, and because it is in this area that some of the most interesting developments in the user-interface have taken place over the past five to ten years, often driving developments in other application domains.

The issues in evaluation discussed in the following sections need to be considered against the background of changes which have been taking place in the systems supporting work in businesses and other enterprises as well as in the home and other contexts. These changes provide an evolving technology and application framework within which both the theory and methods of human-computer interaction need to be further developed in order to be of as much practical value as possible in the context of advanced systems research and development. The key technology trends include movement towards:

- distributed systems running on inter-connected networks

- greater information richness, especially multimedia
- support to autonomous, mobile users who may be physically distant from their organizations
- modular, object-oriented, integrated software
- advanced architectures capable of supporting cooperative work
- greater system intelligence.

These trends and their implications for the evaluation of human-computer interaction are considered in more detail in relation to recent strategic research and development within the IT industry at the end of the chapter. The following sections need to be interpreted within this context of rapid evolution in the technology and application domain.

Theoretical perspectives

The developments outlined above and considered in more detail at the end of the chapter underline the importance of considering different theoretical perspectives in the evaluation of human-computer interaction. Perspectives which may have served well when the systems concerned were at an earlier stage of evolution, where office automation was often seen as more or less synonymous with word processing, need to be considered in relation to alternative perspectives which provide additional insights into the nature of work in modern IT environments. Christie and Gardiner (1990) described four main perspectives, as follows.

- the cognitive
- the social psychological
- the organizational
- the psychophysiological.

As the technology and markets for IT continue to evolve, new kinds of systems are invented. Evaluating these may benefit from drawing on insights offered by additional perspectives. At the time of writing, developments in the area of multimedia and computer support for cooperative work, among others, suggest the potential usefulness of a communications perspective on human-computer interaction, and so we may add to the list above:

- the communications perspective.

These five perspectives on work in IT environments can be summarized as follows.

The cognitive perspective

The cognitive perspective has dominated thinking in HCI (Human-Computer Interaction) for about a decade, although the trends outlined above mean it is increasingly having to give ground to other points of view, especially the social psychological and the organizational.

From the cognitive perspective, human-computer interaction is seen as basically being about the exchange and processing of information within a context of reasoning, problem solving and thinking. At the beginning of the 1980s, there was a widespread view that efficient and effective human-computer interaction could be facilitated by (perhaps even depend upon) building a cognitive model of the user within the computer, so that the computer would 'think like' and 'see things in the same way as' the user; complementing this, it was also felt important (perhaps critical) that the user should 'have' an appropriate 'cognitive model' or 'mental model' of the computer. Much of the research during the 1980s was concerned with these two issues. From this perspective, the key issues in the evaluation of human-computer interaction were to do with the cognitive compatibility or otherwise of the computer system with its human users.

Associated with the emphasis on a cognitive view of human-computer interaction was the belief that much could be achieved by applying the existing science base of research results relating to human cognitive psychology to the design and evaluation of computer systems. This science base had been built up by academic and other researchers working in the field of cognitive psychology for a period of more than thirty years. Whilst their research was aimed at understanding human information processing as it applied to problem solving, thinking and reasoning in general, not specifically in the context of human-computer interaction, it was felt that if the theories were well-founded they should apply well to computer-based work which, the argument went, was an excellent example of human information processing.

A classic example of attempts to apply this science base in a way which would be of practical benefit in the product design and development process is found in the work of Card and colleagues at Xerox Parc (e.g., Card *et al.*, 1983). One of the best known results of their work is the GOMS family of models, loosely based on an interpretation of research results in cognitive psychology. The GOMS models all viewed human-computer interaction in terms of: Goals the user was attempting to achieve; Operators available to the user (including external operators such as pressing a key on a keyboard and internal operators subsumed under the general concept of a mental operator); Methods (sequences of operators suitable for achieving the goals concerned and defined in advance of beginning work on the task); and Selection rules for choosing between alternative methods. The GOMS models provided various methods for predicting the usability of software based on relatively simple and easily acquired data. Usability was defined strictly in terms of time to complete tasks and strictly in terms of expert performance (that is, assuming that the user never made an error in operating the software). This meant that considerations such as ease of learning or the quality of the work done using the system were not encompassed. Despite the deliberate simplifications involved in the models, the work involved in applying them to practical design problems was quite time-consuming in comparison with the somewhat restricted nature of the information they provided.

A contrasting approach to the application of the cognitive psychology science base was taken in Europe by ITT and GEC in the UK in collaboration with Logos Progetti in Italy and a number of academic psychologists, as part of ESPRIT research and development in human-computer interaction (ESPRIT Project 234, see Rogers *et al.*, 1992). An account of this work was published by Marshall *et al.* (1987). Whereas Card *et al.* had deliberately simplified the cognitive psychology research results in order to develop what they called 'engineering models' that could be easily understood and applied by non-psychologists working on product development, Marshall *et al.* deliberately attempted to retain the richness and complexity of the research results in order not to reduce their power in assisting the product design and evaluation team. Instead of simplifying the results, they attempted to explain and clarify them in ways which would make them usable in practical contexts. Marshall *et al.* developed a method for interpreting the research results into a set of design guidelines that could be used to guide the development of user-interfaces and, in the form of checklists and similar instruments, to evaluate prototypes and products. The guidelines were grouped into meaningful categories relating to 'sensitive dimensions' of user-interface design. These were dimensions such as 'navigation' and 'consistency', where it was felt usability would be sensitive to whether the guidelines were applied or not. Whereas other sets of guidelines typically do not indicate how the guidelines were developed or what their scientific basis is, the method described by Marshall *et al.* allows the user to trace any given guideline back to its origin in the research literature and to update it (or add new guidelines) where appropriate in the light of new results. This procedure should minimize the danger of guidelines that may have doubtful basis in research becoming folklore simply by virtue of being endlessly republished in differing guises. The emphasis on explaining the guidelines using practical examples and plain English was intended to facilitate their use by non-psychologists working on product development and evaluation.

As well as interpreting the science base into guidelines, the team also developed a prototype software package which was capable of inspecting a formal description of a user-interface design and evaluating it in terms of a set of usability metrics based on cognitive psychological principles (see Christie and Gardiner, 1987). The work has since been taken further by the MIDAS team at London Guildhall University (see later in this chapter). Complementary work by other research teams, including the MUSIC project team working under ESPRIT, focuses on metrics based on measures of user performance, attitudes and related factors.

One of the key problems in applying guidelines or metrics to the evaluation of human–computer interaction is the effect of the application context. Christie and Gardiner (1990) identified several aspects to this problem, of which the following two will serve as illustration. First, research (e.g., Rabbitt, 1979) strongly suggests that the way that cognitive processes are organized by a person who is new to a task is different from the way they are organized when the person concerned becomes very familiar with the task. Rabbitt found virtually no overlap in the distribution of reaction times obtained from subjects new to a reaction-time task and those who had been practising the task for hundreds of trials; it was not just that the practised subjects reacted faster, it appeared that they seemed to be processing the information in a different way. This makes it difficult to apply any given design guideline to both casual users and well-practised, expert users of computer systems. Secondly, whether well-practised or not, research by Hockey (e.g., Hockey, 1979) suggests that humans can reallocate cognitive resources quite significantly according to the demands of the task being done, particularly between storage (which takes up human memory capacity) and throughput (which enables rapid clearing of internal registers within the cognitive system). This adaptability of the human often means there is no one answer to any given user-interface design problem; it depends on the trade-offs the designer wishes to make in terms of the cognitive processes in which the user will have to engage. Hockey's findings in relation to allocation of cognitive resources provide a link to the psychophysiological perspective considered below.

Further discussion of the cognitive perspective on evaluating human–computer interaction can be found in Gardiner and Christie (1987a).

The social psychological perspective

The social psychological perspective was regarded as very important in the evaluation of IT during the early 1970s when 'IT' (the term was not even used very widely then) meant telecommunications rather than computers. Interest focused on the use of audio and audio–video telecommunications links as alternatives to face–to–face meetings. The launch of Confravision, Inter-City Visual Conferencing and other services stimulated a considerable amount of research around the world, especially in the UK, US, Canada, Sweden and Australia. Much of it was reported by Short *et al.*, (1976) and a summary presented by Christie and de Alberdi (1985). With the passage of time, the results of this research have tended to fade from the collective memory of the IT community but paradoxically, as time has gone on, they have become more relevant. The reason for this is the re-emergence of interest in audio-video person–to–person or group communication, enabled by developments in technology which have become apparent during the 1980s and the last few years in particular. These include the availability of satellite links, developments in local communications networks and, especially, the setting up of ISDN (the Integrated Services Digital Network, which enables text, graphics, audio and video to be transmitted on a common telecommunications network serving businesses and homes alike).

Whilst Confravision-type services, in which people hold meetings in special private or public audio–video studios, are still available, there is in addition growing interest in the concept of desktop conferencing in which video images are presented in windows on the computer screen, with documents, spreadsheets and other computer applications being presented simultaneously in other windows. This type of development brings the telecommunications and computing aspects of IT visibly together at the user-interface and thrusts social psychological issues clearly into the HCI arena.

Social psychology is concerned with how humans interact with one another and the new generation of computer systems is increasingly concerned with providing a technological basis for such interaction. The very term 'HCI' is becoming less appropriate as such systems increasingly support human–human interaction (through a computer/telecommunications system) as well as or instead of human–computer interaction. Key examples of this type of development include Hewlett-Packard's VISION research reported by Gale (1989) and the work by British Telecom and partners in the ESPRIT MIAS project (see Rogers *et al.*, 1992).

One of the key evaluation criteria to have emerged from the research in the 1970s on Confravision-type systems and confirmed in the Hewlett-Packard VISION research as being particularly sensitive to variations in the design of the user-interface is that of Social Presence (sometimes referred to simply as Presence). This is a factor which describes users' perceptions or feelings about a system. It is usually measured using questionnaire techniques and refers to the degree to which users report that they feel they are meeting with people who are 'really there' and that the system feels 'warm' and 'personal' rather than 'cold' and 'impersonal'. The feelings reported are thought to depend on a variety of factors including the extent to which the audio and video channels provide for adequate communication of audible and visual nonverbal cues (e.g., grunts, eye contact and gestures).

This type of evaluation of human-computer interaction is not covered well by cognitive psychological theory and has its basis in social psychological considerations but it is not the only way in which social psychology is becoming increasingly relevant to the evaluation of human-computer interaction. As we move into the latter part of the 1990s and beyond, at least as important as the above examples will be the social psychological aspects of the interaction with the intelligence of the system itself. Even in today's systems, research (e.g., Richards and Underwood, 1984) suggests that the social style of communication adopted by the computer (e.g., formal, polite) can affect the style adopted by the user. Gardiner and Christie (1987b) have discussed how possibilities for multimedia communication of various sorts might aid in avoiding breakdowns in the dialogue between human and computer, and Sheehy and colleagues (e.g., Sheehy, 1987; Sheehy and Chapman, 1987) have examined the nonverbal communication aspects of this in some detail, based in part on research done under ESPRIT. The direction in which technology is advancing and some of the social psychological issues involved are depicted in such television series and films as *2001:A Space Odyssey*, in which the computer HAL can interact socially with humans to the extent of commenting on their drawings and becoming paranoid about their intentions, and *Star Trek*, in which the starship's computer is given a human voice designed to convey a suitably helpful yet respectful and subordinate personality. As advances in artificial intelligence continue, it is not beyond the bounds of possibility that some of the issues described in Greg Bear's novel, *Queen of Angels*, will need to be addressed, including those that arise when the complexity of the artificial system becomes such that the system begins to develop needs of its own which extend beyond those of mechanical information processing.

Further discussion of the social psychological perspective on evaluating human-computer interaction can be found in Christie and Gardiner (1990).

The organizational perspective

The importance of including organizational criteria in the evaluation of human-computer interaction was emphasized by Gardiner as early as 1986 (see also Malone, 1985) and discussed by Christie and Gardiner (1990). The need for this is even more evident now, with increasing interest in groupware and related concepts. One of the central issues in this field is how best to model the organization and how to apply that model to the information processing and work flow requirements of the organization. Models that have been developed to date (e.g., in the ESPRIT PROMINAND project, see Rogers *et al.*, 1992) have emphasized the formal relationships among organizational entities and activities and the need to provide formal structures within which the user can work consistently with organizational practice. Some commercial products such as STAFFWARE have appeared on the market in recent years, mostly aimed at people working from traditional office bases.The European Limitless Office, ELO, demonstrated at the ESPRIT Conference and exhibition in November 1992 and aimed initially at the German insurance market illustrates how it is possible to integrate mobile workers into an electronic office system by providing integrated document, data and voice communications within a structured work environment that guides the worker through the organizational procedures that form the core of his or her work. Gardiner (1986) argued that an important criterion for evaluating human-computer interaction is the level of support given by the system in helping the newcomer to the organization adapt to it. The ELO system meets this criterion to some extent by providing the user with a help and training system integrated into the software supporting the organization's procedures, allowing the newcomer or other user to learn the procedures as well as the system in the course of normal work.

Systems which structure the user's work environment so as to encourage or force correct procedures also need to be evaluated in terms of their social psychological consequences for the individuals and groups concerned. It has been known for well over a decade (see Baird, 1977) that the logical structure of a communication network can have significant effects on group and individual productivity, morale and other factors. Other work going back some twenty years (e.g., Chapanis *et al.*, 1972) has shown that the medium of communication used (e.g., text or voice) can significantly affect the process of communication and task output, in addition to the effects of network structure. Often, these factors vary in opposite directions; for example, a network structure which encourages high group productivity may also tend to lower morale. Some of these issues were considered in ESPRIT research on IT-UPTAKE and have subsequently contributed to the knowledge base for a multimedia training system developed by IMS and partners in the IT-USE project (see Rogers *et al.*, 1992). Project GROUP at London Guildhall University are developing a research framework and demonstration environment in which these issues can be further researched specifically within the context of modern groupware and related types of system.

Further discussion of the organizational perspective on evaluating human-computer interaction can be found in Furnham (1987) and Sloan and Cooper (1987).

The psychophysiological perspective

The different perspectives on human-computer interaction considered above can be largely subsumed under the psychophysiological perspective which emphasizes the human and human-computer system as a system made up of interacting sub-systems. From this point of view, it is possible to regard, for example, the cognitive perspective and the social psychological perspective as relating to cognitive and social sub-systems which interact with each other and with other sub-systems within the human, the human-computer and human-computer-organizational system. In addition, the psychophysiological perspective draws attention to the fact that the human is a biological entity and that aspects of its behaviour are best understood in biological terms. The psychophysiological perspective as explained by Gale (1985) puts particular emphasis on evaluating human-computer interaction in terms of overt behaviour (including specific task performance and broader aspects of behaviour such as absenteeism), subjective experiences (relating to interaction with the particular interface concerned as well as the work situation more generally) and physiological responses (including specific stress responses which may or may not result in physical symptoms).

The physiological aspect of human-computer interaction has played an important part in the evaluation of computer systems for general use and in their evolution during the last fifteen or twenty years. Concern with possible deleterious effects on health, including back problems, eye strain and potential radiation hazards, among other possibilities, has stimulated the industry to adopt improved standards for the design of computer systems. Key examples include the design of keyboards (key travel, pressure required for key closure, keyboard angle and the fact that keyboards—at least on desktop machines— are separate from the base unit so they can be repositioned easily) and displays (refresh rate and contrast ratio and—at least on desktop machines—swivel and tilt). A recent innovation designed to make computer systems even safer for their users has been the low radiation type of monitor. In future years, the ergonomics of displays will be improved even further through the use of low-cost flat panel displays instead of displays based on the CRT (Cathode Ray Tube) technology used today. Flat panel displays are already used in mobile computers.

The integrating framework provided by the psychophysiological perspective provided the theoretical context for the development by Gale and Scane of the CAFE OF EVE concept discussed later in this chapter in connection with the evaluation of networked systems.

Further discussion of the psychophysiological perspective on evaluating human-computer interaction can be found in Gale (1985) and Gale and Christie (1987).

The communications perspective

This perspective on human-computer communication is re-emerging in the context of multimedia and other developments in IT but has its roots in earlier work such as the work by Chapanis and colleagues on natural language communication with computers (e.g., Chapanis *et al.*, 1972) and the work by Alex Reid and colleagues on mediated communication (e.g., Christie and de Alberdi, 1985; Short *et al.*,1976). This perspective focuses on the effects which constraints on communication between human or electronic partners in a task can have on the process and outcome of the work done.

The Theory of HCI Channels, being developed by the authors and their colleagues is a particular model which reflects this perspective. According to the Theory of HCI Channels,

- Interaction between a human and a computer is mediated by one or more communication channels established jointly by the human and the computer.
- Interaction between two or more humans using an electronic (computer-based and/or telecommunication-based) system can be analyzed in terms of the communication channels between each human and the system, those channels being established jointly by the humans and the system.
- At any given point in time, any given HCI channel has a measurable bandwidth. Bandwidth is defined in terms of the maximum number of units of information that can be transmitted through the channel in unit time without loss or error.

Some of the key hypotheses which form part of the theory include the following:

- The total bandwidth available to the human for receiving information is finite and at any given point in time is fixed.
- The total bandwidth available to the human for outputting information is finite and at any given point in time is fixed.
- The human communication system can divide up the total available bandwidth into a number of channels which vary in bandwidth, up to a maximum determined by the sum of their bandwidths which can equal but not exceed the original total. This is analogous to the concept of multiplexing in telecommunications. By analogy with telecommunications theory, there may be some small overhead associated with multiplexing so that the sum of the bandwidths of the individual channels can never quite attain the total overall bandwidth available.
- The bandwidths of the channels created by multiplexing are not determined arbitrarily but reflect a range of standard bandwidths, so that a set of standard types of channel can be identified. An example of a low-bandwidth channel may for convenience be labelled the *text channel*. An example of a high-bandwidth channel may for convenience be labelled the *picture channel*. If a photograph of a scene is transmitted using the picture channel, it may require only a fraction of a second for most of the information about the scene to be received and for the human to be able, for example, to answer questions about it. If the same scene is coded into text and transmitted using the text channel, it may require many seconds or even minutes for most of the information to be transmitted and for the human to be able, for example, to answer many questions correctly.
- The various channels created by multiplexing operate more or less independently of one another with relatively little leakage or interference between them until the maximum limit on the bandwidth available is approached, at which stage errors or a slowing down of processing will be observed.

The Theory of HCI Channels is based on the communications perspective but has interesting links with some of the other perspectives outlined above. For example, the following hypotheses relate it to the psychophysiological perspective:

- The total bandwidth available to the human varies with the human's psychophysiological state of arousal, being at a maximum when the level of arousal is optimum and being reduced as arousal either rises above or falls below the optimum level.
- Bandwidth is affected by other factors such as visual acuity.
- Arousal increases as a monotonic function of the rate of information

received, except that at extremely high rates of information reception arousal may fall.

- Arousal is affected by other factors such as level of alcohol in the blood, semantic content of information received, and other factors.

Special needs

In addition to the theoretical perspectives above, and cutting across them, there is an emerging concern in the IT field with the special needs of some groups of IT users, especially those with special physical needs (e.g., relating to physical infirmity) or special cognitive needs (e.g., relating to limitations in regard to some or all intellectual abilities). Such needs may be more likely to be observed in some demographic groups than others, and the needs of some members of the elderly population are of particular importance in this regard because the elderly represent a large and increasing segment of the general population. In general, the needs of these people are not qualitatively different from the needs of other IT users, but rather represent differences of degree such that there can be significant differences in the emphasis that needs to be put on certain aspects of the user-interface. For example, if programming a video recorder (especially an older model) can be a fairly demanding exercise for a young, healthy person with good eyesight, it can be a serious problem for an elderly person with stiff and painful finger joints and poor eyesight. Both types of user would benefit from controls that are larger, better spaced and more easily identified, but the need is significantly greater for many elderly users. Similar difficulties apply in regard to cognitive factors. If a young, intelligent person sometimes experiences difficulty in following complex operating instructions in a user manual, the problem is much greater for someone who, for whatever reason, may find information processing tasks more demanding.

The increasing importance of the elderly

It could be said that old age is a relatively new phenomenon; people did not always live as long as they tend to nowadays, at least in the developed countries where most IT users are to be found. In the 12-month period to June 1992, 2,353 people in the UK celebrated their 100th birthday, the oldest being 116 years old (Hansard, 1992). People of that sort of age are still rare but thanks largely to improved living conditions and better health care, more people are living longer than ever before and are remaining active for a longer period of their lives. In the European Community alone, there are nine million people over the age of 65 (the statutory retirement age for men in the UK), representing about 17 per cent of the population. A further three million are between ages 60 (the retirement age for women in the UK) and

65, making a total of about twelve million people over the age of 60, almost 20 per cent of the population.

It is an unhappy fact of life that as we grow older, and despite advances in health care, the chances of our developing various physical, intellectual and other impairments increases. Disability and old age tend to go together, so that within a few decades, the developed countries could be facing a major economic and political problem. The Commission of the European Communities estimates there may be between 60 to 80 million disabled and elderly people in the Community by the year 2020, with more than one in four of the population being over the age of 50, and 70 per cent of the disabled population being over the age of 60, with a growth rate in some countries of around 20 per cent per year (TIDE News, 1992).

The Commission of the European Communities has recognized that the trends noted above suggest that the cost to society of caring for the elderly and disabled is likely to increase considerably. At the same time, the continued restructuring of society, including the decline of the extended family, means that an increasing proportion of the costs may fall on the state rather than being absorbed by relatives of those concerned. These factors taken in combination provide an economic imperative (quite apart from any other rationale) for helping the more active and better able members of the ageing population to continue to care for themselves as long as possible, following a lifestyle in which they are integrated into the social and economic activities of society and living fulfilling and independent lives. In recognition of the growing importance of the elderly in the population, the Commission of the European Communities designated 1993 as the European Year of Older People and Solidarity between Generations.

Special needs of the elderly

As Rabbitt (1993) has remarked, 'To consider old age as a disability is not only politically incorrect but also dangerous'. However, as noted above, impairments of various sorts tend to become more common as we grow older, and the very fact of being old rather than young can exacerbate their consequences. As Rabbitt (1993) has put it, 'To be young and blind and deaf is bad enough, but to be old and blind and deaf is worse'.

Types of disability often associated with age include visual, hearing, speech, motor and cognitive impairment. There is frequently a deterioration in visual performance which means that clarity, lack of clutter, size of print and other visual parameters should be of importance to designers and manufacturers of IT or other systems. In visual displays, some older people may tend to find textual instructions easier than icons. It may be unfortunate from this point of view that, according to findings from the TIDE GUIB project (TIDE is explained below), by 1992, 93 per cent of software was being produced on a GUI (Graphical User Interface) platform and of those software houses still producing text-based software, 72 per cent had plans to convert to GUI-based systems.

Other groups

Although the elderly are very important in terms of population size, the designer of IT systems must also take account of the special needs of other groups such as those with physical disabilities or mental handicap.

It is widely accepted (cf Midgley, 1990a, 1990b) that training in IT systems is one of the most useful forms of rehabilitation for many people with physical disabilities, especially in regard to employment prospects. The proliferation of IT in the workplace makes otherwise unattainable jobs available to physically disabled people. IT systems offer a means of facilitating the acquisition of life skills, independence skills and work skills. Professor Stephen Hawkins provides an excellent example. Suffering severely from motor neurone disease which restricts his speech and motor abilities very significantly, he is still able to perform effectively as one of the world's leading physicists thanks to IT. His disease severely affected his ability to speak and eventually he had to have a tracheotomy which removed his speech completely. Thanks to a personal computer, speech synthesizer and communications program, he can now communicate by voice more effectively than before his operation. With the help of his IT communications system, he was able to finish his book, '*A Brief History of Time*', which became a best seller.

Mental handicap and various less significant cognitive impairments may also have special significance in the design of IT systems, and in the training of prospective users of IT systems. The work of centres such as General and Health Information Services in Dublin has shown how, with appropriate training, those with relatively limited cognitive resources can learn to use IT effectively and to enter paid employment using such systems. The possibility of applying multimedia technology to the special IT training needs of such groups of prospective users has been examined by the authors and colleagues in the context of the European TIDE Programme referred to below.

IT in relation to special needs and life skills

For those with special needs to be able to live independent and fulfilling lives, they need to acquire a number of basic life skills, many of which increasingly involve IT. Randall and Gibb (1991) identified the following main areas: shopping, telephoning, mobility, feeding, food-making, dressing, using public transport and communicating needs and wants to others. Much of the equipment used in these various areas is based on IT or uses IT components (e.g., embedded microprocessor systems in washing machine control systems and other appliances). The following four categories provide a number of specific examples:

- *kitchen equipment*, e.g., cookers, microwave ovens and washing machines

- *communications equipment*, e.g., telephones, fax machines and card phones
- *life systems*, e.g., automated teller machines (ATMs), calculators, clocks and watches, vending machines and information systems
- *entertainment systems*, e.g., radios, television sets, video recorders, CD players.

Helping those with special needs to be able to use items such as the above involves both improving the design of the user-interface and providing appropriate training.

Design trade-offs in addressing special needs

Addressing special needs in the design of a user-interface to an IT system is unfortunately not entirely straightforward. Trade-offs will often have to be made, as the following simple example illustrates. An elderly user faced with a visually 'difficult' typeface may tend to read more slowly, but relatively accurately; faced with a visually 'easy' typeface, the same user may tend to read faster, but with a reduction in accuracy (Rabbitt, 1993). The designer must therefore consider not only whether the user is likely to suffer from a particular impairment (such as visual impairment) but must also consider which aspects of performance (e.g. speed or accuracy) are most important in particular situations. Similarly, trade-offs must be made in designing training programmes for IT; providing support in regard to one need may sometimes make learning more difficult for the prospective user in some other respect.

A European research and development programme

As indicated above, recent studies (e.g., Midgley, 1990a, 1990b) support the view that IT and IT training can be of real benefit to people with disabilities. This is recognized in current research and development being done as part of an R&D programme launched by the Commission of the European Communities in March 1991. The programme, TIDE (Technology Initiative for Disabled and Elderly people), supports R&D across a wide range of rehabilitation technologies, including IT, and recognizes the importance of developing appropriate systems, user-interfaces and training systems aimed at the special needs of the populations concerned.

The IT research and development done within the TIDE programme includes work by the authors and their colleagues at London Guildhall University in collaboration with IMS (Dublin) on the ACT-IT project (Application of Compuer-based Training in IT: a multimedia prototype, Project 113). The project was concerned with the application of multimedia technology to the special training needs of the elderly or others with special difficulties in learning IT skills. The project developed a methodology for the analysis of IT skills, an analysis of the special training needs of the elderly and other groups who might have special needs related to cognitive and associated limitations, guidelines for applying multimedia technology and a prototype software system illustrating the functionality that could support authors of multimedia courses aimed at these special groups of prospective IT users (see also Inwood, 1992b).

Special needs and the general population of IT users

The design problems associated with the special needs of groups such as the elderly should not be seen as entirely separate from the needs of the general population of prospective IT users. In the documentation concerning the TIDE programme distributed to prospective participants, the Commission commented that, 'It is important to note that a number of individuals with particular disabilities and nearly all elderly people should not be considered to be a special sub-group whose problems require engineering or human factors solutions which are radically different from those which are helpful to the rest of the population . . . Thus a remarkable number of solutions that enhance access to technology by older people, or by people with disabilities, represent general rather than specific improvements'. This view is consistent with Rabbitt's (1993) comment, 'Improve design for the elderly and you improve design for the young'.

Evaluation in the product life cycle

Each of the theoretical perspectives outlined above contributes to practical methods and tools which can be used in evaluating the human-computer interaction aspects of systems at all stages of their life cycle, from initial research through design and development to use in real work situations.

Each company has its own approach to managing the evolution of a product, from an initial idea to a mature product in practical use on a day to day basis in real work situations, and to some extent each project has its own particular requirements and characteristics. On the other hand, there are recurrent themes and patterns, and a variety of models have been defined which attempt to describe key features of what is involved (e.g., Faehnrich *et al.*, 1987; Foley, 1983, Furner, 1987; Mantei and Teorey, 1988; Meister, 1987; Novara *et al.*, 1986; Roach *et al.*, 1982; Williges, 1987). In broad terms, five main stages can be identified (see also chapter 1):

1. Product development can be considered to start with a basic idea which may be evaluated using market research and other techniques. The idea might be to design a new version of an existing product, or it might be an idea for a very innovative product, different from anything already on the market. The basic idea and general issues relating to functionality and usability are normally evaluated before any significant design work commences. This stage is sometimes called the pre-design stage.

2. The product is designed. The design, or alternative designs, may be evaluated before implementation begins. This stage may involve many iterations and may be combined with Stage 3.
3. The design is implemented. This often involves partial implementation (prototyping) and evaluation of various alternative designs. This stage may be closely associated with Stage 2 and may involve many iterations.
4. The product is evaluated and possibly refined. In the case of software products this may well involve some final debugging before launch.
5. The product is launched and becomes one of the products available on the market. It is evaluated in terms of feedback from the market.

If the product meets with sufficient market success, new versions of it and additions to it may be developed, forming the first stage in a new cycle.

A challenge which has emerged in recent years is a pronounced shortening of the product life cycle. In particular, Stages 2 and 3 have begun to merge. The stimulus for this has been the need to bring products to market faster and faster in order to generate revenue and to compete effectively for market share. The shortening of the life cycle has focused on Stages 2 and 3 because it is in this area that advances in software development environments are making it increasingly possible to develop prototypes rapidly and, where appropriate, incorporate them into the final product. The design, implementation and evaluation aspects of this have become highly iterative and the time scales involved have become increasingly compressed. Whereas in the past, implementation of a design decision might take days or weeks, software developers may now be making decisions on screen at a rate of several a day or, in the case of relatively minor features of the product, several an hour. A plethora of productivity tools, a general move towards object-oriented design and advanced software development environments all contribute to this.

As the life cycle shortens, there is increasing pressure to improve the efficiency with which human-computer interaction aspects can be evaluated or to reduce the amount of such evaluation in order to avoid delaying the process of product development. Methods which might have been appropriate when product life cycles were much longer are becoming less and less acceptable as life cycles shorten.

The following sections consider the aims of evaluation at each of the main stages identified above, the key parties involved and the kinds of information needed.

Defining the objects of evaluation

The concept of evaluation implies there is something to evaluate, and the kind of evaluation that can be carried out in practice depends very much on the nature of what is available. The following conceptualization is largely derived from the definitions suggested by Christie and Gardiner (1990):

1. At this level, aspects of the proposed design may be embodied in a number of vehicles including written specifications, and storyboards.
2. At this level, some aspects of the design have been worked out in depth but there is no real functionality. For example, alternative physical layouts for a keyboard may have been developed as physical shells showing key layouts but the 'keys' do not move, or screens may have been drawn with dummy data included but there are no links to real databases. Some aspects of the design may be illustrated using 'slide shows' incorporating key sequences of screens. 'Interactive slide shows' (usually on a computer screen) may be available to allow the viewer to branch from screen to screen in different ways, following paths a user might be expected to take.
3. Here, some specific aspects of the product have been developed into working prototypes but do not link to one another and there are significant gaps, so that a complete product does not exist. For example, the file management aspect of a word processor may have been implemented in prototype form, allowing the user to locate and retrieve a given document, but there are no facilities implemented for editing documents once retrieved.
4. At this level, the whole product exists as a working prototype but is not yet ready for release. It is known that bugs and design faults exist but it is possible to use the prototype to evaluate many key aspects of the design in real work situations.

These four levels can be termed 'availability levels', referring to the levels at which objects for evaluation are available to the evaluation team. As a general rule of thumb, objects available for evaluation tend to evolve from level 1 to level 4 as the product evolves from an initial idea to a product ready for release. However, this is not the whole story. For example, at the start of the life cycle competing products may be already on the market and may need to be included in the early evaluation work. At later stages, some aspects of the product may develop faster than others, so that different levels of 'evaluation objects' will be available simultaneously. Figure 12.1 suggests what this might look like in practice.

Evaluation tools and techniques

Many of the tools and techniques used to gather the information needed at the various stages outlined above are also used in the evaluation of human work in other contexts and are described in other chapters throughout this book. A classification of general ergonomics methodology with references

Availability Level	Stage of the Product Life Cycle				
	Stage 1: Pre-design	Stage 2: Design	Stage 3: Implementation	Stage 4: Refinement	Stage 5: In real use
Level 4				•	•
Level 3			•	•	
Level 2		•	•	•	
Level 1	•				

Figure 12.1. Typical availability of items for evaluation at different stages of the product life cycle

to particular chapters can be found in chapter 1. Other tools and techniques focus especially on the evaluation of human-computer interaction, or are specific implementations of more general methods.

Various attempts have been made to classify the wide variety of specific evaluation tools and techniques available in HCI. Karat (1988) distinguished between nine main categories, as follows:

- theory-based evaluations
- user-based evaluations
- surveys and questionnaires
- verbal reports
- controlled experiments
- task-based evaluations
- informal design reviews
- formal design analysis
- production system analysis.

The classification has some pragmatic value although in some cases it conflates categories (e.g., formal design analysis and production system analysis). An alternative classification has been offered by Whitefield *et al.* (1991) who group tools and techniques into four main categories based on whether the focus is on the human or the computer and whether the object of evaluation is the real thing (the actual user or the actual computer) or a representation of it. The four categories are:

- *Analytic evaluation methods.* These are based on representations of users and computers and include, for example, the Keystroke Level Model of Card *et al.* (1983). This model and other predictive models can be useful early in the design process for predicting user behaviour given alternative design solutions.
- *Specialist reports.* These are based on human factors or other specialists evaluating prototype designs, drawing on their expertise and on available guidelines.
- *User reports.* These are based on tools and techniques such as question-

naires and interviews, often involving real users but representations (e.g., simulations, outline descriptions, storyboards or other representations) of the computer.

- *Observational methods.* These are based on real users using real computers. (In addition, as explained in chapter 2, observational methods can be used to examine a variety of other aspects of the behaviour of people including users and prospective users of IT systems.)

The distinctions made in the scheme are becoming somewhat less clear as methods of product development evolve. The development of rapid proto-typing tools in particular blurs the distinction between a representation of a planned system and an early prototype of the system itself; even so, the classi-fication is useful in highlighting differences in emphasis between different approaches to evaluation.

The choice of an evaluation tool or technique will depend on a number of factors including, among others, the theoretical perspective or perspectives that are driving the evaluation (see earlier) and the level of availability of the product or system to be evaluated. Figure 12.2 associates some typical evalu-ation tools and techniques with the theoretical perspectives outlined in an earlier section. The tools and techniques listed map onto both the Karat and the Whitefield *et al.* classification schemes. The list is not intended to be comprehensive but highlights some of the more commonly used tools and techniques. They have been grouped according to whether they rely prim-arily on analysis by experts (with little or no involvement of users or prospec-tive users) or analysis of data collected from users or prospective users. The former tend on the whole to be quicker and less expensive; they depend upon the level and domains of expertise of the experts involved. The latter tend to be more time-consuming and more expensive; they depend upon the representativeness of the users or prospective users participating and on the representativeness of the tasks and conditions involved in the evaluations.

There is not sufficient space in this chapter to describe in detail all the tools and techniques in Figure 12.2 and little point in doing so where they have been described elsewhere. The following notes are therefore intended to help guide the reader to appropriate sources of information and to expand on particular points where appropriate.

Expert panel analysis

This can be useful as a form of evaluation in its own right or as a complement to other methods. It consists of a panel of several experts commenting in a structured way on the object of evaluation (e.g., a prototype), usually making recommendations. The experts are often psychologists, human factors special-ists and others with expertise in user-interface design; however, they may also include domain specialists with expertise in particular application areas (e.g., medical applications, technical publishing, financial applications) where they may be able to bring expert knowledge of user requirements and current product offerings to the analysis. Various formats for their reports can be adopted according to the needs of the particular project. The following is based on a format developed by Korte and colleagues at empirica in Bonn and used successfully by them in collaboration with Harmsen and colleagues at the Fraunhofer-Institute for Systems and Innovation Research (FhG-ISI) in Karlsruhe and other colleagues on the ESPRIT ELO (European Limitless Office) Project.

Item: (e.g., 'Labels used for buttons'.)
Problem: (e.g., 'computer-oriented terms are often used where other terms could be used which are more familiar to those working in the application domain concerned.')
Recommendation: (e.g., 'Use terms which are familiar to those in the application domain.' Specific terms would normally be suggested.)
Justification: (e.g., 'The intended users are not well-versed in using computers and cannot be expected to be familiar with the computer terms used. Using terms which they already use will reduce the need for learning new terminology.' Where appropriate, references to the research literature or to industry standards are provided.)

Nielsen and Molich (1990) describe a variant on expert panel analysis based on providing members of the design team with a set of heuristics concerning good design practice (e.g., 'Speak the user's language'; 'Provide feedback') to help them evaluate the design solution as it evolves and to predict possible problems.

Predictive models

These are applied to written specifications, prototypes or other material (depending on the model) in order to predict something important about behaviour, attitudes, feelings or other aspects of human-computer interaction. To the extent that the predictions are valid, they can save time and expense by providing important information early on in the development process, before prototyping is very far advanced, and without the need to run user tests. Some models are based on well-established laws or principles concerned with perceptual-motor performance. For example, Fitts Law, which predicts the time taken to make movements of specified length and accuracy, has been applied to the design and evaluation of user-interface devices such as the mouse (see Gediga and Wolff, 1989; Landauer, 1991). Other models are based on a cognitive perspective of human-computer interaction. A classic example is the GOMS family of models described by Card *et al.*, (1983) and the best-known of the models they developed is the Keystroke Level Model referred to earlier. Their models focused on predicting the time taken to complete simple tasks (such as moving a paragraph of text or setting the margins). Whilst the models have a number of limitations, including the assumption that the user never makes any errors and the fact that other variables such as quality of work done and users' attitudes are not predicted, the models may have practical value in some circumstances and they have been important in stimulating research in this field.

One line of research has been directed towards models which can predict how easy it will be for a new user to learn to use a system. Examples of this include the work by Payne and Green (e.g., 1986) on TAG (Task–Action Grammar), which attempts to model relevant aspects of users' knowledge of the system in terms of grammatical rules, the work by Tanaka *et al.* (1991) on a model called TEM (Text Editing Method) which is concerned with quantification of consistency in user-interface designs, and the production system approach of Kieras and Polson (1985) which is based on the proposition that learning is improved by minimizing the cognitive complexity of the system, modelled in terms of sets of production rules. Kieras and Polson's model makes quantitative predictions for ease of learning and use of a system. Their work has been further developed in recent years (e.g. Bovair *et al.*, 1990).

A second line of research has been characterized by attempts to predict performance in terms of the number of errors made. An assumption underly-

Tool/Technique	Theoretical Perspective					
	Cognitive	Social Psychological	Organisational	Psycho-physiological	Communications	Eclectic
Some tools/techniques based on expert analysis						
Expert Panel Analysis	•	•	•	•	•	•
Predictive models	•	•	•	•	•	•
Audits/guidelines	•	•	•	•	•	•
Objective metrics	•			•		
Dialogue Error Analysis	•					
Some tools/techniques based on data obtained from users or prospective users						
Focus groups						•
Questionnaires						•
Interviews						•
Stakeholder Analysis						•
Observation of users						•
Physiological data				•		
User walkthroughs						•
Controlled tests	•	•	•	•	•	
Field trials	•	•	•	•		
Critical Events						•

Figure 12.2. Evaluation tools and techniques related to theoretical perspectives

ing this approach is that the greater the load on the user, especially on working memory, the greater the likelihood of making errors (see Lerch *et al.*, 1989).

Although much of the work on predictive modelling has focused on a single user working on a single task, multiple cognitive activities may be involved, often occurring in parallel, even in a single task situation and certainly when several tasks are involved. John (1988) has shown how critical path analysis can be used to take account of cognitive activities occurring in parallel.

The GOMS model and later predictive models have been used in the design and evaluation of user-interfaces in a number of application areas, including, for example, help systems (Elkerton and Palmiter, 1991), and text editing (Bovair *et al.*, 1990).

In a somewhat different, more empirical approach to predictive modelling, Williges *et al.* (1992) carried out a series of experimental studies of user performance with an interface and then constructed integrated, empirical models for three dependent variables (the average time needed to locate an information message, the number of user-added keypresses to access the item compared with an error-free search, and the accuracy of transcription). For each variable, they computed a linear regression equation, incorporating values for key independent variables, which best matched the empirical data collected in the experiments. They suggest that the empirical models thus produced can be used by designers to predict user performance under various conditions and so to provide a basis for considering particular trade-offs in the design process.

Further discussion of predictive models can be found in Olson and Olson (1990) and in Landauer (1991).

Audits, guidelines and standards

In this context, an audit instrument is an instrument (usually a paper-based checklist or similar instrument) for assessing the extent to which the object of evaluation (typically a prototype or a competitor's product) meets important criteria. The process of applying the instrument and analyzing the results is the audit itself. The criteria used may be based on industry standards, house style or other considerations but are often based on design guidelines. An example of a published paper-based instrument which could be considered an audit instrument in the present sense can be found in Ravden and Johnson (1989). The method they describe for carrying out the audit using the instrument was developed through experience in two ESPRIT projects (Project 534, 'The Development of an Automated Flexible Assembly Cell and Associated Human Factors Study', and Project 1199/1217, 'Human-Centred CIM Systems'). The criteria on which the audit is based were derived from several sources including the design guidelines presented by Marshall *et al.* (1987).

Some of the difficulties in applying many published guidelines relate to uncertainty about the scientific status of the guidelines. It is often not clear whether they are based on sound research, years of practical experience in a variety of relevant contexts or have weaker foundations. As the links to research are not clear, it is not clear which guidelines need updating when new research results are published. One of the advantages of the guidelines presented by Marshall *et al.* is that the links to research are explicit. Audits typically depend upon the expertise of the evaluator (often a specialist in user-interface design) in assessing the extent to which the prototype or other object of evaluation meets the criteria concerned, although in principle audits can also be carried out by users (for an example, see Maldé, 1992). Several evaluators may be used in order to arrive at an average or consensus evaluation. The format of audit instruments typically allows the evaluator to record the extent to which criteria are met in a simple, standard way, often by means of a rating scale. What is typically not done is to allow for the fact that some criteria may be more important than others. The *Weighted Factors Method*, used for example in the ESPRIT ELO Project (Project 2382, see Rogers *et al.*, 1992), does allow for this. According to the Weighted Factors Method, the evaluator makes two separate judgements for each criterion in the audit: (a) the extent to which the criterion is met, and (b) the importance of that criterion. The first rating can then be weighted by (multiplied by) the second in order to give a weighted score for the prototype in terms of the criterion concerned. The weighted scores can be added over all criteria or all criteria within a particular group in order to provide an overall assessment of the extent to which the prototype or other object of evaluation meets the criteria concerned, taking account of the relative importance of the criteria. By comparing this overall score with the maximum possible score assuming all criteria were met perfectly, it is possible to calculate a *normalized score* which is independent of the number of criteria and which always varies from zero (no criterion is met to any extent) to 100 (all criteria are met perfectly). This allows comparisons to be made between the prototype and other products, between alternative design solutions to a common design problem, or between different aspects of the user-interface. In connection with the latter, Marshall *et al.* (1987) distinguished between 14 'sensitive dimensions' or groups of criteria representing different aspects of the user-interface which affect human-computer interaction and which are sensitive to decisions made by the user-interface design team.

Standards

As the field of human-computer interaction matures, there is increasing emphasis on defining standards at corporate, national and international levels against which to assess products and systems. This movement towards standards may lead to increasing standardization of evaluation tools and methods (cf Reiterer, 1992) but the results of tests in terms of specific standards, no matter how the tests are performed, can be entered into a usability audit which covers those standards as well as other criteria which the evaluation team feel to be important. Stewart (1991) provides a useful directory of HCI standards, published as part of the UK Department of Industry's *Usability Now!* initiative.

Evaluation goals

Setting usability goals or other evaluation goals in advance of developing user–interface design solutions allows the design team to evaluate the extent to which prototypes at different stages of development meet those goals. Where design is based on an iterative process, this procedure in principle can be used to enhance the efficiency of the design process, helping to ensure that effort is not spent unnecessarily on enhancing those aspects of the design which already meet the goals set but is spent instead on improving those aspects which currently fall short. The Weighted Factors Method of Evaluation provides a means of setting evaluation goals within the context of usability or other audits. The following are hypothetical examples of goals that might be set for a product:

'The overall normative score should be at least 95.'

'The score should be at least 90 for at least half of the categories audited.'

'The score should not be lower than 85 in any category.'

Multifaceted evaluation

Audit instruments have typically focused on the concept of usability, with a strong emphasis on the more cognitive and perceptual-motor aspects of human-computer interaction, reflecting the cognitive perspective that has largely dominated work in the field of human-computer interaction until recently. As systems evolve into areas such as multimedia and computer support for cooperative work, it is becoming more important to include other facets in their evaluation. These cover criteria reflecting the social psychological, organizational, psychophysiological and communications perspectives outlined above.

Objective metrics

By objective metrics in this context is meant measurements that can be applied to screen designs, the logic of a dialogue or other aspects of a user-interface without the involvement of users and without relying on the judgement of the evaluator. Such measurements might include, for example, the maximum number of windows that can be open on the screen at any one time, the maximum number of windows that can present motion video simultaneously, the physical size of characters on the screen, and so on. The usefulness of metrics in other fields of human factors has been demonstrated by others (e.g., Pulat and Grice, 1985) and is discussed in some other chapters in this book, including, for example, chapter 19 on anthropometry and the design of workspaces and chapter 20 on computer workspace modelling.

A prototype software package has been developed which was capable of inspecting a formal description of a user-interface design and evaluating it in terms of a set of usability metrics based on cognitive psychological principles (see Christie and Gardiner, 1987). Later work (e.g., Smith, 1990) has taken the research further and has developed software modules which demonstrate the applicability of the approach to actual user-interfaces rather than simply formal descriptions of them. The work at present applies to the older, character-based generations of user-interfaces and additional research is needed to apply the approach to graphical user-interfaces where the problems of data capture are more difficult (Inwood, 1992a).

Dialogue error analysis

Dialogue error analysis focuses on errors that the user can reasonably be expected to make as a result of features of screen layout, dialogue structure or other aspects of the user-interface. By identifying such errors and the factors contributing to them, it is possible to make recommendations for improvements to the design of the interface which should reduce the likelihood of errors occurring and, when they do, provide users with convenient ways of recovering from them. Because operator errors can result from quite small features of the interface, it is necessary to have a rather detailed representation of the interface available, or to have the interface itself. The technique can be useful in analyzing the relatively strong and relatively weak areas of a user-interface in an existing product or a prototype. The technique involves the following main stages:

1. Carry out a hierarchical task analysis or similar analysis of the range of tasks the product is designed to support.
2. For each task or sub-task, identify one or more methods of carrying out that task or sub-task using the product. Each method is defined as a sequence of steps which, if performed correctly, will achieve successful completion of the task or sub-task.
3. For each step in a given method, identify all the errors that a user can reasonably be expected to make, their consequences and the importance of their consequences. This requires expert judgement and depends to some extent on the type of user being considered. There are normnally no more than about four or five reasonable errors per step.
4. For any given error, identify the features of the user-interface that contribute to the error. This requires expert judgement based on an eclectic theoretical perspective.
5. For each feature contributing to a given error, identify a change that

can be made that would improve the user-interface. This requires expert judgement.

6. Taking account of the above, define mechanisms for helping users to recover from errors.

Recommendations stemming from the analysis are grouped according to the errors to which they relate and the importance of the consequences of the errors, and organized according to aspects of the user-interface. Applying the technique systematically is time-consuming and for this reason attention is normally focused on those aspects of the interface with which it is thought the user would normally have to interact frequently and/or where errors would be expected to have especially important consequences. Implementation of recommendations in software or hardware improvements can be expensive and for this reason attention is normally focused on those recommendations which have relatively far-reaching beneficial effects and which relate to especially important aspects of the interface or especially important errors. Sutcliffe and Springett (1992) have shown how it is possible to classify the errors that users make according to categories defined by a theoretical model of user behaviour.

Not all the effects that a user might produce by taking a particular action are necessarily best regarded as errors. Monk (1990) has described a technique, based on 'action-effect rules', which considers both errors and other effects. This technique allows a specification to be evaluated against principles of good design practice, in particular the principles of reversibility and predictability. First, all meaningful actions that a user could make are listed. The action-effect rules are then produced by listing the effects each action could have and by considering the context or conditions determining which effect will occur. The actions, effects and conditions are all described from the user's perspective. Evaluation of the design against the principles is performed by quantifying the number of actions required to reverse the effect if the principle has been violated. Interpretation of the results requires a degree of judgement about the task, the user's context and the practical consequences of violations of principles.

Readers with an interest in techniques related to the above are recommended to read chapter 31 in this book on human reliability assessment.

Focus groups

Focus groups are a form of group interview in which one or two moderators facilitate discussion among about five to ten group members, ensuring that the group focuses on the topic of interest. In the context of IT research, this often includes an introductory phase to the focus group session during which the group is introduced to the concept or product which is to form the focus for the discussion. Other ideas or focal points for discussion may be introduced at further pre-planned points during the session. The scientific status of focus groups in the context of qualitative marketing research has been discussed in a paper by Calder (1977) which was published some years ago but is still well worth reading. A published example of a focus group study in the IT context is provided by O'Donnell *et al.* (1991) in a paper in which they describe a focus group based on discussion of a videotape showing a user's interaction with a product. The comments made by the group during the discussion were content-analyzed and were found to be predictive of actual errors and ratings of task difficulty made by users.

Questionnaires and interviews

Questionnaires and interviews have a long history in research methodology (e.g., see Bouchard, 1976; Moser and Kalton, 1971; and chapter 3 of this book) and are described well in standard texts. The many particular techniques available can be adapted for use in the context of the design and evaluation of IT systems at almost every stage of the research and development process. Some approaches are perhaps worth highlighting. A particular kind of questionnaire is one based on the concept of a usability audit, discussed above. This is effectively a usability audit carried out by users instead of user-interface specialists. An example is provided by Maldé (1992). A particular kind of interview which is used fairly widely at the pre-design stage and other stages of the research and development process is the focus group interview discussed above. A distant cousin of that technique is described by Wright and Monk (1991). Their 'cooperative evaluation' technique involves a user and an interviewer (a member of the design team) engaging in a dialogue with relatively few constraints. The user is given specific tasks to carry out with partial prototypes. Wright and Monk suggest that the technique has the advantage of being natural and easy for the interviewer to carry out, that allowing the interviewer to ask the user questions increases the chances of capturing important information, and the relatively equal status of the user and the interviewer in the situation encourages the user to take a more critical approach to the evaluation. Small-scale questionnaires and interviews are often used in association with other tools and techniques, such as in conjunction with user walkthroughs and controlled tests.

Observation of users

Drury in chapter 2 of this book discusses the role of direct observation in the context of ergonomics evaluations in general. An example of the use of such techniques in the evaluation of user-interface designs in particular is provided by Plaisant and Shneiderman (1992). They describe an evaluation of a touchscreen-driven prototype user-interface to a home automation system in which users were videotaped as they performed a series of representative tasks. The observational data were used in conjunction with other data, including verbal protocol analysis of comments made whilst carrying out the tasks (see chapter 7) and ratings obtained by questionnaire.

Physiological data

Physiological data can be helpful in regard to abnormal conditions such as pain resulting from poor posture, repetitive work or other factors, visual fatigue, headache, stress, and other undesirable conditions, and in regard to the normal range of functioning when working at the interface to an IT system. The role of physiological and related psychophysiological data in evaluating work within an electronic workplace has been discussed in Gale and Christie (1987) and further discussion of particular topics can be found in several chapters in this book, including those by Kilbom (chapter 22) on the measurement and assessment of dynamic work, Corlett (chapter 23) on evaluation of posture, Tracy (chapter 24) on biomechanical methods in posture analysis, Meshkati *et al.* (chapter 25) on mental workload assessment, Cox and Griffiths (chapter 26) on stress, Bullimore *et al.* (chapter 27) on visual performance, and Megaw (chapter 28) on visual fatigue.

User walkthroughs

A user walkthrough is based on a real user or prospective user carrying out one or more designated tasks using a product or prototype, whilst providing comments and/or other data on the process involved. During a user walkthrough, the user is often encouraged to provide a running commentary covering aspects of the interaction of interest to the evaluation team (e.g., the user's expectations, feelings, difficulties, pleasant surprises, and so forth). The commentaries can be analyzed using a variety of techniques and the reader is referred to chapter 7 on verbal protocol analysis for a discussion of these. An example of a user walkthrough is provided by Marshall *et al.* (1990). Their users were encouraged to talk about their interaction with the system as they worked through a planned sequence of task scenarios. Overt behaviour and comments were recorded for subsequent analysis. Polson *et al.* (1992) describe a particular variant which they refer to as a 'cognitive walkthrough' in which attention is focused on the more cognitive aspects of what is involved. The method they describe typically involves presenting a user with an interface or a paper-based representation of an interface and a set of specific questions based on a set of pre-selected tasks. The user is required to explain his or her goals, what actions he or she intends to undertake and what changes to the design he or she feels to be appropriate.

Controlled tests

Controlled test conditions can be set up in order to test theoretical hypotheses, differences between different groups of prospective users, differences between alternative design solutions to a particular design problem, or for a range of other purposes. The principles of experimental design which underpin such tests are described well in other texts and in chapter 5 in this book. Examples of the application of these principles to controlled tests concerned with aspects of user-interface design can be found in the many papers published in this field, of which the following are just a few examples published in recent years: assessment of the usability of icons in a user-interface (Kacmar and Carey, 1991), effects of display size and sentence-splitting on subjects' performance and comprehension (Dillon *et al.*, 1990) and assessing the effects of adding audio and video channels to a network supporting group work (Gale, 1989). (See also chapters 10 and 11.)

Field trials

The complexity of real work situations compared with theoretical models or laboratory situations is such that pilot trials of new products and systems in real work situations are highly desirable. In the IT industry, new products or new releases are typically made available to users on a pilot basis in what are called 'beta tests', so that comments and other feedback can be obtained and 'bugs' or other undesirable features remedied before the product is made generally available. In user organizations developing their own application systems it is also desirable to pilot new features or new applications before making them generally available (see chapter 36). A variety of data can be collected during these trials, based on questionnaires, interviews, focus groups, automatic logging through the system itself, video recording and a variety of other possibilities. An interesting addition to the techniques discussed earlier is the possibility of a 'gripe button'. This is a 'hot button' which the user can press at any time in order to allow him or her to record complaints or other comments into the system. These complaints (or compliments) can be collected automatically through the system itself or off-line at intervals.

Critical events

Some of the most important events that can occur in relation to the use of a system may only occur relatively infrequently or under unusual conditions. For this reason, they may be difficult to explore systematically in a laboratory situation. Such events include accidents but are not restricted to accidents. Some of the issues involved in accident reporting and analysis are relevant to working with IT systems, and are discussed in chapter 32 of this book.

Applicability of tools and techniques

The tools and techniques available to the evaluation team vary in their applicability at the different levels of availability of objects of evaluation as defined earlier in this chapter (cf. Christie and Gardiner, 1990). Figure 12.3 relates some of the key tools and techniques to the levels identified earlier. The number of stars gives a rough indication of relative applicability at the levels shown. It can be seen that many of the tools and techniques can be applied to objects at several or all of the levels identified.

Developing an evaluation plan

The various tools and techniques outlined above contribute to the evaluation team's toolkit. Which particular tools to use from the kit, when and why depend on a variety of factors, but the choice should not be arbitrary, or

Tool/Technique	Availability Level			
	Level 1	Level 2	Level 3	Level 4
Some tools/techniques based on expert analysis				
Expert Panel Analysis	*	**	***	***
Predictive models	*	***	*	*
Audits/guidelines		*	**	***
Objective metrics		*	**	***
Dialogue Error Analysis		*	**	***
Some tools/techniques based on data obtained from users or prospective users				
Focus groups	**	**	*	***
Questionnaires	***	***	***	***
Interviews	***	***	***	***
Stakeholder Analysis	***	**	*	
Observation of users			**	***
Physiological data			**	***
User walkthroughs			**	***
Controlled tests			**	***
Field trials				***
Critical Events			*	***

Figure 12.3. Tools and techniques related to the level of availability of items for evaluation

simply reflect habit or reflect an unswerving allegiance to a particular theoretical perspective no matter what the context. A key aspect of any given evaluation project should be the development of an explicit Evaluation Plan which relates the use of the tools and techniques available to the stages involved in the product development, in a way optimized for the project concerned. The optimization should be with respect to such factors as the confidence one wishes to have in the results of the evaluation at any given stage, how much one is willing or able to pay for the information concerned, the time scales involved, and other factors. An Evaluation Plan needs to be defined at various levels, from high-level planning to detailed design of particular aspects of the evaluation. The higher level planning will typically tend to be done earlier and the more detailed planning later, but there will normally be significant interaction between the different levels and iteration is almost inevitable. Some of the key issues involved in developing an Evaluation Plan are discussed below.

Evaluation goals

The concept of evaluation goals was introduced above in connection with audits and guidelines. The concept can be applied more generally across the whole range of evaluation tools and techniques. An Evaluation Plan should normally specify the aims of the evaluation and define goals in terms of which the evolving user-interface design solution can be assessed. Goals should be defined for each stage in the product life cycle—from pre-design to use in real work situations—for which an evaluation is being conducted. Who sets the goals may itself form part of the Evaluation Plan at the highest level of planning. The relative benefits of involving the hardware and software development team in the setting of evaluation goals compared with the benefits of not doing so needs careful consideration in the context of the particular development. Others who are not necessarily members of the core evaluation team may also need to be involved. These may include representatives of senior management, the marketing department, special interest groups and others.

Selection of tools and techniques

The tools and techniques used must be compatible with the evaluation goals previously defined, but a number of other factors will normally be involved in this aspect of the planning, including the following.

Convergent validity: multiple techniques

An evaluation will only rarely rely on a single tool or technique. Normally several will be involved, looking at the problem from different points of view. Often the recommendations suggested by the results of applying the various techniques will converge on a common core which can be regarded as having a degree of validity that is not dependent upon any single aspect of the evaluation. Other recommendations may complement one another, and in some cases there may be conflicts between recommendations suggested by different techniques. An interesting example in which the results of different evaluation techniques are compared is provided by Desurvire *et al.* (1992) who compared results from a cognitive walkthrough and a heuristic evaluation (see Expert Analysis, above) with those from laboratory-based controlled user tests. They found that the results of the first two methods varied depending upon the expertise and experience of the evaluators and that using them identified substantially fewer problems with the user-interface than did the controlled user tests.

The Evaluation Plan should include some mechanism, often based on expert judgement, for resolving conflicts or differences of emphasis in the results from different evaluation techniques when they arise. For a practical example, the reader may be interested in a paper by Morris *et al.* (1992) who discuss the integration and evaluation of information from three sources (observation of user behaviour by online logging of activity, audio records and video recording) in the context of designing and evaluating a system for computer support of cooperative work.

Resources available

The resource requirements of the various techniques identified as candidates, on the basis of their suitability for providing information relevant to the evaluation goals defined, will need to be compared with the resources available. Resources include financial resources (e.g., for travel expenses, payments to subjects in user tests), hardware, software, rooms (e.g., for user tests), manuals (e.g., of guidelines) and other physical resources, and expertise. The evalution team is likely to be more expert in some areas than in others; their expertise can often be complemented by using outside consultants from the commercial or academic worlds.

The time scale

The time scale for the evaluation needs to be compatible with the time scale for the product development or other activity that is driving the evaluation. This may preclude using some techniques (e.g., controlled tests) which can be time-consuming. It has been noted earlier in this chapter that the problem of planning an evaluation to fit in with the available time scale is becoming increasingly challenging as the time scales for product developments tend to become shorter.

Logistical considerations

Some approaches to evaluation might be ruled out on logistical grounds, e.g., observing users directly may be very difficult where there are security implications or where the work is being carried out in hazardous conditions.

Ethical considerations

The Evaluation Plan needs to be ethical at all levels, from the definition of evaluation goals down to the planning of particular evaluation sessions (e.g., the way in which users in controlled tests or under observation in field trials are briefed). Ethical considerations have a part to play in even the details of statistical analysis, such as the weighting to be given to Type I versus Type II errors. (These refer to the relative probabilities of falsely accepting a chance difference between two conditions as being 'real' compared with the relative probability of dismissing a real difference as 'chance'. The relative weight given to these two types of error is not inconsequential when the two conditions are, for example, the speed of completing a task using a new product compared with using a competitor's.)

Membership of the evaluation team

The traditional model in which the evaluation team, the development team and the users are all separate does not fit all modern product development situations very well. The distinction between the evaluation team and the development team is becoming blurred as productivity tools speed up the development cycle and enable rapid prototyping of alternative software solutions. Where rapid development is of the essence in order to meet market windows, it is not sensible to insist on a separate evaluation team evaluating every design decision at every stage. Even if this were sensible in terms of meeting the market window, it would result in a frustrating situation for the development team who would be forced to dispense with the productivity gains provided by their development tools in order to accommodate the needs of the evaluation team; they would find their creativity significantly impaired by having to wait for feedback from the evaluation team on every little detail of the design. The distinction between the evaluation team and the users is also not always as clear as it might be in some other contexts. Users can, of course, be involved in the evaluation process as subjects in controlled tests, walkthroughs, field trials and so on. However, they can also be involved in other ways. In user organizations implementing IT systems, representatives of various internal user groups may well be actively involved in planning and evaluation at all stages (see chapter 36). In addition, many software products provide facilities for their users to modify aspects of the user-interface according to preferences, so that a user can evaluate the interface as it is and then change it if need be in order to make it more acceptable.

Following experience in a research and development environment within a multinational IT company, Gale and Christie (1987) outlined the concept of a CAFE OF EVE, a Controlled, Adaptive, Flexible Experimental Office of the Future in an Ecologically Valid Environment. The CAFE OF EVE concept is of a research and development environment for the research, design, prototyping and evaluation of advanced office and business systems. The development team, the evaluation team and users participate as equal partners in the process, using a variety of different methods within a physical and organizational environment that combines the best of laboratory-based and field-based approaches. Gale and Christie identified the main benefits of the CAFE OF EVE concept as follows:

- ecological validity
- salience (relevance to user needs)
- symmetrical status of users and developers/evaluators, leading to openness and frankness
- sampling of data
- possibility of long-term, strategic developments
- multiple measures of user behaviour and other variables of interest, contributing to convergent validity of findings and recommendations
- reduced effects of 'Hawthorne' and related effects
- reduced drop-out of subjects from long-term tests
- continuous iterative prototyping
- rapid transfer of laboratory findings to field testing
- interdisciplinary contributions
- public relations, marketing and training value.

Balanced against these potential benefits, there are a number of potential difficulties and problems of which Christie and Gardiner (1990) identified the following:

- high set-up costs
- possible disbenefit in terms of potential delays in development of products
- potential progressive loss of ecological validity as the users become increasingly familiar with the prototypes and concepts being developed and tested and who may therefore develop ways of thinking about the office and skills which are not typical of users in other environments. (Within the framework outlined by Gale and Christie, users work in the CAFE OF EVE on a long-term basis, equivalent to users in other organizational settings.)

Communication of the results of the evaluation

In the context of product development, evaluation is normally not an end in itself but a means to helping in the improvement of the product being developed. For it to be effective in this way, it is essential that the results of the evaluations done are communicated effectively to those who are in a position to use them. Unfortunately, this does not always happen. As an example of what can go wrong, Marshall *et al.* (1990) describe the case of 'Product X' where redeployment of resources within the organization and the priority given to getting the product to market contributed to the results of product evaluations not being used to enhance the product before launch. Following product launch, customer studies indicated that they were experiencing substantially the same problems as had been predicted on the basis of the evaluations carried out during product development.

In order to minimize this kind of problem, evaluation results must be communicated speedily, efficiently and persuasively to those who can use them. At the level of specific communication techniques, it may be noted that some user-interface evaluation teams have found it useful to supplement traditional forms of written and spoken reporting with more technology-based communications. Videotapes have proven particularly useful. They can be used, for example, to provide highlights from focus group discussions, and can be useful as a complete record to which clients, development teams or others can refer. The tapes bring the discussions to life in a way that a written report, even with plenty of 'verbatims' capturing what participants actually said, cannot hope to achieve. Where video recording is not possible, audio recording can be a useful second-best. Videotapes have also been found useful in providing edited highlights of user walkthroughs. The highlights can provide dramatic illustrations of problems (or positive features) which users encounter when trying to use a prototype or product. Multimedia technologies now offer a wider range of possibilities for communicating the results of evaluations effectively.

At a strategic level, it is important for the evaluation team to plan when and what to communicate and to whom. Some attempts have been made to formalize this with reference to structured design methodology. This has in its favour that, where a development team is using a formal structured approach, user-interface evaluations and other ergonomics inputs are likely to be more useful and more readily accepted if they are linked in a logical way to that design methodology. An example is the work of Silcock *et al.* (1990) who outline an approach linked to the Jackson System Development (JSD) method. In a broader context, communication with peer groups, researchers and practitioners in related fields, and others with interests in the development and use of advanced IT systems is important in helping to develop a research and development community that can serve the needs of the industry and its customers within a framework suited to the needs of our evolving society.

Technological developments in the application context

A key aspect of communication and cooperation with those in related fields is the need for the user-interface specialist to work effectively in the context of a rapidly evolving technology and application framework, with associated demands relating to cost-effectiveness, timeliness and other commercial factors. Even academic research which may have longer-term goals needs to take appropriate account of the changes which are occurring if the results are to be pertinent to future developments and inspiring in terms of insights into human behaviour within real-life (rather than purely laboratory) situations.

Implications of the evolution for user-interface evaluation can be considered in relation to the key trends identified by Christie and Stajano (1992) together with the trend towards increasing intelligence in the system. The following discussion of these relates to strategic research and development carried out by the European IT industry during the past decade under the auspices of the Commission of the European Communities as part of the international ESPRIT programme (the European Strategic Programme of Research and development in Information Technology). Whilst the individual projects may not be familiar to the general reader, many of the key product developments during the 1990s and beyond will have used the results of this and subsequent research. The expected continuing collaborative research and development will be among the most significant influences on the future development of evaluation theory and method in human-computer interaction in Europe. (Where specific references are not supplied then the reader is referred to Rogers *et al.* (1992) for further details of ESPRIT projects.)

Connectivity and increasing demands in terms of interoperability

Computing power is becoming decentralized, with increasing emphasis on distributed computing throughout companies and into other emerging application areas including integrated systems in the home. Key projects under ESPRIT include Bull's DCM (Distributed Computing Model), based on the results of eight ESPRIT projects. Other work has led to the formation of a joint venture company, APM Ltd., which has launched ANSAware based on ESPRIT results. ANSAware is used by NASA as the basis for the world's largest distributed computing system, ADS (the NASA Astrophysics Data System), representing a major example of European technology being successful on a world-wide basis. The system means that scientists anywhere in the US, once connected to the network, can have access to computing, database and related resources anywhere on the network, from a microcomputer to a super computer, according to the requirements of the task. Often, a number of different resources cooperate in the carrying out of a single task.

Implications of developments such as these for the evaluation of human-computer interaction are summed up well in a metaphor used by Bull and their collaborators in the COMANDOS project (providing technologies and engineering environments needed to develop distributed computing systems). They suggest that the user of a stand-alone PC (personal computer) can be likened to a musician playing a violin. Simply connecting the PC to a network does not change this much; it would be akin to the musician being able to exchange tape recordings, notes and scores with other musicians and with libraries. In contrast, the user of a distributed system is rather like a conductor of an orchestra. The individual musicians within the orchestra access their data (musical scores), process their data (play their parts of the music) and coordinate their activities with other musicians under the general direction of the orchestral conductor. The work which the conductor does is different from and at a higher level than that of the individual musician. Just as a conductor needs a different kind of 'user-interface' (based on a baton) than a violinist (based on a bow), so the user of a distributed system needs a different user-interface than a user of a PC. In a NASA demonstration of ANSAware, it was explained how the kind of user-interface needed by the NASA scientists using the ADS distributed system was such that it needed the processing power of a workstation rather than a PC. (The term 'workstation' is normally used to refer to a desktop computer that is more powerful than a PC, although today's PCs would have been considered workstations yesterday.) The paradox of this is that whilst distributed systems provide the user with virtually unlimited computing power through the networks, they require more powerful user-interfaces to be able to use that power to greatest effect.

The richness of information handled and user-friendliness

The main changes here have been in the increasing use of graphical user-interfaces, multimedia presentation and multimodal user-system interaction. (These terms are explained below.) Along with these developments has come increasing standardization of various aspects of the user-interface, such as standard ways of using scroll bars to scroll through text or other material presented in a window on the screen. MOTIF, MS Windows and other developments now provide *de facto* industry standards for user-interface design to which an increasing number of products conform. Developments in these areas are likely to reduce the burden of learning new software applications and to accelerate the uptake of IT (Information Technology) within organizations and in everyday life.

The term 'graphical user-interface' refers to user-interfaces which incorporate windows, menus, icons, dialogue boxes and other features drawn on the screen pixel by pixel and is used in contrast to the term 'alphanumeric user-interface' or 'character-based user-interface', older-style interfaces where the designer was constrained to build up an interface from characters rather than pixels. (A pixel is one of the very large number of dots that are illuminated on the screen to form a picture or other display. Text is made up of characters which are organized groups of pixels forming the letters of the alphabet, numerals and other items used over and over again in text, such as punctuation marks.) Many character-based user-interfaces are still in use, associated with applications which have been used for some time, but graphical user-interfaces are becoming the norm for new products.

The term 'multimedia' refers to the use of sound, speech, still images, animation and moving video in addition to text and other media in presenting information to the user. 'Multimodal' refers to the use of several input devices in combination. These include the mouse and other pointing devices for moving a cursor around the screen, touch-sensitive screens which allow the user to instruct the computer by touching the screen (e.g., to select from a menu or indicate a position in a display such as a map), speech input and other forms of input in addition to the keyboard (which was the predominant input device in older-style computer applications in the general office and business environment, although specialist input devices were used in some fields such as Computer Aided Design). The term 'multimedia' is often used in a more general sense to imply the use of graphical user-interfaces and multimodal interaction as well as multimedia information display.

European industry collaboration under ESPRIT has led to some world firsts in the field of multimedia. For example, the portable CD-I player (Compact Disc Interactive player) from Philips is based on concepts developed in the ESPRIT DOMESDAY project; the European industry contribution to the ISO (International Standards Organization) world-standard coding system for video images (ISO standard 11172) was coordinated by the ESPRIT COMIS project.

The implications of developments in multimedia for the evaluation of human-computer interaction revolve around the greater flexibility provided to the user-interface designer and the changes in load that may be put on the user compared with character-based interfaces. Multimedia interfaces usually present a lot more information per unit time to the user and in this sense put greater load on the user. On the other hand, they enable much greater use of pictorial information ('graphics' and 'still and moving images' in the industry jargon); humans are excellent pattern recognizers and users of computer systems can often process pictorial information faster and better than text. ('A picture is worth a thousand words.') This point is discussed in connection with the Theory of HCI Channels.

The autonomy of the individual

Many of the constraints on patterns of working associated with the early days of IT, such as the need to be at a fixed location, are fading as systems become more flexible, providing their users with greater autonomy. On the one hand, the growth of local area, metropolitan area and wide-area telecommunication networks makes it increasingly possible for users to communicate, exchange data and access resources from wherever they happen to be working at the time, provided they have a suitable workstation or other equipment. In parallel, the integration of computing facilities into small, lightweight units that the user can carry around, improvements in battery life and other developments all contribute to providing the user with means of working on the move or at locations where full desktop facilities are not readily available.

Implications for the evaluation of human–computer interaction revolve around the need for a consistent approach to the user-interface across a range of units which vary significantly in size and other characteristics (such as the resolution of display and the availability of colour). At one extreme is the desktop computer with very high resolution, colour display, capable of very fast system response. At the other extreme is the palmtop, a unit which can be held in the palm of one's hand, or a computer badge which can be worn on the lapel (perhaps primarily serving communication functions such as are needed to track the movement of people through public areas such as university buildings or airports, or perhaps including voice communication as depicted in the television series *Star Trek: The Next Generation*). In support of individual autonomy, the challenge is to enable the user to work on any application (within a wide range) on any type of unit according to circumstances, and as part of that to design user-interfaces for the various kinds of unit which are each user-friendly and yet which are consistent with one another in ways which minimize the amount of learning and remembering which the user has to do when moving from one type of unit to another, for instance, when moving from a desktop machine to a notebook computer (light and about A4 size) or palmtop which can be conveniently used when travelling. Some of the issues involved are being addressed in HICOS, an ESPRIT project being led by empirica in Germany.

The types of applications available

The development of software applications has traditionally been a significant bottleneck constraining the rate at which enterprises could exploit the potential benefits of IT. There is increasing adoption of an object-oriented approach, in which software applications are built up in a highly modular way and the individual modules or 'objects' are relatively autonomous units which have enough computer intelligence to know what data to expect from where and then what to do. Modularity, plus products such as C++ and Visual Basic, and advanced software engineering environments such as developed under ESPRIT in ITHACA are all contributing to easing the bottleneck by improving the speed with which high-quality, reliable software can be developed. Further contributions come from the increasing availability of applications which go beyond individual tools, such as word processors or spreadsheets, to offer partial solutions to the business needs of enterprises and which can be tailored to the needs of particular organizations. The work done under ELO, the European Limitless Office, is an example of developments in this area.

Implications for the evaluation of human-computer interaction centre around how well the system and its user-interface have been designed to reflect a common core of user requirements and user-interface features which apply across a range of organizations and work contexts. A number of research and development projects, such as the FAOR and ITHACA projects under ESPRIT, make contributions to this. In ITHACA, for example, part of the work has been to define an application framework for a generic office.

Architectures and the type of work supported

As user-organizations have become more sophisticated in their evaluation of the potential of IT and what they expect of computer-based systems in relation to their business objectives, IT manufacturers have responded by developing systems based on more advanced architectures capable of supporting the kinds of applications which users in the latter part of the 1990s and beyond will expect. (The term 'architecture' is used here to refer to the basic conceptual structure around which the hardware and software of the system are built.) One of the key needs in this area is for systems which can support people working together in teams, departments and other organizational units, including project teams which may involve several organizations working together cooperatively across national boundaries. Terms used to refer to these kinds of applications include 'CSCW' (Computer Support for Cooperative Work) and 'groupware'. Some of the key European work in this field is being done in the ESPRIT project, EUROCOOP.

Implications of these developments for the evaluation of human-computer interaction centre around how well the system provides the user with appropriate functionality and user-interface features which support group dynamics and organizational aspects of the work involved. Considerations include, for example, the suitability of the organizational model presented to the user and the way in which the user can interact with the model, the support for structured interaction among group members (e.g., based on defined organizational procedures) and the support for less structured interactions as well as social psychological aspects of group and organizational behaviour (e.g., the need to feel in real contact with real people and not to be just a 'cog in a machine'). Where some members of the group may be mobile and often away from base, the need for the system to support an adequate level of social interaction in order to maintain corporate spirit and individual morale may be quite important in influencing the success of the system and the organization using it.

Increasing intelligence in the system

In addition to the five trends above, a further trend is likely to become increasingly apparent in future systems, and this is the trend towards greater system intelligence. Systems of the future will take much more initiative for decisions and suggesting solutions; examples include assessing the value of information in databases relevant to current user goals and identifying relevant databases from the vast number of corporate and public databases available through networks. Early indications of this trend include the new generation of 'intelligent' products announced by Microsoft towards the end of 1993.

One way in which intelligence in the system could be manifest is through knowbots. The term 'knowbot' is a registered trademark of the Corporation for National Research Initiatives and is derived from the term 'knowledge robot'. It is an autonomous piece of software that can move from machine to machine on behalf of the user; looking for information, making decisions and carrying out other tasks. Tesler (1991, p. 54) succinctly sums up this aspect of the evolution in computer systems in saying such systems are moving 'from cloistered oracle to personal implement to active assistant'.

Implications for evaluation of human–computer interaction focus on how well the user-interface presents this intelligence to the user and the means it provides for the user to interact with it. One of the most interesting dramatic depictions of this was presented in the television series, *Space Cops*, in which the central character used a portable unit about the size of a walkman which he could fit in his pocket. He interacted with the unit, called 'Box', using spoken natural language and 'Box' would find relevant information for him by searching intelligently through the databases around the world using radio or other over-the-air communications to access the appropriate global communications networks. 'Box' acted as his personal assistant, with virtually unlimited access to information. Although this was yesterday's science fiction, it is clear that the technology is rapidly evolving in the direction of it becoming tomorrow's reality.

The future

The evaluation of human–computer interaction has had a long history but grew rapidly as a distinct field following developments in technological concepts during the early 1980s, especially in the US at the Massachussetts Institute for Technology and other key research and development locations. It was from these developments in particular that the modern type of personal computer user-interface was developed to become a familiar and important part of our technological environment in many spheres. Reflecting the types of system being developed and marketed during the 1980s, theories and methods in the field of human–computer interaction emphasized the user as a human information processing system. Cognitive science and cognitive psychology had a major influence on the way in which researchers looked at problems in human–computer interaction. The technology has moved on considerably since those early days and is continuing to evolve. The electronic systems themselves are becoming more sophisticated in 'social' as well as purely 'cognitive' terms and are increasingly concerned with supporting teams and organizations rather than simply individual human information processors. Theories and methods concerned with the evaluation of human interaction with such systems also need to evolve in order to keep pace with the developments in technology, and to contribute proactively to those developments rather than only reacting to them. The signs are hopeful. Interest in human–computer interaction evaluation, both within industry and within academia, continues to grow.

Acknowledgements

Aspects of this chapter benefit from work done by the authors and their colleagues as part of Office and Business Systems Research at London Guildhall University, especially the work done in collaboration with IMS, Dublin (in association with the Electric Brain Company), and Keith R. Phillips, consultant educational psychologist, on the ACT-IT Project (Project 113) of the TIDE Programme of the Commission of the European Communities and work done on the MIDAS project funded as part of the NAB III Initiative. Aspects of the chapter also benefit from work done by other projects, especially the ELO Project (Project 2382), led by empirica (Germany), part of the ESPRIT Programme of the Commission of the European Communities.

References

Baird, J.E., Jr. (1977). *The Dynamics of Organizational Communication* (London: Harper and Row).

Bouchard, T.J., Jr. (1976). Field research methods: interviewing, question-naires, participant observation, systematic observation, unobtrusive measures. In *Handbook of Industrial and Organisational Psychology*, edited by M.D. Dunnette (Chicago: Rand McNally Publishing Company), pp. 363–413.

Bovair, S., Kieras, D.E. and Polson, P.G. (1990). The acquisition and performance of text-editing skill: a cognitive complexity analysis. *Human-Computer Interaction*, **5**, 1–48.

Calder, B.J. (1977). Focus groups and the nature of qualitative marketing research. *Journal of Marketing Research,* **XIV**, 353–364.

Card, S.K., Moran, T.P. and Newell, A. (1983). *The Psychology of Human-Computer Interaction* (London and Hillsdale, New Jersey: Lawrence Erlbaum).

Central Statistical Office (1992). *Monthly Digest of Statistics.* (London: Her Majesty's Stationery Office), No. 563.

Chapanis, A., Ochsman, R., Parrish, R. and Weeks, G. (1972). Studies in interactive communication: the effects of four communication modes on the behavior of teams during cooperative problem solving. *Human Factors*, **14**, 487–509.

Christie, B. and de Alberdi, M. (1985). Electronic meetings. In *Human Factors of Information Technology in the Office*, edited by B. Christie (Chichester and New York: John Wiley and Sons Ltd), pp. 97–126.

Christie, B. and Gardiner, M.M. (1987). Future directions. In *Applying Cognitive Psychology to User-Interface Design*, edited by M.M. Gardiner and B. Christie (Chichester and New York: John Wiley and Sons Ltd), pp. 315–334.

Christie, B. and Gardiner, M.M. (1990). Evaluation of the human–computer interface. In *Evaluation of Human Work: A Practical Ergonomics Methodology*, edited by J.R. Wilson and E.N. Corlett (London: Taylor and Francis), pp. 271–320.

Christie, B. and Stajano, S. (1992). Advanced business and home systems—peripherals. In *ESPRIT Specific Research and Technological Development Programme in the Field of Information Technology: Results and Progress 1991/92*, edited by S. Rogers, B. Lewendon and M. Vereczkei (Luxembourg: Office for Official Publications of the European Communities), pp. 45–59.

Desurvire, H.W., Kondziela, J.M. and Atwood, M.E. (1992). What is gained and lost when using evaluation methods other than empirical testing. In *People and Computers VII: Proceedings of the HCI '92 Conference, September 1992*, edited by A. Monk, D. Diaper and M.D. Harrison (Cambridge: Cambridge University Press, on behalf of the British Computer Society), pp. 89–102.

Dillon, A., Richardson, J. and McKnight, C. (1990). The effects of display size and text splitting on reading lengthy text from screen. *Behaviour & Information Technology*, **9**, 215–227.

Elkerton, J. and Palmiter, S.l. (1991). Designing help using a GOMS model: an information retrieval evaluation. *Human Factors*, **33**, 185–204.

Faehnrich, K.P., Ziegler, J. and Davies, D. (1987). HUFIT (Human Factors in Information Technology). In *Cognitive Engineering in the Design of Human-Computer Interaction and Expert Systems: Proceedings of the Second International Conference on Human-Computer Interaction, Honolulu, Hawaii, 10–14 August 1987, Volume II*, edited by G. Salvendy (Amsterdam: Elsevier Science), pp. 37–43.

Foley, J.D. (1983). Managing the design of user–computer interfaces. *Computer Graphics World*, December, 47–56.

Furner, S.M. (1987). Practical information about interface operating characteristics for engineering design. In *Proceedings of an IEE Colloquium on Evaluation Techniques for Interactive Systems Design: I Organised by Professional Group C5 (Man-Machine Interaction), 27 March 1987, Savoy Place, London, IEE Digest No. 1987/38*, pp. 4/1–4/6.

Furnham, A. (1987). The social psychology of working situations. In *Psychophysiology and the Electronic Workplace*, edited by A. Gale and B. Christie (Chichester and New York: John Wiley and Sons Ltd), pp. 89–112.

Gale, A. (1985). Assessing product usability—a psychophysiological approach. In *Human Factors of Information Technology in the Office*, edited by B. Christie (Chichester and New York: John Wiley and Sons Ltd), pp. 189–211.

Gale, A. and Christie, B. (Eds.) (1987). *Psychophysiology and the Electronic Workplace* (Chichester and New York: John Wiley and Sons Ltd).

Gale, S. (1989). Adding audio and video to an office environment. In *Proceedings of the First European Conference on Computer Supported Cooperative*

Work, 13–15 September 1989, The Hilton Hotel, Gatwick, London, UK, pp. 121–130.

Gardiner, M.M. (1986). Psychological issues in adaptive interface design. In *Proceedings of an IEE Colloquium on Adaptive Interface Design Organized by Professional Group C5 (Man-Machine Interaction), 4 November 1986, Savoy Place, London, IEE Digest No. 1986/110,* pp. 6/1–6/3.

Gardiner, M.M. and Christie, B. (Eds.) (1987a). *Applying Cognitive Psychology to User-Interface Design* (Chichester and New York: John Wiley and Sons Ltd).

Gardiner, M.M. and Christie, B. (1987b). Communication failure at the person-machine interface: the human factors aspects. In *Communication Failure in Dialogue and Discourse: Detection and Repair Processes,* edited by R.G. Reilly (Amsterdam: Elsevier Science/North-Holland), pp. 309–323.

Gediga, G. and Wolff, P. (1989). On the applicability of three basic laws to human-computer interaction. *Schriftenreihe Ergebnisse des Projekts MBQ, Special Issue 11,* 1–3.

Hansard (1992). *House of Commons Offical Report on Parliamentary Debates,* Monday 8 June 1992, Issue 1590, **209** (23), col 53.

Hockey, R. (1979). Stress and the cognitive components of skilled performance. In *Human Stress and Cognition: An Information Processing Approach,* edited by V. Hamilton and D.M. Warburton (Chichester and New York: John Wiley and Sons Ltd), pp. 141–178.

Inwood, B. (1992a). Developing a useful metric., Poster session at HCI '92: People and Computers, University of York, September 1992.

Inwood, B. (1992b). Supporting the multimedia courseware author: an introduction to the ACT-IT project. In *Proceedings of EDTECH '92, Organised by the Australian Society for Educational Technology in Adelaide, Australia, October 1992.*

John, B.E. (1988). Contributions to engineering models of human-computer interaction. Unpublished doctoral dissertation, Carnegie Mellon University, Pittsburgh.

Kacmar, C.J. and Carey, J.M. (1991). Assessing the usability of icons in user interfaces. *Behaviour & Information Technology,* **10,** 443–457.

Karat, J. (1988). Software evaluation methodologies. In *Handbook of Human-Computer Interaction,* edited by M. Helander (Amsterdam: Elsevier Science Publishers, North Holland), pp. 891–903.

Kieras, D.E. and Polson, P.G. (1985). An approach to the formal analysis of user complexity. *International Journal of Man-Machine Studies,* **22,** 365–394.

Landauer, T.K. (1991). Let's get real: a position paper on the role of cognitive psychology in the design of humanly useful and usable systems. In *Designing Interaction: Psychology at the Human-Computer Interface,* edited by J.M. Carroll (Cambridge: Cambridge University Press), pp. 60–93.

Lerch, F.J., Mantei, M.M. and Olson, J.R. (1989). Translating ideas into action: cognitive analysis of errors in spreadsheet formulas. In *Proceedings of the CHI '89 Conference on Human Factors in Computing Systems* (New York: ACM), pp. 121–126.

Maldé, B. (1992). What price usability audits? The introduction of electronic mail into a user organisation. *Behaviour & Information Technology*, **11**, 345–353.

Malone, T.W. (1985). Designing organisational interfaces. In *Proceedings of CHI '85: Human Factors in Computing Systems, San Francisco, April 1985*, edited by L. Borman and B. Cutris (New York: ACM).

Mantei, M.M. and Teorey, T.J. (1988). Cost/benefit analysis for incorporating human factors in the software lifecycle. *Communications of the ACM*, **31**, 428–439.

Marshall, C., McManus, B. and Prail, A. (1990). Usability of product X – lessons from a real product. *Behaviour & Information Technology*, **9**, 243–253.

Marshall, C., Nelson, C. and Gardiner, M.M. (1987). Design guidelines. In *Applying Cognitive Psychology to User-Interface Design*, edited by M.M. Gardiner and B. Christie (Chichester and New York: John Wiley and Sons Ltd), pp. 221–278.

Meister, D. (1987). Behavioral test and evaluation of expert systems. In *Proceedings of the Second International Conference on Human-Computer Interaction, Honolulu, Hawaii, 10–14 August 1987, Volume II*, edited by G. Salvendy (Amsterdam: Elsevier Science), pp. 539–549.

Midgley, G. (1990a). Developments in IT training for people with disabilities. *Behaviour & Information Technology*, **9**, pp. 397–407.

Midgley, G. (1990b). Functional training in the use of new technologies for people with disabilities. *Behaviour & Information Technology*, **9**, pp. 409–424.

Monk, A. (1990). Action-effect rules: a technique for evaluating an informal specification against principles. *Behaviour & Information Technology*, **9**, 147–155.

Morris, M.E., Plant, T.A. and Hughes, P.T. (1992). CoOpLab: practical experience with evaluating a multi-user system. In *People and Computers VII: Proceedings of the HCI '92 Conference, September 1992*, edited by A. Monk, D. Diaper and M.D. Harrison (Cambridge: Cambridge University Press, on behalf of the British Computer Society), pp. 355–368.

Moser, C.A. and Kalton, G. (1971): *Survey Methods in Social Investigation*, 2nd edition (London: Heinemann Educational Books Ltd).

Nielson, J. and Molich, R. (1990). Heuristic evaluations of user interfaces. In *Proceedings of CHI '90: Human Factors in Computing Systems*, edited by J.C. Chew and J. Whiteside (New York: ACM Press), pp. 249–256.

Novara, F., Bertaggia, N. and Allamanno, N. (1986). Designing adequate human-computer interfaces: methodological issues and considerations on usability. *Paper presented at an ESPRIT Technical Week '86 HUFIT Seminar on Designing Usable IT Products, Brussels, 1 October 1986.*

O'Donnell, P.J., Scobie, G. and Baxter, I. (1991). The use of focus groups as an evaluation technique in HCI: In *People and Computers VI: Proceedings of the HCI '91 Conference, August 1991*, edited by D. Diaper and N. Hammond (Cambridge: Cambridge University Press, on behalf of the British Computer Society), pp. 211–224.

Olson, J.R. and Olson, G.M. (1990). The growth of cognitive modeling in human-computer interaction since GOMS. *Human-Computer Interaction*, **5**, 221–265.

Payne, S.J. and Greene, T.R.G. (1986). Task-action grammars: a model of the mental representation of task languages. *Human-Computer Interaction*, **2**, 93–133.

Plaisant, C. and Shneiderman, B. (1992). Scheduling home control devices: design issues and usability evaluation of four touchscreen devices. *International Journal of Man-Machine Studies*, **36**, 375–393.

Polson, P.G., Lewis, C., Rieman, J. and Wharton, C. (1992). Cognitive walkthroughs: a method for theory-based evaluation of user interfaces. *International Journal of Man-Machine Studies*, **36**, 741–773.

Pulat, B.M. and Grice, A.E. (1985). Computer aided techniques for crew station design: Work-space Organizer - WORG; Workstation Layout Generator-WOLAG. *International Journal of Man-Machine Studies*, **23**, 443–457.

Rabbitt, P. (1979). Current paradigms and models in human information processing. In *Human Stress and Cognition: An Information Processing Approach*, edited by V. Hamilton and D.M. Warburton (Chichester and New York: John Wiley and Sons Ltd), pp. 115–140.

Rabbitt, P. (1993). When do old people find displays difficult: *IEE Colloquium on Special Needs and the Interface*, Digest No. 1993/005 (London: The Institution of Electrical Engineers).

Randall, P. and Gibb, C. (1991). When the chips are down. *Community Care*, **24**, No. 848, pp. 18–19.

Ravden, S. and Johnson, G. (1989). *Evaluating Usability of Human-Computer Interfaces: A Practical Method* (Chichester: Ellis Horwood).

Reiterer, H. (1992). EVADIS II: A new method to evaluate user interfaces. In *People and Computers VII: Proceedings of the HCI '92 Conference, September 1992*, edited by A. Monk, D. Diaper and M.D. Harrison (Cambridge: Cambridge University Press, on behalf of the British Computer Society), pp. 103–115.

Richards, M.A. and Underwood, K.M. (1984). How should people and computers speak to each other? In *Interact '84: First IFIP Conference on Human Computer Interaction*, edited by B. Shackel (Amsterdam: North-Holland), pp. 33–36.

Roach, J., Hartson, H.R., Ehrich, R.W., Yunten, T. and Johnson, D.H. (1982). DMS: a comprehensive system for managing human–computer dialogue. In *Proceedings of the Conference on Human Factors in Computer Systems, Gaithersburg, Maryland, 15–17 March 1982.*

Rogers, S., Vereczkei, M., Richier, A. and Wright, M. (1992). *ESPRIT Advanced Business and Home Systems – Peripherals: The Synopses* (Luxembourg: Office for Official Publications of the European Communities).

Sheehy, N.P. (1987). Nonverbal behaviour in dialogue. In *Communication Failure in Dialogue and Discourse: Detection and Repair Processes*, edited by R.G. Reilly (Amsterdam: Elsevier Science/North-Holland), pp. 325–337.

Sheehy, N.P. and Chapman, A.J. (1987). Nonverbal behaviour at the human-computer interface. In *International Reviews of Ergonomics*, edited by D.J. Oborne (London: Taylor and Francis), pp. 159–172.

Short, J., Williams, E. and Christie, B. (1976). *The Social Psychology of Telecommunications* (Chichester and New York: John Wiley and Sons Ltd).

Silcock, N., Lim, K.Y. and Long, J.B. (1990). Requirements and suggestions for a structured analysis and design (human factors) method to support the integration of human factors with system development. In *Contemporary Ergonomics 1990: Proceedings of the Ergonomics Society Annual Conference*, edited by E.J. Lovesey (London: Taylor and Francis), pp. 425–430.

Sloan, S.J. and Cooper, C.L. (1987). Sources of stress in the modern office. In *Psychophysiology and the Electronic Workplace*, edited by A. Gale and B. Christie (Chichester and New York: John Wiley and Sons Ltd), pp. 113–135.

Smith, H. (1990). Project MIDAS: developing metrics for user interface design. In *Contemporary Ergonomics 1990: Proceedings of the Ergonomics Society's Annual Conference*, (London: Taylor and Francis), pp. 328–333.

Stewart, T. (1991). *Directory of HCI Standards* (London: The Department of Trade and Industry).

Sutcliffe, A.G. and Springett, M.V. (1992). From user's problems to design errors: linking evaluation to improving design practice. In *People and Computers VII: Proceedings of the HCI '92 Conference, September 1992*, edited by A. Monk, D. Diaper and M.D. Harrison (Cambridge: Cambridge University Press, on behalf of the British Computer Society), pp. 117–134.

Tanaka, T., Eberts, R.E. and Salvendy, G. (1991). Consistency of human-computer interface design: quantification and validation. *Human Factors*, **33**, 653–676.

Tesler, L.G. (1991). Networked computing in the 1990s. *Scientific American*, **265**(3), 54–61.

TIDE News (1992). December. (Luxembourg: Office for Official Publications of the European Communities).

Whitefield, A., Wilson, F. and Dowell, J. (1991). A framework for human factors evaluation. *Behaviour & Information Technology*, **10**, 65–79.

Williges, R.C. (1987). Adapting human-computer interfaces for inexperienced users. In *Proceedings of the Second International Conference on Human-Computer Interaction, Honolulu, Hawaii, 10–14 August 1987, Volume II*, edited by G. Salvendy (Amsterdam: Elsevier Science), pp. 21–28.

Williges, R.C., Williges, B.H. and Han, S.H. (1992). Developing quantitative guidelines using integrated data from sequential experiments. *Human Factors*, **34**, 399–408.

Wilson, F. (1993). Including special needs in IT design. *IEE Colloquium on Special Needs and the Interface*. Digest No. 1993/005 (London: The Institution of Electrical Engineers), pp. 2/1–2/4.

Wright, P.C. and Monk, A.F. (1991). A cost-effective evaluation method for use by designers. *International Journal of Man–Machine Studies*, **35**, 891–912.

Chapter 13

Human–machine interfaces for systems control

John R. Wilson and Jane A. Rajan

Introduction

In 1983 a British Airways helicopter crashed into the sea off the Isles of Scilly, with the loss of 20 lives out of 26 on board. At the inquest, the pilot agreed that he had not checked his altimeter properly, but also claimed that he did not see a warning light on the altimeter that should have come on automatically when the helicopter was within 200ft of the surface. He thought that the light might have been partly obscured by the helicopter's control column; 'The stick can cover part of the instruments. It depends on your height and build. In my case it does.' (*The Guardian*, Feb. 24th, 1984).

During maintenance at the BP Oil Refinery at Grangewood in 1987, an explosion and fire occurred $6\frac{1}{2}$ hours after the plant had been put on standby because a temperature cut-out had tripped the plant. One operator was killed. Over-pressurization in a low pressure separator brought about the explosion but the underlying root causes were found to lie in a series of multiple failures in operation, design, procedures and equipment. Amongst these were instrument and display failings which included: inaccurate gauge readings due to the (foreseeable) physical effects of wax and cold; a display which read higher than true because it was offset without the operator's knowledge; displays going off-scale; operators having to estimate flow rates due to lack of displays; controls for different functions being of similar appearance, allowing for inadvertent adjustment; and a breakdown of expectations in the relationship between direction of movement of controls and valve actions (Health and Safety Executive, 1989).

In these cases and countless other examples (see Reason, 1990 for instance) we find inadequate design of the displays and controls making up the human–machine interface; they are particularly inadequate in the way they can lead operators to make errors of omission or commission (see chapters 31 and 32).

At the heart of human–machine systems are the provision of information to the people working there and the control of the system by those people. Indeed, the early years of ergonomics are often characterized by the expression 'knobs and dials ergonomics'. In chapter 1 of this book it was seen that an application-oriented model of ergonomics places the interaction of person and process at the centre of a total systems view, and that displays and controls provide the media for that interaction and for carrying out tasks. It was pointed out in chapter 1 also that, according to the type of system and the view of the analyst, the human–machine interface might physically or conceptually lie between the person and the displays and controls, or between the controls/displays and the machine, or might comprise the controls and displays themselves. In the first of these views the interface is seen as a conceptual entity, in the third as a physical entity, and in the second as either or both.

Controls and displays take a wide variety of forms and this in large part is the reason for the awkward way in which authors (including the present ones) often refer to controls and displays of a 'process', in order to avoid having to refer explicitly to tools, equipment, machines, products or systems. Amongst this variety of interface forms we might find:

- industrial machines which have hand or foot controls (pedals, levers, buttons) and displays of numerical quantities or qualities (states);
- computer systems, which can utilize input devices operated by hand, foot or even head (for the disabled) or where input may be achieved 'directly' through speech, gestures, eye movements, etc., but where information presented to the user is largely by visual display with some auditory feedback;
- simulation and virtual environment systems which aim to give the user some feeling of being in the world which is modelled, where they may have no control, or control via a veridical interface (driving or aviation simulator), or control via novel input devices (dataglove or spaceball in virtual reality for instance);
- essentially display-only systems, usually with visually displayed information (e.g., signs, books) but which also can be auditory (warnings) or tactile (e.g., braille);
- products where the 'control' is built into or is part of the product, and information is displayed back in part through visual confirmation of the control action but also tactilely, kinesthetically or proprioceptively (e.g., toothbrush, razor, screwdriver, hammer, manual gear change);
- transportation where, although traditional type controls and displays exist, the greatest part of information is received by directly seeing the course or track (road or river) and registering the vehicle's state along it.

Definitions of interface forms will depend upon where we draw the 'system' boundary. For instance, instead of a book being a 'read only' display system, we can see reading a book or a set of instructions as a multiple-person system with a *temporal* dimension. One person uses controls—pen, typewriter, wordprocessor, etc.—to produce a display of information which is seen later, on one or several occasions, by themselves or by other people.

This difficulty in clearly defining elements of an interface is found generally, as a quote from the *Guide to Reducing Human Error in Process Operations* shows,

> "It is unfortunate that there is no obvious word which suggests all types of method for communicating between the operator and the plant, or one operator and another, because many human factors recommendations are the same whatever the technology used. For example, most recommendations about the shape of numbers which can be read most quickly and accurately apply whether those numbers appear on conventional instruments, VDUs, controls, or printed reference materials. Questions about how easy controls are to use, apply whether the operators are indicating what action they want to make by turning a knob, pressing a switch, using a list of alternatives on a VDU, or making a telephone call. In the following . . . the word '*display*' refers to all methods of giving information to the operator, the word '*control*' applies to all means by which the operator gives instructions."

(Ball, 1991)

We will follow such a convention in this chapter also, with the interface being the medium through which the two-way exchange of information takes place. A chapter such as this cannot, and should not, address all types of interface; moreover, evaluation of human–computer interfaces is dealt with in detail in chapter 12. Therefore, here we will deal with the sorts of control and display issues generally associated with control panels and control rooms, broadly termed human–machine systems. These include conventional instruments—analogue (quantitative and qualitative) and digital, annunciators, alarms, computer generated displays, information from written or verbal sources, information from the 'state' of a machine, process or environment, and all forms of controlling a process or system. Within our review we will introduce the need for a considered approach, both when analyzing existing human–machine interfaces and when designing and evaluating new ones. Following a brief overview of the human–machine interface and control room issues we discuss at length a set of activities that must be carried out in any analysis and design of interfaces, especially for control rooms. Then some detailed discussion of interface specification, especially for displays, covers choice of display mode, display formats, coding and navigation and structuring of displays. (The increasing role that auditory display of information plays in control rooms is recognized, although not explicitly here.) Finally, although we cannot cover all issues of evaluation we provide a basic control room interface evaluation checklist.

For control-display interfaces and control room ergonomics, other useful sources of information include: Ball, 1991; Goodstein *et al.*, 1988; Ivergård, 1989; Kirwan, 1994; Kragt, 1992; NUREG-0700, 1981; Shirley, 1992; Stanton, 1994; Woods *et al.*, 1987).

The human–machine interface

The classic view of the interface has been to understand the operator as a passive and limited capacity processor of information (Figure 13.1a). In this view, the operator and 'machine' are in a closed loop (although comprising an open system), connected by displays and controls. 'Machine information' is converted into 'operator information' via displays, and controls act as transducers to allow the operator to change a system state. Feedback to the operator comes via the displays and via interaction with the controls (tactile feedback for instance).

In more complex situations, with increased development and use of computer generated information systems, the operator is seen as needing higher level cognitive skills in both normal and abnormal conditions. Skill requirements in perceptual judgement, decision making, problem solving and diagnosis generally have led to more sophisticated models of an operator. Starting about the time of the work of authors in the collections by Edwards and Lees (1974), we can see an expansion of the original simple operator-process loop. The way people interact with systems is modelled as including attributes of the operator such as their mental model, experience, etc., and includes representation of their interaction through formal and informal procedures (Figure 13.1b).

Finally, the nature of work in complex systems now is such that the basic model is frequently conceived of as the 'human as supervisory controller' (Sheridan, 1987). In this view, computer systems mediate between the operators plus their displays and controls on the one hand and the task or process and its sensors and actuators on the other. If we look at Sheridan's ten cause-effect loops in supervisory control (Figure 13.1c), and understand that he defines possible supervisory roles for the operator as planning, teaching, monitoring, intervening and learning, then we can see the need for a structured comprehensive approach to the design of display–control interfaces. If system control can occur in so many ways, and the needs and roles of operators can be so varied, then interface analysis and design must be based on an appreciation of much more than checkpoints for selecting individual instruments.

Over and above the issues implicit and explicit in the above models are influences from outside the direct work context. For instance, the operator mental model might be considered as a product of what Clarke and Mayfield (1977) call 'outside world' knowledge – comprising life experience, social skills etc. – which will affect interaction with the system (social as well as technical) as much as will 'application specific' knowledge gained through

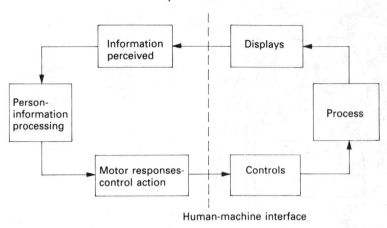

(a)

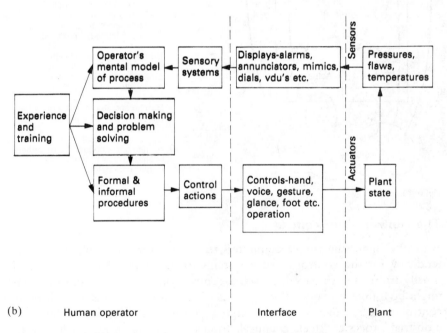

(b)

Figure 13.1 Views of the human–machine interface as simple closed loop (Figure 13.1a), model of human operator in process control (Figure 13.1b) and as supervisory controller (Figure 13.1c – Source Sheridan, 1987)

experience and training. Moreover, although reliable, effective performance will be a product of the work and task related factors of interface, task and training design, there will also be a considerable impact on operator errors from events outside work.

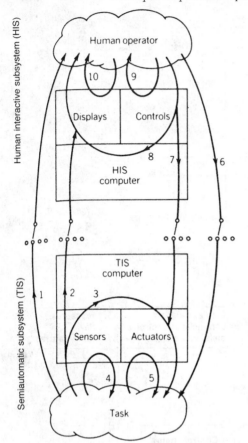

1. Task is observed directly by human operator's own senses.

2. Task is observed indirectly through artificial sensors, computers and displays. This TIS feedback interacts with that from within HIS and is filtered or modified.

3. Task is controlled within TIS automatic mode.

4. Task is affected by the process of being sensed.

5. Task affects actuators and in turn is affected.

6. Human operator directly affects task by manipulation.

7. Human operator affects task indirectly through a controls interface, HIS/TIS computers, and actuators. This control interacts with that from within TIS and is filtered or modified.

8. Human operator gets feedback from within HIS, in editing a program, running a planning model, etc.

9. Human operator orients him- or herself relative to control or adjusts control parameters.

10. Human operator orients him- or herself relative to display or adjusts display parameters.

Figure 13.1c.

The control room context

With the increasing use of computers to control processes and systems the tendency for the control of such systems to be centralized has continued, usually from one primary location, a central control room. Control rooms might be found in many different domains – transport systems (e.g. air traffic control, railways, metros etc.), emergency services (police, fire, ambulance), industrial processes (steel, chemical, food etc.) power plant (nuclear, electricity etc.), security applications (banks, prisons, public or private buildings) etc. Importantly, although the term used generally and in this chapter is 'control room', these may in fact be a series or suite of rooms. 'A *control room* or *control centre* is the place where one or more people, sitting at control desks, conduct control activities. A *control suite* ... is a group of co-located, functionally related rooms, including a control room ... [and] *control complex* is ... a series of functionally related rooms which are on different sites ...' (Wood, 1994).

As Hollnagel (1993) points out the complexity of technological systems has continued to grow over the last two decades and the demands for the safety and efficiency of these systems have grown apace. Increasingly the overall reliability of such systems is limited by the reliability of the people who operate them. As technology has improved and the reliability of the hardware components has increased (although perhaps not as much as engineers might like to think!), so the human operators can become the most unreliable components of a system, unless appropriate ergonomics input is made. It is imperative that the interfaces which provide the means of interaction between the system and the people who operate them, allow the operation of the system to be as efficient and error free as possible. This chapter addresses the issues to be considered in the design of interfaces to support the operation of such complex systems.

In line with the increase in system complexity has come an increase in the amount of information available to the operator to support system operation, and the capability to display this information in a variety of formats. Control room interfaces are now primarily based around several VDU (Visual Display Unit) displays—generally Cathode Ray Tube (CRT's) but also including liquid crystal or other flat panel displays—rather than wall mounted 'panel displays' which use dials and similar display types to present the information to the operator (Figure 13.2). This has had a significant impact on the design of human–machine interfaces. No longer is all the information concerning system parameters available to the operator at the same time. VDU's give far greater flexibility in the design of the information, the same data can be presented in a variety of different formats, but there is a limited amount of information that can be presented on one display screen. To overcome this one operator may have several displays from which the system can be operated. 'How many VDUs do I need?', is one of the questions most often asked by control room designers (Figure 13.3). It must be remembered that, as with any computer interface, the operator has to deal with the complexity of both the plant/process and also the control and surveillance systems. Navigating through a modern VDU-based system requires little manipulative skill but considerable cognitive workload can be involved.

Process for analysis and design of human–machine interfaces

It is possible to define many different processes for design of display/control interfaces. In a sense, this can be regarded as a particular instance, or even as a major part, of human–machine systems design, and thus the types of design process described in chapter 1 will be relevant. Many other processes can be found in the literature, covering interfaces or systems; much of what is said about design and evaluation processes for human–computer interaction in chapter 12 will be relevant, especially given the now almost universal use of distributed computer technology in control rooms as well as the computer-

(a)

(b)

Figure 13.2. Control rooms showing different technologies. Figure 13.2a shows largely floor and wall mounted panel displays at CEGB (courtesy of Tom Mayfield, Rolls Royce and Associates). Figure 13.2b shows a largely VDU-based control room (courtesy of BP Kwinara refinery and Honeywell Control Systems Ltd.)

Figure 13.3. Operator working from a number of VDUs displaying different information in different formats (courtesy of Shell Haven refinery and Honeywell Control Systems Ltd.)

based nature of control-display systems, such as mimics, which have been around for a long time. It is important to emphasize again here that it is a sterile exercise to differentiate 'computer' from 'traditional' interfaces and so this chapter will look at principles and processes germane to design of both. However, we will not cover the sort of territory such as office automation that the reader might expect to find in chapter 12. As far as human–machine system design generally is concerned, the reader is pointed to a vast array of sources of more detailed information, in particular Bailey (1989); Salvendy (1987); Sanders and McCormick (1992); sources specific to control rooms are listed at the end of the introduction to this chapter.

We make no attempt either to define a full process for control-display system design or control room design, which would have integral staged evaluations and iterations embedded within it. Nonetheless, in order that the chapter be given a structure, it is built around a set of activities which, taken in order, might be said to outline an analysis and design process. It is suggested that the procedures and activities shown in Table 13.1 comprise a logically ordered listing of what must be done to develop control/display interfaces for control rooms and human–machine systems generally (for comparison, the proposal from the draft Control Room Ergonomics Standard ISO 11064, part 1, is shown in Appendix 1). It is important to note that the selection of specific interface types and their development or their specification do not

Table 13.1. Human–machine interface design—factors and choices

---

INITIAL ANALYSIS

1. User and task analysis
    Establish tasks and users (operators *and* others)
    Allocation of function
    User needs and constraints; skills analysis
    Task analysis/synthesis
2. Mental model and behaviour assessment
    Consider potential operator mental models
    Assess expected task behaviours
3. Environmental influences and circumstances of use

OUTLINE DESIGN

4. Information analysis
    Consider on-line and off-line information
    Establish information content, sequence (and form)
5. Analysis of potential operator overload/underload
6. Determination of control dynamics
    Analyze or model operator control behaviour
    Assess needs for aiding, quickening, prediction etc.
7. Selection of modality
    Displays
    Controls
8. Prototyping and formative evaluation

DETAIL DESIGN

9. Display instrument specification
    Meet criteria for form then content
    Integrate displays
10. Control specification
11. Control—display integration
12. Detailed evaluation
13. Integration into control room

EVALUATION

14. Evaluation and modification

Source: Author's (JRW) own records.

take place until very late in the process. The implication is that development cannot merely be built around design guidelines and equipment selection rules; before any such guidance can be used the designer must gain a thorough understanding of user, task, system and environment requirements in a detailed analysis.

The remainder of this chapter takes up discussion of the activities defined in Table 13.1. First we will examine in some detail the initial analysis activities—establishing user needs, constraints on design and so on. Second, there is a more cursory look at initial design decisions in terms of the information content and modalities in display and control. Finally, some of the available guidance on interface specification is reviewed; it is obviously impossible to cover all of this and so we have chosen to concentrate upon display systems with particular application in control rooms.

Initial analysis activities

User needs and initial task analysis

As with all ergonomics analysis and design, we need first to establish the requirements for the system, to identify and describe tasks and users, which will involve carrying out a task analysis (see chapter 6 and below). It also involves making some first level decisions on the functions that must be performed and these are described at a gross level as activities that are needed to meet the system objectives (and usually collections of tasks). We need to consider the balance of responsibilities for functions between people and computers or other equipment (allocation of function) and between different people (division of function). The outcomes and decisions made on the basis of such analyses are not irrevocable. They can be returned to, reviewed and revised, in each subsequent stage, on the basis of new or amended information, ideas and opinions. Moreover, the task and user analyses themselves are not once-and-for-all exercises; they will be revised as appropriate in later stages on the basis of new information and decisions.

Where an existing interface is being assessed and analyzed many of the methods, measures and techniques described in the early chapters of this book will be appropriate, particularly verbal protocol analysis (chapter 7) since the work carried out using information interfaces is often not directly observable. At this time it is important to identify any likely significant constraints on the interface from the potential users or their tasks. Factors such as user experience, training and support must be predicted or assessed in terms of the ranges expected to be found, and then accounted for within design. Certain user attributes may determine a major design decision at the outset. Disabled users are an obvious case but other needs may be to design for a variety of cultures and languages, or to understand that military and civilian operators may well behave very differently from each other in circumstances where a technology is being transferred from defence to industrial application.

Often there will be one operator or a team of operators who are highly skilled and knowledgeable about the system, but there may also be a variety of other users of the system, with a wide range of activities, who need to be considered in the design of the interface. Such users include those responsible for maintenance managers or supervisors who require information on system performance, and systems engineers who may require information on the way in which the control system itself is performing. For supervisors, an early decision will be required on the degree to which they are monitoring systems or operators, which will determine whether emphasis should be an equipment position for access, or on workstation configuration for communication. We must appreciate the different levels of operator knowledge involved in working with any particular system, knowledge we may wish to provide, support, enhance or merely to be aware of in designing interfaces. Wirstad (1988) provides a listing of such potential knowledge in process control, including categories of:

- plant layout
- function, construction and capacity of components
- 'manoeuvring' (location and operation of controls and displays)
- system function, construction and flows
- process theory and practice
- identification of disturbances and consequence prediction
- measures to take for disturbances
- procedures for serious incidents
- organization structure
- administrative arrangements
- safety regulations
- supervision.

As well as knowledge required of operators we must also be aware of all the tasks that must be supported by the interfaces, summarized as follows:

Procedural	Tasks that involve following a pre-determined sequence of events
Sensory-motor	For example the physical manipulation of input devices
Communication	The transmission of information between operators without the information being translated into another format (i.e., verbal communication or logs)
Monitoring	The surveillance of the system to identify any change in system status
Fault detection	The identification of an abnormal or unexpected system status
Decision making	The selection between alternative options/actions
Problem solving	The process of resolving uncertainty about system states. A particular kind of problem solving that is especially relevant to this context is fault diagnosis
Prediction	Judgement of likely future system states.

For interface design there is a variety of task analysis methods that can be applied to analyze the information requirements of the users; these need to show who will use the information, when it will be used, what it will be used for and how it will be used (see Table 13.2 later). Such an analysis may also provide a basis for estimation of the potential errors that could occur in performing each task element and their probability of occurrence (see chapter 31 and Kirwan, 1994). We can also use some analyses to determine what the information requirements of each of the tasks are, thus identifying the content of the displays that should be provided to support performance, and first decisions can be made on the display format for this information (see Figure 13.4, for instance). Of course the design of interfaces is dependent on the information available which in turn is dependent on the system hardware and software. Whilst technology may mean that there is a large pool of potential data for display, it is often the case that the display format options are limited by the control system software supplied by the manufacturer of the control system.

The selection of a task analysis method will depend on the interface design application it is to be used for. For example it may be used to identify the frequency of use of the different displays and controls, the content of the information required for VDU-based displays, or the optimal format for displaying task information. Methods which may be used to identify information requirements include Job Process Charts (Tainsh, 1982, 1985 and Figure 13.5) and a modified version of Hierarchical Task Analysis (Piso, 1981; Rajan, 1983) (see Figure 13.6 and chapter 6). Other methods which may be useful in this context include Task Analysis for Knowledge Description (Diaper, 1989) to identify the knowledge requirements of the operator and Operational Sequence Diagrams (Johnson, 1993) to provide information on the sequencing of control and display use. In addition the layout of panel displays can also be analyzed using simple methods such as link analysis or frequency counting (see chapter 2). Where there is no existing interface for comparison, a task synthesis can be carried out using some of the techniques of task analysis.

Task element	Information requirements	Task type	Display format options
Carry out pre-start up checks	Valve positions Pump in a ready state Isolations in place Current maintenance activity	Procedural Checking	Mimic display Sequential Checklist
Initiate start up sequence	Pre-start up checks complete Feedback that sequence has initiated	Operational input Checking	Sequence display
Monitor start up for faults and to check correct sequence	Progress through start up sequence Faults occurring	Monitoring Fault Detection	Alarm display Sequence display

Figure 13.4. Extract from a task analysis to identify information requirements for display design for a pump start up task (source: Rajan, 1993)

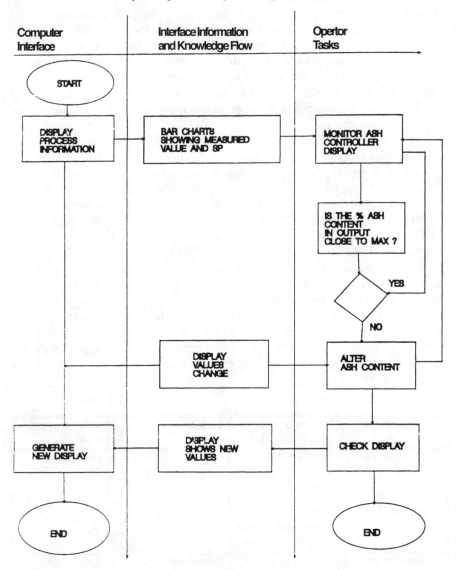

Figure 13.5. Job process chart for ash controller task (Author's own records—JAR)

As complex control tasks are primarily cognitive rather than manual hands-on activities, methods of collecting information for task analysis must reflect this, including verbal protocol analysis (chapter 7), interviews (chapter 3), direct observation (chapter 2), archives (chapter 4), scenario diagnosis and walk-talk throughs (Kirwan and Ainsworth, 1992). The reader should also refer again to chapter 6 on task analysis generally.

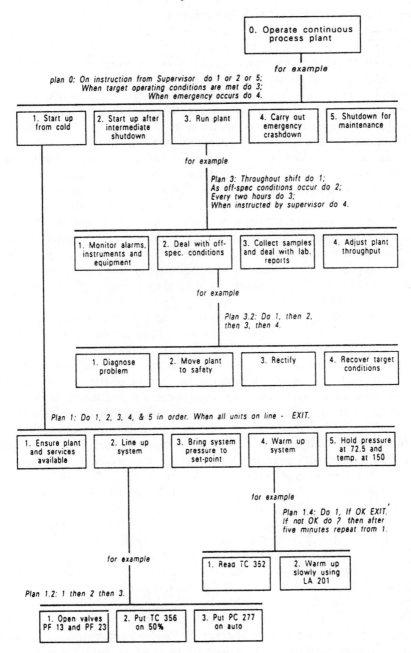

Figure 13.6. Hierarchical task analysis for a continuous process control task (Source: Kirwan and Ainsworth, 1992)

User mental models and task behaviours

User mental models

Imagine someone trying to use a word processing system for the first time, trying to understand control functions in order both to use the system *and* to complete the task whilst employing their previous experience of type-writers: or consider an operator at a remote control task who must carry out a job from a vantage point not used before, trying to get orientated and guide the system through critical operations in the most reliable, safe, effective manner: or the worker at a computerized metal forming machine who must try to understand why parts are showing hairline cracks, and what combination of the many input parameters available would solve the problem, and must do this from process displays as well as by directly viewing the operations: or a self-directed team in electronics assembly, responsible for its own scheduling and trying to work out from a distributed information system why they have a parts delivery bottleneck: or a maintenance engineer on a continuous process plant trying to relate information from system diagnostics to that on the control panel and to that on a giant mimic display. In all these cases, it is reasonable to think of the people involved as having or forming a *mental model* of the system. This mental model (or conceptual model) may be accurate or inaccurate, usable or worthless, but nonetheless the notion of mental models is attractive and useful for ergonomists.

In fact, in many cases the worker will construct and use several mental models: perhaps a symbolic one of the interaction of the variables in the systems—electrical, or chemical, or mechanical; a pictorial one of the form of the system being worked with: and a model of rules governing the operation (correct or faulty) of the system. These models can be relatively concrete or abstract, and may involve representations formed from the system itself, from operating, emergency or maintenance procedures, from instructions and training, or from other systems worked on in the past.

Much debate in the area of mental models has focused on the degree of formality needed in their identification and representation. Within cognitive psychology there is the understanding that the notion has utility only if mental models can be described in computational form (e.g., Johnson-Laird, 1983). On the other hand, within human factors we are usually willing to postpone questions of how people represent and use knowledge in favour of understanding what knowledge is represented and how it is used to make inferences in specific domains (Payne, 1988).

The majority of human factors literature, and especially in human–machine systems, appears to refer to a conceptual (or non-computational) mental model. It is regarded almost as axiomatic that people form some kind of internal representation that constitutes their topographical, structural and functional understanding of a physical system and which allows them to describe, explain, understand and predict system behaviour. It is easiest to conceive such a model as comprising a system simulation that can be reconstituted and run in order to derive or confirm understanding. In ergonomics/human factors there would appear to be some agreement about mental models (Nielsen, 1990; Norman, 1983; Payne, 1988; Rasmussen, 1986; Rutherford and Wilson, 1991; Wilson and Rutherford, 1989) in that they:

- are internal representations of objects, actions, situations, people, etc;
- are built on experience and observation, of the world and of any particular system;
- are simulations run to produce qualitative inferences;
- and constitute topography, structure, function, or operation of the system;
- may contain spatial, causal, contingency relations;
- allow people to describe, predict and explain behaviour;
- underpin people's understanding and behaviour;
- are instantiated each time they are required, and are parsimonious, and therefore are incomplete, unstable and often multiple.

The last point, made by Norman (1983) and others, has high relevance for control/display interface design. For example, the models instantiated by, say, an operator to assist in fault diagnosis at a flexible manufacturing cell will vary each time in type and content. If the problem is product quality related then the cell may be conceptualized in functional form, modelled in terms of the series of transformation processes and the tooling needed to do this. Alternatively, if the problem is to do with hold-ups in components delivery then the cell may be conceptualized physically and spatially in terms of element flows and bottlenecks. In many cases more than one mental model may be formed, which will nonetheless overlap in their content and how they are employed, and which will have gaps or 'inaccuracies' according to operator experience and training.

There are many serious questions about the notion of mental models, including even whether the theoretical notion is necessary or if existing theories of knowledge representation and inference suffice (see, for instance, Bainbridge, 1991). We may be talking of a concept about which there is little agreement on definition, identification, representation and utilization or even adequacy in methodology for their identification (Rutherford and Wilson, 1991). Notwithstanding this, if we can predict or understand even in some fashion what mental models a new operator or user might hold about a system and its relevant domain, and what model they might build through subsequent interaction with the system, then we can improve interface design, training, operating procedures and so on. By understanding the potential users' mental models, and by adapting their own conceptual model accordingly, designers might develop a 'system image' that better matches, sustains and helps develop an appropriate user mental model.

Task behaviours

Extending from the notion of mental models is the consideration of task behaviours found in the work of Rasmussen and his associates (e.g., Goodstein *et al.*, 1988 or Rasmussen, 1986). We can see elsewhere in this book the pervasiveness of Rasmussen's model of skill-, rule- and knowledge-based (S-R-K) behaviour, with its relevance to human reliability and to accident causation (chapters 31 and 32). There is also an intimate relationship between expected task behaviours and the information displayed to support them; in a basic description of how operators perceive information within the S-R-K model, Rasmussen uses a simple display/control system as illustration (Rasmussen, 1986, p. 107).

Skill-based behaviour is what is shown in tasks such as 'in-the-loop' controlling or steering, adjusting and calibrating instrument settings, or assembly tasks. The behaviour is akin to sensory-motor performance—where there is a fairly direct connection between sensory input and motor output with little mediation in cognition. In essence, skill-based behaviour is shown in familiar situations and is where the 'operator' recognizes a 'signal' from the system and understands that this requires a normal routine, then executes a well-learned skilled act more or less automatically. Generally this type of work is undertaken with simple feedback control (comparison of the actual and intended states) which provides error information and thus defines the motor output response. However, there are suggestions that skill-based performance may also be based on feed-forward control and knowledge of the environment, for instance in riding a bicycle.

Rule-based behaviour is said to require a more conscious effort than does skill-based behaviour, and involves following a set of stored rules—in the mind or written down. Here, performance is goal oriented and takes place in familiar but non-routine situations; the operator perceives a 'sign' indicating environmental state(s), and then he or she uses learned rules and procedures once they have recognized certain cues. Control is fed forward through the stored rules. These rules will be derived empirically through experience, or communicated to the operator by colleagues, institutions or training.

The boundary between skill- and rule-based behaviour is indistinct, and may depend in the same situation on training or attention levels. A person working in skill-based mode may not be able to describe how they work, because the behaviour required may have become so automatic to them. The result of working in rule-based behaviour is frequently to co-ordinate and control a sequence of skill-based acts.

At the highest level is knowledge-based behaviour. Here the operator is in an unfamiliar situation with no, few or partial rules available from past experience. A goal is formulated by the operator, based on their perception of the state of the world and on some overall or global aims. Perceiving these 'symbols' allows them to develop a plan, using knowledge, reasoning and experience, and the goal they themselves have formulated helps them work to that plan. The plan itself may be selected and tested through a process of conceptual or physical trial-and-error.

Discussed in outline terms as above, it might seem that Rasmussen's model is fairly simplistic, perhaps naive, and maybe does not capture as much about human behaviour as other models of human information processing. Also, subsequent authors have not always distinguished what they mean by 'skill', 'rules' and 'knowledge' clearly and consistently. On the other hand, the number of lines of important human factors work which have made use of the model is legion.

Within consideration of control-display interfaces, the least that might be said is that any displays must, in general, support all three kinds of behaviour. For instance, the same information display might be needed to guide normal operations, routine maintenance, and fault diagnosis in abnormal pressurized situations, and we have already seen that what starts out as a knowledge-based task may subsequently decompose into a rule- and then a skill-based task. As an example, an in-car navigation and diagnostics system might be used (1) to feed back a continual check on the state of the car sub-systems or position on the road for the driver, (2) to support a routine check by the driver of the brake, suspension, tyre and engine sub-systems, (3) to allow substantial replanning of a route whilst on the move in heavy traffic, or (4) to be used by a garage mechanic in diagnosing fuel injection problems. Not only should the display support all three types of behaviour, it should also allow an operator to move 'up and down' between levels, in terms of information detail and degree of abstraction. Furthermore, in many circumstances it will be valuable for the system itself to support the operator by providing guidance on appropriate behaviour for different situations.

Environment and circumstances of use

Although the person, process, controls and displays constitute a closed-loop sub-system at the heart of human–machine systems, this sub-system is itself open to the environment. Interaction will occur between this task and interface sub-system and all physical, psycho-social and organizational environment factors.

Basic physical constraints are imposed by the location of the interface (Figure 13.7, a and b). If such a location is a control room, then the most basic of these are the architectural constraints of the building in which the control room is located. For example, there may be architectural constraints on the positioning of consoles (e.g., blast proof walls or channelling under raised floors) which reduce the options concerning the number and type of interfaces that can be housed in the control room. There may be environmental constraints (for example the temperature and humidity of the room) affecting suitability of the equipment. The size and shape of the control room will have an impact on the workspace design and consequently on the number and positioning of the VDUs. This will in turn determine how operators

(a)

(b)

Figure 13.7a and b. Location of displays and controls placing physical load on work of operators (courtesy of Tom Mayfield, Rolls Royce and Associates)

will interact together and work as part of an operating team. Social contact should be catered for by grouping operators so that conversation is possible without compromising efficiency, especially important in larger facilities during quieter periods when staffing levels are lower. The interfaces provided in a centralized location such as a control room may be supplemented by interfaces actually on plant or in other locations, to provide information on system status. These may require ruggedization to allow them to conform to safety requirements for equipment to be located on the plant or factory floor. Other specialized requirements are for seismic protection in certain geographical regions or for protection against shock or vibration on a ship or aircraft.

Normal and emergency operation, and planned maintenance

The interface design has in part to predict what the performance of the operator will be; task analysis and consideration of users provide a basis for this to be determined. The interface must take into account the consequences of events and look at different scenarios and modes of operation and predict the information needs of the operators under these different scenarios (Figure 13.8). This will include normal operational scenarios and also abnormal conditions, emergencies and maintenance. In some emergency situations the information for maintaining safe and effective system status may be provided through existing display formats; in other situations special formats may be necessary.

In the case of an abnormal event VDU-based displays are vulnerable to factors such as loss of power. For reasons of reliability and safety, panel-based displays may be provided as back-up to, or instead of VDUs (see later in this chapter). The requirement of an overview of plant state is difficult to meet with VDUs alone, and state boards or hard wired mimics may be needed here. Such displays are also useful for shift changeovers.

It is important to ensure that information concerning critical system states is available, to allow the operator to monitor the system and to bring it to safe status, perhaps in the form of an emergency shut-down panel remote from the control room. In an emergency, automatic safety systems may operate and thus intervention by personnel is prevented or restricted. Information provided to the operators is critical to ensure that they know what is happening, that appropriate action is taken and that personnel are able to anticipate future system states. In emergencies it is imperative that the interface does not add to the workload of staff. Designers of some systems have adopted a strategy to reduce the tendency of people to react in certain inappropriate ways under stress, by ensuring that user intervention in the system in the event of an emergency is prevented for a set period of time to allow the operators to familiarize themselves with the conditions of the system and with information concerning the failures. This avoids the tendency of operators to try and fit the information they have into familiar diagnoses and solutions,

Figure 13.8. Interface design and layout of workstations should allow for all tasks, including ones not directly using interface elements

rather than to explore a range of solutions to the problem in hand (this is shown in availability bias and cognitive lock-up—see Hogarth, 1980 pp. 204–234). An example is the Three Mile Island incident (Reason, 1990 pp. 189–191, 251) where operators and other personnel considered that they had correctly diagnosed the origin and cause of failure and took corrective actions based on this. Additional information which would have indicated that their diagnosis was incorrect was fitted into the original diagnosis or largely ignored, leading to a delay before the correct cause was identified and mitigating action could be taken.

There are a number of factors which impact on interface design under abnormal operating conditions. These include consideration of differences in the tasks to be performed, differences in personnel (including engineers or managers who may be unfamiliar with the displays), differences in workload and thus increased stress from unfamiliar operational conditions, and the fact that during an abnormal scenario the control room is often used as an emergency control centre, which may change needs for access to consoles or for numbers of people in the room and will certainly increase background noise and distractions.

In the design of VDU interfaces for complex systems, maintenance is a critical and often overlooked factor. Studies have shown that, in many cases, errors which are the origins of system failures actually occur during the maintenance phase of the system life cycle (Geyer *et al.*, 1990). Maintenance personnel frequently use the same VDU display formats (or at least the same system) as operating personnel. Even if the information presentation is not common between these two groups of users, information concerning which equipment is undergoing maintenance (both emergency and planned maintenance) is essential to operating personnel, to assure the safety of personnel and equipment during the operational phases of the system. Although it is usual good practice physically to lock-off controls at the panels for equipment undergoing maintenance or repair, it is more difficult to do this reliably working from a VDU. Also, in many situations a shut-down may not be possible, for instance in air traffic control or the emergency services. A further potential problem in safety critical systems, where instrumentation must be checked at intervals as short as 24 hours, is that mistakes in checking can bring about an unforced shut-down of the plant.

Outline design decisions

Information analysis

The purpose of any display at work is to deliver information to someone to allow them to perform a task, whether this be passive (e.g., monitoring) or active (e.g., calibrating or fault diagnosis). In doing this, display design must eliminate or reduce errors of sensing, recognition, perception and decision; in other words, people should be able to recognize the relevant information against its background, distinguish its meaning to understand what is required and use the information to make decisions and perform tasks. The right information must be communicated in the right form to the right person at the right time.

One useful technique to employ in early screening of display designs can be borrowed from the world of Work Study—the systematic questioning process, used in examining a method and process analysis. This can be applied mainly to the information content of displays (but to an extent also to their form), and is a systematic look at *what* information is to be contained in displays before any consideration of *how* such information is to be presented. The questioning procedure shown in Table 13.2 allows a systematic and rigorous look at all parts of the system, whether in analyzing existing displays or assessing needs for new ones.

Table 13.2. Questioning procedure for information analysis

What information is to be displayed?	Why is it necessary?	What else could be displayed?	What should be displayed?
Where is the information to be displayed?	Why there?	Where else could it be displayed?	Where should it be displayed?
When is the information to be accessed/ communicated?	Why then?	When else could it be communicated?	When should it be communicated?
Who is to have/use the information?	Why them?	Who else could have/use it?	Who should have/use it?
How is the information to be presented?	Why that way?	How else could it be presented?	How should it be presented?

Source: Author's (JRW) own records.

Operator underload or overload

Increasingly people at work are interacting with machines or processes almost entirely through an interface rather than by direct sensing and physical actions. As a consequence, the interface will be a major determinant of the load on the operator. Broadly speaking we can talk of physical workload and fatigue, and of mental workload and fatigue. In general we should seek to minimize static or dynamic physical workload (see chapters 22, 23 and 24) imposed on operators by modes of control (e.g., heavy cranks, multiple valve opening, etc.) or by equipment layout (e.g., the position of a display giving a worker neck ache to see it, or controls requiring reach to awkward positions).

For mental load the position is less clear. Despite the theoretical arguments against it (see chapter 26 for instance), the notion of the arousal curve, indicating lower performance or higher errors when people are at high *or* low arousal levels has great face validity for designers and allows general guidance to be given. Stressors from the work (e.g., time pressures), environment (e.g., intrusive noise), and personal factors (e.g., lack of sleep) should be maintained at intermediate levels, certainly not at extremes. Consequences of loads which are too high or too low may be lapses in attention, cognitive lock-up, less co-ordination and timeliness in performance, irritability and so on.

If we take driving as an example almost all of us will have experienced anxious or angry reactions when roads are very busy, we are late, direction signs are confusing and other drivers are behaving erratically, with consequent effect on our driving performance. On the other hand, motorway driving at night can produce very low arousal levels with consequent small and large errors of judgement, and driver fatigue being shown by their fidgeting and being unable to control speed and course simultaneously.

Control dynamics

Although this is in theory a logical place to consider control dynamics in an idealized process for human–machine interface design, in real life most of the considerations discussed above would have been in the light of knowing at least whether the task involved discrete or continuous control. Examples of the former are switch and valve operation to start-up a plant; the latter is when an operator has to keep a plant working with several parameters (e.g., temperatures, flow rates, pressures) held at specified values, and when fluctuations or transients mean that continual adjustment of the variables is needed. Continuous control is also exhibited in most kinds of transport, whether a car, boat, helicopter or submarine. Many modern work tasks require a mixture of continuous and discrete control, and interface design must allow for this. Use of many desktop PCs these days will involve both discrete control—e.g., the sequence of keyboard initiated steps required to edit text, and also continuous control—e.g., use of mouse or space-ball or graphics tablet to input graphics or to 'walkthrough' or around the display. Handwriting is a fundamental form of continuous control.

Amongst the factors to be accounted for in a continuous control interface are:

- what order of control is required? Is the operator making simple step (or zero order) inputs to control distance and position, or is the control of rate (1st order), acceleration (2nd order) or of the much higher orders found in complex systems?
- if the operator is tracking a course, is this pursuit tracking (where target and cursor both move) or compensatory tracking (where change is shown in the discrepancy between target and cursor)? The former mode has control advantages but the latter can make for more effective panel layouts by saving space; whatever the choice, it will have considerable implications for display design;
- would any form of assistance for the operator be of value? We may apply control aiding (e.g., control can be of cursor position or rate of movement acording to whether large distances must be tracked), or display quickening, preview or prediction (all techniques to help when operators might want advance warning of what is to come or of what effects their actions might have).

For a full discussion of human factors issues to do with continuous control, the reader is referred to Kantowitz and Sorkin (1983), Sanders and McCormick (1992) and Wickens (1992).

Selection of control and display modalities

Although controls are most often hand operated and displays are most often visual ones, other modalities are possible. Control may be effected by feet or legs, especially when power must be transmitted, when the hands may be fully occupied and when the control task is one of discrete actuation. Examples are an on-off button for a power press, pedals in most forms of land transport, a foot mouse for a computer or a pedal/shuttle mechanism as on a sewing machine. Novel forms of control may be found with computers, especially for the severely disabled, including use of eye or head movements. Finally, one of the most common forms of input there is, albeit more for human–human systems, is speech. Although considerable technical and human factors difficulties remain, speech is increasingly used for system input even if in a limited and often discrete fashion. Examples are input of commands in a fault inspection task (e.g., 'Pass', 'Red Fault', 'Repeat', 'Scrap'), or in an automated warehousing system (e.g., 'Deliver', 'Transport', 'Rack Three', 'Search').

Selection of mode of input, and especially any decision of whether to use other than hand controls will be on the basis of the task analysis and earlier design decisions. This is true also for display modality selection. Displays may be tactile (e.g., braille or shape coding of control knobs), proprioceptive (e.g., sensing of correct speed and track for cornering in a car), but most will be visual or auditory with emphasis on the former. In general, an auditory display may be chosen when:

- messages are simple and short (even for speech synthesized displays);
- immediate action is required;
- no later referral to the message is needed;
- the information is continually changing;
- the message refers to events in time;
- work is in poor viewing conditions—for instance low illumination or high vibration;
- the recipient is moving around;
- the recipient is receiving a large amount of other visual information;
- the task environment does not have a high degree of background noise.

From the above, it is clear that a principal use of auditory displays is in warnings or other short signals for action. The vast majority of interaction with systems though will be via visual displays and the remainder of this chapter concentrates on this (but see Sorkin (1987) for further discussion of auditory and tactile displays). The reader should, though, bear in mind the role that auditory signals may take as the most important source of information in communication centres, with a VDU providing supporting information, and also the increasing use of CCTV. (Shirley, 1992, provides an excellent fuller look at display issues for industrial – and particularly process control – applications).

Interface specification—display of information

We cannot cover all relevant aspects of interface specification in this chapter. The available research and guidance on even specific parts of this or on particular applications—e.g., computer input devices, or screen coding—is legion. Therefore in this section we will have to be selective.

The interface provides the basis for communication between the system and the user. It provides information to the user on current status and performance of the system. In turn the user can manipulate and monitor the system via the interface. In the design of an interface the consideration of the information needs of the user(s) is probably the single most important element, and it is the display of information we concentrate upon here. Within this, we will first of all compare the two display formats of most applicability to control room design—panel displays and VDUs—before looking at some methods of information coding on displays. The reader should, though, bear in mind the role that auditory signals may take as the most important source of information in communication centres, with a VDU providing supporting information, and also the increasing use of CCTV. (Shirley, 1992, provides an excellent fuller look at display issues for industrial – and particularly process control – applications).

Panel displays and VDUs

One of the principal choices facing the control room interface designer is that between panel and/or visual display unit (VDU) displays, the latter still mainly CRT's. Panel-based displays are usually wall mounted panels in modern control rooms, although they can be found as floor-based consoles. They are usually hardwired and will display the information in one format only, e.g., an analogue dial or light emitting diode (LED), unless information is repeated in other formats to provide redundancy for safety or improved performance (Figure 13.9). For such panel interfaces the critical decisions are

Figure 13.9. Control room detail showing hardwired panel displays and wall-mounted mimics (courtesy of Tom Mayfield, Rolls Royce and Associates)

how the information will be displayed (modality) and where it will be located on the control panel. Controls are also located on the panel amongst the displays and will usually be allocated to individual displays, rather than having the one keyboard with a mouse or joystick that is often the case with VDUs. This is because the different displays require individual controls (e.g., to allocate a set point) and because the size of a panel often means that it would be impractical for several displays to share one control without causing problems of display–control compatibility.

One of the most critical factors in the design of panel-based displays is the layout and grouping of information. Unlike VDU-based displays this formatting cannot be flexible and so must accommodate the full range of tasks that the operator is required to perform. It must also accommodate tasks where two or more operators are required to work side by side. For example, information that is required by more than one operator to carry out different functions concurrently will necessitate special consideration for display placement, or information redundancy through display duplication elsewhere on the panel. Alternatively, VDU overview displays can be projected onto a screen on the wall of the control room, to give a large dynamic overview of key parameters which is accessible to all personnel.

The other important factor in consideration of the design of panel displays is the relationship between controls and displays located on the interface. The location of the control associated with a particular display or range of displays must clearly indicate which display will provide feedback of the effects of that control action. The movement of the control must also conform to the expectations and stereotypes of the user in terms of the effect it will have on the display (see Wickens, 1992 pp. 324–334).

Over the last decade increasing technological complexity has meant that far more demands are placed on the design of the interface to convey relevant and timely information to the operator. The role of the operator is not only one of operation, but through that operation he or she has a major role to assure the safety of personnel and maintain the integrity of the equipment and the environment. A vast number of data points could potentially be displayed to the operator and the importance of displaying the right information at the right time is critical. The use of VDU-based interface technology means that many different formats and combinations of information are possible, so the interfaces can be configured to meet the requirements of a range of different users. However, the move from panel to VDU-technology is not without its disadvantages. The user of a VDU-based system will typically have several screens to access, but even so these screens can only display a limited sample from the range of data available. In addition the operation of the interface to access the information required to operate the system is an additional task which forms part of the operator's workload.

A comparison of panel- and VDU-based displays is given in Table 13.3 (see also Umbers and Gent, 1985).

The number and location of VDUs is dependent on whom they will be

Table 13.3. Panel and VDU displays compared

Advantages of VDU based displays

- They give an increased flexibility in the type and format of data display.
- The same data can be displayed in a variety of formats.
- Data can easily be manipulated on-line (e.g., historical trend displays can be shown over different time axes).
- The displays can more readily be updated to incorporate information about plant modifications and changes.
- CRTs are generally cheaper to install than panel displays.
- CRTs may be more easily maintained and are easily replaced if required.
- CRTs take up less operational space.

Disadvantages of VDU based displays

- Limited resolution (e.g., for the display of alphanumeric characters, although this continues to improve).
- The additional task of accessing data and manipulating the information may increase the time taken to perform tasks.
- It is not possible to display large quantities of detailed process data.
- There may be a time delay in accessing important plant parameters in the event of a fault or incident.
- A back-up system is needed in the event of failure of the electricity supply or the CRT hardware.

Advantages of panel displays

- A large quantity of information is displayed continuously to the operator.
- There is often more physical space for operators to work side by side.
- Operators can respond quickly to alarms and other process problems—as there is no delay in accessing data for either plant information or process input.
- Operators can more readily gain an overview of whole plant status.

Disadvantages of panel displays

- The displays may take up a large area.
- They are custom made and expensive to install.
- The data are displayed in a fixed format.
- Panel displays are difficult and expensive to modify.

---

used by and what they will be used for. The number of VDUs required for a given control room application can be calculated by envisaging a realistic worst case operational scenario, for example a situation where two unrelated failures occur concurrently. In such a situation a minimum of two screens would be needed to display information relating to each failure and a third screen would be needed to display alarm information. In general, most systems require a dedicated screen to display alarm or emergency information, allowing this critical information always to be available to the operator regardless of the current system status information that is being displayed at the interface (Bainbridge, 1991). By implication, the minimum number of VDUs required is three, but the number actually installed should be considered in the light of anticipated normal and worst case scenarios, number and types of users, number of data points for display, and the control room configuration.

In control room contexts, primary operation is often carried out from a CRT-based console arrangement, with panel displays providing an emergency back up, usually to convey information that for safety reasons must be on hardwired displays. Panels may also provide redundancy of information to supplement the VDUs. There may be several parameters critical to operation or safety which would tie up one or more of the display screens if the operators were to display them continuously under normal operational circumstances; one alternative is to present this information on a panel-based display (usually wall mounted)—for example, the fire and gas sensors on an offshore oil rig.

VDUs are not used only as replacement for panel displays, they are also used to replace paper chart recorders, for example for trend information (sometimes the option of a printed version of the displayed data will be available). VDU-based historical data offer advantages over paper media in that storage is more compact and the data can often be manipulated on-line (e.g., the time axis changed or data overlaid with other data). However, without a paper copy the information is not portable for on-line comparison with other information, other than data displayed on adjacent screens, and even then comparison is difficult.

For the display of alarm information, annunciator panels provide a means of instantly recognizing an alarm by its location on the panel, whilst VDU-based alarm lists can be more difficult to read and make it difficult to assimilate information at a glance.

In a context where there are both panel displays and VDUs available, the panel displays are usually used either as a secondary display medium or as back up in the event of failure of the display system or in an emergency. There are also 'hybrid interfaces' where VDUs display some of the system information and are located in a panel with other display media surrounding. These can be problematic interfaces to design from the environmental point of view; for example, lighting should be adjusted to suit both viewing of the VDUs and also the panel displays. Also, a systematic approach must be taken to co-ordinating information format between, say, screen-based and mimic-panel displays; particular attention must be paid to common coding (e.g. colour) between the two media.

Display formats

Within the basic VDU or panel display dichotomy for control room applications there are a number of possible display formats. Some can be applied to both technologies, others are mainly applicable to VDU-based technology.

These formats are: mimic, sequence, loop, alphanumeric, graphical/trend, deviation, object oriented, pictorial, fault and alarm.

Mimic displays

Mimic displays offer a graphical representation of the system (Figure 13.10). This representation can either reflect the functional relationships between elements of the system (for example, a schematic representation of process flows) or the geographical/topographical layout of the elements of the system (for example the physical configuration of a process plant: see Figure 11). Whilst research (Vermeulen, 1987) has indicated that neither is more effective when evaluated in terms of operational performance, many displays that are to be used for more than one task are based on a functional presentation. Such a presentation allows the sequence of system processes to be followed and can help the operator to visualize functional relationships between system elements, thus assisting with problem solving tasks. These display types may have a significant impact on the development of the operators' mental models of system function and so need to be designed carefully to ensure that any inferences an operator may make concerning system functioning are accurate and do not compromise system safety.

Sequence displays

The use of operating procedures to support a wide range of tasks is common practice in many complex systems. Such procedures may be in the form of a written manual or may be computer based, although procedural tasks which follow a predetermined sequence are increasingly automated, leaving the operator free to perform other operational tasks. (It is well recognized, though, that the trend to automate task elements wherever possible does not always enhance the safety or reliability of the system.) For an automated process sequence, the operator will need to know the stage in the process that has been reached, to allow monitoring and early detection of any abnormal events, anticipation of requirements for manual inputs or operator intervention, and, in the event of a fault, whereabouts in the automated sequence it occurred. Sequence displays can make this information explicit. Where procedural tasks are still the operators' responsibility then display requirements are to act as a reminder of, and guide through, the sequence of tasks. Sequence displays may take the form of textual lists (see also alpha–numeric formats), however, a graphical form such as a flow chart or network diagram is preferable (Visick and Simpson, 1985) particularly where the sequences are complex. The displays should provide an overview 'map' of the sequence, indicating operator inputs and giving a positive indication of progress through the sequence.

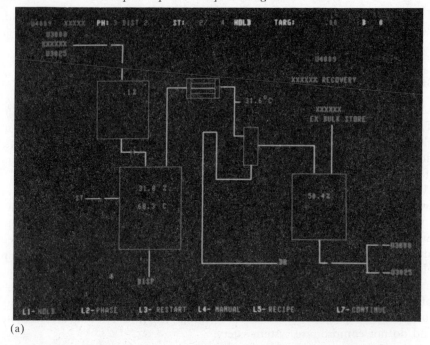

(a)

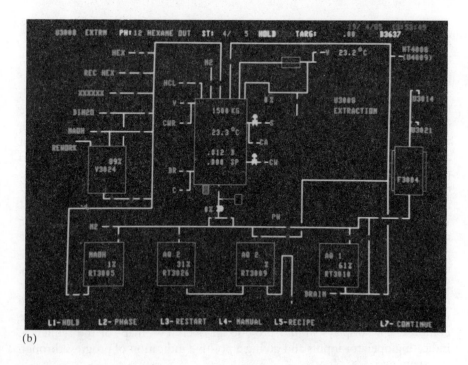

(b)

Figure 13.10 a and b. Examples of mimic displays (Source: The Boots Company plc)

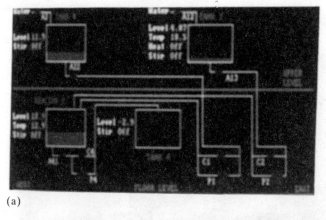

(a)

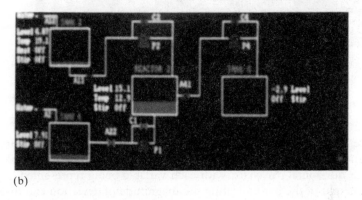

(b)

Figure 13.11. Illustrations of, (a) geographical mimic, (b) topographical mimic (source: ErgonomiQ)

Loop displays

Loop displays are frequently found in process plants as a traditional way of displaying process variable status (Figure 13.12). The information is presented in three values that relate to the controller, i.e., as a setpoint, measured value and output, and usually in the form of a bar chart. The displays may have originally been used to assist the transition to VDU-based technology as they mimic the faceplate displays of three term controllers. On panels the three levels are now most commonly represented using LED-based displays. Loop displays are most useful when there are a limited number of variables for display, detailed information on relative values of set points and measured values are required, and information is required on specific control loops within the system e.g., for maintenance.

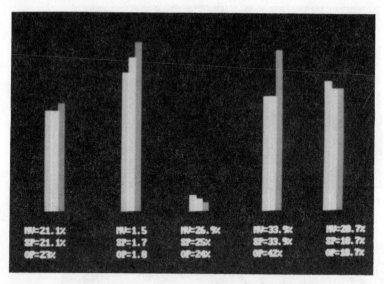

Figure 13.12. Example of a loop display (source: ErgonomiQ)

Deviation displays

These are used primarily for monitoring and fault detection tasks. The displays are used to indicate when a variable deviates outside given threshold values and to present the degree of deviation. The display usually takes the form of a horizontal bar chart, with each variable being represented by a bar and the height of the bar indicating the magnitude of deviation and threshold limits (Figure 13.13). If the bars are placed centrally on the display both

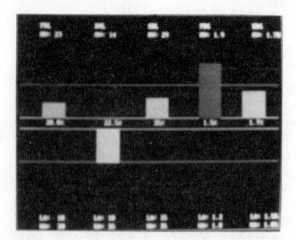

Figure 13.13. Illustration of a deviation display (source: ErgonomiQ)

positive and negative deviations can be indicated. Deviation displays are good for monitoring and fault detection to show movement towards abnormal conditions and to allow a rapid response if required.

Alphanumeric displays

At the most simple level, alphanumeric displays can either be static (e.g., printed instructions on a display) or dynamic (i.e., the values or the written information change over time). On panel-based displays information is usually static with the exception of digital numeric displays (e.g., an LED indicating quantitative information). On VDUs the formats for alphanumeric displays are varied, from lists to codes to natural language. The literature offers a range of guidelines on formatting (e.g., Shneiderman, 1992; Smith and Mosier, 1986), covering issues such as layout, density, and use of abbreviations. In control room contexts alphanumeric displays most commonly take the form of lists, static (e.g., process steps for a batch process) or dynamic (e.g., alarm lists) where the values of the variables or information displayed change according to system state. In addition, there are also dynamic displays in which the data change as a result of operator actions rather than process changes initiated by the system, for example, entering numerals to adjust a set point or entering default values prior to plant start up.

In practice there are many hybrid formats where sections of a display (e.g., overlays or windows) are alphanumeric in style. An example is a mimic display on which items can be selected to bring up a window overlay to give more detailed process information concerning a particular item.

Alphanumeric displays are most useful when there is a requirement for detailed and accurate system or process information, in the diagnosis of faults or reference values for instance.

Graphical/trend displays

In a primarily panel-based control room, trend and other graphically based information will usually be presented on chart recorders, because such information usually requires the provision of an historical record of plant operation and so records are often archived for future reference. VDU-based graphic formats are also used for the display of trend information. They offer the advantage that a large amount of historical data can be stored and readily recalled and that the time axis on historical data can be readily manipulated. Such displays can be used as a diagnostic aid and as a prediction tool to anticipate future plant states. The display of graphs on a VDU is limited by the size and resolution of the display although several graphs can be overlaid for comparison (a maximum of 4–6 is recommended). Line graphs are recommended for ease of use for most tasks (Figure 13.14).

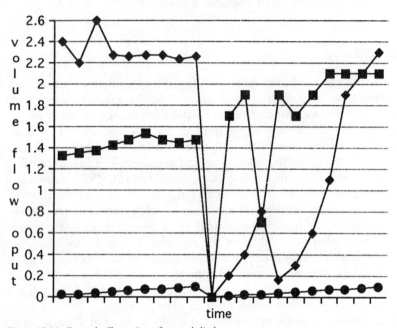

Figure 13.14. Example illustration of a trend display

Polar co-ordinate/object oriented displays

Polar co-ordinate or object oriented displays (also called integrated or shape displays) aim to provide rapid recognition of overall process or system status and can be used as a VDU-based overview display. They commonly take the form of a geometric shape, with the variables to be displayed indicated as points on the geometric object (e.g., at each angle of a hexagon or on the circumference of a circle). The variables are normalized so that when the system is functioning normally the figure will appear in the expected format and deviations in the system lead to deviations in the shape. Operators are able to recognize the shapes formed and so will be able to identify quickly any unusual or unexpected change in system status (Figure 13.15).

The displays are based on the principles of Gestalt psychology, that people will look for form or shape in information and that processing such information takes up less cognitive capacity than processing the individual data elements. The idea of object oriented displays was originally proposed by Coekin (1969) and developed by Goodstein (1981) and Wickens (1986). Whilst any geometric figure could be used, studies have proposed polygons, triangles and rectangles. Wickens reports on some of the tasks that the displays have been used for, which include monitoring, tracking, fault detection and fault diagnosis.

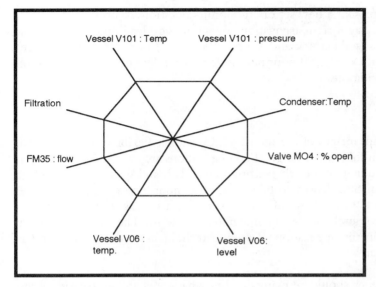

Figure 13.15. Detail from object oriented display format

Pictorial displays

VDUs can offer options for graphics displays that are not possible with panel-based formats, for example high fidelity simulations. This flexibility means that pictorial displays and other high resolution graphics can be used to present information in a way that is easily assimilated and used by the operator. The main advantage of such displays is that they are not constrained by convention and can be tailored to suit the task. As an example, in a paper production process plant a three-dimensional graphic was introduced to show process flows; operators found it easier to see the impact of problems on the overall continuous process than by using the more conventional two-dimensional mimic displays. Other applications where such displays might be useful include provision of a general qualitative overview of a process which can be backed up by detailed quantitative information, a pictorial representation of a particular item of process equipment in a complex plant, which assists the operator to relate operational problems and maintenance issues to the equipment on plant, and generally any information which cannot be easily represented using pre-defined display formats. We are increasingly seeing use of another type of 'pictorial display' in control rooms – close circuit television (CCTV) – for security purposes but also for many other monitoring applications. These include monitoring valves or vessels for steam or gas leaks, observation of personnel working in potentially hazardous areas or in an emergency when access is risky, identifying the location of system problems, or aiding communication in noisy environments or any other situation where there are advantages in giving body signals. Increased use of

CCTV will present a new set of ergonomics questions to do with detection, diagnosis, display system navigation, operator workload and control room environments (see Pethick and Wood, 1989, or Wood, 1992 for instance), although their use on a 24 hour basis is very dependent on associated lighting and plant conditions.

Coding

One of the advantages of the use of VDUs is their flexibility to display data in a variety of formats. A range of coding options are available to the display designer including colour, shape, brightness, flash and spatial coding, which permit information density to be increased without increasing the perceived display density. Careful use of coding can help to break a complex display down into comprehensible but integrated elements. The important factors to consider in the application of coding techniques to the presentation of information are:

- Consistency should be maintained in what codes are used across all displays. To minimize the likelihood of error this includes consistency between all VDU formats and the use of the same code on panel displays and on plant (where appropriate).
- Consistent spatial coding of information fields should be provided (e.g., the menu bar always located at the top of the screen).
- Any code should be easily learnt and follow any conventions and stereotypes of the user population.
- The code should be unambiguous—and items in the code should be readily discriminable from one another.
- Alphanumeric codes should not be too long.
- The code should be unique and not confusable with any other form of coding used within the system.
- Low saturation colours can be used to code 'background' information, that needs to be present but is not immediately important or critical to operation.

Guidelines for visual display coding types are given in Table 13.4 below. Other forms of coding of information received by different sensory channels have an important role in signal re-inforcement and distinction; this includes auditory coding (especially for alarms) and tactile coding of manual controls.

Navigation and structuring of control room VDUs

As already pointed out, the use of VDUs in control applications means that, whilst the operators can access a vast amount of system data, all these data are not displayed simultaneously and continuously. Therefore, they have the additional task of navigating around the display structure and accessing the

Table 13.4. Coding guidelines for visual displays in control rooms

Coding type	Description of code	Applications of code
Size	Use of different sizes of same display element to convey changes in magnitude etc.	Useful way of representing magnitude for levels, temperatures etc. When it is the only code only 3 different sizes are recommended in the code for easy discrimination. Non-linearity of CRTs can make comparative judgements difficult if variations in magnitude of size are small.
Shape	Use of different (usually geometric) shapes to represent categories of items displayed e.g., different classes of marine vessel on a display of a sea channel.	A relatively large number can be used and be readily distinguishable (up to 10). To be effective good resolution (e.g., to ensure a hexagon is not mistaken for a circle) is required and effective contrast, especially if colour is used. It is a useful means of coding for search, counting and comparative type tasks.
Alphanumeric	Use of letters and numbers to form codes to (uniquely) identify elements within a system—often used for labelling purposes e.g., to label valves with a common code on plant and on the display.	Avoid confusability between similar letters, numbers–e.g., X and K, S and 5. They should form a natural category and stand out from other items on the display.
Brightness	The use of different luminance levels to highlight certain items of information or to indicate degrees of magnitude e.g., temperature.	No more than 2 levels should be used to be distinguishable and effective.
Spatial	Use of the location on a display to give a meaning to an item of information e.g., to identify it as a menu item etc.	Should be used consistently throughout the system to indicate: − title pages − information fields − alarms − active and static display areas
Colour	Use of different displayed colours to provide differentiation between items on a display or to impart inherent meaning (colour should only be used as a redundant code.)	Colour has a variety of uses which are task dependent: − To identify and classify information − As a formatting aid − To collate related information across different displays − Reduction of clutter − Visual display structure − Aid to visual search − To connotate danger etc. − To highlight items or status change
Flash	Used to attract attention to items of information on the display.	Up to 4 blink rates are distinguishable—but less is preferable. Rate should be between 1–4 Hz. It is useful for redundant coding. It should be used sparingly. Operators should be able to cancel the blinking.

displays to find the required information. It is important that this structure is clear to the user and accessible in ways which assist, rather than detract from, task performance (Figure 13.16). Users may be offered a variety of options for accessing the different displays within the system, including indirect manipulation of a cursor or other pointing device via a trackerball or joystick, direct manipulation via a touchscreen, or the use of a keyboard with dedicated keys or a coded input. Often more than one option is offered, and for expert users shortcuts to access displays (e.g., to avoid having to progress through the levels of a hierarchy) should be made available. The design of control room VDU formats differs from those of, for example, office automation packages, as the users will be trained to a defined level and will be practised in the use of the system for normal operation. However, the training of new operators will often take place with them sitting by and observing experienced operators, and the structuring of the displays forms an important part of how their mental model of plant or system functioning is developed (see earlier).

The main issues to be addressed in display system structure and navigation are: what is to be presented on a page (the number of data items and the options for accessing other pages, e.g., menu items), the number of levels of pages in a system, the naming and grouping of items on a page, and the method of display selection.

Grouping of items into display pages and the means of navigating around

Figure 13.16. Control room operator utilizing six different types of information, controlled via virtual keyboards (courtesy of Honeywell Control Systems Ltd. and Tom Mayfield of Rolls Royce and Associates)

those display pages are interrelated design issues, dependent on functional relationships and task needs; for example, if the operator controls the system on plant as well as in the control room then topographical layout may be of importance. The naming of groupings is also important but can be problematic when groupings do not have one clear and unambiguous collective description.

Once the potential user population has been defined and the information content and format of the displays determined, then the division of information into display pages for a VDU-based system is the next stage, along with decisions concerning how those pages will be structured and accessed. Hierarchical menus are the most commonly used means of accessing different display formats. There has been a range of research aiming to establish the theoretically optimum structure; the optimal number of menu items on a page is dependent on the way in which the items are grouped, keying and system response times and the number of levels of selections. Snowberry *et al.* (1983) and Paap and Roste-Hofstrand (1986) recommend that a broad and shallow hierarchy be adopted where possible, particularly for experienced operators. Any increase in hierarchy depth can produce slower data access times and lead to an increase in errors in performance (Seppala and Salvendy, 1985). Although support of navigation between pages is commonly achieved by the use of menus, there are alternative mechanisms including: direct interaction (screen item selection), direct selection, tag code access, indexed access, keyword or command access. Navigation may also be supported by use of 'scenario-based' or 'state-related' displays, which have a task orientation; information presented at a point in time is related to current plant conditions and state (Mayfield and Seaton, 1988).

The structure of the displays within a control system context may not be a straightforward application as other influencing factors may apply. Functional or geographical groupings of items into displays may not give even sized and consistent groupings. A larger 'top' level may be required for overview displays. The detail of control information required for the different 'levels' of display may be fixed by organizational constraints.

Where possible the number of pages in a system should be reduced to a minimum, and the information presented on each maximized by the careful use of coding, structuring and labelling. One company has a basic principle that no page in the information system should be more than two control actions (keypresses, touches, selections) away from the current page (Mayfield, 1988). The grouping of items into display 'pages' is critical to effective and safe performance. As information can only be accessed in a limited way, the grouping of items impacts on how individual displays and groups of displays can be used in operation. In grouping (and naming) items for allocation to a display page, important factors include:

- Groupings should be consistent with the functional and causal groupings within the actual system.
- Groupings should be compatible with the operators' mental models of

system functionality. Operators should be able to anticipate which formats will contain which information and the groupings and structure should be meaningful.

- Do not separate directly interacting variables across pages—or where this is unavoidable provide an overview schematic of interacting functionally-related variables.
- Avoid operators having to carry information in memory from one display page to another.
- Displays which may be commonly used together should be checked both for consistency, and the division of information between them.

As a last point, the careful design of window systems is crucial to good control room interface design. They have enormous potential to assist the operator in navigation around the information structure, to underpin intelligent decision support sub-systems and to make the simultaneous display of different data structures or formats more effective. Multimedia interfaces, increasingly in use in control room contexts (Andrews *et al.*, 1973), will almost certainly require good windowed top-level screen displays. Virtual environments also are seen as having a role in control rooms of the future, and usability of these – including navigation – will be just one of a number of consequent ergonomics and technical issues (Wilson *et al.*, 1995).

Evaluation of interfaces

Finally, having carried out our initial analysis and conceptual and detailed design, we must evaluate the resulting interface, or maybe even the whole control room. Evaluations will also be carried out independent of design, for instance as an input to safety audits. Many of the techniques discussed in other chapters of this book must be applied here, especially those of chapters 6, 7, 12, 14, 19, 20, 21, 25, 27, 28, 31, furthermore, we would want to implement our new design in the light of lessons from chapters 35, 36 and 37). Here, though, we will present a set of guidelines written as a checklist, to guide any first level expert evaluation. These guidelines are particular to displays in control rooms, and largely VDUs, but many parts would be relevant in other contexts also. The evaluation checklist is sectioned by topic and is merely a simple guide to the issues involved. There is inevitably some overlap between sections since the whole checklist may not always need to be used.

Displays structure

- Maintain a consistent relationship in the way in which all displays are structured throughout the system.

- Avoid the user needing to maintain anything other than simple items in memory when moving from one display to another.
- Ensure that the display structure is transparent to the user i.e., can the user locate any given item of information within the structure and is the interrelationship between different display pages evident?
- The operator should be able to apply a consistent set of rules for navigation throughout the system.
- Navigation between pages should be simple and the user should be able to enter and exit the structure at any point.
- Ensure the structure of any hierarchical system is appropriate to the user e.g., broad and shallow for expert and frequent users, deeper for novices.
- Ensure the paging structure corresponds to the operator's mental model of the system (or that a coherent mental model can be established through use).
- Provide a page to give an overview of the paging structure.

Information structure

Navigation

- The organization of the display structure should be transparent.
- Moving between screens frequently used in conjunction with each other should be as simple as possible.
- Frequently used displays should be directly accessible (e.g., by dedicated keys).
- Each display format should be labelled with a unique identifier which also shows its place in the navigational structure.

Division and partitioning of information pages

- The operator should be able to perform tasks without having to carry significant information from one display to another in memory.
- Information concerning variables which interact with one another should not be divided across display pages.
- Fields for the presentation of particular types of information (e.g., menus, display titles) should be consistent throughout all displays.
- Simple variable relationships can be divided across pages, but there should be a repeat of information to ensure the relationship is clear.
- Preferred display density is dependent on a range of issues such as the format of information, types of coding, number of dynamic data points, display screen resolution and frequency of use.

Formatting information on the screen

- Displayed information should be standardized in location of particular kinds of information.
- Fields or screen positions for certain types of information e.g., titles, menus, commands and input fields, should be consistent between displays.
- Where possible displays should be symmetrically balanced.
- Important information should be positioned in the upper left/central/ upper right areas of the screen.
- Position and layout of information should provide the user with extra coding about the nature of the information in different parts of the display.
- Data should be grouped to assist with the tasks the user has to perform. This may mean that groupings are on a task-based or functional level.
- Items may be grouped or arranged in a variety of ways dependent on user requirements. For example, Criticality—important items are placed prominently and together; Frequency—frequently used information is placed prominently on the screen, data frequently used together are displayed together; Function—items can be grouped based on function when sequence and frequency are not important; Sequence—items are displayed in the order in which they occur e.g., process flow.
- The use of too many windows or partitions should be avoided.

Display formats

- The formats used should take into full account the task(s) they are used to perform.
- Formats should be matched to user attributes, e.g., polar coordinate displays make use of operators' pattern recognition capabilities, for rapid detection of system deviations.
- Information should be presented in a format which is consistent with the task requirements of the user (e.g., if qualitative state information is required then a digital readout may not assist the user and may actually detract from task performance).

Coding

General coding

- Codes should be used to make information more easily assimilated and to structure displays. In general codes should be: Consistent—both within the display system and with other codes used on plant or in the process; Unambiguous—items in the code should not be confusable; Unique—any coding scheme should be clearly distinguishable from other coding schemes.

Colour

- The use of colour is generally subjectively preferred over monochrome, although it does not always give improvements in performance. Colour can be used to enhance the appearance of a display or as a code. Care must be taken in the use of colour as it can increase the potential for human error.
- Colour should only be used as a redundant code, i.e., all items should be distinguishable without colour, using colour only to enhance the coding.
- Consistent colour coding should be used over all interfaces, in the control room and on plant.
- For accurate discrimination of colours in a code a maximum of seven should be used.
- Colour can be used to relate items that are similar but separated spatially.

Highlighting

- Highlighting should not be overused but used selectively for emphasis and to give feedback.
- Blinking is good for attention getting.
- High brightness has attention getting properties, but less urgency.
- Reverse video should be used in moderation.
- Underlining to highlight text should only be used if it will not contribute to display clutter and if spatial layout permits.
- Do not overuse different fonts and upper case.

Labelling

- Consistency is essential and should be observed in size, font, use of abbreviations and positioning of labels.

Information content of displays

- Ensure that a task analysis and systematic identification of information needs has been applied within system design.
- Check that all potential users of the system have been considered in display design, including operators, management, control system engineers, and maintenance personnel.
- Ensure that the information content of displays matches the tasks it will be used for (i.e., information required for a particular task should be accessed easily and structured in a way that facilitates task performance).
- Ensure that consistency in the presentation of information is maintained wherever possible between plant and control room.

Acknowledgements

Considerable efforts were made in commenting on an earlier version of this chapter by Tom Mayfield and John Wood, and it is greatly improved as a result. Any remaining deficiencies are the authors' own.

References

Andrews, D.M., Smith, P.A., Wilson, J.R. and Mayfield, T.F. (1993). The use of multimedia in training process control operators. Presented at the Annual Conference of the Ergonomics Society, Herriot Watt University, April (available from the JOE, University of Nottingham.

Bailey, R.W. (1989). *Human Performance Engineering: A Guide for System Designers,* 2nd edition (London: Prentice Hall).

Bainbridge, E.A. (1991). Multiplexed VDT display systems: a framework for good practice. In *Human Computer Interaction and Complex Systems,* edited by G. Weir and J. Alty (London: Academic Press), pp. 189–210.

Ball, P.W. (Ed.) (1991). *The Guide to Reducing Human Error in Process Operations.* Report SRDA-R3 of The Human Factors in Reliability Group (Warrington: The SRD Association).

Clarke, A.A. and Mayfield, T.F. (1977). Ships bridge and wheelhouse ergonomics design study. Proceedings of the Human Factors in the Design and Operation of Ships Conference. Gothenburg.

Coekin, J. (1969). A versatile presentation of parameters for rapid recognition of total state. In *Manned System Design,* edited by J. Moraal and K.-F. Kraiss (New York: Plenum), pp. 153–179.

Diaper, D. (1989). Task analysis for knowledge description; the method and an example. In *Task Analysis for Human Computer Interaction,* edited by D. Diaper (London: Ellis Horwood).

Edwards, E. and Lees, F.P. (Eds) (1974). *The Human Operator in Process Control.* (London: Taylor and Francis).

Geyer, T.A.W., Bellamy, L.J., Astley, J.A. and Hurst, N.W. (1990). Prevent pipe failures due to human errors. *Chemical Engineering Progress* **86** (11), pp. 66–69.

Goodstein, L.P. (1981). Discriminative display support for process operators. In *Human Detection and Diagnosis of System Failures,* edited by J. Rasmussen and W.B. Rouse (New York: Plenum), pp. 433–449.

Goodstein, L.P., Andersen, H.B. and Olsen, S.E. (1988). *Tasks, Errors and Mental Models* (London: Taylor and Francis).

Health and Safety Executive (1989). *The Fire and Explosion at BP Oil (Grangemouth) Refinery Ltd* (London: HMSO), pp. 15–35.

Hogarth, R. (1980). *Judgement and Choice* (Chichester: John Wiley and Sons).

Hollnagel, E. (1993). Requirements for dynamic modelling of man–machine interaction. Proceedings of ANS Seminar, November 4–5, 1992, Kyoto, Japan; to be published in *Nuclear Engineering and Design,* 1993.

Ivergård, T. (1989). *Handbook of Control Room Design and Ergonomics* (London: Taylor and Francis).

Johnson, G.I. (1993). Spatial operational sequence diagrams in usability investigations. In *Contemporary Ergonomics 1993,* edited by E.J. Lovesey (London: Taylor and Francis).

Johnson-Laird, P.N. (1983). *Mental Models*, (Cambridge: Cambridge University Press).

Kantowitz, B.H. and Sorkin, R.D. (1983). *Human Factors: Understanding People-System Relationships* (New York: John Wiley).

Kinkade, R.G. and Anderson, J. (Eds) (1984). *Human Factors Guide for Nuclear Power Plant Control Room Development*, EPRI Report NP-3659 (Palo Alto, CA: Electric Power Research Institute).

Kirwan, B. (1994). *A Guide to Practical Human Reliability Assessment* (London: Taylor and Francis).

Kirwan, B. and Ainsworth, L.K. (1992). *A Guide to Task Analysis* (London: Taylor and Francis).

Kragt, H. (1992). *Enhancing Industrial Performance* (London: Taylor & Francis).

Kurke, M.I. (1961). Operation sequence diagrams in systems design. *Human Factors*, **3**, 66–73.

Mayfield, T.F. (1988). Processor based displays: the flexible control panel. Proceedings of the SARS '88 Conference Human Factors and Decision Making: Their Influence on Safety and Reliability. Manchester.

Nielsen, J. (1990). A meta-model for interacting with computers. *Interacting with Computers*, **2**, 147–160.

Norman, D.A. (1983). Some observations on mental models. In: *Mental Models* edited by D. Gentner and A. Stevens (Hillsdale, NJ: Erlbaum), pp. 7–14.

NUREG-0700 (1981). *Guidelines for Control Room Design Reviews* (US Nuclear Regulatory Commission).

Paap, K.R. and Roske-Hofstrand, R.J. (1986). The optimal number of menus per panel. *Human Factors*, **28**, 377–385.

Payne, S.J. (1988). Methods and mental models in theories of cognitive skill. In: *Artificial Intelligence and Human Learning*, edited by J. Self (London: Chapman and Hall).

Pethick, A.J. and Wood, J. (1989). Closed circuit television and user needs. In: *Contemporary Ergonomics* 1989, (ed: E.D. Megaw). London: Taylor and Francis, 450–455.

Piso, E. (1981). Task analysis for process control tasks. *Journal of Occupational Psychology*, **54**, 247–254.

Rajan, J.A. (1993). Human factors in control room design: reducing risk and maximising safety. *Loss Control Newsletter*, **4**, 20–22 (London: Sedgwith Energy).

Rasmussen, J. (1986). *Information processing and human–machine interaction* (Amsterdam: North Holland).

Reason, J. (1990). *Human Error* (Cambridge: Cambridge University Press).

Rutherford, A. and Wilson, J.R. Searching for the mental model in human–machine systems. In: *Models in the Mind: Perspectives, Theory and Application*, edited by Y. Rogers, A. Rutherford and P. Bibby (London: Academic Press), pp. 195–223.

Salvendy, G. (Ed.) (1987). *Handbook of Human Factors* (New York: John Wiley).

Sanders, M.S. and McCormick, E.J. (1992). *Human Factors in Engineering and Design*, 7th edition (New York: McGraw-Hill).

Seppala, P. and Salvendy, G. (1985). Impact of depth of menu hierarchy on performance effectiveness in a supervisory task: computerised flexible manufacturing system. *Human Factors*, **27**, 713–722.

Sheridan, T.B. (1987). Supervisory control. In *Handbook of Human Factors,* edited by G. Salvendy (New York: John Wiley & Sons), pp. 1243–1268.

Shirley, R.S. (1992). Computer Graphics for Industrial Applications. Englewood Cliffs, N.J.: Prentice Hall.

Shneiderman, B. (1992). *Designing the User Interface: Strategies for Effective Human–Computer Interaction,* 2nd edition (New York: Addison-Wesley).

Smith, S.L. and Mosier, J.N. (1986). *Guidelines for Designing the User Interface* (Mitre Corporation).

Snowberry, K., Parkinson, S.R. and Sisson, N. (1983). Computer display menus. *Ergonomics,* **26**, 669–712.

Sorkin, R.D. (1987) Design of auditory and tactile displays. In: *Handbook of Human Factors* (ed: G. Salvendy). New York: John Wiley, 549–576.

Stanton, N. (ed.) (1994). *Human Factors in Alarm Design* (London: Taylor and Francis).

Tainsh, M. (1982). On man computer dialogues with alpha-numeric status displays for naval command systems. *Ergonomics,* **25**, 683–703.

Tainsh, M. (1985). Job Process Charts and man computer interaction within naval command systems. *Ergonomics,* **28**, 555–565.

The Guardian (1984). Helicopter pilot admits piloting error. Feb. 24th.

Umbers, I.G. and Gent, E. (1985). Warren Spring Laboratory DTI: Stevenage Report No. CR 2758 CON.

Vermeulen, J. (1987). Effects of functional or topographically presented process schemes on operator performance. *Human Factors,* **29**, 383–395.

Visick, D.S. and Simpson, M.P. (1985). *A Study of Current and Potential Methods of Displaying Sequence Information in Process Plants,* Report No. CR 2764 (Stevenage: Warren Spring Laboratory).

Wickens, C.D. (1986). *'The Object Display': Principles and a Review of Experimental Findings,* Technical Report No. CPL 86-6 (Champaign-Urbana, IL: University of Illinois).

Wickens, C.D. (1992). *Engineering Psychology and Human Performance,* 2nd edition (New York: HarperCollins).

Wilson, J.R., Brown, D.J., Cobb, S.V. D'Cruz, M.D. and Eastgate, R.M. (1995). Manufacturing operations in virtual environments. PRESENCE (in press).

Wilson, J.R. and Rutherford, A. (1989). Mental models: theory and application in human factors. *Human Factors,* **31,** 617–634.

Wirstad, J. (1988). On knowledge structures for process operators. In: *Tasks, Errors and Mental Models,* edited by L. Goodstein, H. Andersen and S. Olsen (London: Taylor and Francis), Ch. 3.

Woods, D.D., O'Brien, J.F. and Hanes, L.F. (1987). Human factors challenges in process control: the case of nuclear power plants. In *Handbook of Human Factors,* edited by G. Salvendy (New York: John Wiley and Sons), pp. 1724–1770.

Wood, J. (1992). The human factor: special feature CCTV. Security Management Today, April, 28–30.

Wood, J. (1994). The developing international standard on control room ergonomics, ISO 11064. Proceedings of the 12th Congress of the International Ergonomics Association, Toronto, 15–19 August.

Appendix

Proposed ergonomic design process for control centres. Draft Control Room Ergonomics Standard ISO 11064, Part 1

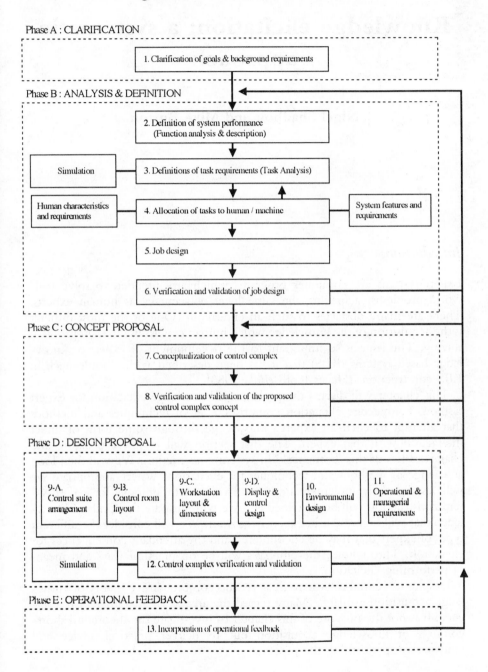

Phase A : CLARIFICATION

1. Clarification of goals & background requirements

Phase B : ANALYSIS & DEFINITION

2. Definition of system performance (Function analysis & description)

Simulation

3. Definitions of task requirements (Task Analysis)

Human characteristics and requirements

4. Allocation of tasks to human / machine

System features and requirements

5. Job design

6. Verification and validation of job design

Phase C : CONCEPT PROPOSAL

7. Conceptualization of control complex

8. Verification and validation of the proposed control complex concept

Phase D : DESIGN PROPOSAL

9-A. Control suite arrangement

9-B. Control room layout

9-C. Workstation layout & dimensions

9-D. Display & control design

10. Environmental design

11. Operational & managerial requirements

Simulation

12. Control complex verification and validation

Phase E : OPERATIONAL FEEDBACK

13. Incorporation of operational feedback

Chapter 14

Knowledge elicitation: a systematic approach

Nigel Shadbolt and Mike Burton

Introduction

Expert systems are computer programs which are intended to solve real-world problems, achieving the same level of accuracy as human experts. There are many obstacles to such an endeavour. One of the most difficult is the acquisition of the knowledge which human experts use in their problem solving. The issue is so important to the development of expert or knowledge-based systems (KBS) that it has been described as the 'bottle-neck in KBS construction' (Hayes-Roth *et al.*, 1983).

This chapter will discuss the problem of knowledge elicitation for expert systems. Knowledge elicitation comprises a set of techniques and methods that attempt to elicit an expert's knowledge through some form of direct interaction with that expert. The first section will review the nature and characteristics of this 'bottleneck' in system construction. We will then look at a range of methods and techniques for elicitation. Where appropriate we will describe their implementation in software. Methodologies for expert system construction will be described and we will illustrate the kinds of knowledge that will be present in expert behaviour. We will consider the different types of expert that may be encountered and the attendant consequences for elicitation. Throughout, the emphasis will be on practical ways and means of performing elicitation.

Despite its central role there is, as yet, no comprehensive theory of knowledge acquisition available. Many regard the area as an art rather than a science. It is not the purpose of this chapter to investigate the theoretical shortcomings of knowledge acquisition but to deliver practical advice and guidance on performing the process.

Expert systems

In the early days of Artificial Intelligence much effort went into attempts to discover general principles of intelligent behaviour. Newell and Simon's (1963) General Problem Solver exemplified this approach. They were interested in uncovering a general problem solving strategy which could be used for any human task.

In the early 1970s this position was challenged. A new slogan came to prominence—'in the knowledge lies the power'. A leading exponent of this view was Edward Feigenbaum of SRI. He observed that experts are experts by virtue of domain specific problem solving strategies together with a great deal of domain specific knowledge. It was the attempt to incorporate these various sorts of domain knowledge which resulted in the class of programs called Expert or Knowledge-Based Systems.

Throughout this article we will be assuming that current commercially available expert system software will be the implementation vehicle for the programs. Thus the form in which the knowledge will be implemented is likely to be standard *production rules* with a *structured object* facility such as *frames*. For a review of the major types of expert system architecture see Jackson (1990) and of different knowledge representation formalisms see Shadbolt (1989) and Rich and Knight (1991).

The problem of elicitation

The people who build expert systems, the knowledge engineers, are typically not people with a deep knowledge of the application domain. However, it is the knowledge engineers who must gather the domain knowledge and then implement it in a form that the machine can use.

In the simplest case, the knowledge engineer may be able to gather information from a variety of non-human resources: text books, technical manuals, case studies and so on. However, in most cases one needs actually to consult a practising expert. This may be because there is not the documentation available, or because real expertise derives from practical experience in the domain, rather than from a reading of standard texts. The task of gathering information generally, from whatever source, is called *knowledge acquisition* (KA). The sub-task of gathering information from the expert is called *knowledge elicitation* (KE). In this chapter we will be concentrating on KE. Few systems are ever built without recourse to experts at some stage. Those systems not informed by actual expert understanding and practice are often the poorer for it.

Many problems arise before elicitation of the detailed domain knowledge for an expert system is ever conducted. There are failures in appreciating what it is realistic to build. Sometimes the failure is in formulating the role of the system. On other occasions there is an inadequate understanding of the task environment. Very often the effort and resources required to build systems are underestimated: this occurs in both the development and maintenance of systems. A particularly nasty situation arises when the knowledge engineer is expected to conjure up knowledge for areas in which no evidence of systematic practice exists at all. Knowledge engineers seem to be expected to provide theories for domains where there is no theory. Providing we can avoid all of these obstacles then we get down to detailed issues of KE.

Two questions dominate in KE. How do we get experts to tell us, or else show us, what they do? How do we determine what constitutes their problem solving competence? The task is enormous, particularly in the context of large expert systems. There is a variety of circumstances which contrive to make the problem even harder. Much of the power of human expertise lies in laid-down experience, gathered over a number of years, and represented as heuristics*. Often the expertise has become so routinized that experts no longer know what they do or why.

There are obviously clear commercial reasons to try to make KE an effective process. We would like to be able to use techniques that will minimize the effort spent in gathering, transcribing and analyzing an expert's knowledge. We would like to minimize the time spent with expensive and scarce experts. And, of course, we would like to maximize the yield of usable knowledge.

There are also sound engineering reasons why we would like to make KE a systematic process. We would like the procedures of KE to become common practice and conform to clear standards. This will help ensure that the results are robust, that they can be used on various experts in a wide range of contexts by any competent knowledge engineer. We also hope to make our techniques reliable. This will mean that they can be applied with the same expected utility by different knowledge engineers. Placing elicitation on such a systematic footing will also be important in the development of methodologies that direct the process of expert system specification, construction and maintenance.

We will begin by describing, in sufficient detail for the reader to apply them, examples of major KE methods. We will mention other techniques and where the reader can find out more about them. We will then review aspects of expertise and cognition that are likely to affect the KE process directly. We will also indicate where software support is available for some of these KE methods. Finally, we will describe the methodologies for acquisition and the construction of programmes of acquisition that are beginning to emerge.

Elicitation techniques

The techniques we will describe are methods that we have found in our previous work to be both useful and complementary to one another. We

---

*An heuristic is defined as a rule of thumb or generally proven method to obtain a result given particular information.

can subdivide them into natural and contrived methods. The distinction is a simple one. A method is described as natural if it is one an expert might informally adopt when expressing or displaying expertise. Such techniques include interviews or observing actual problem solving. There are other methods we will describe in which the expert undertakes a contrived task. The task elicits expertise in ways that are not usually familiar to an expert. The first two categories of elicitation method are both natural under this definition and are varieties of interview and protocol analysis.

The structured interview

Almost everyone starts in KE by determining to use an interview. The interview is the most commonly used knowledge elicitation technique and takes many forms, from the completely *unstructured* interview to the formally-planned, *structured* interview. The structured interview is a formal version of the interview in which the knowledge engineer plans and directs the session. The structured interview has the advantage that it provides structured transcripts that are easier to analyze than unstructured chat. The formal interview which we have specified here constrains the expert-elicitor dialogue to the general principles of the domain. Experts do not work through a particular scenario extracted from the domain by the elicitor; rather the experts generate their own scenarios as the interview progresses. The structure of the interview is as follows:

1. Ask the expert to give a brief (10 minute) outline of the target task, including the following information:
   (a) An outline of the task, including a description of the possible solutions or outcomes of the task;
   (b) A description of the variables which affect the choice of solutions or outcomes;
   (c) A list of major rules which connect the variables to the solutions or outcomes.
2. Take each rule elicited in Stage 1, ask when it is appropriate and when it is not. The aim is to reveal the scope (generality and specificity) of each existing rule, and hopefully generate some new rules.
3. Repeat Stage 2 until it is clear that the expert will not produce any additional information.

It is important in this technique to be specific about how to perform stage 2. We have found that it is helpful to constrain the elicitor's interventions to a specific set of *probes,* each with a specific function. Here is a list of probes (P) and functions (F) which will help in stage 2:

P1 Why would you do that?
F1 Converts an assertion into a rule

P2   How would you do that?
F2   Generates *lower order* rules

P3   When would you do that?
     Is ⟨the rule⟩ always the case?
F3   Reveals the generality of the rule and may generate other rules

P4   What alternatives to ⟨the prescribed action/decision⟩ are there?
F4   Generates more rules

P5   What if it were not the case that ⟨currently true condition⟩?
F5   Generates rules for when current condition does not apply

P6   Can you tell me more about ⟨any subject already mentioned⟩
F6   Used to generate further dialogue if expert dries up.

The idea here is that the elicitor engages in a type of slot/filler dialogue. Listening out for relevant concepts and relations imposes a large cognitive load on the elicitor. The provision of fixed linguistic forms within which to ask questions about concepts, relations, attributes and values makes the elicitor's job very much easier. It also provides sharply focused transcripts which facilitate the process of extracting usable knowledge. Of course, there will be instances when none of the above probes are appropriate (such as the case when the elicitor wants the expert to clarify something). However, you should try to keep these interjections to a minimum. The point of specifying such a fixed set of linguistic probes is to constrain the expert to giving you all, and only, the information you want.

The sample of dialogue below is taken from a real interview of this kind. It is the transcript of an interview by a knowledge engineer (KE) with an expert (EX) on VDU fault diagnosis*.

EX   I actually checked the port of the computer
KE   Why did you check the port?
EX   If it's been lightning recently then it's a good idea to check the port
     + because lightning tends to damage the ports
KE   Are there any alternatives to that problem?
EX   Yes, that ought to be prefaced by saying do that if it was several keys
     with odd effects + not necessarily all of them, but more than two
KE   Why does it have to be more than two?
EX   Well if it was only one or two keys doing funny things then the thing
     to do is check they're closing properly + speed would affect all keys,
     parity would affect about half the keys

This is quite a rich piece of dialogue. From this section of the interview alone we can extract the following rules.

        IF   there has been recent lightning

---

*In the transcripts we use the symbol + to represent a pause in the dialogue.

THEN	check port for damage
IF	there are two or fewer malfunctioning keys
THEN	check the key contacts
IF	about half the keyboard is malfunctioning
THEN	check the parity
IF	the whole keyboard is malfunctioning
THEN	check the speed

Of course these rules may need refining in later elicitation sessions, but the text of the dialogue shows how the use of the specific probes has revealed a well-structured response from the expert★.

Semi-structured interviews

Techniques exist to impose a lesser amount of structure on an interview. We mention two examples here. One of these is the Knowledge Acquisition Grid (LaFrance, 1987). This is a matrix of knowledge types and forms: examples of knowledge forms are *layouts* and *stories*; examples of question types are *grand tour* and *cross–checking*. A grand tour involves such things as distinguishing domain boundaries and organization of goals; cross–checking involves the engineer attempting to validate the acquired knowledge by, for example, playing devil's advocate.

Secondly, there is the teachback technique of Johnson and Johnson (1987). This involves creating an intermediate representation of the knowledge acquired, which is then 'taught back' to the expert, who can then check or, when necessary, amend the information.

Unstructured interviews

Unstructured interviews have no agenda (or, at least, no *detailed* agenda) set either by the knowledge engineer or by the expert. Of course, this does not mean that the knowledge engineer has no goals for the interview, but it does mean that he or she has considerable scope for proceeding; there are few constraints. The advantages of this approach stem from this lack of constraints. Firstly, the approach can be used whenever one of the goals of the interview is that the expert and the knowledge engineer establish a rapport. There are no formal barriers to the discussion ranging as either participant sees fit. Secondly, the engineer can get a broad view of the topic easily; he or she can 'fill in the gaps' in his or her own perceived knowledge of the domain, thereby making himself or herself more 'comfortable' with his or her mental model (see chapter 13). Thirdly, the expert can describe the domain in a

---

★In fact, a possible second-phase elicitation technique would be to present these rules back to the expert and ask about their truthfulness, scope and so forth.

way with which he or she is familiar, discussing topics that he or she considers important and ignoring those he or she considers uninteresting.

The disadvantages are clear enough. The lack of structure can lead to inefficiency. The expert may be unnecessarily verbose. He or she may concentrate on topics whose importance he or she exaggerates. The coverage of the domain may be too patchy. The data acquired may be difficult to integrate, either because they do not form a unity, or because there are inconsistencies. This last will be an even more likely occurrence if the information provided by several experts is to be collated.

In all the interview techniques (and in some of the other generic techniques as well) there exist a number of dangers that have become familiar to knowledge engineers. One problem is that experts will only produce what they can verbalize. If there are nonverbalizable aspects to the domain, the interview will not recover them. This can arise from two causes. It may be that the knowledge was never explicitly represented or articulated in terms of language (consider, for example, pattern recognition expertise). Then there is the situation where the knowledge was originally learnt explicitly in a propositional or language-like form. However, in the course of experience it has become routinized or automatized*. This can happen to such an extent that experts may regard the complex decisions they make as based on hunches or intuitions. Nevertheless, these decisions are based upon large amounts of remembered data and experience, and the continual application of strategies. In this situation they tend to give *black box* replies 'I don't know how I do that . . .', 'It is obviously the right thing to do . . . '.

Another problem arises from the observation that people (and experts in particular) often seek to justify their decisions in any way they can. It is a common experience of the knowledge engineer to get a perfectly valid decision from an expert, and then to be given a spurious justification. For these and other reasons we have to supplement interviews with additional methods of elicitation. Elicitation should always consist of a programme of techniques and methods. This brings us on to consider another technique much favoured by knowledge engineers, protocol analysis.

Protocol analysis

Protocol analysis (PA) is a generic term for a number of different ways of performing some form of analysis of the expert(s) actually solving problems in the domain. In all cases the engineer takes a record of what the expert does—preferably by video or audio tape—or at least by written notes. Protocols are then made from these records and the knowledge engineer tries to extract meaningful structure and rules from the protocols.

We can distinguish two general types of PA—*on-line* and *off-line*. In on-

---

*We often use computing analogy to refer to this situation and speak of the expert as having *compiled* the knowledge.

line PA the expert is being recorded solving a problem, and concurrently a commentary is made. The nature of this commentary specifies two sub-types of the on-line method. The expert performing the task may be describing what he or she is doing as problem solving proceeds. This is called *self-report*. A variant on this is to have another expert provide a running commentary on what the expert peforming the task is doing. This is called *shadowing*.

Off-line PA allows the expert(s) to comment retrospectively on the problem solving session—usually by being shown an audio-visual record of it. This may take the form of retrospective self report by the expert who actually solved the problem, it could be a critical retrospective report by other experts, or there could be group discussion of the protocol by a number of experts including its originator. In the case in which only a behavioural protocol is obtained then obviously some form of retrospective verbalization of the problem solving episode is required.

Before PA sessions can be held, a number of pre-conditions should be satisfied. The first of these is that the knowledge engineer is sufficiently acquainted with the domain to understand the expert's tasks. Without this the elicitor may completely fail to record or take note of important parts of the expert's behaviour.

A second requirement is the careful selection of problems for PA. The sampling of problems is crucial. PA sessions may take a relatively long time, only a few problems can be addressed (Shadbolt and Burton, 1989). Therefore, the selection of problems should be guided by how representative they are. Asking experts to sort problems into some form of order (Chi *et al.*, 1982a) may give an insight into the classification of types of problem and help in the selection of suitable problems for PA (see also the next two sections on concept sorts and laddering).

A further condition for effective PA is that experts should not feel embarrassed about describing their expertise in detail. It is preferable for them to have experience in thinking aloud. Uninhibited thinking aloud has to be learned in the same way as talking to an audience. One or two short training sessions may be useful, in which a simple task is used as an example. This puts the experts at ease and familiarizes them with the task of talking about their problem solving.

Where a verbal or behavioural transcript has been obtained we next have to contemplate its analysis. Analysis might include the encoding of the transcript into 'chunks' of knowledge (which might be actions, assertions, propositions, key words, etc.), and should result in a rich domain representation with many elicited domain features together with a number of specified links between those features.

There are a number of principles that can guide the protocol analysis. For example, analysis of the verbalization resulting in the protocol can distinguish between information that is attended to during problem solving, and that which is used implicitly. A distinction can be made between information brought out of memory (such as a recollection of a similar problem solved in the past), and information that is produced 'on the spot' by inference. The knowledge chunks referred to above can be analyzed by using the expert's syntax, or the pauses he or she takes, or other linguistic cues. Syntactical categories (e.g., use of nouns, verbs, etc.) can help distinguish between domain features and problem solving actions, etc.

In trying to decide when it is appropriate to use PA bear in mind that it is alleged that different KE techniques differentially elicit certain kinds of information. With PA it is claimed that the sorts of knowledge elicited include the 'when' and 'how' of using specific knowledge. It can reveal the problem solving and reasoning strategies, evaluation procedures and evaluation criteria used by the expert, and procedural knowledge about how tasks and sub-tasks are decomposed. A PA gives you a complete episode of problem solving. It can be useful as a verification method to check that what people say is what they do. It can take you deep into a particular problem. However, it is intrinsically a narrow method since usually one can only run a relatively small number of problems from the domain.

When actually conducting a PA the following is a useful set of tips to help enhance its effectiveness. Present problems and data in a realistic way. The way problems and data are presented should be as close as possible to a real situation. Transcribe the protocols as soon as possible. The meaning of many expressions is soon lost, particularly if the protocols are not recorded. In almost all cases an audio recording is sufficient, but video recordings have the advantage of containing additional and disambiguating information. Avoid long self report sessions. Because of the need to perform a double task the process of thinking aloud is significantly more tiring for the expert than is being interviewed. This is one reason why shadowing is sometimes preferred. In general, the presence of the knowledge engineer is required in a PA session. Although the knowledge engineer adopts a background role, his very presence suggests a listener to the interviewee, and lends meaning to the talking aloud process. Therefore, comments on audibility or even silence by the knowledge engineer are quite acceptable.

Protocol analyses share with the unstructured interview the problem that they may deliver unstructured transcripts which are hard to analyze. Moreover, they focus on particular problem cases and so the scope of the knowledge produced may be very restricted. It is difficult to derive general domain principles from a limited number of protocols. These are practical disadvantages of protocol analysis, but there are more subtle problems.

Two actions, which look exactly the same to the knowledge engineer, may be the result of two quite different sets of considerations. This is a problem of impoverished interpretation by the knowledge engineer. He simply does not know enough to discriminate the actions. The obverse to this problem can arise in shadowing and the retrospective analyses of protocols by experts. Here the experts may simply wrongly attribute a set of considerations to an action after the event. This is analogous to the problems of misattribution in interviewing.

A particular problem with self report, apart from being tiring, is the possibility that verbalization may interfere with performance. The classic demonstration of this is for a driver to attend to all the actions involved in driving a car. If one consciously monitors such parameters as engine revs, current gear, speed, visibility, steering wheel position and so forth, the driving invariably gets worse. Such skill is shown to its best effect when performed automatically. This is also the case with certain types of expertise. By asking the expert to verbalize, one is in some sense destroying the point of doing protocol analysis—to access procedural, real-world knowledge.

Having pointed to these disadvantages, it is also worth remembering that context is sometimes important for memory—and hence for problem solving. For most nonverbalizable knowledge, and even for some verbalizable knowledge, it may be essential to observe the expert performing the task. For it may be that this is the only situation in which the expert is actually able to perform it.

Finally, when performing PA it is useful to have a set of conventions for the actual interpretation and analysis of the resultant data. Ericsson and Simon (1984) provide the classic exposition of protocol analysis although it is oriented towards cognitive psychology. Useful additional references are Belkin *et al.* (1987), Firlej and Hellens (1991), Kuipers and Kassirer (1983), and McGraw and Harbison-Briggs (1989). A full discussion of verbal protocol analysis—on-line self reporting—is provided by Bainbridge and Sanderson in chapter 7.

The techniques discussed so far are *natural* and intuitively easy to understand. Experts are used to expressing their knowledge in these sorts of ways. The techniques that follow are what we have termed *contrived* and permit the expression of knowledge in ways that are likely to be unfamiliar to the expert.

Concept sorting

Concept sorting is a technique that is useful when we wish to uncover the different ways an expert sees relationships between a fixed set of concepts. In the version discussed here an expert is presented with a number of cards on each of which a concept word is printed. The cards are shuffled and the expert is asked to sort the cards into either a fixed number of piles or else to sort them into any number of piles the expert finds appropriate. This process is repeated many times. Using this task one attempts to get multiple views of the structural organization of knowledge by asking the expert to do the same task over and over again. Each time the expert sorts the cards he should create at least one pile that differs in some way from previous sorts. The expert should also provide a name or category label for each pile on each different sort. Performing a card sort requires the elicitor to be not entirely naive about the domain. Cards have to be made with the appropriate labels before the session. However, no great familiarity is required as the expert provides all the substantial knowledge in the process of the sort. We now provide an example from our VDU domain to show the detailed mechanics of a sort. The concepts printed on the cards were faults and corrective actions drawn from a structured interview with the expert. He had outlined 20 faults and outcomes:

A	damaged VDU port	B	faulty key contacts
C	incorrect parity setting	D	program is busy
E	damaged tube	F	terminal not switched on
G	mains fuse blown	H	fuse in VDU blown
I	loose connection or break in line	J	press control-c
K	press control-q	L	press control-z
M	VDU power supply fault	N	software fault
O	video section fault on VDU	P	incorrect terminal speed setting
Q	terminal requires reset	R	VDU transmission fault
S	transmit line fault	T	receive line fault.

The expert was shown possible ways of sorting cards in a *toy* domain, as part of the briefing session, and then asked to sort the real elements in the same way.

The dimensions/piles (P) which the expert used for the various sorts (S) were as follows:

S1 P1 hardware fault P2 software fault
S2 P1 easy to clear P2 difficult to clear
S3 P1 no component damage P2 component damage usually as symptom of another outcome
    P3 component damage as an outcome
S4 P1 communication fault P2 not a communication fault
S5 P1 no output P2 no response
    P3 garbled output P4 combination of no response and garbled output.

Table 14.1 shows the pile of each sort for each element. You will see that many of the elements are distinguishable from one another—even with these few sorts.

Using this information we can attempt to extract decision rules directly. An example of a rule extracted from the sorting is:

IF	it is a hardware fault	(sort 1/pile 1)
AND	it is easy to clear	(sort 2/pile 1)
AND	it is component damage	(sort 3/pile 3)
AND	it is NOT a communication fault	(sort 4/pile 2)
AND	there is garbled output	(sort 5/pile 3)
THEN	there are faulty key contacts	(outcome B)

As you can see from the example rule such sorts produce long and cumber-

Table 14.1. Tabulated results from the card sort

Card	Pile of Sort 1	Pile of Sort 2	Pile of Sort 3	Pile of Sort 4	Pile of Sort 5
A	1	2	3	1	4
B	1	1	3	2	3
C	2	1	1	1	3
D	2	1	1	2	1
E	1	2	3	2	1
F	2	1	1	2	1
G	1	1	2	2	1
H	1	1	2	2	1
I	1	2	3	1	4
J	2	1	1	2	2
K	2	1	1	2	2
L	2	1	1	2	2
M	1	2	3	2	1
N	2	2	1	1	2
O	1	2	3	2	1
P	2	1	1	1	3
Q	2	1	1	1	4
R	1	2	1	1	4
S	2	2	3	2	4
T	1	2	1	1	4

some rules. In fact many of the clauses may be redundant—once you have established that the fault is a hardware one, there is no need to check whether it is component damage. However, the utility of this technique does not reside solely in the production of decision rules. We can use it, as we have said, to explore the general interrelationships between concepts in the domain. We are trying to make explicit the implicit structure that experts impose on their expertise.

When using any of these KE methods knowledge engineers should beware a type of semantic mindset whereby the expert or elicitor focuses on only one type of knowledge element. In concept sorting the cards can name knowledge elements of any type—objects in a domain, tasks, goals, actions, etc. The restriction is that in any sorting session the cards should be of the same knowledge type.

Variants of the simple sort are different forms of *hierarchical* sort. One such version is to ask the expert to proceed by producing first two piles, on the second sort three, then four and so on. Finally we ask if any two piles have anything in common. If so you have isolated a higher order concept that can be used as a basis for future elicitation.

The advantages of concept sorting can be characterized as follows. It is fast to apply and easy to analyze. It forces into an explicit format the constructs which underlie an expert's understanding. In fact it is often instructive to the expert. A sort can lead the expert to see structure in his or her view of the domain which he or she has not consciously articulated before. Finally, in domains where the concepts are perceptual in nature (i.e., X-rays, layouts and pictures of various kinds) then the cards can be used as a means of presenting these images and attempting to elicit names for the categories and relationships that might link them.

There are, of course, features to be wary of with this sort of technique. Experts can often confound dimensions by not consistently applying the same semantic distinctions throughout an elicitation session. Alternatively, they may over simplify the categorization of elements, missing out important caveats.

An important tip with all of the contrived techniques we are reviewing is always to audio tape these sessions. An expert makes many asides, comments and qualifications when sorting or ranking and so on. In fact one may choose to use the contrived methods as means to carry out auxiliary structured interviews. The structure this time is centred around the activity of the technique. It is worth noting that we have found (Schweikert *et al.*, 1987) that an expert's own opinion of the worth of a technique is no guide to its real value. In methods such as sorting we have a situation in which the expert is trying to demonstrate expertise in a non-natural or contrived manner. He or she might be quite used to chatting about his or her field of expertise, but sorting is different and experts are suspicious of it. Experts may in fact feel they are performing badly with such methods. However, on analysis one finds that the yield of knowledge is as good and sometimes better than for non-contrived techniques (Shadbolt and Burton, 1990).

Laddered grids

Once again this is a somewhat contrived technique, and you will need to explain it fully to the expert before starting. The expert and the knowledge engineer construct a graphical representation of the domain in terms of the relations between domain or problem solving elements. The result is a qualitative, two-dimensional graph where nodes are connected by labelled arcs. No extra elicitation method is used here, but expert and elicitor construct the graph together by negotiation.

In using the technique the elicitor enters the conceptual map at some point and then attempts to move around it with the expert. A formal specification of how we use the technique is shown below together with an example of its use.

Start the expert off with a seed item

Move around the domain map using the following prompts

To move DOWN the expert's domain knowledge:

Can you give examples of ⟨ITEM⟩?

To move ACROSS the expert's domain knowledge:

What alternative examples of ⟨CLASS⟩ are there to ⟨ITEM⟩?

To move UP the expert's domain knowledge:

What have ⟨SAME LEVEL ITEMS⟩ got in common?

What are ⟨SAME LEVEL ITEMS⟩ examples of?

To elicit essential properties of an item:

How can you tell it is ⟨ITEM⟩?

To discriminate items:

What is the key difference between ⟨ITEM 1⟩ and ⟨ITEM 2⟩?

The elicitor may move around the domain map of knowledge in any order which seems convenient. As the elicitation session progresses, the elicitor keeps track of the elicited knowledge by drawing up a network on a large piece of paper. This representation allows the elicitor to make decisions (or ask questions) about what constitutes higher or lower order elements in the domain, what differences exist between elements in the network. In order to give the reader the flavour of the technique, there follows an extract from a laddered grid elicitation session. Once again, the knowledge domain is VDU fault diagnosis.

KE: What examples of a 'no response' problem are there?
EX: It's a software problem, unless a small set of keys is affected.
KE: Can you give me an example of what you do if there's a small set of keys affected?
EX: Yes. You check the key contacts.
KE: Can you give me examples of actions for software problems?
EX: Yes. You can intervene or don't intervene. You don't intervene if there's a user error where they misinterpret a slow program as this fault.
KE: Can you give me an example of when you intervene?
EX: When there's an editor problem. Then you use control keys.
KE: Can you give me an example of using control keys?
EX: Press control-c, control-q and control-z.
KE: What is the difference between control-c and control-q?
EX: Control-q gets you out of control-s and control-c gets you out of the program. Control-q should be used before control-c.
KE: What is the difference between control-c and control-z?
EX: Control-c gets you out of the program and should be used before control-z. Control-z quits the system.

In the course of this laddered grid interview the elicitor drew up a hierarchical representation of the domain as shown in Figure 14.1.

This hierarchy gives rise to the following set of rules which could be included in the knowledge base of an expert system for VDU fault finding.

IF there is no response
AND it is a software problem

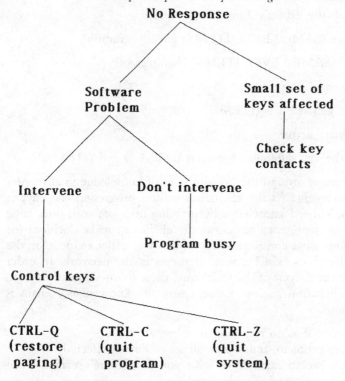

Figure 14.1. Laddered grid in the VDU domain

AND	intervention is needed
THEN	press control-q
IF	control-q doesn't work
THEN	press control-c
IF	control-c doesn't work
THEN	press control-z

As with the previous contrived method it is important to keep an audio record of the session for future review or transcription. Also it is a technique that can be used on a variety of knowledge types; objects, actions, tasks, goals, etc. We have found that this form of knowledge elicitation is very powerful for structured domains. As with other contrived techniques we have found that whilst an expert may think this technique is revealing little of interest, subsequent analysis provides good quality rules.

The limited information task

A technique which does not provide a spatial representation of the domain, but rather a set of hints or suggestions which may prove useful in expert system construction is a technique called the *limited information task* (Hoffman, 1987) or *20 questions* (Grover, 1983). The expert is provided with little or no information about a particular problem to be solved. The expert must then ask the elicitor for specific information which will be required to solve the problem. The information which is requested, along with the order in which it is requested, provides the knowledge engineer with an insight into the expert's problem solving strategy. One difficulty with this method is that the knowledge engineer needs a good understanding of the domain in order to make sense of the expert's questions, and to provide meaningful responses. The elicitor should have forearmed himself with a problem from the domain together with a *crib sheet* of appropriate responses to the questions.

In one of the versions of the limited information task which we use we tell the expert that the elicitor has a scenario in mind and the expert must determine what it is. The scenario might represent a problem, a solution or a problem context. The expert is told that they may ask the elicitor for more information, though what the elicitor gives back is terse and does not go much beyond what was asked for in the question. The expert may be asked to explain why each of the questions was asked.

An example of the kind of interaction produced by this technique is shown below. Here the problem domain is in the construction of lighting systems for the inspection of industrial products and processes.

EX: Is this in the manufacturing industry?
KE: Yes
EX: So we've ruled out things like fruit, vegetables, cows?
KE: Yes
EX: Is it the metal industry?
KE: The material is wood
EX: So we could be dealing with a large object here like a chair or table
KE: The object is large
EX: It's likely to be a 3-D object, you've got to pick it up and turn it over
KE: That's right
EX: So what I need now are the dimensions of this object in terms of the cube that will enclose it
KE: It would have similar dimensions to the table top
EX: Do I inspect one surface or all the surfaces?
KE: All of them
EX: Is the inspector looking for one or many faults?
KE: One particular fault
EX: Can you describe it for me?
KE: It's pencil marks about half an inch long
EX: What colour is the wood?
KE: Dark unfinished wood
EX: We've got a contrast problem here. At this point I'd go and look at the job + to see if the graphite pencil marks reflect light + sometimes

it does, but it depends on the wood + if it does you can select the light to increase the contrast between the fault and the background

$$\vdots$$

EX: I'd be doing this in three phases: first a general lighting, then specific for surface lighting, and then some directional light [expert then gives technical specifications for these types of light]

This interview gives us an interesting insight into the natural line of enquiry of an expert in this domain. Often expert systems gather the right data but the order in which they are gathered and used can be remote from how an expert works. This can decrease the acceptability of the system if other experts are to use it, and it also has consequences for the intelligibility of any explanations the system offers in terms of a retrace of its steps to a solution.

It will be seen that we can once again extract decision rules directly from the dialogue:

> IF   fault colour is black
> AND   object colour is dark
> THEN   contrast is a problem

The drawbacks to this technique are that the elicitor needs to have constructed plausible scenarios and the elicitor has to be able to cope with questions asked of him. The experts themselves are sometimes uncomfortable with this technique, this may well have to do with the fact that, as with other contrived techniques, it is not a natural means of manifesting expertise. Whilst a few scenarios may reveal some of the general rules in a domain the elicitation is very case specific. In order to get the range of knowledge for a sweep of situations many scenarios would need to be constructed and used.

An interesting variation on the method which we have used successfully is a form of telephone consultancy. Here we take two domain experts and place them at opposite ends of a table and ask them to imagine that one is a 'client' who is ringing up the other, a 'consultant', to ask for advice concerning a particular problem. They then engage in a conversation in which the consultant tries to elicit the nature and context of the problem, and finally attempts to offer appropriate advice. In this variation of the limited information task you can rely on one of the experts to generate interesting cases. In addition, the expert 'role playing' as the client can provide appropriate responses to the consultant's enquiries. The only drawback is that sometimes experts fabricate extremely difficult cases for each other in order to test each other's mettle!

A taxonomy of KE techniques

We have sampled some of the major approaches to elicitation and where appropriate given a detailed description of techniques that are likely to be of use. There are many variants on the methods we have described. In Table 14.2 we have provided a taxonomy of methods with which we are familiar together with a primary reference for each one.

Having discussed the principal methods of elicitation we should spend a little time reflecting on the nature of two other major components of the KE enterprise, namely—the experts and the expertise they possess.

On experts

Experts come in all shapes and sizes. Ignoring the nature of your expert is another potential pitfall in KE. A coarse guide to a typology of experts might make the issues clearer. Let us take three categories we shall refer to as *academics, practitioners, samurai* (in practice experts may embody elements of all three types). Each of these types of expert differs along a number of dimensions. These include: the outcome of their expert deliberations, the problem solving environment they work in, the state of the knowledge they possess (both its internal structure and its external manifestation), their status and responsibilities, their source of information, the nature of their training.

How are we to distinguish these different types of expert? The academic type regards their domain as having a logically organized structure. Generalizations over the laws and behaviour of the domain are important to them. Theoretical understanding is prized. Part of the function of such experts may

Table 14.2. A taxonomy of elicitation methods

Non-contrived
   Interviews
      Structured
         Fixed Probe (Shadbolt and Burton, this volume)
         Focused Interviews (Hart, 1986)
         Forward Scenario Simulation (Grover, 1983)
      Semi-Structured
         Knowledge Acquisition Grid (LaFrance, 1987)
         Teach Back (Johnson and Johnson, 1987)
      Unstructured (Weiss and Kulikowski, 1984)
   Protocol Analysis
      Verbal
         On-line (Johnson *et al.*, 1987)
         Off-line (Elstein *et al.*, 1978)
         Shadowing (Clarke, 1987)
      Behavioural (Ericsson and Simon, 1984)
Contrived
   Conceptual Mapping
      Sorting and Rating (Gammack, 1987a, 1987b)
      Repertory Grid (Shaw and Gaines, 1987a)
      Pathfinder (Schvaneveldt *et al.*, 1985)
   Goal Decomposition
      Laddered Grid (Hinkle, 1965)
      Limited-Information Task (Grover, 1983; Hoffman, 1987)

be to explicate, clarify and teach others. Thus they talk a lot about their domains. They may feel an obligation to present a consistent story both for pedagogic and professional reasons. Their knowledge is likely to be well structured and accessible. These experts may suppose that the outcome of their deliberations should be the correct solution of a problem. They believe that the problem can be solved by the appropriate application of theory. They may, however, be remote from every day problem solving.

The practitioner class on the other hand are engaged in constant day to day problem solving in the domain. For them, specific problems and events are the reality. Their practice may often be implicit and what they desire as an outcome is a decision that works within the constraints and resource limitations in which they are working. It may be that the generalized theory of the academic is poorly articulated in the practitioner. For the practitioner heuristics may dominate and theory is sometimes thin on the ground.

The samurai is a pure performance expert—the only reality is the performance of action to secure an optimal performance. Practice is often the only training and responses are often automatic.

One can see this sort of division in any complex domain. Consider for example medical domains where we have professors of the subject, busy housemen working the wards, and medical ancillary staff performing many important but repetitive clinical activities.

The knowledge engineer must be alert to these differences because the various types of expert will perform very differently in KE situations. The academic will be concerned to demonstrate mastery of the theory. They will devote much effort to characterizing the scope and limitations of the domain theory. Practitioners, on the other hand, are driven by the cases they are solving from day to day. They have often *compiled* or *routinized* any declarative descriptions of the theory that supposedly underlies their problem solving. The performance samurai will more often than not turn any KE interaction into a concrete performance of the task—simply exhibiting their skill.

But there is more to say about the nature of experts and this is rooted in general principles of human information processing★. Psychology has demonstrated the limitations, biases and prejudices that pervade all human decision making—expert or novice. To illustrate consider the following facts, all potentially crucial to the enterprise of KE.

It has been shown repeatedly that the context in which one encodes information is the best one for recall. It is possible then, that experts may not have access to the same information when in a KE interview, as they do when actually performing the task. So there are good psychological reasons to use techniques which involve observing the expert actually solving problems in the normal setting. In short, protocol analysis techniques may be necessary, but will not be sufficient for effective knowledge elicitation.

Consider now the issue of biases in human cognition. One well known

---

★An excellent review of the psychology of expertise is Chi *et al.* (1988).

problem is that humans are poor at manipulating uncertain or probabilistic evidence. This may be important in KE for those domains which require a representation of uncertainty. Consider the rule:

IF   the engine will not turn over
AND   the lights do not come on
THEN   the battery is flat with probability X

This seems like a reasonable rule, but what is the value of X, should it be 0·9, 0·95, 0·79? The value which is finally decided upon will have important consequences for the working of the system, but it is very difficult to decide upon it in the first place. Medical diagnosis is a domain full of such probabilistic rules, but even expert physicians cannot accurately assess the probability values.

In fact there are a number of documented biases in human cognition which lie at the heart of this problem (see for example Kahneman *et al.*, 1982). People are known to undervalue prior probabilities, to use the ends and middle of the probability scale rather than the full range, and to *anchor* their responses around an initial guess. Cleaves (1987) lists a number of cognitive biases likely to be found in knowledge elicitation, and makes suggestions about how to avoid them. However, many knowledge engineers prefer to avoid the use of uncertainty wherever possible.

Cognitive bias is not limited to the manipulation of probability. A series of experiments has shown that systematic patterns of error occur across a number of apparently simple logical operations. For example, *Modus Tollens* states that if 'A implies B' is true, and 'not B' is true, then 'not A' must be true. However people, whether expert in a domain or not, make errors on this rule. This is in part due to an inability to reason with contrapositive statements. Also in part it depends on what A and B actually represent. In other words, they are affected by the content. This means that one cannot rely on the veracity of experts' (or indeed anyone's) reasoning.

All this evidence suggests that human reasoning, memory and knowledge representation is rather more subtle than might be thought at first sight. The knowledge engineer should be alert to some of the basic findings emanating from cognitive psychology. Whilst no text is perfect as a review of bias in problem solving the book by Meyer and Booker (1991) is reasonably comprehensive.

On expertise

Clearly the expertise embodied by experts is not of a homogeneous type. In constructing expert systems it is likely that very different types of knowledge will be uncovered which will have very different roles in the system. There are a number of analyses available of the *epistemology* of expertise. Our analysis is based to a large extent on that of Breuker (1987).

Firstly, we can distinguish what is called *domain level* knowledge. This term is being used in the narrow sense of knowledge that describes the concepts and elements in the domain and relations between them. This sort of knowledge is sometimes called *declarative*, it describes what is known about things in the domain. The propositions below can both be seen as domain level knowledge in this sense.

damaged VDU is a hardware fault
transmit line fault is a software fault

Extract A: Analysis of a laddered grid obtained from an expert VDU technician

There are also knowledge and expertise which have to do with what we might call the *inference level*. This is knowledge about how the components of expertise are to be organized and used in the overall system. It tells us the type of inferences that will be made and what role knowledge will play in those inferences. This is quite a high level description of expert behaviour and may often be implicit in expert practice. The following is a description of knowledge about part of an inference level structure called *systematic diagnosis*.

To perform systematic diagnosis we will have knowledge about a complaint, and knowledge about observables from the patient or object. We select some aspect of the complaint and using a model of how the system should be performing normally we look to see if a particular parameter of the system is within normal bounds.

Extract B: Analysis of verbal and behavioural protocols obtained from an abdominal pain expert

Another type of expert knowledge is the *task level*. This is sometimes called *procedural* knowledge, to do with how goals and sub-goals, tasks and subtasks should be performed. Thus in a classification task there may exist a number of tasks to perform in a particular order so as to utilize the domain level knowledge appropriately. This type of knowledge is present in the following extract.

First of all perform a general inspection of the object. Next examine the sample with a hand lens. Next use a prepared thin-section and examine that under a cross-polarizing microscope.

Extract C: Analysis of a verbal protocol obtained from an expert geologist

Finally, there is a level of expert knowledge referred to as *strategic* knowledge. This is information that monitors and controls the overall problem solving. This can have to do with the way resources are used. What to do if the proposed solution fails or is found to be inappropriate in some way? What to do when faced with incomplete or insufficient data? Such information is contained in the following extract from an interview.

> If I had time I would always check the video driver board. If its a ... machine I'd always check that because they are notorious for going wrong.

> Extract D: Part of a structured interview transcript obtained from an expert VDU technician

Any field of expertise is likely to contain these various sorts of knowledge to greater or lesser extents. At any particular knowledge level the information may be explicit or implicit in an expert's behaviour. Thus in some domains the experts may have no real notion of the strategic knowledge they are following whilst in others this knowledge is very much in the forefront of their deliberations. Also, of course, the requirements on a system about how far it needs to implement these various levels will vary. It is almost universally acknowledged that significant reasoning about problem domains requires more than just modelling simple relationships between concepts in the domains. It may require causal models of how objects influence and affect one another, models of the processes in which objects participate. This is a hard problem. And often the limitations of first generation expert systems mean that sophisticated domain models cannot be supported.

This brings us to a final important feature of KE. Often the process of acquisition yields much more knowledge than can be implemented. In this case the knowledge engineer ought to think about laying down the knowledge in a format that will allow it to be used when more powerful implementation systems arrive. Putting knowledge into deep freeze in this way requires using an expressive and unambiguous intermediate representation of the knowledge to be stored. A number of candidates exist for this; Young and Gammack (1987) provide a brief review of the alternatives.

Since knowledge elicitation is such a time consuming and expensive business, the results of which cannot all be immediately used, there is an increasing interest in developing ways of storing, archiving and retrieving knowledge that makes the best use of the elicitation investment (Neches *et al.*, 1991). The key to this lies in a change in our way of thinking about the content of knowledge-based systems. This has already been outlined earlier in this section. It is called the *knowledge level* view and was originally conceived by Newell (1982).

An important version of the knowledge level thesis was put forward by Clancey (1985) on *heuristic classification*. The structure, shown in Figure 14.2, was the result of a rational reconstruction of a number of existing expert systems. His claim was that the knowledge bases of many systems were for the most part *undifferentiated*. The knowledge bases of these systems had been built with little regard as to how the knowledge was used. His analysis

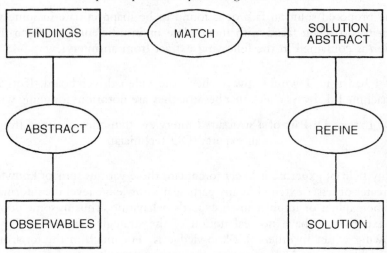

Figure 14.2. Heuristic classification (from Clancey, 1985)

uncovered what he saw as an important type of problem solving system. Not all systems would contain and use knowledge in this way. Not all systems would be examples of heuristic classification, but some important ones were.

One system which Clancey characterized as heuristic classification was MYCIN (Shortliffe, 1979). We shall illustrate his approach using MYCIN-like knowledge. Figure 14.2 is to be understood as a structure of what we have termed earlier as the inference level. It tells us what kinds of inferences are performed in this domain and the type of knowledge used by these inferences. The rectangles should be seen as types of data and the ellipses as types of inference.

Let us take the left hand side of this structure that contains a process called *abstraction*. This is the process by which *observations* or data are transformed into abstract observations or *findings*. The process of abstraction can be effected by a number of means. One of these is *qualitative* abstraction. Examples are shown below—here we move from quantitative observations to qualitative findings.

if	patient has white blood cell count <2500
then	patient has low white blood cell count

if	patient has temperature >101
then	patient has fever

What we have provided above is the actual *domain level* knowledge that plays the inference layer role of *abstraction*. The idea is to see much of the knowledge in MYCIN's knowledge base as to do with this type of knowledge level process—the process of abstraction, moving from a quantitative description of data to a qualitative one.

The top part of Figure 14.2 involves inferences from *findings* to *abstract solutions*. This type of inference is called *matching*—and is understood as a type of association knowledge. An example of such knowledge might be a rule such as:

> if   patient has fever
> and   patient has low white blood cell count
> then   patient has gram negative infection

Notice that the concept of gram negative infection might be viewed as a diagnosis or solution but it is not a very specific one. What sort of gram negative infection is it? The right hand side of the heuristic classification structure deals with inference types which refine general to specific solutions. Such knowledge might consist of hierarchical typologies containing knowledge such as

> gram negative infection
> has sub-types
> e. coli infection
> :
> etc.

To establish a particular infection the system is likely to use knowledge that discriminates the sub-types.

Notice that in this account although we have talked around the inference structure from left to right no explicit control knowledge has been given. We might have stipulated that the system start with patient data and reason forward to a possible solution. We might have stated that the system should hypothesize a solution and see if there was evidence from the patient's condition to support the hypothesis. Or else a mixture of these ways of moving around the structure of Figure 14.2 might have been adopted. This additional knowledge is, of course, the *task layer* we mentioned earlier. Sometimes the standard method of moving around the inference structure is modified. This might arise if, when a particular disease is suspected, then one immediately looks for a particular piece of supporting observational data. Such knowledge comprises the *strategies layer*.

What we have described was exemplified via a MYCIN-like example. But heuristic classification as a type of problem solving might apply to many domains; financial assessment of an individual's credit worthiness, the likelihood of finding a mineral resource at a particular location, etc. It is this generality that is the power of these knowledge level approaches. If we know what kind of application we are building we can use the models to indicate the type of knowledge we need to acquire, how we might structure the knowledge base, and how we can archive and index knowledge for future use.

These knowledge level models have formed the basis for a number of important methodologies that aim to support the knowledge engineering process.

Methodologies and programmes of KE

Are there any guidelines as to how KE techniques should be assembled to form a programme of acquisition? In other words, when should we use the various techniques? The choice may depend on the characteristics of the domain, of the expert, and of the required system. Furthermore, it is clear that some techniques are going to be more costly in terms of time with the expert, or with the effort required for subsequent analysis of transcripts.

There are a number of articles and books available on 'how to do knowledge elicitation'. These often contain advice of the most general kind, and emphasize the pragmatic considerations of expert system development. General reviews can be found in Firlej and Hellens (1991), Hart (1986), Hoffman (1987), Kidd (1987), McGraw and Harbison-Briggs (1989), and Welbank (1983). While these reviews are based on experience of the general kind, there have also been a number of attempts to make formal recommendations.

Gammack and Young (1985) offer a mapping of knowledge elicitation techniques onto domain type. Their analysis requires that domain knowledge be separated into different categories, and provides suggestions about the particular techniques which are most likely to be effective in each category. Although the analysis alerts one to the fact that there are different types of knowledge with any domain, it does not provide engineers with guidelines about how to identify each type.

The most thorough attempt to integrate KE procedure is provided by KADS—Knowledge Acquisition and Domain Structuring (see Wielinga *et al.*, 1992 and Schreiber *et al.*, 1993 for an overview of KADS). KADS embodies seven principles for the elicitation of knowledge and construction of a system. These principles are:

1. The knowledge and expertise should be analyzed before the design and implementation starts.
2. The analysis should be model-driven as early as possible.
3. The content of the model should be expressed at the knowledge level.
4. The analysis should include the functionality of the prospective system.
5. The analysis should proceed in an incremental way.
6. New data should be elicited only when previously collected data have been analyzed.
7. Collected data and interpretations should be documented.

Principle 1 is quite straightforward and requires no further explanation. Principle 2 requires that one should bring to bear a model of how the knowledge is structured early on in the process, and use it to interpret subsequent data. Principle 3 means that one should use an appropriate intermediate level knowledge representation device, and try to characterize knowledge in terms of its use and functioning. Principle 4 is a reminder that a complete analysis includes an understanding of how the system is to work—e.g., who will use it, and in what situation. One cannot gain full understanding of the problem simply by trying to map out an expert's knowledge without regard to how it will be used. Principle 5 emphasizes the fact that there is a wide variety of related topics within a domain. This means that construction of a model should be 'breadth-first', embodying all aspects at once, rather than attempting fully to represent one sub-part after another. Principles 6 and 7 are once again straightforward. Like many of the best recommendations the utility of these statements is most apparent when they are not adhered to. As KADS develops it is attempting to impose more structure on the general knowledge engineering process.

One of the most obvious aspects of this development is the characterization of increasing numbers of knowledge level models of the sort described in the previous section. The identification of an appropriate model for one's application becomes an important knowledge elicitation exercise. The model selection process exploits a classification tree of knowledge level models, such as shown in Figure 14.3. The way that the tree is used in KA is as follows. The first task for the knowledge engineer in KA is model selection. What

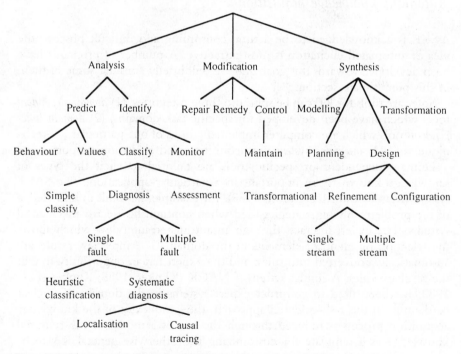

Figure 14.3. The taxonomic hierarchy of KADS-I problem solving models

he or she has to do is to traverse this tree until a leaf node is reached; he or she then has his or her model. The traversal of the tree is given by a series of questions about the domain, each of which corresponds to a node of the tree. The answers to the questions are multiple choice, and each option corresponds to a move down a particular branch of the tree.

Various libraries and configurations of such models exist or are under development. The main examples of these are KADS (Wielinga *et al.*, 1992), generic tasks (Chandrasekaran and Johnson, 1993), problem solving methods (Puerta *et al.*, 1992), and Steels' components of expertise (Steels, 1990).

All of these approaches claim that knowledge acquisition should be viewed under the metaphor of model building, rather than the mining of information (Wielinga *et al.*, 1992). The discussion above emphasizes this point—in short, KE is not an independent part of constructing an expert system, and viewing it as such will lead to inefficient practice. Whether or not the adoption of these various methodologies makes for more efficient and effective KE is a moot point as these claims have not been formally evaluated. There is no doubt that between them they are winning many converts. In so far as they impose a discipline on the KE process they are likely to be very important.

Automatic knowledge acquisition

As KE is acknowledged to be a time-consuming and difficult process, the idea of automated elicitation is most attractive. A number of programs have been developed towards this goal, and we will briefly consider some of them in this penultimate section.

Software tools for KE can be split into three categories: (1) *domain dependent* tools which have been developed for specific task domains; (2) *domain independent* tools which are computer implementations of one particular KE technique; and (3) *integrated systems* for acquisition and elicitation support.

Domain dependent or specific tools are tailored to elicit the types of knowledge known to be important in a particular application. The SALT system (Marcus and McDermott, 1989), for example, has built in knowledge of the problem solving strategies used when configuring electro-mechanical systems. The expert interacts through an automated interview which allows the selection by menu of elements in the domain. The interview results are automatically converted into rules, and the expert has an opportunity to edit the resultant rules. A similar system, KNACK (Klinker, 1988; Klinker *et al.*, 1989) has been used to construct expert systems in the domain of nuclear hardening. In this task-oriented approach, the complexity of the knowledge acquisition process is reduced through the use of a model of the required knowledge as a template for customizing the otherwise general KA techniques and tools. These systems are by definition restricted to particular tasks and domains. Moreover, most of these domain-oriented tools have remained research prototypes.

The second approach to automation of the KA process is to focus on one particular technique and support its use. Of those systems which implement individual standard techniques, many of the most successful are based on the *repertory grid*. This technique has its roots in the psychology of personality (Kelly, 1955) and is designed to reveal a conceptual map of a domain, in a similar fashion to the card sort as discussed above (see Shaw and Gaines, 1987a, for a full discussion). The technique as developed in the 1950s was very time-consuming to administer and analyze by hand. This naturally suggested that an implemented version would be useful.

There are several repertory grid programs on the market. One of the earliest and best known programs was ETS (Boose, 1985) although this was developed primarily as a research tool. KSS0 (Gaines, 1990) which we illustrate below formed the basis for a number of commercial products.

Briefly, subjects are presented with a range of domain elements and asked to choose three, such that two are similar and different from the third. Suppose we were trying to uncover an astronomer's understanding of the planets. We might present him with a set of planets, and he might choose Mercury and Venus as the two similar elements, and Jupiter as different from the other two. The subject is then asked for their reason for differentiating these elements, and this dimension is known as a construct. In our example 'size' would be a suitable construct. The remaining domain elements are then rated on this construct.

This process continues with different triads of elements until the expert can think of no further discriminating constructs. The result is a matrix of similarity ratings, relating elements and constructs. This is analyzed using a statistical technique called *cluster analysis*. In KE, as in clinical psychology, the technique can reveal clusters of concepts and elements which the expert may not have articulated in an interview.

The automated versions are run in such a way that the repertory grid is built-up interactively, and the expert is shown the resultant knowledge. Experts have the opportunity to refine this knowledge during the elicitation process. In Figure 14.4 we can see that the expert has so far generated seven constructs along which the planets vary. In this case a nine point rating scale has been used and in the case of the construct size the smallest planet, Mercury, has been given a rating 1 and the largest, Jupiter, a rating of 9. The other planets have been rated in a comparative manner along the size construct. The analysis has already revealed clusters of both constructs and elements. Thus Jupiter and Saturn are clustered together at around 82% similarity, Neptune and Uranus at around 88%, and these two pairs are clustered at around 80%. An astronomer might well observe that this group of four planets constitute the gas giants. A new concept has been uncovered. Similarly, constructs can be clustered. We see that the constructs relating to tem-

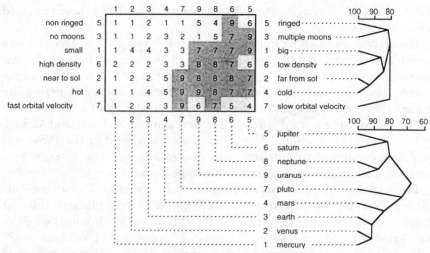

Figure 14.4. Knowledge elicited using KSS0. Heavier shading indicates higher values for elements in a construct

perature and distance from the sun are clustered. Such associations can reveal causal or other law-like relations in the domain.

Given the type of analysis shown in Figure 14.4, KSS0 is also able to output a set of machine-executable rules. An example from our sample domain is shown below.

Induct Analysis of planets   26-Nov-90 10:34:08
  Induce Unknown Significance 80·00%
  Target Attribute 7 Based On Entities 1 Through 9

Induct Rules from planets   26-Nov-90 10:34:09
A5=non ringed → A1=small 60·0% 21·0%
A2=far from sol and A5=non ringed → A1=small 100% 33·3%

A6=low density → A1=big 100% 8·78%
A3=multiple moons → A1=big 100% 19·8%

A1=small and A2=far from sol → A7=slow orbital velocity 100% 22·2%
A6=low density → A7=slow orbital velocity 33·33% 52·9%

A2=near to sol → A7=fast orbital velocity 100% 8·78%
A1=small → A7=fast orbital velocity 66·67% 41·7%

These systems by-pass the human elicitor altogether and have been used with considerable success in domains where solutions can be comfortably enumerated (see Boose, 1986 for a list of successful applications of ETS). Indeed, they can be found a place in any programme of elicitation.

A second automated acquisition technique not discussed previously is induction-based machine learning. Machine induction is the process whereby a program is presented with many solved problems (or other appropriate data sets) from the domain, and uses statistical regularities to infer underlying rules. An introductory overview to this technique can be found in Hart (1986). The best known algorithm for this is ID3 (Quinlan, 1979, 1983), and versions of this are now available in some small-scale shells. Indeed a variety of induction was used to generate rules from the repertory grid data set shown above.

To illustrate the standard use of such induction methods imagine you are presented with hundreds of cases of patient symptoms alongside decisions about the diagnosis. It may be that a doctor cannot readily articulate a rule relating a set of symptoms to a particular diagnosis. If they occur reliably together with the diagnosis, the program can induce a rule relating them.

Although this technique sounds attractive, it has several associated problems. Rules may be induced which do not have any basis in the domain. This can arise according to quite arbitrary decisions such as the order in which the learning examples are input. Furthermore, there is the problem of cause and effect. The fact that a particular set of symptoms co-occurs with a diagnosis on the learning set does not necessarily imply a causal relation. Research continues on more sophisticated machine learning techniques for KE (e.g., Michalski, 1983; Mitchell, 1982). Clearly though, induction programs do not provide a complete solution to automatic elicitation.

Although programs continue to be written to support single KA techniques there is an increasing trend towards the third type of approach mentioned at the beginning of this section. This approach is to integrate several KA tools. The idea is that the whole is greater than the parts. One such system, ProtoKEW (Shadbolt, 1992; Reichgelt and Shadbolt, 1992), incorporates repertory grids, laddering, card sorts and machine induction. ProtoKEW has facilities for translating the results of elicitation from one tool to another. The knowledge engineer is also able to integrate and execute the results of different tools. The toolset was developed as part of a large ESPRIT project* and runs on SUN workstations under UNIX. Though primarily a research tool ProtoKEW is available by licence from the first author.

A development of this work was supported by funding to the first author from the UK Science and Engineering Research Council and Department of Trade and Industry. This was to develop a KE toolset running on laptop PCs. The system is called SET (Structured Elicitation Toolkit) and is shortly to be launched as a commercial product.

The tools built into SET include: protocol editing (enabling text and documents to be annotated and analyzed), concept and attribute laddering, card sorts and repertory grids. In addition, a powerful machine learning method is included for detecting patterns and rules amongst large amounts of data. The results of elicitation are stored in a persistent object-oriented database. All the tools are able to access this database providing a means of transferring knowledge between the various tools. The entire package is interfaced

---

*ESPRIT project P2567 known by the acronym ACKnowledge. See Anjewierden et al., 1992 for a fuller description of this project.

through an intuitive direct manipulation interface. We have provided a flexible hypertext system able to annotate objects in the database together with concise documentation and tutorial material.

There is in prospect the emergence of even more strongly integrated workbenches. In systems such as KEW (Anjewierden *et al.*, 1992) and VITAL (Shadbolt *et al.*, 1993; Motta *et al.*, 1993) we see the bginnings of attempts to integrate both sets of KA tools and model based methodologies of the sort discussed in the previous section. These workbenches contain libraries of knowledge level models. The expert system builder is guided through the process of KA and KE by the use of structured methodologies supported by the workbench. The ambition is a natural one. It is to build knowledge-based systems for knowledge elicitation and acquisition. Such systems indicate the shape and form of the next generation of knowledge engineering tools.

Conclusions

The problem of knowledge elicitation is a subtle and complex one. This chapter has described some of the techniques that are used in this enterprise and indicated where software support for the process is becoming available. But we have also sought to provide an indication of the difficulties inherent in doing this kind of work. At present knowledge elicitation is itself a form of expertise. Experienced knowledge engineers come to recognize the subtleties of expert thinking. They develop skills that allow them to capture an expert's knowledge despite the many obstacles they face. Recently, methodologies have begun to emerge that seek to structure and manage the acquisition process. These are still under development and even when mature will still require the skills of the knowledge engineer.

As expert systems and other knowledge intensive technology becomes more widely available, more people will have to face the issue of knowledge elicitation. In order to provide support for this enterprise, research is underway in areas as disparate as software engineering, artificial intelligence and psychology. Experience to date has identified many interesting and difficult problems that together form the research agenda for the future. Knowledge engineering will continue to be a challenging discipline.

References

Anjewierden, A., Wielinga, B. and Shadbolt, N.R. (1992). Supporting knowledge acquisition: The ACKnowledge project. In *ESPRIT-92 Knowledge Engineering,* edited by L. Steels *et al.* (Brussels: CEC).
Belkin, N.J., Brooks, H.M. and Daniels, P.J. (1987). Knowledge elicitation using discourse analysis. *International Journal of Man–Machine Studies,* **27**, 127–144.

Boose, J.H. (1985). A knowledge acquisition program for expert systems based on personal construct psychology. *International Journal of Man–Machine Studies,* **23**, 495–525.

Boose, J.H. (1986). *Expertise Transfer for Expert System Design* (New York: Elsevier).

Boose, J.H. and Bradshaw, J.M. (1987). Expertise transfer and complex problems: using AQUINAS as a knowledge-acquisition workbench for knowledge-based systems. *International Journal of Man–Machine Studies,* **26**, 3–28.

Breuker, J., Wielinga, B., van Someren, M., De Hoog, R., Screiber, G., de Greef, P., Bredeweg, B., Wielemaker, J., Billeaut, J.P., Davoodi, M., and Hayward, S. (1987). Model-Driven Knowledge Acquisition: Interpretation Models (KADS deliverable D1, University of Amsterdam, Dept of Social Science Informatics).

Burton, A.M., Shadbolt, N.R., Hedgecock, A.P. and Rugg, G. (1987). A formal evaluation of knowledge elicitation techniques for expert systems: domain 1. In *Research and Development in Expert Systems IV,* edited by D.S. Moralee (Cambridge: Cambridge University Press).

Burton, A.M., Shadbolt, N.R., Rugg, G. and Hedgecock, A.P. (1988). Knowledge elicitation techniques in classification domains. In *ECAI-88: Proceedings of the 8th European Conference on Artificial Intelligence,* pp. 85–93.

Chandrasekaran, B. and Johnson, T.R. (1993). Generic Tasks and Task Structures: History, Critique and New Directions, in *Second Generation Expert Systems,* edited by J.-M. David, J.-P. Krivine and R. Simmons. (Berlin: Springer) pp. 232–72.

Chi, M., Feltovitch, P. and Glasser, R. (1982a). Categorization of experts and novices. In *The Handbook of Intelligence,* edited by R. Sternberg (Hillsdale, NJ: Lawrence Erlbaum).

Chi, M.T.H., Glasser, R. and Rees, E. (1982b). Expertise in problem solving. In *Advances in the Psychology of Human Intelligence: 1,* edited by R.J. Sternberg (Hillsdale, NJ: Lawrence Erlbaum).

Chi, M.T.H., Glasser, R. and Farr, M.J. (1988). *The Nature of Expertise* (Hillsdale, NJ: Lawrence Erlbaum).

Clancey, W.J. (1985). Heuristic classification. *Artificial Intelligence,* **27**, 289–350.

Clarke, B. (1987). Knowledge acquisition for real time knowledge based systems. In *Proceedings of the First European Workshop on Knowledge Acquisition for Knowledge Based Systems.* Reading University, UK.

Cleaves, D.A. (1987). Cognitive biases and corrective techniques: proposals for improving elicitation procedures for knowledge based systems. *International Journal of Man–Machine Studies,* **27**, 155–166.

Cooke, N.M. and McDonald, J.E. (1986). A formal methodology for acquiring and representing expert knowledge. *Proceedings of the IEEE,* **74**, 1422–1430.

Diederich, J., Ruhmann, I. and May, M. (1987). KRITON: a knowledge-acquisition tools for expert systems. *International Journal of Man–Machine Studies,* **26**, 29–40.

Elstein, A.S., Shulman, L.S. and Sprafka, S.A. (1978). *Medical Problem Solving:*

An Analysis of Clinical Reasoning (Cambridge, Mass: Harvard University Press), Cited in Kuipers and Kassirer.

Ericsson, K.A. and Simon, H.A. (1984). *Protocol Analysis: Verbal Reports on Data* (Cambridge, Mass: MIT Press).

Firlej, M. and Hellens, D. (1991). *Knowledge Elicitation: A Practical Handbook.* (Englewood Cliffs, NJ: Prentice Hall International).

Gaines, B. (1990). An architecture for integrated knowledge acquisition systems. *Proceedings of the 5th AAAI Knowledge Acquisition for Knowledge-Based Systems Workshop, Banff, Canada.*

Gaines, B.R. (1988). Knowledge acquisition systems for rapid prototyping of expert systems. *INFOR, 26,* 256–285.

Gaines, B.R. and Linster, M. (1990). Development of second generation knowledge acquisition systems. In *Current Trends in Knowledge Acquisition,* edited by B. Wielinga *et al.* (Amsterdam: IOS Press).

Gammack, J.G. (1987a). Formalising implicit domain structure. In *Proceedings of the SERC Workshop on Knowledge Acquisition for Engineering Applications. SERC report RAL-87-055.*

Gammack, J.G. (1987b). Different techniques and different aspects of declarative knowledge. In *Knowledge Acquisition for Expert Systems: A Practical Handbook,* edited by A.L. Kidd (New York: Plenum Press).

Gammack, J.G. and Young, R.M. (1985). Psychological techniques for eliciting expert knowledge. In *Research and Development in Expert Systems,* edited by M. Bramer (Cambridge: Cambridge University Press).

Grover, M.D. (1983). A pragmatic knowledge acquisition methodology. In *IJCAI-83: Proceedings 8th International Joint Conference on Artifical Intelligence.*

Hart, A. (1986). *Knowledge Acquisition for Expert Systems* (London: Kogan Page).

Hayes-Roth, F., Waterman, D.A. and Lenat, D.B. (1983). *Building Expert Systems* (Reading, Mass: Addison-Wesley).

Hinkle, D.N. (1965). The change of personal constructs from the viewpoint of a theory of implications. Unpublished Ph.D. thesis, University of Ohio.

Hoffman, R.R. (1987). The problem of extracting the knowledge of experts from the perspective of experimental psychology. *AI Magazine, 8,* 53–66.

Jackson, P. (1990). *An Introduction to Expert Systems,* 2nd edition (Reading, Mass: Addison Wesley).

Johnson, L. and Johnson, N. (1987). Knowledge elicitation involving teach-back interviewing. In *Knowledge Elicitation for Expert Systems: A Practical Handbook,* edited by A. Kidd (New York: Plenum Press).

Johnson, P.E., Zualkernan, I. and Garber, S. (1987). Specification of expertise. *International Journal of Man–Machine Studies, 26,* 161–181.

Kahneman, D., Slovic, P. and Tversky, A. (Eds.) (1982). *Judgement under Uncertainty: Heuristics and Biases* (New York: Cambridge University Press).

Kelly, G.A. (1955). *The Psychology of Personal Constructs* (New York: Norton).

Kidd, A.L. (Ed.) (1987). *Knowledge Acquisition for Expert Systems: A Practical Handbook* (New York: Plenum Press).

Klinker, G. (1988). KNACK: Sample-driven knowledge acquisition for reporting systems. In *Automating Knowledge Acquisition for Expert Systems,* edited by S. Marcus (Boston, Mass: Kluwer Academic).

Klinker, G., Boyd, C., Dong, D., Maiman, J., McDermott, J. and Schnellbach, R. (1989). Building expert systems with KNACK. *Journal of Knowledge Acquisition,* **1**, 299–320.

Kuipers, B.and Kassirer, J.P. (1983). How to discover a knowledge representation for causal reasoning by studying an expert physician. In *IJCAI-83: Proceedings of the 8th International Conference on Artificial Intelligence.*

LaFrance, M. (1987). The Knowledge Acquisition Grid: a method for training knowledge engineers. *International Journal of Man–Machine Studies,* **27**.

Marcus, S. and McDermott, J. (1989). SALT: a knowledge acquisition language for propose-and-revise systems. *Artificial Intelligence,* **39**, 1–38.

Marcus, S., McDermott, J. and Wang, T. (1985). Knowledge acquisition for constructive systems. In *IJCAI-85: Proceedings of the Ninth International Joint Conference on Artificial Intelligence.*

McGraw, K.L. and Harbison-Briggs, K. (1989). *Knowledge Acquisition: Principles and Guidelines* (Englewood Cliffs, NJ: Prentice-Hall International).

Meyer, M. and Booker, J. (1991). *Eliciting and Analysing Expert Judgement: A Practical Guide* (London: Academic Press).

Michalski, R. (1983). A theory and methodology of inductive learning. In *Machine Learning: An Artificial Intelligence Approach,* edited by R. Michalski, J. Carbonell and T. Mitchell (Palo Alto: Tioga).

Michalski, R., Carbonell, J. and Mitchell, T. (Eds.) (1986). *Machine learning: An AI Approach, Volume 2,* (Los Altos, CA: Morgan Kaufman).

Mitchell, T. (1982). Generalization as search. *Artificial Intelligence,* **18**, 203–226.

Motta, E., O'Hara, K. and Shadbolt, N. (1993). Grounding GDMs: a structured case study. *Journal of Knowledge Acquisition,* **5**(4).

Neches, R., Fikes, R., Finin, T., Gruber, T., Patil, R., Senator, T. and Swartout, W. (1991). Enabling technology for knowledge sharing. *AI Magazine,* **12**(3), 37–56.

Newell, A. (1982). The knowledge level. *Artificial Intelligence,* **18**, 87–127.

Newell, A. and Simon, H.A. (1963). GPS, a program that simulates human thought. In *Computers and Thought,* edited by E. Feigenbaum and J. Feldman (New York: McGraw-Hill).

Puerta, A.R., Tu, S.W. and Musen, M.A. (1992). Modelling Tasks with Mechanisms (Knowledge Systems Laboratory Technical Report KSL-REPORT 92–30).

Quinlan, J.R. (1979). Rules by induction from large collections of examples. In *Expert Systems in the Micro-Electronic Age,* edited by D. Michie (Edinburgh: Edinburgh University Press).

Quinlan, J.R. (1983). Learning efficient classification procedures and their application to chess endgames. In *Machine Learning: An Artificial Intelligence Approach,* edited by R. Michalski, J. Carbonell and T. Mitchell (Palo Alto: Tioga).

Reichgelt, H. and Shadbolt, N.R. (1992). ProtoKEW: a knowledge-based system for knowledge acquisition. In *Research Advances in Cognitive Sci-*

ence, Volume 5: *Artificial Intelligence,* edited by D. Sleeman and N.O. Bernsen (Hillsdale, NJ: Lawrence Erlbaum).

Rich, E. and Knight, K. (1991). *Artificial Intelligence,* 2nd edition (New York: McGraw-Hill).

Schreiber, A. *et al.* (1993). *KADS: A Principled Approach to Knowledge-Based System Development* (London: Academic Press).

Schvaneveldt, R.W., Durso, F.T., Goldsmith, T.E., Breen, T.J., Cooke, N.M., Tucker, R.G. and De Maio, J.C. (1985). Measuring the structure of expertise. *International Journal of Man–Machine Studies,* **23**, 699–728.

Schweickert, R., Burton, A.M., Taylor, E.N., Shadbolt, N.R. and Hedgecock, A.P. (1987). Comparing knowledge elicitation techniques: a case study. *Artificial Intelligence Review,* **1**, 245–253.

Shadbolt, N.R. (1989). Knowledge representation in man and machine. In *Expert Systems: Principles and Case Studies,* 2nd Edition, edited by R. Forsyth (London: Chapman and Hall).

Shadbolt, N.R. (1992). Facts, fantasies and frameworks: the design of a knowledge acquisition workbench. In *Contemporary Knowledge Engineering and Cognition,* edited by F. Schmalhofer, G. Strube and Th. Wetter (Heidelberg: Springer-Verlag).

Shadbolt, N.R. and Burton, A.M. (1989). Empirical studies in knowledge elicitation. *ACM-SIGART Special Issue on Knowledge Acquisition,* 108.

Shadbolt, N., Motta, E. and Rouge, A. (1993). Constructing knowledge-based systems. *IEEE Software,* November, 34–39.

Shapiro, A. (1987). *Structured Induction in Expert Systems* (New York: Addison-Wesley).

Shaw, M.L.G. and Gaines, B.R. (1987a). An interactive knowledge elicitation technique using personal construct technology. In *Knowledge Acquisition for Expert Systems: A Practical Handbook,* edited by A.L. Kidd (New York: Plenum Press).

Shaw, M.L.G. and Gaines, B.R. (1987b). KITTEN: knowledge initiation and transfer tools for experts and novices. *International Journal of Man–Machine Studies,* **27**, 251–280.

Shortliffe, E.H. (1979). *Computer-Based Medical Consultations: Mycin* (New York: American Elsevier).

Steels, L. (1990). Components of expertise. *AI Magazine,* **11**(2), 30–49.

Weiss, S. and Kulikowski, C. (1984). *A Practical Guide to Designing Expert Systems* (Towata, NJ: Rowman and Allanheld).

Welbank, M.A. (1983). A review of knowledge acquisition techniques for expert systems. British Telecom Research, Martlesham Heath.

Wielinga, B.J. *et al.* (1992). KADS: a modelling approach to knowledge engineering. *Knowledge Acquisition Journal,* **4**(1).

Young, R.M. and Gammack, J. (1987). Role of psychological techniques and intermediate representations in knowledge elicitation. In *Proceedings of the First European Workshop on Knowledge Acquisition for Knowledge Based Systems,* Reading University, UK.

Part IV

Assessment and design of the physical workplace

The physical environment comprises what must be the most widely explored of any group of ergonomics variables. These variables are components of most situations we investigate and affect people and their performance in a multitude of ways.

In spite of the long history of environmental assessments, the methods and techniques are still developing. This must be, in part, because of the impact of computers and information technology in general (see chapter 9 computerized data collection), but it is also because our understanding of the psychophysiological and physiological effects of environmental stressors is increasing.

Moreover, the array of methods used reflects these various effects. Different aspects of the physical environment have been said to have effects—positive and negative—on all the facets of individual well-being described earlier in chapter 1. All environments at the extreme will affect workers' health; what we need are data to tell us at what levels such effects as noise-induced hearing loss, or heat stress, or musculo-skeletal diseases might be found. Here, of course, we need to know not just the physical stimulus measurements but also exposure times, and also what alleviating effect different job, task or equipment designs, or personnel selection, might have.

Ideally, of course, we are concerned with keeping harmful environmental variables at well below (or sometimes above) the levels where health effects are found. Therefore we would wish, for instance, to eliminate causes of annoyance, discomfort or dissatisfaction, such as perceived flicker from fluorescent lights, discomfort glare on a VDU, irritating air draughts, noise annoyance, cramped workspaces etc.

The third type of possible effects of the physical environment is upon performance. This can occur through the first two, health problems or discomfort and dissatisfaction, of course, but may also be produced more directly. Examples are disability glare on a VDU, speech interference noise levels, cold conditions leading to loss of dexterity, or reduced vigilance through distractions from poor seating. Possible performance decrement has been proposed in terms of output and errors, for mental and physical tasks, and with respect to both direct task-related measures and also systemic ones such as absenteeism or labour turnover. It must be said that, in general, results on performance effects of the environment are more equivocal than those on health or discomfort/dissatisfaction.

The final proposed outcome of working environments, good or poor, is the attitudes that such conditions might engender in the workforce. This might be seen in, for instance, such opinions (spoken or unspoken) as: 'If management think so little of us to give us such conditions to work in then why should we co-operate with what they want'. Consequences will be resistance to change, lack of innovation or dynamism, a tendency to cure rather than prevention, and generally an unwillingness to give of skills and time. Working environments perceived as good on the other hand may evoke opposite reactions.

This section opens with three chapters reviewing all the effects described above in the context of the visual, climatic and auditory environments; consequently methods of assessment are described which will allow us to evaluate working conditions and to initiate the correct remedial action where necessary. Howarth (chapter 15), Parsons (chapter 16), and Haslegrave (chapter 17) explain, for the unwary, the multiplicity of physical characteristics and units of measurement of environmental stimuli, the different human responses of relevance, and the consequently diverse methods and techniques of measurement. They also describe evaluation criteria such that assessments can be made, and provide examples of practical working environment assessment.

Vibration, covered by Bonney in chapter 18, differs somewhat from the first three types of physical work environment; it will be a factor only in certain working conditions. It is relatively recently that hand-arm vibration has had attention from a major body of researchers, probably as a result of the widespread use of chain saws in forestry in the last twenty years. Measuring the extent of Vibration White Finger still does not embrace an agreed practice, the argument lying primarily between clinical judgement and the use of psychophysical threshold measures of spatial or vibratory perception. For some situations, e.g., chain saws or hand grinders, the assessment of vibration levels is quite structured and anti-vibration mountings can bring the levels down to those defined by ISO, keeping energy input to the hands such that damage is unlikely. Other work situations, e.g., riding motorcycles or hand fettling of castings, are less tractable and require changes to the process to overcome the problems.

Whole body vibration covers a range of experiences from seasickness to helicopter flying. Here the major area of motor vehicle riding has been emphasized since many of the techniques can be extended in either direction along the vibration spectrum. The use of mathematical models and computer simulations seems likely to extend in whole body vibration studies, at least on the design side. A greater understanding of the effects on people, however, will still require experimentation. Since only low stress levels are acceptable in research studies the importance of exploiting field evidence will be obvious.

The last three chapters in this section have close links, all having anthropometry as a major component. Pheasant (chapter 19) outlines the subject, the sources of data and the methods for applying them to real problems. Superficially the procedures are simple, but in fact there are many poor applications of anthropometry which arise to a great extent through insufficient recognition of the reasons for the choice of the various dimensions or percentiles. Pheasant gives clear guidance in this respect.

Computer graphics have made major strides in recent years and permit the simulation of workplaces at much less cost than physical mock-ups (Porter *et al.*, chapter 20). As is emphasized in the chapter, this does not obviate the need for physically testing the design, but it does bring a good design more quickly to realization and allows the rapid exploration of a greater number of different designs before investing in mock-ups or prototypes. As computer technology advances we can expect even more sophistication in these methods, including the calculation of environmental physiological and psychological aspects in addition to the physical layouts.

The last chapter (21) by Corlett deals with a particular design or testing application, industrial seating, which has important anthropometric components. Many people believe they are competent to design seats, and the wide variety of seats available does, in part, support their view. The comments of seat users, particularly where the seat is a necessary part of a workplace, gives clear evidence that a lot of their confidence is misplaced.

The seat evaluation chapter brings into study the utility and acceptability of the seat, as well as its dimensions. In fact these two factors are relevant for any workplace, and variations on the methods in this last chapter could be used to test their effectiveness in these reports. One of the reasons for incorporating this chapter within the group is to point out these broader aspects, and to round off what might otherwise have been seen as primarily technical matters with a reminder that the ergonomist is interested more broadly than just in the relationships between physical dimensions.

Chapter 15

Assessment of the visual environment

Peter A. Howarth

Introduction

The scope of this chapter

Vision provides us with more information than all of our other senses combined. As a consequence, the environmental conditions necessary to optimize the eyes' performance are of paramount importance.

While the visual sense is in some ways exquisitely fine, in other ways it is dreadfully misleading. Although we can detect an offset in a line of as small as 5″ of arc (the width of a pencil viewed at a distance of 300 m), when indoors under artificial lights we can think that two pieces of cloth are the same colour—to discover on going outside that they are quite different. In the first of these examples, visual performance is dependent upon the *quantity* of light, whereas in the second it is the *quality* (the spectral characteristics) of the light which limits performance.

There are many different aspects of the visual environment to consider when trying to produce the conditions necessary for good visual performance, not all of them immediately obvious. Consider an example: visual displays incorporating liquid crystal alphanumeric characters are to be installed in a self-service petroleum pump. What visual considerations are there? Some spring to mind immediately, such as character readability and the positioning of the display so that it can be seen by drivers of different heights, sitting or standing. But the relevant elements of the visual environment are not simply those related directly to the visual task in isolation from its surroundings. Other questions need to be asked: Is the display going to be used both during the day and night? If so, how is it going to be lit, and will supplementary lighting be needed? What is known of the spectral characteristics of the light that will fall on the display? Will it be lit by monochromatic sodium road lighting, and what account needs to be taken of this in the use of colour in the design? If supplementary lighting is going to be used, how should it be positioned to avoid producing veiling reflections in the display, and will discomfort be produced from glare? Is the display going to be covered with either glass or plastic, and if so what reflections will be produced in this cover by artificial lights, by car headlights, or by the sun as it crosses the sky? Clearly the issues involved in assessing the visual environment are not restricted to the ambient light levels!

This chapter discusses conditions of the visual environment that may be of concern to the ergonomist, and indicates why and how they may need to be assessed. Of necessity, the chapter is general in nature: the aim is to provide the reader with sufficient information to enable him or her to approach a wide range of problems with an adequate understanding of the key issues. With the background information provided here, texts dealing with specific issues—such as the Chartered Institute of Building Services (CIBS) Code of Interior Lighting (CIBS, 1984)—should be more readily understood. Finally, further information on the issues considered here, along with full references, can be found in standard texts such as Boyce (1988).

The nature of vision

It is often said that vision runs on light. While it is true that the eye's photoreceptors respond to light, they do not signal absolute light level—adaptation removes this information at the earliest stage of the visual process. Because of its adaptation systems the human eye has a tremendous total operating range, and the illuminance at midday on a Florida beach may be $10°$ times higher than on a dark and cloudy night on the Scottish moors. However, although the eye can operate in either place, at any one time it is restricted to a small part of this range. It is this adaptation that makes us so poor at *absolute* visual judgements; on the other hand we are usually extremely good at *relative* judgements. Rather than responding to light level *per se*, the visual system responds to *variation* (temporal, spectral or spatial) in light. It is more appropriate to say that vision runs on contrast—a luminance or a chromatic change in either space or time.

Contrast describes a variation in light, and this chapter begins with a review of the nature of light and how it affects the visual system. This is followed by a description of the characteristics of natural and artificial light sources, and a discussion of which aspects of the visual environment deserve attention. Finally, those parameters which it is realistic and useful to assess in a practical situation are considered.

How light affects the visual system

Light is a small portion of the electromagnetic radiation spectrum. This spectrum includes radio waves, microwaves, ultraviolet radiation, infra-red radiation, and X-rays. What is special about light is that we can see it. While this seems an obvious statement, there is an important fact to be gleaned from it—namely, 'light' is defined by the human visual system, and not by the light source. If you cannot see it, it's not light. The sun, for example, radiates a wide range of the electromagnetic spectrum, some of which passes through the atmosphere and some of which is filtered out by it. However, the limits of what we call light are those of the human eye [approximately 380–760 nm (10^{-9} m)] not those of the sun. The spectral sensitivity of the eye, termed the Vλ function, is shown as a solid line in Figure 15.1. A photometer measuring *light* will, in effect, measure the total amount of radiation present over the spectrum as weighted by this function. This is done by passing the radiation through filters, so that the combination of the filter and the radiation sensor has the same response characteristics as the human eye.

The effectiveness of electromagnetic radiation in producing the sensation of vision depends upon how sensitive the eye is to the wavelengths present.

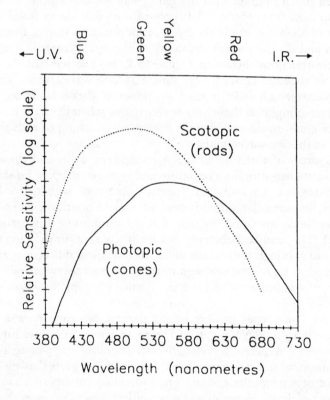

Figure 15.1. The spectral sensitivity of the eye for daytime light levels (solid line), known as the Vλ function, and the sensitivity for nightime light levels (dashed line), known as the V'λ function. The colour names describe the appearance of each portion of the spectrum (c.f., a rainbow)

The *spectral characteristics* of the light (i.e., the relative amount of radiation at each wavelength) give it its characteristic appearance (i.e., its colour), and the weighted sum of the radiation present at each wavelength determines the *amount of light* present.

A potentially troublesome point to note here is that the spectral sensitivity of the eye is not totally independent of the light level. The eye has two kinds of photoreceptor, which have different operating ranges. One kind are called 'cones', of which there are three types (to give us colour vision), the other kind are called 'rods'. At normal light levels the cones are the active photoreceptors, and the rods are essentially saturated (their photopigment is highly bleached). However at lower light levels, e.g., around dusk, there is not enough light for cones to operate, and rods become the active photoreceptors. The spectral sensitivity of the rod system differs from that of the normal cone system, and the change in sensitivity of the eye as the light level changes from photopic (cone) to scotopic (rod) levels is termed the 'Purkinje shift'. Purkinje, a Czech physician, noticed in 1825 that while red and blue paint on signposts looked the same brightness during daylight, at night the blue looked much brighter than the red. There are two aspects of the Purkinje shift to note from Figure 15.1: going from cone vision (solid curve) to rod vision (dashed curve) (1) the peak of the function moves from 555 nm to 507 nm and (2) the overall sensitivity of the eye increases.

From the functions shown in Figure 15.1, we can see that the absolute sensitivity of rods and cones is very similar at long wavelengths. This shows why long-wavelength light is used to 'preserve' dark-adaptation, e.g., on ship's bridges at night: at these long wavelengths, when the light level is high enough for cones to operate there is much less bleaching of rods than there would be at shorter wavelengths.

Other sources of variation in spectral sensitivity exist, and these include differences between stimulus conditions and between people. To standardize light measurement, a number of 'standard observers' with defined spectral sensitivities for particular stimulus field sizes, field positions, and adaptation levels have been specified by the CIE (Commission Internationale de l'Eclairage). The 'standard observer' is a useful concept for defining units and standards and although individuals differ from it, these differences are generally too small to be of practical significance. Unless otherwise indicated, the calibration of a light meter will use the normal photopic function shown in Figure 15.1.

The three cone types present in the normal eye each have a different spectral sensitivity (the $V\lambda$ graph of Figure 15.1 is a composite function for the whole eye). A given wavelength will stimulate each cone type by a different amount, and it is the *relative* stimulation of each that gives rise to the percept of a particular colour. The individual colours of a rainbow are seen because each wavelength produces a different ratio of cones stimulation. However, the same ratio (and hence the same colour appearance) can be produced by different combinations of wavelengths. A mixture of long-wavelength (red) light and short–middle wavelength (green) light can provide the same relative stimulation as a middle wavelength (yellow) light, and in appearance the two will be indistinguishable★. However, the colour rendering properties of these two light sources will be very different: under the mixture a rose will look red and a leaf will look green, whereas under the single light both would look yellow.

The *temporal characteristics* of the light source may be of interest when artificial lights are used, either because flicker is noticeable in the environment or because of interactions between the light source and the equipment being used. The visual system integrates light over a finite period of time, and a light flickering faster than the integration time will be perceived as steady rather than flickering. This integration time varies with light level, and so modern films, for example, appear to be continuous rather than flickering, even though their frequency is below the maximum that humans can detect under optimal conditions.

The *spatial characteristics* of tasks and lighting are the final aspect of the visual environment considered here. Luminance and chromatic variation can occur across a room, across a workplace, and at different heights from the ceiling. Directional lighting can facilitate tasks such as inspection. Also, lighting may be deliberately arranged within a room so that features stand out; this can be for safety (highlighting fire extinguishers and exits) as well as for aesthetic reasons.

Spatial variation is not always advantageous, however. At any given moment the eye has a limited operating range, and a large range of light levels within the environment is to be avoided because visual performance will be degraded at the extremes. This effect can be seen in large rooms when sunlight shines through a window, providing lots of light for the areas near the window but leaving the areas away from the window relatively gloomy. Here, without supplementary lighting the details in the shadows may be below threshold, and cannot be seen.

Light sources

It is appropriate here to consider light sources as being either natural or artificial.

Natural light

The sun is the main source of natural light. Direct sunlight, light which has been scattered by the atmosphere to become skylight, and light which has

---

★This is a *metameric* match: two objects (in this case the light sources) appear to be identical in colour even though their spectral composition differs.

been bounced off the moon all originate from the sun. Three aspects of light from the sun are considered here.

First, the wavelength spectrum of light from the sun is broad, and contains no large peaks. The spectrum of an overcast northern sky is illustrated in Figure 15.2(a). At different times of the day, and at different places in the sky, the relative amount of energy at each wavelength changes. The most noticeable example of this occurs late in the day, when the emission spectrum of the western sky is predominantly long-wavelength. For colour rendering purposes, daytime skylight has been accepted as 'normal' light, with a clear northern sky providing the spectral standard.

Second, the sun is giving off energy in a manner which can be considered to be continuous. There is essentially no rapid temporal variation in sunlight,

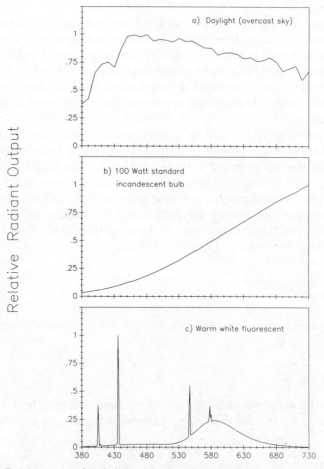

Figure 15.2. Emission spectra of (a) daylight, and two artificial sources: (b) an incandescent bulb, and (c) a fluorescent tube. Each graph has been normalized so that its peak equals 1

with slow natural variations being caused by clouds, by the earth's rotation, and by eclipses.

Third, the sun as a source is very small in angular terms, while the sky as a source is very large, unless walls and ceilings block it out. Although direct sunlight is a lot more intense than skylight, because of this size difference *the major component of natural lighting in buildings comes from the sky* and not from the sun, even on a cloudless day. This is particularly true in those parts of the world where the sun is reputed never to shine!

Incandescent light

When an object is heated, it gives off electromagnetic radiation. If you could take a metal ball and paint it perfectly matt black, then at normal room temperature it would neither reflect nor emit any light. These properties are what is meant by the term 'black body'. If you started heating the ball, it would eventually change colour and start to glow, i.e., it would emit radiation within the visible spectrum. An electric element of a cooker is like this—turn the cooker on and the black element starts glowing red (and also gives off lots of long-wavelength radiation as heat). The metal ball would vary in colour as its temperature changed, and at different temperatures it would have a characteristic (broad band) emission spectrum and, consequently, a characteristic colour. Because of this known variation in colour with temperature, light sources are often specified in terms of their 'colour temperature', which is the temperature at which a black body would have the same colour appearance as the source.

In the same way that a cooker element radiates light as well as heat, a thin piece of wire heated by passing electricity through it will give off light. Again, the emission spectrum of the light will depend upon the temperature to which the filament has been heated. An incandescent light is simply a thin piece of tungsten wire, surrounded by inert gas within a glass container, heated to make it glow.

Figure 15.2(b) shows the spectral emission of a typical, commercial light bulb, which is yellow–white in appearance. The hotter the wire, the more it glows and the relatively greater amount of light it emits—particularly at short wavelengths (with a consequent rise in its colour temperature). However, the hotter the wire, the more evaporation there is from its surface and the shorter its life expectancy. So there is a trade-off between lamp temperature (and hence light output) and lamp life. A recent improvement in this trade-off has come with the use of halogen gases within the lamp. These combine with the evaporating tungsten during the lamp's operation, and the evaporated tungsten does not become deposited on (and blacken) the inner wall of the bulb. Also, after evaporating into the halogen gas the tungsten is deposited back onto the filament. This increases the lamp's life expectancy, and it can then be run at a higher temperature with both a higher luminous efficacy (i.e., it gives off more light per unit of electricity) and also a spectrum more closely approaching that of skylight.

What about the temporal characteristics of incandescent lamps? An incandescent bulb run from the mains supply (50 Hz in the UK, Australia and Europe, 60 Hz in the USA) will show a temporal variation in its light output. However, Figure 15.3(a) shows that this variation is small—there is low luminance modulation. Although the filament will alter its temperature with the variation in the electricity, once the light is switched on the filament remains at a high temperature and never has a chance to cool down. Because the electrical variation produces only a small temperature change, there is little (<5%) variation in the light output over each cycle and so flicker from incandescent lights is not a problem.

Discharge lamps

The common fluorescent tube is an example of a discharge lamp. These work on a very different principle from that used in an incandescent lamp. Electrons emitted from a cathode at one end of the tube pass through a pressurized gas to an anode at the other end of the tube. Some are captured by atoms of the gas, thereby raising the atoms' energy level. However, like many high-energy objects these atoms are unstable. They subsequently release electromagnetic energy, and if this energy is within the visible spectrum, it

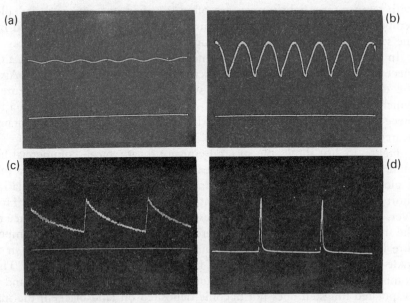

Figure 15.3. Temporal variation of light output from four different sources: (a) incandescent bulb, (b) fluorescent tube, (c) green phosphor (P31) VDU, (d) white phosphor (P4) VDU. In each case the height above the baseline indicates intensity. Each trace shows variation over 1/20th second: the mains frequency was 60 Hz and the VDUs were driven at 50 Hz

is called light. Unlike the light from incandescent lamps, the emission from discharge lamps is usually at discrete wavelengths. The use of different gases, and different gas pressures, produces a variety of emission spectra.

The two gases most commonly used in commercial discharge lamps are mercury, which is found in the ubiquitous fluorescent tube, and sodium, found in road lights. Because discharge lamps generally emit at only a few discrete wavelengths, this emission by itself does not give satisfactory colour appearance. To improve the colour-rendering properties of fluorescent tubes, the inside of the tube is coated with a phosphor. This coating absorbs energy emitted by the lamp at discrete wavelengths (including the invisible ultraviolet) and re-emits light across a broad band of wavelengths within the visible spectrum. In Figure 15.2(c) the peaks are caused by the gas emission while the lower, continuous, background is produced by the phosphor. The choice of phosphor will determine the appearance of the emission spectrum, and lamp manufacturers are capable of producing green, blue, and even red fluorescent tubes. Normally tubes are produced with a warm appearance (yellowish-white) or with a cool appearance (blue-white), the use of which depends on the application and taste of the user. For example, to improve the product appearance, store meat counters are often lit by a 'warmer' redder, light than would be acceptable for normal interior applications.

Low pressure sodium lights have long been used for roadway lighting. These produce 'sodium yellow' light—almost monochromatic light of a wavelength just below 600 nm. These lamps are now being superceded by high pressure sodium lights which, although they have a broader spectral output, are still spectrally centred around the sodium lines, and appear yellow. These high pressure lamps have a higher luminous efficacy than normal incandescent bulbs, and can provide over 100 lm W^{-1} as opposed to 17 lm W^{-1}. For economic reasons they are being used increasingly in applications where accurate colour rendition is not necessary. However, because of the limited spectral output of these lamps they may not always be subjectively satisfactory when used as a sole light source. This situation may be remedied by providing both yellow-white high pressure sodium and blue-white cool fluorescent lamps, which together provide a broader overall spectrum. The combination can be subjectively very pleasant, mimicking as it does the combination of yellow sunlight and bluer skylight.

Discharge lamps may also differ from incandescent lamps in their temporal characteristics. As seen in Figure 15.3, the modulation of a discharge lamp is much higher (generally about 50%) than that of an incandescent bulb. With alternating current of 50 Hz the electrical signal driving the discharge lamp is, in effect, rectified when light is produced, and the light level rises and falls 100 times per second. This frequency is well above the maximum detectable visually, which for large fields is usually no higher than 60 Hz and, therefore, fluorescent light flicker should not be a problem. Experience tells us otherwise: before the introduction of Visual Display Units (VDUs) into the workplace, fluorescent tubes were the principal topic of visual environmental complaints amongst office workers. However, one reason for these complaints is that sometimes other frequencies are also present in the emission. A mains frequency flicker can occur at the ends of the tubes; this can be remedied with appropriate shielding. Also, after thousands of hours of use, sub-harmonic flicker may be seen over the whole of the tube. This is remedied by tube replacement.

One solution of the problem of fluorescent tube flicker is to stagger electronically the phase of a *bank* of tubes. When some are fully on others are not, and this strategy reduces the overall modulation. However, this solution is being superceded, and recent developments in ballast (the circuitry driving the tube) technology have enabled tubes to be driven at much higher frequencies (e.g., 20 kHz) than that of the mains, effectively eliminating all flicker. As well as reducing operating costs (by increasing the luminous efficacy of the lights), these high-frequency lamps have been reported to reduce complaints of headaches in offices lit by fluorescent lamps (Wilkins *et al.*, 1988).

The other main difference between incandescent lamps and fluorescent tubes is their spatial extent. Although not all discharge lamps are large (some are the size of a normal light bulb) long tubes are commonly found in domestic, industrial and commercial applications. The amount of light given off by any point on the tube is far less than that given off by a point on a naked incandescent bulb providing the same illuminance. For this reason, we might expect fluorescent tubes to give rise to *fewer* complaints about glare (see p. 467). However, because it appears less bright a fluorescent tube will often be left exposed while an incandescent bulb will usually be incorporated into a glare-reducing fitting, and so the comparison is not a fair one.

Self-luminous tasks

Traditionally, the visual environment has been considered as consisting of two elements: the objects being viewed (such as paper copy, the road, instruments and dials) and their lighting (such as daylight, room lights or task-specific lights). However, the increasing use of self-luminous sources, such as VDUs, televisions, and LEDs, has led to a reconsideration of this viewpoint because the visual characteristics of the object are far less dependent upon the illuminating light than previously. As an illustration, VDUs are discussed briefly below.

VDUs essentially use the same technology as televisions and other cathode ray devices, where an electron beam scans rows of phosphor dots. The beam is turned on and off, and the phosphor glows or does not glow. Again, the spectral, temporal, and spatial characteristics of the display should be considered.

Spectral

The main monochrome phosphors in present use are amber (Phosphor P134), white (Phosphor P4), and green (Phosphor P31 or P138). If luminance-matched, none has an intrinsic spectral advantage over the others*.

After the phosphor has been excited by the electron beam it emits radiation within the visible spectrum. The VDU phosphor, like that of fluorescent lights, does not decay instantaneously. Rather, it glows for a while, with ever decreasing intensity, after the beam has passed over it. The time course of this persistence varies between phosphors, so the temporal characteristics of all screens are not identical. For example, green ⟨P31⟩ phosphor has a slower decay than white ⟨P4⟩. Whether or not this is preferable depends upon the application and upon personal choice. Figures 15.3(c) and (d) show the change in luminance over time of a single character on a green and on a white VDU, respectively. In each case, the discrete event of the phosphor excitation and subsequent decay is apparent. The greater persistence of the green phosphor [the long tail of Figure 15.3(c)] results in a lower average luminance variation over time, or modulation, when compared with the white phosphor. For an application where characters remain for a while in the same place on the screen this reduced modulation may be preferable because screen flicker will be less apparent. However, for an application where characters are being moved frequently the longer-persistence phosphor is often considered annoying, because of the 'ghost image' left briefly behind.

The *type* of ambient illumination, incandescent or discharge, has no spectral effect on the performance of screen-based tasks. (However, there may be spatial differences between them because the larger the area of a lamp (e.g., a fluorescent tube) the more likely it is to be reflected in a screen.)

Temporal

Apparent flicker (see later) is the major identifiable complaint arising from the temporal variation of VDUs. Often, you can see flicker from a VDU or a TV when looking to the side of the screen, but the flicker disappears when you look directly at it. This is because the eye is generally more sensitive to flicker in its periphery than in the centre of its visual field. Also, generally the larger the stimulus the more sensitive you are to flicker†, and so it is more apparent on a screen with dark characters and a bright background than on a screen with lit characters and a dark background. The flicker may be less obvious if the screen luminance is reduced, however this may be at the cost of decreasing the visibility of the screen characters, with a consequent

---

*The claim that because green is closer to the peak of the human spectral sensitivity curve it is preferable from a user's viewpoint is incorrect if the screens are equiluminant. The advantage of having a phosphor which peaks where the eye is most sensitive is that less radiation is needed to produce a given luminance.

†Sensitivity to flicker is context-dependent, and while frequencies as high as 90 Hz can be detected under optimum conditions, 60 Hz is generally taken as the maximum detectable under normal lighting conditions.

decrease in visual performance. The long-term solution to this flicker problem is for the screen refresh rate to be raised.

As well as inherent flicker, older screens often show two problems which are sometimes considered as spatial rather than temporal. These are 'jitter' and 'swim', and each is well-described by its name. The first is a fast oscillatory movement of a character around a central point, the second is a slow drift of the characters, as if the screen was being viewed through moving water. Both are caused by circuit instabilities and inadequate filtering.

Spatial

The main issue with VDUs, like other self-luminous tasks, is that there may be conflicting requirements of the ambient lighting. Generally, the higher the ambient light level, the lower the contrast on a screen between a task and its background: the luminance of each will be raised by the same numerical amount, and so the ratio between them will be lowered. However if an operator is reading paper documents, these will require a higher illuminance than is generally recommended for VDU work. If a problem exists, and a reasonable compromise cannot be achieved, the solution could be to provide specific task lighting on the documents. (This solution will also meet the aim of having a higher illumination level on the immediate tasks, with a lowering of illumination level on the surrounding areas.)

Evaluation

Whilst it is possible to evaluate VDU problems objectively, a subjective assessment of the severity of the problem is generally a more realistic approach. There is no need here for intricate measurements, objective or subjective, as the only real issue to be faced is whether the screen is subjectively satisfactory for the user. This information may be determined from the operator by interview or questionnaire, and a satisfactory investigation of the issue could include both these user responses and an 'expert evaluation' of the severity of the problem. Of course, the ultimate solution to any substantial problem is to replace the screens.

Parameters of interest and their assessment

There are three main reasons why measurement of the visual environment is usually undertaken.

1. *Health and safety.* The lighting of an area should be adequate to ensure that people can live safely, and it should not in itself be a health hazard. Measurement of the visual environment can provide information as to whether or not these criteria are met.

2. *Visual performance.* There is a large body of knowledge telling us how visual performance is affected by variables in the visual environment, such as illuminance and contrast. Assessment of these variables can provide information about the expected performance in the location considered.

3. *Aesthetic reasons.* A pleasant environment is conducive to well-being, and will usually result in less stress and better task performance.

Laws, standards, and guidelines exist to provide information about the requirements for the first two reasons. The third is often a matter of personal preference, although many guidelines and recommendations for aesthetically pleasing lighting seem to have good general support.

The following section discusses the various parameters which might be relevant, and provides background information about each; the *taking* of measurements is discussed further in the final section.

Amount of light

In the recent past, measurement and specification of light has been complicated by the use of different units in different countries, and by different groups of people within the same country. The text below uses currently accepted SI units. Because alternative units will still occasionally be encountered, for example in instructions for older instruments, and in older texts, descriptions of some previous units, and their conversion factors, have been included here.

Because of the adaptability of the human visual system we are very poor at making *absolute* judgements (but very much better at making *relative* judgements) and the eye itself is not a particularly trustworthy meter. Hence, light levels are usually measured objectively (although their subjective appearance should not be ignored), and Figure 15.4 illustrates the different concepts involved. There are two distinct aspects to consider. The first is the question of how much light there is present to 'light things up'—how much light is falling on this book, or how much light goes into your eye. This is termed *illuminance*. The second is how much light is being given off by something, and this is termed *luminance* (or luminous intensity if the source is small). The difference between illuminance and luminance can be illustrated by closing this book (after reading the next sentence!). The amount of light falling on the book is constant—the illuminance doesn't change—but overall the page reflects more light than the book cover, and so the page has a higher average luminance.

Illuminance

If we took a photocell which faithfully recorded every photon which landed on it and placed it on a desk, then the photocell output would depend upon

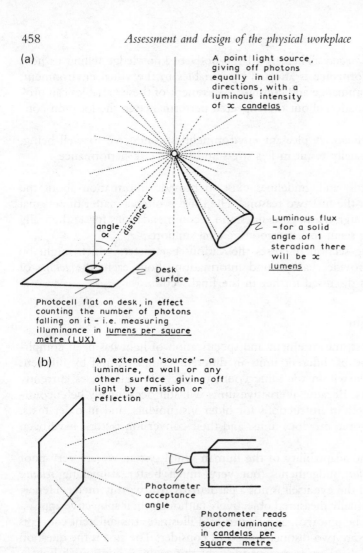

Figure 15.4. The measurement of light: (a) illuminance, (b) luminance

the number of photons falling on it, in other words the *illuminance* on the desk surface. Illuminance measures tell you how much light is falling within the photocell catchment area, and nothing else. This amount of light could have been produced by a single bulb, a bank of fluorescent tubes, or by daylight—the source is irrelevant here, the measurement simply tells you how much light there is.

Illuminance varies with the position of the source. Imagine you had a very small light bulb hanging from a wire above a desk. If the bulb was radiating equally in all directions (and no light was reflected from the walls or ceiling) the number of photons falling on the photocell would depend upon three factors:

1. The area of the photocell's collecting surface.
2. The distance from the source to the photocell.
3. The angle between the desk surface and the light source.

Consider these in turn.

1. The larger the photocell area, the greater the number of photons falling on it; make the catchment area twice as big and twice the number of photons are captured. However, the number of photons captured *per unit area* would be constant. Unless there is variation across the desk surface, the size of the collecting area of the photocell is irrelevant. We can think of an illuminance measurement as being a measure of how many photons per unit area are falling on the desk, and the unit used for measurement ('lumens per square metre' which for convenience is termed 'lux') reflects this fact.

2. The further away the photocell is from the source the fewer photons will fall on it, and we can easily determine the quantitative relationship between the source distance and the illuminance value. Imagine that the photocell had an area of 1 cm^2 and was directly below the tiny light bulb at a distance of 1 m. Because the source radiates equally in all directions the amount of light falling on the photocell is proportional to the 'solid angle' made by the photocell at the source. The solid angle is the area of the photocell divided by the square of the distance from it to the source, and is measured in steradians (sr). So here the 1 cm^2 photocell at 100 cm distance would subtend $1/100^2 = 1 \times 10^{-4}$ sr.

The point to note here is that the solid angle varies inversely with the *square* of the distance, and from the previous deduction this means that the number of photons falling on the photocell varies inversely in proportion to the *square* of the distance from it to the source. Because the illumination on a surface is expressed in terms of the amount of light falling on it per unit area, the inverse square law follows: the illuminance on the surface varies with the square of the distance between the surface and the source.

To illustrate that *illuminance* depends on the distance from the source to the object being illuminated, imagine a torchlight with a diverging beam shone directly at a wall from a couple of inches away. If you walked away from the wall two things would happen: the circle of light on the wall would increase, and the amount of light falling at any point on the wall would decrease. Double the distance from the torch to the wall and you increase by four the area of wall illuminated, decreasing by four the illuminance (i.e., the amount of light per unit area of the wall) falling on the lit portion of the wall.

3. The above examples have assumed that the photocell was positioned perpendicularly to the source. What if the photocell were angled—which is what would happen if it was placed on an inclined document holder rather than flat on the desk—with the light source directly above it? Here the amount of light per unit area falling on the photocell would be less than if it were perpendicular to the source, and the illuminance varies with the angle

that the photocell makes with the perpendicular. If we start with the photocell directly perpendicular to the source the illuminance will be at a maximum value. If we now rotate the photocell so that its face is no longer exactly perpendicular to the source, the illuminance on the face will be reduced by the *cosine* of the angle of rotation. For example, when the photocell is rotated 60° the illuminance will be reduced by a factor of two, as $\cos 60° = 0·5$.

ILLUMINANCE UNITS

The amount of light falling on a given surface area, the luminous flux per unit area, is the illuminance at a point on the surface. The SI unit is the *lux*, which is one lumen per square metre ($lm\ m^{-2}$). That is to say, when luminous flux (see below) of 1 lumen is spread over a surface area of 1 m², the illuminance is 1 lux. An alternative name, not in common use, for the lux is the 'metre candle', while the 'phot' is the illuminance when 1 lumen is spread over 1 cm².

The *foot-candle* is a unit still frequently encountered, especially on old illuminance meters; it is an illuminance of $1\ lm\ ft^{-2}$, and is equal to 10·76 lux.

The *troland* is a unit used in vision research, when an eye with a 1 mm² entrance pupil views a surface with a luminance of $1\ cd\ m^{-2}$ (see below) then the retinal illuminance is 1 troland.

For illustration, Table 15.1 provides some representative indoor illuminance values recommended by the CIBS.

Luminous intensity (point sources)

In the previous section we considered the light source to be very small, and to give off light equally in all directions. If this source was so small that it

Table 15.1. Examples of recommended illuminance values (from the CIBS Code, 1984)

Condition	Recommended value (lux)
Rarely visited locations, with limited perception of detail required, e.g., railway platforms	50
Continuously occupied areas, with limited perception of detail required e.g., waiting rooms	200
General offices / Airport ticket counter	500
Drawing boards / Bench and machine work (fine detail)	750
Cloth inspection / Assembly work (fine detail, e.g., watchmaking)	1500
Inspection of extremely fine detail (e.g., small instruments)	2000

effectively had no area, it would be called a 'point source'. In fact, this is a theoretical concept as all real sources have a finite area to them. However, the idea of radiation being emitted from a single point equally in all directions is a useful one in explaining the concepts of *luminous intensity* and *luminous flux*. The more intense a source is, the more light it gives off in all directions: luminous intensity is an attribute of the light source. Now consider a cone with a point source at its centre: the more intense the source the more photons there would be within this cone; in other words, the greater 'luminous flux' there would be within the cone, see Figure 15.4(a). So luminous flux describes the light itself, and not the source.

An ideal point source gives off radiation equally in all directions. Although real sources can sometimes be assumed to have negligible area (consider the angular subtense of a motorcycle headlight viewed from a mile away) the property of directionality does not generally hold. Interior and exterior light fittings are often designed with reflectors or shades to produce a certain pattern of light, and the 'hot spots' of a headlight are a good example of this.

Given that the luminous flux varies with direction, it follows that the luminous intensity of a real (as opposed to a theoretical) source also varies with direction.

LUMINOUS INTENSITY UNITS
Luminous intensity is a measure which describes the amount of light emitted by a point source. The SI unit is the *candela* (cd). This basic unit of light is actually defined in terms of the emission of a black body at the freezing point of platinum, 2040 K.

LUMINOUS FLUX UNITS
When light is emitted from a point source, the 'density of photons' within a cone centred at the source is termed the luminous flux. The number of photons within the cone will be a function of both the source intensity and the cone solid angle, and the SI unit of luminous flux is the *lumen* (lm). This is defined as the luminous flux emitted through a solid angle of 1 sr from a point source of intensity 1 cd. A point source with a luminance of 1 cd will emit overall a total of 12·57 (i.e., 4π) lm.

Luminance (extended sources)

In practice, one generally needs to consider extended sources rather than point sources. Usually a fluorescent tube, a light fitting, or some other source such as the sky subtends an appreciable angle at the eye, and cannot be considered to be infinitesimally small. Hence, luminance is specified in candelas per square metre of the source rather than in candelas. To take the measurement, a light meter with a small acceptance angle is used which will collect light within, e.g., 1°, 0·5°, or 1′, depending on the instrument. It makes no difference how far away the source is, as long as it is larger than the meter's acceptance angle. Consider a luminance meter pointed at a wall, collecting and recording all of the light within its acceptance angle. If we move the meter closer to the wall more photons will be collected by the photocell from any given point. But because the acceptance angle of the meter is fixed, the portion of the wall from which photons are being collected by the photocell is also reduced. Because the increased number of photons collected from each point is balanced by the decreased area of the wall within the meter's acceptance angle the luminance reading remains constant. So luminance, unlike illuminance, is *independent* of the measurement distance.

LUMINANCE UNITS

Luminance describes how much light is emitted from a surface, and the SI unit is the *candela per square metre* (cd m^{-2}). (This unit in the past has been called a 'nit' but, to the chagrin of schoolchildren everywhere, this name is no longer in use.)

The other unit that may be encountered is the *stilb*, which is 1 cd cm^{-2}.

Reflectance

Illuminance and luminance are closely linked. The amount of light falling on a point on the wall is its illuminance, and the amount of reflected light coming back from the wall is its luminance. If all of the light that fell on the wall was reflected, then the values of the illuminance and the luminance would be the same, using appropriate units. If some of the light was absorbed, then the values would differ. The *reflectance* of the wall may be found by comparing the illuminance and the luminance values, as explained below.

An important point to consider before going further is the *directionality* of reflections. A perfect matt white surface will reflect light equally in all directions, irrespective of the direction of the ambient light, and is called a perfectly *diffuse* reflecting surface (Figure 15.5(a)). On the other hand, a highly faithful reflector, like a mirror, will only reflect light at the same angle to the normal that the ambient light makes (Figure 15.5(b)). Most surfaces lie somewhere between these two extremes, with a higher luminous flux occurring at the angle of reflection than elsewhere (Figure 15.5(c)). (Try moving this book and see if you can detect specular reflection on the paper from any light sources.) Finally, if a surface reflects light back along its own path it is termed a retro–reflector.

The greater the proportion of unwanted specular reflection from a surface, the more likely a person is to experience annoyance, discomfort and degraded visual performance. Where specular reflection is likely to be a problem, the positioning of light fittings must take this into account: areas of relatively high luminance should be avoided at places where they might be reflected into the eye. Where new equipment is used in a previously-designed room, such as new VDUs in an old office, the aim should be to allow for adjustability in the equipment so that specular reflections from the screen can be reduced.

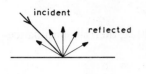

a) A perfect diffuse reflector
(also known as a Lambert
or as a cosine reflector)
reflects equal luminous flux
in all directions.

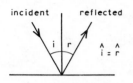

b) A perfect specular reflector
reflects all light at an
angle equal to the angle
of incidence

c) Most surfaces reflect light
in all directions, but with
a greater proportion at an
angle equal to the angle
of incidence.

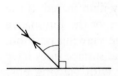

d) A retro-reflector reflects
light back along exactly
the same path.

Figure 15.5. The four types of reflection: (a) diffuse, (b) specular, (c) mixed, (d) retro-reflection

REFLECTANCE UNITS

Irrespective of its angle of incidence, all of the light falling on a *perfect diffuse* reflecting surface is emitted equally in all directions. No light is lost, and it is intuitively appealing to have units which are defined in such a way that the values for luminance and illuminance are identical here. This is the rationale behind the old (now deprecated) luminance units 'apostilb' and 'foot-lambert'. When the illuminance falling on a perfect diffuse reflector is 1 lux, the luminance of the surface will be 1 *apostilb* (asb). It follows that if the surface does not reflect all of the light, then the luminance in apostilbs will be equal to the reflectance of the surface, multiplied by the illuminance in lux. The same rationale applies for imperial units, and with an illuminance of one foot-candle (i.e., $1 \, \text{lm ft}^{-2}$), the luminance of a perfect diffuse reflector will be one *foot-lambert*. (It follows that $10.76 \, \text{asb} = 1$ foot-lambert.)

The accepted SI luminance unit is the candela per square metre, and for a perfectly diffuse reflecting surface there is a simple relationship between it and the apostilb:

$$1 \text{ cd m}^{-2} = \pi \text{ asb.}$$

So if the illuminance on a wall of reflectance 0·8 is 220 lx, then the luminance of the wall will be:

$$(220 \times 0·8) = 176 \text{ asb} \tag{1}$$

or

$$\frac{(220 \times 0·8)}{3·14} = 56 \text{ cd m}^{-2}. \tag{2}$$

The first calculation is more straightforward, so why is the use of the apostilb deprecated? The reason is that in practice most surfaces are *not* perfectly diffuse, but rather have some specular component. Because of this the wall luminance will vary, and the measured value will depend upon the observer's position.

Instead of using the reflectance of the wall to determine the wall luminance, a quantity called the 'luminance factor' is used. The luminance factor is the ratio of the luminance of a reflecting surface, viewed in a given direction, to that of a perfect white diffusing surface identically illuminated. If the reflecting surface is itself a perfect diffuser, then the value of the luminance factor is the same as the reflectance, is independent of the viewing position, and cannot be greater than one. On the other hand, if the surface does have specular reflections, then the luminance factor will vary with viewing position, and at the angle of reflection it could be greater than one. Hence:

$$\text{Luminance (cd m}^{-2}) = \frac{\text{illuminance (lux)} \times \text{luminance factor}}{\pi}.$$

The reflectance value of a surface is normally found for 'light', i.e., the whole of the visible spectrum. It is also possible (although not normally needed in practice) to determine the reflectance values for different wavelengths. If this were done, then it would be seen that different coloured walls had different reflectance spectra. For example, a wall which looks red might reflect a large proportion of long-wavelength light, but little medium- or short-wavelength light, while one which looks green might reflect a large proportion of medium-wavelength light, but little long- or short-wavelength light.

Flicker

Flicker, noticeable rapid fluctuations in light level, can be a serious problem in artificial environments. Unfortunately, objective measurement of flicker is not simple because it requires rapid-response equipment, normally available only to a lighting specialist.

Subjective assessment of flicker is, however, much more feasible and both the area of noticeable flicker and the degree of noticeable flicker can be adequately assessed by descriptive means. Also, because the periphery of the eye is more sensitive than the central area to flicker, subjective assessment may actually be a more relevant method. Hence, when dealing with flicker the precise circumstances under which it is seen, such as the luminance and the position of the source in the visual field, should be noted.

In considering the subjective assessment of flicker, we should take a lesson from noise assessment (see chapter 17 of this book). In the same way that the psychological effect of noise can be independent of the sound level (think about the dripping of a tap) flicker can have an annoyance or a distractive effect out of all proportion to its physical magnitude. A subjective assessment of the flicker should not only consider the physical aspects of the stimulus, such as the perceived flicker strength, but should also evaluate the psychological effect that the flicker is having on the person. The positive side to flicker is that because it is very attention-getting, its use is a very good visual method of conveying warning information.

Two further aspects need to be considered. First, some people are especially sensitive to flicker. Epileptics are an extreme example, and for them flicker (particularly at frequencies around 10 Hz) can provoke seizures. Second, flicker which is not visually detectable may still affect parts of the visual system. The human retina responds to flicker at high frequencies (over 100 Hz) even though the light appears steady, and no flicker is seen. It remains to be determined whether other parts of the human visual system are affected by high flicker frequencies, and whether performance or comfort are affected.

Colour

Again, a full spectral assessment of the environment requires specialized equipment. However, this assessment is very rarely needed.

Different light sources have different colour rendering properties, and the colour of an object is determined both by the spectral composition of the light source, and by the spectral reflectance properties of the object and its surround. In this way two objects, like a shirt and tie, can appear identical in colour under one set of lights (a metameric match) but can be noticeably different from each other under another set of lights. The eye is easily fooled under these circumstances. However, while changes in colour appearance like this do occur, the adaptability of the visual system is such that most objects will retain their correct appearance under a wide variety of light sources. This 'colour constancy' can be seen outdoors when a cloud passes over the sun—although the spectral composition of the light falling on the grass and the trees changes drastically, the colours of the objects do not seem to change.

When considering the chromatic aspects of the environment, both the colour rendering properties of the light sources and the pleasantness (or otherwise) of the lighting and the environmental colours must be included. The first of these is likely to have been considered by the lighting designer in an environment where it is important, e.g., where colour discrimination or colour matching are included in the tasks performed in that location. This occurs more often than casual thought would suggest, for example, text and instructions are often colour coded. There is an important point to stress here: colour coding is valueless if the ambient light does not reveal the colour differences. Consider sodium light, which is monochromatic. In this light the three cone types are always stimulated in the same ratio. All objects appear to be the same colour, yellow, and depending on how much of the light they reflect, some are darker than others. Hence, the colour coding of road signs has to be revealed at night by means of supplemental lighting.

Good colour rendering is not always necessary, and other criteria may take precedence. In some situations (such as industrial exteriors and warehouse interiors) the lights may have a simple safety function. Good colour discrimination is not required, for example, if the lighting only has to reveal the presence or absence of objects. Here the cost of running the lights may be a more important criterion than their pleasantness.

In an environment where people have to spend a large proportion of their working time the colour rendering properties of the lights, or combination of the lights, plays an increasingly important role. Lamps with poor colour rendering are generally considered to provide a less pleasant environment than those with good rendering. Of particular importance here is the appearance of skin tones: the better the appearance of skin, the more subjectively preferable is the light. Although 'warm' lamps are generally considered preferable to those with a 'cool' appearance, lamps with a narrow spectral band (such as high pressure sodium) generally provide an unacceptable environment for long-term working. Also of interest here, and a factor where little research has been performed, is how people performing different occupations prefer different spectral combinations. Draughtspeople and designers, for example, may tend to use cool, blue-white light, while office work is more commonly performed under warmer, yellower light.

Daylight coming from a northern sky is broad-band (see Figure 15.2) and is the reference illuminant used for colour-vision testing. Different lamps are compared with this standard as far as their colour rendering properties are concerned, and the CIE have devised a 100 point scale, the 'colour rendering index' (CRI) for lamps. As a generality, the higher the value of the CRI the better the lamp performs (e.g., incandescent lamps may have a CRI of 99, an artificial daylight fluorescent lamp has a CRI of 93, while a white fluorescent lamp has a CRI of 56 (Boyce, 1988)). However, these are *overall* values for the lamps, and a lamp with a high score does not necessarily perform well over all of the spectrum, although it should give good rendering of most colours.

Colours will often have specific culturally-based meaning associated with them (e.g., red = stop, green = go)★. Here assessment of the use of colour involves measuring the performance associated with the colour rather than the colour itself. Also, visual performance can sometimes be enhanced by using colour to provide a contrast between a task and its surround. Unfortunately, while the concept of colour contrast has some intuitive meaning, (most people would consider the colour contrast between red and blue to be greater than that between yellow and green), as yet there is no widely-accepted objective metric by which it can be evaluated. At present, this assessment has to be made on a subjective basis.

Glare

A number of quite distinct visual problems have been grouped together under the heading of 'glare'. These problems, such as discomfort and reduced visual performance, have in common that they are all associated with light levels that are *relatively* high when compared with the ambient light levels. Although different forms of glare may occur simultaneously, they are essentially independent because they do not have the same underlying physiological mechanism. It is not surprising, then, that it is possible to have discomfort without disability, and vice versa, even though both will be often found occurring together.

Discomfort glare

When a portion of the visual field has a much higher luminance than its surround, a feeling of discomfort may occur around the eyes and brow. While this effect has been studied empirically for many years, the physiological mechanisms involved are still unknown. The facial and the iris muscles have been suggested as the site of the discomfort, but possibly neither are involved. When a glare source is turned on, people tend to wrinkle their brow and partially close their eyelids; concurrently the iris sphincter muscle contracts and the pupil constricts. Whether these are simply *responses* to the increased light and the associated discomfort, or whether they *cause* the discomfort is not known.

Although the mechanism of discomfort glare is unknown, the conditions

---

★Nature has developed colour coding, as seen in the colouring of wasps and bees: yellow and black hoops warn of a stinging insect (or an insect impersonating one). While 'yellow' is a human percept (the insect itself is not yellow/black, but rather we perceive it that way) we can infer from the colour coding that whatever might otherwise prey on these insects have some form of colour vision. Similarly, the use of colour in camouflage depends upon the colour vision of the observer: two objects may be metamerically matched to a person with normal colour vision, yet to someone with a colour vision defect (or a normal person looking through a coloured filter) they will look different. Because of this, people with colour vision defects have been used in the past to 'spot' camouflaged objects.

under which discomfort occurs have been well established for a number of years. The early work of Hopkinson, Petherbridge, and Guth (Hopkinson and Collins, 1970) revealed that:

Discomfort *increases* with:
    an increase in the luminance of the glare source,
    an increase in the angular size of the glare source at the eye.
Discomfort *decreases* with:
    an increase in the luminance of the background,
    an increase in the angular position of the glare source relative to the line of sight.

By definition discomfort is subjective, and discomfort glare is not easily quantified. A given physical configuration of lights will not only give rise to different reported amounts of discomfort from different people, but also to different reported amounts of discomfort from the *same* person on different occasions. Subjective assessment in this situation is not particularly reliable. On the other hand, the *physical* parameters of different lighting configurations are, in theory, easy to determine. If it is known how the above four factors (glare source luminance, size and position, and background luminance) interact, it should be possible to measure aspects of the environment and determine on an objective scale how good or how bad the environment is. This is the rationale behind the various glare indices established in different countries—they say little about how an individual will respond, but they do allow an objective evaluation to be made of the lighting configuration. Calculation of the Glare Index involves determining the luminance, size and position of the glare source(s). Full details for the calculations involved may be found in CIBS publications (1984 and further information list at the end of this chapter).

Despite the apparent validity of these studies in carefully-controlled experimental conditions, the evaluation of the glare indices in practice is complicated. While the luminance and size of the glare source(s) are easily measured, there are real difficulties in determining the value of both the background luminance and the glare source position. First, the luminance of different walls and different parts of the ceiling will vary, and the problem arises of what value constitutes the 'true' background luminance. In practice, an 'average' value has to be estimated. Second, there is not one unique glare index for one room position, but rather one glare index for each position in the room *and* each position of the eyes.

As someone looks around a room, the glare index will vary. One of the parameters mentioned above is the angle between the line of sight and the glare source. So while a head and eye position may be specified for measurement purposes, e.g., the person may be assumed to be looking down at their desk, or perhaps to be looking straight ahead, each position is arbitrarily chosen and is not one which the person necessarily adopts or has problems with.

Not surprisingly, there is a poor correlation between subjective reports of discomfort and the glare index. The advantage of the index is that it does provide an objective description of the environment. The disadvantage of the index is that this figure does not in itself describe well the subjective discomfort of an individual subjected to that glare.

An alternative way of evaluating glare is to use an 'expert observer' approach, where an individual experienced in glare assessment can evaluate the degree of discomfort they feel in a certain situation. The simplest way of initially determining whether there is a problem is to shield the suspected glare source(s) and see if there is an improvement in comfort. If so, then you can safely assume that there was some discomfort present in the first place. This can then be evaluated more precisely. A number of descriptive scales have been produced, perhaps the most useful of which is that proposed by Hopkinson (Hopkinson and Collins, 1970). Four criterion points were used, each of which describes a transitory position of subjective feeling:

1. Just perceptible.
2. Just acceptable.
3. Just uncomfortable.
4. Just intolerable.

If necessary, the scale can be extended by including end points (imperceptible, and intolerable) and by allowing descriptions between the criterion points (e.g., not quite acceptable, but not yet uncomfortable). This will produce a nine point scale. An alternative method is simply to rate the amount of discomfort on a rating scale, with defined anchor points having specific, known, rating values, e.g., a rating scale where imperceptible = 0, just uncomfortable = 5, and intolerable = 10. (See chapter 3 for general discussion of rating scales.)

There are a number of ways to reduce discomfort glare. Consider in turn the four factors mentioned previously. Reducing the glare source luminance will reduce the discomfort. This solution may not always be feasible, because the source could be providing illumination for a different part of the room. Increasing the angle between the viewer's line of sight and the glare source, by moving either the source or the person, will decrease the discomfort. Another, less obvious, solution is to *increase* the background luminance. It seems counterintuitive that increasing the amount of light present reduces the discomfort, but this manoeuvre reduces the contrast between the source and the background. A practical way of doing this is by painting the walls and ceilings so that they reflect more light, thereby increasing the ambient light level. Finally, the solid angle of the glare source can be reduced by shielding the source. Changing the luminaire fitting to one with a narrower cut-off angle will decrease the visible extent of the source.

These solutions are not necessarily independent. For example, by reducing the size of the glare source the illuminance it provides for some other purpose may be decreased, necessitating an increase in the source luminance. In the same way, a bare light bulb or a luminaire with a small fitting or shade may be improved by increasing the size of the fitting. Although the glare source solid angle is increased, the source luminance is lowered, with a consequent reduction in discomfort. These are all solutions which are imposed on the person by outside manipulation of the light sources. Where feasible, it is preferable to allow individuals control over their immediate visual environment, with directional local lighting, and in this way enable them to position the light sources in such a way that they are comfortable.

As a final point, the long-term effects of small amounts of discomfort glare are unknown. It is reasonable to suppose that someone working all day under conditions where glare is just perceptible will have a build-up of discomfort over the day. The need for improvement here is greater than in the situation where someone is briefly exposed to 'just uncomfortable' glare once or twice per day. Similarly, someone's idea of how tolerable glare is will depend upon what they are doing, and how interested and involved they are in it. A person watching television may choose to have a low ambient light level but a bright screen—environmental conditions which in other circumstances would be considered intolerable. The point to remember here is that the glare assessment cannot be considered in isolation from the context in which it is made.

Disability glare

While the physiological mechanisms involved in discomfort are not understood, the ways in which an extraneous light source can affect visual performance are quite clear. All involve contrast reduction. The degradation of visual performance occurs in one of two ways: a glare source can act *directly* by reducing the contrast between an object and its background, or *indirectly* by affecting the eye.

'Direct' disability glare can occur because of a discrete reflection, such as the specular reflection of a light source from the surface of a VDU screen. Here the luminance of both the object (the characters being viewed) and the background (the surrounding screen) are raised by the addition of the extra light but the contrast is reduced. It can also occur because of a diffuse reflecting veil over the whole of the task, as is seen, for example, when a car windscreen mists up. The whole of the scene looks grey and washed out, and both luminance contrast and colour contrast are diminished. These two examples have in common that the contrast between the object and the background is decreased, with a consequent reduction in object visibility. Hence to reduce the disability one should raise the contrast between the task and the background.

'Indirect' disability glare affects the eye and not the visual task. It is seen, for example, when a car approaches at night with its headlights on full beam and your eyes get dazzled. The disability in this situation has two sources:

(a)   there is scatter within the eye reducing the retinal image contrast, and

(b) the adaptation level of the eye is raised as the car approaches.

After the car has passed it takes a little while for the eye to re-adapt to the ambient light level.

The fact that scatter within the eye decreases visual performance is particularly pertinent when one considers age-related changes to the eye (see chapter 27). The media within the eye generally become more opaque with age, and in particular the lens tends to go 'cloudy'. Although these changes will increase the effect of direct disability glare to a certain extent, the real problems are seen in the presence of indirect disability glare. This is because for the older person opacities within the media will scatter the light to a much greater extent than is found for a younger person. Hence an older person who is able to perform perfectly well under normal lighting conditions may be *severely* disabled in the presence of indirect disability glare—and the reason for this disability may not be visually apparent to a younger person.

Both sources of disability are present in other situations, such as when an object is seen against or near a bright window. They have been termed 'indirect' here to emphasize that they act by affecting the eye rather than affecting the scene. Indirect disability glare is usually found in circumstances where discomfort will also be present because in both cases the eye itself is affected by the glare. Like discomfort, the disability is often reduced by *raising* the light level. On a city street a car headlight, even when it is not dipped, causes less disability than the same light on an unlit road because the extra ambient light raises the eyes' adaptation level. In the city, the headlight is less bright in comparison with its surroundings, and we can take this reasoning further by considering that the same headlight would produce virtually no disability during daylight.

There is a problem encountered less often where performance is reduced, which can still be considered as glare, and this is where there is an area of *low* light as compared with the ambient light level to which the eye is adapted. The problem faced here by the visual system is exactly the same as that faced by a photographer encountering deep shadows. If the exposure is set manually for the ambient light the details in the shadows are lost, whereas if the camera is set to record these details everything else in the photograph will be overexposed. The eye's adaptation system acts like an automatic camera, but it isn't possible to vary it in the way you can with a manual override camera. The automatic setting the eye will adopt is appropriate for the ambient light, and not for the shadows, and so details in the shadows are lost to the visual system.

Contrast rendering, and directionality of light

The effectiveness of a lighting configuration is not dependent only upon whether any noticeable disability is present. Despite this, quantification of lighting effectiveness is an area in which laboratory-based research has been applied in the past largely to lighting design rather than to environmental assessment.

The relative positions of the light source, the visual task, and the observer determine how effectively the task contrast is rendered, and recently a measure of lighting effectiveness, the Contrast Rendering Factor (CRF) has been devised. The CRF has been used mainly in regard to paper-based tasks, which is where it is at its most useful. If the task lighting in an office is suspected to be deficient, then measurement of the CRF would be an appropriate way to investigate the problem. Ideally, the CRF is measured by comparing the contrast of the object under the ambient lighting with its contrast under reference lighting (completely diffuse, unpolarized illumination). It is important to realize that the CRF is specific to a particular target, a particular location, and a particular observer position and is not a measure which describes the lighting alone. So the CRF of writing on matt paper will be different from that of writing on glossy paper under otherwise identical conditions. The CIE Publication 19/2 (and Boyce, 1988) provides details of assessing the CRF if access to reference lighting is available.

If reference lighting is not available, do not despair, as two approximate methods are described in the CIBS code (CIBS, 1984). For simplicity, these methods assume 'standard tasks', and, interestingly, because the task is no longer a variable, they *do* both provide a quantitative description of the lighting configuration. In the first method, the illuminance at different planes at the workplace is measured, and the value of the CRF is found from a calibration graph. In the second method, a CRF gauge is used at the workplace, and the CRF value is read directly.

Generally, the higher the CRF, the more acceptable the visual performance. This might lead one to suppose that the reference lighting conditions, uniform diffuse illuminance, could be considered as 'ideal'. However, this is not the case as CRF is not the only criterion by which the directionality of lighting is judged. Uniform, shadow-free lighting gives an extremely bland appearance to an interior, with the solidity of objects being less readily apparent because of the lack of relief. For example, facial features are far more pleasant when seen under directional lighting than under flat lighting.

The directionality and modelling of the lighting at any point in a room may be quantified by combining two measures. One is the scalar illuminance, the other is the illumination vector. These may be measured by suspending a cube (at any orientation) at the position of interest and measuring the illuminance value of each face. Incidentally, in the absence of a cube, this measurement may still be taken: improvise by imagining that you have a cube, and then hold the illuminance meter against each imaginary face. Averaging the values of the six cube faces gives the scalar illuminance. The illumination vector is then calculated by constructing a vector diagram of the differences within each of the three pairs of opposing faces, as illustrated in Figure 15.6.

Combining the illumination vector value with the scalar value gives the

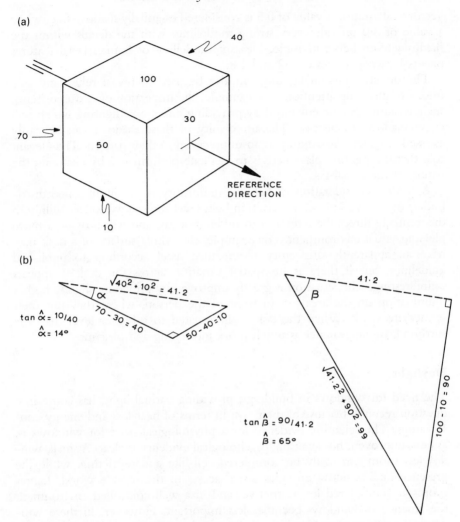

Figure 15.6. Calculation of the scalar and vector illuminances: (a) The illuminance values are measured on each face of a cube suspended in space, and the scalar illuminance is their arithmetic average (50 lux). (b) A vector diagram can be constructed by first determining the difference between each pair of faces (90, 40, and 10 lux). Then the angle of the illumination vector in the horizontal meridian *relative to the reference direction shown in* (a) is determined as shown in the left-hand portion of (b); tan α = 10/40 so α = 14°. The magnitude of this vector is the square root of $40^2 + 10^2$ (the two measurements taken in the horizontal plane) which equals 41·2. The angle in the vertical meridian, *relative to the direction just determined,* can now be found, as shown in the right-hand portion of (b). Tan β = 90/41·2 and so β = 65°. The *magnitude* of the illuminance vector is the square root of the sum of the squares of these two sides: the square root of ($41·2^2 + 90^2$) which equals 99 lux. ANSWER: In figure (a) the illumination vector has a magnitude of 99 lux, in a direction 14° to the left of the reference direction, and 65° below this direction. The vector : scalar ratio = 99/50

vector/scalar ratio. A value of 0·5 is considered essentially shadow-free, while a value of 3·0 provides very strong modelling, with the details within the shadows often being invisible. Pleasant modelling of human facial features occurs between values of 1·2 and 1·8.

The directionality of lighting can also be used to reveal relief, and as a means of directing attention. For example, the inspection of textured items, such as cloth, may be enhanced by providing directional lighting which will cause shadows to be cast. The uniformity of these shadows can be easily assessed—defects showing up as an imprecision in their pattern. The elegant role that the lighting plays here is to facilitate performance by changing the nature of the visual task.

A further consideration in lighting uniformity is the illuminance distribution over a workplace. Variation in light level across a workplace will both differentially direct the attention to different areas, and will provide a more pleasing visual environment. For example, the whole surface of a desk may have an apparently-satisfactory illuminance level according to published guidelines, but if there is no spatial variation across the desk it appears uninteresting. This situation is greatly improved by providing slightly higher illumination on the area just in front of the person, where they put their immediate work. When the task is illuminated to a higher level than the surround, the appearance is much more interesting and pleasant.

Daylight

The need for windows in buildings, providing natural light, has come into question recently because of their cost in terms of heat-loss and energy conservation. The scientific evidence for a physiological need for windows is, at best, unproven, however the psychological evidence is clear. A small, windowless room can easily be considered cell-like and restricting, while the presence of a window provides visual access to the outside world. Larger rooms are considered less restrictive, and in a well-controlled environment the absence of windows becomes less important. However, in these windowless environments the information, such as the time of day, and the variety provided by the changing outside light is still absent.

As well as the illuminance variation over the day, the provision of daylight can also provide spatial and spectral variation within a room, again decreasing the monotony of the environment. The illumination variation may be quantified as a change in the 'daylight factor' across a room. The daylight factor is the ratio of the illuminance from the skylight measured on a horizontal surface within the room to the illuminance from the skylight (*not* direct sunlight) measured on a horizontal plane which has an unobstructed access to the hemisphere of the sky. At different positions within the room the daylight factor will vary, and if required this variation may be assessed over a room. For interiors where the daylight factor variation is not large, such as rooms with skylights, or rooms which are not too deep, when the average daylight factor is 5% or greater an interior will appear generally to be well day-lit. Also, if the illuminance from the sky is not known, the relative daylight factor at different positions within the room will give information about the spread of daylight within the environment.

The pleasantness of variation in spectral content and luminance level is not restricted to natural lighting. On the contrary, in some parts of the world the design of mood lighting provides a considerable source of revenue for interior designers and fixture designers alike.

Overview of measuring the visual environment

Standards and guidelines

Lighting guidelines and standards come in two forms, either statutory, where certain legal requirements have to be met, or recommendatory, where the recommendations provided are given as examples of 'good current practice'. In neither case should the values given be assumed to be 'ideal'. For example, in the case of a legal minimum requirement for a certain illuminance at a workplace, the *minimum* required is not necessarily the *optimum* value. Illuminance values which were considered 'good practice' 20 years ago are now, with the advent of more efficient lamps, often considered inadequate.

Most countries have their own standards and guidelines, and specific values will not be presented here. Rather, the reference and further information sections contain details of documents which contain the necessary information.

What measures are needed?

An assessment of the visual environment is going to be needed in one of two situations. The first is a design situation where the concepts discussed previously will have to be considered in a somewhat abstract manner. The example given in the introduction, that of designing a petroleum pump display, is a case in point. The second situation is a more immediately practical case, where an environment already exists and has to be evaluated in some way. Generally the assessment will be needed because problems are known to be present, and the appropriateness of the measures to be taken depends upon what these problems are.

The preceding sections have discussed the variety of factors of concern in the visual environment. Outside of the laboratory, many of these lend themselves to subjective assessment, and flicker is a good example. Objective physical quantification of the frequency and modulation of flicker is complicated, however it is usually unnecessary. The appropriate question may be 'Is there unpleasant flicker present?'—which has to be answered on subjective rather than on objective grounds.

In an assessment of an existing environment, the people who live and work in that environment are an important resource. As well as providing information about what environmental problems exist, their severity, and the appropriateness of the measures being taken to alleviate them, people familiar with the environment can also provide information which reveals problems which are not immediately obvious. Problems which may occur at different times of the day, or only under certain circumstances (such as when an infrequently used piece of equipment is operating) fall into this category. Once the problem has been established, then the relevant measures can be taken.

An example

Suppose you are approached by a client who feels that the average illuminance within an office is too low at the end of the day, and that some of the desks are inadequately lit. You are asked to make an assessment of the environment, and to suggest improvements. Off you go to the office armed with a luminance meter, an illuminance meter, and a tape measure. What do you then do?

Consider this problem in the following four stages:

Average illuminance within the room

The first issue to be considered in determining the average illuminance within the room is 'how many measurement points are needed, and where should the measurements be taken? The answers to these questions depend upon the size and dimensions of the room, and the accuracy required. The following method will give accuracy to within 10%.

The first step is to determine the dimensions of the room, including the ceiling height, and to sketch a scale plan of a cross-section of the room. The positions and dimensions of the luminaires should be marked on this sketch. Next, the 'Room Index' (an overall size-measure of the room parameters) is found from the ceiling height and the length of one wall. For a square room, it is numerically half the ratio of the wall length to the ceiling height. If the room is almost square, then take the measurement for the largest wall in determining the wall:ceiling ratio. The minimum number of measurement points can then be found from Table 15.2.

Table 15.2. Minimum number of measurement points needed for different sized rooms

Ratio wall length:ceiling height	Room index	Minimum number of points
<2	<1	4
2–4	1–2	9
4–6	2–3	16
>6	>3	25

If the room cannot be considered as square, an extra calculation is involved. First measure the *smaller* wall length, and the ceiling height. From these figures you can determine the minimum number of points needed within the square defined by the dimension of the smaller wall. This then tells you how far apart the measurements points need to be. By applying this value to the room *as a whole* the required number of measurement points may be calculated. Take the example of a room which is 20 m × 30 m with a 4 m ceiling. If the room were 20 m × 20 m then the wall:ceiling ratio would be 5, and from the above table a minimum of 16 measurement points would be needed. So in the 20 m × 30 m room, which is 1·5 times as large, 24 measurement points would be needed to maintain the distance between measurement points.

Greater accuracy than 10% can be achieved by increasing the number of measurement points, and when time allows the more points taken the better. Doubling the number of points halves the error, and accuracy to within 5% can then be claimed.

Once you have determined the number of measurement points, the next step is to divide the plan of the room into a grid consisting of a number of squares each equal in area. If necessary, the number of measurement points should be increased to ensure that the grid is symmetrical. The idea is to take illuminance measurements at the centre of each of these squares. However, before doing this check from the plan that the measurements are not going to be biased by having a disproportionate number of measurements taken directly underneath luminaires. The illuminance measurements should be taken in the horizontal meridian, at around desk height, without shadows being cast on the meter and the values should be recorded directly onto the sketch plan of the office. In this way, you will also have a visual indication of the illuminance variation over the room to use in conjunction with the arithmetic average for the room as a whole.

The desk illumination

Again, it is a good idea to sketch a plan of what is being measured, in this case each desk. While doing this, note down what measurements need to be determined, and under what conditions they need to be taken. One position on the desk may be identified as being the location where the primary tasks are performed, with other areas being either where objects used less are placed, or simply storage space. If objects on the desk are inclined, in the way that a document holder may be, then the horizontal plane may not be the appropriate place to take the measurement and this should be noted on the sketch. Measurements should be taken with any occupants in place, so that any shadows usually cast by the person are present.

As well as the illuminance falling on the desk, the luminance of the various points of interest can be determined, as measured from the occupant's usual position. Knowing the illuminance on an object, and its luminance, the 'luminance factor' may be calculated (see earlier).

What other measurements are needed?

Depending on the circumstances, further measures could be appropriate. In the present example, the luminances and reflectances of the walls, floor and ceiling *should* be determined. Wall reflectance is usually important to the lighting of small rooms, but less so for large rooms where the area immediately adjacent to the wall is the only part of the room affected by light reflected from it. On the other hand, the larger the room the greater the importance of the ceiling reflectance. While a small room can have a satisfactory appearance with a darker ceiling, a large room needs to have a large proportion of the light incident on the ceiling reflected back into the room.

Optionally, a subjective 'expert observer' assessment of discomfort glare could be taken from each workplace; the relative daylight factor could be measured at each grid point; the contrast rendering factor could be established; measurements could be repeated to determine the variation over the day, and so on. While these measures might all be outside the original brief, where time and enthusiasm permits, they enable a much more complete description of the workplace to be made and provide a sounder basis on which to make recommendations about appropriate changes.

Assessment of the environment

Now that the measurements have been taken, the values obtained can be compared with values known to represent good practice, as detailed in the CIBS Code. Here the recommendation for standard service illuminance (the mean illuminance over the lifetime of the lamps) for a general office is 500 lux, with a minimum of 200 lux for any working surface. The Code also gives recommendations which are applicable here for illuminance ratios:

(a) The ratio of the minimum illuminance to the average illuminance over the task area should not be less than 0·8. While a greater degree of non-uniformity is acceptable between task areas and areas of the room where no tasks are performed, the illuminance on areas adjacent to the tasks should still not be less than 0·3 of the task illuminance.

(b) In an interior with general lighting, the ratio of the average illuminance on the ceiling to the average illuminance on the horizontal working plane should be within the range 0·3 to 0·9.

(c) In an interior with general lighting, the ratio of the average illuminance of any wall to the average illuminance on the horizontal working plane should be within the range of 0·5 to 0·8.

(d) In an interior with localized or local lighting, the ratio of the illuminance on the task area to the illuminance around the task should be not more than 3:1.

By comparing the measured values with the above recommendations, the extent of any problem present in the office may be evaluated, and remedial action may be suggested. Here the extra measurements suggested may be helpful in differentiating between the available solutions. For example, if the wall and ceiling reflectances are acceptable, then the solution may lie in modification of the luminaires, whereas if these reflectances are low the situation could be improved by painting the walls and ceiling.

This example is, of necessity, short and addresses a specific and relatively standard issue. Assessment of other visual environments will involve different circumstances and problems. For example, the criteria by which the design of visual aspects of a car interior is judged will be rather different from those applicable in an office. With this in mind, the next section contains a checklist to indicate the range of problems which might be present in *any* visual environment, and the questions that may need to be asked.

Checklist

What are the problems?
What measures are needed?
What guidelines are available?

Consider each of the following if appropriate:
Light levels
Luminance and illuminance:
  Where is the light needed?
  What variation is there?
    across the room?
    across the workplace?
  Is supplementary light needed anywhere?
    at any particular time of the day?
    at any particular time of the year?
    for particular purposes related to the immediate task?
    for purposes unrelated to the tasks, e.g., safety lighting.
Surfaces
  What are the reflectances of the various surfaces?
    the walls,
    the ceiling,
    the floor.
  What are the reflectance ratios between them?
Glare
  Discomfort glare:
    Is it a problem?

Subjective or objective assessment needed?

Can it be relieved by modifying the environment?

Disability glare:

Is it affecting performance?

Can it be relieved by moving or shielding the lights?

Temporal aspects

Is flicker apparent:

in the visual task itself, e.g., a VDU?

in the environment, e.g., from fluorescent tubes?

Chromatic considerations

Is the colour rendering of concern?

Is the colour rendering of the lights acceptable?

Is the colour appearance of the lights acceptable?

Is colour discrimination a factor for concern?

How good is the colour rendition of the present lights?

Is supplementary light with better colour rendition needed for the task, to enhance colour discrimination?

Is colour coding present, and if so are the lights adequate for the colours to be discriminated?

Spatial considerations

Directionality:

Is there a specific need for directional lighting?

What variation is there over the room?

What variation is there over the day? How is the daylight supplemented by artificial light, and how does this vary over the day?

Are shadows a problem under working conditions?

Highlighting:

Are there any features that need special consideration to increase their conspicuity?

Are there any special-purpose needs, such as safety lighting with a back-up power supply?

Users

Are there particular user groups with specific requirements?

Non-visual considerations of the visual environment

How can the environment be maintained under the desired conditions: what maintenance is needed?

What non-visual effects occur, e.g., heat from luminaires?

References

Boyce, P.R. (1988). *Human Factors in Lighting* (London: Macmillan).

CIBS (1984). *Code for Interior Lighting* (London: Chartered Institution of Building Services).

Hopkinson, R.G. and Collins, J.B. (1970). *The Ergonomics of Lighting* (London: Macdonald).

Wilkins, A.J., Nimmo-Smith, I., Slater, I.A. and Bedocs, L. (1988). Fluorescent lighting, headaches and eye-strain. Presented at the *National Lighting Conference,* Cambridge, March 1988.

Further information

CIBS publications: These cover a wide variety of issues, e.g., glare, lighting for VDUs, lighting guides for various applications such as engineering, libraries, hospitals. Information can be obtained from: Chartered Institution of Building Services, 222 Balham High Road, London SW12 9BS.

CIE (Commission Internationale de l'Eclairage) publications: These also cover a wide range of issues, such as colour rendering, light measurement, and visual performance. Information on their availability can be obtained from lighting organizations of most countries (e.g., CIBS, IES).

Cronly-Dillon, J., Rosen, E.S. and Marshall, J. (1985). *Hazards of Light: Myths and Realities, Eye and Skin* (Oxford: Pergamon Press).

Egan, M.D. (1983). *Concepts in Architectural Lighting* (New York: McGraw Hill). A comprehensive book covering vision as well as lighting in an easily understood volume.

Galer, I. (Ed) (1987). *Applied Ergonomics Handbook: Chapter 9, The Environment—Vision and Lighting* (London: Butterworths). Although short, this is a useful reference chapter dealing with the principles of good lighting, and practical design considerations.

IES Lighting Handbook. (New York: Illuminating Engineering Society of North America). This handbook comes in two volumes, a Reference volume (1984) and an Applications volume (1987). The Society can be contacted at: Illuminating Engineering Society of North America, 345 East 7th Street, New York, NY 10017.

Interior Lighting Design Handbook (London: The Lighting Industry Federation). The Federation is a useful source of current information, and may be contacted at 207 Balham High Road, London SW17 7BQ.

Overington, I. (1976). *Vision and Acquisition* (London: Pentech Press). A reference book about visual performance. Not easy reading, and no longer completely up to date, but a comprehensive collection of information.

Pritchard, D.C. (1985). *Lighting,* 3rd Edition (London and New York: Longman). This inexpensive paperback is fairly up to date, and is highly recommended.

Wyszecki, G. and Stiles, W.S. (1982). *Color Science: Concepts and Methods, Quantitative Data and Formulae,* 2nd edition (New York: John Wiley).

A standard reference work on colour: comprehensive, but not easy reading.

As well as the above sources, information can be obtained from the bodies which produce standards in individual countries (e.g., British Standards, American ANSI standards, German DIN standards). Also, International Standards have been established in many areas by the International Standards Organisation (ISO).

Chapter 16

Ergonomics assessment of thermal environments

Kenneth C. Parsons

Introduction

An integral part of applied ergonomics methodology is to consider how people will be affected by their environment. Traditionally, visual, acoustic (noise and vibration) and thermal environments are considered as well as spatial, psychosocial and chemical (e.g., air quality) environments. It is usual to consider effects in terms of separate components; however, it should always be remembered that people work in 'total' environments and interactions of environmental components may be important in some applications. Environmental effects are often considered in terms of workers' health, comfort and performance and this can contribute to human systems design and analysis in terms of overall system performance, worker safety, workstation design and so on.

Thermal environments can be divided conveniently into hot, neutral (or moderate) and cold conditions. Applied ergonomics methods for assessing thermal environments include objective methods, subjective methods and methods using mathematical (usually computer) models. Objective methods include measuring the physiological response of people to the environment. Responses in terms of sweat rate, internal body temperature, skin temperatures and heart rate are useful measures of body strain (see chapter 22). Performance measures at simulated or actual tasks can also be useful. Subjective measures are particularly helpful when assessing psychological factors such as thermal comfort and satisfaction. They can also be useful in quantifying the effects of moderate cold or moderate heat stress. Mathematical models have become popular in recent years because, although often complex, they can be easily used in practical applications, employing digital computers. Some of the more sophisticated rational (or causal) models involve an analysis of the heat exchange between people and their environment and also include dynamic models of the human thermoregulatory system. Empirical models can provide useful mathematical equations which 'fit' data obtained from exposing human subjects to thermal conditions (see also chapter 8).

The aim of this chapter is to present the principles behind practical methods for assessing human response to hot, moderate and cold environments and to present a practical approach to assessing thermal environments with respect to human occupancy.

The principles

The following brief discussion provides the underlying principles behind assessing human response to thermal environments. For a fuller discussion and references the reader is referred to a standard text such as Parsons (1993).

Relevant measures

It is now generally accepted that there are six important factors that affect how people respond to thermal environments. These are air temperature, air velocity, radiant temperature, humidity and the clothing worn by, and the activity of, the human occupants of the environment. In any practical assessment, instruments and methods for quantifying these factors may be used.

Thermoregulation

People are homeotherms, that is they react to thermal environmental stimuli in a manner which attempts to preserve their internal body ('core') temperature within an optimal range (around 37°C). If the body becomes too hot, vasodilation (blood vessel expansion) allows blood to flow to the skin surface (body 'shell') providing greater heat loss. If vasodilation is an insufficient measure for maintenance of the internal body temperature then sweating occurs resulting in increased heat loss by evaporation. If the body loses heat then vasoconstriction reduces blood flow to the skin surface and hence reduces heat loss to the environment. Shivering will increase metabolic heat production and can help maintain internal body temperature.

The physiological reaction of the body to thermal stress can have practical consequences. A rise or fall in internal body temperature can lead to confusion, collapse and even death in humans. Vasoconstriction can lead to a reduction in skin temperature and complaints of cold discomfort and a drop in manual performance. Sweating can cause 'stickiness' and warmth discomfort and 'mild heat' can provide a drop in arousal, and so on.

Thermal indices

A useful tool for describing, designing and assessing thermal environments is the thermal index. The principle is that factors that influence human response to thermal environments are integrated to provide a single index value. The aim is that the single value varies as human response varies and can be used to predict the effects of the environment. A thermal comfort index for example would provide a single number which is related to the thermal comfort of the occupants of an environment. It may be that two different thermal environments (i.e., with different combinations of various factors such as air temperature, air velocity, humidity and activity of the occupants) have the same thermal comfort index value. Although they are different environments, for an ideal index identical index values would produce identical thermal comfort responses of the occupants. Hence environments can be designed and compared using the comfort index.

A useful idea is that of the standard environment. Here the thermal index is the temperature of a standard environment that would provide the 'equivalent effect' on a subject as would the actual environment. Methods of determining equivalent effect have been developed. One of the first indices using this approach was the effective temperature (ET) index (Houghten and Yaglou, 1923). The ET index was in effect the temperature of a standard environment (air temperature equal to radiant temperature, still air, 100% relative humidity for the activity and clothing of interest) which would provide the same sensation of warmth or cold felt by the human body as would the actual environment under consideration.

Heat balance

The principle of heat balance has been used widely in methods for assessing human responses to hot, neutral and cold environments. If a body is to remain at a constant temperature then the heat inputs to the body are balanced by the heat outputs. Heat transfer can take place by conduction (K), convection (C), radiation (R) and evaporation (E). In the case of the human body an additional heat input to the system is the metabolic heat production (M) due to the burning of oxygen by the body. Using the above, the following body heat equation can be proposed:

$$M \pm C \pm R \pm K - E = S \qquad (1)$$

If the net heat storage (S) is zero then the body can be said to be in heat balance and hence internal body temperature can be maintained. The analysis requires the values represented in equation (1) to be calculated from a knowledge of, for example, the physical environment, clothing and activity.

Rational indices

Rational thermal indices use heat transfer equations (and sometimes mathematical representations of the human thermoregulatory system) to 'predict' human response to thermal environments. In hot environments the heat balance equation (1) can be rearranged to provide the required evaporation rate (E_{req}) for heat balance ($S = 0$) to be achieved, e.g.,

$$E_{req} = (M - W) + C + R. \tag{2}$$

(K can often be ignored and W is the amount of metabolic energy that produces physical work.) Because sweating is the body's major method of control against heat stress, E_{req} provides a good heat stress index. A useful index related to this is to determine how wet the skin is; this is termed skin wettedness (w) where:

$$w = \frac{E}{E_{max}} = \frac{\text{actual evaporation rate}}{\text{maximum evaporation rate}}. \tag{3}$$
$$\text{possible in that environment}$$

In cold environments the clothing insulation required (IREQ) for heat balance can be a useful cold stress index based upon heat transfer equations.

Heat balance is not a sufficient condition for thermal comfort. In warm environments sweating (or skin wettedness) must be within limits for thermal comfort and in cold environments skin temperature must be within limits for thermal comfort. Rational predictions of the body's physiological state can be used with empirical equations which relate skin temperature, sweat rate and skin wettedness to comfort.

Empirical indices

Empirical thermal indices are based upon data collected from human subjects who have been exposed to a range of environmental conditions. In hot environments curves can be 'fitted' to sweat rates measured on individuals exposed to a range of hot conditions. There has been little research of this kind for cold conditions, however a wind chill index was developed based upon the cooling of cylinders of water in outdoor conditions. Wind chill provides the 'trade-off' between air temperature and air velocity. Comfort indices have also been developed entirely empirically from subjective assessments over a range of environmental conditions.

Direct indices

Direct indices are measurements taken on a simple instrument which responds to environmental components similar to those to which humans respond. For example a wet, black globe with a thermometer placed at its centre will respond to air temperature, radiant temperature, air velocity and humidity. The temperature of the globe will therefore provide a simple thermal index which, with experience of use, can provide a method of assessment of hot environments. Other instruments of this type include the temperature of a heated ellipse and the integrated value of unaspirated wet bulb temperature, air temperature and black globe temperature (WBGT).

Measuring instruments

Air temperature is traditionally measured using a mercury in glass thermometer, although more recently thermocouples and thermistors have been used. An advantage of electronic instrumentation is that values can be continuously recorded and fed into digital computers for later analysis. The dry bulb of a whirling hygrometer gives a value of air temperature. If there is a large radiant heat component in the environment then it will be necessary to shield the air temperature transducer (e.g., using a wide mouthed vacuum flask). Air humidity can be found from the wet and dry bulb of a whirling hygrometer. Other methods include capacitance devices and hair hygrometers.

Radiant temperature is usually quantified in the first analysis by measuring black globe (usually 150 mm diameter) temperature. Correcting the globe temperature for air temperature and air velocity allows a calculation of mean radiant temperature. If more detailed analysis is required then instruments for measuring plane radiant temperatures in different directions should be used. Correction factors may be necessary to allow for the shape of the human body; however use of a globe thermometer provides a satisfactory initial measurement method.

Air velocity should be measured down to about $0.1 \mathrm{~m~s}^{-1}$ in indoor environments. Generally cup or vein anemometers (i.e., masses rotated by moving air) will not measure down to such low air speeds. Suitable instruments are hot wire anemometers, where the cooling power of moving air over a hot wire, corrected for air temperature, provides air velocity, or Kata thermometers, where air movement cools a thermometer and cooling time is related to air velocity. Kata thermometers are however cumbersome and time consuming to use in practical applications.

In more recent years integrating systems have been developed which detect all four environmental parameters and integrate measurements into thermal indices which predict, for example, thermal comfort. The instruments are usually simple to use and provide practical solutions for the non-expert. They are often relatively expensive however. Another development has been the use of transducers connected to digital storage devices, to allow recording of environmental conditions over long periods of time. The devices usually allow easy interfacing with digital computers where sophisticated analysis can be performed.

Subjective methods

Thermal comfort is usually defined as 'that condition of mind which expresses satisfaction with the thermal environment'. The reference to 'mind' emphasizes that this is a psychological phenomenon and hence the importance of subjective assessment methods. Subjective methods range from simple thermal sensation votes to more complex techniques where semantic or cognitive models of human perception of thermal environments can be determined. In a simple practical assessment of thermal environments two types of scale are generally used. One type is concerned with thermal sensation and the other is concerned with acceptability (i.e., a value judgement). Specific questions regarding general satisfaction, draughts, dryness and open-ended questions asking for other comments provide a useful brief subjective form, especially for measuring the acceptability and comfort of an environment. There are biases and errors which can occur in taking subjective measures, but it should not be forgotten that the best judges of their thermal comfort are the human occupants themselves. (See chapter 3 on indirect observation.)

Human performance

Despite a number of studies having been carried out on the effects of thermal environments on human performance there is no specific information of practical value. Human performance can be considered in terms of physical (e.g., manual dexterity) and psychological (e.g., behavioural, cognitive) effects. If heat or cold stress is sufficiently severe that internal temperatures pass beyond limits at which major physiological effects occur (e.g., causing collapse or hallucinations) then clearly performance will be impaired. Within such limits, effects are influenced by factors such as motivation, level of proficiency at the task and individual differences.

The state of practical knowledge is such that it is not yet possible to predict reliably effects on manual or cognitive performance in hot environments or effects on cognitive performance in cold environments. Major effects do occur, however, on manual dexterity, cutaneous sensitivity and strength in cold environments. Hand skin temperature can be used to predict effects and it is generally agreed that to maintain manual dexterity the hands should be kept warm. The distraction effects of thermal stress can also degrade performance, particularly in cold environments.

Clothing

Clothing can be worn for protection against environmental hazards and for aesthetic reasons as well as for thermal insulation. In thermal terms a microclimate is produced between the human body and the clothing surface. The applied ergonomist should ensure that the microclimate allows the body to achieve desirable physiological and psychophysical objectives. Physiological objectives may include the maintenance of heat balance for the body, and the preservation of skin temperatures and sweating at levels which allow for comfort. An interaction between thermal aspects and material type should also be considered (e.g., the effect of 'scratchy' materials being exacerbated by sweating).

The dry thermal insulation of clothing is greatly affected by how much air is trapped within clothing layers as well as within clothing. There are two common units which quantify dry clothing insulation; these are the CLO and the TOG. The CLO value is a clothing insulation value which is intended to be a 'relative unit', compared to a normal everyday costume necessary for thermal comfort in an indoor environment. For example, a 'typical' business suit (including underclothes, shirt, etc.) is often quoted as having a thermal insulation of 1·0 CLO. A nude person has zero CLO. In terms of thermal insulation 1 CLO is said to have a value of $0·155 \ m^2 \ °C \ W^{-1}$. The TOG value is a unit of thermal resistance and is a property of the material. It can be measured on a heated flat plate, for example, in terms of heat transfer. It does not necessarily relate to the thermal insulation provided to a clothed person. For comparison, 1 TOG is equal to $0·1 \ m^2 \ °C \ W^{-1}$. When clothing becomes wet, due to sweating or external conditions, then the clothing insulation is altered (usually greatly reduced). There are methods which estimate the thermal properties of wet clothing; however, these are crude and thermal insulation values for wet clothing are not well documented. Additional factors which can greatly affect the thermal insulation of clothing include pumping effects due to body movement, ventilation and wind penetration.

In application it is not usually necessary to have a detailed description of clothing insulation properties. The important point is whether the clothing achieves its objectives. The objectives for clothing will be determined in an overall ergonomics system analysis, involving a description of the objectives of the organization, task analysis, allocation of functions, job design and so on. This will include the objectives in terms of thermal insulation which will also involve the design of the clothing (to include pockets to keep hands warm, devices to keep workers cool in hot environments (e.g., ice jackets), etc.).

User tests and trials (described generally in chapter 10) will provide important information about whether clothing meets its objectives. Objective measures such as sweat loss, skin temperatures and internal body temperatures or subjective measures of thermal sensation, comfort or stickiness can be of great value. Performance measures at actual or simulated tasks can also be used to evaluate whether clothing has achieved its thermal and other objectives.

Safe surface temperatures

The applied ergonomist is also interested in the effect there will be on the body from physical contact between the human skin and surfaces in a workplace; for example, what sensation is caused by bare feet on a 'cold' floor or whether brief contact with a domestic product (e.g., a cooker, oven door, knobs or a kettle) will result in pain or burns. There are many factors involved in determining human response. These include the type and duration of contact, the material and condition of the surface and the condition of the human skin. However, a simple model based upon the heat transfer between two semi-infinite slabs of material in perfect thermal contact can provide a practical method of assessment.

The principal point is that there is a contact temperature which exists between the skin and the material surface which is dependent upon the physical properties of the skin and the material and which will influence the effect on the person. For example, if one touched a metal slab at 100°C then the contact temperature would be of the order of 98°C and the metal would be felt to be extremely hot. If, however, one touched a cork slab at 100°C then the contact temperature would be around 46°C and the cork would feel much less hot than the metal. Contact temperature can therefore be used to predict effects on the body, and can be calculated from the following equation:

$$T_{con} = (b_1 T_1 + b_2 T_2)/(b_1 + b_2) \tag{4}$$

where T_1 and T_2 are the initial surface temperatures (°C); T_{con} is the contact temperature (°C); and b_1 and b_2 are thermal penetration coefficients calculated from the following equation:

$$b = (K\rho c)^{1/2} \, JS^{-1/2} \, m^{-2} \, °C^{-1}$$

where K is the thermal conductivity; ρ is density; and c is specific heat. To calculate T_{con} the values of b for human skin and for different materials are required. These are provided in Table 16.1.

McIntyre (1980) argues that if one assumes a skin temperature of 34°C and a simple reaction time for an individual of 0·25 s then, from the work of Bull (1963), whose data suggest a partial burn at a skin temperature of about 85°C for 0·25 s contact, one can estimate temperatures which would

Table 16.1. Thermal penetration coefficients (b)

Material	b ($JS^{-1/2} \, M^{-2} \, °C^{-1}$)	Momentary contact surface temperature for burn threshold (°C)
Human skin	1000	
Foam	30	—
Cork	140	450
Wood	500	187
Brick	1000	136
Glass	1400	121
Metals	>10000	90

produce partial burns. Some of these temperatures are provided in Table 16.1 (from McIntyre, 1980).

A practical approach to providing maximum surface temperatures for heated domestic equipment is provided in British Standard 4086 (BSI, 1966). The Standard considers three categories of material type and three types of contact duration. A summary of the limits is provided in Table 16.2.

Because of the complexity of the problem and ethical considerations regarding experimentation involving pain and burns on human subjects, knowledge is incomplete in this area. In addition, specifications for safety of manufacturing products often involves other costs and benefits to be considered (see chapter 34 in this book). The limits provided in BS 4086 (BSI, 1966) have recently been reviewed and an additional document, BS PD 6504 (BSI, 1983), has been produced which provides background medical information and data in terms of discomfort, pain and burns. There is, however, little information concerning the inter- and intrasubject variation in response, effects of skin condition and effects on different populations (e.g., the aged, children) and other variables important for practical application.

International standards

There have been many national and international standards concerned with thermal comfort, heat stress and cold stress, and because of renewed interest in indoor and outdoor environments there is increasing activity in this area. Thermal comfort standards can define conditions for thermal comfort and indicate the likely degree of discomfort of occupants of thermal environments. Standards for heat stress and cold stress attempt to specify conditions that will preserve health and often comfort and performance. Standards can also provide guidance on environmental design and control, they standardize methods to allow comparison and they contribute to assessment and evaluation.

Influential institutions throughout the world that produce standards or guidelines include the American Society of Heating, Refrigerating and Air Conditioning Engineers (ASHRAE) in the USA, the Chartered Institute of Building Services Engineers (CIBSE) in the UK, National standards bodies,

Table 16.2. Maximum surface temperatures (°C) for heated domestic equipment (BSI, 1966)

	Handles (Kettles, pans, etc.)	Knobs (not gripped)	Momentary Contact
Metals	55	60	105
Vitreous enamelled steel and similar surfaces	65	70	120
Plastics, rubber or wood	75	85	125

the ISO and also European Standardization under CEN, the World Health Organization (WHO), the American Conference of Governmental Industrial Hygienists (ACGIH), the International Labour Organization (ILO) and more. Many standards have defined limits or at least methods in terms of rational or empirical thermal indices or physiological condition (e.g., body 'core' temperature). There are also standards that provide techniques and methods (e.g., physiological and subjective assessment methods).

A significant contribution over the last twenty years has been the co-ordinated development of ISO standards for the ergonomics assessment of thermal environments. These are influential and it is likely that in the future they will be adopted in similar form as European Standards.

ISO standards

The collection of ISO (International Organisation for Standardisation) standards and documents, concerned with the ergonomics of the thermal environment, can be used in a complementary way to provide an assessment methodology. The subject is divided into three principal areas—hot, moderate and cold environments—and remaining standards are divided into human reaction to contact with solid surfaces and supporting standards.

For the assessment of hot environments a simple method based on the WBGT (wet bulb globe temperature) index is provided in ISO 7243. If the WBGT reference value is exceeded a more detailed analysis can be made (ISO 7933) involving calculation, from the heat balance equation, of sweating required in a hot environment. If the responses of individuals, or of specific groups, are required (for example in extremely hot environments) then physiological strain should be measured. Methods of measuring mean skin temperature, heart rate, internal body ('core') temperature and mass loss are all described in ISO 9886.

ISO 7730 provides on analytical method for assessing moderate environments and is based on the Predicted Mean Vote and Predicted Percentage of Dissatisfied (PMV/PPD) index and on criteria for local thermal discomfort, in particular drafts. If the responses of individuals or specific groups are required, then subjective measures should be used (ISO DIS 10551).

ISO TR 11079 (Technical report) provides an analytical method for assessing cold environments involving calculation of the clothing insulation required (IREQ) from a heat balance equation. This can be used as a thermal index or as a guide to selecting clothing.

ISO work on contact with solid surfaces is divided into hot, moderate and cold surfaces and standards are in initial stages of development. Ergonomics data to establish temperature limit values for touchable surfaces of machines are provided in proposed European standard EN 563 (1994). Supporting standards include an introductory standard (ISO DIS 11399) and standards for estimating the thermal properties of clothing (ISO 9920) and metabolic heat production (ISO 8996). Other standards consider instruments and measurement methods (ISO 7726) and medical supervision of individuals exposed to hot or cold environments (ISO CD 12894). Standards still in working document form are those concerned with symbols and units, proposals to consider the responses of disabled persons, and the thermal comfort performance of buildings over long periods.

ISO DIS 11399 (1993) presents a description of principles and methods of application of the series of ISO standards and should be consulted for an initial overview. The ISO working system showing how the collection of standards can be used in practice is presented in Figure 16.1. The objectives and status of the standards are shown in Table 16.3.

Thermal models and expert systems

The increase in knowledge of the human response to thermal environments and methods of modelling the response, and the development of the digital computer, have allowed complex but useful methods to be used easily in

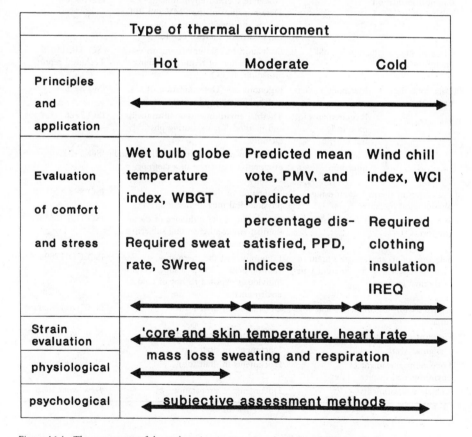

Figure 16.1. The assessment of thermal environments using the series of ISO standards

Table 16.3. Aims, title and status of ISO standards concerned with the ergonomics of the thermal environment

Aims of the standard		Title of the document	Status
General presentation of the set of standards in terms of principles and application		Ergonomics of the thermal environment: principles and application of International Standards	ISO DIS 11399
Standardization of quantities symbols and units used in the standards		Ergonomics of the thermal environment: definitions symbols and units	New work item
Comfort and thermal stress evaluation			
Thermal stress evaluation in hot environments	Analytical method	Hot environments — Analytical determination and interpretation of thermal stress using calculation of required sweat	ISO 7933
	Diagnostic method	Hot environments — Estimation of the heat stress on working person, based on the WBGT-index (wet bulb globe temperature)	ISO 7243
Comfort evaluation		Moderate thermal environments — Determination of the PMV and PPD indices and specification of the conditions for thermal comfort	ISO 7730
Thermal stress evaluation in cold environments		Evaluation of cold environments — Determination of required clothing insulation, IREQ	ISO TR 11079 Technical report
Data collection standards	Metabolic rate	Ergonomics — Determination of metabolic heat production	ISO 8996
	Requirements for measuring instruments	Thermal environments — Instruments and methods for measuring physical quantities	ISO 7726
	Clothing insulation	Estimation of the thermal insulation and evaporative resistance of a clothing ensemble	ISO 9920
Evaluation of thermal strain using physiological measures		Evaluation of thermal strain by physiological measurements	ISO 9886
Subjective assessment of thermal comfort		Assessment of the influence of the thermal environment using subjective judgement scales	ISO DIS 10551
Selection of an appropriate system of medical supervision for different types of thermal exposure		Ergonomics of the thermal environment — Medical supervision of individuals exposed to hot or cold environments	ISO CD 12894
Contact with hot, moderate and cold surfaces		Documents in draft form	New work items
Comfort of the disabled		Documents in draft form	New work item
Design of work for cold		Documents in draft form	New work item
Long-term assessment of environmental quality		Documents in draft form	New work item
Vehicle environments		Documents in preparation	New work item

practical application. Practical models which simulate how people respond to hot, moderate or cold environments can be used to assess and design thermal environments. They can also be integrated into larger computer-based expert and knowledge-based systems for use by the ergonomics practitioner.

Examples of models which simulate the human response to thermal environments are provided by Haslam and Parsons (1987). Models of the human thermoregulatory system controlling a passive body (e.g., made up of cylinders and a sphere with thermal properties similar to those of the human body) can be used to predict changes in temperature within and over different parts of a clothed body. These models can then simulate how persons could respond in terms of heat stress or cold stress in outdoor environments or in terms of thermal comfort indoors. Investigations of the nature of expertise used in assessing the human response to thermal environments, coupled with the requirements of ergonomics practitioners and simulations using such models as are described here, can provide the input to expert systems. An example of the structure of such a system as described by Smith and Parsons (1987) is provided in Figure 16.2. It is probable that such systems will become valuable tools in integrating the principles and knowledge involved in the assessment of human response to thermal environments for practical application. Techniques for the elicitation of knowledge in general for expert systems are discussed elsewhere in this book in chapter 14.

The practice

Despite the lack of some information the principles mentioned above can provide a practical methodology which can be used to assess thermal environments with respect to effects on their human occupants. Almost by definition practical assessments will have factors specific to particular applications and one universal method is therefore difficult to provide. A general description of methods for assessing extreme environments is given later. More detailed guidance is also provided for the assessment of moderate environments.

Practical assessment of hot environments

The assessment of hot environments is particularly important as danger to health can occur rapidly. For environments where experience has been gained in monitoring workers, working practices can be developed and conditions monitored using simple thermal indices (e.g., wet bulb globe temperature (WBGT) index, wet globe temperature (WGT)).

The WBGT index is used in ISO 7243 (ISO, 1989) as a simple method for assessing hot environments. For conditions inside buildings and outside buildings without solar load:

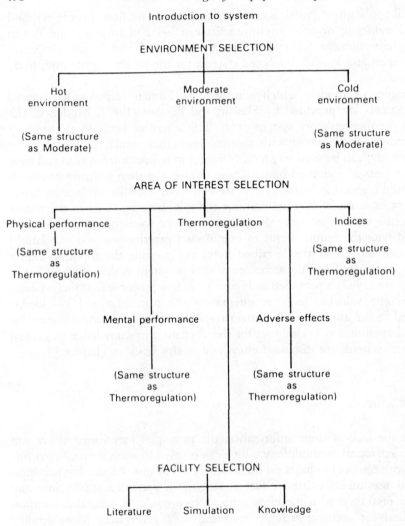

Figure 16.2. System tree diagram

$$\text{WBGT} = 0 \cdot 7 \, t_{nw} + 0 \cdot 3 \, t_g \tag{5}$$

and outside buildings with solar load:

$$\text{WBGT} = 0 \cdot 7 \, t_{nw} + 0 \cdot 2 \, t_g + 0 \cdot 1 \, t_a \tag{6}$$

where t_{nw} is the natural wet bulb temperature (i.e., not 'whirled'); t_g is 150 mm diameter black globe temperature; and t_a is the air temperature.

Acclimatization programmes, before workers begin work, are useful. It is particularly important that workers do not become unacceptably dehydrated (e.g., greater than 4% of body weight lost in sweat) or have an unacceptably elevated internal body temperature (e.g., greater than 38·0–38·5°C). A more detailed analysis can be provided by using rational assessments of the environment. Allowable exposure times based on such factors as predicted elevated internal body temperature or dehydration can be provided. It is important to remember however that there are individual differences in workers and that knowledge of heat transfer for the human body is incomplete. Experience is therefore required in the use of rational models.

If individuals are exposed to extremely hot environments then individual physiological measures of heart rate, internal body temperature and sweat loss should be taken and each worker observed closely.

Practical assessment of cold environments

Similar general guidelines apply to cold environments as were described for hot environments. A simple index such as the wind chill index can be used when experience has been gained with its use, where:

$$WCI = (10 \sqrt{V} + 10 \cdot 45 - V) (33 - t_a) \tag{7}$$

where V is the air velocity (m s^{-1}); and t_a is the air temperature (°C). The effects associated with different values of the wind chill index (from Parsons, 1993) are:

WCI	Effect
200	Pleasant
400	Cool
1000	Cold
1200	Bitterly cold
1400	Exposed flesh freezes
2500	Intolerable

When the WCI value is calculated it is often useful to calculate the t_a value which would provide the same wind chill in 'calm' air ($V = 1 \cdot 8$ ms^{-1}).

Clothing is important and a compromise must be reached between thermal insulation and clothing design to reduce effects on worker performance and safety. Of particular interest is the temperature of the body's extremities (hands and feet). Prolonged exposure may lead to thermal injury but severe discomfort distraction and loss of manual dexterity are the most commonly occurring effects. Although there is some debate, 'back of hand' temperatures above 20–25°C should maintain some comfort and performance. Hand temperatures of less than 10–15°C are usually unsatisfactory, although they should not produce injury. Performance effects will depend upon duration of exposure. Cold can produce severe discomfort and this has behavioural and distractive effects. Distraction may reduce manual and cognitive workload capacity.

Rational indices can be used to predict the required clothing insulation for heat balance and thermal comfort and also allowable exposure times based upon a drop in body 'core' temperature. Physiological measures of heart rate, mean skin temperature and body core temperature should be used if assessing individuals in cold environments. Any drop in body core temperature is unsatisfactory, but 36°C is a lower working limit. A core temperature of below 35°C is defined as hypothermia. Medical screening of subjects should take place before exposure to either hot or cold environments.

Practical assessment of moderate environments

A common request to the environmental ergonomist is to assess an indoor climate such as found in an office. The practical method used in an actual case is outlined here, although some adjustment to the results has been made to illustrate points.

The workers in a large office were complaining that their thermal environment was unacceptable. The ergonomist was asked to assess the environment, quantify the problem and make recommendations for improvement if necessary. One day was allowed for the assessment and a total of four days for the whole project, including both analysis and final report, a not unusual time restriction.

Worker relations

Complaints about working environments can be stimulated by other work related problems and it is important for the ergonomist to gain an impression of the physical, social and organizational environment in general. In addition it is useful to have the co-operation and understanding of management and workers. The worker representative was therefore contacted and the ergonomist introduced. It was explained that the ergonomist was attempting to improve the thermal environment conditions. The physical and subjective measures which were to be taken were also demonstrated to the workers' representative who then passed on the information to the office occupants.

Where, when and what to measure

The question of where and when to measure is a question of statistical sampling. The thermal environmental conditions will vary throughout a space and also with time (during the day, night and seasonal variations). The more measuring points in the room and the more measuring times, in general, the more accurately the environment can be quantified. This then is a question of resources. Only one day was allowed for measurement so a plan of the office was obtained and individual workplaces identified. Measurements should be taken at the positions of the workers. Ankle, chest and head heights were chosen as measuring points at each workplace. Ten workplaces were chosen as the sample, 'evenly spread' throughout the office. Measurements were taken over a 3-h period under what had been established as 'typical' conditions throughout the morning, a time when complaints had been received. The ventilation systems were identified and set to normal working. Outside weather conditions were noted.

A 150 mm diameter globe thermometer was placed at each workplace (only two were available so they had to be moved around) for at least 20 min before readings were taken. Using a hot wire anemometer, air velocity and air temperature were measured at ankle, chest and head height of the worker. A whirling hygrometer was used at chest height to measure wet and dry bulb temperatures (dry bulb was used as a cross check for air temperature with the air temperature sensor on the hot wire anemometer). The workers' clothing and activity were noted, and movements throughout the room were also noted.

Subjective assessment forms (see Appendix) were handed to each worker and collected centrally (i.e., the working position was noted but a degree of anonymity was maintained). The subjective forms allowed some information to be collected regarding time variations (i.e., outside the survey time) and general satisfaction.

Analysis

PHYSICAL MEASURES

Analysis of physical measurements takes place in two parts. The first part is to obtain for each measurement point air temperature, mean radiant temperature, relative humidity and air velocity from the instrument measures and also to determine metabolic heat production and clothing insulation values. The second part is to predict the degree of discomfort. The subjective measures are analyzed separately and complement the physical measures.

The air temperature and air velocity were measured directly using the hot-wire anemometer. The mean radiant temperature (t_r) is obtained from globe temperature (t_g) corrected for air temperature (t_a) and air velocity (V). If the mean radiant temperature is within a few degrees of room temperature then McIntyre (1980) suggests that:

$$t_r = t_g + 2 \cdot 44 \sqrt{V(t_g - t_a)} \qquad (8)$$

where temperatures are in °C and air velocity in m s^{-1}. Relative humidity is calculated from the dry bulb (air temperature) and aspirated (whirled) wet bulb of the whirling hygrometer. Table 16.4 provides typical values. Table 16.5 provides metabolic heat production values for typical activities and Table 16.6 provides clothing insulation values for typical clothing. Useful infor-

Table 16.4. Relative humidity (%) from dry bulb and aspirated wet bulb temperature

Dry bulb temperature (°C)	Aspirated wet bulb temperature (°C)									
	12	14	16	18	20	22	24	26	28	30
12	100									
14	79	100								
16	62	81	100							
18	49	64	82	100						
20	37	51	66	83	100					
22	28	40	54	68	83	100				
24	20	31	43	56	69	84	100			
26	14	24	34	45	58	71	85	100		
28	9	18	27	37	48	59	72	85	100	
30	5	12	21	30	39	50	61	73	86	100

Table 16.5. Estimates of typical metabolic heat production values

Activity	Metabolic heat production($W m^{-2}$)
Seated, at rest	58
Standing, relaxed	70
Standing, light arm work	100
VDU operation	70
Driving	70–100

Table 16.6. Estimates of typical clothing insulation values (1 CLO = $0 \cdot 155$ m² °C W^{-1})

Type of clothing	Clothing insulation (CLO)
None	0
Light summer clothing (briefs, shorts, short sleeved shirt, light socks, light shoes)	$0 \cdot 3$
Light work clothing (light underwear, cotton long sleeved workshirt, light long trousers, socks, shoes)	$0 \cdot 65$
Light business suit (including underclothing etc.)	$1 \cdot 0$
Heavy business suit (including underclothing etc.)	$1 \cdot 5$

mation is provided by presenting all physical data in a table, or on the office plan, in the final report.

Table 16.7. PMV values for air temperature, clothing and activity (assume: mean radiant temperature = air temperature, air velocity = $0\cdot15$ m s^{-1} and relative humidity = 50%)

Clothing (CLO)	Activity (W m^{-2})	Air temperature (°C)						
		16	18	20	22	24	26	28
0·65	58	—	−2·7	−2·0	−1·3	−0·6	0·0	0·8
1·0	58	−2·1	−1·6	−1·1	−0·5	0·0	0·6	1·2
1·5	58	−1·1	−0·7	−0·3	0·2	0·6	1·1	1·5
0·65	70	−2·2	−1·7	−1·2	−0·6	0·0	0·5	1·0
1·0	70	−1·3	−0·9	−0·5	0·0	0·4	0·9	1·3
1·5	70	−0·5	−0·2	0·2	0·5	0·9	1·2	1·6
0·65	100	−0·9	−0·5	−0·1	0·3	0·6	1·0	1·4
1·0	100	−0·3	0·0	0·3	0·6	1·0	1·3	1·6
1·5	100	0·3	0·5	0·7	1·0	1·3	1·5	1·8

PREDICTION OF WHOLE-BODY THERMAL DISCOMFORT

Despite some theoretical limitations, one of the most useful thermal comfort indexes is the predicted mean vote (PMV) of Fanger (1970) which is used in ISO 7730 (ISO, 1993). Air temperature, mean radiant temperature, air velocity, humidity, clothing and activity values can be integrated to predict the mean thermal sensation vote of a large group of people on a seven point thermal sensation scale (as used on the subjective assessment form in the Appendix). The values range from PMV = 3 (hot) through PMV = 0 (neutral) to PMV = −3 (cold). PMV = 0 provides comfort conditions. From the PMV value a predicted percentage of dissatisfied (PPD) value can be calculated. This is related to the percentage of people likely to complain about the thermal conditions. Values of PMV and PPD are presented in Tables 16.7 and 16.8 for typical environmental conditions.

The thermal discomfort results for all ten workplaces will not be presented here; however the PMV and PPD values for each workplace were calculated and labelled on a copy of the plan of the office, for the final report. This showed the predicted whole-body thermal sensation (comfort) pattern over the office and those areas of likely complaint.

The following is an example of calculations for one workplace; the physical measurements were: t_a = 18°C; t_r = 18°C; V = 0·15 m s^{-1}; relative humidity = 50%; clothing insulation = 0·65 CLO; metabolic rate = 70 W m^{-2}.

Table 16.8. Interpretation of PMV values in terms of thermal sensation and Predicted Percentage of Dissatisfied (PPD)

Sensation	Cold	Cool	Slightly cool	Neutral	Slightly warm	Warm	Hot
PMV	−3	−2	−1	0	1	2	3
PPD (%)	—	75	25	5	25	75	—

Using Tables 16.7 and 16.8

$$PMV = -1\cdot7 \text{ and } PPD = 62\%.$$

Therefore the prediction is that, on average, a person will be between slightly cool and cool at this position. Also it can be seen that for all other conditions remaining the same an increase in air temperature from 18 to 24°C will provide a PMV value of 0 required for comfort. This could be a recommendation, or a recommendation could be made in terms of increased CLO value, etc.

LOCAL THERMAL DISCOMFORT
As well as overall or whole-body thermal sensation thermal conditions can produce effects on local areas of the body. For example, cold air moving around the workers ankles may cause a draught. The most common forms of local discomfort are caused by cooling due to air movement, heat losses due to asymmetric radiation (e.g., a radiant draught caused by workers sitting next to cold walls or windows) and thermal gradients. There is some debate about conditions which produce discomfort, but cool air movements (especially, if fluctuating) should be avoided above $0\cdot15 \text{ m s}^{-1}$, and particularly for exposed skin areas and if the subject is already cool. Radiant asymmetry should not exceed 10°C (less in the case of heated ceilings) and vertical temperature gradients should not be greater than 3°C. General observation of the workplaces, air velocity measures at the three heights (ankle, chest and head), and mean radiant temperatures will provide an indication of possible local thermal discomfort.

Dryness is probably related to air velocity, humidity and air and radiant temperatures, and is usually due to the evaporation of fluids from the eyes, nose and mouth which can lead to various problems, for example, with contact lenses. Local discomfort and other factors such as dryness and overall satisfaction should also be examined using subjective methods.

SUBJECTIVE RESPONSES
Analysis of subjective responses involves determining the average of, and variation in, response. The responses of how workers felt at the time of measurement can be compared with predicted responses. In general subjects in the office example used, gave a wider range on the scale than the predicted measures. The subjective measures were also presented on a plan of the office in the final report. On average workers were between slightly cool and cool with some subjects cold and some neutral. Draughts were reported in some areas. Most workers wished to be warmer. Responses regarding general sensation at work were similar to responses made about the conditions when they were measured. Most people were generally dissatisfied with the thermal environment.

Concluding remarks and recommendations

The above measurement and analysis allowed recommendations to be made in a final report which were related to the original objectives. An average increase in air temperature was recommended with some specific recommendations about draughts for particular workstations. It was also noted that the high level of dissatisfaction indicated may be due to general work or workplace dissatisfaction and not simply related to thermal conditions.

References

B.S.I. (1966). *Recommendations for Maximum Surface Temperatures of Heated Domestic Equipment.* (London: British Standards Institution).

B.S.I. (1983). *Medical Information on Human Reaction to Skin Contact with Hot Surfaces.* (London: British Standards Institution).

Bull, J.P. (1963). Burns. *Postgraduate Medical Journal,* **39**, 717–723.

Fanger, P.O. (1970). *Thermal Comfort.* (Copenhagen: Danish Technical Press).

Haslam, R.A. and Parsons, K.C. (1987). A comparison of models for predicting human response to hot and cold environments. *Ergonomics,* **30**, 1599–1614.

Houghten, F.C. and Yaglou, C.P. (1923). Determining equal comfort lines. *Journal of American Society of Heating and Ventilation Engineering,* **29**, 165–176.

ISO 7726: 1985. Thermal environments—Instruments and methods for measuring physical quantities. (Geneva: International Standards Organisation).

ISO 7243: 1989. Hot environments—Estimation of the heat stress on working man, based on the WBGT-index (wet bulb globe temperature). (Geneva: International Standards Organisation).

ISO 7933: 1989. Hot environments—Analytical determination and interpretation of thermal stress using calculation of required sweat rate. (Geneva: International Standards Organisation).

ISO 8996: 1990. Ergonomics—Determination of metabolic heat production. (Geneva: International Standards Organisation).

ISO 9886: 1992. Evaluation of thermal strain by physiological measurements. (Geneva: International Standards Organisation).

ISO 7730: 1993. Moderate thermal environments—Determination of the PMV and PPD indices and specification of the conditions for thermal comfort. (Geneva: International Standards Organisation).

ISO 9920: 1993. Estimation of the thermal insulation and evaporative resistance of a clothing ensemble. (Geneva: International Standards Organisation).

ISO CD 12894: 1993. Ergonomics of the thermal environment—Medical supervision of individuals exposed to hot or cold environments. (Geneva: International Standards Organisation).

ISO DIS 10551: 1993. Assessment of the influence of the thermal environment using subjective judgement scales. (Geneva: International Standards Organisation).

ISO DIS 11399: 1993. Ergonomics of the thermal environment: Principles and application of International Standards. (Geneva: International Standards Organisation).

ISO TR 11079 (Technical Report): 1993. Evaluation of cold environments—Determination of required clothing insulation, IREQ. (Geneva: International Standards Organisation).

McIntyre, D.A. (1980). *Indoor Climate*. (London: Applied Science Publishers).

Parsons, K.C. (1993). *Human Thermal Environments*. (London: Taylor and Francis).

Smith, T.A. and Parsons, K.C. (1987). The design, development and evaluation of a climatic ergonomics knowledge based system. In: *Contemporary Ergonomics 1987*, edited by E.D. Megaw (London: Taylor and Francis), pp. 257–262.

Appendix

The following subjective form was used in a moderate office environment where workers had been complaining about general working conditions. Various details about workers' characteristics and location were collected separately. The form was handed to workers for completion at their workplace. Question 1 determines the workers' sensation vote on the ASHRAE/ISO scale. Note that this can be compared directly with the measured PMV. Question 2 provides an evaluation judgement. For example, question 1 determines subject's sensation (e.g., warm). Question 2 compares this sensation with how the subject would like to be. Questions 3 and 4 provide information about how workers generally find their thermal environment. This is useful where it is not practical to survey the environment for long durations. Questions 5 and 6 are catch-all questions about workers' satisfaction and any other comments. Answers to these questions will provide information about whether more detailed investigation is required. Answers will also indicate factors which are obvious to the workers but not obvious to the investigator.

Please answer the following questions concerned with YOUR THER-
MAL COMFORT.

1. Indicate on the scale below how you feel NOW.
   Hot
   Warm
   Slightly warm
   Neutral
   Slightly cool
   Cool
   Cold
2. Please indicate how you would like to be NOW
   Warmer   No change   Cooler
3. Please indicate how you GENERALLY feel at work:
   Hot
   Warm
   Slightly warm
   Neutral
   Slightly cool
   Cool
   Cold
4. Please indicate how you would GENERALLY like to be at work:
   Warmer   No change   Cooler
5. Are you generally satisfied with your thermal environment at work?
   Yes   No
6. Please give any additional information or comments which you think
   are relevant to the assessment of your thermal environment at work
   (e.g., draughts, dryness, suggested improvements, etc.).

Chapter 17

Auditory environment and noise assessment

Christine M. Haslegrave

Introduction

Sound in our environment can be generated by transport of all forms, by equipment and machinery, and by people themselves. In the community, this is mainly due to traffic noise, to neighbours, and to radios and other domestic equipment. In the working environment, office noise comes from people, typewriters, printers and telephones. In manufacturing industry, it comes mostly from motors, fans, pumps and compressors, moving machinery, contact between tool and workpiece, and resonant plates or housings. Traffic and machinery can also set up vibrations in the structure of the building. All these have to be considered when identifying the sources of noise and investigating ways of reducing sound levels.

It is rare to experience silence in the modern world and our auditory environment is very complex. Sound can bring pleasure and information, but unnecessary sound or too much sound is annoying, distracting and possibly harmful. Also, it is not always possible to separate these effects. It is quite possible for a sound to be wanted by one hearer and unwanted by many others, and conflicts may arise between differing interests in both living and working environments. For this reason, we tend to distinguish between sound and noise. Kryter (1985) has defined noise in terms of its effects on people as 'audible acoustic energy that adversely affects the physiological or psychological well-being'. As he says, this is consistent with the usual definition of noise as 'unwanted sound'.

Ergonomists therefore need to evaluate auditory environments which may include sounds which are unwanted although they are not so loud as to be harmful, or sounds which are enjoyed by one person while annoying to another. They need to measure noise levels to assess whether there is a problem in terms of safety, working performance, comfort or annoyance. Further than this, they may need to identify the source and specify the action or protection which is needed.

In order to do this, they must be equipped to measure hearing ability, noise or sound levels of individual sources and of the environment, and to be able to assess the risk of damage to hearing. Some typical applications are:

1. Hearing assessment and screening.
2. Industrial health and safety.
3. Workplace (re)design.
4. Design and evaluation of the effectiveness of communication signals.
5. Design and performance evaluation of protective equipment.
6. Community noise assessment.
7. Noise suppression and shielding.
8. Machine design and testing.

Thus, the complexity of our auditory environment, and our highly subjective and personal responses to it, require an equally wide range of measures for use in its evaluation. This chapter presents the methodology for measuring sound and outlines some of the range of measures or criteria which have been proposed for assessing noise. The aim is to indicate some of the most useful techniques which are currently used by ergonomists, but without attempting to cover either specialist techniques or the concerns of acoustic engineers who deal with noise control, architectural acoustics or design of communication systems. Similarly there is no intention of covering the physics of sound or physiology of hearing, other than to introduce the necessary concepts and terms.

For a more detailed study of the ergonomic aspects introduced in this chapter the reader can consult texts dealing with hearing, perception of sound and the effects of noise on work and health, such as Jones and Chapman (1984), Kryter (1985), Loeb (1986) or Sanders and McCormick (1992). Guidance on acoustic treatments and techniques for reducing noise levels can be found in publications such as those by Brüel and Kjaer (1987) and the Health and Safety Executive (1983), as well as in noise control handbooks.

Units of measurement of sound

Sound is a variation of pressure in the air (or in any other elastic medium) which the human sense of hearing detects. It is transmitted in the form of pressure waves as a series of compressions and rarefractions travelling outwards from the source of the sound. The speed of the wave depends on the medium, but in air at 20°C sound travels at a velocity of 344 m s^{-1}. The amplitude of the sound pressure wave is the fluctuation above or below the ambient air pressure. Our sensation of loudness however does not come diretly from the pressure, but from the intensity of the sound wave which is the rate at which energy is transmitted by the wave. This is defined in terms of the energy passing through a unit area in unit time or sound power per unit area. Pressure can be measured directly, but it is more difficult to measure sound power. Sound pressure and sound power are however closely related, as will be shown later. It is therefore usual to measure sound levels in terms of sound pressure and to use this to calculate sound power. The units used for noise measurements in fact normally take account of this.

Auditory stimuli at the ear result from the combined signals received from sound sources in the environment. The waveforms (or variation in intensity/amplitude with time) of typical sound signals are shown in Figure 17.1.

A pure tone (which might be obtained from a tuning fork or a vibrating string) vibrates at a single frequency and can be represented as a sine wave, but most sounds are made up of complex tones containing many frequencies. Complex signals may be analyzed into their component sine waves (with characteristic frequency, amplitude and phase) to understand the content of the signal. This is known as frequency or Fourier analysis. A frequency spectrum indicates the principal frequencies contained within the signal, and their relative intensities (energies). When a sound is made up of frequencies covering most of the sound spectrum, it is known as white or broad-band noise. An impulsive signal (such as a door slamming or a hammer blow) is a single pressure pulse with a very fast rise time (around 25 ms or less) to the initial peak amplitude, followed by small pressure oscillations decaying over about 1 s.

The audible range of sound levels is enormous, from a quiet whisper up to the level of a warning siren (which is around the pain threshold), representing a range in power of the sound signal of over one to one billion as shown in Table 17.1. Similarly, the ear is sensitive to a large range of frequencies—the audible range is approximately 20–20000 Hz, with greatest sensitivity between 2–5 kHz. Most speech is between 300 and 700 Hz, with all vowel sounds below 1000 Hz, but sibilant consonants may be higher than 5000 Hz. Low frequency sounds below 20 Hz are usually termed infrasound. The effects of infrasound on human listeners are not yet well understood, and this range is not usually considered when assessing the auditory environment.

Definition of units

Several units are used for sound measurement, and it is helpful to give some brief definitions to show the relationships between the various basic measures. A fuller description of the measures can be found in texts such as Kohler (1984).

Sound is described by intensity and by frequency. (The corresponding sensations in human hearing are termed loudness and pitch.) Intensity or power of the oscillations in the air is measured in watts per metre^2 (W m^{-2}), while pressure is measured in Newtons per metre^2 (N m^{-2}); frequency is

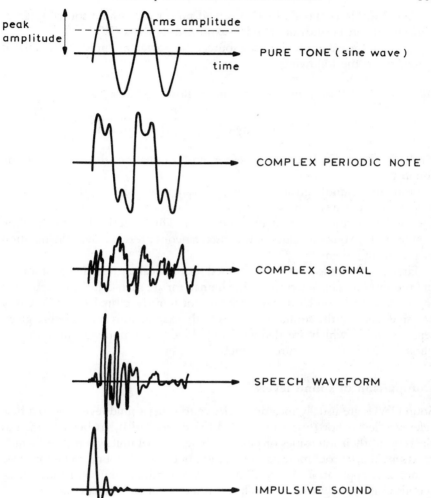

Figure 17.1. Waveforms of various sound signals

Table 17.1. Range of intensity and pressure of audible sound

Intensity (W m^{-2})	Pressure (N m^{-2})	Decibel level	
10^{-12}	0·00002	0	Hearing threshold
3 × 10^{-6}	0·04	65	Conversation
10^{-4}	0·2	80	Town traffic
10^{-2}	2	100	Workshop
3	36	125	Jet at take-off (60 m away)
100	200	140	Pain threshold

measured in Hertz (Hz). The total sound power of a sound source (such as a machine) can be measured and is expressed in watts (W).

In a sound wave emitted from a source, the power and pressure are related according to the following equation:

$$\text{Power/unit area} = \text{energy flow/unit area} = \frac{\text{pressure}^2}{\text{density of air} \times \text{speed of sound}}.$$

The sound power level is therefore proportional to the square of the sound pressure.

From the sound signals in Figure 17.1, it may be seen that the sound amplitude fluctuates rapidly over time, so the root mean square value (rms) pressure is the measurement normally used. This is derived from the time average of the squared values of the instantaneous pressures over the duration of the measurement.

These are definitions of the units used for physical measurements of sound, but the auditory characteristics of the human ear are such that the corresponding sensations of loudness and pitch are not linearly related to the intensity and frequency of the sound. Measures of these sensations are therefore given separate units (which are defined later): loudness is normally measured in phons, and pitch is measured in mels.

Comparison of sound levels

Sound levels are usually measured relative to other sound levels, or to a base reference level. For this, a unit called the decibel (dB) has been defined as the ratio of their intensities or powers. (The original unit was the bel, which represented a 10-fold increase in intensity, but this was found to be too large in practical applications.) The decibel is a logarithmic unit and thus corresponds quite closely to the non-linear response of the human ear.

The base reference sound intensity is chosen as the power of a standard vibration in the air which is just on the threshold of hearing, or one billionth of a watt per square metre (10^{-12} W m^{-2}). Thus, a sound level of N dB is related to its intensity of I W m^{-2} by

$$N \text{ dB} = 10 \log_{10} \frac{I}{I_0}$$

where I_0 is the reference level intensity at threshold of hearing (10^{-12} W m^{-2}).

Given the relationship between sound intensity and pressure,

$$N \text{ dB} = 10 \log_{10} \frac{I}{I_0} = 10 \log_{10} \frac{p^2}{p_0^2} = 20 \log_{10} \frac{p}{p_0}$$

where p is the sound pressure level and p_0 is the reference level amplitude at threshold of hearing (2×10^{-5} N m^{-2}).

The values of intensity, pressure and decibels for some typical sounds within the audible range are shown in Table 17.1. The relationships between the three scales are shown in Table 17.2. A sound which is 10 times louder than another has an intensity level of 10 dB relative to it. Since the decibel scale is logarithmic, a sound 100 times louder is said to be 20 dB louder. The smallest change in loudness that the human ear can discriminate is 1 dB, but in practice the minimum difference needed to recognize a sound above the background noise level is 3 dB, which represents a doubling in intensity or loudness. A doubling of the sound pressure level is equivalent to an increase of 6 dB in sound level.

Since the various units are related in this way (with loudness or sound power proportional to the square of sound pressure) a useful guide to changes in sound levels is given by:

$$+ \quad 3 \text{ dB} = 1\cdot4 \times \text{sound pressure level} = 2 \times \text{sound power level,}$$

$$+ \quad 6 \text{ db} = 2 \times \text{sound pressure level} = 4 \times \text{sound power level,}$$

$$+ 20 \text{ db} = 10 \times \text{sound pressure level} = 100 \times \text{sound power level.}$$

Effects of several sources of sound

What then happens when two sounds are heard at the same time? How are their effects added? Sound levels are not directly additive when they are measured in logarithmic dB units. If two sounds are received at the ear simul-

Table 17.2. Decibel scale

Ratio of intensities (powers) of two sounds		Ratio of pressures of two sounds	Relationship in decibels	
1		1	0	
10		3·16	10	
100		10	20	
	200			23
	400			26
	600			28
	800			29
1000		31·6	30	
10000		100	40	
100000		316	50	
1000000		1000	60	
1		1	0	
1/10		1/3	−10	
1/100		1/10	−20	

taneously, the resultant is the sum of the energies $(I_1 + I_2)$ in the two sounds and so the combined sound power level becomes:

$$\text{combined sound power level (W m}^{-2}) = I_1 + I_2.$$

Thus, the decibel level is:

$$\text{combined sound level (dB)} = 10 \log \frac{I_1 + I_2}{I_0}.$$

The combined loudness of sounds from two or more sources can therefore be calculated by using this formula, but a simpler rule of thumb (accurate to 0·5 dB) is also given in Table 17.3.

If the two sounds are of equal intensity (difference 0 dB), their combined intensity is 3 dB higher than the intensity of either. If the difference is 5 dB, the combined sound is 1 dB higher than the greater. If the difference is 10 dB, the combined sound is approximately equal to the louder sound and there is no noticeable difference in the intensity of the resultant sound. For example:

two sounds of 75 dB combine to give a sound level of 78 dB,
two sounds of 75 dB and 77 dB combine to give a sound level of 79 dB,
two sounds of 75 dB and 85 dB combine to give a sound level of 85 dB.

As a practical consequence of this, the sound of a machine or piece of equipment can be measured even in the presence of background noise, provided that the background noise level is at least 10 dB below the machine noise level. If the background noise level is higher than this, a correction factor for the measured machine noise level can still be calculated as shown previously. It is also useful to note that the ratio of two sound levels is calculated by subtracting the levels in decibels: this gives the signal-to-noise ratio as the difference in decibel levels of the desired signal and the unwanted noise.

Table 17.3. Addition of two sound levels in decibels

Difference (dB) between sounds	Add to higher (dB)
0	3
1	2·5
2	2
4	1·5
6	1
8	0·5
10	0

Measures of noise

It is obvious from this brief review that measurements of sound or noise levels will be influenced by the temporal, intensity and spatial characteristics of the signals. The temporal characteristics can include both the frequency spectrum and fluctuations in the overall sound level with time. A variety of measures have therefore been developed for specific purposes, and the most important are discussed below.

Subjective measures

Since the response characteristics of the ear are non-linear, both frequency and intensity affect our perception of loudness of a sound. We do not perceive a sound arriving at the ear to be equally loud at 20 Hz and at 10 kHz— a sound at 20 Hz appears to be very much quieter. Curves of equal subjective loudness can be plotted, giving the sound levels which provide constant perceived loudness at various frequencies and indicating the combinations which appear equally loud to a human listener. This shows, for example, that a 50 Hz tone must have a loudness of about 85 dB to give the same subjective loudness as a 1000 Hz tone at 50 dB.

Subjective loudness is measured in a unit called a phon. This unit is equivalent to the decibel at 1000 Hz and is also logarithmic. While the phon measures the subjective equality of sounds, a unit called the sone was defined to measure the relative loudness of sounds. One sone is defined as the loudness of a 1000 Hz tone of 40 dB (40 phons). A sound of 2 sones is one judged to be twice as loud. The phon and sone scales are related by the following formula:

$$\text{phons} = 40 + 10 \log_2 \text{sones}.$$

Every increase of 10 phons then doubles the loudness in sones, so that, for example, 50 phons is equivalent to 2 sones.

There is also a subjective measure of pitch, which corresponds to frequency for a pure sinusoidal tone and to the fundamental frequency for complex waveforms. Perceived pitch is given a unit called a mel—defined by a pure tone of frequency 1000 Hz at a sound pressure level of 60 dB, which is said to have a pitch of 1000 mels. Any two tones separated by a given number of mels appear equally far apart in pitch, regardless of their frequency.

Weighted measures

In view of the non-linearity of hearing response, most instruments are designed to measure sound levels on a decibel scale which is weighted to match the characteristic of the ear. Several scales have been developed for different purposes and their response characteristics have been standardized internationally. The A-, B-, and C-weighted scales were developed to match the responses for sounds of low, moderate and high intensity. The most commonly used is the dBA scale, which gives the best correlation with subjective tests of perceived loudness, and also with ratings of noise annoyance. The A-scale was in fact designed to match the 40 phon equal loudness contour.

This means that the actual measurement of the sound pressure level is converted to a weighted dBA sound level, in accordance with the response characteristic shown in Figure 17.2 which takes account of the middle range of frequencies to which the human ear is most sensitive. The other scales shown in Figure 17.2 are less commonly used. The B-scale was designed to match the equal loudness contour at 70 dB, and the C-scale for a flat rating across the frequency range. There are also D-weighted scales which were designed as equal noisiness scales for assessing aircraft noise. The unweighted sound levels are normally only used in frequency analysis.

Empirical measures

In most everyday situations noise contains sounds from various sources, of differing frequencies and intensities, and also extending over different periods of time. In order to describe the noise environment, various statistical distribution measures are used. These are based on noise levels measured on the dBA scale. The most important are defined as follows:

Equivalent level of sustained noise (L_{eq}). This is the average level of sound

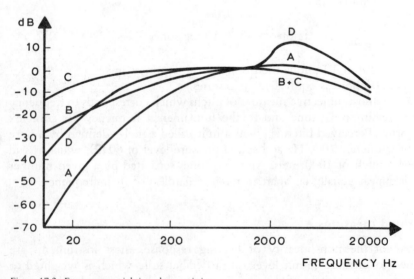

Figure 17.2. Frequency weighting characteristics

energy over a given period of time, which integrates all the fluctuating noises to represent them as an average steady level. It therefore takes account of short but high peaks.

Median noise level (L$_{50}$). This is the noise level which is exceeded for 50% of the time period.

Background noise level (L$_{90}$). This is the 10th percentile level—the level that is exceeded for 90% of the time.

Peak noise level (L$_1$ or L$_{10}$). This is the 99th or 90th percentile level—the level that is exceeded for 1% or 10% of the time.

All these indices can be used to give a measure of the total environment in, say, an office or a factory. The last three give a feel for the range of the noise—the 'average' level, the low background level and the highest levels heard over a period of time. They are measured by taking a continuous recording of the dBA level over a known time period and then performing a statistical analysis of the record.

Day–night equivalent level (L$_{dn}$). This is a 24-h L$_{eq}$ used for community noise exposure, where the value of L$_{eq}$ is measured over 24 h but the night-time readings between 22.00 and 06.00 hours are weighted with an increase of 10 dB.

Sound exposure level (SEL or L$_{AE}$). This is useful for comparing unrelated noise events in terms of their total acoustic energies: the energy is integrated over the duration of the event and expressed as the equivalent level over 1 s. It is often used to describe the noise energy of a single event such as a passing car.

Choice of instrumentation

In order to measure the auditory environment, instruments are needed to measure sound levels within the range of 0–150 dB and over the frequency range up to 20 kHz. A variety of types of instrument are used and the more generally used are described briefly below. Further information on measurement, calibration and analysis techniques can be found by consulting manufacturers' handbooks or by reference to texts such as Kohler (1984), Hassall and Zaveri (1979) or Peterson and Gross (1978).

Sound level meters

Sound level meters are the most commonly used instruments for measuring the acoustic environment and most general purpose meters contain several signal measures which are suitable for different applications. The basic measuring system is shown diagrammatically in Figure 17.3. At its simplest, the sound level meter consists of a microphone which converts pressure variations into electrical signals. These are then amplified by an electronic network and displayed in some form, either on a digital or analogue instrument, or as a

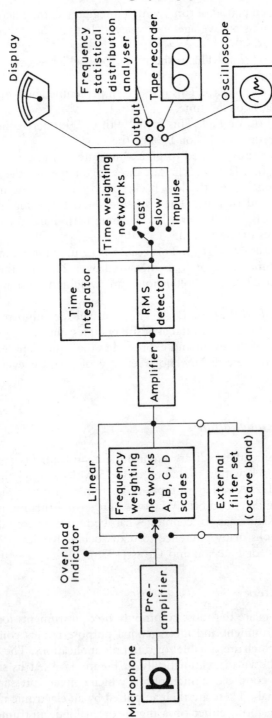

Figure 17.3. Diagram of a typical sound level meter

continuous reading which is recorded on tape, computer or hard copy. The sound level displayed is usually the instantaneous rms value of the signal in decibels.

Sound pressure level (SPL) meters normally have an appropriate frequency weighting network (with a response matching the A-, B-, C- or D-scale) in addition to providing the unweighted (linear) decibel level. When used with the A-weighted network, the response of the instrument is similar to that of the human ear. Portable filter sets giving octave band or 1/3 octave band frequency analysis are commonly used in conjunction with an SPL meter to give the frequency content of the sound. These record the sound level in each filter band. Figure 17.4 shows a sound level meter in use coupled to a filter set. In laboratory applications, wave analyzers may be used for more detailed narrow band frequency analysis.

Integrating sound level meters are capable of measuring over a longer time period and provide the average sound level and L_{eq} value. These integrate and average the sound over a period determined by the response characteristic of the instrument: for 200 ms on the FAST setting or 500 ms on the SLOW setting. The SLOW setting allows the user to read the overall sound level displayed even when the signal is fluctuating rapidly. The measured level will depend on the time weighting setting, which should of course be appropriate to the type of source and purpose of the measurement. Additional analysis of the temporal characteristics over a longer period of time may

Figure 17.4. Measurement of machine noise in the field

require an external statistical distribution analyzer. The noise signals may also be recorded on tape for further analysis at a later date, but in this case a reference signal level must also be recorded for calibration during analysis.

Impulsive sounds cannot be measured accurately on most normal SPL meters because the response time of the instrument is not sufficiently fast. On some meters a 'hold' facility permits a display of either peak level or maximum rms level, which is useful for measuring impulsive noise. Some meters may have impulse time weighting with a time constant of 35 ms, instead of peak hold. Detailed analysis of the waveform of impulsive sound would normally be made using an oscilloscope or external analyzers.

Sound level meters should also have an input amplifier overload indicator, for use when measuring on the weighted scales, because the difference between the weighted and unweighted levels can be several decibels. The reading displayed on the weighted scale (say dBA) may be below the limiting amplitude while the unweighted decibel level exceeds the maximum limit, and it is therefore possible to overload and clip the peak amplitude of the input signal.

Sound level meters require calibration before use, either by means of built-in calibrating networks or with an external acoustic calibrator held over the microphone. The calibrator is a reference sound source of accurately specified sound level. It is useful as a reference level when recording noise for later analysis.

Dosemeters

Dosemeters are small integrating sound level meters which are used to measure the total personal exposure to noise during a period such as a working day, and are usually carried in a pocket or attached close to the wearer's ear. These give the total A-weighted sound energy received during the measurement period, and may be used to calculate the L_{eq} value. The dose is often displayed as the proportion of the maximum permitted 8-h dose (usually 90 dBA). The instrument may also indicate whether a standard peak noise level has been exceeded.

Probe microphones

Probe microphones are used for measuring sound in the ear or other small spaces, for instance for evaluating the effect of hearing protectors. These are coupled to the sound level meter by means of a small probe tube, and it is necessary to apply a correction for loss of sound pressure in the tube. An alternative method of measuring the sound in earphones is to use an acoustic coupler (artificial ear), which is a standard cavity having an acoustic impedance similar to that of the ear.

Sound generation

Noise sources are often required in experimental work, for instance in the measurement of hearing ability or in testing hearing protectors. Recordings of speech and other sounds may be used, but other noise generators are available to produce single tones, continuous or intermittent signals. These can be played either through earphones or in free-field using loudspeakers. Pure tones can be generated using a system incorporating an oscillator, amplifier and attenuator. Complex tones can be generated by waveform generators. The frequency of the signal is controlled by an oscillator and the intensity by an attenuator, while the frequency range may be limited by the use of filters. Several types of noise are commonly used:

1. *Wide-band noise* containing a wide range of frequencies (within the bandwidth limits specified).
2. *Narrow-band noise* containing only a small range around a chosen frequency.
3. *White noise* containing all the audible frequencies and with an essentially flat spectrum.
4. *Pink noise* also containing the full range of audible frequencies, but weighted towards the lower frequencies so that the sound pressure spectral density is inversely proportional to frequency.

Impulse or impact noise can be generated by, for instance, using high voltage spark plugs or solenoid activated hammers. It is more difficult to produce impulsive sounds with specified characteristics. If recordings are used, the speaker is likely to 'ring', which extends the duration of the signal, while the signal itself may be distorted by limitations in the dynamic range of the speaker.

Analysis techniques

Detailed analysis of noise signals and environments can be complex and theoretical. A short review of the techniques can be found in Kohler (1984), and noise control handbooks should be consulted for the more technical details.

Amplitude analysis

A continuous reading from a SPL meter can be analyzed to give an amplitude–time plot, showing the variation in intensity over a long period. This may be used to compute statistical distribution parameters, such as median or background noise levels. The waveform of a signal over a short duration can be displayed by means of an oscilloscope.

Frequency analysis

Frequency analysis is used when detailed information is needed about a complex sound signal. It can help to isolate possible sources of noise in machinery or in a complex auditory environment. It is also used to evaluate the relative contributions of different frequency components when assessing the risk of damage to hearing. The most usual analysis is the plot of a frequency or power spectrum, which displays the sound energy across the range of frequencies. The frequency range can be split into frequency bands (usually one octave or one third octave wide) which are analyzed separately.

An octave is an interval which represents a doubling in frequency: the upper frequency bound is twice the lower frequency bound and the centre frequency of the band is taken as the geometric mean of the two bounds. The octave bandwidths used for industrial sound analysis are normally similar to those forming the musical scale, but the two sets of octave bands do not coincide. The centre frequencies of the two sets are:

Musical octaves   32   64   128 256 512 1024 2048 4096 Hz.

Industrial sound octaves
 37·5 75   150   300   600   1200 2400 4800 9600 19200 Hz.

When more accurate frequency analysis is required, one third octave bands can be used by logarithmically dividing each octave band into three. Narrow band analysis (using a spectrum analyzer) gives more detail. The bandwidth is usually defined as a fixed percentage of the frequency to be analyzed. For example, a 6% frequency band at centre frequency 500 Hz would have a bandwidth of 30 Hz covering the range between 485 Hz and 515 Hz. Filter sets are used in conjunction with SPL meters for octave band analysis in field use, but narrow band analysis would normally be performed using a recording of the noise.

Further analysis may be used to investigate the temporal characteristics of the noise. For instance, a spectrogram or sonogram is a three-dimensional representation of the frequency, intensity and duration of a signal. This is a plot of frequency against time for a signal or sample of relatively short duration (e.g., speech), in which the density (darkness) of the lines represents the intensity of the sound at each frequency.

Noise measurement procedures

Noise of individual machines or products may be measured in the laboratory or in an anechoic chamber, if precise measurements are required. For most purposes, however, this is not necessary and it is sufficient to use a quiet period in the workplace itself. Environmental measures obviously have to be carried out in the field and under normal conditions. The procedures adopted for measuring noise levels under these different conditions will be described separately.

Since the intensity of sound waves emitted from a noise source is attenuated with distance, noise level decreases with the square of the distance from the source. A measurement of sound level is therefore meaningless unless the location of the measurement is specified. This is normally chosen as the position or positions at which people are likely to be present.

Field surveys

A useful checklist of the procedures which need to be adopted in a field survey is given in Beranek (1971). There are obviously problems in measuring individual noise sources in a workplace or in a town environment, due to the presence of other noise sources. It may be possible to choose a quiet period at night or at the weekend, when very little other machinery is working, but it is unusual in modern environments to have no background sources of noise. Even at night there is likely to be traffic noise, and in the built environment there is frequently central heating, air conditioning or other plant operating. However, reference to Table 17.3 shows that this will have a small or negligible effect on the noise measurement, provided the noise level is at least 10 dB above the background level.

If the difference is less than 3 dB (in any frequency band), or if the source noise level is below that of the background, it cannot be measured reliably. If the difference is between 3 dB and 10 dB, the effect of the noise source may usually be calculated to an acceptable degree of accuracy by comparison with the background level measured with the source switched off. This is more difficult when background noises are intermittent or fluctuating, and in any field survey care has to be taken to monitor these during the measurement period.

Laboratory measurements

Background noise is likely to be less of a problem in the laboratory, but measurements there are liable to be affected by the structure of the enclosed space (by walls or other objects in the acoustic field). Sound may be reflected off hard surfaces (such as steel or concrete), or alternatively may be absorbed by fabric surfaces or acoustic tiles. When sound is reflected, sound level distribution in the space will depend on the phase relationships of the incident and reflected waves. This may increase noise levels or interfere with speech and other signals. In rooms, sound may be reflected several times from the walls and cause reverberation. (Reverberation time is the time taken for the resultant sound to decay.) A discussion of the effects of sound fields can be found in Beranek (1971) or Peterson and Gross (1978).

Briefly, in enclosed spaces, there are three distinct regions around a noise source:

1. *Near field* where there may be interference between the emitted sound wave and reflected waves, so that the sound level varies with slight changes in meter location. The near field extends over a distance approximately equal to the wavelength of the lowest frequency emitted or to twice the greatest dimension of the source machine (whichever is the greater).

2. *Far field,* which approximates to free-field conditions. The noise intensity decreases according to the inverse square law. This region can be identified by noting whether the sound level measurements obey this law.

3. *Reverberant field* close to reflecting surfaces, where the reflected noise is diffuse and levels are high. In this case, measurements of the source will not be accurate as the noise levels depend on the room geometry and absorption properties of the surfaces.

For accuracy, measurements should if possible be made in the far field region.

Two extreme conditions are used for laboratory measurements: the anechoic chamber and the reverberation chamber. In an anechoic chamber (such as the one shown in Figure 17.5), the surfaces are covered in highly sound absorbent material, so that there are no reflections or echoes, simulating the free-field conditions experienced outdoors. A reverberant room is one in which the walls completely reflect sound energy and where no walls or surfaces are parallel. This gives a diffuse sound field in which sound energy is

Figure 17.5. Anechoic chamber measurement of the noise level of a ship's whistle (Reproduced with permission from the Motor Industry Research Association, Nuneaton, UK)

uniformly distributed. The sound pressure at any point is an average value due to the many reflections. This is most suitable for measuring total power output of a noise source. Both the anechoic chamber and the reverberation chamber can be used to determine the sound power of a source, but an anechoic chamber can also be used to determine the reflectivity of surfaces.

Most measurements are in fact made in a semi-reverberant room—typical of most normal working conditions—where there is a mixture of direct and reflected sound, and it is necessary to ensure that the measurements are not made in the near field.

Location of microphone

The microphone and observer can both interfere with the acoustic field being measured by blocking or reflecting sound waves, causing significant errors which may be as large as 6 dB. If possible, the sound level meter should be left mounted on a stand. If it is handheld, the observer should hold it at arm's length.

In general, whether outdoors or in an anechoic chamber, a directional (free-field or frontal incidence) microphone is used and should be pointed towards the noise source. The alternative random-incidence microphones are designed to record sound from all directions and are normally used in reverberant or diffuse sound fields. However, it should be noted some test standards prescribe different procedures to these. If a random-incidence microphone is used in a free-field, its readings are most accurate when it is orientated at an angle of 70–80° to the source.

The microphone is normally located in a position representative of a hearer's ear, but without the person present. If the hearer is standing, ear height is usually assumed to be 1·5 m. In seated workplaces, the measurement might be taken at the height of the individual person's ear. The measurements must be made at all positions at which people may be exposed to the noise, and also take account of normal operating conditions and working practices. For instance, measurements of the noise levels of machinery should include locations used during activities such as maintenance.

In the working environment, it is sometimes necessary for the machine operator to be present while the noise measurements are made. The microphone must then be positioned as close to the operator's head as possible, while avoiding reflections from the head or body or absorption in clothing. This means that the microphone should be at least 50 mm from the side of the head (and preferably further away). Where miniature instruments are attached to the hearer's collar or helmet, or inside noise protectors, it is necessary to take account of such factors and to make corrections to the measured values.

Noise mapping

Noise mapping is useful for an initial survey of a complex environment such as a machine workshop. The noise levels in a working area or around a machine are measured systematically and plotted in the form of a noise map, as shown in Figure 17.6. Maps of this form can be used to identify zones of noise danger (particularly where people may be moving around), and to indicate where preventive measures are required. They can also be used in planning measures to isolate noise sources. Such noise maps are the auditory equivalent of the room illuminance assessment method for the visual environment, as described in chapter 15 of this book.

The noise levels should be measured at as many locations as are necessary to give accurate noise contours and plotted on a sketch of the layout of the machinery and workplace. The measurements should be taken at regular intervals, either as a grid covering the area of interest or around the circumference of a machine. If the sound field is complex, it will be necessary to take measurements at closely spaced intervals.

Accuracy of measurements

Regular calibration of instruments is essential to ensure the accuracy of measurements. A discussion of some of the methods used is given in Kohler (1984).

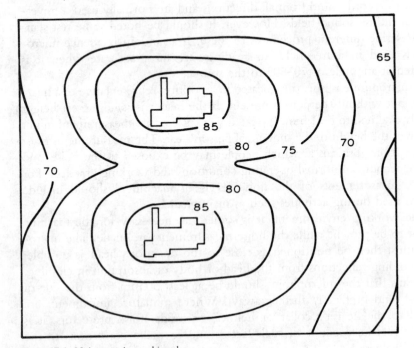

Figure 17.6. Noise map in machine shop

Outdoors several factors may affect the measurements. Wind is the most common problem, since this causes air turbulence and low frequency noise at the microphone. It is usual to shield the microphone with open-cell foam when taking measurements outside, although this is not fully effective at higher wind speeds. Wind, temperature and humidity can also affect the actual sound levels, as they change attenuation over distance and can create shadow zones. For a discussion of these atmospheric effects, and the attenuation produced by surroundings and barriers or walls, see Beranek (1971).

Identifying noise sources

Various methods can be used to identify or isolate specific sources of noise in a complex environment. Where possible the different components of the machine or environmental noise should be switched on or off separately and investigated in turn. Sound intensity analyzers are available with highly directional probe microphones capable of determining the direction of sound propagation and identifying noise sources, but simpler methods can also be used.

Potential sources can be isolated by surrounding them with a suitable absorbent material, such as lead sheet, which can easily be shaped around components. Alternatively, they can be shielded by barriers or temporary walls of attenuating materials such as boards sandwiched with acoustic foam.

In isolating noise sources it may be necessary to consider whether reflection or reverberation are contributing to the noise level. In some cases housing and panels may act as a sounding box, either through vibration or by reflecting noise from other sources. It may help in investigating these effects to use acoustic materials to fill the air spaces and damp the reverberations. Weights may be added to increase damping and test the effects of different components.

Standard test methods

The general principles of noise measurement have been outlined in the preceding part of this chapter, but standardized test procedures also have been developed to ensure accurate and comparable measurements for specific applications. In addition to international standards (ISO), there may be national and industrial standards.

ISO 2204 (ISO, 1979) is a guide to noise problems, covering the general procedures used in the measurement of noise and the evaluation of its effects on human beings. This document lists ISO standards which are applicable to specific problems. Current ISO standards dealing with acoustic measurements are also listed in Brüel and Kjaer (1987), which contains guidance on arrangements of instruments which are suitable for testing to these standards.

Procedures for measurement of vehicle or traffic noise are given in ISO

5130 (ISO, 1982a) and also in ISO 1996 (ISO, 1982b) which deals more generally with community response to noise. There is a British Standard BS 4142 (BSI, 1990) dealing with the measurement of environmental noise. Other applications of measurements of communication and annoyance effects are covered later.

The procedures adopted for measurement of noise emitted by machinery or other equipment depend on the purpose of the measurements, and particularly on whether this is to determine the sound power output of the machine or the sound pressure levels to which operators are exposed in the workplace. ISO 3740 (ISO, 1980) is a guide to the measurement of sound pressure levels of machines and equipment, which helps in deciding which methods should be used for different test environments (the more detailed methods being given in the subsequent series of standards ISO 3741 to ISO 3746, which are identical to those of BS 4196 (BSI, 1981)).

ISO 3744 (ISO, 1981a) covers testing in free-field conditions over a reflecting surface, which corresponds to many industrial workplaces. By this method, measurements are made over a hypothetical surface (defined as either a hemisphere or a rectangular parallelepiped) which envelops the noise source under investigation. These are used to calculate the sound power level of the source, and may also be used to compare machines or to rate equipment on its sound power output, or for prediction of the sound pressure level at any given location around the machine/source.

If the workplace is not free-field (for instance where it is close to a wall or surrounded by other machinery), the survey method of ISO 3746 (ISO, 1975) should be used to determine the weighted sound pressure level (SPL) at prescribed microphone positions close to the sound source. This may also be used to calculate the sound power level of the source, and is useful for rating the sound output of a source producing steady noise when it cannot be moved to an acoustic chamber for more accurate measurements.

Both ISO methods require large numbers of measurements, and some prescribed locations may be difficult to access on larger machines, especially in crowded machine shops. British Standard 4813 gives a simpler method for measuring the sound power level from machine tools (BSI, 1972). However, it cannot be used to estimate the total power output of a machine, and is less accurate for identifying highly directional noise sources. Following BS 4813, noise levels are measured at a height of 1·5 m and at intervals around the machine at a distance of 1 m from its surface. The microphone positions should not be more than 1·5 m apart, to obtain sufficiently detailed analysis, with a minimum of five measurements one of which should be at the point of highest sound level. These measurements can be used to plot a noise map of the environment, but can also be used to calculate the mean noise level at the 1 m distance. In addition a measurement should be taken at the operator's position, and at any other positions occupied by personnel for a significant period of the working day.

Measurement of hearing

A person's hearing ability is measured in terms of the minimum threshold of perception, using the normal techniques for threshold measurements such as the 'method of limits'. The measurements are usually made with an instrument called an audiometer, and the measurements presented in the form of an audiogram. 'Normal' thresholds have been established (see for instance ISO, 1990) and loss or impairment of hearing is usually defined as a minimum change of 10 or 15 dB in the threshold. Hearing level is taken as the amount by which the average threshold is raised in an individual, so that a positive hearing level represents hearing ability worse than the defined 'normal' level at any frequency. In an audiogram, the reference level of 0 dB represents the normal threshold of hearing (i.e., for people who have no hearing disability or age deterioration) at each frequency over the auditory range.

Audiometer

An audiometer consists of a calibrated oscillator and amplifier which presents pure tones over the range of audible frequencies. At each frequency, the amplifier can be scanned through the range of intensities when measuring the hearing threshold. The tones are normally presented through headphones, and the signal can be directed to one or other of the pair of earphones, so that the two ears can be tested separately. Two forms of audiometer are used: either manual or automatic (Bekesey), as described later.

Measuring hearing thresholds

The audiometer is operated by the experimenter who alters signal frequency and sound level. The hearing threshold is measured at discrete frequencies across the hearing range, usually as the average of the ascending and descending thresholds. The absolute threshold for a sound is the minimum level which can be detected by the subject, when presented in the absence of other sounds, so that measurements should take place in a very quiet room. The audiometer and operator should both be screened from the direct view of the subject, so that no cues are given to the presentation of the test signal. In order to avoid any influence from other sounds on the measurements of the hearing threshold, no words should be spoken to the subject and the subject should be asked to indicate the presence or absence of a noise by silent signal such as the movement of a finger.

Bekesey audiometers permit a semi-automatic version of the procedure, similarly based on the 'method of limits'. The sound level is varied automatically, decreasing while the subject holds down a switch (which controls the direction of the motor driving the attenuator), and increasing when it is released. The subject is asked to hold the switch pressed down only as long as the sound is audible, and therefore maintains it at threshold level. A continuous trace of the hearing level is recorded at each frequency, and the mean value indicates the auditory threshold. Frequency is scanned across the range in a programmed sequence so that the tone is maintained at each frequency for a short period to establish the threshold.

Whichever technique is used, a period should be allowed before the measurements are made for the subject to adapt to the low noise level and recover from any temporary threshold shift due to previous exposure to noise. The US regulations on occupational noise exposure (OSHA, 1983) specify that the first baseline audiograms taken in a hearing conservation programme should be taken after at least 14 h without exposure to workplace noise, although subsequent annual audiograms are permitted at any time during the working day.

Sounds may be presented through speakers instead of earphones, but there may be some differences in the measured threshold (perhaps up to 6 dB) due to the differences in the acoustic field. According to Loeb (1986), the threshold in free-field conditions is likely to be lower than when presented through earphones, and the binaural threshold is likely to be lower than the monaural threshold.

Measuring masking thresholds

Although absolute hearing threshold is measured in a quiet environment, it is often important to know the hearing threshold for a signal or for speech in a noisy environment. This is termed the masked threshold, which defines the threshold of detection (not intelligibility) against the background level of noise. The effect is measured by determining the absolute threshold of the sound when presented alone, then measuring it in the presence of the masking sound. The amount of masking is then defined as the difference in decibel level by which the threshold of audibility is raised above the absolute threshold.

The probability that a signal will be detected increases with the level of the signal above the background. So the masked threshold is sometimes defined as the level at which there is a given probability (say 75%) of correct detection of the signal. This is not an absolute measure since it depends on factors such as the level of expectancy of the hearer and the rise time, duration and temporal shape of the signal. For a detailed discussion of the factors involved, the reader may consult texts such as Sorkin (1987).

Assessing the risk of hearing damage

Although a single violent sound can cause damage to the ear-drum, this is rare and hearing loss is commonly caused by long-term exposure to noise. The exposure does not need to be continuous, since the effects of intermittent exposure are cumulative. The main indicator used to assess hearing damage is a change in the hearing threshold of an individual. If this is measured before and after exposure to loud noise the threshold shift may be determined by the difference in the two thresholds. More usually hearing loss is assessed by the drop below the population norm. In industrial hearing conservation programmes, the operators at risk are assessed at the start of their employment and changes in hearing are monitored by annual tests (OSHA, 1983).

Temporary threshold shift (TTS) is a short-term and reversible change experienced after exposure to loud noise, which may persist for minutes or hours depending on the exposure. Since recovery starts as soon as the noise ceases, it is necessary to specify the time at which the TTS is determined. This is normally measured 2 min after exposure.

Noise-induced permanent threshold shift (NIPTS) is the long-term effect of exposure to noise and is not reversible.

The mechanisms of TTS and NIPTS are not necessarily identical, but Kryter (1985) has suggested that the TTS of a group of workers after 8-h exposure can be used as a criterion for the risk of long-term damage.

Risk of hearing damage has to be assessed in terms of duration of exposure as well as of noise level. ISO Standard 1999 (ISO, 1990) gives a risk table in relation to age, duration of exposure and intensity of noise, which shows that intensities above 90 dBA seriously risk hearing damage.

Since industrial workers are exposed to noise which may vary considerably over a working period, it is important to calculate the noise levels over the whole period of exposure (usually over an 8 h working day). The noise dose is calculated from the measured L_{eq} value (equivalent level of sustained noise). In measuring this, it should be remembered that operating conditions, especially communication signals, may contribute to the total noise dose. When this occurs in occupations, such as aircraft pilots, where the signals are presented through headphones, the noise levels can be recorded at the ear, if necessary using a miniature or probe microphone, and the noise dose can be obtained by later analysis of the recording (Glen, 1976).

According to Davies and Jones (1982), most standards apply the 'equal-energy principle', which suggests that exposure to higher intensities can be permitted for short periods providing that the total energy within an 8 h period does not exceed the normally permitted dose (as in ISO 1999 (1990)). This allows for quiet periods during the working day, but does not take account of recovery which may occur in these periods. For example, the equal energy principle assumes that a continuous 4 h exposure followed by 4 h quiet has the same effects as four 1 h exposures to the same level of noise separated by quiet periods of 1 h. On this assumption, halving the duration of exposure permits the sound level to be increased by 3 dB. The American OSHA Standard (OSHA, 1983) uses a less conservative 5 dB halving rule.

For a discussion of assessment of occupational noise-induced hearing loss and standards for noise exposure limits, the reader should consult Kryter (1985) or Davies and Jones (1982). Kryter includes comments on the methodology of audiometry and an extensive review of the evidence for the relative influence of presbycusis, sociocusis and nosocusis (physiological ageing, exposure through activities of everyday living, and damage through disease or trauma, respectively) on hearing thresholds.

The risks from the complex effects of impulsive noises are reviewed in both Loeb (1986) and Kryter (1985), who quote damage risk contours giving an indication of acceptable maximum peak SPL values for impulsive noises of different durations and repetition frequencies.

Measurement of effects of noise on comfort and performance

People vary enormously in their responses to noise, being influenced by factors such as intrusion into privacy or whether the sound is intermittent or unexpected, and probabilistic or statistical measures are needed to make any assessment of the annoyance and performance effects of noise. Some of the reasons for individual differences are discussed by Jones and Davies (1984). People can adapt well to a continuous background of sound, and experiments therefore have to be designed carefully to include an adequate degree of realism in the experimental conditions. A corollary to this is that measurements of the level of annoyance may be due to factors other than the noise levels which are the subject of the investigation: noise annoyance may just be a symptom of poor morale or stress.

Annoyance

Many different measures are used to assess noisiness, annoyance and intrusiveness, some specific to environmental, community or transport noise. It is important to realize that loudness or noisiness cannot be equated with annoyance and investigators have found large differences in assessments of noise sources judged on these criteria (Kryter, 1985). Various aspects of perceptions of noisiness have been studied: judgements of noisiness or loudness of the neighbourhood, dissatisfaction with present noise level, frequency with which annoyance is felt, degree of annoyance, and interference with activities (Jones and Davies, 1984). These should be carefully distinguished in any survey which is conducted, since measures taken simply to reduce noise levels may not be appropriate to reduce the level of annoyance.

The annoyance level of noise is frequently measured by means of surveys and questionnaires, but it must be remembered that the wording of questions or of instructions given to the subjects can have a considerable influence on the responses and ratings. Advice on the design of appropriate rating scales may be found in Loeb (1986) (and in chapter 3 of this book in general terms).

Noise annoyance is obviously a multi-factorial problem. For example, at least thirteen primary and derived measures have been used to assess community noise (Sanders and McCormick, 1992). A few of the evaluation methods which can be used are discussed below.

Many indices have been used to quantify disturbance caused by fluctuating noise (e.g., noise rating (NR), noise and number index (NNI), traffic noise index (TNI), noise pollution index (NPI)), and these are discussed in Loeb (1986) and Kryter (1985). These indices may often be recorded on noise level analyzers, which perform statistical analyses of noise over long duration measurement periods.

Kryter (1985) developed scales of annoyance or noisiness (in units called noys) which have been used for measures of aircraft and traffic noise. These were based on experimentally derived equal annoyance curves (pN dB) analogous to the equal loudness curves (phons). However, annoyance has additional psychological, social and economic dimensions so that Kryter noted that the threshold of perceived annoyance can vary with previous exposure, location indoors or outdoors, time of day, and the impulsive or startle characteristics of the noise.

Schultz (1978) produced a dosage response curve as a predictive tool. This relates noise exposure (in terms of a day–night average sound level L_{dn} in dBA) to the level of community annoyance, based on large scale surveys from various countries (mostly related to transport noise). Others suggest that short duration noises can be particularly annoying and have proposed measures using peak levels of noise events, as for aircraft overflights (Fidell, 1984). Other special measures have been developed to evaluate aircraft noise exposure (ISO, 1978), and a discussion of some of these may be found in Ollerhead (1973).

Preferred noise criteria (PNC) curves were introduced as design criteria for background noise in offices, rooms or halls (Beranek, 1971). They were based on the speech interference level (SIL), which is a measure of speech intelligibility and is described in the next section. The noise criteria curves are a set of arbitrary sound spectra for steady noises, serving as a reference for rating noise environments. They are also useful in deciding where in the frequency spectrum the greatest benefits could be obtained by noise reduction treatments. PNC curves are based on the need for acceptable speech communication, and specify that the loudness in phons should not exceed SIL by more than 22 units. The spectrum of the measured background noise is plotted over the noise criteria curves, and the noise is rated by the number of the curve which equals or just exceeds the noise spectrum at any point. This rating is then compared to a table of recommended ratings to evaluate the suitability of the noise level for the particular room environment.

It is not only loud noises that can be annoying. The effects of intrusiveness of low-level (or infrequent) noise exposure have been investigated by Fidell and others (Fidell *et al.*, 1979; Fidell and Teffeteller, 1981). They found that the L_{eq} measure is insensitive to these types of noises and have developed methods of predicting the level of intrusion or annoyance from measures of signal detectability.

Speech intelligibility

Noise can mask speech, but the degree of masking depends on various factors since there is a large measure of redundany in speech, especially when the hearer is familiar with the subject matter. Thus, the assessment of speech intelligibility is not a simple question of measuring the signal-to-noise ratio. Speech intelligibility can be measured most directly by listening tests, presenting different types of standardized material, such as nonsense syllables, phonetically balanced word lists which contain all speech sounds (or phonemes), and sentences. The A-weighted decibel scale is normally used to measure sound level in these applications.

Several methods have been developed to predict speech intelligibility from physical measures of the speech signal, noise environment and task parameters (e.g., speaker–hearer distance). Three of the methods are given here, and discussions of their relative merits can be found in Kryter (1985) and other textbooks.

1. *The Articulation Index* (ANSI, 1969) predicts the likelihood of difficulties with speech communication, and is calculated from the differences between the SPLs of the speech and masking noise in various frequency bands. These are weighted according to their importance in the intelligibility of speech. Originally, 20 bands were used between 250 and 7000 Hz, but the articulation index is now calculated from octave or one third octave analysis. The signal and noise levels are measured in each band and the differences weighted and summed to give the articulation index.

The relationship between the articulation index and intelligibility has been determined experimentally for different types of spoken material, and is expressed in a series of graphs in the ANSI Standard which indicate the percentage of syllables, words or sentences which will be correctly understood. Corrections are given in the standard to take account of factors such as reverberation or noise interruption. An indication of the difficulty of speech communication is given from the following values of articulation index:

< 0·4	difficulties likely
0·4–0·7	some difficulties may occur
> 0·7	good speech communication possible

2. *Speech Interference Level* (SIL) is a simpler method (also specified in ANSI (1969)) which does not require direct measurement of the speech level. The SIL is calculated from the mean of noise SPL for octave bands centred at 500, 1000, 2000 and 4000 Hz. Again, this value is compared with tables relating it to speech intelligibility for different voice efforts (normal, raised, shouting) and for varying distances between speaker and listener. Preferred speech interference level is a similar measure which is related to the maximum distance over which speech communication is possible (ISO, 1974).

3. *Direct Measurements of SPL* (in dBA) have also been used as an index of speech interference, predicting SIL. Loeb (1986) suggests a relationship of

SIL = SPL − (9 or 10 dB).

Webster (1979) produced a chart relating dBA, voice effort and speaker–hearer distance to quality of communication, so that either SPL or SIL could be used in assessing this.

Design of warning and information signals for safety and audibility

The effects of masking on warning and alarm signals are discussed in Webster (1984), who gives a method for predicting the level at which a pure tone signal will be audible. However, additional criteria are needed to ensure that the signal is also attention-getting and recognizable. The perception of masked signals varies with frequency, signal-to-noise ratio and absolute level of background noise.

Four main factors need to be taken into account in designing auditory warnings and signals: audibility, noise dose, startle and discriminability (Coleman *et al.*, 1984). Reports from Patterson and Milroy (1979) and Coleman *et al.* (1984) describe procedures which have been used to design the appropriate sound level for auditory warnings.

The first stage in this process is the measurement or prediction of the masked threshold (and modelling techniques for this are mentioned later). As described by Coleman *et al.*, 'The method employed . . . involved the tape recording of the particular background noise and a detailed narrow-band spectral analysis followed by computerized calculation of masked thresholds, incorporation of a 97·5th percentile absolute threshold criterion, and a graphical output showing the complete design window with its constituents.'

This 'design window' defines the upper and lower boundaries of frequency and intensity within which the signal components must lie in order both to be audible and to attract attention without causing startle or risk of hearing damage. Figure 17.7 shows how several factors are considered in specifying the 'design window' for a warning signal. This is constructed from knowledge of the hearing abilities of the workforce, the nature of the masked threshold due to the background noise and the levels above threshold which are necessary for the signal to be audible without causing startle. It is also possible in this to take account of the attenuation effects when using hearing protectors.

As seen in Figure 17.7, the thresholds may also be set for different population criteria: the maximum level to avoid startle is determined for people with normal hearing, while the minimum effective signal level is set for the people in the population who have the lowest hearing ability.

Standards and guidelines for noise exposure

International and industry-wide standards have been drawn up for assessing noise levels in different environments. The effects of noise on performance in different types of task are difficult to define, since they are task specific,

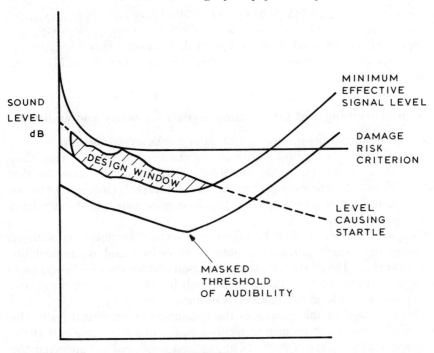

Figure 17.7. 'Design window' method for specifying signals to be used in noisy environments (after Coleman *et al.*, 1984)

and must often be assessed by experimentation. However, general guidelines can be found in most ergonomics textbooks.

Evidence of the risk of hearing damage was presented in ISO (1990), showing clearly that exposure to noise of over 90 dB for 8 hours a day will produce significant deafness, and that the potential for damage increases with the frequency of the noise source. It is now generally recognized that a risk of hearing damage exists at lower levels of noise and most countries have established regulations on maximum acceptable levels for industrial workplaces. The 1986 Directive of the European Community (Commission of the European Communities, 1986) aimed to reduce noise levels to below 80 dBA by 1994. The Directive defined three action levels at which employers must introduce progressive administrative and engineering controls to protect their employees. The Occupational Safety and Health Administration (OSHA, 1981, 1983) in America has established noise exposure limits for both continuous and intermittent noise at workplaces, in terms of a noise dose related to the time-weighted average sound level over an 8 hour working day. The requirements are again based on three action levels.

Some standards set an upper limit for peak SPL (140 dB is often used for impulsive noise exposure), and in America exposure to any steady noise level is not permitted above 115 dBA (OSHA, 1983). However, there is less consensus on standards for exposure to impulsive sound or to infrasonic and ultrasonic frequencies. A review of the research is given in Kryter (1985).

Assessment of personal protectors

Hearing protectors have to be assessed for their effect on the audibility of communication signals, as well as for the attenuation they provide.

Evaluating attenuation

Noise attenuation curves are normally measured by the manufacturers and are supplied with hearing protection equipment. The attenuation may be measured by methods such as ANSI S3.19-1974 (ANSI, 1974). The American Environmental Protection Agency (1979) requires manufacturers to test hearing protectors to this standard and to label them with a noise reduction rating (NRR). The NRR is calculated from 1/3 octave band analysis of measured attenuation. This can then be used to estimate the noise exposure (in dBA) for the wearer: either as the workplace sound level in dBC less NRR, or as the sound level in dBA plus 7 dB less NRR (OSHA, 1983).

Sutton and Robinson (1981) review various other procedures which can be used to estimate the level of protection given to a wearer (usually defined as the reduction in dBA at the ear) from a knowledge of the frequency response of the protector and the frequency spectrum of background noise. The most accurate method of calculating the protection provided is given in ISO 1999 (ISO, 1990), where the attenuation (minus a variance correction) is subtracted from the workplace sound levels measured with octave band analysis. It should be noted, however, that these methods indicate the protection provided under optimum conditions and do not allow for poor fitting of the protector or noise leakage due to hair or other factors. Various studies have shown that the methods overestimate the actual protection when measured in workplace conditions (Berger, 1983; Lempert, 1984).

A subjective method of measuring attenuation (by threshold shift) can also be used, as outlined in ISO 4869 (ISO, 1981b). A minimum of 10 subjects should be tested (using normal audiometric techniques), although more should be used if possible. The subjects must have a hearing threshold level in either ear which is not worse than 15 dB at frequencies below 2 kHz and not worse than 25 dB at frequencies above 2 kHz. The test signals specified are pink noise filtered through 1/3 octave bands from 63–8000 Hz. This method can be used to compare or rank different models of protectors, and to evaluate design features which may affect performance.

The Health and Safety Executive (1990) Noise Guide then gives a method of calculating the 'assumed protection' from the attenuation values obtained from testing a group of subjects. The 'assumed protection' is taken as the mean attenuation at each frequency minus the standard deviation over the tests (probably of the order of 5 dB), in order to allow for the variation in protection between wearers.

Dummy heads and miniature microphones can also be used to measure attenuation, although the results may be expected to differ from subjective measurements because of the effects of bone conduction leakage. A semi-objective technique, in which a miniature microphone is attached to a subject's ear, can be used to investigate the effects of hair, glasses or helmets with different models of hearing protector.

Evaluating audibility of communications

Speech communication needs to be considered when assessing the effectiveness of hearing protectors. This can be done in terms of the articulation index, speech interference level or by direct experimental testing of intelligibility. Although the articulation index generally assumes normal hearing, it can be calculated against reduced hearing thresholds, and a measure of AIIHA (articulation index incorporating hearing ability) is suggested (Coleman *et al.*, 1984). Coleman *et al.* also propose a technique for selecting protectors, which takes into account the interaction between the attenuation characteristics and hearing ability, the range of noise and the speech conditions.

Simulation and modelling of the acoustic environment

A few mathematical models of the auditory environment have been developed to assist ergonomics analyses. Three such simulations are described here related to the design of auditory warning signals, the reduction of noise levels in factories and the acceptability of traffic noise in the urban environment. (The reader should also refer to chapter 8 on modelling and simulation.)

Models of the auditory filter have been used to demonstrate the effects of masking of signals in a noisy environment, and to predict the audibility and discriminability of signals used in the workplace (Patterson *et al.*, 1982; Coleman *et al.*, 1984). The models assume that the ear acts as a set of bandpass filters, and predict the masked thresholds of signals from the measured spectrum of the noise environment by modelling the characteristic of the auditory filter centred on the signal frequency to be heard, then integrating the noise power within the filter passband.

Shield (1980) describes a computer model called Noiseshield, which was developed to predict noise levels during the design of a factory. This uses information on the layout of the area (dimensions of the room and construction materials of walls, floor and roof), and location and sound power output of all machines and other noise sources. The package can also be used to evaluate noise problems in existing factories, by investigating the effects of treatments such as erecting barriers or enclosures around machines, or the introduction of sound absorbing materials on floor or walls.

Two approaches to modelling road noise have been reported by Clayden *et al.* (1975), who described the early development of a model to predict traffic noise levels in the urban environment given a plan of the buildings and the locations of vehicles, and by Nelson (1973) who used a model to predict the temporal distribution of noise from freely flowing traffic (using speed, traffic mix, and distance from the road).

Concluding remarks

It is apparent from the wide variety of techniques introduced in this chapter that one of the most important considerations in any noise assessment is the choice of appropriate criteria for the evaluation. It is rarely sufficient simply to measure the sound pressure level or acoustic power at a single location, since the effects of noise on hearers are considerably more complex. Both perception and individual responses or preferences are usually influenced by the social or occupational context in which the sound is heard as well as by the auditory characteristics of the human ear.

When assessing a noisy environment, ergonomists may need to consider the effects on safety, distraction, transmission of information, annoyance or pleasure for different groups of hearers. These are evaluated in different ways and a range of criteria may be needed to assess the environment. It is hoped that the brief discussion in this chapter will give some guidance on the most appropriate methods to choose for assessments of community, transport and industrial environments.

References

ANSI (1969). *American National Standard for Calculation of the Articulation Index.* ANSI S3.5. (New York: American National Standards Institute).

ANSI (1974). *American National Standard Method for the Measurement of Real-ear Protection of Hearing Protectors and Physical Attenuation of Ear-muffs.* ANSI S3.19-1974. (New York: American National Standards Institute).

Beranek, L.L. (1971). *Noise and Vibration Control.* (New York: McGraw-Hill).

Berger, E. (1983). Using the NRR to estimate the real world performance of hearing protectors. *Sound and Vibration,* **18**, 5, 26–39.

Brüel and Kjaer (1986). *Noise Control, Principles and Practice.* (Naerum, Denmark: Brüel and Kjaer).

Brüel and Kjaer (1987). *Acoustic Measurements According to ISO Standards and Recommendations.* (Naerum, Denmark: Brüel and Kjaer).

BSI (1972). *Method of Measuring Noise from Machine Tools Excluding Testing in Anechoic Chambers.* BS 4813. (London: British Standards Institution).

BSI (1981). *Sound Power Levels of Noise Sources.* BS 4196. (London: British Standards Institution).

BSI (1990). *Method for Rating Industrial Noise Affecting Mixed Residential and Industrial Areas.* BS 4142. (London: British Standards Institution).

Clayden, A.D., Culley, R.W.D. and Marsh, P.S. (1975). Modelling traffic noise mathematically. *Applied Acoustics,* **8,** 1–12.

Coleman, G.J., Graves, R.J., Collier, S.G., Golding, D., Nicholl, A.G.McK., Simpson, G.C., Sweetland, K.F. and Talbot, C.F. (1984). *Communications in Noisy Environments.* Technical Memorandum TM/84/1 (EUR P.74). (Edinburgh: Institute of Occupational Medicine).

Commission of the European Communities (1986). *Council Directive of 12 May 1986 on the Protection of Workers from the Risks Related to Exposure to Noise at Work.* Council Directive 86/188/EEC. Official Journal of the European Communities, NO L 137/29, 24 May 1986.

Davies, D.R. and Jones, D.M. (1982). Hearing and noise. In *The Body at Work,* edited by W.T. Singleton (Cambridge: Cambridge University Press), pp. 365–413.

Environmental Protection Agency (EPA) (1979). Noise labeling requirements for hearing protectors. *Federal Register,* **42**(190), 56139–56147.

Fidell, S. (1984). Community response to noise. In *Noise and Society,* edited by D.M. Jones and A.J. Chapman (Chichester: John Wiley), pp. 247–277.

Fidell, S. and Teffeteller, S.R. (1981). Scaling the annoyance of intrusive sounds. *Journal of Sound and Vibration,* **78,** 291–298.

Fidell, S., Teffeteller, S.R., Horonjeff, R.D. and Green, D.M. (1979). Predicting annoyance from detectability of low-level sounds. *Journal of the Acoustical Society of America,* **66,** 1427–1434.

Glen, M.C. (1976). The contribution of communications signals to noise exposure. *Applied Ergonomics,* **7,** 197–200.

Hassall, J.R. and Zaveri, K. (1979). *Acoustic Noise Measurements* (Naerum, Denmark: Brüel and Kjaer).

Health and Safety Executive (1983). *100 Practical Applications of Noise Reduction Methods* (London: HMSO).

Health and Safety Executive (1990). *Noise at Work—Noise Assessment, Information and Control. Noise Guide No. 5 Types and Selection of Personal Ear Protectors* (London: HMSO).

ISO (1974). *Assessment of Noise with Respect to Its Effect on the Intelligibility of Speech.* ISO Technical Report TR 3352-1974 (Geneva: International Standards Organisation).

ISO (1975). *Acoustics—Determination of Sound Power Levels of Noise Sources—Survey Method.* ISO 3746 (Geneva: International Standards Organisation).

ISO (1978). *Acoustics—Procedure for Describing Aircraft Noise Heard on the Ground.* ISO 3891 (Geneva: International Standards Organisation).

ISO (1979). *Guide to International Standards on the Measurement of Airborne Acoustical Noise and Evaluation of its Effects on Human Beings.* ISO 2204 (Geneva: International Standards Organisation).

ISO (1980). *Acoustics—Determination of Sound Power Levels of Noise Sources—*

Guidelines for the Use of Basic Standards and for the Preparation of Noise Test Codes. ISO 3740 (Geneva: International Standards Organisation).

ISO (1981a). *Acoustics—Determination of Sound Power Levels of Noise Sources— Engineering Methods for Free Field Conditions over a Reflecting Plane.* ISO 3744 (Geneva: International Standards Organisation).

ISO (1981b). *Acoustics—Measurement of Sound Attenuation of Hearing Protectors—Subjective Method.* ISO 4869 (Geneva: International Standards Organisation).

ISO (1982a). *Acoustics—Measurement of Noise Emitted by Stationary Road Vehicles—Survey Method.* ISO 5130 (Geneva: International Standards Organisation).

ISO (1982b). *Acoustics—Description and Measurement of Environmental Noise.* ISO 1996 (Geneva: International Standards Organisation).

ISO (1990). *Acoustics—Determination of Occupational Noise Exposure and Estimation of Noise-induced Hearing Impairment.* ISO 1999 (Geneva: International Standards Organisation).

Jones, D.M. and Chapman, A.J. (1984). *Noise and Society* (Chichester: John Wiley).

Jones, D.M. and Davies, D.R. (1984). Individual and group differences in the response to noise. In *Noise and Society,* edited by D.M. Jones and A.J. Chapman (Chichester: John Wiley), pp. 125–153.

Kohler, H.K. (1984). The description and measurement of sound. In *Noise and Society,* edited by D.M. Jones and A.J. Chapman (Chichester: John Wiley), pp. 35–76.

Kryter, K.D. (1985). *The Effects of Noise on Man,* 2nd edition (London: Academic Press).

Lempert, B.L. (1984). Compendium of hearing protection devices. *Sound and Vibration,* **18**, 26–39.

Loeb, M. (1986). *Noise and Human Efficiency* (Chichester: John Wiley).

Nelson, P.M. (1973). *A Computer Model for Determining the Temporal Distribution of Noise from Road Traffic.* TRRL Laboratory Report 611 (Crowthorne: Transport and Road Research Laboratory).

Ollerhead, J.B. (1973). Noise: how can the nuisance be controlled? *Applied Ergonomics,* **4**, 130–138.

OSHA (Occupational Safety and Health Administration) (1981). Occupational noise exposure; hearing conservation amendment. *Federal Register,* **46**, 4078–4179.

OSHA (Occupational Safety and Health Administration) (1983). Occupational noise exposure; hearing conservation amendment; final rule. *Federal Register,* **48**, 9738–9783.

Patterson, R.D. and Milroy, R. (1979). *Existing and Recommended Levels for Auditory Warnings on Civil Aircraft.* Civil Aviation Authority Contract Report (Contract No 7D/S/0142). (Cambridge: Medical Research Council, Applied Psychology Unit).

Patterson, R.D., Nimmo-Smith, I., Weber, D.L. and Milroy, R. (1982). The deterioration of hearing with age: frequency selectivity, the critical ratio, the audiogram, and speech threshold. *Journal of the Acoustical Society of America,* **72**, 1788–1803.

Peterson, A.P.G. and Gross, E.E. (1978). *Handbook of Noise Measurement,* 8th edition (Concord, MA: General Radio Company).

Sanders, M.S. and McCormick, E.J. (1992). *Human Factors in Engineering and Design,* 7th edition (New York: McGraw-Hill).

Schultz, J. (1978). Synthesis of social surveys on noise annoyance. *Journal of the Acoustical Society of America,* **64,** 377–405.

Shield, B.M. (1980). A computer model for the prediction of factory noise. *Applied Acoustics,* **13,** 471–486.

Sorkin, R.D. (1987). Design of auditory and tactile displays. In *Handbook of Human Factors,* edited by G. Salvendy (New York: John Wiley), pp. 549–576.

Sutton, G.J. and Robinson, D.W. (1981). An appraisal of methods for estimating effectiveness of hearing protectors. *Journal of Sound and Vibration,* **77,** 79–91.

Webster, J.C. (1979). Effects of noise on speech. In *Handbook of Noise Control,* 2nd edition, edited by C.M. Harris (New York: McGraw-Hill).

Webster, J.C. (1984). Noise and communication. In *Noise and Society,* edited by D.M. Jones and A.J. Chapman (Chichester: John Wiley), pp. 185–220.

Chapter 18

Human responses to vibration: principles and methods

Rosemary A. Bonney

Introduction

Vibration can affect people's health, comfort and performance. The methods used to assess the effects of vibration include objective measurement, subjective measurement and mathematical modelling. This chapter defines the principles of hand/arm and whole body vibration as well as outlining the methods used to evaluate the effects they may have on the individual. It is not intended as a full and deep review of the subject; for those requiring more information the reader is referred to detailed texts such as Brammer and Taylor (1982), Dupuis and Zerlett (1986) and Wasserman (1987).

General principles

Vibrations are mechanical oscillations produced by regular or irregular movements of a body about its resting position. The direction in which this motion occurs must be defined in terms of its three orthogonal components:

'x'—front to back,
'y'—side to side,
'z'—up and down.

Both hand/arm and whole body vibration measurements are obtained with respect to internationally agreed biodynamic coordinate systems, as defined by ISO 2631 (ISO, 1985). (See Figure 18.1.)

Once the direction of the motion has been defined, its other characteristics, in terms of repetition rate or frequency (number of cycles per unit time) and amplitude (distance displaced from resting position) must also be specified (see Figure 18.2). The frequency (in hertz) is the number of times a body vibrates in a specific time period, normally 1 s. Amplitude is the maximum

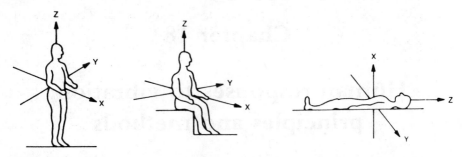

Figure 18.1. International vibration coordinate system for a standing, sitting and lying person (ISO 2631, 1985)

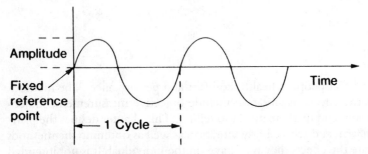

Figure 18.2. Frequency (number of cycles/unit time) and amplitude for sinusoidal vibration

amount a point of interest is displaced from its central position. Since achievement of a given amplitude at a chosen frequency requires a certain acceleration (which, of course, implies a certain maximum velocity), a waveform can be defined in terms of frequency and acceleration alone. Acceleration is more often used in place of amplitude to define a waveform, not least because the acceleration can be measured more readily than amplitude or velocity in most practical cases.

Many structures which have elastic properties will tend to vibrate freely at a particular frequency, called the natural frequency. If vibration is applied to a structure at or near this frequency then it will resonate, or vibrate at an amplitude greater than that applied to it. At other frequencies the opposite of resonance occurs so that the body absorbs or reduces the input intensity. This is known as damping or attenuation.

The human body is extremely complex, composed of organs, bones, joints and muscles. Each of these parts can be affected in the ways described above. At some frequencies therefore, they might vibrate at greater amplitudes than the vibration applied to them and at others they may absorb and attenuate the inputs. It is evident that input vibrations are rarely pure sinusoids, but can have complex and varying wave forms, where a number of waves are superimposed on each other. Due to the variations in natural frequency and damping of various parts of the body, the transmission of such inputs to regions remote from where they are initiated will add a further layer of complexity. Nevertheless, resonance can have such a powerful effect that the resonant characteristics of a body system can be measured. This is done by comparing the vibration intensity of the system at the point of stimulation and at a point of exit. Resonance is said to occur when the output/input amplitude ratio is greater than unity. At the same time, the phase shift between the input and output waveforms can give some indications of possible loadings on the intervening structures.

The two areas where the effects of vibration on the human body have been given most attention are hand/arm vibration and whole body vibration. When a person is exposed to either, changes in health, comfort and performance will be noticed as a reaction to the stress imposed. However the principles and methods used to identify these effects are very different due to the nature of the different body parts affected.

Hand/arm vibration

Principles

Vibration of hand and arm body segments is caused primarily by the use of vibrating hand tools. Blood circulation in the fingers can be affected, causing bone and joint deformities and muscle and nerve problems.

The most widely documented symptom is called *vibration white finger* (VWF) or Reynaud's phenomenon. VWF can first be identified by a slight tingling or numbness in the fingers. Later the tips of one or two fingers may suffer from attacks of blanching particularly in the morning and in cold conditions. The fingers affected and distribution of damage are initially dependent on the tool type and the method by which it is held. VWF attacks usually last less than 1 h and are ended by a red flush and pain, when the blood rushes back to the finger tips. During an attack, sensitivity to touch, pain and temperature are reduced.

The time taken for VWF to appear depends on the level of exposure for the individual. For very high levels of vibration this may be only a few months, however for most occupations it is usually 5–10 years. The period elapsing before symptoms appear is called the *latent* interval. VWF as well as being extremely painful, can reduce finger sensitivity and manual dexterity. Thus work requiring fine finger movement may no longer be possible. The symptoms can arise sooner where the exposure to vibration occurs in the cold, and can be exacerbated by smoking, hence any investigation should control for these factors. The symptoms can be brought on by cold, which makes it socially handicapping and many outside activities can be unpleasant and difficult. Employees may be forced to change their jobs as a result of the disease and at the moment there is no universal cure. Some remission from early stages of the disease has been noted on people who have been withdrawn from exposure (Riddle and Taylor, 1982).

Of the population, 10% exhibit VWF symptoms, even though they are not actually exposed to a vibrating environment (Griffin, 1982). This makes control and diagnosis of the disease more difficult. Recommendations as to the acceptable limits of vibration exposure below which the disease is unlikely to occur, are provided in ISO 5349 (ISO, 1986) (see Figure 18.3).

Equipment and processes which are regularly associated with the causes of VWF are:

(a) hand fettling of castings and forgings,
(b) grinders or other hand-held rotary tools,
(c) pneumatic hammers, drills and other oscillating tools,
(d) chain-saws, and
(e) motor cycles.

These equipments vibrate at frequencies between 5–1000 Hz, the range of frequencies within which VWF is most likely to occur. The actual risk resulting from use of these tools depends on the length of exposure, the

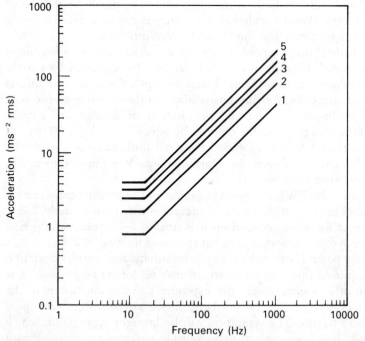

Figure 18.3. Acceptable limits of vibration exposure of the hand (ISO 5349, 1986). Curves 1 to 5 refer to multiplying factors, associated with exposures per 8 hour shift of 4–8, 2–4, 1–2, ½–1, and up to ½ hour respectively

thermal environment in which they are used, and how well the tools have been maintained, as well as on the physical state of the user.

Practical assessment techniques for vibrating hand tools

Where vibrating hand tools are being used the ergonomist may be called in to identify whether there is a risk of VWF, quantify its severity and make recommendations for design improvements if necessary.

The first requirement is to use a reliable method for assessing the incidence of VWF amongst the work-force. Two observational methods will be described here, and for further information the reader should refer to Griffin (1990, chapter 14).

One of the established methods in widest use is that of Taylor and Pelmear. Several variants have been proposed by them, and others, over the years but the sequence of stages given in Table 18.1 remains substantially the same. The method indicates the stages through which the disease passes, and the values for a given individual are obtained by questioning rather than direct observation. The values set down are based on the conditions of the fingers as given in the left hand side of the table, complications which may exist are not incorporated. The right hand side of the table lists probable effects, and they are given as an aid to correct stage identification.

A later proposal is that by Griffin (1982) whose categorization system deals solely with symptoms experienced in the subject's hands (Figure 18.4). Although the concept was to record, and score, finger blanching, as shown in Figure 18.4, Griffin (1990, p. 574) points out that the same scoring system

Table 18.1. Stage assessment for vibration-induced white finger (Taylor and Pelmear, 1975)

Stage	Condition of Digits	Work and Social Interference
00	No tingling, numbness or blanching of digits.	No complaints.
0T	Intermittent tingling.	No interference with activities.
0N	Intermittent numbness.	No interference with activities.
0TN	Intermittent tingling and numbness.	No interference with activities.
01	Blanching of one or more fingertips with or without tingling and numbness.	No interference with activities.
02	Blanching of fingers beyond tips, usually confined to winter.	Slight interference with home and social activities. No interference at work.
03	Extensive blanching of digits. Frequent episodes summer as well as winter.	Definite interference at work, at home and with social activities. Restriction of hobbies.
04	Extensive blanching, most fingers. Frequent episodes summer and winter.	Occupational changed to avoid further vibration exposure because of severity of signs and symptoms.

Note: Complications are not used in this grading.

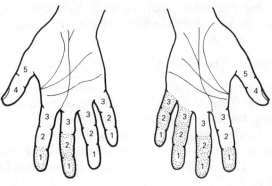

Right Hand

Digit	Th.	1	2	3	4
Possible Score	4+5	1+2+3	1+2+3	1+2+3	1+2+3
*Actual Score	0	1	3	0	0
Total Score	4/33				

Left Hand

Digit	Th.	1	2	3	4
Possible Score	4+5	1+2+3	1+2+3	1+2+3	1+2+3
*Actual Score	0	1	3	6	6
Total Score	16/33				

(Note T (tingling) or N (numbness) may be inserted as appropriate
where score = 0)
*Example   TOTAL SCORE : 4_R, 16_L

Maximum Score on
Either Hand   *Possible Action*

1   Vibration-exposed person to be regularly warned
           of problem and informed of possible
           consequences

3   Vibration-exposed person to be advised not to
           continue with vibration work

5   Vibration-exposed person to be removed from
           vibration work

9   Compensation of vibration-exposed worker

Possible Use of Proposed Categorisation of Vibration-induced White Finger
(The action appropriate to each score should be defined by the
appropriate authorities: the actions listed are only intended to
illustrate the possible use of the proposed method of categorisation)

Figure 18.4. Categorization system proposed for vibration-induced white finger, with actions of worker warning and protection (Griffin, 1982)

can be used to record the extent of tingling, numbness, etc. The progress or regression of symptoms can be recorded by this method, and it can easily be fitted into computerized records. Griffin recommends that this technique should be used only by an occupational physician.

The text below the diagram illustrates a hypothetical example of how the scores may be used to define appropriate initial remedial action. Whether the proposed actions are appropriate for the given scores must await further field studies. It must also be emphasized that these methods, and those which follow, are designed to identify the presence of hazardous vibrations and are not clinical diagnoses which define treatment. This latter activity is for the medical practitioner, who should also be involved where any case of VWF is suspected. The aim of the ergonomist is to move forward from the identification of the hazard to its reduction or elimination.

An assessment by Taylor and Pelmear (1975) of a number of clinical tests for VWF proposed that the Renfrew Ridge Aesthesiometer gave the most reliable separation between people with VWF and those not exposed to vibration. A development of the device, using a rising ridge around a disc was developed by Corlett *et al.* (1981) (see Figure 18.5). This provided closer control over experimental errors. Studies by Marsh (1986) suggest that increased discrimination can be achieved by using as a measure the degree of rotation of the disc between the just noticeable appearance of the ridge and the 'just noticeable disappearance' of the ridge.

The other most generally well-considered method suitable for field use is a vibration aesthesiometer. This has its detractors, in that concern is expressed about the level of adaptation experienced by the fingers which may bias the measurements.

Apart from these objective measures, discussion with the workers concerned will provide information on symptoms and when they occur, which will give additional evidence for identification. Methods of working, characteristics of tools and equipment and the circumstances under which they are used, as well as the various uses to which they are put, will give the investigator knowledge to help in recognizing the likely paths for the vibration transmission. Such directions of enquiry will increase the opportunity to understand the particular situation and help to reduce the effects of precon-

Figure 18.5. Adapted Renfrew Ridge Aesthesiometer used for detecting VWF

ceptions and over-simplifications, such as can all too frequently occur if the investigator is unfamiliar with the practical working activities.

Recording the vibration from tools is not just a matter of screwing accelerometers onto the handle. Many tools create shocks during use, presenting overloads which can cause a DC shift in the record, and which can be misinterpreted as a low-frequency vibration. It is customary to interpose a resiliant pad between accelerometer and tool, which acts as a mechanical filter to vibrations over 2000–3000 Hz. Care must also be taken to see that the mounting does not have its own natural frequency which superimposes onto the tool frequencies.

Piezoelectric accelerometers are most commonly used for measuring tool vibration, together with the appropriate recording equipment. Where tape recording of the results is being used, the response of the recorder must be sufficient for the purpose, and this normally means using a high quality one. If piezoresistive accelerometers are used, it is necessary to employ damped accelerometers; the undamped version is too easily damaged by shock. Some workers mount the accelerometer on a plate which is gripped by the operator, seeking to record the vibrations transmitted to the hand. Rasmussen (1982) reported studies of various forms of such mountings which gave transmissibility values close to unity, i.e., the transfer function was flat over the range of frequencies of interest for hand/arm vibration.

The intensity of a worker's grip on the tool is an important aspect; grip is almost always necessary for control and the accelerations should be recorded at least in the direction in which the grip by the fingers is being exerted. It is the transmission of vibration to the fingers which is of major importance, and both frequency and acceleration are required. The measure of acceleration is related to the energy present in the system.

Apart from measuring the vibration characteristics, the extent of exposure to vibration must also be recorded. Activity sampling (discussed as 'occurrence sampling' in chapter 2 on direct observation) to give the proportion of the working period over which the person is exposed may be sufficient, but it may be desirable also to record the lengths of work and rest periods. Especially where working sessions are long and breaks short, it is advisable to maintain the record of actual work and rest intervals, since a general statement of percentage exposure per day could be misleading.

When investigating hand/arm vibration the objective is usually to determine if the levels to which the person is exposed will induce VWF during a working lifetime. Current practice is to specify how to assess the energy input to the hand rather than just limiting values of frequency and acceleration. The method is described in an appendix to ISO 5349 (ISO, 1986).

Where exposures cannot be reduced to the recommended levels and engineering changes cannot reduce the severity of the vibration, recourse is usually made to limiting total operating times. Brammer (1982a,b) proposed a method for estimating the latent interval, and standard deviation of that interval, for a given level of the frequency weighted rms. component acceleration. This latter value is the one defined in the ISO appendix (ISO, 1986) and represents the energy input to the hand. The method does provide some measure which is of use in the design of new products as well as being useful for those concerned with the improvement of current practices.

For test purposes various ways of mounting vibration recording equipment, particularly on chain-saws, have been proposed, which may be specified for standards purposes. These methods endeavour to record vibration without any confounding by the variability introduced by the human operator, and they are of particular value when assessing the effects of various designs of vibration isolators on, for example, chain-saw handles (Reynolds and Wilson, 1982).

Results obtained through using the techniques described provide the data for assessing the level of exposure against the appropriate standard, which in this case is ISO 5349 (ISO, 1986).

Recommendations to reduce the effects of hand/arm vibration

Some general (non-medical) measures to reduce the effects of hand/arm vibration are recommended in ISO 5349 (ISO, 1986), which is summarized as follows:

Technical recommendations

Measure vibration in all three orthogonal directions on the handles (or gripped parts of tool or component) and compare with ISO recommended levels. If they exceed these levels:

1. Reduce the vibration by anti-vibration handle/mountings, or where appropriate, eliminate any out-of-balance components in the equipment.
2. Where grinding on pedestal grinders is concerned, improve the wheel dressing to reduce wheel irregularities and introduce softer wheels.
3. Improve the maintenance of equipment.
4. Ensure that the tools *are* appropriate for the task.
5. Where rotary hand-held grinders are in use, provide flexible suspension to reduce the forces needed to manipulate the tool.

Advice to the employer

1. Ensure that adequate training and instruction are provided in the proper use of the tool.
2. The worker should let the tool do the work, grasping it as lightly as is possible consistent with safe work practice and tool control.

3. Vibration hazards are reduced when continuous vibration exposure over long periods is avoided. Therefore work schedules with adequate rest breaks are essential.
4. The worker should use the tools only when absolutely necessary.
5. Investigate for the incidence of VWF at least annually.

Advice to employee

1. Wear adequate clothing to keep body temperature at an acceptable level, and wear gloves whenever practically possible during use of the tools.
2. Before starting a job warm the hands, and thereafter keep them as warm as possible.
3. Reduce smoking when using vibrating hand tools; smoking can act as a vasoconstrictor, i.e., it reduces blood supply to the fingers.
4. Should signs of tingling, numbness or blueness in the fingers occur, see a physician and possibly seek a change of job.

Whole body vibration

Principles

Unlike hand/arm vibration there is no particular injury which is due specifically to *whole body vibration* (WBV). However, it is recognized that exposure can be detrimental to health, comfort and performance and that methods should therefore be sought to minimize WBV effects. Vibration can be sensed in different parts of the body across a very wide range of frequencies, from 0·1–10000 Hz (see Figure 18.6). However, it is generally agreed that human sensitivity to WBV is greatest around 4–8 Hz in the 'z' (up and down) direction and 1–2 Hz in the 'x' (front to back) and 'y' (side to side) directions. Furthermore, there is strong epidemiological evidence to suggest that there is an increased level of low-back pain and gastrointestinal problems amongst people who are exposed to WBV at these frequencies for long periods of time (Heliovaara, 1987; Hulshof and van Zanten, 1987; Kelsey, 1975; Rosegger and Rosegger, 1960).

Human reaction to WBV has been studied extensively, mainly in relation to transportation industries; these studies have been aimed primarily at improved comfort and performance for crew and passengers. Other studies have investigated the actual amount of vibration reaching the operator through the seat surface for land vehicles, aircraft and ships (e.g., Coermann, 1960; Lovesey, 1975).

It is difficult to establish simple and acceptable principles for measuring exposure to vibration since the acceptable level of exposure is very much dependent on the environment within which the operator is exposed. Fur-

Frequency (Hz)

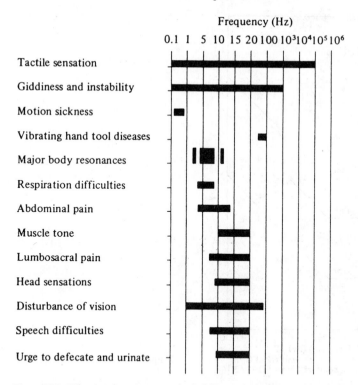

Figure 18.6. Vibration frequencies at which different physiological effects occur

thermore, most criteria which seek to limit vibration exposure are based on studies using sinusoidal vibration. ISO 2631 (ISO, 1985) defines the limits of exposure to vibration in both the 'z' and 'y' directions (1–80 Hz range) in terms of three criteria:

(a) preservation of health, *exposure level* (EL),
(b) working efficiency, *frequency decreased proficiency boundary* (FDPB), and
(c) comfort, *reduced comfort boundary* (RCB).

The levels for each criterion are defined in terms of the maximum time for which an operator should be exposed to vibration. The levels range from 1 min to 24 h. For any exposure time:

$$EL = 2FDPB \text{ and}$$

$$RCB = FDPB/3 \cdot 14 \text{ (see Figure 18.7).}$$

Oborne (1983) has criticized the 1978 version of ISO 2631 (which is very similar to the 1985 amended version) and claimed that the experimental evi-

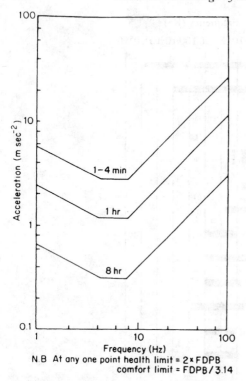

Figure 18.7. ISO acceptable limits of whole body vibration exposure for working efficiency

dence for it is lacking, both in quantity and quality. The standard however is in wide use, giving guidelines for exposure limits which should not be exceeded.

Practical assessment techniques for whole body vibration

Most situations of exposure to WBV occur whilst people are in a seated position. As with hand/arm vibration it is necessary to record the amplitudes and accelerations of the vibrations to which the person is exposed, together with the periods of exposure. It is also important to note the posture of the person, since it has been found that the way the person sits in, say, a driving seat can affect the amount of vibration reaching the upper part of the body (Wilder *et al.*, 1982). Posture changes and time spent in different postures should be noted during observation.

Major groups who are exposed to WBV, and on whom much research attention has been focused, are vehicle drivers and pilots. In the following discussion the vehicle driver will be used as the example.

As the seat itself may be flexible, it is necessary to measure its transmissibility. This is the ratio of the acceleration appearing at the seat surface to that imposed on the floor to which the seat is attached. By running a series of tests using accelerometers attached to the floor and to a plate on which the subject sits, and gathering data from a sequence of frequencies, the natural frequency of the loaded seat will be found. This is the imposed frequency for which the transmissibility ratio is greatest. Where the ratio is less than unity, the seat is providing damping.

As the body is a complex organism, the various segments of which have their own natural frequency and damping characteristics, it is not sufficient to assume that the frequency at the seat surface is experienced all over the body. If the head under vibration is being investigated, a lightweight bite bar can be used, carrying a triaxial accelerometer.

Basic instrumentation for obtaining vibration measurement is:

an accelerometer,
a preamplifier,
a multichannel FM tape recorder, and
a Fourier analyzer.

Vibration will be measured using accelerometers, which produce a voltage directly proportional to the acceleration value of the motion. Piezoelectric accelerometers are the most commonly used because they are small, light and have a wide operating frequency range from about 0·2 Hz to several kilohertz. Most accelerometers are designed to measure vibration in one direction only; thus one accelerometer is required for each of the directions of motion being measured. For the seated subject of our example, to measure the vibration at the seat surface, a triaxial accelerometer can be used embedded in the centre of a hard rubber disc and positioned where the operator's buttocks would be located. Before recording, each accelerometer must first be calibrated.

Once the accelerometers have been attached they must then be connected to a preamplifier so that the vibration signal can be recorded with a magnetic tape recorder. The data can then be analyzed using commercially available systems (for example, as supplied by Brüel and Kjaer) designed to display the predominant frequencies and amplitudes of the particular waveform. This method of separating a vibration signal into a number of individual components is called *Fourier* or *spectral analysis*. By comparing the results to ISO 2631 it should be possible to identify whether the level of vibration to which the subject is being exposed is too high. It should be emphasized that to obtain the frequencies and amplitudes of vibrations at different points on a human body is by no means as easy as these simple descriptions might make it appear, and the experimental controls are complex. In addition the equipment required for vibration analysis is costly and sophisticated. Further advice should be sought before embarking upon such experiments and analyses. A discussion of the measurements related particularly to heavy transport vehicles will be found in SAE J1013 (SAE, 1980).

In addition to measuring the actual amount of vibration to which the operators are exposed, it is also important to question them about any discomfort they may feel as a result. Use of subjective measures can provide useful information about the environment which would not be available if only objective measures were used. Whitham and Griffin's (1978) semantic scaling technique may be used as an indicator to describe those areas of the body which are most severely affected by the vibration environment. Rating scales, reviewed by Oborne (1978), can be used to find acceptable tolerance levels within the constraints of the system to which the operator, in this case the driver, is exposed. The response to the subjective questions, however, depends on the interviewee's expectations; for example, an army tank driver would obviously have different expectations of what is comfortable compared with a driver of a family saloon car.

It is known that poor postures and heavy lifting, as well as vibration, are related to back injury problems. Thus in any vibration assessment it is important to consider not only the level of vibration to which the drivers are exposed but also the constraints placed on them within the confines of their equipment, as well as the tasks carried out before and following driving. Only in this way will important parameters relating to the problem be properly identified.

Recommendations to reduce the effects of whole body vibration

1. Where feasible, for example, for rail travel, reduce vibration at source by minimizing the undulations of the surface over which the vehicle must travel.

2. Reduce the transmission of vibration to the driver by improving vehicle suspension, and altering the position of the seat within the vehicle.

3. Decrease the amount of vibration to which the driver is exposed by reducing the speed of travel, minimizing the exposure period, and increasing the recovery time between exposures.

4. Modify the seat and control positions to reduce the incidence of forward or sideways leaning of the trunk; maintain back rest support. Eliminate awkward postures due to difficulty of seeing displays or reaching controls.

Methods for evaluating the effects of vibration on performance have not been mentioned. Body parts tend to vibrate in sympathy with vibrating machinery; vibration can therefore reduce motor control, producing hand unsteadiness, or can cause oscillations of the eyeballs resulting in difficulty in focusing. There is little evidence to suggest that vibration can affect central information processes.

Conclusions

The overall objectives of this chapter have been: first, to summarize the principles and methods for evaluating the effects of vibration in situations where exposure to the human operator is an unnecessary and unexpected problem, and secondly, to provide some recommendations as to how the effects might be reduced. This has been done for both hand/arm and for whole body vibration.

Methods employed to assess vibration environments are dependent upon whether we are interested in effects in terms of whole body or hand/arm vibration. For both situations a major problem with exposure to vibration is that often neither work-force nor management realizes the potential dangers and they do not take the recommended precautions at the right time. The overall aim of vibration assessment should be to determine the extent of vibration exposure, and to compare this with recommended limits, but in any case to redesign equipment and jobs such that exposure is minimized in terms of degree and time.

Acknowledgements

Thanks are due to Malcolm Page and Ken Parsons for their comments on this chapter.

References

Brammer, A.J. (1982a). Relations between vibration exposure and the development of vibration syndrome. In *Vibration Effects on the Hand and Arm in Industry,* edited by A.J. Brammer and W. Taylor (New York: John Wiley), pp. 283–290.

Brammer, A.J. (1982b). Threshold limit for hand–arm vibration exposure throughout the work day. In *Vibration Effects on the Hand and Arm in Industry,* edited by A.J. Brammer and W. Taylor (New York: John Wiley), pp. 291–301.

Brammer, A.J. and Taylor, W. (Eds) (1982). *Vibration Effects on the Hand and Arm in Industry* (New York: John Wiley).

Coermann, R.R. (1960). The passive mechanical properties of the human thorax-abdomen system and of the whole body system. *Journal of Aerospace Medicine,* **31**, 915–924.

Coermann, R.R. (1962). The mechanical impedance of the human body in sitting and standing position of low frequencies. *Human Factors,* **4**, 227–253.

Corlett, E.N., Akinmayoa, N.K. and Sivayoganathan, K. (1981). A new aesthesiometer for investigating Vibration White Finger (VWF). *Ergonomics,* **24**, 603–630.

Dupuis, H. and Zerlett, G. (1986). *The Effects of Whole-Body Vibration* (Berlin: Springer-Verlag).

Griffin, M.J. (1982). *The Effects of Vibration on Health.* Memorandum 632. (University of Southampton: Institute for Sound and Vibration Research).

Griffin, M.J. (1990). *Handbook of Human Vibration* (London: Academic Press).

Heliovaara, M. (1987). Occupation and risk of herniated lumbar intervertebral disc or sciatica leading to hospitalization. *Journal of Chronic Diseases*, **40**, 259–264.

Hulshof, C. and van Zanten, B.V. (1987). Whole-body vibration and low back pain: a review of epidemiological studies. *International Archives of Occupational and Environmental Health*, **59**, 205–220.

ISO (1985). *Guide for the Evaluation of Human Exposure to Whole Body Vibration*. ISO 2631. (Geneva: International Standards Organisation).

ISO (1986). *Guidelines for the Measurement and Assessment of Human Exposure to Hand Transmitted Vibration*. ISO 5349 (Geneva: International Standards Organisation).

Kelsey, J.L. (1975). An epidemiological study of acute herniated lumbar intervertebral discs. *Rheumatic Rehabilitation*, **14**, 144–159.

Lovesey, T. (1975). The helicopter: some ergonomic factors. *Applied Ergonomics*, **6**, 139–149.

Marsh, D.R. (1986). Use of a wheel aesthesiometer for testing sensibility in the hand. *Journal of Hand Surgery*, **11-B/2**, 182–186.

Oborne, D.J. (1978). The stability of equal sensation contours for whole body vibration. *Ergonomics*, **21**, 651–658.

Oborne, D.J. (1983). Whole body vibration and International Standard ISO 2631: a critique. *Human Factors*, **25**, 55–69.

Rasmussen, G. (1982). Measurement of vibration coupled to the hand–arm system. In *Vibration Effects on the Hand and Arm in Industry,* edited by A.S. Brammer and W. Taylor (New York: John Wiley), pp. 89–96.

Reynolds, D.D. and Wilson, F.L. (1982). Mechanical test stand for measuring the vibration of chain saw handles during cutting operation. In *Vibration Effects on the Hand and Arm in Industry,* edited by A.S. Brammer and W. Taylor (New York: John Wiley), pp. 211–224.

Riddle, H.F.V. and Taylor, W. (1982). Vibration-induced white finger among chain sawyers nine years after the introduction of anti-vibration measures. In *Vibration Effects on the Hand and Arm in Industry,* edited by A.S. Brammer and W. Taylor (New York: John Wiley), pp. 169–172.

Rosegger, R. and Rosegger, S. (1960). Health effects of tractor driving. *Journal of Agricultural Engineering Research*, **5**, 241–275.

SAE (1980). *Measurement of Whole Body Vibration of the Seated Operator of Off Highway Work Machines*. SAE J1013. SAE Handbook Volume 4 (Warrendale, PA: Society of Automotive Engineers, Inc.).

Taylor, W. and Pelmear, P.L. (Eds) (1975). *Vibration White Finger in Industry*. (London: Academic Press).

Wasserman, D.E. (1987). *Advances in Human Factors/Ergonomics,* Volume 8, *Human Aspects of Occupational Vibration*. (Amsterdam: Elsevier).

Whitham, E.M. and Griffin, M.J. (1978). The effects of vibration frequency and direction on the location of areas of discomfort caused by whole body vibration. *Applied Ergonomics*, **9**, 231–239.

Wilder, D.G., Woodworth, B.B., Frymoyer, J.W. and Pope, M.H. (1982). Vibration and the human spine. *Spine*, **7**, 243–254.

Chapter 19

Anthropometry and the design of workspaces

Stephen T. Pheasant

Introduction

Anthropometry is the branch of the human sciences which deals with body measurements—particularly those of size, shape and body composition. Biomechanics is the application of mechanical principles to the study of the structure and function of the human body. As applied to ergonomics, the two are closely linked, since the science of biomechanics commonly provides the *criteria* for the application of anthropometric data to the problems of design.

This chapter is concerned with anthropometry and the methods for using anthropometric data. For the latter it will be necessary to consider criteria, but a fuller discussion of biomechanics will be found in chapter 24. For the wider applications of these subjects to design, the reader is referred to Chaffin and Anderson (1992), Clark and Corlett (1984) and Pheasant (1986).

Anthropometric data

The variability of most bodily dimensions may be described, to a tolerable degree of accuracy, by a mathematical function known as the *normal distribution*. This name does not imply that it describes the distribution of 'normal people'—whatever that might mean anyway—but rather as meaning 'the distribution which you will find most useful in practical affairs'. To deal with this slight semantic difficulty, it is sometimes referred to as the Gaussian distribution, after Johann Gauss (1777–1855), the German mathematician and physicist who first explored its mathematical properties in detail. Gauss's concern in this respect was with random errors in physical measurement. The fact that bodily characteristics such as stature (i.e., standing height) are also normally distributed is an empirical observation due to the English anthropologist and geneticist Sir Francis Galton (1822–1911). Geneticists have in fact contrived certain plausible mathematical models to account for the normal distribution of anthropometric characteristics, but the details of these need not concern us here.

The distribution of stature in adult British men is shown in Figure 19.1. Frequency is plotted vertically and stature is plotted horizontally. Beneath the horizontal axis is a second scale showing percentiles of stature. In any particular characteristic, *n*% of the population concerned are smaller than the *n*th percentile, i.e., the *n*th percentile is the value which is exceeded in (100 − *n*)% of cases. Note that the percentiles are close together in the centre of the distribution and widely spread in the tails.

The curve is symmetrical about the mid-point. The 50th percentile value is also the most common value and the frequency declines systematically as you enter the tails of the distribution.

Figure 19.2 shows exactly the same data, plotted in a slightly different way. This is the cumulative form of the normal distribution, sometimes known (because of its shape) as the *normal ogive*. Plotted horizontally is stature; plotted vertically are percentiles of stature. This is a particularly useful way of presenting the data for our present purposes because it allows us to read off directly the percentage of people who will be *accommodated* (i.e., satisfied) with respect to a particular anthropometric criterion. To take a slightly trivial example, Figure 19.2 tells us the percentage of men who would be able to pass under a doorway of a given height, without running the risk of banging their heads.

The normal distribution is completely described by two parameters, the mean and the standard deviation. The mean is the same thing as the familiar

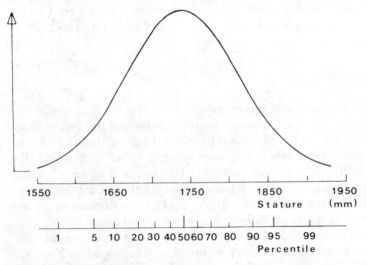

Figure 19.1. The normal distribution of the stature of British men. After Pheasant (1986)

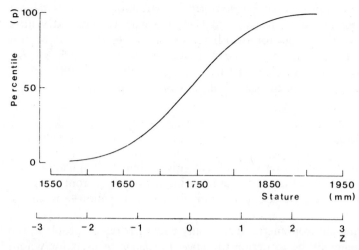

Figure 19.2. The normal distribution of the stature of British men, plotted in cumulative form. After Pheasant (1986)

arithmetical average and for normally distributed variables it is equal to the 50th percentile. The standard deviation is a measure of dispersion; it describes the extent to which an individual might be expected to differ from the mean. So we might say, for example, that the mean stature of a carefully selected sample of high-jumpers was greater than the mean stature of the population at large, but that their standard deviation was less.

If the mean (*m*) and the standard deviation (*s*) of a normally distributed variable are known, then any percentile (X_p) which we might happen to require may be calculated from the following equation:

$$X_p = m + s\,z$$

where *z* (the standard normal deviate) is a factor for the percentile concerned. Values of *z* for some commonly used percentiles (*p*) are given in Table 19.1

Table 19.1. Selected values of *z* and *p*

p	*z*	*p*	*z*
1	−2·33	99	2·33
2·5	−1·96	97·5	1·96
5	−1·64	95	1·64
10	−1·28	90	1·28
15	−1·04	85	1·04
20	−0·84	80	0·84
25	−0·67	75	0·67
30	−0·52	70	0·52
40	−0·25	60	0·25
50	0·00	50	0·00

which is the cumulative version of the normal distribution plotted in tabular form. British men have a mean stature of 1740 mm with a standard deviation of 70 mm, whilst the values for British women are $m = 1610$, $s = 62$; for American men $m = 1755$, $s = 71$; and American women $m = 1625$, $s = 64$. More complete versions of Table 19.1 are given in Pheasant (1986, 1990), as is a further discussion of the statistical basis of anthropometrics.

User populations

The anthropometric characteristics of any given human population will depend upon a number of factors. The most important ones, from the point of view of ergonomics, are sex, age, ethnicity and occupation—usually in that order.

When applying anthropometric data to any particular design problem, the first step will generally be to define the *target population* of users for whom the product (workstation, environment) is intended, and to locate a source of anthropometric data for the population concerned (or for one which resembles it as closely as possible in relevant respects). The consequences of using inappropriate data will be that fewer people will be satisfied than intended—sometimes quite drastically so—or even worse, the performance of the eventual users will be impaired.

In most adult populations a difference of about 7% between the average heights of men and women (and a somewhat larger difference in their standard deviations) will be found. On average, men will also be larger in most other respects, although the magnitude of the difference will vary from dimension to dimension. The most important exception to this rule is hip breadth. In addition to the more obvious differences in shape between men and women, it is worth noting that men have proportionally greater limb lengths. That is, if we were to compare a man and a woman of *equal stature*, we should expect the man to have longer arms and legs, bigger hands and feet, and so on. There will also be proportional differences in dimensions which have a substantial soft tissue component. This is partly because men have (on average) greater muscle bulk whereas women have greater bodily fat; and it is partly due to sex differences in fat distribution.

People also change in shape as they get older. In our society at least, they tend to put on weight; and after the age of about 55, muscle bulk begins to decrease and the spine begins to shorten due to changes in the properties of the intervertebral discs.

Anthropometric differences due to the ageing process itself are confounded with differences due to long-term historical processes known as secular trends. In Britain, Europe and North America, people have been getting steadily taller. For the last century or so, there has been an upward trend in adult stature at the rate of about 10 mm/decade (an inch/generation). This trend was first noticed by Galton in the latter part of the last century. In

Britain and in North America though there is some evidence that the trend has now come to a halt, but in Japan it is still under way at a very rapid rate.

The ethnic groups of the world differ in both size and shape. Figure 19.3 shows sitting height (i.e., the distance from the seat to the crown of the head) plotted against stature. The oblique lines on the chart show sitting height divided by stature. Each of the major ethnic divisions of the world include both tall and short ethnic groups, but they tend to have characteristic body proportions. Black Africans have relatively long legs for their stature; far eastern peoples have relatively short legs (particularly so the Japanese). Europeans and Indo-Mediterraneans (who together make up the so-called Caucasoid division of mankind) are somewhere between the two extremes.

A compilation of basic anthropometric data for the British population, with some advice on its use, will be found in Pheasant (1990), whilst a wider range of information, including adult dimensions for certain other nationalities, is given in Pheasant (1986). The largest collection of anthropometric data published in any one place at present is held by NASA (1978) although a number of international organizations are seeking to draw together an international data set appropriate for ergonomics applications.

Criteria

Before pursuing some of the methods for using anthropometric data we must look at how we decide whether what we do will be adequate for the purpose. We need criteria both to guide us in our applications and to provide a framework against which we can test our decisions.

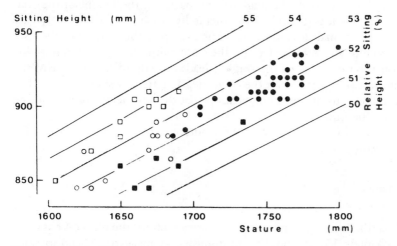

Figure 19.3. Ethnic differences in the relationship between average sitting height and average stature in samples of adult men. (● = European, ○ = Indo-Mediterranean, □ = Far Eastern, ■ = African). After Pheasant (1986)

In the context of ergonomics, a criterion is a standard of judgement which defines the extent to which a particular product (workstation, environment) is appropriately matched to its human users. For a criterion to have any practical (or scientific) value, it must be possible to specify the operation which an investigator would have to perform, in order to determine whether the criterion had indeed been satisfied in any particular case. A criterion which may be defined in this way is known as an *operational criterion*. There is not much point in saying that a product should be 'easily usable' unless we can define usability in terms of how it could be measured. Some product standards work in this way; BS 6652, *Packages Resistant to Opening by Children*, defines a set of procedures for conducting an experiment to determine how easily people are able to open a particular container. An experiment of this kind is called a *user trial* (see chapter 10 in this book for a full review of user trials). An ergonomics criterion of this kind indicates whether an existing product is to be regarded as satisfactory but it does not reveal much about how the product should be designed. This has both advantages and disadvantages.

Design criteria are hierarchical. At the highest level there are very general concepts: comfort, efficiency, safety, usability and so on. In themselves, these are not easy to define in operational terms—at least not in ways which are directly useful to the designer. To get round this problem, we need to break down these high-level criteria into subordinate criteria at successively lower levels in the hierarchy. Consider the criteria which might define an 'ergonomically-designed chair'. One of these would obviously be comfort. Subordinate to this would be more specific design principles, such as the provision of adequate postural support and the avoidance of pressure hot-spots on supporting surfaces. These in turn could be broken down into component parts dealing, for example, with the angle of the backrest, the height of the seat, and so on. At this lower level of the hierarchy, it should be relatively easy to provide operational design criteria. For example, the criterion, 'the user should be able to sit with the feet on the floor without experiencing undue pressure on the underside of the thighs', leads directly to 'the height of the seat should not be greater than the lower leg length of a short user' and in turn leads directly to the recommendation that the seat height should not be greater than 400 mm.

The criteria which define a successful outcome to the design process fall into three main groups:

comfort,
performance,
health and safety.

(We will see that these groupings recur in assessment of the physical environment, in chapters 15, 16 and 17 and discussed as measures of performance and stress in chapter 5 also.)

Case studies show that those ergonomics measures which increase comfort and well-being are also likely to improve productivity (and vice versa). For example, Ong (1984) studied a group of data entry operators at an airline computer centre in Singapore, before and after ergonomic improvements to the design of their workstations. He found both an increase in productivity and a dramatic reduction in the symptoms of visual and muscular fatigue and discomfort. Performance as measured by keystrokes per hour increased by 25%—but at the same time, the error rate dropped from an overall 1·5% (1 character in 66 incorrect) to 0·1% (1 character in 1000 incorrect). Dainoff and Dainoff (1986) describe experiments on data-entry operators in which comparisons were made between a workstation which was designed according to commonly accepted ergonomics guidelines and one which deliberately broke most of the rules. (The workstations differed with respect to seat design, keyboard height, screen location, task lighting, glare, etc.) Performance was 25% better at the ergonomically-designed workstaton. When the differences in lighting were eliminated there was still a performance difference of 17·5%. The subjects also expressed a preference for the ergonomically-designed workstation and experienced less back and shoulder pain. Aside from any humanitarian considerations involved, performance differences of this magnitude amply justify the costs of the ergonomics intervention (see also chapters 1 and 34).

In many problems of workstation design, the immediate objective will be to achieve appropriate muscular efforts for the performance of a given task. We may reasonably assume that this will have both short- and long-term benefits. In the short-term it will reduce fatigue—and by so doing, improve both performance and subjective comfort. In the long-term, it will reduce the incidence of conditions such as back pain, neck pain and repetitive strain injuries. The sickness absence which results from these common musculoskeletal disorders is economically costly, both for the organization concerned and for society as a whole.

Physiological and biomechanical principles

To achieve the desirable outcomes described earlier we need some rational procedures and principles. To help us to decide the broad arrangements of equipments, displays and controls, and layouts in general, there are four principles, stated by McCormick and Sanders (1987, p. 343), which might be seen as a normalization of common sense; because they are so evidently ignored in many situations they should not be underrated.

1. *Importance principle.* Those components which are most essential to safe and efficient operation should be in the most accessible positions.

2. *Frequency of use principle.* Those components which are used most frequently should be in the most accessible positions.

3. *Function principle.* Components with closely related functions should be located close to each other.

4. *Sequence of use principle*. Components which are often used in sequence should be located close to each other and their layout should relate logically to the sequence of operation.

Note that the term 'accessible' relates not only to physical accessibility (such as ease of reach) but also to visibility and to other more abstract characteristics.

Further sets of principles are needed to deal with the direct relationships between the person and the workplace. These will arise from consideration of the physiological, biomechanic and sensory (information transfer) relationships required if the people concerned are to maintain the comfort, performance and health aspects which were stated earlier to be the basis of our design requirements. Reference to ergonomics texts, appropriate to the aspects being designed, will give information on the requirements to be met if good performance is to be possible.

Within industry it has been common to rely on concepts such as the principles of easy movement, still listed in many methods texts, for decisions on workspace arrangements. These principles are limited in how they accord with the needs of human physiology and psychology, and Corlett (1978) proposed a set which was more in line with current knowledge (Table 19.2). These provide more specific guidance for the choice of anthropometric dimensions to meet our basic principles.

Table 19.2. New principles for workplace layout

1.	The worker should be able to maintain an upright and forward facing posture during work.
2.	Where vision is a requirement of the task, the necessary work points must be adequately visible with the head and trunk upright or with just the head inclined slightly forward.
3.	All work activities should permit the worker to adopt several different, but equally healthy and safe, postures without reducing capability to do the work.
4.	Work should be arranged so that it may be done, at the worker's choice, in either a seated or standing position. When seated, the worker should be able to use the backrest of the chair at will, without necessitating a change of movements.
5.	The weight of the body, when standing, should be carried equally on both feet, and foot pedals designed accordingly.
6.	Work activities should be performed with the joints at about the mid-point of their range of movement. This applies particularly to the head, trunk and upper limbs.
7.	Where muscular force has to be exerted it should be by the largest appropriate muscle groups available and in a direction co-linear with the limbs concerned.
8.	Work should not be performed consistently at or above the level of the heart; even the occasional performance where force is exerted above heart level should be avoided. Where light hand work must be performed above heart level, rests for the upper arms are a requirement.
9.	Where a force has to be exerted repeatedly, it should be possible to exert it with either of the arms, or either of the legs, without adjustment to the equipment.
10.	Rest pauses should allow for all loads experienced at work, including environmental and information loads, and the length of the work period between successive rest periods.

Application of anthropometric data and criteria

Design limits

The slope of the normal ogive is steepest at the mean and it decreases steadily in the tails of the distribution. This has an extremely important consequence for ergonomics in that it is increasingly difficult to accommodate extreme individuals. Let us consider a specific example: the desirable range of adjustment for the height of a working chair. For the purposes of argument we shall ignore things like the table with which the chair will be used and the task which will be performed (both of which are, in practice, very important) and we shall consider the chair taken in isolation. A seat which is too high causes undue pressure on the undersides of the thighs; one which is too low makes standing up and sitting down needlessly difficult and encourages a slumped position of the spine. Most ergonomics books would recommend therefore that the height of a chair should be a little below the popliteal height of its user. (Popliteal height is the vertical distance from the floor to the crease at the back of the knee.) It so happens that the popliteal height of British adults (ignoring sex differences) has a mean of 455 mm and a standard deviation of 30 mm. If we assume the optimal height of a seat to be 40 mm less than popliteal height, then the distribution of optimal heights will have a mean of $455 - 40 = 415$ mm, with a standard deviation of 30 mm. By calculating percentiles of this distribution, we can calculate the percentage of the target population who would be matched by any given range of height adjustment. From Table 19.1 we find that the 25th and 75th percentiles are 0·67 standard deviations below and above the mean, respectively. Hence to satisfy the 50% of users who are between these limits, we would need to make our chair adjustable by $0·67 \times 30 = 20$ mm, on either side of the mean value, i.e., from 395 to 435 mm. Repeating this calculation for other percentiles, the following results are obtained:

Percentage satisfied	50	60	70	80	90	95	98
Millimetres adjustment	40	50	62	77	98	118	140

We are in a situation of diminishing returns in which each additional unit of adjustment yields less benefit in terms of the percentage of satisfied users.

In practical terms we have to set limits on our attempts to satisfy the largest possible number of users. Where should we set these *design limits*? By convention we usually choose to accommodate (i.e., satisfy) the range of users who are between the 5th and 95th percentile of whatever characteristic we are dealing with. In some problems (like the one we have just considered) this will accommodate 90% of people; in others (as we shall see shortly) it will accommodate 95%. This is an arbitrary choice based on expediency. Beyond these limits the situation of diminishing returns begins to tell against us rather dramatically. In some cases however (such as when dealing with health and safety issues) we will need to set wider limits. The rule of thumb for making such decisions is to consider the worst possible consequences of a mismatch.

Fitting trials

A fitting trial is an experimental investigation of the relationships between the dimensions of a product (workstation, environment) and the dimensions of its users. In general, the experimental subjects will be asked to try out an adjustable mock-up of the product concerned. Critical dimensions of the product will be adjusted through a range of values, and the subjects will be asked to express their preferences with respect to comfort, ease of use, and so on. A fitting trial is therefore a special kind of psychophysical experiment. It is important that the subjects in the experiment should be a representative sample of users of the product—both with respect to their body dimensions and with respect to their general fitness and anything else which might be relevant.

Suppose we wished to conduct an experiment to determine the narrowest gap between two obstructions, of a given height, that subjects could pass through without experiencing undue inconvenience. We could do the experiment simply by shifting furniture around, so as to create gaps of different widths. We could start with a gap which was so narrow that no one could get through; and then systematically widen it (perhaps in 100 mm increments) until we had a gap which everyone could get through. If we plotted a graph of gap width (on the horizontal axis) against the percentage of people who could get through (on the vertical axis), we should expect to get a relationship which looked something like Figure 19.2, i.e., the cumulative form of the normal distribution. We could make it look exactly like Figure 19.2 by calculating the mean and standard deviation of the minimum gap widths through which each individual subject could pass, and then use these parameters to calculate percentiles, using the z values in Table 19.1. (In essence we are smoothing our data by fitting a normal distribution.) The experiment could be made more sophisticated by observing whether the subjects passed through crab-wise or head on, or by asking them whether it was easy or difficult, and so on. In fact, many different criteria could be used, and if the data were plotted for each of these criteria in turn, a series of normal distributions, spaced out along the horizontal axis, would be expected.

To conduct a fitting trial to determine the optimal height for a control (like a door handle or a light switch) or for a working surface would be slightly more complicated. A gap between obstacles can be too narrow for convenience, but it cannot reasonably be too wide; whereas a door handle or a work bench can be too low, too high or just right. We could start with a position which was too low for everyone; then as we raised the level of the object, we would expect the percentage of subjects who judged it to be too low to decrease steadily. As it did so, the percentage of people saying it was just right would climb to a peak value, before beginning to fall again—

as an increasing number of people began to say that it was too high. The characteristic form which the results of such experiments are generally found to take, is shown in Figure 19.4.

The latter experiment has a considerably greater element of subjective judgement than the former (which is much more of a go/no-go situation, in which you can either get through the gap or else you cannot). This has a number of important consequences. Characteristically, it is found that the range of optima reported in ascending trials (i.e., from low to high) is placed higher than the range reported in descending trials (i.e., from high to low). This is true of psychophysical experiments in general.

If we conduct our fitting trial with each subject running through the range of possible positions of the door handle, from low to high and back to low again, and plot the results, a graph similar to that of Figure 19.5 would result. Subjects have experienced all the positions from 'too low' to 'too high', and a section in between which they say is satisfactory. The test is conducted, for each person, with the height changing up and down, to incorporate the effects of ascending and descending trials as described above. We use subjects who are at the extremes of the body dimensions which are relevant to the dimension of the object which we are testing; in the example of Figure 19.5

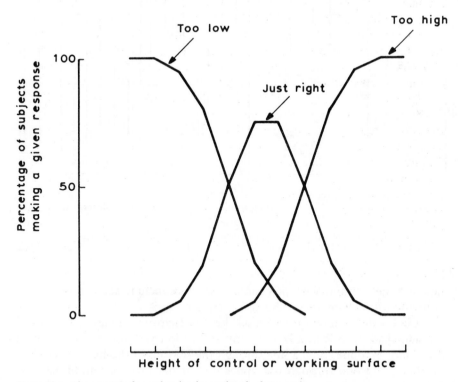

Figure 19.4. Characteristic form taken by the results of a fitting trial

SUBJECT

DIMENSION (cm)	5th				50th				95th			
	M		F		M		F		M		F	
	1	2	3	4	5	6	7	8	9	10	11	12
110												
107.5												
105												
102.5												
100												
97.5												
95												
92.5												
90												
87.5												
85												
82.5												
80												
77.5												
75												
72.5												
70												
67.5												
65												
62.5												
60												

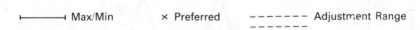

├────────────┤ Max/Min   × Preferred   – – – – – – – Adjustment Range

Figure 19.5. Graphical display of the results of a fitting trial, from which the extent of adjustment to the dimension may be deduced. 12 subjects were used, 6 male and 6 female, in the 3 percentile groups shown

these are the 5th and 95th percentiles. We may include some others if we wish, and 50th percentiles were also used here.

If a line can be drawn through all the 'satisfactory' dimensions, then this indicates that there is one level of this dimension which suits everyone. If this cannot be done, then a line through the bottom of the highest 'satisfactory line', and one through the upper end of the lowest, demonstrate the minimum range of adjustment needed to suit the population concerned. The method due to Jones (1963), finally needs all the separately selected dimensions putting together and checking with the sample of subjects, since dimensions may interact with each other, which would only be revealed by their final grouping and checking.

It is also worth noting that the form of results shown in Figure 19.4 also turns up in quite different areas of ergonomics, for example, in studies of thermal comfort (Fanger, 1973; Grandjean, 1988; and see chapter 16). In fact it is likely to be relevant to studies of subjective preference in general. The lines on the graph could represent all sorts of things: too soft and too hard; too cool and too warm; too light and too dark; too young and too old, etc.

The anthropometric method of limits

The process by which we establish final design recommendations with respect to anthropometric and biomechanical criteria, is known as the *method of limits* (e.g., Woodworth and Schlosberg, 1954). The term is borrowed from psychophysics. The anthropometric method of limits is essentially a model or analogue of the fitting trial, in which the anthropometric data stand as substitutes for the experimental subjects. You could say that, in applying this method, we are attempting to predict, using pencil and paper methods, what the result of a fitting trial would be if we were to perform one. (Note that the fitting trial is a special case of the psychophysical method of limits.)

In applying this method to any particular design problem, we are attempting to establish those boundary conditions, which make an object 'too big', 'too small', and so on, with respect to certain design criteria. In the simplest cases we may do this by inspection. This requires us to identify the *limiting user*—that is a hypothetical individual, who by virtue of his or her extreme bodily characteristics, is particularly difficult to accommodate with respect to the criteria concerned. If the limiting user is accommodated, it necessarily follows that the majority of the population, who are less demanding in their requirements, will be accommodated as well.

Anthropometric criteria fall into three principal categories: clearance, reach, and posture.

Clearance

Clearance problems include those which relate to head room, knee room, elbow room and so on, as well as those concerned with access through passageways, around and between equipment and into equipment for maintenance purposes. These are amongst the most important issues in workspace design since mismatches in these respects may be particularly hazardous. Unresolved clearance problems may also have knock-on effects in terms of unsatisfactory working postures, which the user is forced to adopt. For example, if the distance between the surface of the seat and the underside of a table does not provide adequate clearance for the thighs and knees of the users, then they may have to adapt to the situation by pushing the chair backwards (and leaning forward excessively to perform the task) or by perching on the front of the seat (and hence losing the support of the backrest). This is quite a common problem at service counters and cash tills.

Clearance should be adequate for the largest user. For practical purposes it will usually be expedient to set the design limits at the 95th percentile hence by definition accommodating 95% of the target population.

As an example, suppose we wish to find the minimum clearance required between the arms of a chair. We look up a suitable table of anthropometric data. Searching through this table we find the dimension 'hip breadth', which for our target population (British adults) has 95th percentile values of 405 mm for men and 435 mm for women. We adopt the larger figure. This would be an exact fit for the hips of the limiting user, but it would not allow for her clothing (since anthropometric data are usually quoted for unclad people) nor would it allow her any leeway. Making a commonsense correction for these we arrive at a round figure of 500 mm. (We could, if we wished, formalize this last stage, by specifying exactly what clothing ensemble she is likely to be wearing, and how much leeway we wish to give, but in reality (and for this problem) this degree of precision is rarely realistic.)

The process we have gone through could be summarized as follows:

Criterion: seat breadth ≥ hip breadth.
Limiting user: 95th percentile woman = 435 mm hip breadth.
Corrections: clothing and leeway = 65 mm.
Design recommendation: seat breadth ≥ 500 mm.

This application of the method of limits is in many respects equivalent to the first fitting trial we discussed in the previous section; and if we plotted a graph of the seat breadth (horizontally) against the percentage of users accommodated (vertically) we should again get a normal ogive, as shown in Figure 19.2.

Reach

Reach problems include those which are concerned with the location of controls in the workspace as well as things like the height of a seat (where it is necessary for the feet to reach the floor) and the height of visual obstructions. Seat depth also falls into this category since it is necessary for the user to 'reach' the backrest without undue pressure on the backs of the knees.

The procedure for dealing with reach problems is the same as the one used for clearance problems, except in this case the limiting user will be a small member of the target population, usually a person who is 5th percentile in the relevant characteristic.

Note that both the clearance and the reach criteria impose limits in one direction only. The clearance criterion indicates when an object is too small, but not when it is too large—and conversely for reach. Both clearance and reach therefore are *one-tailed constraints*. There may of course be other constraints acting in the opposite direction. In the case of clearance and access problems, these might include economy of space, or the reduction of distances travelled. In laying out working areas, we often find that clearance problems (e.g., elbow room) interact with reach problems (e.g., the accessibility of controls). The interaction of two opposing sets of constraints, creates a situation which is equivalent to the one which is described in Figure 19.4.

Note also that there is an important class of design problems (concerned with the safeguarding of machinery) in which the conventional criteria of clearance and reach are reversed. In these cases we actively wish to *prevent* access to hazardous areas, or to place the hazards *out of reach*. So the limiting user might, for example, be a person with long slender limbs who could reach the furthest distance through the smallest aperture.

Posture

Posture problems are inherently more complicated. For example, a working surface which is too high is just as undesirable as one which is too low. This limits the design in two directions to give a *two-tailed constraint* of the kind shown in Figure 19.4. In this situation there are two options open to us. We either have to provide an adjustable workstation, so that each user may set it to his or her own optimum dimensions (as discussed previously under 'Design limits'); or else we have to settle on a single overall compromise value which will maximize the number of users who are accommodated and minimize the inconvenience suffered by the remainder. Supposing our problem concerns the height of a work bench to be used by a standing person for performing a certain manipulative task. (It is assumed that the bench will be used by men and women.) A suitable height is between 50 and 100 mm below elbow height. It follows from the shape of the normal distribution that the single overall compromise value which will accommodate the greatest number of people is 75 mm below the average (or 50th percentile) elbow height. Looking up an appropriate table of data, we find the 50th percentile elbow height is 1090 mm for men and 1005 mm for women. This gives an overall average (i.e., for both sexes) of 1048 mm, to which we should add 25 mm for shoes, giving 1073 mm. Subtracting the 75 mm and rounding up, gives us a final recommendation of exactly 1000 mm. We may express this formally as follows:

> *Criterion:* (elbow height − 100 mm) < working surface height < (elbow height − 50 mm).
> *Optimal compromise:* 50th percentile elbow height − 75 mm = 973 mm.
> *Corrections:* 25 mm for shoes.
> *Design recommendation:* working surface height = 998 mm, rounded 1000 mm.

At this point it will be pertinent to consider how many users will be mismatched with respect to the criteria and how seriously inconvenienced they will be. This will enable us to decide whether a single compromise height will indeed be satisfactory, or whether an adjustable workstation will be necessary.

For a further discussion of the method of limits and further examples of its application see Pheasant (1986, 1990). For a compilation of design standards and guidelines concerned with anthropometric and ergonomic issues, see Pheasant (1987).

Conclusions

In this chapter some of the basic uses of anthropometric data have been described, together with an introduction to sources. When choosing anthropometric data it is necessary to read the small print. Who have been measured? What ages? Both sexes? How many in each sample? How old are the data? Listings of anthropometric data and their diagrams look very impressive, but this is no substitute for reliability and statistical validity.

When preparing a design for a particular situation, e.g., in a factory or office, and you are not sure whether or not tabulated data are suitable, it is valuable to measure some major dimensions for a sample of potential users (e.g., their stature). By comparing the mean and standard deviations of relevant dimensions for the situation you are concerned with, to the tabulated dimensions in published data, simple statistical tests will confirm whether or not they are from the same populations. If so, then you can use the tabulated values with confidence.

It must be emphasized that ergonomics is not synonymous with anthropometrics; it is an unfortunate belief in some design offices that this is the case, and that an ergonomic situation is achieved by choosing dimensions in conjunction with anthropometric tables. Anthropometrics are a necessary, but by no means sufficient, contribution to ergonomic design and evaluation. If the reader keeps in mind that we are dealing with a whole person, with needs for comfort, performance, interest and all those other things which make people so fascinating, then we are unlikely to reduce them to a simple matter of linear dimensions, no matter how vital the statistics.

References

Chaffin, D.B. and Anderson, G.B.J. (1992). *Occupational Biomechanics,* 2nd edition (New York: John Wiley and Sons).

Clark, T.S. and Corlett, E.N. (1984). *The Ergonomics of Workspaces and Machines: A Design Manual* (London: Taylor and Francis).

Corlett, E.N. (1978). The human body at work: new principles for designing workspaces and methods. *Management Services,* May, 20–52.

Dainoff, M.J. and Dainoff, M.H. (1986). *People and Productivity—A Manager's*

Guide to Ergonomics in the Modern Office (Toronto: Holt, Rinehardt and Winston).

Fanger, P.O. (1973). *Thermal Comfort* (New York: McGraw-Hill).

Grandjean, E. (1988). *Fitting the Task to the Man—An Ergonomic Approach,* 4th edition (London: Taylor and Francis).

Jones, J.C. (1963). Fitting trials. Architects Journal Information Library, February p. 321.

McCormick, E.S. and Sanders, M.S. (1987). *Human Factors Engineering and Design,* 5th edition (New York: McGraw-Hill).

NASA (1978). *Anthropometric Source Book.* US National Aeronautics and Space Administration.

Ong, C.N. (1984). VDT workplace design and physical fatigue: a case study in Singapore. In *Ergonomics and Health in Modern Offices,* edited by E. Grandjean. (London: Taylor and Francis), pp. 484–494.

Pheasant, S.T. (1986). *Bodyspace—Anthropometry, Ergonomics and Design* (London: Taylor and Francis).

Pheasant, S.T. (1987). *Ergonomics—Standards and Guidelines for Designers* (London: British Standards Institution).

Pheasant, S.T. (1990). *Anthropometrics–An Introduction,* 2nd edition (London: British Standards Institution).

Woodworth, R.S. and Schlosberg, H. (1954). *Experimental Psychology,* 3rd edition (London: Methuen and Co).

Chapter 20

Computer aided ergonomics and workspace design

J. Mark Porter, Martin Freer, Keith Case and Maurice C. Bonney

Man-modelling CAD systems*

Computer aided design (CAD) methods are becoming very popular with engineers as they provide considerably more flexibility than conventional techniques. Although they are now commonplace in manufacturing industries the great majority of CAD systems completely ignore the most important component of the human–machine system being designed—humans themselves.

The importance of an ergonomics input to a design is now recognized by many industries as being essential. The increasing complexity of modern systems and the social, economic and legislative pressures for good design have led to the demand for the ergonomics input to be made available as early as possible in the design programme, starting preferably at the concept stage. Traditionally, ergonomists have had to wait until the mock-up stage before being able to perform a detailed evaluation of a prototype design. This delay has several consequences, which will be discussed later in this chapter, all of which are detrimental to the design process.

Clearly, the optimum solution is to provide a means of supplying the ergonomics input in a complementary fashion to the engineering input; the logical conclusion being to develop CAD systems with facilities to model both equipment and people. Recognizing the potential of this solution, in some cases as early as the late 1960s, several research teams have developed man-modelling CAD systems. These have met with varying degrees of success but, essentially, they are design tools which enable evaluations of postural comfort and the assessment of clearances, reach and vision to be conducted

---

Editors' note: 'Man' is used in this chapter as a generic term in preference to human or people in the context of modelling systems, since this is the terminology employed in the area.

on the earliest designs, and even from sketches. In order to achieve these predictions, the systems need: three-dimensional modelling of equipment and workplaces which can be displayed on a computer graphics screen; three-dimensional man models (representations of the human form which can be varied in size, shape and posture for a variety of populations); evaluative techniques, based around the man model, to assess reach, vision, fit and posture; and a highly interactive user interface which allows the user to tailor design evaluations to their own requirements.

Existing systems

Existing man-modelling CAD systems show considerable differences in the extent to which the above facilities have been developed, and the breadth of their potential applications. Brief descriptions of the most established, or most recently developed man-modelling systems are given below:

ANYBODY

This is a three-dimensional ergonomics template or stencil of the human form which is used in conjunction with the CADKEY workplace modelling system which can be run on an IBM AT (see Figure 20.1). The system is believed to have been developed originally as 'OSCAR' by the Hungarian Design Council in Budapest, but in its current form is marketed by IST GmbH in Germany. The templates are available for men and women in a variety of sizes (females: 5th and 50th percentiles; males 50th and 95th

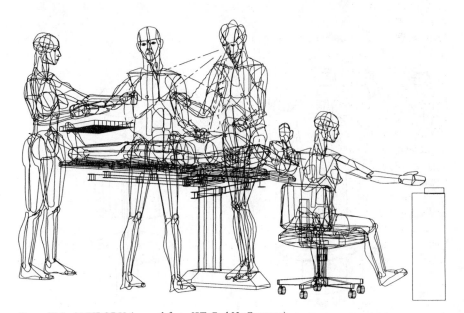

Figure 20.1. ANYBODY (sourced from IST GmbH, Germany)

percentiles), and shapes (ectomorph, mesomorph and endomorph). The anthropometric data are taken from DIN 33402 part 2 which presents information for the German public, although 50th percentile models are available for other databases such as those presented in *Bodyspace* (Pheasant, 1986). The developers suggest that these templates can be linearly scaled to represent other body dimensions. More recently IST have released a system called ANTHROPOS (apparently a development of ANYBODY) which includes some animation features. No published studies on the development or application of either system have been found to date.

APOLIN

Stands for APproximation by Ovals and LINes (APOLIN) technique, which uses parameter based, universal geometrical elements for modelling of human body segments (see Figure 20.2). The basic concept of the system is similar to that of SAMMIE (Grobelny *et al.*, 1992). The system is based upon primitive objects developed using the constructive solid geometry (CSG) method and is capable of using objects generated externally (e.g., AutoCAD or RoboSolid) using DXF file format. The man-model has 15 links with 14 joints, the anthropometry covers 1st to 99th percentile adults and joint constraints can be actively modified. Vision, posture and reach can be evaluated in a 3D environment which allows the building of models in complex structural hierarchies. At the time of writing the commercial availability of this system is unclear.

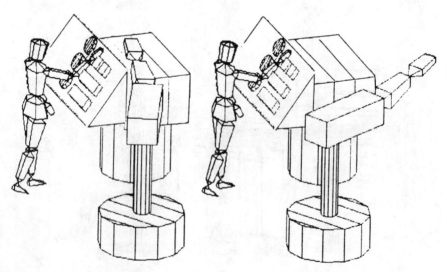

Figure 20.2. Design of robotic workstation using APOLIN (reproduced from Grobelny *et al.*, 1992)

BOEMAN

Developed by the Boeing Corporation, Washington in 1969 for use in checking cockpit layout (see Figure 20.3), the system was complex to use and it was not designed for interactive use as graphics terminals were not commonly in use at that time.

BUFORD

Developed by Rockwell International, California (see Figure 20.4), it offers a simple model of an astronaut, with or without a space suit. Body segments can be selected separately and assembled to construct any desired model, although these segments must be moved individually to simulate working postures. The model does not predict reach but a reach envelope of two-handed functional reach can be defined and displayed around the arms. It is not generally available.

CAR (*Crew Assessment of Reach*)

Developed by Boeing Aerospace Corporation for use by the Naval Air Development Centre in the USA, the system is designed to estimate the percentage of users (i.e., aircrew) who will be able to be accommodated physically in a particular workstation. The analysis is purely mathematical and the system has no graphical display. It is only available for in-house assessment of reach in aircraft crew stations.

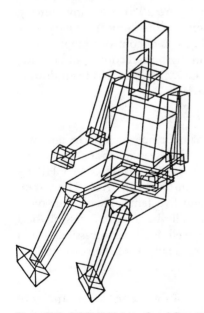

Figure 20.3. BOEMAN (reproduced from Dooley, 1982)

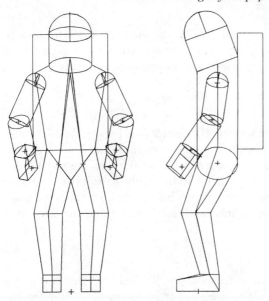

Figure 20.4. BUFORD (reproduced from Dooley, 1982)

COMBIMAN (*Computerized Biomechanical Man* Model)
Developed by the Armstrong Aerospace Medical Research Laboratory to evaluate the physical accommodation of pilots in aircraft crewstations, the pilot model is constructed using an array of interconnected triangles which can be reduced to just a profile view (as in Figure 20.5) from any viewing angle. COMBIMAN can produce pilot visibility plots to meet military standards, reach tests can be conducted for various types of control taking account of the clothing and harnessing being used. Strength predictions can be made for seated pilots based upon empirical data. COMBIMAN has been distributed to the major aerospace industries since 1978.

CREW CHIEF
This is a three-dimensional model of a maintenance technician and was developed by the Armstrong Aerospace Medical Research Laboratory and the Human Resources Laboratory (see Figure 20.6). Much of CREW CHIEF's functionality is based upon that incorporated in COMBIMAN. The system can generate 10 sizes of model (five male and five female) with four types of clothing and 12 initial postures. It can usefully be used to assess physical access for reaching into confined spaces as well as visual accessibility and strength analysis. CREW CHIEF has been generally available since 1988.

CYBERMAN (*Cybernetic Man* Model)
Developed by the Chrysler Corporation in 1974 for use in design studies of car interiors (see Figure 20.7). There are no constraints on the choice of joint

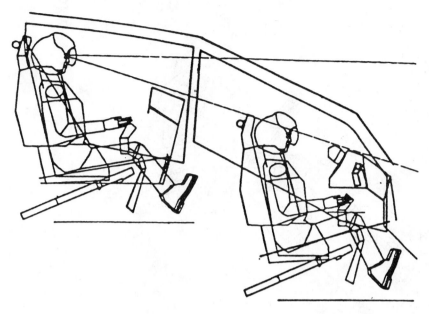

Figure 20.5. COMBIMAN. The plot shows a side view of a Helicopter crewstation (reproduced from McDaniel, 1990)

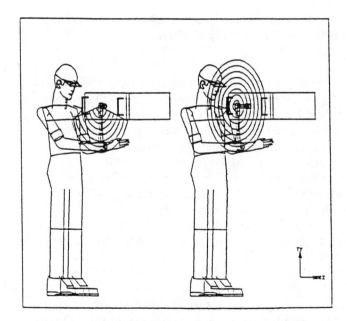

Figure 20.6. CREW CHIEF: Rotation of ratchet wrench is limited by handles on the box (left plot). The use of an extension rod between the ratchet and socket results in unobstructed rotation (right plot) (reproduced from McDaniel, 1990)

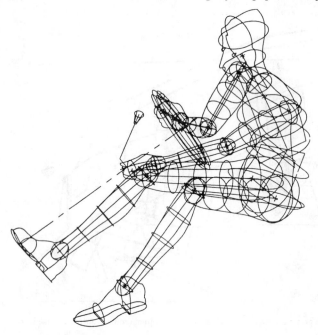

Figure 20.7. CYBERMAN (reproduced from Dooley, 1982)

angles so the man–model's usefulness for in-depth ergonomics evaluations is rather limited. This system is not generally available.

ERGOMAN

A relatively new man-modelling facility forms a constituent part of the ERG-ODATA system (see Figure 20.8). It is based upon the ELUCID CAD system software. The man-model has a 20 link body segment architecture, with 22 joints. Anthropometry is mainly controlled by varying body segment link length (joint to joint distance) with some ability to define depths and circumferences. The main man-modelling functions include: choice of morphotypes, joint angle limits, simulation of zero gravity, clothing simulation, upper limb reach areas, fields of vision, tool handling, motion and collision detection. Included with the ERGODATA system are a number of databases covering anthropometry, biomechanics, strength, a computerized human movement catalogue and a general ergonomics bibliography. For more details see Coblentz *et al.* (1991).

ERGOSHAPE

This system developed at the Institute of Occupational Health in Helsinki offers a two-dimensional manikin which runs within the AutoCAD system (see Figure 20.9). The models can be viewed from four viewpoints (left, right, top and front). The manikin can be constructed from up to nine seg-

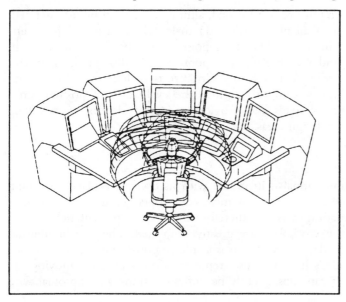

Figure 20.8. ERGOMAN. Reach capability assessment from the maximal reach areas for upper arms (reproduced from Mollard *et al.*, 1992)

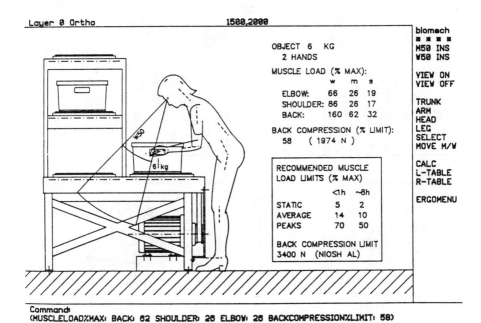

Figure 20.9. ErgoSHAPE. The figure shows the screen of the AutoCAD system when the biomechanical calculation of the ErgoSHAPE system is used (reproduced from Launis and Lehtelä, 1992)

ments, a set of basic postures is provided, although they can be user defined, and the manikins are available in male and female, 5th, 50th and 95th percentiles, or a user specified size. The anthropometric database includes Finnish, North European and North American populations although the manikins can be linearly scaled as required. In addition to two-dimensional reach and vision evaluation the system also permits the evaluation of postural stress resulting from vertical loads and provides recommendation charts giving guidance in various design areas.

ErgoSPACE

This is a three-dimensional man-model with its own workplace modelling facilities (see Figure 20.10). The system was developed within the restrictions imposed by microcomputers and therefore the graphic presentations of the man-model and the workplace are greatly simplified. The man-model has 17 joints and, in order to attain a reasonable response time, a stick model (i.e., the man-model's link structure) representation is used for moving the model. The model can subsequently be enfleshed using an ellipsoidal wire frame.

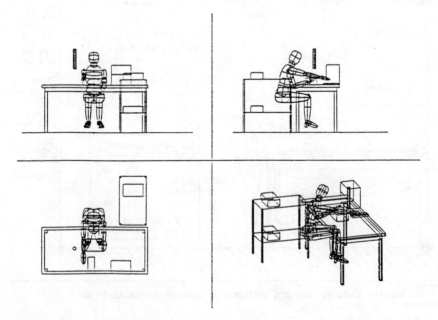

Figure 20.10. ErgoSPACE. Four views (front, plan, side and perspective) of a workstation evaluation (reproduced from Launis and Lehtelä, 1990)

FRANKY

Developed by Gesellschaft für Ingenieur-Tecnick (GIT) mbH in Essen (see Figure 20.11), it has a very similar (and comprehensive) suite of facilities to SAMMIE (which is described below). However FRANKY is not presently commercially available following the closure of GIT in 1987.

JACK

Developed by the University of Pennsylvania with extensive funding from NASA and the US Army Research Office, amongst others (see Figure 20.12). This man-model has 71 segments and 70 joints, including a 17-vertebrae spine and fully articulated hands. The default human figure is based upon data from the Society of Automotive Engineers for the 50th percentile male, although anthropometric data from a variety of populations can be entered and manipulated through the Spread Sheet Anthropometric Scaling System (SASS). Physical dimensions, joint limits, moments of inertia, centres of mass and strength data can be entered. The complexity of the man-model's flesh shape and structure can be controlled by the user. The system enables the evaluation of reach, fit, vision, posture, and torque load on joints and has been used to simulate human performance in the Apache Helicopter and the US Space Station. The system also allows the creation of human-like motion via an animation package, and includes rendering facilities enabling colour surface shading, reflections, shadows and textures (for clothing portrayal). Its availability is currently limited to Silicon Graphics workstation computers. For more information see Badler (1990).

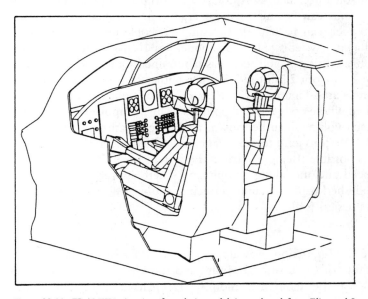

Figure 20.11. FRANKY. An aircraft cockpit model (reproduced from Elias and Lux, 1986)

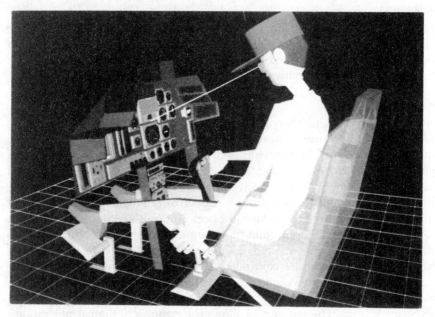

Figure 20.12. JACK. Colour shaded image of a cockpit model

MANNEQUIN

A recent PC based man-model from HUMANCAD, which, although it includes a basic workplace modeller, is primarily designed to be used with other graphics software, such as AutoCAD. Workplace models created in other CAD systems can be imported for ergonomics evaluations or Mannequin man-models can be exported as simple 'people pictures' for use in other CAD drawing systems. The man-model has 16 major body segments as well as articulated fingers and toes, has constrained joints and comes in 5 different body sizes from 2·5th to 97·5th percentile, created from a database of 10 nationalities (see Figure 20.13). Posture can be controlled by use of a range of standard whole body or hand postures or by manipulation of individual limbs. Reach volumes can be shown, as can sight paths and views from the man-model. Various joint torques can be calculated and a limited amount of animation is possible. Being PC based the system has limitations in terms of graphics speed and modelling complexity. There appears to be little in the way of published papers detailing principal exploitation of the system as an ergonomics design tool.

MINTAC

*M*an Machine *INT*er*A*C*tion was developed in 1984–5 by the Kuopio Regional Institute of Occupational Health and the University of Oulu (see Figure 20.14) for the Computervision CAD/CAM system. The three-dimensional man-model is based upon the anthropometric database published

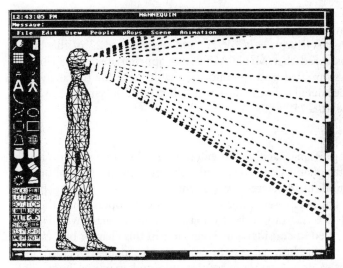

Figure 20.13. MANNEQUIN. A screen image showing field of view analysis and the system's iconic interface (sourced from HUMANCAD, 1991)

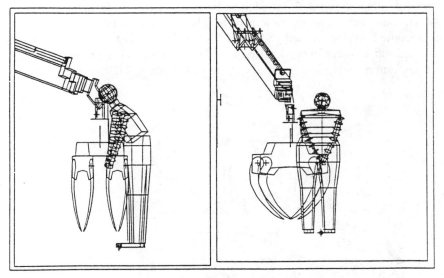

Figure 20.14. MINTAC. An operator is shown lubricating grease nipples of a tractor's grabbing mechanism (reproduced from Kuusisto and Mattila, 1990)

by Dreyfuss (1967) for the American civilian population, although the model is adjusted to simulate the wearing of winter clothes in order to evaluate difficult working postures encountered in Finnish agriculture and forestry. The simple man-model contains six links: lower links (one rigid block which can be selected in a choice of 13 postures), back, upper arms and forearms. The man-model was designed to be compatible with the OWAS working posture analysis system (Karhu *et al.*, 1977, and see chapter 23 of this book) although it is considered that MINTAC is not appropriate for widespread use because of its simplified posture, and is suitable only for the analysis of heavy work. Further details of this, and some other Finnish systems, are given in Kuusisto and Mattila (1990).

SAMMIE

System for Aiding Man-Machine Interaction Evaluation was originally developed at Nottingham University and subsequently at Loughborough University with funding generated by commercial consultancy (see Figure 20.15). SAMMIE has been used extensively by its developers since the mid 1970s and SAMMIE CAD Ltd currently market their software world-wide. This system is described in considerable detail later in this chapter but, briefly, it is a versatile three-dimensional system comprising a man-model of completely variable anthropometry, with comfortable and maximal joint angle constraints for each of its joints, together with its own workplace modeller. The general purpose nature of the system makes it suitable for a wide range of applications and special or logical relationships between model components can be defined allowing the models to be functional, for example, the operational movements and limitations of pedals, doors, seats or levers can easily be specified and executed. SAMMIE provides sophisticated ergonomics facilities and a powerful workplace modelling system, across a range of different computer systems (e.g., SUN, Silicon Graphics, HP/Apollo). Recent

Figure 20.15. SAMMIE. An adjustable computer workstation model

descriptions of the SAMMIE system include Case *et al.* (1990a), Case *et al.* (1990b) and Porter *et al.* (1990).

TADAPS

*T*wente *A*nthropometric *D*esign *A*ssessment *P*rogram *S*ystem was developed by the University of Twente in The Netherlands to run on VAX computers (see Figure 20.16). The system is based upon ADAPS, developed by Delft University for the PDP-11 computer in the late 1970s (see Post and Smeets, 1981). The basic man-model consists of 24 segments although the amount

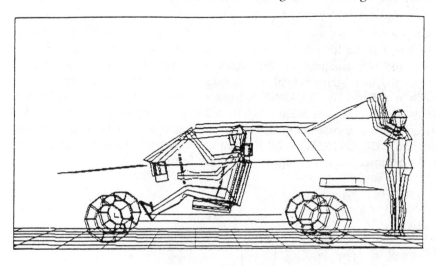

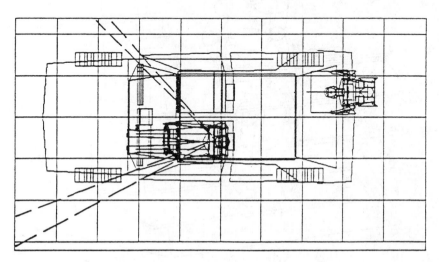

Figure 20.16. TADAPS. The two views show an analysis of reach (top) and vision (bottom) (reproduced from Westernik *et al.*, 1990)

can be reduced or extended to suit the intended application. The system includes its own workplace modeller. The anthropometric database comprises Dutch men, women and 4-year-old boys as well as American pilots and it is relatively easy to create models of other populations. All percentiles can be chosen although the man-model is linearly scaled from the 50th percentile proportions and it is not clear whether individual body segments can be set to different percentile values. TADAPS offers a prediction of the compression and shear force of the intervertebral disc L5-S1 for various postures and external loads.

WERNER

Developed at the Institute of Ocupational Health at the University of Dortmund and implemented on the Astari ST personal computer (see Figure 20.17), the three-dimensional man-model consists of 19 segments, each of which is defined by simple solids most of which are ellipsoids. A convex hull is constructed over these solids to define a silhouette of the man-model. The model's anthropometry appears to be based on the German National Standard DIN 43116. WERNER communicates with AutoCAD to provide its three-dimensional workspace modelling features.

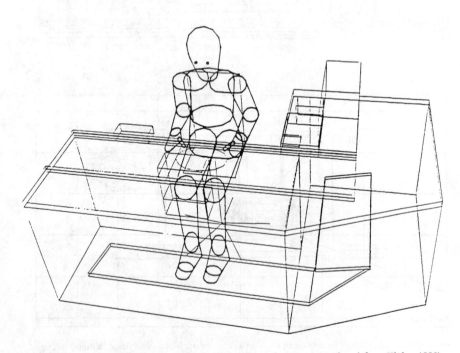

Figure 20.17. WERNER. An evaluation of a cash desk workstation (reproduced from Kloke, 1990)

Further information

Dooley (1982) and Rothwell (1985) and more recently Porter *et al.* (1993) present surveys of man-modelling systems, including many of the above systems. Other sources include: Elias and Lux (1986) (FRANKY); Lippmann (1986) (OSCAR); McDaniel (1976) (COMBIMAN); Porter (1992) and Porter and Freer (1987) (SAMMIE).

A man-modelling CAD system

The SAMMIE system will now be described in more detail to demonstrate how a man-modelling CAD system can be used as an effective ergonomics tool; the facilities discussed are indicative of what is possible with such methods.

Equipment and workplace modelling

The workplace modelling system is used to generate full-size 3D geometric representations of a working environment and specific items of equipment. A boundary representation form of solid modelling is used to enable the system to be highly interactive whilst maintaining a sufficiently accurate 3D model. This method requires that solid shapes are constructed from a description of the location of their vertices, a knowledge of which vertices are joined together to form edges and which edges form plane polygon faces.

Models of considerable complexity can be built quickly from a range of parametrically defined primitive shapes such as cuboids, polyprisms, and cones (see Figure 20.18). These are constructed interactively from a primitives menu, requiring very brief specification, such as object name, depth, width,

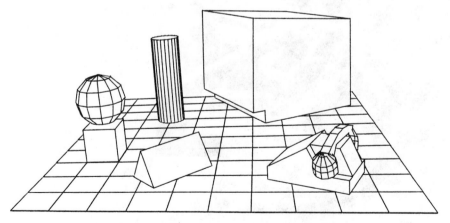

Figure 20.18. Examples of simple model types available in the SAMMIE system. The telephone is an example of how models are formed from these basic types

etc. Complex non-regular solids can be developed by describing vertex locations together with edge and face definitions and solids of revolution can be created by defining an axis of revolution and the desired profile. Although truly curved surfaces cannot be built this is rarely a cause for concern as sufficient accuracy can be obtained from a suitably configured faceted model (see Figure 20.19). A reflection facility is also available so that mirror images of solids can be constructed automatically; for example, only one side of a car needs to be defined manually.

Specification of logical or functional relationships between items in the model is achieved using a hierarchical data structure, an example of which is shown in Figure 20.20. This hierarchy allows the designer to move the whole car as one unit or to open individual doors or the boot (see Figure 20.21), to rotate the steering wheel or to adjust the tilt of the driver's seat cushion. To achieve this selectivity, users need to travel across and up or down the data structure until they reach the level which will control the particular item(s) to be adjusted.

The data describing the 3D workplace model can be taken directly from engineering drawings or sketches and entered via the interactive 'primitive' modelling menu, modelled off-line in data definition format, or it can be imported via IGES or DXF format files from other CAD systems (see Figure 20.22). Another important feature of the system is the interactive geometric editing facility which allows model modifications in ways relevant to various design situations. Importantly the structural validity of the model is main-

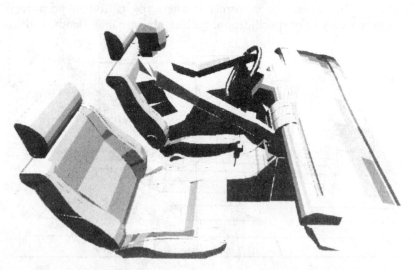

Figure 20.19. A wide angle perspective view of a car interior showing complex surface detail using 'faceted' objects to imitate curves

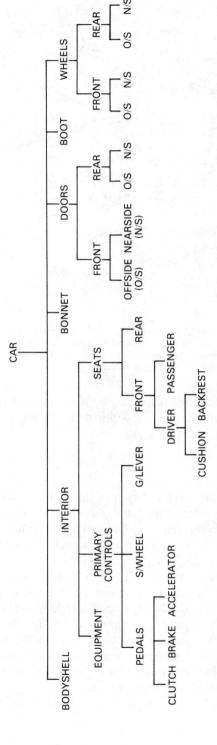

Figure 20.20. An example of the hierarchical data structure which allows the model to be functional as shown in Figure 20.21

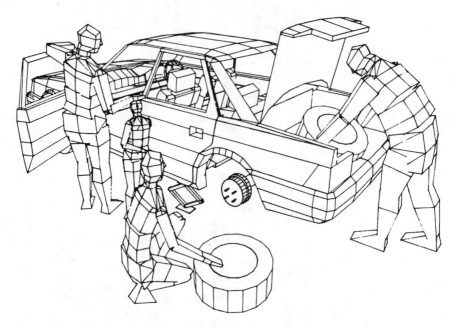

Figure 20.21. A complex car model. A hierarchical data structure enables functional as well as geometric relationships to be modelled, thus all moving parts of the model can be made to function. For example the car's doors, bonnet and boot can be made to open and close. Inside the car it is possible to adjust the seat and steering wheel within the design specification

tained during modifications, for example, if a table was modelled as a top and four legs, then increasing the width of the table would automatically reposition the legs to maintain a valid model.

Man-modelling

The man model is a 3D representation of the human body with articulation at all the major body joints. Limits to joint movement can be specified and the dimensions and body shape of the man-model can be varied to reflect the ranges of size and shape in the relevant national and/or occupational populations (see Figure 20.23).

The man-model has a set of 18 joints and 21 straight rigid links structured hierarchically to represent the major points of articulation and the body segment dimensions (see Figure 20.24). Optional hands can be introduced consisting of 16 links with 18 joints (see Figure 20.25). The hierarchical structure is similar to that shown for the car, so that when the man-model's right upper arm is raised, then the right forearm and hand follow accordingly. By dropping down the hierarchy users can control just the forearm and hand together or just the hand, at their discretion.

The size, shape and range of postures permitted are a function of the

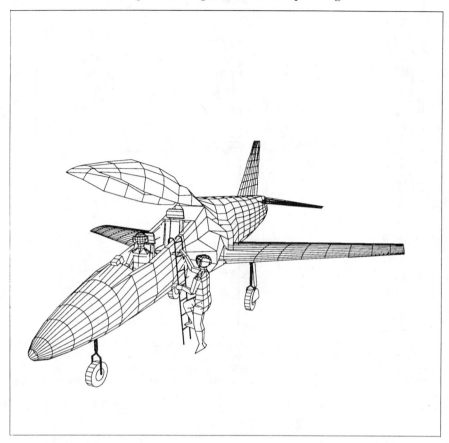

Figure 20.22. An aircraft model imported to SAMMIE from the UNIRAS system

anthropometric and biomechanical databases chosen by the user. The data required consist of the linear dimensions between adjacent joints (e.g., from elbow to wrist), the body segment parameters of weight and centre of gravity, and the absolute and 'normal' limits for each joint in each of the three degrees of freedom (i.e., flexion–extension, abduction–adduction and medial–lateral rotation).

The limb length data can be stored as either a set of mean dimensions together with standard deviations, or as a set of dimensions explicitly defining the anthropometry of an individual. The displayed man-model can be interactively amended by changing the overall body percentile, individual link percentile, explicit link dimension and the use of correlation equations to relate internal link dimensions to external anthropometric dimensions.

The anthropometric database provided varies according to need, and users can incorporate any anthropometric database suitable for their chosen application, either to define a population data set or to build a specific individual.

The flesh shape is controlled by a classification system known as somato-

Figure 20.23. Shown, from left to right, are male models of 95th, 50th and 5th percentile stature from a chosen population. The system also enables changes to be made to individual limbs allowing representation of specific users or groups of users. The shape of the models' flesh envelope can be varied in accordance with somatotypes providing a useful evaluative technique for situations involving work in confined spaces

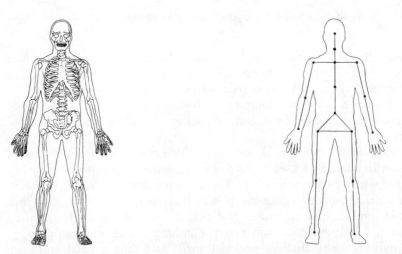

Figure 20.24. The link structure of the man-model is a simplification of the human skeletal frame, with pin-joints suitably constrained to simulate human movement capabilities. The rigid links between the joint centres are defined by use of anthropometric data and are usually displayed with 3D flesh shapes

typing (Sheldon, 1940) which enables the extent of endomorphy, (plumpness), mesomorphy (muscularity) and ectomorphy (leanness) to be specified; the somatotype number and the height and weight enables 17 body dimensions to be obtained from Sheldon's experimental data.

The joint constraints prevent the man-model being positioned in an unattainable posture. For example, it is impossible to abduct the elbow. The system indicates whether a selected joint angle is within the 'normal' range of movement, within the maximum range, or infeasible. The limb dimensions and somatotype can be interactively altered to construct 3D man-models to the user's unique specification if desired. Additionally, the joint constraints can be actively edited by the user to suit particular design situations, for example, joint movement range might be limited to represent disability, joint angle comfort ranges, the effects of bulky clothing or unusual working conditions, for example, where high gravity forces may severely limit arm movement.

Variable anthropometry allows the evaluation of body clearances (fit) and reach. In addition, the 'man's view' facility allows the user to display the man-model's field of view on the graphics screen. These facilities allow the user to predict the likely work postures that a given design will enforce. For example, a tall and fat model of a driver might be shown to adopt a slouched posture to gain sufficient headroom with arms at full stretch to the steering wheel under which the thighs are trapped. The view to the main driving displays may be obscured by the steering wheel, causing the driver to slouch to an even greater extent. This posture can be visualized by the designer and specified in terms of joint angles which can be compared with recommended angles in the literature (e.g., Rebiffe, 1966 for the driving task). The ergonomist would be able to comment upon such a posture saying that tall drivers of that particular car would suffer considerable discomfort in the neck, shoulders, lower back and thighs. Furthermore, the design can then be interactively modified by lowering the seat or raising the roof-line, and re-positioning or providing adjustment to the steering wheel.

Ergonomics facilities

The system has several facilities to help the user assess the ergonomics of a particular design.

A clasher routine

This facility automatically detects whether two solids are intersecting and, if this is the case, it flashes the appropriate items to attract the user's attention. This feature can be used to check clearances with the man-model set to an appropriate size and shape, say 99th percentile limb lengths and an extreme endomorph. Alternatively, visual inspection from a variety of angles will achieve the same result.

Reach algorithms

Reach can be assessed simply by positioning the arms or legs so that the hands or feet either contact, or fail to contact, a specified control or point in space (see Figure 20.23). This method could become tedious for a large number of controls so an algorithm has been developed which predicts a feasible posture for the arms or legs given a specified model item or co-ordinates to be reached. Generally, there will be a large number of feasible postures for any successful reach attempt. The algorithm selects the limb posture to be displayed by attempting to minimize the extension of the joints away from their neutral positions and by preferring the greater extension of distal links to those that are more proximal. This feature does not ensure that the displayed limb posture is the likely posture adopted by a human, but it does confirm whether or not the reach attempt will be successful. If a reach attempt fails, the system displays this fact together with the distance by which it failed.

There are two other automated methods to define reach: reach areas and reach volumes. Both methods are especially suited to concept design as they are generated without specifying control locations or co-ordinates. The first method enables envelopes of reach areas to be overlaid on any surface of the design as an aid to assessing suitable positions for control locations. The second method is an extension of this whereby reach is assessed over a number of imaginary surfaces parallel to either the frontal, sagittal or transverse planes of the man-model. An example of a reach volume in the transverse plane is shown in Figure 20.26; such information is particularly useful for locating controls above head height. A major study was conducted using this facility to determine both hand and foot reach zones for drivers of agricultural tractors (Reid *et al.*, 1985).

Vision tests

The view 'seen' by the man-model (man's view) is under the full control of the user (see Figure 20.27). For example, one can select left, right or a mean eye position, 60 or 120° cone of vision and specify the angle of vision using the eyes and/or head as appropriate. Constraints limit the maximum angles of vision from the eyes. As with reach, the testing of vision can be achieved manually by directing the head and eyes or else the user can specify the model item or co-ordinates to be viewed; the resulting view, together with the visual angle and viewing distance, will be displayed automatically. Sight lines can be attached to the man-model to show preferred, acceptable or maximal visual angles and distances based on any set of recommendations appropriate to a particular design scenario.

Vision can further be evaluated by using 2D visibility plots showing vision across a given surface (e.g., checking vision of the facia in a vehicle and, in particular, through the steering wheel) and 3D visibility maps which enable

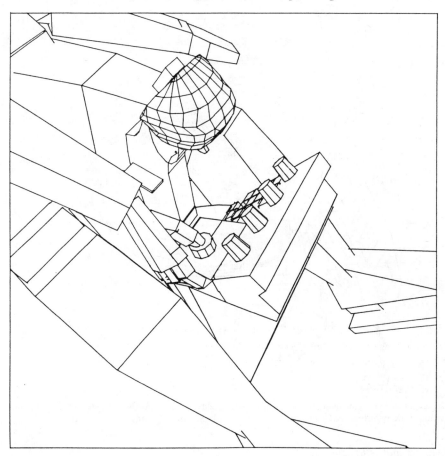

Figure 20.25. A close up view of a cockpit model showing the pilot's pedestal controls and joystick. A functional hand model is being used to demonstrate alternative postures for operating a push button control

the visual field to be described in terms of areas or volumes that are obscured from view by workplace structures (e.g., checking external visibility from a vehicle through the windows). Simple calculations allow one to determine the maximal visibility in any plane (vertical or horizontal) so, for example, the user can check whether a tall car driver would be able to see signposts without leaning forward or whether a train driver can see track side signals without having to move out of his seat. Aitoff equal area projections can be taken giving a full 360° field of view from a single view point (see Figure 20.28), particularly useful in aircraft and helicopter evaluation where clear fields of view need to be described in terms of visual angles.

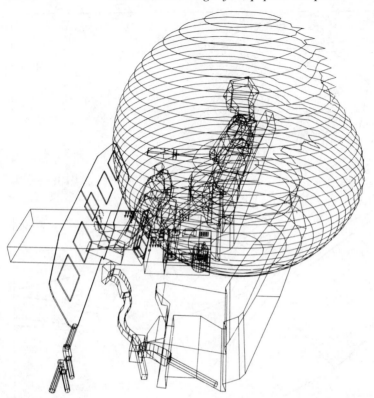

Figure 20.26. This plot illustrates volumetric reach facility (available for both hands and feet). In this example the right hand reach for a 50th percentile male helicopter pilot is being assessed

Mirrors and reflections

The mirror modelling facility can be used to design mirrors for vehicles (see Figure 20.27) or to determine whether reflections will be a problem in windscreens or computer screens. The mirror parameters of focal length, convexity/concavity, size and orientation are all variable and can be inter-actively adjusted to provide the required field of view displayed on the mirror surface, as seen by the man-model.

Saving postures

Having selected an appropriate size and shape of man-model and adjusted his or her posture to suit the task demands and physical contraints of the workplace, it is important that this posture can be stored and recalled at a later date. This facility exists and it enables the user to run through a sequence of typical work postures in rapid succession, for example driving forwards, depressing the clutch and engaging first gear, depressing the clutch, engaging reverse gear and looking rearwards (see Figure 20.29).

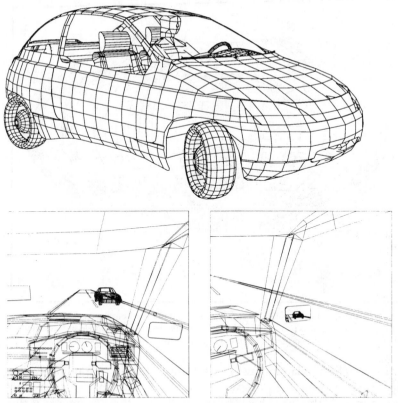

Figure 20.27. SAMMIE's viewing facilities enable the evaluation of the visual field for the full range of operator sizes. Left shows a 95th percentile male driver's view of the car controls and displays and the road through the windscreen. Right shows a 95th percentile male driver's view in his off-side exterior mirror. Reflections in the windscreen at night due to unshielded illuminated displays can also be identified at an early stage in the design

User dialogue

The system is highly interactive and allows designers to proceed through the design process in a manner determined by their own requirements rather than in a predetermined manner. The system is operated via an easy to use, graphical, menu based dialogue using either a mouse driven cursor to select menu options, by direct entry of command abbreviations from the keyboard or by using a MACRO command processor (described later). Each menu, of which there are nearly 40, contains logically named commands grouped according to their functions. A brief description of some of the main menus is given below.

View menu

The status of the graphics display is governed by four main parameters. The first is the centre of interest, basically what the user is looking at, either

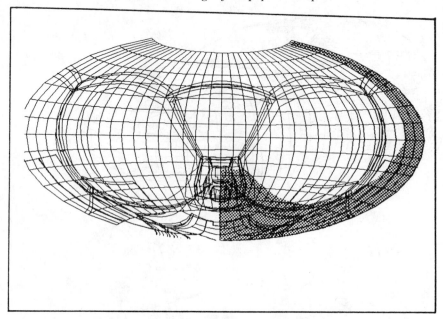

Figure 20.28. An Aitoff Equal Area projection showing 360° field of view from the rear seat in a tandem aircraft cockpit. The shaded area is part of an overlay indicating the legislative clear field of view requirements for this particular aircraft type (note that the non-shaded area is the clear field of vision)

directly or through the man's view. The second is the viewing point, which can be set at the man-model's eyes or any other point in 3D space around or inside the models that have been constructed. The third parameter is the choice between displaying view in plane parallel projection (e.g., engineering drawing style) or in perspective and the fourth is the size of the displayed model, which is set by the scale factor in plane parallel projection and by the acceptance angle (i.e., the viewing angle) in perspective. The 'view menu' contains a variety of ways of interactively changing these parameters and it also provides a director of 'saved views' which the user has set up for future use.

Workplace menu

These commands allow the interactive positioning of models or component parts of models in the workplace. Items can be shifted or rotated about either their own (local) axis system or the global axis system. An example of this important distinction is illustrated in Figure 20.30.

A commonly used alternative to specifying the shift distance in millimetres is to 'drag' the chosen item(s) to a desired location on the screen using the light pen, keyboard cursor keys or mouse. This method can be faster because the location can be changed in two axes simultaneously and the accuracy can be maintained by increasing the scale of the model.

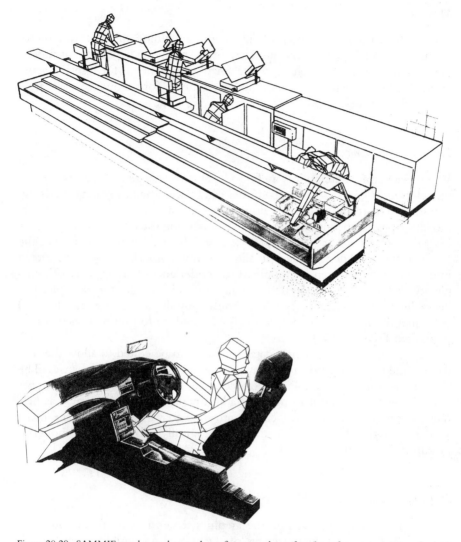

Figure 20.29. SAMMIE can be used to evaluate fit, postural comfort (by reference to joint angle data) and reach to controls or other important workplace items, in a delicatessen, for example (top) or in a car (bottom). As well as the appraisal of static reach and comfort it is also possible to consider sequences

Display menu

Complex models take longer to be drawn on the graphics screen and sometimes these models appear confusing. The 'display menu' allows the user to select which items need to be displayed as required.

Man menu

This menu contains a variety of sub-menus including the 'anthropometry menu' for changing the anthropometry of the man-model, the 'joint movement menu' for postural changes, the 'man's view menu' for displaying the view seen by the man-model and the 'reach menu' for producing reach areas and reach volumes.

Workplace editor menu

When evaluating a design it is useful to be able to change the size or shape of model items. Objects and group entities can be modified by scaling, shearing, extruding and re-dimensioning along a variety of axis systems. Additionally model shapes can be interactively changed by using the cursor to 'drag' vertices, edges or faces of objects on the screen. This feature is invaluable at the concept stage in design since it allows simple models to be constructed initially which can then be modified to model an increasing level of complexity as the design progresses. A wrap around console, for example, can begin life as a number of simple cuboids roughly arranged as needed and subsequently shapes can be interactively 'dragged' to form neat angled corner joints (see Figure 20.31).

Additionally the hierarchical data structure can be edited to allow the user to redefine logical relationships between model items as the need arises. The man-model can, for example, be attached to a seat such that he or she moves with the seat throughout its range of adjustment, or equipment such as helmets, boots, back packs, etc., can be attached to various parts of the man-model such that they move and remain logically related to the model as posture is changed.

Rendering menu

Models are usually displayed on the graphics screen in wire frame form (see Figure 20.26) so that all the edges of the model are visible, even though some in reality would be totally or partially obscured by solid objects. For extra clarity, the 'hidden lines' can be automatically removed (e.g., Figure 20.29), or the model can be colour surface rendered (e.g. Figure 20.19).

Plot menu

The end result of a design and/or evaluation will usually be in the form of a variety of views taken of the model. These are output in a wide variety of plot formats which can be sent to a wide variety of output devices, or to graphics packages for further manipulation or rendering, etc.

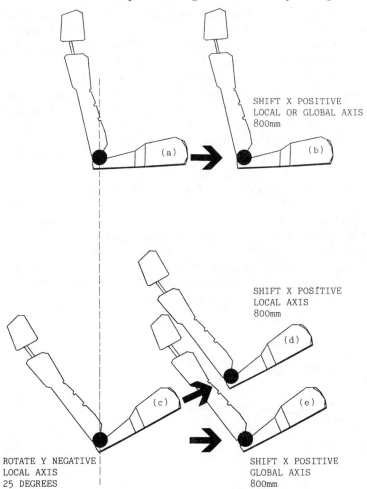

SHIFT X POSITIVE
LOCAL OR GLOBAL AXIS
800mm

SHIFT X POSITIVE
LOCAL AXIS
800mm

ROTATE Y NEGATIVE
LOCAL AXIS
25 DEGREES

SHIFT X POSITIVE
GLOBAL AXIS
800mm

Figure 20.30. An example of the use of the local and global axis systems available in the SAMMIE system. In some orientations these axis systems are identical, as shown in (a) and (b) where the seat is shifted 800 mm along the global or local X axis. In (c) the car seat has been rotated about its local Y axis to produce seat tilt. (If it had been rotated about the global Y axis then it would have pivoted around the centre of the available workspace.) Examples (d) and (e) show how a subsequent 800 mm shift along the local and global axis systems, respectively, can produce different results. If the intended movement is to simulate fore and aft adjustment of the seat, then only (e) is appropriate

Macro command processor

A MACRO command processor enables users to generate files of commands outside of SAMMIE (using any appropriate text editor). In their simplest form these files would contain sequences of commands in a form identical to that which could be entered through the normal user interface, usefully allowing users to construct and retain sets of commands which suit their needs in particular situations. A more powerful aspect of the command processor is the ability it has to include variables and programming logic to control the issuing of commands thus providing users with a mechanism with which to customize their usage. Another important use of command files is the development of evaluation programs. A command file could contain a defined set of reach and vision tests, for example, which could be automatically applied while also varying the size and shape of the man-model, in effect processing a large number of different users through the design. In this way an assessment of the percentage of the target population accommodated by a design can be conducted with minimal effort. The range of user sizes could include the modelling of all the individuals recorded in a relevant anthropometric survey, using the command file to run them all through the test set, identifying all individuals who fail to complete any test successfully. Alternatively, where the full survey data are not available, the Monte Carlo method (see Churchill, 1978) might be used to generate a wide range of users from a knowledge of means, standard deviations and correlation coefficients for the survey data. Lastly, the command files are used to generate sets of standard ergonomics evaluations that can be applied to different designs. Thus a standard set of evaluations could be defined that would commonly be applicable to cockpit models, for example, where issues such as

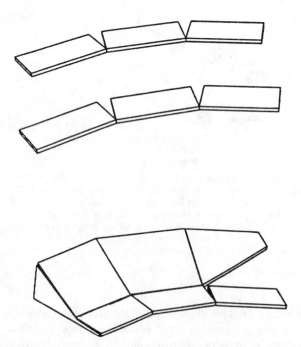

Figure 20.31. A group of six simple cuboids primitives (top) are shown laid out horizontally in the approximate form of a console. By interactively 'dragging' the corners and edges of the cuboids a wrap-around console can be quickly built (bottom). Note that the right hand side of the console is shown partially completed

vision to specific displays, reach to joystick, knee room, etc., are common concerns.

Case studies using SAMMIE

Two short projects carried out using SAMMIE are described here, to give the reader an insight into the way in which such systems are used.

Computer workstation design

The aim of this project was to design an integrated workstation to be used in the computer aided design of printed circuit boards. The original workstation was purely a grouping together of the hardware needed to perform the required functions, which resulted in a three-sided configuration, comprising an alphanumeric VDT on the left, an A0 digitizer board in the centre and a graphics VDT on the right. Not unexpectedly, this arrangement was far from satisfactory with a high incidence of physical discomfort reported by the users. The manufacturers then designed two prototype integrated workstations where the graphics VDT and a much reduced digitizer, which was sunk into the worksurface, were placed directly in front of the user. However, both these designs were found to cause problems for the user for several reasons, including lack of thigh clearance, forward leaning over the worksurface, difficult reach to the keyboard and an excessive viewing distance to the graphics VDT. The manufacturers were both surprised and disappointed when these problems came to light within the first few days of testing, as they had invested considerable time and expense to produce the prototypes. However, most of their attention had been directed at the engineering problems and the interface design had suffered as a consequence.

Following initial discussions with the manufacturers, it was decided to develop three alternative designs using SAMMIE, covering a range of manufacturing costs. These designs are illustrated in Figure 20.32 and are now briefly described:

(a) This was the cheapest design with all the components free standing on the fixed height worksurface. Whilst this option may appear satisfactory as a paper specification, the visualization of the workstation clearly shows its shortcomings, such as the lack of space for paperwork, the likely wrist and arm discomfort arising from the raised digitizer board, and the generally clumsy layout.

(b) This was the most expensive design as it offered both worksurface height and tilt adjustment. The digitizer was sunk into the worksurface and the workstation could be set up for either left- or right-handed use as it was divided into two modules; this feature also made it considerably more portable.

(c) This was the medium cost design which had all of the features of (b) above except the adjustable tilt angle. The VDTs were adjustable.

These designs were presented to the manufacturers in the form of slides, as reproduced here. The plots were visually enhanced by an industrial designer who was closely involved in the project. The manufacturers were able to visualize accurately the concept workstations knowing that the system had been used to evaluate the designs in terms of fit, reach, vision and posture. The chosen design was (c) because of several factors, namely its aesthetic appeal, ease of manufacture, cost and sound ergonomics. This workstation was manufactured successfully and the product was nominated for a design award the following year. Further details of this project can be found in Porter (1981).

Tram driver's workstation

This project was conducted by SAMMIE CAD Ltd. in co-operation with Design Triangle of Cambridge and concerned the design of the STIB Tramway 2000 in Brussels. SAMMIE was used to investigate and propose design solutions to a number of human factors problem areas identified from an initial assessment of the proposed tram cab design. The cab design was complicated by the need for a wraparound console providing sufficient surface area, within a limited cab space, for the required controls and displays which would allow the driver to sit comfortably facing forward when driving, and also swivel around to face back down the tram when selling tickets to passengers. Furthermore, the placement of an electrical equipment cupboard in the rear wall bulkhead placed severe width restrictions upon the driver's cab entrance space.

A full functional model was built of the entire tram from engineering drawings (see Figure 20.33). STIB provided anthropometric data which were used to generate a range of man-models for the drivers, with passengers being derived from European data from Pheasant (1986).

An appraisal of drivers entering and leaving the cab showed that drivers would experience difficulties by being required to twist and bend due to the narrow door width and low ceiling height. It was also found that a high entrance step and severly limited standing space just inside the cab door served to compound the difficulty drivers would experience. Drivers were forced to adopt uncomfortable, unstable and somewhat contorted postures when entering the cab, especially when carrying their log books and personal equipment bag (see Figure 20.34). As a consequence potential modifications to the ceiling height, floor panels, side door gear boxes and the rear wall cupboard were discussed between the designers and customer with a view to substantially improving cab access.

A range of man-models was used to determine a best possible driver package based upon a wraparound console design. The end result was a packaging

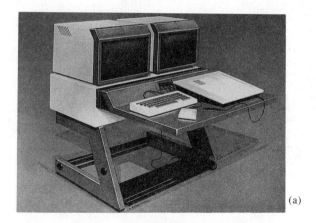

(a)

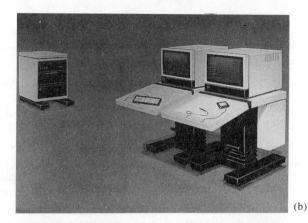

(b)

(c)

Figure 20.32. Three alternative designs of computer workstations: (a), (b) and (c) were the low cost, expensive and medium cost alternatives respectively. The plots were enhanced by an industrial designer

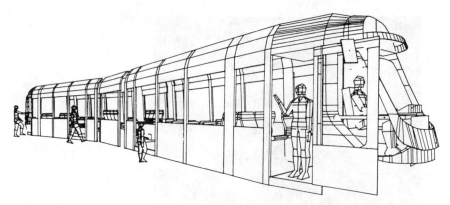

Figure 20.33. The model of the STIB Tramway 2000 vehicle. A variety of passenger sizes, including a child, are shown

specification that identified a range of seat position and angle adjustments, the console height, positions for the main driving controls and displays and the external visual field for a range of drivers from small females through to large males. A number of packaging problems were identified. Firstly it was shown that there was a need for an adjustable foot rest, since seat height adjustment alone was insufficient to allow all drivers to operate the main driving controls, with ease and in comfort, at a fixed height console while resting their feet on the floor. Secondly the specified seat was shown to have too long a seat cushion for many smaller drivers. Thirdly the combination of a high seat position (for optimal external vision) and the console height (limited by the need to ensure it does not limit downward external vision in town traffic) and structural elements of the cab walls and console supports meant that knee and leg space under the console was severely limited causing problems when adjusting the original seat swivel mechanism. Lastly there was shown to be a conflict between the two main driver tasks, i.e., driving the tram and selling tickets to the passengers. A package designed to allow the best possible ease of operation, vision and comfort when driving was found to be severely compromised by the requirement to have the driver swivel around and sell tickets through the cab back wall (a reasonably fre-quent task). Changes to the seat position were needed to allow the driver to swivel the chair fully and to be able to remain seated while operating the ticketing equipment (there was insufficient space to consider standing operation). In conjunction with the rest of design team a variety of new seat swivel mechanisms and rotation points were investigated (see Figure 20.35). A mechanism was identified that allowed the seat to move and swivel such that both main driver tasks could be easily accomplished, without the need to make major structural changes to the cab or console, and which was both mechanically feasible and cost effective.

SAMMIE's mirror surface facility was used to identify acceptable external

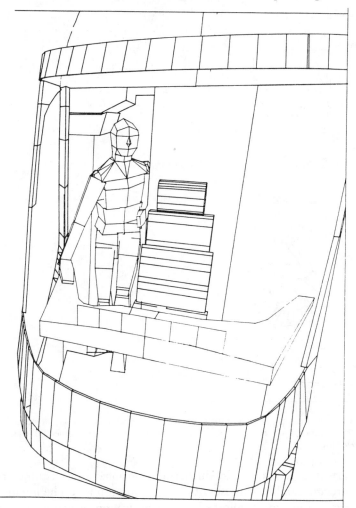

Figure 20.34. A view through the front of the cab showing a large operator twisting and bending as he enters the cab through the door as originally designed

mirror locations and to demonstrate that the full range of drivers could obtain adequate views of passengers entering and leaving the tram and of vehicular traffic (see Figure 20.36). This involved testing across a number of compromise solutions, involving engineering, cost, production feasibility and passenger safety considerations, as well as optimum viewing.

Man-modelling in CAD enabled both the identification and quantification of ergonomics problems with the design and, being graphical, enabled these problems to be communicated easily to other members of the design team and to the customer. Importantly, good communication, leading to a clear understanding of the problems, allowed the design team rapidly to generate

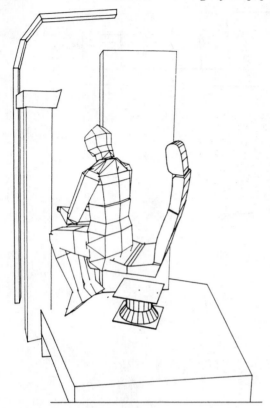

Figure 20.35. A view, from the front of the cab showing the driver swivelled around on one of the seat options to sell tickets to passengers. Note that much of the cab structure has been 'turned off' to show the situation more clearly

a range of alternative solutions and furthermore enabled the customer to weigh up the strengths and weaknesses of various compromises. The iterative nature of the evaluation process allowed the quantification of various design alternatives and the eventual development of a best possible ergonomic tram design. The tram will eventually be put into service in Brussels. Projects such as this highlight the value of computer-based man-modelling techniques in the evaluation and design of complex working environments. Other SAM-MIE projects have been described in Bonney *et al.* (1979a,b), Case and Porter (1980), Levis *et al.* (1980), Porter and Porter (1987) and Porter and Case (1980), and see Figures 20.29, 20.37, 20.38 and 20.39.

The advantages of using CAD

There are several important advantages to using 3D man-modelling CAD systems in design and these are now discussed briefly.

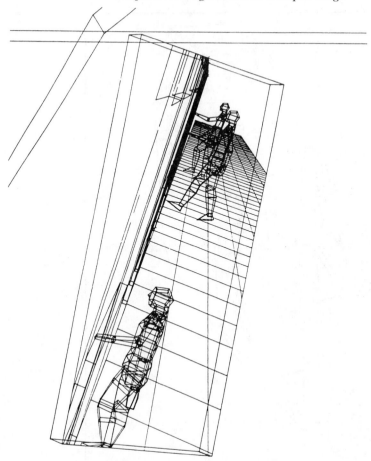

Figure 20.36. A driver's eye view of an external mirror showing passengers boarding the tram from ground level. The figure in the foreground is a 50th percentile 10 year old. The mirror positions, angles and focal length were varied to obtain the optimum field of view

Reduced timescale

This clearly can be a major factor and it may often decide whether or not the project receives any ergonomics input at all. Time can be saved in several areas, for example, the construction of a computer-based mock-up might take between 1 (simple) to 5 (complex and large) days compared with as many months using wood, glass fibre or other materials. Subject selection can be a time-consuming process when conducting user trials, whereas the anthropometric database of the computer system can be used to select the required man-models in seconds. For example, when designing driving packages it is important to consider people with long legs and short arms because they will have a personal conflict between positioning the seat rearwards for good leg posture, whilst having the steering wheel at full stretch, or having

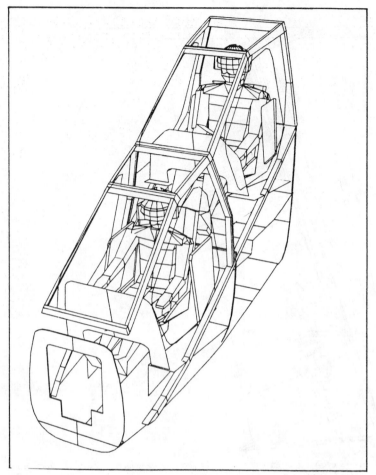

Figure 20.37. A tandem cockpit helicopter model

the seat further forwards for good arm posture at the sacrifice of leg posture. The best solution is to provide steering wheel adjustment but this require-ment may not be apparent if user trials are rushed using only a small handful of subjects who may have similar percentile reach with their hands and feet. Another saving is made at the evaluation stage as only a few man-models are examined compared with 20–30 subjects, with the ensuing lengthy data analysis.

Early input of ergonomics expertise

Because of the rapid modelling facilities it is possible to start the ergonomics input right at the beginning of the project. This is particularly necessary as engineers are using CAD systems themselves and the design might be virtually

Figure 20.38. An evaluation of a concept design for an advanced maritime control centre

finished from their point of view by the time the first full size mock-ups are ready for traditional user trials.

Iterative design

Early commencement coupled with reduced timescale make it very easy to establish an iterative design programme and to promote the exploration of a wide range of design solutions. Compromises are an essential feature of design and the above features are important ingredients in developing the optimum trade-off between, for example, cost and the ergonomics specification.

3D analysis

Apart from user trials, other traditional techniques involve using anthropometric data or 2D manikins. Both of these methods are unsatisfactory for complex tasks, for example, driving a tractor and ploughing a field (see Figure 20.40). The driver will have both feet operating foot controls, one hand will be on the steering wheel and the other will be on a hydraulic control lever to adjust the height of the plough. The driver will be looking both in front and, twisting the spine, over the right shoulder to the furrows behind. This posture cannot be assessed without using 3D analysis.

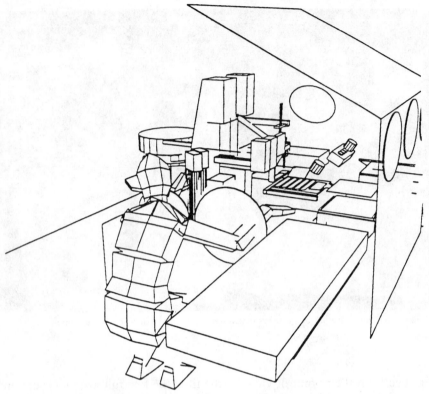

Figure 20.39. An operator performing maintenance on a machine through glove ports. The nature of the machine's process requires that it remains sterile. Note that much of the glass shielding between the operator and the machine has been 'turned off' for clarity

Improved communication

Computer graphics provide an excellent means of presenting ergonomics input to design committees. The visual impact of the ergonomics specifications is far stronger and easier to grasp than numerous recommendations in a report. Additional realism can easily be supplied using the services of an industrial designer or stylist (see Figure 20.41) and this collaboration improves communication within the design team.

Cost effective ergonomics

The use of CAD is cost effective because of the advantages described above. If the ergonomics input lags behind the engineering, then the end result is often last minute modifications which take time and money to implement or a product that does not meet the full ergonomics specification. Both of these are undesirable; the first because it increases the development and pro-

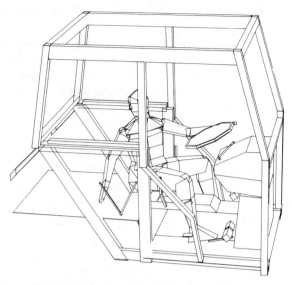

Figure 20.40. Being three-dimensional, the man models can assume complex postures. For example, the tractor driver shown above must be able to reach the hydraulic control and watch the plough as well as operating the normal driving controls

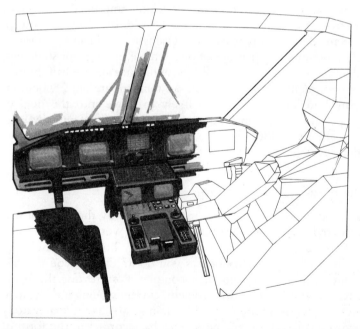

Figure 20.41. A concept model of a helicopter cockpit interior. The combined strengths of the ergonomist and stylist are clearly shown in the above photograph. The ergonomics contribution to the design can be communicated powerfully using 3D graphics

duction costs, whilst the second is likely to reduce the success of the product or service.

There are few disadvantages, and these are more to do with restricting the potential advantages. One problem is that CAD is a powerful tool and, like any tool, it can be dangerous in the wrong hands. The selection of relevant and accurate databases and decisions concerning workstation design and posture require the skills of an experienced ergonomist or a designer/engineer with suitable training. The systems are designed to supplement an ergonomist's skills, not replace them. It would be short-sighted to think that such systems can replace totally user trials; they should only be used to explore alternative designs, to eliminate the poor ones and select and, if possible, improve upon, the promising ones. The results of the CAD evaluation should lead straight to an in-depth user trial with working prototypes, especially if the tasks are complex and performed under adverse conditions.

Future developments

The future of man-modelling CAD systems looks very promising as manufacturing organizations are always looking for ways to reduce development times and costs, whilst producing high quality design for the increasingly 'design aware' public. Useful developments might variously include: increasingly sophisticated and realistic dynamic strength modelling, improved methods of man-model control (data gloves, whole body co-ordinate measuring suits, etc.), expert systems that support design and evaluation, human behavioural modelling (e.g., functional control operations, fatigue, movement strategies, etc.), animated human movement and collision avoidance and perhaps human response to environmental variables such as temperature and vibration. Increasingly man-modelling would seem likely to move into the field of virtual reality, especially for looking at dynamic human movement in systems and animated visualization. With regard to SAMMIE, the following enhancements are under development.

Control of the man-model's posture

The current methods for setting the posture are limited by the fact that it is often difficult to predict the actual posture that people would adopt in some circumstances. For example, could you specify exactly how you would get out of a car without taking mental notes as you do it? Even if you do this, it would be quite tedious to set up such complex postures for the man-model. One interesting solution to this problem, currently being investigated, is the use of a catsuit worn by the user with strain gauges at the major body joints. This device enables the user's posture to be recorded in the form of voltages which could be linked directly to the control of the man-model's posture. Another use of the catsuit would be to collect postural data from a sample of people performing a variety of tasks and use the findings as a database.

Anthropometric database

Very few anthropometric surveys take sufficient measurements to define an accurate 3D model of people. In addition both external dimensions and the location of joint centres, including ranges of movement, are required. It has been suggested (Bonney *et al.*, 1980) that surveys should take into account these requirements and take more comprehensive measurements to maximize the potential applications of their data. The major problem with this request is the time and cost required. However, new developments in recording methods allow the automated collection of thousands of measurements that define points across almost the entire body surface in a matter of seconds. Such data would allow the direct definition of man-model body shapes and forms in a much more complex and realistic manner than is the case in current systems. A man-model desribed by this sort of data is currently under development, the major effort being the development of manageable ways of controlling the complex curved surfaces used to represent the man-model such that users can still make use of more conventional data sources, covering a wider number of subjects and nationalities, and the identification of joint centres within the flesh envelope. A major enhancement that this kind of body surface model would provide is the ability to have actively deforming flesh shapes (e.g., the buttocks change shape when the man-model is moved from a standing to a sitting posture). Apart from a more realistic and visually attractive model this facility will allow the modelling of other non-solid surfaces such as seat cushions, and importantly it will allow man-models to be placed on seats in a highly realistic manner, where both the man-model and seat will deform. This is not possible in any man-modelling system at present. Currently users of these systems rely on H-points, best estimates or, if the proposed seat is available, seat compression tests.

Other development areas under investigation include an expert system design shell, more complex constraint modelling (e.g., the ability to define multiple constraints such as fixing a foot to a pedal and the hips to a seat and causing knee and ankle angles to remain within comfort tolerance as either the anthropometry of the leg is changed or the seat is moved) and a dynamic strength modeller which takes account of the complex interactions between muscle groups in different postures (e.g., the forearm can exert a greater lifting force with the hand facing upward than with the hand facing downward).

References

Badler, N. (1990). *Human Factors Simulation Research at the University of*

Pennsylvania. Report MS-CIS-90-67, Department of Computer and Information Science, University of Pennsylvania, USA.

Bonney, M.C., Blunsdon, C.A., Case, K. and Porter, J.M. (1979a). Man–machine Interaction in Work Systems. *International Journal of Production Research*, **17**, 619–629.

Bonney, M.C., Case, K., Porter, J.M. and Levis, J.A. (1979b). Design of mirror systems for commercial vehicles. *Applied Ergonomics*, **11**, 199–206.

Bonney, M.C., Case, K. and Porter, J.M. (1980). User needs in computerised man models. In *Anthropometry and Biomechanics: Theory and Application,* edited by R. Easterby, K.H.E. Kroemer and D.B. Chaffin (New York: Plenum Press), pp. 97–101.

Case, K. and Porter, J.M. (1980). SAMMIE: a computer aided ergonomics design system. *Engineering*, **220**, 21–25.

Case, K., Bonney, M.C., Porter, J.M. and Freer, M.T. (1990a). Applications of the SAMMIE CAD system in workplace design. In *Work Design in Practice,* edited by C.M. Haslegrave, J.R. Wilson, E.N. Corlett and I. Manenica (London: Taylor and Francis), pp. 119–127.

Case, K., Porter, J.M. and Bonney, M.C. (1990b). SAMMIE: a man and workplace modelling system. In *Computer-Aided Ergonomics,* edited by W. Karwowski, A. Genaidy and S.S. Asfour (London: Taylor and Francis), pp. 31–56.

Churchill, E. (1978). Statistical considerations in man-machine designs. In: *Anthropometric Source Book*, Volume 1: *Anthropometry for Designers*, NASA Scientific and Technical Information Office, NASA Reference Publication 1024, Chapter IX.

Coblentz, A., Mollard, R.and Renaud, C. (1991). Ergoman: 3-D representations of human operator and man–machine systems. *International Journal of Human Factors in Manufacturing*, pp. 167–178.

Dooley, M. (1982). Anthropometric modelling programmes—a survey. *IEEE Computer Graphics and Applications*, **2**, 17–25.

Dreyfuss, H. (1967). *The Measure of Man: Human Factors in Design,* 2nd edition (New York: Whitney Library of Design).

Elias, H.J. and Lux, C. (1986). Gestatung ergonomisch optimierter Arbeitsplatze und Produkte mit Franky und CAD. [The Design of ergonomically optimized workstations and products using Franky and CAD.] *REFA Nachrichten*, **3**, 5–12.

Grobelny, J., Cysewski, P., Karwowski, W. and Zurada, J. (1992). APOLIN: a 3-dimensional ergonomic design and analysis system. In *Computer Applications in Ergonomics, Occupational Safety and Health,* edited by M. Mattila and W. Karwowski (Amsterdam: Elsevier Science Publishers BV), pp. 129–135.

HUMANCAD (1991). *Mannequin User Guide* (New York: Humancad).

Humanscale (1978). *Humanscale Data Portfolios* by Diffrient, N., Tilley, A.R. and Harman, D., for Henry Dreyfuss Associates, MIT Press, USA.

Karhu, O., Kansi, P. and Kuorinka, I. (1977). Correcting working postures in industry: a practical method for analysis. *Applied Ergonomics*, **8**, 199–201.

Kloke, W.B. (1990). WERNER: a personal computer implementation of an extensive anthropometric workplace design tool. In *Computer-Aided*

Ergonomics, edited by W. Karwowski, A.M. Genaidy and S.S. Asfour (London: Taylor and Francis), pp. 57–67.

Kuusisto, A. and Mattila, M. (1990). Anthropometric and biomechanical man models in computer-aided ergonomic design structure and experiences of some programs. In *Computer-Aided Ergonomics,* edited by W. Karwowski, A.M. Genaidy and S.S. Asfour (London: Taylor and Francis), pp. 104–114.

Launis, M. and Lehtelä, J. (1990). Man models in the ergonomic design of workplaces with the microcomputer. In *Computer-Aided Ergonomics* edited by W. Karwowski, A.M. Genaidy and S.S. Asfour (London: Taylor and Francis), pp. 68–79.

Launis, M. and Lehtelä, J. (1992). ergoSHAPE—a design oriented ergonomic tool for AutoCAD. In *Computer Applications in Ergonomics, Occupational Safety and Health,* edited by M. Mattila and W. Karwowski (Amsterdam: Elsevier Science Publishers BV), pp. 121–128.

Levis, J.A., Smith, J.P., Porter, J.M. and Case, K. (1980). The impact of computer aided design on pre-concept package design and evaluation. In *Human Factors in Transport Research,* Volume 1, edited by D.A. Oborne and J.A. Levis (London: Academic Press), pp. 356–364.

Lippmann, R. (1986). Arbeitsplatzgestaltung mit Hilfe von CAD. [Workstation design with help from CAD.] *REFA Nachrichten,* **3,** 13–16.

McDaniel, J.W. (1976). Computerized biomechanical man-model. *Proceedings of the 6th Congress of the International Ergonomics Association and the 20th Annual Meeting of the Human Factors Society,* pp. 384–389.

McDaniel, J.W. (1990). Models for ergonomic analysis and design: COMBI-MAN and CREW CHIEF. In *Computer-Aided Ergonomics,* edited by W. Karwowski, A.M. Genaidy and S.S. Asfour (London: Taylor and Francis), pp. 138–156.

Mollard, R., Ledunois, S., Ignazi, G. and Coblentz, A. (1992). Researches and developments on postures and movements using CAD techniques and ERGODATA. In *Computer Applications in Ergonomics, Occupational Safety and Health,* edited by M. Mattila and W. Karwowski (Amsterdam: Elsevier Science Publishers BV), pp. 337–343.

Pheasant, S. (1986). *Bodyspace* (London: Taylor and Francis).

Porter, J.M. (1981). *Ergonomic Aspects of CAD Workstations.* Report No. CAS.027. (Loughborough: SAMMIE CAD Ltd.).

Porter, J.M. (1992). Man Models and Computer-aided Ergonomics. In *Computer Applications in Ergonomics, Occupational Safety and Health,* edited by M. Mattila and W. Karwowski (Amsterdam: Elsevier Science Publishers BV), pp. 13–20.

Porter, J.M. and Case, K. (1980). SAMMIE can cut out the prototypes in ergonomics design. *Control and Instrumentation,* **12,** 28–29.

Porter, J.M. and Freer, M.T. (1987). *The SAMMIE System, Information Booklet,* 5th edition (Loughborough: SAMMIE CAD Ltd.).

Porter, J.M. and Porter, C.S. (1987). *An Ergonomics Study of the C69 Stock Cab.* Unpublished report for London Underground Ltd.

Porter, J.M., Case, K. and Bonney, M.C. (1980). Computer generated three-dimensional visibility chart. In *Human Factors in Transport Research,* Volume 1, edited by D.J. Oborne and J.A. Levis (London: Academic Press), pp. 365–373.

Porter, J.M., Case, K. and Bonney, M.C. (1990). Computer workspace modelling. In *Evaluation of Human Work: A Practical Ergonomics Methodology,* edited by J.R. Wilson and E.N. Corlett (London: Taylor and Francis), pp. 472–499.

Porter, J.M., Case, K., Freer, M.T. and Bonney, M.C. (1993). Computer-aided ergonomics design of automobiles. In *Automotive Ergonomics,* edited by B. Peacock and W. Karwowski (London: Taylor and Francis), pp. 43–77.

Post, F.H. and Smeets, J.W. (1981). ADAPS: computer-aided anthropometrical design. *Tijdschrift voor Ergonomic* **6**(4), 11–18 (in Dutch).

Rebiffe, R. (1966). An ergonomic study of the arrangement of the driving position. *Ergonomics and Safety in Motor Car Design, London Symposium,* 27 September, pp. 26–33.

Reid, C.J., Gibson, S.A., Bonney, M.C. and Bottoms, D. (1985). Computer simulation of reach zones for the agricultural driver. In *Proceedings of the 9th International Congress of the International Ergonomics Association,* edited by I.D. Brown, R. Goldsmith, K. Coombes and M.A. Sinclair (London: Taylor and Francis), pp. 646–648.

Rothwell, P.L. (1985). Use of man-modelling CAD systems by the ergonomist. In *People & Computers: Designing the Interface,* edited by P. Johnson and S. Cook (Cambridge: Cambridge University Press), pp. 199–208.

Sheldon, W.H. (1940). *The Varieties of Human Physique* (New York: Harper and Bros.).

Westerink, J., Tragter, H., Van Der Star, A. and Rookmaaker, D.P. (1990). TADAPS: a three-dimensional CAD man model. In *Computer-Aided Ergonomics,* edited by W. Karwowski, A.M. Genaidy and S.S. Asfour (London: Taylor and Francis), pp. 90–103.

Chapter 21

The evaluation of industrial seating

E. Nigel Corlett

Introduction

From an ergonomics perspective it is necessary to recognize that the work seat is as much a tool to achieve the work objectives as any other piece of equipment in the workplace. Its design and functioning will be influenced by the tasks to be done, the other equipment to be used, the environment and, of course, the individual human differences. It will be evident that there is unlikely to be one seat suitable for all jobs and the concept of an 'ergonomic chair' independent of the tasks is not possible. Since what we try to do affects our postures, e.g., what we look at or reach for, then it is evident that the contributions of the seat to comfort and support should be developed in relation to these activities.

Seat requirements

From such considerations as these, Table 21.1 provides some of the important requirements for a work seat. They are based on the seat as a full body support, rather than as a temporary perch. There is utility in a perch, where most of the body weight is still on the feet, but it is not part of the present discussion.

Maintaining one third or less of the body weight on the feet was shown by Eklund *et al.* (1982) to be necessary if people were not to complain of leg discomfort. More than this amount on the legs requires continuous muscular activity for body support, a major source of discomfort. The requirement to allow changes of posture is a good ergonomic one and also a practical necessity in many jobs. The use of a high seat at, for example, a supermarket checkout may increase the sitter's reach by allowing movement of the legs, whilst still being fully supported by the seat.

In many cases a work seat has to resist other forces than just the body weight of the sitter. Where movements of the arms are needed, or forces exerted by arms or legs, these forces must be transmitted through the body and the seat to the ground. It is evident that a backrest is a channel for such

Table 21.1. Functional factors in sitting

The task
  Seeing
  Reaching
  Exerting forces

The sitter
  Support weight
  Resist accelerations
  Under-thigh clearance
  Trunk–thigh angle
  Leg loading
  Spinal loading
  Neck/arms loading
  Abdominal discomforts
  Stability
  Postural changes
  Long-term use
  Acceptability
  Comfort

The seat
  Seat height
  Seat shape
  Backrest shape
  Stability
  Lumbar support
  Adjustment range
  Ingress/egress

---

forces on many occasions, otherwise the musculature of the trunk must be tensed to provide a semi-rigid path for the force transmission. The generated muscle tensions will increase the load on the spinal column, particularly in the lumbar spine, which is the major channel for load transmission from the upper to the lower part of the body. A backrest can also reduce loads on the lumbar spine by transmitting part of the gravity forces due to the head, arms and upper trunk through the seat structure rather than the spine (Corlett and Eklund, 1984).

The evaluation of seating

A chair is for a sitter, not for itself, and there is no doubt that this is a case where form must follow function. Thus the priorities in any evaluation must include the responses of sitters, both in their behaviour and in their subjective judgements. It also follows, as noted in the literature, that to evaluate a chair on its dimensions alone, and how they relate to anthropometric data, is inadequate; it should not be necessary to repeat this, but unfortunately there is still a widely held view that anthropometric data are sufficient for seating selection, and even workspace design (see also Pheasant's strictures on this in chapter 19).

It might be as well to begin this section with almost the last words from a paper by Shackel *et al.* (1969): 'Seating comfort is still a very complex problem and the only valid approach is the experimental method'. This does not rule out the need for dimensional criteria, but certainly it calls for a better understanding of them than we currently possess. The Shackel *et al.* paper demonstrates no significant correlations between the British Standards Institution recommended dimensions and the reported comfort. In a study by Langdon (1965) of a group of key punch operators, average stature 64 inches, 85% reported their seat as being comfortable, even though the average seat height was 19·2 inches. This highlights a point made by Branton (1969) and others, that chair comfort has many dimensions, including the task and possibly the appearance. These are discussed by Shackel *et al.* (1969) and further by Lueder (1983) who comments, after an extensive review 'little insight is available into the meaning of comfort'.

Table 21.2 indicates what methods have been used to assess many of the sitting aspects which are important for chair users. It is evident that people can be asked their views on all the sitting aspects listed. The choice of methods will, to some extent, depend on the aims of the investigations and hence whether or not laboratory equipment and conditions are appropriate. In what follows some examples of practical evaluations are presented, demonstrating the selections of test methods which have been found useful. These are based around the seat model of Table 21.3 which links functions, effects and the required measures.

A more comprehensive overview of seating evaluations would extend the 'methods' setion of Table 21.3 to include research techniques, including X-rays, optical posture recording systems and a wider range of physiological areas of study. Much seating research, complementary to the areas discussed here, is concerned with understanding the reasons why discomforts arise, spinal discs suffer damage or particular postures are adopted. This fundamental work, on which the practical evaluations of working seats are based, requires its own review but is not further discussed here.

Dimensional evaluations

Although body sizes are not sufficient data for the design of seating, it is obvious that they are essential. Pheasant (1984) has provided up-to-date data for the UK population, together with useful notes on their use. The clearances which are desirable in fixing the length and height of the seat have been described by Akerblom (1954), Floyd and Roberts (1958) and Murrell (1965). Briefly, for a chair in use at a table, the seat length must clear the calf of a 5th percentile female user, and the height allow light contact under the thighs when the sitter's feet are flat on the floor. This latter requirement makes some adjustment necessary if a 95th percentile male and a 5th percentile female are to use the same chair.

Associated with the question of dimensions is the requirement for adjust-

Table 21.2. Some of the methods used in assessing the functional qualities of industrial seating

Functional factors	Dimensional measurements	Fitting trials	Force or pressure measurements	Bio-mechanics calculations	Observations or timing of behaviours	Subjective judgements overall	Subjective judgements body parts	Check-lists	Cross-modality	Reach/force/stability	Stature changes
Seeing	✓	✓				✓					✓
Reaching	✓	✓				✓				✓	✓
Seat	✓	✓				✓	✓	✓		✓	✓
Backrest	✓	✓				✓	✓	✓		✓	✓
Adjustment	✓	✓				✓				✓	
Ingress/egress	✓			✓	✓	✓			✓		
Stability			✓	✓	✓	✓				✓	
Support weight			✓	✓		✓	✓				
Under-thigh clearance	✓		✓	✓	✓	✓			✓		✓
Trunk–thigh angle	✓			✓				✓			
Leg load			✓	✓	✓	✓	✓				✓
Spinal load			✓	✓		✓	✓				
Neck/arm load					✓	✓	✓				
Posture changes			✓		✓	✓					
Long-term use			✓		✓	✓	✓				✓
Acceptability			✓		✓	✓		✓	✓		
Comfort			✓	✓		✓	✓			✓	✓
Lumbar support	✓	✓				✓	✓				

Methods

Table 21.3. Seating model for assessment of industrial seats (adapted from Eklund, 1986)

Functional factors		Responses and effects	
		Initial	Subsequent
The task	(detailed items	Postures	Discomfort
The sitter	as in Table	Loads	Pain
The seat	20.1)	Pressures	Disease
		Influences on blood flow	Reduction in performance
		Discomfort	
		Preferences	
Measures			
Workplace dimensions		Biomechanical load	Rating
Work weights		EMG	Ranking
Work forces		Stature change	Clinical examination
Work reaches		Rating	Epidemiological studies
Work time patterns		Ranking	Performance
Anthropometry		Dilations of body parts	
Strength		Linear measurements	
		Posture	

ment. Jones (1963) has put forward a simple and effective fitting trials procedure, using subjects from the extremes of the population. (See Pheasant's chapter 19 on anthropometry and the design of workspaces.) In running such trials we might note the comments of Branton (1969) that people older than 30–35 years are likely to be more sensitive to discomfort than younger people. Again, the evidence is in favour of experimentation since, particularly for work chairs the variability in the tasks will require evidence for the facility to see, to reach, to exert forces effectively and to maintain a stable position on the seat.

By using the fitting trials procedure, direct evidence is obtained for the need for adjustments on the chair, and the necessary extent of these adjustments. LeCarpentier (1969) noted the variability in dimensions recorded by the same subjects on different days when studying comfort. Although these findings are important for recognizing the flexibility inherent in dimensions for comfort, the same variability might not exist for some constraining work activities, and there is clearly need for further investigation into the precision to be expected from fitting trials conducted to assess a seat for various functions.

Ingress and egress

These factors are especially important in two cases in particular, seating for the elderly and sit-stand seats. The former is not part of this chapter, but studies by Shipley et al. (1969) showed that accessibility of chairs for the elderly was best assessed by the times taken to get into and out of them, coupled with observations of the subjects' behaviours during these manoeuvres. However, for people without disability, time might not be such a good means of separating seats.

For sit-stand seats the advantages of a saddle are reduced when their use is observed. For women in skirts they are awkward. The times of use for what is often a hard seat surface which supports only a part of the buttocks, as well as observations of the manoeuvres to get on and off, are probably the best methods. Added to this must be stability requirements, which are discussed later. As will be a constant refrain in this chapter, the experiences of the users must also be part of the evaluations.

Observational methods

It is an obvious point that the investigator should watch what the users do, the other side of the coin to gathering their experiences of using the chair. Kember (1976) utilized an adaptation of Benesh notation, a choreographer's tool, to record in detail the postures and movements of a chair user. It is capable of recording on a time base and, providing the three months' intensive training suggested by Kember is not prohibitive, will record sitting activities in great detail.

For most studies however this investment in training will prove excessive, for there is still the need, after recording, for a detailed analysis.

Most workers have used simpler methods, of which that by Branton and Grayson (1967) is typical. It may be modified to include actions, and so on, relevant to the problem under investigation with little difficulty. For their easy-chair comfort study they selected positions of the head, trunk, arms and legs, with separate numbers in each body part for the important postures (see Table 21.4). If, for example, the head was also supported by the hands, or

Table 21.4. Coding of sitting postures (Branton and Grayson, 1967)

Head		
	Free of support	1.
	Against headrest	2.
	Against side wing	3.
	Supported by hands	4.
Trunk		
	Free from backrest	1.
	Against backrest	2.
	Lounging/slumped back	3.
Arms (one or both)		
	Free from armrest	1.
	Upon armrest	2.
Legs		
	Free, both feet on floor	1.
	Crossed at knee	2.
	Crossed at ankle	3.
	Stretched forward	4.

the legs were stretched forward, the relevant number on the recording sheet was also ringed. The resulting records are easy to enter into, and analyze by, computer and the coding is easy to learn.

Clearly the whole range of observational techniques can be drawn on and modified to suit the investigators. Analysis, however, usually takes longer than recording, and even activity sampling methods (Grandjean, 1980; Branton, 1969), which are very useful to record the postural behaviour over several hours and are simple to apply in many cases, can condemn the investigator to many hours of tedious analysis. If body markers can be used, electronic analysis methods are becoming more practical and in the laboratory are well used (Corlett *et al.*, 1986). In the field, however, body markers are not often practical and video recording not always welcomed.

The length of time, and time of day are both important in observational studies. The length of time must be relevant to the task. Branton and Grayson (1967) used 4–5 h, since that was the length of a long train journey, and hence the period of use of the seat. Shipley *et al.* (1969) studied people in homes for the elderly over the whole day. On the other hand, where work imposes a variety of activities during the day, as for many office workers, randomly chosen periods of 1 or 2 h, or half a day may be appropriate. The samples should be taken at intervals over the whole day, however. Both Branton and Grayson and Shipley *et al.* report significant differences between morning and afternoon sitting behaviours. There can also be differences between men and women in the postures adopted (Branton and Grayson, 1967) and in the length of the sitting periods (Shipley *et al.*, 1969), both factors which may apply in situations other than those studied by these workers.

Subjective methods

This heading embraces a wide range of methods, requiring an understanding of psychophysics. Comfort is one of the variables which is most often examined by ranking or rating scales (and the reader is referred to chapter 3 for general discussion of these). Allen and Bennett (1958), for an air pilot's seat, used a forced choice method of ranking all body areas in decreasing levels of comfort. Corlett and Bishop (1976) preferred to focus on discomfort, asking subjects either to rank, or to rate, body areas perceived as suffering discomfort. There was no requirement to cover all body areas. (See chapter 23 on static muscle loading in this book.)

The recognition of discomfort appears to change with circumstances. Branton (1969) proposed a model in which people traded off stability against relaxation, indicating that no posture will remain comfortable for long periods. Shipley (1980) suggests that discomfort varies with arousal and attention; periods when attention 'turns inward towards the condition of the self' cause what has been ignored to receive attention, producing fluctuations in subjective discomfort.

Several workers echo the view of Habsburg and Mittendorf (1980) that people seem to attempt to produce a general view of comfort unless asked to consider 'is it for me/not for me?' This question focuses the response more sharply and the instruction to subjects to consider how it is for them at the time the questions are asked, rather than to leave this aspect of the survey implicit, would appear to be good practice.

A *general comfort rating* scale, developed by Shackel *et al.* (1969), is given in Figure 21.1. It is fine enough to evaluate even small differences in discomfort, and was administered by Drury and Coury (1982) every half hour of their trials. Shackel *et al.* (1969) provide details of the range of comfort of the 10 chairs they tested, and Drury and Coury (1982) also show their chairs on the same metric. It is thus possible to compare results from this scale with other workers, and get some feel for the relative effectiveness of the chair under test.

The question of focus is important, Wachsler and Learner (1960) found that the only factors which correlated highly with feelings of comfort in their aircrew study was comfort of the buttocks and back. If these were accommodated, subjects were willing to overlook other discomforts when rating a seat. In later studies the use of a 'body map' to obtain *body part discomfort* (BPD) has been widely used, often based on Corlett and Bishop (1976).

The use of BPD scales, either with or without scales looking at somatic conditions, gives 'before and after' comparisons based on personal data which are very compelling evidence of changes and their effects, providing the controls or intervening variables are maintained. Shackel *et al.* (1969) are not the only researchers to draw attention, for example, to the effects of appearance on judgements of seat quality. Extended periods of use were employed by Bendix *et al.* (1988) prior to investigating their subjects' responses to two chairs under comparison. Drury and Coury (1982) report a comparison between several hours' use and a rapid evaluation procedure; the two studies

General comfort rating

Please rate the chair on your feelings *now*

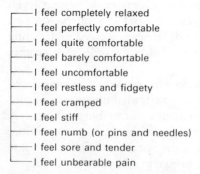

    I feel completely relaxed
    I feel perfectly comfortable
    I feel quite comfortable
    I feel barely comfortable
    I feel uncomfortable
    I feel restless and fidgety
    I feel cramped
    I feel stiff
    I feel numb (or pins and needles)
    I feel sore and tender
    I feel unbearable pain

Figure 21.1. General comfort rating scale of Shackel *et al.* (1969)

shared closely similar results. These workers used small groups (12 subjects). Jones (1969) has suggested that economies can be obtained by training testers, quoting that teaching testers to discriminate different levels of discomfort can give highly repeatable results. These however are within-testers; the between-testers comparison gave considerable differences.

Apart from focusing attention on body parts, subjects can be asked to focus attention on chair details, such as the *chair feature check list* (CFCL) (Shackel *et al.*, 1969; Drury and Coury, 1982) or on aspects of the task, which could be rated for difficulty. This last presupposes a good task analysis, as called for by Lueder (1983), which is not often done.

The CFCL shown in Figure 21.2, is again by Shackel *et al.* (1969), but modified by Drury and Coury (1982). It provides an opportunity to get the mean and distribution of the effects of the various aspects of a chair as experienced by the sample of sitters. Typically, those dimensions which the subjects can adjust prior to the test will give a smaller range, and a mean closer to the optimum, than the scores on other features.

When evaluating seating, the length of time during which the subjects are exposed to the conditions is important. Jones (1969) illustrated examples where certain boundary levels of discomfort, just distinguishable by untrained

Chair feature checklist

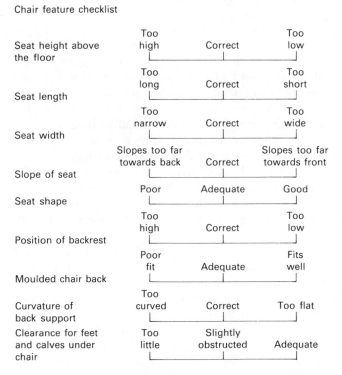

Figure 21.2. Chair feature checklist of Shackel *et al.* (1969), modified by Drury and Coury (1982)

subjects, did not appear until after 3 h or more. Branton (1966) identified a pattern of sitting behaviour in less time than this, whilst LeCarpentier (1969) reported that after 20 min he found no changes in preferred easy-chair sittings even though the subjects sat for 2 h.

Shackel *et al.* (1969) used brief (5 min) periods of sitting, under carefully controlled conditions, and a ranking procedure, to separate a range of chairs so that they were suitable for a group of people. Drury and Coury (1982) used a short period for adjustment of the chair by each subject to an initial position and, in common with several other studies, then evaluated its effectiveness against longer periods of sitting.

Cross–modality matching (CMM)

Discomfort has been related to pressure distribution measured on the seat and backrest (Wachsler and Learner, 1960; Habsburg and Mittendorf, 1980). Lueder (1983) comments that 'the lack of an accepted measure of comfort [. . . has] frequently caused comfort to be relegated to a low priority in comfort (sic) decision making'. Whilst accepting the first point, chair researchers appear to have put comfort high on their list of criteria, as well as giving it extensive discussion in their papers.

In the discussion on comfort by Branton (1969), he reports an attempt to use a hand dynamometer in a cross-modality match with feelings of bodily tension. Although unsuccessful in his trials to relate the responses to seat features, he considered the method had some potential, and that it would facilitate the subject's ability to evaluate responses to the seat.

One attempt to quantify perceptions of pressure and their links with discomfort is reported by Gregg (personal communication). Using a sphygmometer cuff as the other modality, he asked subjects to use magnitude estimation to estimate perceived pressure, and demonstrated a highly significant reliability between repeated trials for estimating perceived pressure. With this information Gregg used the pressure cuff to determine the pressure values which characterize a discomfort scale with five points lying between 'no noticeable discomfort' and 'extremely uncomfortable'. He found repeatability to be good.

Extending these studies he then compared the results of BPD scales, and pressure recordings from a cuff on the subject's arm, in a number of trials, in which he compared seat heights, pressure on the ischii and discomfort experienced in various body parts, whilst sitting for 2 h. He demonstrated close agreement between the two methods. By combining a CFCL with the body part assessments by cross–modality matching (CMM) using the cuff, a recognition of some of the causes of discomfort was obtained. In particular it was found possible to identify pressure distribution across a seat surface by requiring the pressure judgements to be made via the comparison with a cuff.

Posture changes and stability

An aspect of behaviour which researchers have sought to use as an indicator of seat comfort is the amount of posture change which ocurs during sitting. The argument is that if the subject changes position frequently (fidgets), then the seat is not comfortable. Difficulties have been found with using the concept; for example, it is generally agreed that some changes in posture are desirable, and of course some are necessary due to the task's demands. In looking at automobile seating, Rieck (1969) found that although a questionnaire separated one seat from the other four, the measurement of small movements in the seats, using a force platform, did not. Branton (1966) found distinct changes of posture in relation to seat shapes, although he was looking more at gross changes than 'fidgets'.

The requirement for stability has usually been discussed in terms of the chair tipping over during use, or of its resistance to movement during work activities. Branton (1969) raised the point of the stability of the sitter, in the seat, in the context of easy-chair comfort. He pointed out that 'if the seat does not allow postures which are both stable and relaxed, the need for stability seems to dominate that for relaxation', requiring muscular activity to secure stability. It may appear self evident that people will not relax on a seat if, by so doing, they would fall out of it, but some proposals for forward-sloping seats have certainly increased the efforts needed to maintain stability at the expense of relaxation. This is not to say that, in some cases, the trade-off has not been satisfactory; it is just necessary to note that muscular effort is a concomitant of the design.

Changes in stature

Having devised a seating model which, *inter alia*, proposed that one requirement of a good industrial seat was that it should reduce the load on the spine, Eklund (1986) examined several methods for evaluating a seat in these terms. Biomechanical analysis, in conjunction with a force platform (Eklund *et al.*, 1983) or an instrumented chair, (Eklund *et al.*, 1987) could assess the loads on the spine, but provided no direct evidence of the validity of the values, or their actual effects on the sitter.

Eklund and Corlett (1986) reported a study to compare BPD, biomechanical analysis and stature change measures for industrial seats. The study was conducted in a laboratory, and involved a force production task, a sideways viewing task (as in some fork-lift truck driving) and a sit-stand seat. They noted that the three methods were all effective but contributed different areas of information. The theoretical evaluation of the load, using biomechanics, was quick and inexpensive. The stature change method needed at least 30 min exposure by each subject, may need several subjects but was better for major loading situations. It had the advantage of giving a numerical measure of the actual effects of the load on the spine. Discomfort assessment was inexpensive, sensitive and suitable for field work but it, too, required long exposure by the subjects. Trials with the precision stadiometer had demonstrated its practicability in an industrial setting (see also chapter 23 for discussion of the stadiometer).

Comprehensive evaluation procedure

1. Starting with the need to evaluate a range of chairs for their utility in various situations, and no comprehensive assessment procedure available, the procedure of Shackel *et al.* (1969) involved:

    (a) ranking of the chairs for preference, whilst sitting on them but not being allowed to see or touch them;
    (b) long-term sitting whilst working, with regular completion of a general comfort rating and body part discomfort (BPD) ranking; and
    (c) completion of a chair feature check list (CFCL) at the end of the session.

This procedure showed where chairs were inadequate, either in dimensions or in relation to tasks, and which chairs were preferred.

2. Where a single chair has to be tested, for example when a prototype has been built or a company is exploring the purchase of a particular chair, Drury and Coury (1982) proposed a variation on this set of tests. They utilized the data from Shackel *et al.* (1969) as a basis for part of their evaluation, also bringing in a modified general comfort questionnaire and the CFCL from their methods. They used LeCarpentier's (1969) concept of initial adjustments and the BPD procedure of Corlett and Bishop (1976).

The procedures adopted, in summary, were:

    (a) comparison with anthropometric data, standards and principles;
    (b) opportunity to adjust the chair until the feeling of comfort is maximized for the sitter, typically taking about 5 min;
    (c) a sitting and working period of 2·5 h, with general comfort and BPD scales given every half hour; and
    (d) CFCL given at the end of the sitting session.

The results of the *general comfort* and CFCL measures were compared with Shackel *et al.* (1969), to identify how the chair lay in the general field. They confirmed by further studies the reliability of the battery of methods, and demonstrated its sensitivity by noting the discrimination which occurred, and the opportunities available for the interpretations of differences between the different measures.

Drury and Coury (1982) report that the method enabled a single chair to be evaluated both for its use in given jobs and to provide information to the maker on its weak points. It would certainly be desirable to have a broader base for comparison than the 10 chairs from the Shackel *et al.* (1969) study, but even without this section of the procedure, the method is probably the most economic procedure available which covers the major features in sitting which are important to the user.

3. Recognizing the complexity of any seating assessment, Yu *et al.* (1988) used a fractional factorial experiment to evaluate seven variables in seat design for a sewing task. These were seat height, seat angle, whether or not the seat rocked, whether or not it swivelled, backrest distance, backrest height and backrest angle. The statistical procedure is one that minimizes experimental costs and demonstrated the practicality of this form of experimental design. Their dependent variables were general discomfort, BPD, stature change and electromyography (EMG).

Two females, spanning 90% of female stature, were used as subjects. In a summary table of results the use of BPD and stature change provided most information, and their results supported each other. The other two methods did not contribute additional information so far as the selection criteria for the chair were concerned.

4. To develop a seat for a university auditorium, Wotzka *et al.* (1969) first did an activity sampling study of student activities in four large auditoria, accompanied by a questionnaire to the students about the seat and desk characteristics. From this they deduced a range for the variables of importance and built five seats to test their effectiveness. A number of subjects attended one test session of 20–30 min. First, the seat was adjusted for each subject 'until the most comfortable posture had been achieved', when each subject was given a questionnaire to assess six body parts for 'uncomfortable, medium or comfortable'. From this trial, a set of seats, including some modified as a result of the previous experiment, were ranked by paired comparisons, followed by the use of the same body part questionnaire. The final test was in an auditorium, using the same seats as in the paired comparison study. Here, again the same questionnaire was used.

This study linked the questionnaire with observed behaviours and body part judgements in a long study, which required a large number of subjects. This has the advantage of increasing the confidence one may have in the seat's acceptability, but increases the time and expense of the study.

The study is typical of those which associate questionnaires with other measures. Hünting and Grandjean (1976) have used activity sampling and discomfort bi-polar questionnaires for selected body parts in the comparison of seats. They have deduced preferences, and desirable changes, from the results, although subject numbers have been large.

Conclusions

A list of the functional factors relevant to seat use is given in Table 21.3. When evaluating a seat, those factors relevant to the purposes of the seat must be studied. To define which are relevant may need a task analysis, as Lueder (1983) has pointed out. An activity sampling study of postures linked with the task analysis, will give some insight into the ways that people are coping with the job, helped by, or in spite of, the chairs provided.

From such a study, combined perhaps with measurements of the seat and workplace and backed by occupational health evidence and/or questionnaire or BPD study, the present state of a situation can be documented. This would provide data for a focused attack on the seating and its associated factors, as well as a baseline for comparing the results of any changes. These data would come particularly from the combination of measurements, observations and BPD measures. But then, having made the design decisions and the related tests which appear to be relevant, what must be done about the testing of the final results?

A two-stage procedure seems appropriate. With only one or a few chairs, but with data derived from previous studies, the methodology used by Drury and Coury (1982) is appropriate. Briefly, this is a dimensional comparison with standards or other validated data, a short self-adjustment period for each subject, then a long (2 h plus) exposure when the seat is used for its designed purposes, during which BPD are gathered, combined at the same sessions with a general comfort scale if this is seen as desirable. At the end of the test period, a CFCL is administered.

This test procedure locates the chair within a spectrum of other chairs used for similar purposes and identifies weaknesses in the new design, as subjectively recognized by users. However, if certain additional requirements are needed, e.g., a given reach envelope, reduced back load compared with existing chairs, or similar, then additional measures must be added to the above procedure. In the light of Yu *et al.* (1988), measures of stature change, the BPD and the CFCL may well give all the necessary data plus clear evidence of spinal loading.

The wide variation in chair users, as well as the wide range of uses to which a chair will be put, require a further stage after the installation of a number of the modified chairs. It is at this final stage that the activity sampling and general questionnaire are useful. The procedure is not necessarily expensive but it does allow the modes of use to be assessed against what was expected in the design and in the first experimental stage. It also assesses the acceptability of the design, giving a comparison with the initial study and indications of any gains made.

The reader will recognize that there is still much to do in chair research and application. Our knowledge of comfort factors is still small and our ability to provide an effective working seat has much room for improvement. This is in no way to discredit the studies which have demonstrated good designs for particular cases. But a walk around an office, factory or supermarket, or a discussion with an orthopaedic surgeon, would reveal how far there is to go. If one thing is in the ergonomist's favour, it is the increasing recognition of the health and safety aspects of seating. A major weakness in the computerized office is the operator's back, and it is this 'discovery' that has again put seating firmly on the ergonomist's agenda.

Note

This chapter is largely based upon the Ergonomics Society's Lecture 1989, reproduced in *Ergonomics*, **32**, 257–269.

References

Akerblom, B. (1954). Chairs and sitting. In *Symposium on Human Factors in Equipment Design,* edited by W.F. Floyd and A.T. Welford (London: H.K. Lewis).

Allen, P.S. and Bennett, E.M. (1958). *Forced Choice Ranking as a Method of Evaluating Psychological Feelings.* Technical Report No. 58 (USAF, WADC).

Bendix, A., Jensen, C.V. and Bendix, T. (1988). Posture, acceptability and energy consumption on a tiltable and knee support chair. *Clinical Biomechanics*, **3**, 66–73.

Branton, P. (1966). *The Comfort of Easy Chairs.* Interim report. Furniture Industry Research Association, UK.

Branton, P. (1969). Behaviour, body mechanics and discomfort. *Ergonomics*, **12**, 316–327.

Branton, P. and Grayson, G. (1967). An evaluation of train seats by observation of sitting behaviour. *Ergonomics*, **10**, 35–51.

Corlett, E.N. and Bishop, R.P. (1976). A technique for assessing postural discomfort. *Ergonomics*, **19**, 175–182.

Corlett, E.N. and Eklund, J.A.E. (1984). How does a backrest work? *Applied Ergonomics*, **15**, 111–114.

Corlett, E.N., Wilson, J. and Manenica, I. (Eds) (1986). *The Ergonomics of Working Postures, Section 5: Seats and Sitting.* (London: Taylor and Francis).

Drury, C.G. and Coury, B.G. (1982). A methodology for chair evaluations. *Applied Ergonomics*, **13**, 195–202.

Eklund, J.A.E. (1986). Industrial seating and spinal loading. Ph.D. thesis University of Nottingham. Distributed by Dept of Industrial Ergonomics. University of Technology, Linköping, Sweden.

Eklund, J.A.E. and Corlett, E.N. (1986). Experimental and biomechanical analysis of seating. In *The Ergonomics of Working Postures,* edited by E.N. Corlett, J. Wilson and I. Manenica (London: Taylor and Francis).

Eklund, J.A.E., Corlett, E.N. and Johnson, F. (1983). A method for measuring the load imposed on the back of a sitting person. *Ergonomics*, **26**, 1063–1076.

Eklund, J.A.E., Houghton, C.S. and Corlett, E.N. (1982). *Industrial Seating, A Report of Some Pilot Studies*. Internal report. University of Nottingham, published in Eklund (1986).

Eklund, J.A.E., Örtengren, R. and Corlett, E.N. (1987). A biomechanical model for evaluation of spinal load in seated work tasks. In *Biomechanics XB*, edited by B. Jonsson (Champaign, IL: Human Kinetics Publishers).

Floyd, W.F. and Roberts, D.F. (1958). Anatomical and physiological principles in chair and table design. *Ergonomics*, 2, 1–16.

Grandjean, E. (1980). *Fitting the Task to the Man* (London: Taylor and Francis).

Habsburg, S. and Mittendorf, L. (1980). Calibrating comfort: systematic studies of human responses to seating. In *Human Factors in Transport Research*, edited by D.J. Oborne and T.A. Levis (New York: Academic Press).

Hünting, W. and Grandjean, E. (1976). Sitzverhalten und subjektives Wohlbefinden auf Schwenkbaren und fixierten Formisitzen. *Zeitschrift für Arbeitswissenschaft*, 30, 161–164. (Quoted in report by E. Grandjean and W. Hünting from Federal Institute of Technology, Zurich of 20.10.1976.)

Jones, J.C. (1963). Fitting trials. *Architects Journal*, 137, 321–325.

Jones, J.C. (1969). Methods and results of seating research. *Ergonomics*, 12, 171–181.

Kember, P. (1976). The Benesh movement notation used to study sitting behaviour. *Applied Ergonomics*, 7, 133–136.

Langdon, F.J. (1965). The design of card punches and the seating of operators. *Ergonomics*, 8, 61–65.

LeCarpentier, E.F. (1969). Easy chair dimensions for comfort—a subjective approach. *Ergonomics*, 12, 328–337.

Lueder, R.K. (1983). Seat comfort: a review of the construct in the office environment. *Human Factors*, 25, 701–711.

Murrell, K.F.H. (1965). *Ergonomics* (London: Chapman and Hall).

Pheasant, S.T. (1984). *Anthropometry, an Introduction for Schools and Colleges*, BSI Education No. PP7310. (London: British Standards Institution).

Rieck, A. (1969). Über die Messung des Sitzkomforts von Autositzen. *Ergonomics*, 12, 206–211.

Shackel, B., Chidsey, K.D. and Shipley, P. (1969). The assessment of chair comfort. *Ergonomics*, 12, 269–306.

Shipley, P. (1980). Chair comfort for the elderly and infirm. *Nursing*, supplement 20, *Sleep and comfort*.

Shipley, P., Haywood, J., Furness, W. and Rose, J. (1969). *Testing Easy Chairs for the Elderly*. Report to the Research Institute for Consumer Affairs, London.

Wachsler, R.A. and Learner, D.B. (1960). An analysis of some factors influencing seat comfort. *Ergonomics*, 3, 315–320.

Wotzka, G., Grandjean, E., Burandt, U., Kretschmar, H. and Leonhard, T. (1969). Investigations for the development of an auditorium seat. *Ergonomics*, 12, 182–197.

Yu, C.-Y., Keyserling, W.M. and Chaffin, D.B. (1988). Development of a work seat for industrial sewing operations: results of a laboratory study. *Ergonomics*, 31, 1765–1786.

Part V

Analysis of work activities

Having examined and explained the methods and techniques basic to all areas of ergonomics investigation, looked in more depth at techniques developed or adapted to study some particular design and evaluation applications, and then devoted a number of chapters to the broad area of physical environment assessment, we turn now to analysis techniques for work activities. Each of the chapters in this section takes either a specific, if broad, area of work activity, such as visual performance, or a specific consequence of certain types of work, as in stress assessment. The individual contributions draw upon any number of the methods discussed earlier in the book in more basic and general form.

Work involves both physical and mental activities and this section considers both of them. The physiological costs of work are a reality to many, and are not necessarily alleviated by modern technology. Hence it is proper that we should be informed about the major methods for evaluating physical effort, including simple methods of field investigation to recognize when this effort is high or excessive.

Although dynamic work (Kilbom, chapter 22) is obvious when it is perfor-med, static and postural load (Corlett, chapter 23) is often overlooked. It is easy to assume that if someone is sitting down there is no load of any signifi-cance, and forget that muscles are still active holding a posture. If the job restricts the opportunity for change between muscles, then the situation is exacerbated. People will often discount the discomfort of postural loading, expecting to 'feel tired after work'. However these manifestations give guid-ance to the ergonomist on where to look for the sources of problems, and methods to assess the effects on the person are important in very many situ-ations. It will have been noted, in the chapter on seating at the end of the last section, how subjective methods were utilized, whilst many previous chapters all use people's responses as primary data for analysis.

When risks, and boundaries for levels of exposure, have to be assessed, we need numbers. Chapter 24 by Tracy on biomechanics is cautious about the acceptance of figures but strongly in favour of biomechanic analyses for 'before and after' comparisons. If figures are to be used as absolute measures and matched against criteria, the availability of a range, or measure of the likely spread of results, together with the proposals for using 'worst-case' values, contributes to the assurance that the decisions taken will be conserva-tive.

If the assessment of physical workload is difficult and contentious, then the situation for mental workload (Meshkati *et al.*, chapter 25) is no different. The methods used do parallel some of those for physical workload— especially performance (or primary task measures), subjective assessment and physiological measurement. Even the fourth group of techniques–secondary task measures—can be used for physical workload too, for instance checking the effects of extended keyboard use by testing subsequent performance on one of the many tests of manipulative ability.

Of course, one way in which mental workload assessment differs from physical workload assessment is in the less overt and visible nature of behav-iour involved. Direct observation of task performance or worker response can often at least confirm or illuminate investigations of physical work. This is less easy for mental work, and techniques such as protocol analysis (see Bainbridge and Sanderson; chapter 7) may be needed. A similar problem, i.e., that the causative factors are not easily or directly observable, is also found in the assessment of stress. In chapter 26, Cox and Griffiths discuss the various models or definitions of stress—as with many topics in this book the model or definition chosen can affect our selection of methods as well as of preventive strategies. They then address the identification of stress potential, and recognition and measurement of stress in individuals.

The next two chapters in this section are complementary reviews of the assessment and consequences of performing visual work, appropriate since over 90 per cent of work task-related input is visual. Bullimore *et al.* (chapter 27) are concerned with the performance of visual tasks, whereas Megaw (chapter 28) concentrates upon one possible outcome or group of outcomes of visual performance, namely visual fatigue. As well as the types of method that may be adapted for use in almost any situation, visual processes and their consequences may be assessed by use of some particular techniques also, including specific performance tests (e.g., for colour vision) and observed behavioural measures such as eye movements. Indeed, visual task performance is a very rich topic as far as methods and techniques development is concerned. This is enlarged upon in chapter 29 in which Kumashiro gives a view from Japan of measurement of workload which concentrates upon a measure of visual function (critical flicker frequency), one of attention and performance (concentration maintenance function) and one of physiological function (heart rate variability).

In conclusion, if there were one theme which connects all the work activity analysis in this section, it is that measurement and assessment of effects on people are made doubly difficult because of the problems in defining, in unambiguous terms, the phenomena concerned. For instance, workload, physical or mental, is made harder to understand and evaluate if we are unsure of its causes; disagreement about what visual fatigue is does not make the task of assessing it any easier; and so on.

Chapter 22

Measurement and assessment of dynamic work

Åsa Kilbom

Introduction

The human body is continuously required to perform physical work. Three main types of demand must be met:

(a) moving the body or its parts, e.g., in walking and running,
(b) transporting or moving other objects, e.g., in carrying, lifting, hitting, cranking, and
(c) maintaining the body posture, e.g., in forward stoop of the body, twisted trunk, raised arms.

When exposed to these demands the human body responds with a complex series of events, leading to the performance of muscular exercise. Thus the muscle contraction is the end point of events taking place in the sensory organs, the brain, nervous system, lungs, heart and blood vessels and musculo-skeletal systems (see Figure 22.1).

The term *physical stress* is often used to describe the demands, while *strain* is used to describe the response in the human body (although see discussion of different stress models in chapter 26). The assessment of these physical stresses and strains is an important component of ergonomics. It is used to identify excessive physical stresses and to design external demands so that they fit the capacity of the workers. This chapter will deal with the quantitative evaluation of (1) physical stress related to demands of the cardiovascular system from dynamic work, and (2) physiological strain on the cardiovascular and pulmonary systems, predominantly during dynamic work.

Sometimes physical stress can be accurately predicted. For example, the energy and the power needed to transport the body up a ladder can be calculated, knowing the weight of the body, the vertical distance and the time available. In a similar way the force or torque necessary to maintain body

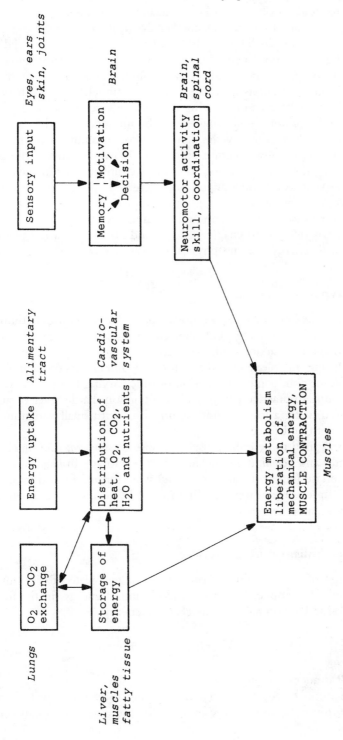

Figure 22.1. Pathways leading to muscular contractions

posture or to hold an object can be calculated. (Evaluation of the resulting biomechanical strain is covered in chapter 24 whilst evaluation of postures and of static muscular loading are discussed in chapter 23.) In mixed tasks, encompassing body movements, exertion of forces and maintaining body posture in a complex time sequence, it is often impossible to predict the demands, although attempts have been made to model them. In these cases the demands must be measured, using the methods described below.

Obviously the response—*the strain*—will be influenced by the capacity of the individual and not only by the demands. For optimum performance all systems of the body must function efficiently. However, any of the organs participating in the events leading to muscle contractions can have a low functional capacity or small dimensions, thereby limiting the capacity for muscular work. Those systems that most commonly limit the rate of physical work are the cardiovascular system and the muscles.

The reader is referred to chapters 23, 24 and 30 also for a full picture of issues related to energy use and physical stresses at work.

Different types of exercise

During *dynamic* exercise muscles are shortening and lengthening rhythmically, e.g., in running and walking. The shortening phase is also called *concentric* exercise, while a contraction with simultaneous lengthening is called *eccentric* exercise. During *static* exercise the muscles maintain a contraction with unchanged force and length for a period of time varying from a few seconds to several hours. In practice purely static contractions hardly ever occur; the most common being a low intensity contraction with small variations both in muscle length and force.

Other terms used to describe muscle exercise are *isometric* (i.e., unchanged muscle length) and *isotonic* (i.e., unchanged muscle force). In *intermittent* work muscular exercise (either dynamic or static) is performed for a duration of a few seconds up to several minutes, interrupted by rest, then resumed, again followed by a pause and so on.

Muscle metabolism during exercise

The discharge of a nerve impulse on the motor end plate of the muscle is the signal for a fast conversion of chemically bound energy, in the form of ATP (adenosine triphosphate), to mechanical energy:

$$ATP \rightleftarrows ADP + phosphate + energy$$

where ADP is adenosine diphosphate.

Available stores of ATP are very limited, and therefore they have to be built up continuously from energy obtained by the oxidation of glucose and fatty acids. Protein is also metabolized but at a much lower rate, its main function being to provide material for tissue repair and growth.

The metabolism can take place either aerobically (with oxygen) or anaerobically (without oxygen). Oxygen is transported to the muscles by the circulation. If enough oxygen is available the aerobic pathway is chosen, because it is more efficient as it leads to a more complete metabolism of energy-rich nutrients and gives less fatigue. The local fatigue perceived in conjunction with anaerobic muscle metabolism is probably caused by the lowering of pH that takes place when lactate is produced.

$$\text{aerobic pathway} \begin{cases} \text{free fatty acids} \\ \text{glucose} \end{cases} + O_2 \rightarrow CO_2 + H_2O + \text{energy}$$

$$\text{anaerobic pathway} \qquad \text{glucose} \rightarrow \text{lactate} + \text{energy}$$

Note that the breakdown of ATP is also anaerobic, but no lactate is produced.

Role of the cardiovascular system during exercise

The main role of the cardiovascular system during exercise is to transport:

(a) heat from exercising muscles to the body surface,
(b) nutrients (fatty acids and glucose) from their stores in liver and fatty tissue to exercising muscles,
(c) oxygen from lungs to muscles,
(d) CO_2, H_2O and lactate from muscles to lungs, liver and kidneys for excretion and metabolism.

In order to meet these demands the circulation can increase the transporting capacity by a factor of 100 within a few minutes after the onset of exercise. This is achieved by:

(a) distribution of more blood to exercising muscles,
(b) a more efficient uptake of O_2 and excretion of CO_2,
(c) increasing the total blood flow, by increasing the cardiac stroke volume and heart rate.

Physical working capacity

The ability of any individual to perform physical work varies within very wide limits. These variations are mainly due to genetic factors which influence both body dimensions and functional capacity. The ability to perform aerobic work is best evaluated through measurements of the *maximum aerobic power*, measured as the highest uptake of oxygen per minute (max $\dot{V}O_2$) that can be achieved during dynamic exercise with large muscle groups. Physical inactivity, e.g., a few days in bed, leads to a fast reduction of the maximal $\dot{V}O_2$, whereas training gives nearly as fast an increase. Thus the aerobic power achieved in an individual is the effect of an adaptation process. Variations in average male and female values by age are presented in Figure 22.2 (adapted from Astrand, 1960).

Fatigue and recovery

Anaerobic metabolism leading to lactate accumulation and muscular fatigue takes place:

(a) at the onset of dynamic exercise,
(b) during heavy dynamic exercise, i.e., when the energetic demands exceed 50% of the individual maximal aerobic power,
(c) during static exercise exceeding about 10% of the maximal muscle strength.

Some of the lactate produced during exercise can be metabolized in the muscles and/or removed. However, during very heavy dynamic work or during static exercise blood circulation cannot keep up with the demands on oxygen supply and lactate removal, and this leads to lactate accumulation,

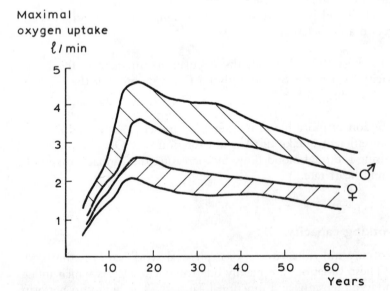

Figure 22.2. Maximal aerobic power in physically active men and women by age. The solid lines indicate mean values, the upper shaded area represents +2 S.D. for men and the lower shaded area represents −2 S.D. for women. Modified from Astrand (1960)

lowered pH, perception of fatigue and reduced endurance. The endurance during static contractions decreases as the intensity of the contraction increases. For a relatively unlimited endurance, without subjectively perceived fatigue, contraction intensities of 5–10% of maximal muscle strength must not be exceeded.

The cycle of fatigue and recovery is demonstrated in Figure 22.3. The rate of recovery after fatiguing exercise depends on:

(a) the duration of exercise,
(b) the intensity of exercise, and
(c) the physical fitness of the individual.

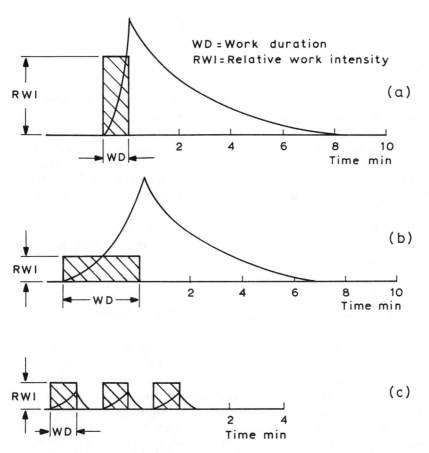

Figure 22.3. Schematic representation of build-up of fatigue, and recovery in three types of exercise. (a) Very high intensity exercise during 1 min, followed by recovery which in this case took around 9 min. (b) Exercise at ⅓ of previous intensity, performed during 3 min, followed by recovery which took around 7 min. (c) Intermittent exercise consisting of three exercise periods at ⅓ of initial intensity, each performed during 1 min with 1-min pauses interspersed. Note the brief recovery periods

The rate of recovery depends to a large extent on the ability of the blood circulation to supply oxygen to tissues (repayment of the 'oxygen debt'). Therefore recovery has usually been studied by measuring heart rate after exercise.

Recent research indicates that recovery of muscle function may not be complete even though all circulatory variables have returned to pre-exercise levels. After static contractions heart rate may normalize, but on repeated contractions muscle strength and endurance may still be reduced. After prolonged static or eccentric exercise to exhaustion full recovery may take several hours and maybe even days. The most likely cause of this reduced performance capacity is damage to the structure of the muscles (including ruptures and oedema formation) and maybe also depletion of energy-rich nutrients in the muscles.

Dynamic work measurements

Evaluating energy expenditure

Energy expenditure during dynamic work is expressed in kilojoules (kJ) or in litres of oxygen consumed (VO_2). Usually energy expenditure is expressed per time unit, i.e., as power, in kilojoules per second (kJ/s) or in litres of oxygen per minute ($\dot{V}O_2$). The rationale for measuring oxygen uptake is that the amount of oxygen consumed during aerobic exercise is directly proportional to the amount of energy produced within the body. Thus, oxygen uptake is an indirect measure of the demands on work output, i.e., it is a measure of physical stress.

Of the energy produced a large part is in the form of heat. The remaining energy production during exercise is used for the work performed, i.e., pedalling, walking, etc.

$$\text{Mechanical efficiency} = \frac{\text{external work produced}}{\text{total energy production}}.$$

The mechanical efficiency during different kinds of activities varies from 0 to 50–60%. During static exercise, for example, no external work is performed as no movements take place, and therefore the mechanical efficiency is 0%. In most everyday activities mechanical efficiency varies between 0 and 20%. In the basic constituents of physical work—cycling, walking, cranking, hitting, carrying—the oxygen uptake varies very little between individuals who perform the same type of exercise. In work tasks where one's own body is carried (e.g., walking) there will be some differences between individuals due to variations in body weight. Unless these differences are extreme, measurements obtained in a small group of subjects can be used as a measure of the demand, the physical stress, in that situation. The more complex and skill-demanding a physical performance is, the more oxygen uptake varies between different performers. Consider, for example, the difference in oxygen uptake between horse riders. A skilled rider uses less oxygen per minute to master the horse than an unskilled one, who uses additional muscle contractions for balancing on horseback and in preparation for unexpected actions from the horse.

For most purposes, including manual labour, oxygen uptake is a good indicator of physical stress. Important exceptions are tasks which induce a heavy heat stress, tasks with a large static component, and other activities which demand a large proportion of anaerobic metabolism. For such tasks oxygen uptake gives valuable information about the aerobic component, but the measurements must be supplemented with others (see later under heart rate and body temperature measurements). For example, short-lasting, very demanding athletic tasks (e.g., short distance running) should not be evaluated only on the basis of oxygen uptake. However, if the oxygen uptake is measured over the entire *exercise plus recovery* period the total demands on energy consumption can be calculated. Because occupational activities usually go on for several hours, it is very uncommon that they contain an appreciable degree of anaerobic exercise. Exceptions may be all-out life saving operations (fire-fighting, diving, emergency maintenance) but in such cases the tasks are usually short lasting and time is given afterwards for recovery. Such situations can be potentially hazardous especially when combined with high heat load (see later). Performers of such tasks should therefore be selected for good physical performance capacity, and the task time may have to be controlled in order to avoid heat stroke or accidents caused by fatigue.

Thus measurement of oxygen uptake during work aims at assessing the physical stress during a work operation. It is used in order to:

(a) identify the most demanding tasks in an occupation. Such tasks often exclude the weaker members of the work force from taking a certain job. The most strenuous tasks can be redesigned to make them less demanding, job rotation or pauses can be introduced, or the task can be reduced in pace (see Figure 22.4),

(b) compare the demands of alternative ways of performing a task, for example with improved tools or work routines (see Figure 22.5),

(c) evaluate the component of dynamic exercise in a complex work situation. Many jobs have components of static and dynamic exercise as well as heat stress. Strain measures like heart rate (see later) and rectal temperature indicate the total physiological response to a number of different stressors. Oxygen uptake however increases appreciably only as an effect of dynamic exercise (see Figure 22.4).

The task demands can also be expressed in relation to the maximal aerobic power of the individual. Consider Table 22.1, which gives oxygen uptake during a number of occupational or everyday tasks. By tradition tasks

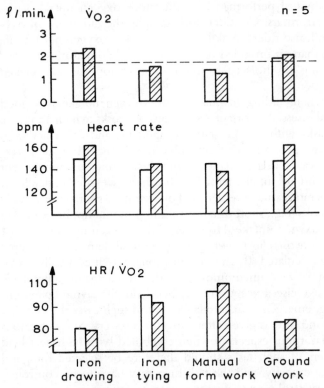

Figure 22.4. Oxygen uptake and heart rate during four common tasks in the building industry. All tasks were performed at a normal (unfilled columns) and an impeded (filled columns) pace. The level for 'very heavy work' = $\dot{V}O_2 > 1.75$ l/min, has been marked. Observe the high $HR/\dot{V}O_2$ quotient in some tasks, probably indicating static components. Source. Kilbom *et al.*, 1989

demanding oxygen uptake over 2·0 l/min are considered extremely heavy, those demanding 1·5–2·0 l/min are considered very heavy, those demanding 1.0–1.5 l/min are considered heavy, those demanding 0·5–1·0 l/min are considered moderate, and those demanding less than 0·5 l/min are classified as light. However, the physiological response, the strain, will obviously vary markedly between different individuals when performing the same task. For an individual with a maximal oxygen uptake of 1·75 l/min, tasks with an oxygen uptake of 1 l/min demand 60% of his or her maximal capacity, while the same task requires only 30% of the maximal capacity of an individual with a maximum of 3·0 l/min. Thus the aerobic strain, expressed in relation to the individual's maximal aerobic power, may vary considerably between individuals doing the same task.

Many investigations have demonstrated that the demands during an 8-h work day should not exceed about 30–40% of the individual maximal capacity. For shorter time periods a somewhat higher proportion can be used. For very short-lasting tasks—a few minutes—close to maximal demands can

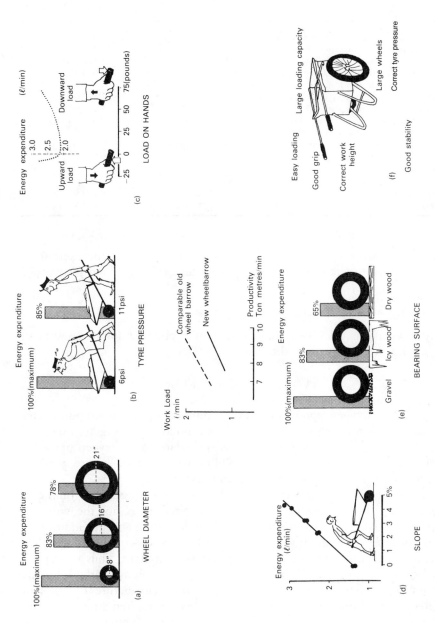

Figure 22.5. Oxygen uptake and productivity in wheelbarrowing. Large differences were obtained with changes in wheel diameter, tyre pressure, positioning and grip of handles, slope and structure of the terrain. The optimal design improved productivity by 40% with an unchaged energy expenditure. From Hansson, 1970

Table 22.1. Oxygen uptake during various physical tasks and activities

Activity	Oxygen uptake (l/min)
Running, skiing, swimming (male elite)	>5·0
Running, skiing, swimming (female elite)	>4·0
Cycle ergometer, external load 200 W	2·8–2·9
Fire-fighting, manual work in forestry and mining	2·0–3·0
Cycle ergometer, external load 150 W	2·1–2·2
Heavy industrial work, heavy gardening and agriculture	1·5–2·0
Cycle ergometer, external load 100 W	1·5–1·6
Heavy cleaning and manufacturing, fast walking or slow running	0·8–1·5
Cycle ergometer, external load 50 W	0·9–1·0
Walking 4–5 km/h, nursing, catering, light manufacturing, changing between sitting, standing and walking	0·6–1·0
Passive standing	0·4–0·5
Sitting assembly work, driving, office work	0·3–0·6
Passive sitting	0·2–0·4
Supine	0·2–0·3

be met provided a sufficiently long rest period is given after the exercise to allow for lactate metabolism (see also earlier under 'Fatigue and recovery').

Measuring and analyzing oxygen uptake

Oxygen uptake is calculated after analysis of expired air for its content of oxygen and (usually) carbon dioxide.

$$\dot{V}O_2 = \dot{V}_E (CO_2i - CO_2e)$$

where $\dot{V}O_2$ is oxygen uptake in litres per minute, $\dot{V}_E$ is expired pulmonary ventilation in litres per minute, CO_2i is concentration of oxygen in inspired air and CO_2e is concentration of oxygen in expired air. Oxygen uptake is usually expressed in litres per minute, STPD (*s*tandard *t*emperature, *p*ressure and saturation, i.e., 0°C, 760 mm Hg and *d*ry).

Expired air volume is either measured, or calculated from the volume of inspired air. The volume of inspired and expired air is not the same, as expired air volume is slightly expanded through heating and the addition of carbon dioxide and humidity. Therefore CO_2i must be corrected for differences in temperature, humidity and carbon dioxide content. Thus a complete calculation of oxygen uptake includes the analysis of volume, carbon dioxide and oxygen concentrations of expired air, and temperature of ambient inspired air. Carbon dioxide and oxygen content of inspired air are constant (0·03 and 20·94%, respectively), and the temperature of expired air is assumed to be 37°C.

In order to convert oxygen uptake to energy consumption, an energy coefficient of 20·2 kJ at rest and 20·6 kJ at exercise˙ should be used. Thus, the combustion of one litre of oxygen with a mixture of carbohydrates, fat and protein yields on the average 20·2 kJ at rest. (Previously energy consumption was often expressed in calories, where 1 kcal = 1000 cal = 4·186 kJ.)

Measurements of oxygen uptake require relatively expensive and complex instrumentation. For the laboratory, fully automated systems which measure ventilatory air flow and temperature, current barometric pressure and humidity, ventilation rate and concentrations of oxygen and carbon dioxide in expired air are available. Large errors in the calculation of oxygen uptake can be introduced if the volume of expired air is not correctly measured, or if the air sample for oxygen analysis is mixed with ambient air. Therefore the system for collection and analysis must fit tightly to the subject's mouth via a mouth piece, while a nose clip is used to eliminate leakage from the nose. A whole face mask can also be used but it is less reliable   due to risk of leakage (especially for bearded persons) and a larger 'dead space'.

For field studies simplified automated instruments are used. They usually analyze only inspired air flow and expired oxygen concentration, making assumptions for other factors. Thus a small error is introduced, but within the most common range of oxygen consumption values this error is usually less than 5%.

Instead of this complex procedure a simple measurement of pulmonary ventilation can be used to predict oxygen uptake. For light to moderately heavy exercise intensities pulmonary ventilation is closely related to oxygen uptake. During heavy exercise (oxygen uptake over 2·0 l/min and pulmonary ventilation above 40–50 l/min) pulmonary ventilation is an unreliable measure of energy expenditure.

Measuring physical working capacity

The maximal aerobic power (max $\dot{V}O_2$) can be accurately measured during dynamic exercise with large muscle groups, e.g., cycling or running. Other physical activities which can produce maximal or near maximal oxygen uptakes are skiing, rowing and swimming, but such activities are for practical reasons not often used for testing purposes.

Choice of cycle ergometer vs. treadmill

The advantages of cycle ergometers are:

(a)   the subject is more stationary than on the treadmill, and therefore all mesurements are performed more easily and accurately,

(b)   the mechanical efficiency during cycling varies little between individuals. This means that a given workload on the cycle ergometer can be expected to yield very similar values for oxygen uptake in a group of subjects. On a treadmill, however, running at a certain speed will give a larger range of oxygen uptakes for a group of subjects, because of their varying body weights,

(c) elderly subjects usually find maximal exertion on the cycle ergometer less frightening than on the treadmill.

The advantage of treadmill exercise is that maximal oxygen uptake is up to 8% higher than on the cycle ergometer. Moreover, treadmill tests are usually preferred by young and fit subjects.

Test protocol for cycle ergometer tests

The subject is usually tested on two to three submaximal and one maximal workload (Figure 22.6). The submaximal workloads are chosen to correspond to 35–40, 50–60 and 70–80% of the individual's expected maximal aerobic power. In practice this often coincides with a power output of 50, 75 and 100 W for female subjects and 50, 100 and 150 W for males. For small or old individuals and for well-trained persons, the test loads must be adjusted. The testing on submaximal workloads serves several purposes:

1. The investigator can observe the subject's reaction to exercise, register pulmonary ventilation, heart rate (see below) and subjective responses

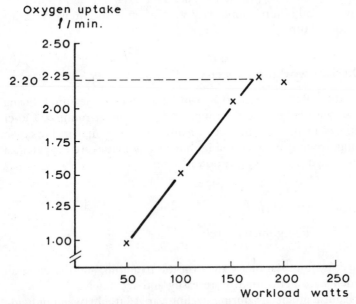

Figure 22.6. Relation between workload and oxygen uptake during submaximal and maximal exercise on the cycle ergometer. The maximal oxygen uptake was in this case reached at a workload of 175 W and it was 2·20 l/min. On another day the subject was also tested on a workload of 200 W, but could not further increase the oxygen uptake. The latter phenomenon is referred to as levelling off of oxygen uptake

and thereby more easily estimate which maximal workload should be used.

2. Contraindications to further work stress can be identified, i.e., symptoms and physiological signs indicating cardiovascular, pulmonary or musculoskeletal disease.
3. Submaximal workloads serve as a warm-up, i.e., they increase muscle temperature. Thereby exchange of oxygen and carbon dioxide is facilitated, and the risk of strains and sprains is decreased through improved muscle co-ordination.

Exercise on each submaximal workload is maintained for 5–6 min. Oxygen uptake should not be measured until during the last 1–2 min of this exercise period, as it takes 1–3 min after the start of exercise for oxygen uptake to stabilize at a steady state.

The maximal workload is chosen so as to exhaust the subjects completely within 3–7 min. Usually oxygen uptake is measured continuously after 2 min and the highest value obtained thereafter is used as max $\dot{V}O_2$.

Test protocol for treadmill

The workload is gradually increased through increments of slope and speed of the treadmill. Usually subjects are tested for 2–3 min on each workload without pauses between, and oxygen uptake is measured during the last minute of each workload. As the subject approaches exhaustion, continuous measurements are made. A number of different test protocols for treadmill exercise have been described. If oxygen uptake is not measured during the test, it can be estimated from the number of metabolic units, METS (1 MET = oxygen uptake at rest, i.e., 3.5 ml $\times$ kg $\times$ min) required for a certain increment in slope and speed of the treadmill.

An alternative to cycle ergometer and treadmill exercise tests is the so-called *step test*, which may be suitable in field studies, where sometimes no other testing equipment is available. In the Harvard step test, the physical working capacity is estimated by letting the subject step up and down a stool at a given pace. In modified step tests the height of the stool may be varied to suit the subject's estimated capacity. Gradational tests, using submaximal and maximal intensities, have also been described. Step tests, however, are difficult to standardize, measurements of, for instance, heart rate are difficult to perform, and the tests often lead to muscle soreness.

For further reading on exercise testing the reader is referred to Åstrand and Rodahl (1986, pp. 354–390).

Contraindications to exercise testing

Before the test the subject must be asked about cardiovascular or pulmonary disease, ongoing infections and medications, and exercise habits. Ongoing infections and chest pain are absolute contraindications, whereas previous cardiac insufficiency, angina pectoris, myocardial infarctions and hypertension are relative contraindications. In such cases exercise testing must only be done under electrocardiograph (ECG) surveillance, by trained medical staff and with resuscitation equipment available. Symptoms like chest pain, excessive breathlessness and certain arrhythmias that occur during exercise are indications to stop the test immediately. In subjects above 40 years of age it is advisable to register ECG before, during and after the exercise test. With the above precautions the risks of exercise testing are exceedingly small. A further reduction of risk is obtained by doing a submaximal work test (see 'Heart rate during exercise' below) using the heart rates at submaximal workloads to predict the maximal oxygen uptake.

Heart rate during exercise

Heart rate increases during both static and dynamic exercise, during heat exposure, and as an effect of psychological stress. Thus a heart rate increase is an *unspecific* cardiovascular strain response, and the interpretation of heart rate recordings must always be made against a background knowledge of the circumstances of the recording. Other factors that can influence heart rate are tobacco smoking, certain types of medication, ongoing infections, and so on.

During moderate dynamic exercise at a constant workload (e.g., on a cycle ergometer) heart rate increases during the first 1–3 min and then reaches a steady state. Steady-state heart rates are linearly related to workload or oxygen uptake (Figure 22.7). A very unfit individual will have his or her heart rate–oxygen uptake relationships shifted to the left, and a well-trained individual will have it shifted to the right. During static exercise steady-state levels are usually not reached.

Submaximal exercise testing

The relationship between heart rate and oxygen uptake is the basis for the method to predict maximal oxygen uptake from submaximal heart rates. By measuring heart rate at, say, 3 submaximal workloads, the maximal oxygen uptake can be predicted, either by extrapolation to the maximal heart rate or by using nomograms. These methods require knowledge of the maximal heart rate, which gradually decreases with age, from around 220/min in young children to around 160/min at age 60. However, there is also a relatively large variation in maximal heart rate between different individuals, even at the same age. Thereby an error of about 10% is introduced in the prediction of maximal oxygen uptake, compared with direct measurements.

The same safety precautions as during maximal testing must be used, i.e., a brief medical history regarding cardiovascular and pulmonary disease and infections must be taken, and the subjects must be instructed to report any

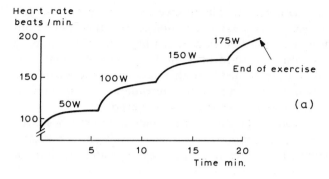

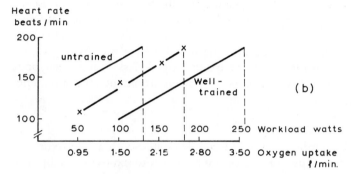

Figure 22.7. Heart rate during a standardized cycle ergometer test. (a) Heart rate increase and steady-state levels during exercise on increasing workloads. Note that no steady state was reached at the maximal workload. (b) Crosses indicate steady-state heart rate levels from Figure 22.7(a). The maximal heart rate, 180 beats/min, was reached at workload 175 W, at an oxygen uptake of 2·5 l/min. Heart rate—oxygen uptake relationships for an untrained and a well-trained individual with their respective maximal oxygen uptakes, are also indicated

symptoms apart from normal breathlessness and general fatigue. Usually the test is interrupted when heart rates 30 beats/min below the expected maximal heart rate are reached, i.e., at heart rate 170/min for a 20-year-old and 130/min for a 60-year-old. In order to avoid some other factors which influence heart rate, testing should not be performed within 1 h after smoking or after a large meal, and the air temperature in the laboratory should be around 18 °C.

Heart rate as a strain measure during occupational work

The continuous measurement of heart rate during work is a common method to evaluate cardiovascular strain. The measurements are relatively simple to perform, and the results are usually reliable. The most commonly used system is telemetry, i.e., the ECG impulse is registered via chest electrodes and trans-mitted to a receiver. The receiver identifies the R-waves of the ECG signal and stores them in a microprocessor, where the number of beats is counted for given time periods (usually 1 min). The receiver can be positioned either on the wrist of the subject or in an external station, where the signals from several transmitters, using different frequencies, can be stored. The circuitry of the receiver can be constructed to identify the R-wave with good accuracy, so that no artefacts are recorded. Another method is to record the ECG signal continuously using miniaturized tape recorders. This also permits analysis of arrhythmias and is often used for clinical purposes.

If no measurement system is available, heart rate can nevertheless be recorded manually with reasonably good accuracy. Usually the work tasks to be studied are interrupted regularly at 1–5 min intervals and the time for 10 beats is recorded with a stopwatch.

As already emphasized heart rate is an unspecific measure of cardiovascular strain. Therefore measurements must be supplemented with activity recordings, i.e., the type of physical activity, psychologically stressful situations, simultaneous heat exposure, etc., must be noted. Often the investigation is supplemented with measurements of oxygen uptake, skin and deep body temperatures, subjective ratings of exertion and environmental measurements.

Analysis of heart rate recordings during work

The occupational recordings should be supplemented with a submaximal or maximal exercise test. The purpose of this testing is to expose the subject to a standardized activity for reference, and also to measure the maximal capacity. At least some of the exercise workloads should be designed so that they resemble the type of exercise performed occupationally. For example, if the occupational tasks studied are mainly performed walking, the exercise test should preferably include testing at one or two walking speeds on the treadmill. Or if the occupational tasks are mainly performed by the upper body, an exercise test including arm cranking may be most appropriate.

The analysis includes (see Figure 22.8):

1. Calculation of average heart rate. This average value is related to the oxygen uptake (or workload) at the same heart rate during the standardized exercise test. This will permit an estimation of the proportion of the maximal oxygen uptake used.
2. Identification of the most demanding tasks.
3. Analysis of the mean heart rate for each of the different activities performed.
4. A comparison between the distribution of heart rates for the studied job (or work task) and that of other jobs.

These analyses can be used to estimate the severity of the work tasks, both in terms of calculated proportion of maximal oxygen uptake, and in terms of cardiovascular strain. Heart rates during prolonged work up to 90 beats per minute are considered to indicate light cardiovascular strain, 90–110

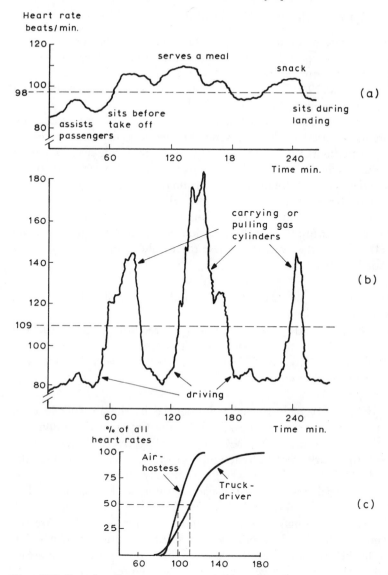

Figure 22.8. Examples of heart rate curves recorded during occupational work. (a) Air-hostess. (b) Truck driver delivering heavy gas cylinders to customers. (c) Cumulative heart rate curves. Note the very steep slope and narrow range of heart rates in the air-hostess and the much wider range of heart rates in the truck driver

moderate, 110–130 heavy, 130–150 very heavy and 150–170 extremely heavy strain. Moreover, work tasks requiring ergonomic interventions can be identified.

Recovery heart rate: the Brouha method

If no automatic system for heart rate measurement is available, and if the work cannot be interrupted for manual recordings, the cardiovascular strain can nevertheless be estimated using recovery heart rate measurements done according to Brouha (1960). This method has even been claimed to be more reliable than measurements during exercise.

Immediately after the termination of a work task, the subject is seated and the heart rate is measured for three minutes after cessation of work. Recovery heart rate measurements can be performed repeatedly during a work day in order to evaluate whether the workload is too high, and whether recovery is incomplete between the different tasks. Brouha (1960, pp. 107–8) has outlined the practical procedure for assessing work and recovery. Heart rate, HR, during the first, second and third minutes after exercise is obtained by counting the beats during the last 30 s of each of the minutes, and doubling the value. $HR_{AV1,2,3}$ is the average of these three values, and is highly correlated with total cardiac cost.

1. If $HR_1 - HR_3 \geqslant 10$, or if HR_1, HR_2 and HR_3 are all below 90, then recovery is normal.
2. If the *average* of HR_1 over a number of recordings is $\leqslant 110$, and $HR_1 - HR_3 \geqslant 10$, the workload is not excessive.
3. If $HR_1 - HR_3 < 10$, and if $HR_3 > 90$, then recovery is inadequate for the task requirements.

The rate of recovery of heart rate is influenced by the absolute level of heart rate at the interruption of work, and by the fitness of the individual. In addition, heat exposure influences the rate of recovery. If recovery is unsatisfactory, the work must be redesigned in such a way as to reduce the physical stress. The intensity of exercise is usually difficult to influence, especially in industrial tasks where the pace is often set by machines. The fitness of the individual too is difficult to influence, and even in well designed, intensive training experiments, improvements of 10–20% are the best that can be achieved. The best way of securing sufficient recovery is usually to limit the duration of each task, or, if this is not possible, to increase the duration of the pauses between tasks (see also Figure 22.3).

Rating of perceived exertion

There is a well-known, curvilinear relationship between the intensity of a range of physical stimuli and our perception of their intensity. A positively accelerating relationship has been found between physical workload and perceived exertion. Thus, in a given individual, there is a highly reproducible relation between, e.g., the workload on a cycle ergometer and the perceived exertion. In the so-called Borg, or RPE (rating of perceived exertion), scale

(Borg, 1985), the scale steps have been adjusted so that the ratings, from 6 to 20, are linearly related to the heart rate divided by ten. The scale (Figure 22.9) is presented to the subject before the start of the exercise test and the 'endpoints'—6 and 20—are thoroughly defined. The scale is then shown to the subject at the end of each exercise intensity and he or she is asked for a rating. The verbal explanations are used as support information. Recently a non-linear scale, which also has ratio properties, was developed.

These scales can be used to supplement physiological measurements during exercise testing and occupational work tasks. They often provide valuable additional information about subjective responses, especially in cases where the heart rate response is unreliable (e.g., patients with atrial fibrillation, or medication which influences heart rate). Similar scales have also been developed to quantify intensity of pain.

The RPE scales have been used to a limited extent in industry. The idea of substituting physiological measurements by subjective ratings is attractive, as ratings do not require any instrumentation. However, recent findings in industry and in industrial tasks suggest that ratings are influenced not only by the overall perception of exertion, but also by previous experience and motivation of the subjects. Thus highly motivated subjects tend to underestimate their exertion.

Body temperature during heat exposure and exercise

Body temperature increases during exercise, since a large proportion of the energy produced is converted to heat. To some extent the exercise perform-

6	NO EXERTION AT ALL
7	EXTREMELY LIGHT
8	
9	VERY LIGHT
10	
11	LIGHT
12	
13	SOMEWHAT HARD
14	
15	HARD (HEAVY)
16	
17	VERY HARD
18	
19	EXTREMELY HARD
20	MAXIMAL EXERTION

Figure 22.9. Borg scale for rating of perceived exertion

ance benefits, since the metabolic processes work faster in higher temperatures. Thus deep body temperature is adjusted in relation to the relative workload, and although this adjustment is slow and takes at least 30 min, it is highly accurate. For example, during exercise corresponding to 50% of maximal aerobic power, deep body temperature is 38·0°C.

However, human tissues have a limited range of temperature tolerance, so the major part of the heat produced must be dissipated. In low environmental temperatures radiation and convection are the main ways of heat dissipation, whereas sweating is the only possibility in high environmental temperatures. With an ambient air temperature above 37°C, or in intense radiant heat, external heat is transferred to the body and must also be dissipated through sweating. Since sweating demands increased blood circulation to the skin, heat exposure puts large demands on the cardiovascular system. Hence exercise in hot environments induces increments in heart rate far above those obtained at the same exercise intensity in a cold environment (Figure 22.10).

Prolonged heavy sweating also leads to loss of fluid, which can severely reduce the blood volume and thereby further increase the cardiovascular strain. A loss of 1% body weight through sweating leads to a deterioration in the physical working capacity and reduced orthostatic tolerance. The corresponding heart rate increase is around 10 beats/min. The World Health Organisation recommends that sweat production should not exceed 4 l in an 8-h work period. Thus physical work in hot environments imposes large stresses on the human physiology. If uptake and production of heat cannot be balanced by heat dissipation, deep body temperature increases with a concomitant risk of heat exhaustion and heat stroke. Deep body temperatures above 38·0°C indicate that this balance may be upset.

Through environmental measurements of wet bulb temperature, radiant heat and air velocity, an index, WBGT, can be calculated. This index can be used to predict the risk of heat exhaustion during exercise at different intensities. A more accurate prediction can only be obtained through physio-

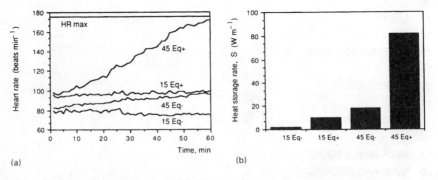

Figure 22.10. Heart rates (a) and heat storage (b) during standard work at increasing workloads at 15°C and 45°C, with (Eq+) and without (Eq−) heavy equipment. From Sköldström, 1987

logical measurements of heart rate, deep body and skin temperatures, and fluid balance. A more extensive discussion on the measurement of the body's responses to heat will be found in chapter 16 in this book.

Conclusions

Physiological measures of physical activity have a long history of contributions to ergonomics. They provide readily measured indications of the dynamic efforts people are exerting, and well recognized and reliable limits can be used to ensure that people are not overloaded. As pointed out at the end of the chapter, there are confounding factors which must be watched for. Emotional stresses, including anxiety, can add to the physiological costs of human effort, whilst adverse thermal environments will detract in a major way from the available work capacity of a person.

Subjective measures, epitomized by the Borg scale, give a cross comparison to the physical measures, and allow the investigator to explore how the job feels to the job holder. In situations where a single extreme stressor is acting, a one-dimensional measure could be appropriate. In most situations though, we need more than one measure in order to understand how best to modify, to design, or to assess a situation.

References

Åstrand, I. (1960). Aerobic work capacity of men and women with special reference to age. *Acta Physiologica Scandinavica* (Suppl. 169) **49**.

Åstrand, P.O. and Rodahl, K. (1986). *Textbook of Work Physiology* (New York: McGraw-Hill).

Borg, G. (1985). *An introduction to Borg's RPE-Scale* (Ithaca, NY: Movement Publications).

Brouha, L. (1960). *Physiology in Industry* (Oxford: Pergamon).

Hansson, J.F. (1970). *Ergonomics in the Building Industry*. Research Report No. 8. Byggforskningen, State Council of the Building Industry.

Kilborn, Å., Jörgensen, K. and Fallentin, N. (1989). Recording workload in occupational work – a comparison between observation methods, physiological measurements and subjective estimates. National Board of Occupational Health and Safety, Sweden, Arbeteoch Hälsa, no. 38.

Sköldström, B. (1987). Physiological responses of fire fighters to workload and thermal stress. *Ergonomics*, **30**, 1589–97.

Chapter 23

The evaluation of posture and its effects

E. Nigel Corlett

Introduction

The maintenance of postures and the support of loads are particular examples of the performance of static work. Although these are both quite common, what can be overlooked, presenting particular difficulties for the analyst, are those cases where postures *and* other physical activities intermingle. If one or the other situation predominates it may be sufficient to assess the whole situation on the basis of its worst aspect, but if serious levels of effort are required whilst posture is maintained, this procedure is unsatisfactory.

In general we can say that, whilst the limitations on dynamic physical activities felt by a person would be high heart rate and shortage of breath, the limits to static work will be the experience of muscular pain. As described in chapter 22 by Kilbom, this arises particularly from the anaerobic metabolic activity of the muscles, whose blood supply is restricted due to the increased intramuscular pressure. A consequence of this pressure is that the heart rate does not represent the static effort involved, although post-effort heart rate can be an important indicator of the existence of static load (see Brouha, 1960).

If we are to evaluate posture and static work, therefore, we must have some overview of the major contributors to static workloads. Five dimensions relevant to the definition of a posture are given below, from which, by suitable measures, we can deduce key components which contribute to the loads experienced:

  (i)   The angular relationship between body parts
 (ii)   The distribution of the masses of the body parts
(iii)   The forces exerted on the environment during the posture
 (iv)   The length of time that the posture is held

(v) The effects on the person of maintaining the posture.

Measurements of some, or all, of the above five dimensions help us to define the following:

(i) The stability of the person in that posture
(ii) Estimates of muscle loads and joint torques
(iii) Estimates of fatigue levels and recovery times (which may require some additional measures)
(iv) Comparisons of conditions against criteria, e.g., to assess hazards in load handling.

Although this chapter recognizes the above structure, as with so much else in this book it cannot be too strongly emphasized that what is chosen for measurement from the first list depends on what it is that we want to know from the second list. Thus for those wishing to use the information given here, a review of the whole chapter is desirable before selecting any particular methods. It should also not need emphasis that the physical and psychophysical measurements to be described are not all that contribute to people's experience of workloads. When undertaking fieldwork to identify sources of load, chapters 30 and 35 also describe a number of other important measures with brief reasons why they are important.

Methods for direct measurement of the effort involved in holding a posture, as well as its effects, are less common than for dynamic work. The range of methods embraces estimation techniques (biomechanics and estimates from maximum voluntary contraction); measures of muscular activity (analysis of the electromyographic—EMG—signal); measures of the resultant effects (e.g., spinal shrinkage); subjective measures (e.g., discomfort recordings) and a wide range of interpretive methods. These last range from epidemiological studies to estimates of likely effects from the use of posture recordings (using such as OWAS, NIOSH, Nordic Questionnaire, posture targetting) and posture measurements (e.g., goniometers, SELSPOT or CODA). The measurement of posture and its effects is more extensively discussed in Corlett *et al.* (1986).

Posture recording

As a first thought, it may be proposed that photographs or videos, perhaps in pairs of views orthogonal to each other, would be enough to record postures. It is true that for some situations this is adequate, for example, just for a visual record or if some particular angles are to be measured. In other cases, however, it will be realized that, although the posture may have been recorded using a video or by still photography, the data for analysis still have to be retrieved from the recorded images. Where accuracy is needed, problems of parallax in a plane transverse to the optical axis can be considerable.

However, rarely is a record of posture of use on its own; it is necessary also to have data on task activities, the loads moved, conditions of the people concerned, and the workplace. It is fitting, therefore, before continuing with a number of methods for recording postures, to comment on the wider aspects of postural loading. It is also appropriate to note that postural data are frequently used in biomechanical analysis, and chapter 24 on biomechanics should be consulted to assess what data are required, so that the method most appropriate for the need is selected.

The longer-term effects of posture and work activities, whilst under less control by the investigator, are of great importance. Epidemiological methods are outside the scope of this text but the ergonomist has need of data from epidemiologists and occupational health professionals and is sometimes in a position to contribute data to them.

Direct observation methods

The measurement of the angles between body parts, or their angles in relation to the environment, are frequently required. For biomechanical analysis accuracy should, in general, be high. There are several procedures available which assess postures in relation to their contributions to discomfort, strain, stability or force exertions and for these a wider margin of error can often be acceptable.

OWAS

Although work study (methods engineering) offers several charts and symbol systems for recording work activities, these do not embrace the whole posture. One of the earliest whole posture coding systems for industrial use was developed in Finland, to investigate the working postures in a steelworks. The company Ovako Oy, in conjunction with the Finnish Institute of Occupational Health (1992) developed the OWAS method. Postures are observed, and recorded as shown in Figure 23.1a. An accompanying assessment sheet, similar to Figure 23.1b, enables each posture to be assessed for acceptability or else for appropriate remedial action.

The OWAS code for a posture comprises a record of the posture itself in the first three figures, the load or force used is indicated by the fourth figure and a record of the stage in the cycle or task is recorded in the fifth figure. The procedure is to glance at the work to take in the posture, force and work phase, then to look away and record it. Thus the work activities can be sampled, and from these samples estimates can be made of the proportions of time during which forces are exerted or postures held.

The assessment sheet of Figure 23.1b allows the assessment of the likely musculoskeletal load even for a single posture combination of back, arms and

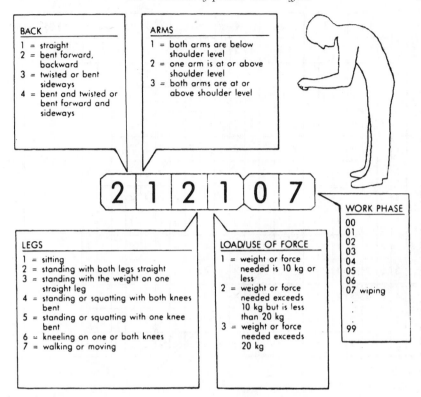

Figure 23.1(a). Items of the OWAS method and an example code for a particular task

legs, with the action categories. Where the activity is frequent, though the load is light, the sampling procedure permits the estimation of the proportions of time the limbs or back are spent in the various working postures. These postures may then be evaluated for adequacy by using the table of Figure 23.1c, where action categories for the various postures in relation to their estimated times of use during the working day are given.

It is evident that these measures could be taken from video recordings, and it is recommended that video should always support direct observation. The usual precautions when using activity sampling should be followed, such as not using a sampling period which is a factor of the task cycle time, although for non-cyclic work equal intervals of thirty or sixty seconds can be suitable. Observers require training to a standard which ensures that their recordings are consistent and that their observation errors are at an agreed level; under 10% is desirable. Regular testing of users is also desirable if confidence in the system is to be maintained.

Procedures such as OWAS have some limitations. The postural assessment is broad and the force estimations, too, have broad categories. Their great strength is the facility they provide for the rapid identification of most of

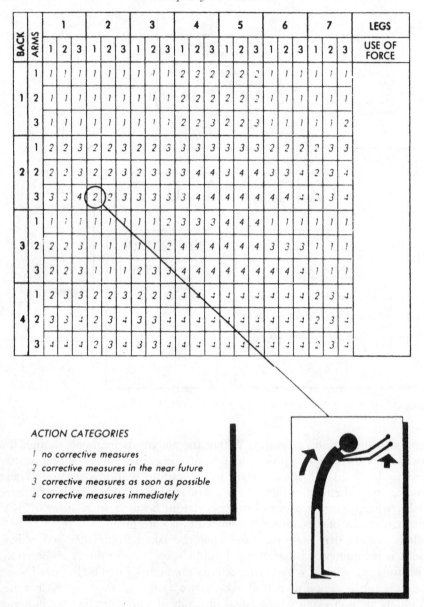

BACK	ARMS	1			2			3			4			5			6			7			LEGS
		1	2	3	1	2	3	1	2	3	1	2	3	1	2	3	1	2	3	1	2	3	USE OF FORCE
1	1	1	1	1	1	1	1	1	1	1	1	2	2	2	2	2	2	1	1	1	1	1	1	
	2	1	1	1	1	1	1	1	1	1	1	2	2	2	2	2	2	1	1	1	1	1	1	
	3	1	1	1	1	1	1	1	1	1	1	2	2	3	2	2	3	1	1	1	1	1	2	
2	1	2	2	3	2	2	3	2	2	3	3	3	3	3	3	3	3	2	2	2	2	3	3	
	2	2	2	3	2	2	3	2	3	3	3	4	4	3	4	4	3	3	4	2	3	4		
	3	3	3	4	2	2	3	3	3	3	3	4	4	4	4	4	4	4	4	2	3	4		
3	1	1	1	1	1	1	1	1	1	2	3	3	3	4	4	4	1	1	1	1	1	1	
	2	2	2	3	1	1	1	1	1	2	4	4	4	4	4	4	3	3	3	1	1	1	
	3	2	2	3	1	1	1	2	3	3	4	4	4	4	4	4	4	4	4	1	1	1	
4	1	2	3	3	2	2	3	2	2	3	4	4	4	4	4	4	4	4	4	2	3	4	
	2	3	3	4	2	3	4	3	3	4	4	4	4	4	4	4	4	4	4	2	3	4	
	3	4	4	4	2	3	4	3	3	4	4	4	4	4	4	4	4	4	4	2	3	4	

ACTION CATEGORIES

1 *no corrective measures*
2 *corrective measures in the near future*
3 *corrective measures as soon as possible*
4 *corrective measures immediately*

Figure 23.1(b). Action categories in the OWAS method for work posture combinations

the major inadequate postures. Furthermore, as they are easy to learn and use, they can be employed by a wide spectrum of the workforce, and then alert people to those aspects of activity which may be hazardous.

It will be evident that OWAS seeks to identify postures which put the body in positions where force exertions can be dangerous. Balanced and

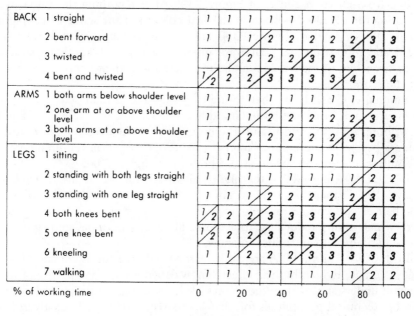

	% of working time → 0	20	40	60	80	100
BACK 1 straight	1 1 1	1 1	1 1	1 1	1	
2 bent forward	1 1 1	2 2	2 2	2 3	3	
3 twisted	1 1 2	2 2	3 3	3 3	3	
4 bent and twisted	1/2 2 2	3 3	3 3	4 4	4	
ARMS 1 both arms below shoulder level	1 1 1	1 1	1 1	1 1	1	
2 one arm at or above shoulder level	1 1 1	2 2	2 2	2 3	3	
3 both arms at or above shoulder level	1 1 2	2 2	2 2	3 3	3	
LEGS 1 sitting	1 1 1	1 1	1 1	1 1	2	
2 standing with both legs straight	1 1 1	1 1	1 1	1 2	2	
3 standing with one leg straight	1 1 1	2 2	2 2	2 3	3	
4 both knees bent	1/2 2 2	3 3	3 3	4 4	4	
5 one knee bent	1/2 2 2	3 3	3 3	4 4	4	
6 kneeling	1 1 2	2 2	3 3	3 3	3	
7 walking	1 1 1	1 1	1 1	1 2	2	

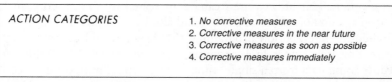

ACTION CATEGORIES

1. *No corrective measures*
2. *Corrective measures in the near future*
3. *Corrective measures as soon as possible*
4. *Corrective measures immediately*

Figure 23.1(c). Action categories in the OWAS method for work postures according to percentage of use over the work period

symmetrical postures are the ones which are, in general, acceptable. Pushing, pulling or moving loads when the person is twisted, or the body is in other ways asymmetrically loaded, are recommended for change. Results can be presented as diagrams, showing the proportions of harmful (category 3) and very harmful (category 4) work activities in the various phases of the person's work, which illustrate forcefully the conditions of the job. Measurements after changes have taken place, and subsequently, will demonstrate the benefits, and monitoring of the jobs over the years is facilitated by the relative simplicity of the process.

RULA

A procedure analogous to OWAS was developed by McAtamney and Corlett (1993) to assess the exposure of people to postures, forces and muscle activities known to contribute to Upper Limb Disorders (ULD). This Rapid Upper Limb Assessment (RULA) technique uses observations of postures adopted by the upper limbs, the neck, back and legs, recording the values drawn from charts A and B of Figure 23.2. After recording the values representing the observed posture in the first column of the score sheet (Figure 23.3), tables A and B (Figure 23.4) are used to obtain a posture score for the A and B body groups. Values for muscle use and loads are then extracted from the tables of Figure 23.5 and also entered into their appropriate spaces on the score sheet.

The scores C and D are then found by adding the separate scores as shown in Figure 23.3. From these the Grand Score is found from table C of Figure 23.6. Enter scores C and D into the boundaries of the diagram and note the value where row and column intersect. Appropriate action requirements for the different scores are given at the bottom of Figure 23.6.

In RULA, as in OWAS, the higher the code number, at any stage of the analysis, the further does the part concerned depart from a desirable posture. So changes concentrate on reducing the magnitudes of the individual numbers, which in turn reduce the total score. Thus the analyst has some guidance regarding where to introduce changes.

Both OWAS and RULA are easily learned and give consistent and reasonable accuracy. Evidently their effectiveness depends on the expertise of their users in understanding the ergonomics of the working situations under study and the technical possibilities for change. As with other methods given later in this chapter, their effectiveness can be increased by sound work analysis (see chapters 2 and 6, and Drury 1987) and also by incorporation into a more comprehensive intervention programme (McAtamney and Corlett, 1992).

For investigating upper-limb activities, Armstrong *et al.* (1982) illustrate a procedure using data from a single camera. The upper limb posture is coded, according to the diagram of Figure 23.7a, at equal, short, intervals through a number of work cycles. Hand grip forces should be obtained where these are noted as forceful. The observations can be checked off in columns, one for each posture, and then plotted on a time base, as in Figure 23.7b. Right and left arms, or values before and after a change, can be shown on the one plot, thus providing comparisons in a convenient form. Such a detailed analysis will demonstrate where adverse postures and forces are maintained or frequently repeated. A repeated analysis after changes have been introduced will then show the improvements to the postural loadings and any other benefits from the intervention.

As with the other methods for sampling postures, sampling at appropriate intervals can build up a picture of joint activities and forceful exertions from which interpretations of body loadings can be derived. Inspection of the video and discussion with the job holders in relation to identified extreme or frequent loadings can lead to reductions in overall workloads and levels of hazard.

Notation

Detailed records of postures have been used to record ballet for some two centuries. Two popular methods in wide use are the Benesh Notation

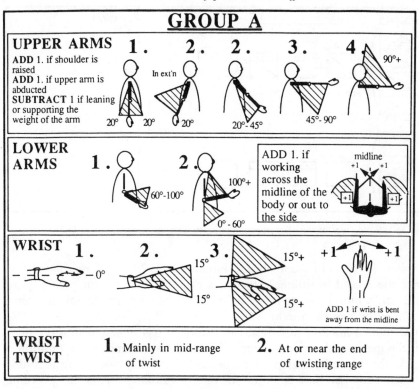

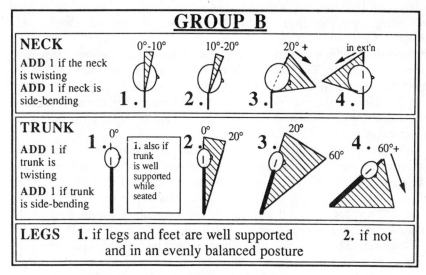

Figure 23.2. Items for assessment when using the RULA method

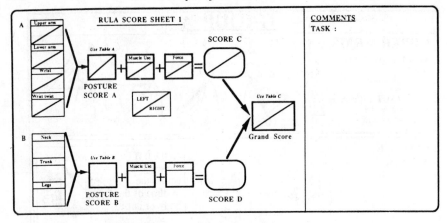

Figure 23.3. Score sheet for recording observations, using diagrams shown in Figure 23.2

(Kember, 1976) and Labanotation (Hutchinson, 1970). Both methods incorporate the record of timing, which is a major advantage, but each requires about three months of training in order to become proficient in even a simplified form of the methods.

The trained notator, however, can record in real time and in more detail than the methods described above. This detail can extend to the position of fingers, for example, as would be needed if a ballet was being recorded. Although proposals have been put forward to include effort in the recording, this has been based on the appearance of the performer as would be needed in recording a piece of choreography and not on any measure which would be suitable for work analysis.

A procedure, which does not incorporate timing data, is posture targetting (Corlett *et al*, 1979). This makes use of a diagram (Figure 23.8), which has 'targets' located alongside each of the major limb segments, and on head and trunk. As shown in the posture target record of Figure 23.9, the same position of the targets, but without the human shape, can be used, saving space on the recording form.

Marking the form requires the user to take the posture of Figure 23.8 as the normal or zero position. For a person in this position a mark would be made on the centre of each target. Movements forward by limbs, trunk or head (in the sagittal plane) would require a mark along the vertical axis of the target ('up' for forward) and positioned on that axis according to the estimated angle of the displacement. Each concentric circle marks off a 45° step. Displacements to the side of the body would be marked on the horizontal axes, again in a position according to the estimated angle. Directions (in a horizontal plane) are marked along the appropriate radius or between appropriate radii. An example of the use of the system is shown in Figure 23.9.

There are cases where the trunk may be twisted rather than bent, or where

TABLE A   Upper Limb Posture Score

UPPER ARM	LOWER ARM	WRIST POSTURE SCORE							
		1		2		3		4	
		TWIST 1	2	TWIST 1	2	TWIST 1	2	TWIST 1	2
1	1	1	2	2	2	2	3	3	3
	2	2	2	2	2	3	3	3	3
	3	2	3	3	3	3	3	4	4
2	1	2	3	3	3	3	4	4	4
	2	3	3	3	3	3	4	4	4
	3	3	4	4	4	4	4	5	5
3	1	3	3	4	4	4	4	5	5
	2	3	4	4	4	4	4	5	5
	3	4	4	4	4	4	5	5	5
4	1	4	4	4	4	4	5	5	5
	2	4	4	4	4	4	5	5	5
	3	4	4	4	5	5	5	6	6
5	1	5	5	5	5	5	6	6	7
	2	5	6	6	6	6	6	7	7
	3	6	6	6	7	7	7	7	8
6	1	7	7	7	7	7	8	8	9
	2	8	8	8	8	8	9	9	9
	3	9	9	9	9	9	9	9	9

TABLE B   Neck, Trunk, Legs Posture Score

NECK POSTURE SCORE	TRUNK POSTURE SCORE											
	1		2		3		4		5		6	
	LEGS 1	2	LEGS 1	2	LEGS 1	2	LEGS 1	2	LEGS 1	2	LEGS 1	2
1	1	3	2	3	3	4	5	5	6	6	7	7
2	2	3	2	3	4	5	5	5	6	7	7	7
3	3	3	3	4	4	5	5	6	6	7	7	7
4	5	5	5	6	6	7	7	7	7	7	8	8
5	7	7	7	7	7	8	8	8	8	8	8	8
6	8	8	8	8	8	8	8	9	9	9	9	9

Figure 23.4. Tables for evaluating posture scores A and B, to be entered on score sheet of Figure 23.3

MUSCLE USE SCORE

Give a score of 1 if the posture is;

mainly static, e.g. held for longer than 1 minute
repeated more than 4 times/minute

FORCES OR LOAD SCORE

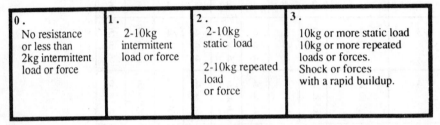

0.	1.	2.	3.
No resistance or less than 2kg intermittent load or force	2-10kg intermittent load or force	2-10kg static load 2-10kg repeated load or force	10kg or more static load 10kg or more repeated loads or forces. Shock or forces with a rapid buildup.

Figure 23.5. Muscle and load score tables; values to be entered into score sheet of Figure 23.3

a bend and a twist occur together. In most of these the recording of the positions of arms and legs will result in the trunk adopting the appropriate position. Where it is felt desirable to record the angle of twist of the trunk with respect to the hips, the arc between the head and trunk targets—marked 'trunk twist' on Figure 23.9—may be marked at the appropriate point, using the same angular scale as for the targets.

With only a modest amount of practice this procedure can be learnt, and its accuracy can be tested with goniometers (see below) during the learning period. The posture can be reconstructed from the diagram, and with simple mathemetics the angles of all segments can be related to each other, to permit the whole posture to be redrawn, or input to a computer. With measures of body weight, body sizes and exerted forces, a biomechanics program can then be used to calculate required torques and loads.

The biomechanics program given in chapter 24 has been modified to use posture targetting as input (Tracy and Corlett, 1991). The transfer of the targeted point directly from the record form to the computer screen, which displays a similar target to that on the recording form, avoids errors in measurement and interpretation of the record. The computer converts the target representation to angles and then pursues the analysis, giving quicker analysis than if interpretation or measurement had to be used. With the increasing power of laptop technology it will, no doubt, soon be possible to undertake direct input in the field (see chapter 9)

SCORE D (NECK, TRUNK, LEGS)

	1	2	3	4	5	6	7+
1	1	2	3	3	4	5	5
2	2	2	3	4	4	5	5
3	3	3	3	4	4	5	6
4	3	3	3	4	5	6	6
5	4	4	4	5	6	7	7
6	4	4	5	6	6	7	7
7	5	5	6	6	7	7	7
8	5	5	6	7	7	7	7

SCORE C (UPPER LIMB) — row labels 1 to 8

TABLE C Grand Score Table

ACTION LEVEL 1   A score of one or two indicates that posture is acceptable if it is not maintained or repeated for long periods.

ACTION LEVEL 2   A score of three or four indicates further investigation is needed and changes may be required.

ACTION LEVEL 3   A score of five or six indicates investigation and changes are required soon.

ACTION LEVEL 4   A score of seven or more indicates investigation and changes are required immediately.

Figure 23.6. Table to determine action levels, using scores C and D from score sheet of Figure 23.3

A comment on observational methods

All observational methods have a deceptive appearance of simplicity, giving the potential user the impression that their use is easy and their results simple to determine and conclusive. Unfortunately, this is not so, and potential users should be aware of the need for training in the method, monitoring of its use and supporting knowledge for the effective application of its results.

Those with work study/methods engineering experience will not be surprised by these comments. The estimation of postural angles requires practice, and should be subject to practice on refresher courses to estimate the errors for the different body angles being assessed. Confidence in the results can

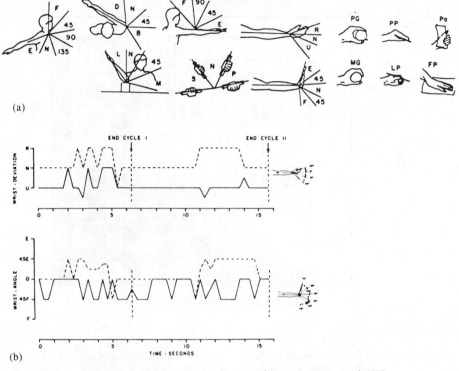

(a)

(b)

Figure 23.7. Recording of investigation, using coding according to Armstrong *et al.* 1982

then be based on measures of error, rather than being based on the confidence of the user. In any case, as in any measurement procedure, enough repeated measures should be taken to allow a calculation of variance, where this is possible.

Direct measurement

Goniometers

Individual angles can be measured using a simple goniometer, of the types shown in Figure 23.10, which will give angles to the vertical, or the angle between adjacent body segments. Accuracy is perfectly adequate for most purposes. The pendulum goniometer gives quick measures and is very simple to use. Where spinal measures are being made, care is needed since flexion of the spine introduces a pronounced curvature. The goniometer can be used at about L3 to L5, to get an indication of the angle at the lumbar-sacrum junction, and on the lower part of the thoracic spine for an approximate estimation of the spinal angle.

A combination of instruments in conjunction with a flexible rod, to record

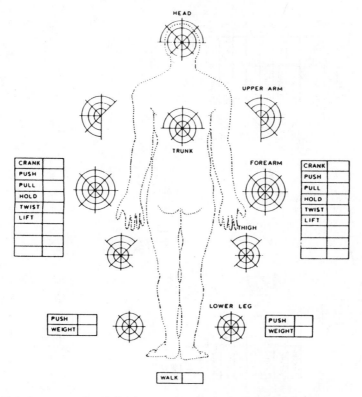

Figure 23.8. Posture targets, with the datum position shown by the dotted figure

spinal shape, can be used for a better estimate (Burton, 1986). The changes in spinal flexibility have been well documented by use of such a system, and it could be used to demonstrate the difficulties in performance faced by some workers with limited mobility, due either to their condition or to workplace restrictions.

Recordings for segment or whole-body postures are possible. The ease of recording from small electronic pendulum potentiometers can allow several to be mounted at selected points on a subject, sufficient to reconstruct from the recordings the postures adopted. Since the recording could cover a whole waking period, gathering the data as analogue signals on magnetic tape, computerized analysis is readily possible (see chapter 9 on computer data collection and analysis).

Instrumentation for the recording of spinal motion over a period of time has been developed by many workers, including Marras *et al.* (1990).★ These devices strap to the subject's body around the chest and the hips. The space between, along the line of the spine, is linked by instrumentation, and the

★The Lumbar Motion Monitor, Chattanooga Group Inc, PO Box 489, Hixson, TN37343, USA.

Analysis of work activities

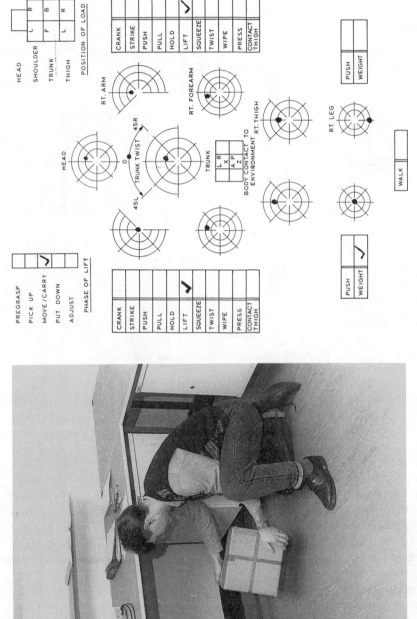

Figure 23.9. Posture target records taken from a photograph

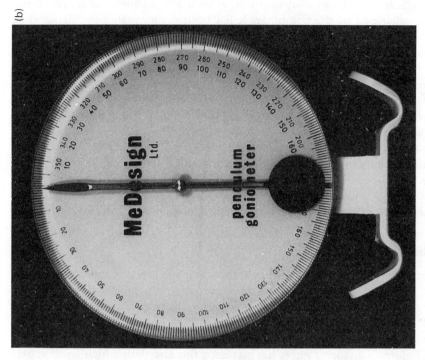

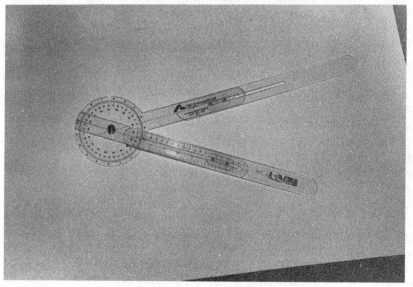

Figure 23.10. Goniometers. The radiating arms can be set either side of a joint, or used to assess displacement from a datum direction. The circular goniometers operate by gravity on the pendulum

result of deflecting the spine in any direction is recorded on-line by a computer. Software allows all deflections, as well as velocities and accelerations, to be calculated by the computer. It has been demonstrated by Marras *et al.* (1990) that differentiation between conditions, and investigation of the level of hazard, in materials handling jobs, can be better distinguished by including the velocity and acceleration data in the analysis of lumbar movement than by relying on displacement alone.

A different approach has been to use flexible lightweight tubes containing strain-gauged strips, which are strapped to the joint, or along the spine. Each allows joint movement to be measured in two planes simultaneously and the results can be stored on unobtrusive recorders for direct reading or later transfer. The procedure gives consistent results, and is commercially available.* It has been used in clinical studies as well as in ergonomics investigations, e.g. spinal flexion of the neck (Parsons and Thompson, 1990) and lumbar movement by motor mechanics (Boocock *et al.*, 1994).

Indirect observations

There are several commerical methods which use three or more video cameras, together with software, to track markers on a subject moving in the cameras' fields of view. Advances in computer software are making it increasingly possible for laboratories to develop their own systems using 'frame-grabbing' techniques, for example, and taking data from video recordings.

The use of a video camera has increased the ease with which work activities may be recorded for later analysis. As yet, however, the analysis is still a long and tedious process. Unless care is taken to record truly representative samples, to do a proper work analysis and to record from several positions so that joint angles can be estimated accurately, the analyst is probably better using simpler, direct measurements at the workplace itself.

An example of the use of a simple video camera, combined with work analysis and other techniques has been described by Drury (1987). The problem was to evaluate the hazard of upper limb disorders due to posture and repetition. Drury recorded at least two representative cycles from five different camera positions: 1) a front three-quarter view from above the operator's head, 2) a direct left side view, 3) a direct right side view, 4) a direct front view and 5) a direct rear view. From the task analysis the sub-tasks were decided upon, so that a sequence of single motions was identified. Each was defined, using the video record, onto an analysis form (Figure 23.11). For the purpose of the study, the joint angle values were then coded, using the classification of Figure 23.12, and recording the class numbers on another copy of the form. The coding gives higher numbers to more serious deviations from the neutral position and the resultant sheet of coded values demonstrates where serious problems are to be expected. By further calculations,

---

*Penny and Giles Biometrics Ltd, Blackwood, Gwent, NP2 2YD, UK

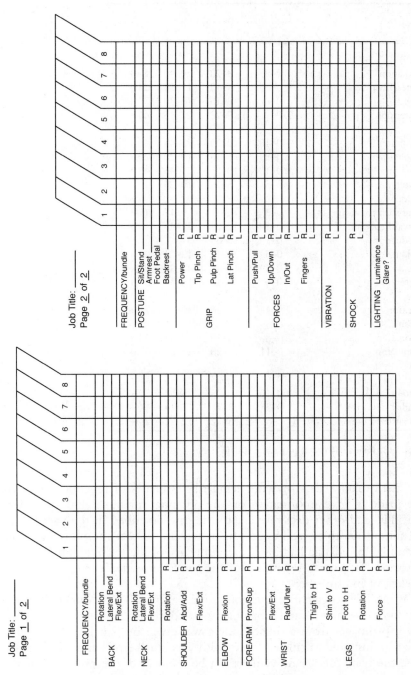

Figure 23.11. Posture analysis form, using data from video recordings (Drury 1987)

Zones and Joint Angles

	Zone, degrees			
	0	1	2	3
Neck				
Rotation	0–8	8–20	20–40	40 +
Lateral bend	0–5	5–12	12–24	24 +
Flexion	0–6	6–15	15–30	30 +
Extension	0–9	9–22	22–45	45 +
Back				
Rotation	0–10	10–25	25–45	45 +
Lateral bend	0–5	5–10	10–20	20 +
Flexion	0–10	10–25	25–45	45 +
Extension	0–5	5–10	10–20	20 +
Shoulder				
Outward rotation	0–3	3–9	9–17	17 +
Inward rotation	0–10	10–24	24–49	49 +
Abduction	0–13	13–34	34–67	67 +
Adduction	0–5	5–12	12–24	24 +
Flexion	0–19	19–47	47–94	94 +
Extension	0–6	6–15	15–31	31 +
Elbow				
Flexion	0–14	14–36	36–71	71 +
Forearm				
Pronation	0–8	8–19	19–39	39 +
Supination	0–11	11–28	28–57	57 +
Wrist				
Flexion	0–9	9–23	23–45	45 +
Extension	0–10	10–25	25–50	50 +
Radial deviation	0–3	3–7	7–14	14 +
Ulnar deviation	0–5	5–12	12–24	24 +

Figure 23.12. Joint angle codes for posture analysis (Drury 1987)

using performance data or work standards, estimates of repetitions per day were made to give information on the potential for upper limb disorders. From this, and other data relevant to the case described in the paper, a set of requirements to be addressed was drawn up, so that the working group responsible for change had a succinct summary of the relative importance of the various factors. A repetition of the procedure after change gave clear evidence of improvement.

It is possible to calculate true joint angles using a single camera, but the position of the cameras and subject must be controlled. Computation for the analysis of images from a single camera was given by Jian Li, *et al.* (1990). They demonstrated calculated angles with very small errors and consider that where the plane of the movement can be controlled, the procedure is not difcult. This constraint arises, in practice, where industrial tasks are frequently repeated in a single workspace, such as in assembly line work.

Video methods often use markers on the body to provide measurement points. These can be stripes placed, for example, along the long axes of the limbs, reflective spots on appropriate body parts (e.g., over the joints of limbs) or small lights similarly mounted. An optical system which used reflective markers is the Cartesian Optoelectronic Dynamic Anthropometer (CODA)* device. This equipment is usable on-line to a computer. It scans a number (up to 12) of retroreflective pyramidal markers with light from three segmented mirrors at a rate of 300 times/s. One mirror rotates in the vertical plane centrally between two horizontally rotating mirrors which are about 1m apart. The system, which is illustrated in Figure 23.13 will record the position of each marker (which is identified by its colour) up to ±0·1mm in the plane parallel to the mirrors and ±0·3mm perpendicular to this plane, within a workspace of a four metre cube.

Figure 23.14 shows the use of an optical pointer to locate and record points outside the vision of the instrument. Two different markers are spaced a known distance apart and at a known distance from the point of the rod. The computer is programmed to recognize the markers when both are visible, identify their positions and calculate the position in space of the point of the rod from this information. In use, one marker is covered and the point of the rod located on the required anatomical point. The concealed marker is then revealed to CODA, the computer giving a 'bleep' when the recording is complete. To cancel a false reading, the computer may be programmed so that when a marker of a different colour is displayed, the previous reading is cancelled. This device overcomes a major restriction of all vision systems, that they cannot see the other side of the subject (many use multiple cameras to compensate for this). The optical pointer is usable only for stationary postures. However, by the use of this technique a posture can be introduced into the computer. Force data from strain gauged instruments can also be input, and calculations done virtually on-line. Figure 23.15 shows a stick man displayed on a screen, developed from the CODA readings, with body load data presented alongside. Applications of this method are discussed in chapter 24 on biomechanics.

Effects on the person of maintaining a posture

Maximum voluntary contraction

Perhaps the earliest scientific approach to estimating the appropriateness of static loads was to evaluate, experimentally, the holding times for various loads, expressing the result as the times for which a person could hold proportions of the maximum load. The force required to achieve maximum load is referred to as maximum voluntary contraction (MVC).

Whilst MVC may be measured quite simply (in many cases using a spring

---

*Charnwood Dynamics Ltd, Loughborough, Leicestershire, LE12 8LY, UK.

(a)

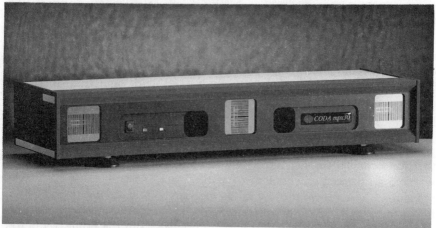

(b)

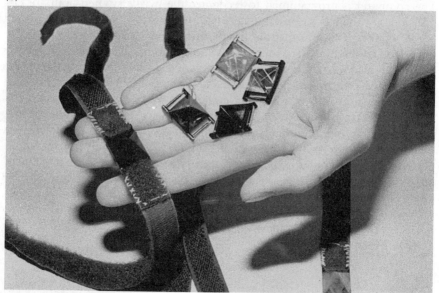

Figure 23.13. The CODA mpx 30 optical posture recording instrument (a), with examples of the retroreflective pyramidal markers (b)

balance) there are some essential controls. An impulse force is not required
for the measurement; the instruction to a subject is usually of the form 'build
up your maximum force gradually, over a period of 2-3 s, and hold it for
3 s'. The value used is the mean force over the last, relatively constant, period.

The relationship between force and holding time, demonstrated by Monod
and Scherrer (1965) and by Rohmert (e.g. 1973a,b) is a logarithmic one
(Figure 23.16). Today it is accepted that a long-term *constant* static effort

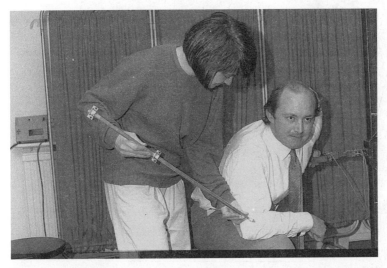

Figure 23.14. The optical pointer in use, providing direct input for reference points, even when the point itself is invisible to CODA

Figure 23.15. On-line display of the posture via a 'stick man', and immediate calculation of forces and torques at the desired joints

greater than 2–3% of MVC is unacceptable, although at one time 15% was believed to be possible. Knowledge of the force holding-time relationship, which appears to hold for most skeletal muscle, does allow us to estimate the effects of some postures, and provide guidance as to their appropriateness. The maximum holding time for a posture is not, by itself, a very useful measure, since we wish to know the frequency with which the posture may be held, and the consequent likelihood of damage. Hence recovery from

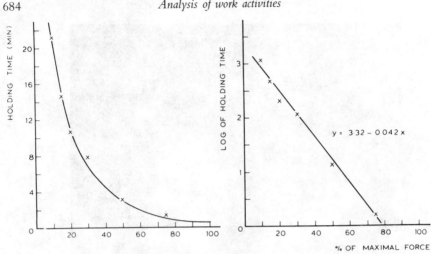

FORCE-TIME RELATIONSHIP FOR PULL CONDITION

Figure 23.16. Subjects held various percentages of their maximum force, exerted by pulling at shoulder level, for as long as they could. The log-normal relationship gives a means for calculation of intermediate values

static workloads is of interest. In experimental work, evidence of recovery has been taken as being when the same posture can be held again for the same maximum time. In gathering such data we must seek to achieve the same level of motivation for each test and treat subjects as their own controls. Thus we calculate the forces as a percentage of the subject's own MVC, and provide rest periods as a proportion of each subject's own holding time. A typical recovery curve is shown in Figure 23.17 (Milner *et al.* 1986). It arises from a number of different forward bending postures held for as long as possible (T_1), after which the subject had a rest interval equal to 12 times T_1 and then repeated the posture again (T_2). An important consequence of the relationship shown in this curve is that if the maximum holding time is considerable, e.g., the posture is such that discomfort builds up over a long period, then the recovery also takes a very long time.

If we are seeking MVC for a particular situation it is important that the posture adopted, including foot positions and any constraints due to the workplace, are repeated during the tests, so that, as nearly as possible, the same muscle groups are used. Few *practical* force exertions are undertaken by just a single group of muscles, so, if a number of different muscles is recruited to do the task, unless the posture is identical, the group of muscles could be operating differently, or else other groups of muscles may be called into play. It is more usual to find static work interspersed with rest pauses, and in practical situations the requirement is often to estimate the feasibility of a given work-rest cycle.

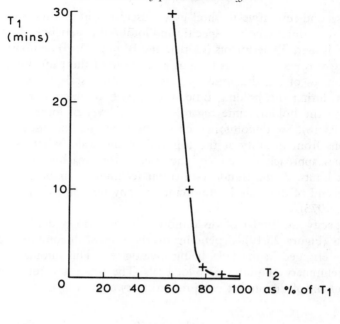

Graph of $T_2 = T_1{}^{0.854}\,e^{\frac{-0.152}{I}}$

For $I = 1200\ \% \ T_1$

Forward bending posture

Figure 23.17. The graph shows the recovery (T_2) after a rest pause I equal to twelve times the first holding time (T_1). Data for the formula came from 42 subjects experiencing five different postures and five rest intervals

Subjective methods

Two subjective methods can contribute to the evaluation of static work. Borg's scale has already been described, together with its rationale, in Kilbom's chapter 22 on dynamic work. Although its use in static work is not valid in terms of the relationship of the numbers to the person's heart rate, the judgements of severity do give important information. An example of this arose from a demonstration of the difficulties involved in butchery. A rig was available which permitted subjects to exert single-handed forces on a set of knife handles, pulling them in directions across and down the body in the vertical plane and exerting as much force as possible. Although differences in the forces were found when working at different heights, they were relatively modest. However, when subjects used Borg's scale to judge the difficulty of pulling in each of the directions, large differences were identified in the ease of the knife movement in each of the directions, which separated them more clearly than the analysis of the imposed forces could do.

The second subjective method uses muscular pain as a measure. Because 'pain' is sometimes seen as a specific and localized experience, the term 'discomfort' is used. Experiments (Corlett and Bishop, 1976) demonstrated that, if a force was exerted for as long as possible until the pain was unbearable and estimates of the discomfort levels made on a scale (5 or 7 points) at intervals during the holding time, the growth of discomfort was linearly related to the holding time regardless of the level of force being exerted (Figure 23.18). So, discomfort itself can be used as a linear scale. There are deviations from linearity at the top end of the scale, where subjects will sometimes approach the end of the scale before reaching their own discomfort limits. A magnitude estimation technique can be adopted if the extreme end of the scale is important (see any text on psychophysics, e.g., Stevens, 1975).

To specify the site(s) of discomfort, a body map is used, divided into segments (Figure 23.19), depending on the sites of discomfort experienced by those engaged in the tasks being investigated. This infomation is found from preliminary enquiries or pilot trials. The procedure for mapping the development of discomfort can proceed in two ways:

1. At intervals during the whole working day, people are asked to point to the site(s) of current discomfort on the body map. Then they are asked to rate the intensity of discomfort at each identified site on a 5 or 7 point scale, preferably by marking a paper scale which is 'anchored' at the 0 and 5 (7) points by 'no discomfort' and 'extreme discomfort', respectively. These scores are plotted against time of day for each body site, dividing the scale during analysis at the 1/2 points as well to give

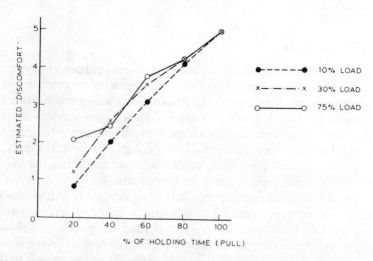

Figure 23.18. Mean values for overall discomfort ratings when different percentages of MVC are exerted for as long as possible

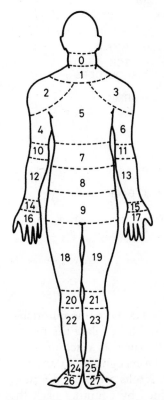

Figure 23.19. The body map for evaluating body part discomfort, either by rating or ranking

an effective 10 (or 14) point scale. Since it is likely that differences in body size or person-equipment relationships will cause changes in the distribution of discomfort around the body, the effect of adding together the scores from several subjects should be considered carefully; it is usually unwise.

The reason for urging the collection of data throughout the working day is that recovery from static load is slow (see above), and a lunch break is often not sufficient to achieve full recovery. A study of engravers (Figure 23.20), where four workers were studied and the average result taken, illustrates the point forcibly. It is evident from the high discomfort levels reached that the posture is extreme, and the curves representing the most heavily loaded body parts appear as one curve spread across the whole day, rather than the morning curve repeated in the afternoon.

2. The second way in which the body-mapping procedure can be used is recommended when rating of individual sites will take up too much of the subject's time. The person can be asked to point out the site(s)

Analysis of work activities

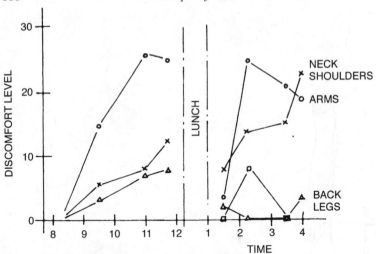

Figure 23.20. Discomfort scores, rating on a 7-point scale by four workers. The relative continuity of the neck and arms graphs over the lunch period will be noted

which are most uncomfortable, then those next most uncomfortable and so on until no more sites are reported. We have asked the person to identify a sequence of just-noticeable differences in discomfort, so it is unlikely that more than five or six will be recognized, as any text on psychophysics will confirm. Again these results are plotted against time of day, but a numerical value is obtained by counting back the number of levels of discomfort reported and numbering them, using the no-discomfort sites as zero, the last reported sites as 1, and so on. Although less detailed than the previous method, it is much quicker, involves less explanation to the subjects and reveals the most heavily loaded body parts equally as well.

Of course, all the other aspects of experimental control apply for this method as well.

Postural load data

As a major contribution to the faster gathering of data on musculoskeletal problems, Institutes of Occupational Health in the Nordic countries have designed the Nordic Questionnaire (NMQ) (Kuorinka *et al.*, 1987). This provides a standard format for gathering data on musculoskeletal problems. Increased information about the incidence and epidemiology of these complaints is very necessary. Where data are needed for a particular investigation, such a questionnaire can be supplemented by additional questions, but its use will enable data from different studies to be compared, and the large data pool arising from its use in the Nordic countries can be used for comparative purposes also (see also chapter 30).

The NMQ was very attractive to the UK Health and Safety Executive and in consequence it was evaluated for widespread use by that authority. Small modifications were made, to be sure that it was unambiguous for native UK English speakers (Dickinson *et al.* 1992). This modified questionnaire is included as an appendix at the end of the chapter, by kind permission of the Health and Safety Executive. It follows a format as given in chapter 3 on questionnaires, with a prologue, the core questions, some classification questions and an epilogue. The musculoskeletal disorders survey begins, after the personal details section, with a general survey to give estimates of prevalence and disability. Following this are four sections seeking more specific information for each of four body areas. These seek to establish the severity of any disorder. The final sections include a general health questionnaire and an opportunity to give more details of the work of the respondent.

The example shown was for a survey of supermarket checkout operators. Ten supermarkets, and 481 operators, were used in the initial evaluations. Their results in terms of reliability of responses were similar to those of Kuorinka *et al.* (1987) but they emphasized that a minimum 80% return is necessary if prevalence rates are to be realistic. The HSE experience was that to bring respondents together in groups to answer the questionnaire provided a better return rate than just issuing the forms and asking for their return. Numbers suffering some disability appeared to be less affected by the way the form was administered, those suffering presumably having more reason to see it to be to their advantage to reply.

Another tool developed at the Swedish National Board of Occupational Safety and Health is the single sheet analysis for identifying musculoskeletal stress factors (see Figure 23.21). This is self-explanatory and uses the site of discomfort or injury to focus attention on a number of possible workplace faults which could be their causes. The list of possible causes is equally applicable to the body–mapping procedure described earlier, enabling a direct link to be made to the sources of the problems. After changes have been introduced, it is clear that these same methods can be used to demonstrate any improvements which have been achieved.

Electromyography

Electromyography (EMG), the recording of myoelectric signals which occur when a muscle is in use, can be used to assess the level of activity occurring over a period of time. It can also be used to show the presence of muscle fatigue, a state when a skeletal muscle is unable to maintain a required force of contraction (Hagberg, 1981). A high correlation has been shown between EMG activity and muscular force, for both static and dynamic activities (Hagberg, 1981). This relationship was once thought to be linear, but is now proposed as exponential (Lind and Petrofsky, 1979; Hagberg, 1981).

When a muscle begins to fatigue, there is an increase in the amplitude in the low frequency range of EMG activity (Petrofsky *et al.*, 1982). There is

Kemmlert, K. Kilbom, A. (1986) National Board of Occupational Safety and Health, Research Department, Work Physiology Unit, 171 84 Solna, Sweden

1. Is the walking surface uneven, sloping, slippery or nonresilient?
2. Is the space too limited for work movements or work materials?
3. Are tools and equipment unsuitably designed for the worker or the task?
4. Is the working height incorrectly adjusted?
5. Is the working chair poorly designed or incorrectly adjusted?
6. (If the work is performed whilst standing): Is there no possibility to sit and rest?
7. Is fatiguing foot-pedal work performed?
8. Is fatiguing leg work performed eg:
   a) repeated stepping up on stool, step etc.?
   b) repeated jumps, prolonged squatting or kneeling?
   c) one leg being used more often in supporting the body?
9. Is repeated or sustained work performed when the back is:
   a) flexed forward, more than 20°?
   b) severely flexed forward, more than 60°?
   c) bent sideways or twisted, more than 15°?
   d) severely twisted, more than 45°?
10. Is repeated or sustained work performed when the neck is:
   a) flexed forward, more than 15°?
   b) bent sideways or twisted, more than 15°?
   c) severely twisted, more than 45°?
   d) extended backwards?
11. Are loads lifted manually? Notice factors of importance as:
   a) periods of repetitive lifting
   b) weight of load
   c) awkward grasping of load
   d) awkward location of load at onset or end of lifting
   e) handling beyond forearm length
   f) handling below knee height
   g) handling above shoulder height
12. Is repeated, sustained or uncomfortable carrying, pushing or pulling of loads performed?
13. Is sustained work performed when one arm reaches forward or to the side without support?
14. Is there repetition of:
   a) similar work movements?
   b) similar work movements beyond comfortable reaching distance?
15. Is repeated or sustained manual work performed? Notice factors of importance as:
   a) weight of working materials or tools
   b) awkward grasping of working materials or tools
16. Are there high demands on visual capacity?
17. Is repeated work, with forearm and hand, performed with:
   a) twisting movements?
   b) forceful movements?
   c) uncomfortable hand positions?
   d) switches or keyboards?

Method of application.

* Find the injured body region
* Follow while fields to the right
* Do the work tasks contain any of the factors discribed?
* If so, tick where appropriate

Also take these factors into consideration:
a) the possibility to take breaks and pauses
b) the possibility to choose order and type of work tasks or pace of work
c) if the job is performed under time demanded or psychological stress
d) if the work can have unusual or unexpected situations
e) presence of cold, heat, draught, noise or troublesome visual conditions
f) presence of jerks, shakes and vibrations

Column headers (body regions): neck shoulders, upper part of back | elbows, forearms hands | feet | knees and hips | low back

Figure 23.21. Method for the identification of musculoskeletal stress factors which may have injurious effects

also a shift in the frequencies towards the lower end of the spectrum as fatigue occurs.

Although needle electrodes, entering specific muscles, are used for medical research, occupational EMG records are usually taken from surface electrodes. These are stuck over the central part of the muscle and leads taken, via pre-amplifiers, to amplification and recording equipment. Telemetering can be done, but the existence of miniaturized circuitry and recorders has rendered this less necessary. Skin preparation is necessary for reliable recordings. This involves the use of fine sandpaper to remove layers of dead skin prior to fixing the electrodes. Many modern electrodes, usually small silver discs, have a central hole into which electrode jelly, a saline grease, may be inserted with a syringe after the electrode is in place. The placement is generally over the central part of the muscle, where most of the active fibres will lie, and the electrodes will be from 3–5 cm apart. An 'earthing' electrode is sometimes included, placed away from the muscle being recorded and where other muscular activity is unlikely to be picked up. Signals from the electrodes should be pre-amplified as close to the electrodes as possible, to increase the signal-to-noise ratio, before passing them via a low pass filter to a recorder or analyzer. As opposed to clinical EMG, where the quality of the signal is of importance, in occupational EMG the quantity of the signal is usually the important factor. The signal is analyzed with respect to its frequency or amplitude, and the amplitude is usually analyzed with the signal given as a percentage of that from a standard MVC taken prior to the investigation. This conversion permits comparison across different tasks and different people.

There are three major methods for EMG analysis: the integrated EMG (IEMG), Fourier analysis and amplitude probability distribution function (APDF) analysis.

IEMG

The integrated EMG gives a measure of the power in the signal. Integrating circuits accumulate the root mean square (rms) values of the signal, recording when a certain selected total value has been reached and starting the addition again. The visual record shows a series of triangular waves, with equal peaks but spaced more closely where the EMG signal was greater. Counting the peaks per unit time or calculating the rms value, again per unit time, provides values representative of the muscular activity.

Fourier analysis

The speed of response of muscle fibres varies, and they are conventionally divided into fast and slow twitch fibres. As a muscle is used, slow twitch activity becomes more evident, and is taken to be a sign of increased muscular fatigue. The frequencies bound up in a raw EMG signal are identified by Fourier analysis, a procedure which breaks down the signal into its component sine waves. Usually the analysis is done by taking successive short samples of the EMG, analyzing them and presenting the results as a frequency spectrum, showing the frequency and amplitude of the component waves.

Amplitude Probability Distribution Function

Work by Jonsson (1976) and Hagberg (1979) demonstrated the utility of analyzing EMG in terms of the amplitudes present in the signal. Each excursion of the signal represents the innervation of muscle fibre(s) to exert the force. Large excursions are related to the exertion of external force or rapid movement; small and frequent excursions can be interpreted as indicating the maintenance of static work, e.g., for holding a position. The analysis is relatively simple in concept. Again, short successive samples of the signal are taken and the amplitudes of all the peaks counted and grouped. They are plotted as an amplitude spectrum or as a cumulative amplitude distribution function. If the latter plot is adopted, the 10th decile is proposed as the level which demonstrates the static work load.

Although close correlations have been reported between EMG analyses and force, Hagberg (1981) has noted some sources of variance in experiments. The relationship will change with temperature (such as might arise from high levels of work activity), with fatigue, with whether the contraction is concentric or eccentric, and with changes in velocity of contraction. For much occupational work these factors may not be serious influences on results, but should be considered in relation to the quoted literature for any extensive studies. Where defined test contractions are easy to apply and the muscle action substantially isometric, the APDF can be a useful measure of the muscle performance.

Comment on EMG

Each method has its uses, depending on the problems under study. As will be recognized, all methods may be used from the same EMG recordings. The amount of data collected in occupational EMG is usually very large as longer recording periods are required. This creates difficulties in storage and analysis. Also artefacts are more common, such as movement artefacts, changes in recording due to temperature changes or muscle isolation.

Occupational EMG is a good technique for assessing which muscles are used in a task but is of more limited use in accurately assessing the fatigue process. It is, however, of value to establish the changes in the amplitude and frequency domains, with respect to time, to gain some understanding of how a muscle is operating. It is a tool which should be used in conjunction with other assessment techniques for a good understanding of a work situation.

Spinal loading

Although in the investigation of postural loading, where loads are of short duration, a biomechanical analysis may be suitable, exposure over a longer period can introduce changes which alter the spinal characteristics. The effect of exposures over a long period, or loads which are frequently repeated, can be assessed by measuring changes in total stature (Eklund and Corlett, 1984).

The forces imposed on the lumbar spine during the day are the gravity loadings from thorax, head and arms, together with the components of forces exerted by the arms which have to be transmitted to the pelvis. These loads cause a reduction in stature over the day of 15mm or more due to a slow decrease in spinal disc height; this will vary with age. Such a shrinkage is recovered when lying down. To compare the effects of different workplaces, postures or work regimes on spinal loading, the use of a precision stadiometer is required (Figure 23.22). By close control of the experiment and measure-

Figure 23.22. Precision stadiometer

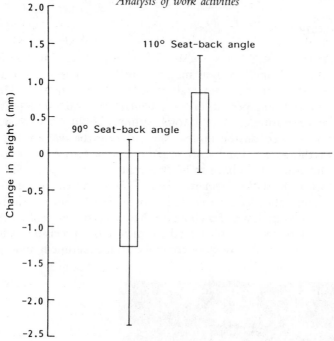

Figure 23.23. Changes in height during simulated driving, for two different seat-back angles

ment protocol, changes in stature of about 0.5mm can be identified. As with many biological response measures, it is preferable to use subjects as their own controls, since averaging across subjects—due to differences in responding—increases the variance considerably.

There are several points to note when using the technique, which has been used in the workplace as well as in the laboratory. As with all precision measurements, tight experimental control is required. If repeated tests are to be made, the time of day and prior activities of the subjects should be consistent. As recovery of height loss is quite rapid, the periods between exposure to load and subsequent measurement should be kept short and, again, consistent. Rest pauses between any sequence of measurements should also be controlled so that no extra increase or major decrease of load arises, say from a major change in posture. Thus if, at one point in an experimental sequence, a subject lay down, the resultant change in disc condition would make the effects of a subsequent test condition very different from earlier trials (Abu Amin et al., 1988).

Some instruction to subjects is needed on how to position themselves on the stadiometer, to help in maintenance of a consistent posture. The experimenter must also check and control weight distribution between heels and soles, location of the spine on the micro-switched pads set into the backboard of the instrument, head position (which is assessed by the nose marker), and ensure that the subject has folded arms. About five recordings over 2 or

3 s are taken. The subject then steps off, and back on the stadiometer immediately, is repositioned and more readings are taken. The repetition gives measures for the estimation of error as well as evidence for the consistency of the measuring posture.

For any study it is desirable to differentiate stature changes due to load from those which would normally arise under gravity loadings in the postures associated with the situation under investigation. Thus it will often be advisable to run a number of preliminary measures to establish the rate of shrinkage prior to loading, following with the trials. Extrapolation of the initial measures to the time of the final measures under load will allow assessment of a loading effect which will be independent of the expected rate of shrinkage at that time of the day. Recent work by Althoff *et al.* (1992) has demonstrated the importance of this point, as well as the difficulty of establishing it in some cases.

A study by Foreman (1989) noted that compressibility of the heel is a major influence on total stature. He showed changes ranging between 2 and 6mm in a group of 20 subjects over a period of about 1·5min. Hence it is desirable that, where experimental conditions reduce the load on the heel, some control is exercised to take this change into account. This might be done by recording initial calibration curves for heel compression and correcting results in relation to the time from standing on the stadiometer. The work quoted earlier, by Althoff *et al.* (1992) describes techniques to exercise control over heel pad shrinkage and regular diurnal variations, as well as noting the effects on shrinkage of lumbar disc diameter. They note that, where close control is exercised, an accurate and reliable measurement of stature change, and hence an estimation of spinal stress, is possible in practical situations.

Some results from the use of the method are shown in Figure 23.23. Comparison was made between seat-back angles for drivers of a car. Subjects 'drove' a video driving game for 1-h periods. The changes in stature for seat geometries support the work of Andersson and Örtengren (1974), who showed that spinal load was least for a seat–back angle of about 110°.

The technique has been used to study nursing activities (Foreman and Troup, 1987), design of working seats (Corlett and Eklund, 1983), the effects of circuit weight training and running (Leatt *et al.* 1986), the effects of equipment arrangement and vibration on vehicle driving (Bonney, 1988) and the effects of overhead working (Burton and Tillotson, 1991).

Recent work by Jafry (1993) suggests that, where subjects are seated during experimental studies, a stadiometer designed for use on a seated subject is desirable. When changing from sitting to standing the change in spinal posture may allow a relatively rapid change in disc height which confounds the effects of the experimental loadings.

Advances in posture evaluation

In the introduction we set out five dimensions relevant to the investigation of postural loads and static work. However consideration of some of the measures discussed will show that more than these factors have been brought into the analyses. Although research techniques can be relatively specific and focus on one or two measures, the field worker must look more widely. This wider view is one of the features of field investigations which makes causation such a problem to define in a court case, but at the same time helps to elucidate apparently conflicting field results.

This problem is taken up more extensively in chapter 30, for instance, where approaches are sought which may not pin down in a quantitative manner the contributing share from each of a number of variables in an investigation, yet still a logical and satisfactory explanation for the observations resulting from a study can be obtained. This is a natural order of things; the complexities of nature do not reveal themselves immediately. We gradually recognize the ways in which apparently disparate components interact. Only when they are recognized as relevant and contributory can we begin to measure their contributions. A qualitative understanding is where we start.

Even in the rather restricted—yet still complicated—field of posture evaluation it is inevitable that the list in the introduction will be extended. This is clear from the four key problem areas which succeed our five components; what we know about them already tells us that there is, for example, more to fatigue than just muscle metabolism and biomechanics. Ways of quantifying these others influences will arise, be built into the tool kit and become normal practice. This is one challenging area for methods development. It is also one of the ways in which practical studies, illuminated by the broad ergonomics base of physiological, psychological, environmental and social scientific knowledge, contribute to the more reliable understanding of the real world. The iterative interplay between science and practice, conducted by a wondering mind, is the process through which reliable knowledge is consolidated into a more comprehensive understanding. Readers will realize that the research literature is where they must look for such developments in methods. That is where they are tested and their contributions assessed, after which their utility can be explored in a wide range of field studies. We can expect, therefore, and should welcome, many changes to ergonomics methods, including those for posture evaluation, over time.

References

Abu Amin, A., Corlett, E.N. and Bonney, R.A. (1988). Does wearing a seat belt alter the load on the back whilst driving. In *Contemporary Ergonomics*, edited by E.D. Megaw (London: Taylor and Francis).

Althoff, I., Brinkman, P., Frobin, W., Sandover, J. and Burton, K. (1992). An improved method of stature measurement for quantitative determination of spinal loading. *Spine,* **17**, 682–693.

Andersson, B.J.G., and Örtengren, R. (1974). Lumbar disc pressure and myoelectric back muscle activity during sitting. **II**. Studies on an office chair. *Scandinavian Journal of Rehabilitation Medicine,* **6**, 115–121.

Armstrong, T.J. Foulke, J.A. Joseph, B.S. and Goldstein, S.A. (1982). Investigation of cumulative trauma disorders in a poultry processing plant. *Journal of the American Industrial Hygene Association,* **43**, 103–115.

Bonney, R.A. (1988). Some effects on the spine from driving. *Clinical Biomechanics,* **3**, 236–240.

Boocock, M.G., Jackson, J.A., Burton, A. and Tillotson, K.M. (1994). Continuous measurement of lumbar posture using flexible electrogoniometers. *Ergonomics,* **37**, 175–185.

Brouha, L. (1960). *Physiology in Industry* (Oxford: Pergamon).

Burton, K. (1986). Measurement of regional lumbar sagittal mobility and posture by means of a flexible curve. In *The Ergonomics of Working Postures,* edited by E.N. Corlett, J.R. Wilson and I. Manenica (London: Taylor and Francis).

Burton, A.K. and Tillotson, K.M. (1991). Measurement of spinal strain to estimate loads on the spine in overhead working postures. Report of the Spinal Research Unit, School of Human and Health Sciences, Huddersfield University, Huddersfield, UK.

Corlett, E.N. and Bishop, R.P. (1976). A technique for assessing postural discomfort. *Ergonomics,* **19**, 175–182.

Corlett, E.N. and Eklund, J.A.E. (1983). The measurement of spinal load arising from work seats. *Proceedings of the Human Factors Society 27th Annual Meeting,* Santa Monica, CA.

Corlett, E.N., Madeley, S. and Manenica, I. (1979). Posture targetting: a technique for recording working postures. *Ergonomics,* **22**, 357–366.

Corlett, E.N., Wilson, J.R. and Manenica, I. (1986). *The Ergonomcis of Working Postures* (London: Taylor and Francis).

Dickinson, C.E., Campion, K., Foster, A.F., Newman, S.J., O'Rourke, A.M.T. and Thomas, P.G. (1992). Questionnaire development: an examination of the Nordic Musculoskeletal Questionnaire. *Applied Ergonomics,* **23**, 197–201.

Drury, C.G. (1987). A biomechanical evaluation of the repetitive motion injury potential of industrial jobs. *Seminars in Occupational Medicine,* **2**, 1, 41–50 (New York: Thieme Medical Publishers Inc.).

Eklund, J.A.E. and Corlett, E.N. (1984). Shrinkage as a measure of the effect of loads on the spine. *Spine,* **9**, 189–194.

Finnish Institute of Occupational Health (1992). OWAS, a method for the evaluation of postural load during work. Publication office, Topeliuksenkatu 41 aA, SF 00250 Helsinki, Finland.

Foreman, T.K. (1989). Low back pain prevalence, work activity analyses and spinal shrinkage. Unpublished Ph.D. thesis, University of Liverpool.

Foreman, T.K. and Troup, J.D.G. (1987). Diurnal variations in spinal loading and the effects on stature. *Clinical Biomechanics*, **2**, 48–54.

Hagberg, M. (1979). The amplitude distribution of surface EMG in static and intermittent static muscular exercise. *European Journal of Applied Physiology*, **40**, 265–272.

Hagberg, M. (1981). An evaluation of local muscular load and fatigue by electromyography. *Arbete och Hälsa*, **24**, Solna, Sweden.

Hutchinson, A. (1970). *Labanotation* (London: Oxford University Press).

Jafry, T. (1993). The effects of vibration, posture and operating foot pedals on spinal loading. PhD thesis, University of Nottingham.

Jian Li, Bryant, J.T. and Stevenson, J.M. (1990). Single camera photogrammetric technique for restricted 3D motion analysis. *Journal of Biomedical Engineering*, **12**, 69–74.

Jonsson, B. (1976). Evaluation of the myoelectric signal in long-term vocational electromyography. In *Biomechanics V*, edited by A.P.V. Komi (Baltimore: University Park Press), pp. 509–514.

Kember, P.A. (1976). The Benesh movement notation used to study sitting behaviour. *Applied Ergonomics*, **7**, 133–136.

Kuorinka, I., Jonsson, B., Kilbom, Å., Vinterberg, H., Biering-Sørenson, F., Andersson, G. and Jorgensen, K. (1987). Standardized Nordic questionnaires for the analysis of musculoskeletal symptoms. *Applied Ergonomics*, **18**, 233–237.

Leatt, P., Reilly, T. and Troup, J.D.G. (1986). Spinal load during circuit weight training and running. *British Journal of Sports Medicine*, **20**, 119–124.

Lind, A. and Petrofsky, J.S. (1979). Amplitude of the surface EMG in fatiguing isometric contractions. *Muscle and Nerve*, **2**, 257–264.

McAtamney, L. and Corlett, E.N. (1992). Reducing the risks of work related upper limb disorders: a guide and methods. Institute for Occupational Ergonomics, University of Nottingham.

McAtamney, L. and Corlett, E.N. (1993). RULA: a survey method for the investigation of work-related upper limb disorders. *Applied Ergonomics*, **24**, 91–99.

Marras, W.S., Ferguson, S.A. and Simon, S.R. (1990). Three dimensional dynamic motor performance of the normal trunk. *International Journal of Industrial Ergonomics*, **6**, 211–214.

Milner, N.P., Corlett, E.N. and O'Brien, C.O. (1986). A model to predict recovery from maximal and submaximal isometric exercise. ch. 13 in: *The Ergonomics of Working Postures*, edited by E.N. Corlett, J.R. Wilson and I. Manenica (London: Taylor and Francis Ltd).

Monod, H. and Scherrer, J. (1965). The work capacity of a synergic muscle group. *Ergonomics*, **8**, 329–338.

Parsons, C.A. and Thompson, D. (1990). Comparison of cervical flexion in shop assistants and data input VDT operators. In *Contemporary Ergonomics* (London: Taylor and Francis).

Petrofsky, J.S., Glaser, R.M. and Phillips, C.A. (1982). Evaluation of the amplitude and frequency components of the surface EMG as an index of muscle fatigue. *Ergonomics*, **25**, 213–223.

Rohmert, W. (1973a). Problems in determining rest allowances. 1. Use of modern methods to evaluate stress and strain in static work. *Applied Ergonomics*, **4**, 91–95.

Rohmert, W. (1973b). Problems in determining rest allowances. 2. Determining rest allowances in different tasks. *Applied Ergonomics*, **4**, 158–162.

Stevens, S.S. (1975). *Psychophysics* (Chichester: John Wiley and Sons Ltd).

Tracy, M.F. and Corlett, E.N. (1991). Loads on the body during static tasks: software including the posture targetting method. *Applied Ergonomics*, **6**, 362–366.

Appendix★

Surv No. `2050` `0` `1`

MUSCULOSKELETAL DISORDERS SURVEY OF SUPERMARKET CHECKOUT OPERATORS

Dear Sir / Madam,

With the co-operation of your employer and trade unions we are conducting a survey to find out the extent to which muscle and joint aches and pains are experienced by employees in retail occupations.

We are interested in mild and severe problems affecting muscles, ligaments, nerves, tendons, joints and bones suffered both at work and away from work. This could mean sprains, strains, inflammations, irritations and dislocation. For the purpose of this survey we are not interested in any injuries to the skin.

We would like you to complete this questionnaire about your health. All answers will be treated as **strictly confidential** and individual answers will not be made known to anyone other than the survey team.

The more questionnaires that are completed, the greater will be the accuracy and usefulness of the findings, the better to help us improve health and safety at work.

Thank you for your help.

Claire Dickinson

Claire Dickinson

HOW TO ANSWER THE QUESTIONNAIRE

Please complete this questionnaire by answering ALL questions as fully as possible. Some of the questions require a written answer, for others you need only tick a box. ☑

Please do not write in the margin.

PERSONAL DETAILS

1 Today's date   Day Month Year

2 Sex   Male ₁☐   Female ₂☐

3 Date of birth   Day Month Year

4 What is your weight?
   stones ☐   pounds ☐ or   kg ☐

5 What is your height?
   feet ☐   inches ☐ or   cm ☐

6 Are you right or left handed?
   right ₁☐   left ₂☐   able to use both hands equally ₃☐

1

★The Nordic Musculoskeletal Questionnaire (NMQ) as modified by the UK Health and Safety Executive, and applied to a study of supermarket checkout operators.

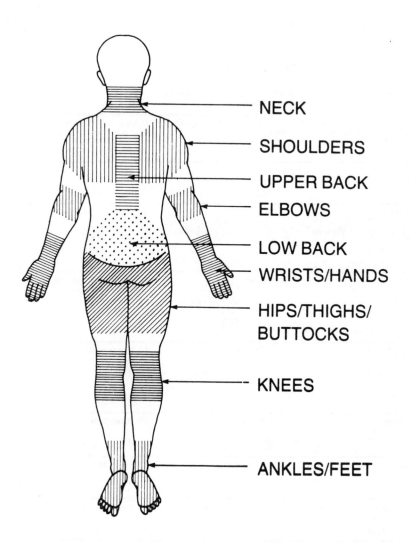

NECK

SHOULDERS

UPPER BACK

ELBOWS

LOW BACK

WRISTS/HANDS

HIPS/THIGHS/
BUTTOCKS

KNEES

ANKLES/FEET

This picture shows how the body has been divided. Please answer the three questions shown opposite for each body area.

Body sections are not sharply defined and certain parts overlap. You should decide for yourself which part (if any) is or has been affected.

2

MUSCULOSKELETAL DISORDERS

Rec 0 2

Please answer by using the tick boxes ☑
– *one tick for each question*

Please note that this part of the questionnaire should be answered, even
if you have never had trouble in any parts of your body.

Have you at any time during the **last 12 months** had **trouble (such as ache, pain, discomfort, numbness)** in:	Have you had **trouble** during the **last 7 days**:	During the **last 12 months** have you been **prevented** from carrying out normal activities (eg. job, housework, hobbies) because of this trouble:
1 Neck No ☐1 Yes ☐2	**2 Neck** No ☐1 Yes ☐2	**3 Neck** No ☐1 Yes ☐2
4 Shoulders No ☐1 Yes ☐2 in the right shoulder ☐3 in the left shoulder ☐4 in both shoulders	**5 Shoulders** No ☐1 Yes ☐2 in the right shoulder ☐3 in the left shoulder ☐4 in both shoulders	**6 Shoulders (both/either)** No ☐1 Yes ☐2
7 Elbows No ☐1 Yes ☐2 in the right elbow ☐3 in the left elbow ☐4 in both elbows	**8 Elbows** No ☐1 Yes ☐2 in the right elbow ☐3 in the left elbow ☐4 in both elbows	**9 Elbows (both/either)** No ☐1 Yes ☐2
10 Wrists/hands No ☐1 Yes ☐2 in the right wrist/hand ☐3 in the left wrist/hand ☐4 in both wrists/hands	**11 Wrists/hands** No ☐1 Yes ☐2 in the right wrist/hand ☐3 in the left wrist/hand ☐4 in both wrists/hands	**12 Wrists/hands (both/either)** No ☐1 Yes ☐2
13 Upper back No ☐1 Yes ☐2	**14 Upper back** No ☐1 Yes ☐2	**15 Upper back** No ☐1 Yes ☐2
16 Lower back (small of the back) No ☐1 Yes ☐2	**17 Lower back** No ☐1 Yes ☐2	**18 Lower back** No ☐1 Yes ☐2
19 One or both hips/thighs/buttocks No ☐1 Yes ☐2	**20 Hips/thighs/buttocks** No ☐1 Yes ☐2	**21 Hips/thighs/buttocks** No ☐1 Yes ☐2
22 One or both knees No ☐1 Yes ☐2	**23 Knees** No ☐1 Yes ☐2	**24 Knees** No ☐1 Yes ☐2
25 One or both ankles/feet No ☐1 Yes ☐2	**26 Ankles/feet** No ☐1 Yes ☐2	**27 Ankles/feet** No ☐1 Yes ☐2

3

Neck trouble   Rec $\boxed{0 \mid 3}$

How to answer the questionnaire:

By neck trouble we mean pain, ache or discomfort in the shaded area only.

Please answer by using the tick boxes $\boxed{\checkmark}$ – **one tick for each answer.**

---

1 Have you **ever** had any neck trouble (ache, pain, numbness or discomfort)?

   Yes   No   If you have answered **NO** to this question, do not answer questions

  1 ☐   2 ☐   2-12 but please go to the section on **shoulder trouble** page 6.

---

2 Have you **ever** hurt your neck in an **accident**?

   Yes   No

  1 ☐   2 ☐

  If the answer is **NO**, please go on to Question 3.

  If **YES**:

  2a Was the accident at work?

     Yes   No

    1 ☐   2 ☐

                            Month Year

  2b What was the approximate **date** of the accident? ☐☐☐☐

---

3 Have you **ever** had to **change duties** or **jobs** because of neck trouble?

   Yes   No

  1 ☐   2 ☐

---

4 What do you think brought on this problem with your neck?

  1 Accident ☐ 2 Sporting Activity ☐ 3 Activity at Home ☐

  4 Activity at Work ☐ 5 Other ☐ **(please specify)** [   ]

5a What year did you **first** have neck trouble? [19   ]

5b What year was your **worst** neck trouble? [19   ]

                                   Mild   Severe   Very, Very Severe

6 How bad was the pain during the **worst** episode? 1 ☐   2 ☐   3 ☐

---

7 Have you **ever** been absent from work because of neck trouble?

   Yes   No

  1 ☐   2 ☐

  If the answer is **NO**, please go on to Question 8.

  If **YES**:

  How many **times**?

7a [   ]

  How many **days** have you been absent from work with neck trouble **in total?**

7b [   ] days

  How many **days** have you been absent from work with neck trouble in the **last 12 months?**

7c [   ] days

4

8 How **often** do you get or have you had neck trouble?

daily	1 ☐
one or more times a week	2 ☐
one or more times a month	3 ☐
one or more times a year	4 ☐
one or more times every few years	5 ☐
one episode of trouble only	6 ☐

9 What is the **total length of time** that you have had neck trouble during the **last 12 months**?

0 days	1 ☐
1 - 7 days	2 ☐
8 - 30 days	3 ☐
More than 30 days, but not every day	4 ☐
Every day	5 ☐

10 Has neck trouble caused you to **reduce** your activity during the **last 12 months**?

10a Work activity (at home or away from home)

Yes   No

1 ☐   2 ☐

10b Leisure activity

Yes   No

1 ☐   2 ☐

11 What is the **total length of time** that neck trouble has prevented you from doing your **normal work** (at home or away from home) during the **last 12 months**?

0 days	1 ☐
1 - 7 days	2 ☐
8 - 30 days	3 ☐
More than 30 days	4 ☐

12 Have you been seen by a doctor, physiotherapist, chiropractor, or other such person because of neck trouble during the **last 12 months**?

Yes   No

1 ☐   2 ☐

If the answer is **NO,** please go on to the next section.

If **YES:**

12a Where? (more than one box can be ticked)

Medical centre at work	1 ☐
GP	2 ☐
Hospital	3 ☐
Private doctor	4 ☐
Osteopath or chiropractor	5 ☐
Other*	6 ☐

* If you have ticked *Other* please give details [   ]

5

Shoulder trouble   Rec 0 | 4

How to answer the questionnaire:

By shoulder trouble we mean pain, ache or discomfort in the shaded area only.

Please answer by using the tick boxes ☑ – **one tick for each answer.**

---

1 Have you **ever** had any shoulder trouble (ache, pain, numbness or discomfort)?

Yes   No   If you have answered **NO** to this question, do not answer questions
1 ☐   2 ☐   2-12 but please go to the section on **low back trouble** page 8.

---

2 Have you **ever** hurt your shoulder in an **accident**?

No   Yes
1 ☐   2 ☐ my right shoulder
         3 ☐ my left shoulder
         4 ☐ both shoulders

If the answer is **NO**, please go on to Question 3.

If **YES**:

2a Was the accident at work?

Yes   No
1 ☐   2 ☐

2b What was the approximate **date** of the accident?   Month Year ☐☐☐☐

---

3 Have you **ever** had to **change duties** or **jobs** because of shoulder trouble?

Yes   No
1 ☐   2 ☐

---

4 What do you think brought on this problem with your shoulder?

1 Accident ☐   2 Sporting Activity ☐   3 Activity at Home ☐
4 Activity at Work ☐   5 Other ☐ **(please specify)** [_____]

5a What year did you **first** have shoulder trouble?   19 [____]
5b What year was your **worst** shoulder trouble?   19 [____]

                                              Mild   Severe   Very. Very Severe
6 How bad was the pain during the **worst** episode?   1 ☐   2 ☐   3 ☐

---

7 Have you **ever** been absent from work because of shoulder trouble?

Yes   No
1 ☐   2 ☐

If the answer is **NO**, please go on to Question 8.

If **YES**:

How many **times**?
7a [_____]

How many **days** have you been absent from work with shoulder trouble **in total?**
7b [_____] days

How many **days** have you been absent from work with shoulder trouble in the **last 12 months?**
7c [_____] days

6

8 How **often** do you get or have you had shoulder trouble?

daily	1 ☐
one or more times a week	2 ☐
one or more times a month	3 ☐
one or more times a year	4 ☐
one or more times every few years	5 ☐
one episode of trouble only	6 ☐

9 What is the **total length of time** that you have had shoulder trouble during the **last 12 months**?

0 days	1 ☐
1 - 7 days	2 ☐
8 - 30 days	3 ☐
More than 30 days, but not every day	4 ☐
Every day	5 ☐

10 Has shoulder trouble caused you to **reduce** your activity during the **last 12 months**?

10a Work activity (at home or away from home)

Yes   No

1 ☐   2 ☐

10b Leisure activity

Yes   No

1 ☐   2 ☐

11 What is the **total length of time** that shoulder trouble has prevented you from doing your **normal work** (at home or away from home) during the **last 12 months**?

0 days	1 ☐
1 - 7 days	2 ☐
8 - 30 days	3 ☐
More than 30 days	4 ☐

12 Have you been seen by a doctor, physiotherapist, chiropractor, or other such person because of shoulder trouble during the **last 12 months**?

Yes   No

1 ☐   2 ☐

If the answer is **NO,** please go on to the next section.

If **YES:**

12a Where? (more than one box can be ticked)

Medical centre at work	1 ☐
GP	2 ☐
Hospital	3 ☐
Private doctor	4 ☐
Osteopath or chiropractor	5 ☐
Other*	6 ☐

* If you have ticked *Other* please give details [   ]

7

Low back trouble   Rec [0][5]

How to answer the questionnaire:

By low back trouble we mean pain, ache or discomfort in the shaded area whether or not it extends from there to one or both legs (sciatica).

Please answer by using the tick boxes ☑ – **one tick for each answer.**

---

1 Have you **ever** had any low back trouble (ache, pain, numbness or discomfort)?

Yes No   If you have answered **NO** to this question, do not answer questions
1☐   2☐   2-12 but please go to the section on **wrist/hand trouble** page 10.

---

2 Have you **ever** hurt your low back in an **accident**?

Yes No
1☐   2☐

If the answer is **NO**, please go on to Question 3.

If **YES**:

2a   Was the accident at work?

Yes No
1☐   2☐

2b   What was the approximate **date** of the accident?   Month Year
                                                             ☐☐☐☐

---

3 Have you **ever** had to **change duties** or **jobs** because of low back trouble?

Yes No
1☐   2☐

---

4 What do you think brought on this problem with your back?

1 Accident☐   2 Sporting Activity☐   3 Activity at Home☐
4 Activity at Work☐   5 Other☐ **(please specify)**☐

5a What year did you **first** have low back trouble?   19☐
5b What year was your **worst** low back trouble?   19☐

---

6 How bad was the pain during the **worst** episode?   Mild   Severe   Very, Very Severe
                                                          1☐   2☐   3☐

---

7 Have you **ever** been absent from work with low back trouble?

Yes No
1☐   2☐

If the answer is **NO**, please go on to Question 8.

If **YES**:

How many **times**?
7a   ☐

How many **days** have you been absent from work with low back trouble **in total?**
7b   ☐ days

How many **days** have you been absent from work with low back trouble in the **last 12 months?**
7c   ☐ days

8

8 How **often** do you get or have you had low back trouble?

daily	1☐
one or more times a week	2☐
one or more times a month	3☐
one or more times a year	4☐
one or more times every few years	5☐
one episode of trouble only	6☐

9 What is the **total length of time** that you have had low back trouble during the **last 12 months**?

0 days	1☐
1 - 7 days	2☐
8 - 30 days	3☐
More than 30 days, but not every day	4☐
Every day	5☐

10 Has low back trouble caused you to **reduce** your activity during the **last 12 months**?

10a Work activity (at home or away from home)

Yes   No
1☐   2☐

10b Leisure activity

Yes   No
1☐   2☐

11 What is the **total length of time** that low back trouble has prevented you from doing your **normal work** (at home or away from home) during the **last 12 months**?

0 days	1☐
1 - 7 days	2☐
8 - 30 days	3☐
More than 30 days	4☐

12 Have you been seen by a doctor, physiotherapist, chiropractor, or other such person because of low back trouble during the **last 12 months**?

Yes   No
1☐   2☐

If the answer is **NO**, please go on to the next section.

If **YES:**

12a Where? (more than one box can be ticked)

Medical centre at work	1☐
GP	2☐
Hospital	3☐
Private doctor	4☐
Osteopath or chiropractor	5☐
Other*	6☐

* If you have ticked *Other* please give details [_____]

9

Wrist or hand trouble   Rec [0][6]

How to answer the questionnaire:

By wrist or hand trouble we mean pain, ache or discomfort in the shaded area only.

Please answer by using the tick boxes ☑ – **one tick for each answer.**

---

1. Have you **ever** had any wrist or hand trouble (ache, pain, numbness or discomfort) ?

Yes   No   If you have answered **NO** to this question, do not answer questions
1.☐   2.☐   2-12 but please go to **General health questionnaire** on page 12.

---

2 Have you **ever** hurt your wrist or hand in an **accident**?

No   Yes
1☐   2☐ my right wrist or hand
       3☐ my left wrist or hand
       4☐ both wrists or hands

If the answer is **NO,** please go on to Question 3.

If **YES**:

2a   Was the accident at work?

Yes   No
1☐   2☐

2b   What was the approximate **date** of the accident?   Month Year ☐☐☐☐

---

3 Have you **ever** had to **change duties** or **jobs** because of wrist or hand trouble?

Yes   No
1☐   2☐

---

4 What do you think brought on this problem with your wrists or hands?

1 Accident☐   2 Sporting Activity☐   3 Activity at Home☐
4 Activity at Work☐   5 Other☐ **(please specify)** ☐

5a What year did you **first** have wrist or hand trouble?   [19   ]
5b What year was your **worst** wrist or hand trouble?   [19   ]

---

6 How bad was the pain during the **worst** episode?   Mild 1☐   Severe 2☐   Very, Very Severe 3☐

---

7 Have you **ever** been absent from work with wrist or hand trouble?

Yes   No
1☐   2☐

If the answer is **NO,** please go on to Question 8.

If **YES**:

How many **times**?
7a   ☐☐

How many **days** have you been absent from work with wrist or hand trouble **in total**?
7b   ☐☐ days

How many **days** have you been absent from work with wrist or hand trouble in the **last 12 months**?
7c   ☐☐ days

8 How **often** do you get or have you had wrist or hand trouble?

daily	1 ☐
one or more times a week	2 ☐
one or more times a month	3 ☐
one or more times a year	4 ☐
one or more times every few years	5 ☐
one episode of trouble only	6 ☐

9 What is the **total length of time** that you have had wrist or hand trouble during the **last 12 months**?

0 days	1 ☐
1 - 7 days	2 ☐
8 - 30 days	3 ☐
More than 30 days, but not every day	4 ☐
Every day	5 ☐

10 Has wrist or hand trouble caused you to **reduce** your activity during the **last 12 months?**

10a Work activity (at home or away from home)

Yes   No
1 ☐   2 ☐

10b Leisure activity

Yes   No
1 ☐   2 ☐

11 What is the **total length of time** that wrist or hand trouble has prevented you from doing your **normal work** (at home or away from home) during the **last 12 months**?

0 days	1 ☐
1 - 7 days	2 ☐
8 - 30 days	3 ☐
More than 30 days	4 ☐

12 Have you been seen by a doctor, physiotherapist, chiropractor, or other such person because of wrist or hand trouble during the **last 12 months**?

Yes   No
1 ☐   2 ☐

If the answer is **NO,** please go on to the next section.

If **YES:**

12a Where? (more than one box can be ticked)

Medical centre at work	1 ☐
GP	2 ☐
Hospital	3 ☐
Private doctor	4 ☐
Osteopath or chiropractor	5 ☐
Other*	6 ☐

* If you have ticked *Other* please give details [＿＿＿＿＿＿＿＿＿＿＿]

11

GENERAL HEALTH QUESTIONNAIRE

Rec $\boxed{0\ 7}$

We should like to know how your health has been in general, **OVER THE PAST FEW WEEKS.**
Please circle the answer which you think most nearly applies to you.

HAVE YOU RECENTLY:

1	been able to concentrate on whatever you're doing?	Better than usual	Same as usual	Less than usual	Much less than usual
2	lost much sleep over worry?	Not at all	No more than usual	Rather more than usual	Much more than usual
3	felt that you are playing a useful part in things?	More so than usual	Same as usual	Less useful than usual	Much less useful
4	felt capable of making decisions about things?	More so than usual	Same as usual	Less useful than usual	Much less useful
5	felt constantly under strain?	Not at all	No more than usual	Rather more than usual	Much more than usual
6	felt that you couldn't overcome your difficulties?	Not at all	No more than usual	Rather more than usual	Much more than usual
7	been able to enjoy your normal day-to-day activities?	More so than usual	Same as usual	Less so than usual	Much less than usual
8	been able to face up to your problems?	More so than usual	Same as usual	Less able than usual	Much less able
9	been feeling unhappy and depressed?	Not at all	No more than usual	Rather more than usual	Much more than usual
10	been losing confidence in yourself?	Not at all	No more than usual	Rather more than usual	Much more than usual
11	been thinking of yourself as a worthless person?	Not at all	No more than usual	Rather more than usual	Much more than usual
12	been feeling reasonably happy, all things considered?	More so than usual	About same as usual	Less so than usual	Much less than usual

SCORE
0011

13 How often do you experience any of the following symptoms during or after work?
For each symptom, put a tick in the appropriate box.

	Frequently	Sometimes	Rarely	Never
Fatigue	1 ☐	2 ☐	3 ☐	4 ☐
Headaches	1 ☐	2 ☐	3 ☐	4 ☐
Disturbed vision	1 ☐	2 ☐	3 ☐	4 ☐

14 Do you wear spectacles or contact lenses whilst working a check-out?
Yes No
1 ☐ 2 ☐

12

INFORMATION ABOUT YOUR JOB

Rec [0|8]

1 How many years and months have you been doing your **present type of work** at this supermarket?

Years + Months ☐ ☐

If less than one month, how many weeks? ☐

2 Have you worked in other supermarkets?

No 1☐ Yes 2☐

2.1 If yes, what is the total length of time you worked on checkouts elsewhere, before working at this supermarket?

Years + Months ☐ ☐

If less than one month, how many weeks? ☐

3 Do you have any other paid job other than at this supermarket?

Yes 1☐ No 2☐

4 On average, how many hours a week do you work at this supermarket? (including overtime but excluding the main meal break)

Hours ☐

5 How many of these hours are spent working on a check-out?

Hours ☐

6 Do you rotate or change your duties regularly during the day?

Yes 1☐ No 2☐

If **YES,**

6a How often?

Changing once every hour 1☐
Changing once about every 2 hours 2☐
Changing once about every 2-4 hours 3☐
Other 4☐

If you have ticked *Other* please say how often ________________

7   On average how many breaks do you have each working day?

[   ]

---

8   Ignoring your lunch-break, how long is each of your breaks on average?

minutes

[   ]

---

9   Do you experience any difficulty in operating the following equipment?

	Yes 1	No 2	Don't Use It 3
Laser scanning			
Electronic cash register			
Weighing scales			

		Yes 1	No 2
10a	**Do you** adjust the backrest of your seat?		
10b	**Do you** adjust the footrest to your seat?		
10c	**Do you** adjust the height of the seat?		
10d	**Do you** move the seat to or from the desk?		

Chapter 24

Biomechanical methods in posture analysis

Moira F. Tracy

Introduction

Biomechanics is the study of forces on the human body. It is used in the ergonomics field most often to assess manual handling tasks. This chapter outlines the types of task for which biomechanical analysis is or is not an appropriate method, details the tools required to carry out an analysis, and describes various methods and models used in the field, with their advantages and disadvantages. When interpreting results of a biomechanical analysis we come to the question of how reliable are the simplifications and approximations made in the analysis. And finally, what are our criteria for 'safe' levels of force and how reliable are they?

Relevants tasks

Biomechanics is a useful tool in most manual handling situations, whether people are lifting, pushing, pulling, or even when no load is handled but the body's own weight is creating postural stress. The human body is complex and biomechanics cannot at this stage give very fine detail. For instance it is, to our knowledge, not possible to show through calculations which of two backrest shapes would be superior: the posture or force from the backrest would need to change quite noticeably for calculated results of body loadings to show differences.

Biomechanics is best used as a comparative method because, as will be shown throughout this chapter, results rest on simplifications and approximations. The method is particularly successful at demonstrating possible improvements obtained from redesigning a task. It can also be used to identify the most stressful parts of a job.

Interpretation of results

Having calculated forces at joints such as elbows or shoulders, and forces on the spine or within trunk muscles, results for various tasks or designs can be compared. It is also possible to assess the feasibility and safety of tasks even at the design stage, by comparing these forces with recommended limits. However, as will be discussed in this chapter, these limits are not absolute guarantees of safety, so results of a biomechanical analysis should not be used in isolation, but combined with other assessment methods. Direct observation, discomfort charts or questionnaires may identify sources of discomfort which force calculations cannot. For repetitive tasks a physiological assessment is often necessary as well. Injury statistics will help to identify problems for which biomechanics can give relevant answers and help towards an assessment of the cost-effectiveness of redesign.

Drury *et al.*'s (1983) evaluation of a palletizing aid provides a good example of the use of biomechanics with other methods. Very simple calculations based on video recordings showed to what extent the load on the lumbar spine was reduced when a palletizing aid was used. The heart rate was also shown to go down, and the authors used these factors and injury statistics to evaluate the cost effectiveness of introducing the aids. The interesting point is that if this study had based itself only on accepted 'safe' limits of spinal load, it would have found with this particular biomechanical model that they were quite acceptable without an aid. The use of several methods to evaluate the task ensured that problems were not overlooked and biomechanics demonstrated the benefits of introducing an aid.

How detailed should the analysis be?

The simplest biomechanical calculations are those relating to a static posture and force in the sagittal plane, i.e., where there is neither twisting nor lateral bending. The problem is purely two-dimensional (2D), and there are no extra forces caused by accelerations and inertia. It is easy to record the posture in that one plane and all the calculations can be done with a small calculator.

If the task is characterized by much lateral bending or twisting, problems or possible improvements would be overlooked with a 2D analysis. Recording posture in three dimensions (3D) requires special methods outlined later. Calculations are quite lengthy and a computer program is usually necessary. A 3D analysis is therefore quite time-consuming unless the posture-recording method and computer program have already been set up.

If the task is not static but dynamic, extra forces resulting from the accelerations have to be added to the calculations. The analysis now requires continuous monitoring of posture, along with the value and direction of the acceleration of each limb. This may restrict dynamic analyses to the laboratory and to our knowledge this has not yet been done in 3D. In 2D it is possible sometimes to avoid recording posture and acceleration continuously by using simulations. This has been done for lifting, the accuracy of the results depending on how similar the lifting technique is to the one used as a database by the computer program. Finally it is quite common to analyze dynamic tasks statically, freezing at the beginning, middle, and end of the task, for instance. If the task was done slowly and smoothly, this is a good approximation. Otherwise this may underestimate forces during the acceleration phase up to two or three times (Garg *et al.*, 1982).

Equipment required

A biomechanical analysis requires measurements of posture, hand-force (and any other forces, such as the force from a backrest) and, in the case of dynamic tasks, accelerations. Hand-forces can be measured with a spring balance, but in some cases a tool equipped with strain gauges may have to be constructed. This can be interfaced to a computer for on-line recordings. One quick-and-ready method is to press onto bathroom scales. Sometimes it is practical to place subjects on bathroom scales or a force plate and take one reading when they are applying a force and one reading when they are not. The difference is due to the force applied.

The choice of a posture recording method will depend on whether the analysis is 2D or 3D, in the field or in the laboratory. It may not be necessary to record posture when using some 2D computer programs: the posture is input via a stick-man on the screen. However it is best to base this on a photograph in order to input a realistic posture. Photographs or videos can be digitized for 2D or 3D posture recordings. This is time consuming in 3D but often used as a simple field method in 2D. Another simple method for static 2D tasks is to take measurements from the subject with a tape measure and plumb bob (Schultz *et al.*, 1983). This is more of a laboratory method as it would interrupt real work in many cases.

For 3D analyses or dynamic tasks, scanners like Selspot, Vicon, Coda or Elite are practical when on-line to a computer, though costly and usually not very portable (Tracy *et al.*, 1987). A method for field work in the future could be to interface a posture-recording suit based on light-emitting diodes (Samuelson *et al.*, 1987) or strain gauges. Also for field work, a pen and paper method such as Corlett *et al.*'s (1979) posture targetting, or the Labanotation used in dance recording, may also be of interest as posture inputs to a computer evaluation (see chapter 23).

Human variability

Biomechanics is sometimes used with the aim of determining safe limits for most of the population. At this stage, various aspects of human variability must be borne in mind. The most obvious one is the large range in strength found in the population. Men are usually stronger than women, but there is considerable overlap, some women exceeding some men. (Strength testing though does not necessarily identify people who will be the most capable of doing a task, if the test requires the use of different muscle groups from those used in the task (Keyserling *et al.*, 1980).)

Human variability in body weight and stature also comes into biomechanics. The posture people choose to adopt for a given task is another factor, but given a particular posture, the loads on the body are greatest for larger body weights and limb lengths. It is therefore recommended to use 95th percentile body weight and stature in order to err on the side of safety.

Finally, when results of biomechanical calculations are evaluated against 'safe' limits, it must be borne in mind that people differ in their susceptibility to back pain or injury. Hutton and Adams (1982) have shown that as a rule women's spines are more susceptible to fracture than men's, and age also is a weakening factor. Troup *et al.* (1987) found that people who had experienced back pain chose to lift lower loads. The mechanism of back pain is still uncertain in many cases: tests on cadaveric spines do not necessarily produce the same effects as those observed *in vivo* (Brinckmann, 1986). The best predictor of susceptibility to back pain seems, at the moment, to be a history of previous low-back trouble (MacDonald, 1984).

Principles of biomechanical calculations

The aim in this section is to demonstrate how loads at any body joint are calculated. These results can be used in a comparative way, for instance to quantify the improvements obtained from redesigning a task. On the other hand they can be compared with population data on maximum strength capabilities, in order to assess how strenuous the task is.★ The present section will also demonstrate how forces within the low back can be evaluated with a simple 2D model.

The calculation of forces rests on the principle that all forces must balance each other if the body is to be in equilibrium. If there is a resultant force in any direction, the body will move in that direction. The calculation of loads at body joints which follows is in fact the calculation of moments, or turning forces around a point. Moments must also balance, so that the sum of moments around any point is zero, if there is to be no rotation.

Moments and lever arms

The moment, or torque, of a force about a point, is a measure of the turning force round the point. For instance holding a weight in the hand creates a moment around the elbow, tending to make it extend. Muscles spanning the

---

★This last aspect is discussed in the next main section. Readers who do not need to know the details of calculations can turn directly to that section after reading the paragraph on moments and lever arms.

elbow provide the opposite moment by contracting, so that the elbow is able to support the weight. The greater the weight, the larger the elbow moment.

In general terms, the moment of a force about a point is the product of the force and the perpendicular distance between the point and the line of action of the force. This is shown in Figure 24.1. A weight of 100 N* held in the hand creates a moment of $100 \times 0 \cdot 20 = 20$ Nm (Newton metres) for the position shown. If the weight was held with the arm hanging down, the lever arm of the force would be zero, and so the moment about the elbow would be zero. There would, of course, still be a force at the elbow, resisting the downward pull of the 100 N weight.

So far we have not taken into account another force exerting a moment round the elbow: the weight of the hand and forearm. The location of the centre of gravity of the hand and forearm, and their weight, can be estimated from tables, presented later in the chapter (see section 'Inputs to biomechanical calculations').

A simple example with a 2D low-back model

To evaluate forces in the lumbar region, the moment around a point of the low back is calculated in the same way as has been demonstrated for the elbow. Then a model of the muscular layout is used to calculate how much force the back muscles need to exert to counteract this moment. This enables the compression force on the spine itself to be evaluated, which is a much used criterion for safety.

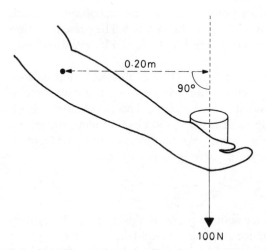

Figure 24.1. The moment at the elbow due to the 100 N weight held at the hand is the product of the force (100 N) and the perpendicular distance (0·20m) through which it acts

---

*Newtons (N) are units of force or weight. A 1 kg mass weighs approximately 10 N. More on this and other units appears at the end of this section.

What follows is a simple example in 2D (see Figure 24.2): the posture and hand-force are in the sagittal plane. Calculations in 3D, for lateral bending and twisting, will follow after that.

A subject is depicted holding a 100 N weight and moments are calculated around the point indicated by a star, situated on a point of the lumbar spine. The weight of the part of the body above this point is 400 N in this example, acting with a lever arm of 0·20 m. It creates a moment of 80 Nm. Add to this the effect of the weight at the hands, acting with a 0·60 m lever arm, and the total moment created around the starred point is 140 Nm. Thus the weight at the hands and the subject's own body weight tend to flex the trunk: trunk muscles and ligaments must counteract this so that the posture is held.

Many studies have been carried out to record the trunk extension moment capabilities of subjects. We will return to this later in the chapter but note that a range of 100–700 Nm has been found in the literature, for male subjects without back pain. The subjects whose maximum strength was 100 Nm would therefore not be capable of exerting the 140 Nm of our example. For the strongest subjects, those capable of 700 Nm, it would be an easy task.

Forces within the trunk can be evaluated at this stage, using a model of

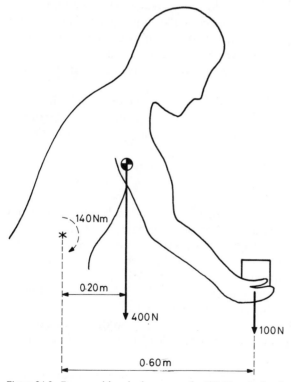

Figure 24.2. Forces and low-back moment for 100 N at the hands

the low back. Such models vary in detail and complexity, as will be seen later, but the principle can be shown here. In a simple 2D model, the 140 Nm trunk flexion moment is resisted by back muscles alone. The greater the leverage those muscles have from the spine, the smaller the force needed from them. Let us suppose the line of action of the back muscles is 5.8 cm posterior to the spine (this is the average from a recent study: more information is given in the section 'Inputs to biomechanical calculations'). The force these muscles need to exert is $140/0.058 = 2414$ N. This is much larger than the body weight or the hand-force, because the muscles are balancing the moments through a very small lever arm.

As the back muscles pull to counteract moments, they compress the lumbar spine. The weight of the body above the lumbar spine and the weight at the hands also compress it, so finally the total compression force is the sum of all these components: $100 + 400 + 2414 = 2914$ N. This again may seem large, but compression tests have shown that the spine can, in general, withstand this type of force. Further discussion on this is given in the section 'Forces on the low back'.

Figure 24.3 shows all the forces involved in this simple problem: the sum of all these forces is zero, and so is the sum of all moments around any point. The point indicated by a star on the lumbar spine was chosen only to simplify calculations, as the compression force does not create a moment round this point. In this example the centre of mass of the upper part of the body is shown as being 0.20 m from the lumbar spine. In fact this would not be known in an ordinary problem—what is estimated is the location of the centre of mass of each body segment, and so the moments created by each segment can be summed.★

We will now go on to calculations in 3D static or dynamic tasks. However, in many cases very valuable information can be obtained from the simple calculations described so far (for instance Drury *et al.*'s (1983) evaluation of palletizing aids discussed earlier).

Moments in 3D space

The following, on moments in 3D space, contains details which need not be read by users who will not actually be performing calculations.

Previous examples were restricted to postures and forces in one plane, the sagittal plane. For asymmetric postures, or forces not contained in this plane, the calculation of the moments can be made by the following method. Figure 24.4 shows a force **F** (bold type indicates a vector) acting at the hand, with **r** the vector running from the elbow to the hand. The simplest way to determine the moment, **M**, around the elbow is to record the x, y, z components of vectors **r** and **F**, along a set of perpendicular axes. Most measuring

---

★The next section, 'Moments in 3D space', shows in detail how this is done. Calculations are simplified because in 2D situations, $r_y = 0$, $F_y = 0$, $M_x = 0$, $M_z = 0$.

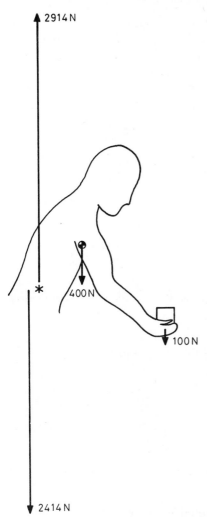

2914 N

*

400 N

100 N

2414 N

Figure 24.3. Equilibrium of forces when holding a 100 N weight

equipment will allow this. Thus the components of **r** are r_x, r_y, r_z along the x, y, z axes, and those of **F** are F_x, F_y, F_z. The moment **M** around the elbow is also a vector and it is simple to calculate its components M_x, M_y, M_z. M_x is the turning force in the (y, z) plane, M_y is the turning force in the (z, x) plane, and M_z is the turning force in the (x, y) plane.

The resultant of M_x, M_y, M_z, is the size of the vector **M**. From Pythagoras' theorem:

$$M^2 = M_x^2 + M_y^2 + M_z^2.$$

Analysis of work activities

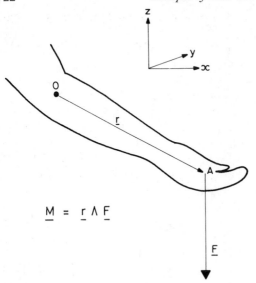

$$M = r \wedge F$$

Figure 24.4. The moment of a force about O is the vector product (noted $\wedge$) of the vector r running from O to A, with the force vector F

M_x, M_y and M_z are obtained through the following equations:

$$M_x = r_y F_z - r_z F_y$$
$$M_y = r_z F_x - r_x F_z$$
$$M_z = r_x F_y - r_y F_x.$$

A short-hand notation for these is:

$$\mathbf{M} = \mathbf{r} \wedge \mathbf{F}$$

$\mathbf{M}$ is described as the vector product (noted $\wedge$) of $\mathbf{r}$ and $\mathbf{F}$ (the order is important).

For calculations of moments in a 2D situation we have seen that the moment was the product of the force and its perpendicular distance (lever arm). This gives the same result as the equations above, so the method chosen depends on which is easiest to record: the lever arm, or components along the x, y and z axes.

Each component of the moments, M_x, M_y, M_z, represents the turning force round the x, y, or z axis. For instance in Figure 24.4 the y axis is directed into the paper, and M_y is the moment round that axis, and represents the flexion/extension moment about the elbow.

This is the only moment in the case of Figure 24.4: the reader can verify from the above equations that M_x and M_z are zero. (The only component of $\mathbf{F}$ is along the z axis and $\mathbf{r}$ and $\mathbf{F}$ are in one plane: $r_y = 0$, $F_x = 0$, $F_y = 0$.)

This means that there are no twisting or abduction/adduction requirements on the elbow.

The sign (positive or negative) of a moment indicates the direction of the turning force. In the example of Figure 24.4, F_z is negative, so from the equations, M_y is positive. This represents the extension effect force **F** has on the elbow. A negative M_y would represent a flexion effect.

One method of working out the meaning of a positive or negative moment is as follows: stick out your right-hand thumb in the direction of the selected axis, y in this case. Your other fingers naturally curl round in the direction of the rotation corresponding to a positive M_y, in this case, elbow extension.

For this system, a 'right-handed' set of axes is needed, i.e., y should go into the paper, as in Figure 24.4, not out of it. An easy trick to ensure the axes are right-handed is to point the right-hand thumb along the z axis. The fingers curl round to indicate the direction from x to y (Figure 24.5). All force and posture recording should be done in a right-handed set of axes for consistency.

So far we have looked at the moment in 3D created by one force, **F**. When calculating the moment round a point in the low back, for instance, the weights of several body segments have to be taken into account. One method is to add up the moment that each of these forces creates round the low back. This is summarized by the equation:

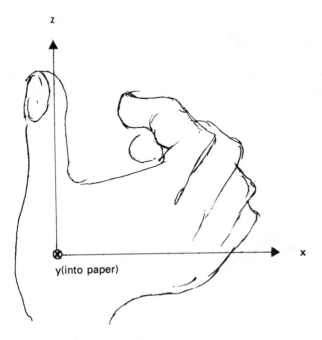

Figure 24.5. Conventions for a system of axes

$$\mathbf{M} = \mathbf{r} \wedge \mathbf{F} + \mathbf{r}_1 \wedge m_1\mathbf{g} + \mathbf{r}_2 \wedge m_2\mathbf{g} + \ldots$$

where $\mathbf{F}$ is an external force acting on the body, such as a weight at the hands; $\mathbf{r}$ is the vector running from the low back to the point of application of $\mathbf{F}$; m_1, m_2 are the masses of body segments (kg); $\mathbf{r}_1$, $\mathbf{r}_2$ are the vectors running from the low back to the centres of mass of body segments; and $\mathbf{g}$ is the acceleration due to gravity ($9 \cdot 81$ m/s², downwards).

A second method to calculate $\mathbf{M}$, which is useful when moments at several joints of the body are already known, is to calculate the moment round the wrist, then use this result and add the effect of the forearm weight to work up to the elbow, and so on to the shoulder, till the low back is reached. This second method, although it may seem less immediate, is more economical if the moments at the wrist, elbow and so on were required anyway. The following equation is used (symbols are shown on Figure 24.6).

$$\mathbf{M} = \mathbf{r}_{cm} \wedge m\mathbf{g} + \mathbf{M}_{adj} + \mathbf{r}_{adj} \wedge \mathbf{R}_{adj}$$

where $\mathbf{M}$ is the moment at the selected joint; $\mathbf{M}_{adj}$ is the moment at the adjacent joint; $\mathbf{r}_{adj}$ is the vector running from the selected joint to the adjacent one; m is the mass of the segment between these two joints; $\mathbf{r}_{cm}$ is the vector running from the selected joint to the centre of mass of the segment; and $\mathbf{R}_{adj}$ is the resultant force calculated at the adjacent joint.

For example:

$\mathbf{R}_{adj}$ at the wrist is $\mathbf{F} + m_{hand}\mathbf{g}$
$\mathbf{R}_{adj}$ at the elbow is $\mathbf{R}_{wrist} + m_{forearm}\mathbf{g}$

and so on, if $\mathbf{F}$ is a force on the hand.

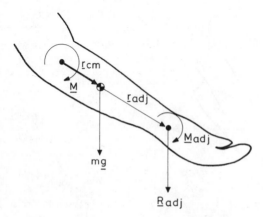

Figure 24.6. The moment M_{adj} at the wrist can be used to calculate the moment M at the elbow

Moments in dynamic tasks

Previous sections have focused on calculations for static tasks, but if body segments are going through accelerations this will add the effect of inertial forces to the moments at the joints. The main difficulty in performing an analysis of a dynamic task is recording instantaneous accelerations throughout the task. The problem has therefore been tackled mainly for lifting in the sagittal plane, where acceleration data are obtained from video recordings, say. Ayoub and El Bassoussi (1978) have included a prediction of accelerations in a dynamic computer model: it is approximated as a function of the angle of each limb at the end of the lift and the duration of the lift.

The most common simplification is to ignore accelerations, and treat the problem as a static one. This may lead to errors if the task is performed quickly. McGill and Norman (1985) have tested this on lifting tasks: they carried out both static and dynamic evaluations of the load on the L4–L5 intervertebral joint. Results with the dynamic analyses were on average 19% higher than with the static approximation, and could go up to 52% higher. A similar investigation by Garg *et al.* (1982) resulted in dynamic evaluations two to three times higher than static ones. These differences are probably due to differences in lifting speed, method, and weight lifted. McGill and Norman (1985) proposed a quasi-dynamic model, in which the only dynamic component was the acceleration of the object lifted. Tsuang *et al.* (1992) have shown that differences between static, quasi-dynamic and dynamic analyses increase mainly with the speed of the lift.

Still related to lifting in the sagittal plane, Garg *et al.* (1982) have found that the peak low-back moment found throughout the whole dynamic lift can be approximated by a static evaluation, if the posture selected is the one at the initiation of the lift. However the validity of this approximation may depend on the type of lift analyzed.

In summary, analyzing dynamic tasks with a static or quasi-dynamic model saves experimental and computer effort, but is an approximation liable to considerable error.

In order to interpret results obtained from a dynamic analysis, more research is needed on voluntary dynamic moments, and on responses of tissue and intervertebral disc material to high loads applied for a short instant. At the moment, the greatest value of a dynamic analysis is for comparative purposes. The rest of this section will describe the calculations required for a dynamic analysis.

The moment at a particular joint varies throughout the motion due to changes both in lever arms and in accelerations. At any instant, the moment depends on the value of the linear acceleration of the centre of mass of each segment, on the direction of this acceleration, and also on the angular acceleration of each segment. The resistance to rotation that an object has depends on its mass and shape and is described by its moment of inertia. This parameter differs for different axes of rotation, but some moments of inertia for movement in the sagittal plane can be found in the literature (Winter, 1979).

The general equation for the moment **M** at a joint at a particular instant is:

$$\mathbf{M} = \mathbf{r}_{cm} \wedge m\mathbf{g} + \mathbf{M}_{adj} + \mathbf{r}_{adj} \wedge \mathbf{R}_{adj} + \mathbf{r}_{cm} \wedge m\mathbf{a} + \mathbf{I} \wedge \ddot{\boldsymbol{\theta}}$$

The first three terms have already been described (see Figure 24.6) in the previous section on static moments. Symbols used in the additional terms are: **a** is the linear acceleration of the centre of mass of the segment (ms^{-2}); $\ddot{\boldsymbol{\theta}}$ is the angular acceleration of the segment about its centre of mass (degrees s^{-2}); and **I** is the moment of inertia of the segment about its centre of mass ($kg\,m^2$).

For 3D tasks, Ito *et al.* (1980) provide detailed equations for a dynamic analysis of a man-model made of several links, each with a given position, orientation, velocity and acceleration.

A note on units for moments

In the previous example, the lever arm was expressed in metres (m), the force in Newtons (N) and so the moment was obtained in Newton-metres (Nm). These are SI units and therefore recommended; however the following units also can be found in the literature.

The kg-force or kilopond is the force exerted by a mass of 1 kg due to gravity; it is equal to 9·81 N. Thus if a 10 kg object is held at the hand, it exerts a force of 98·1 N. Its weight is 98·1 N, while its mass is 10 kg.

One can also come across the pound (lb). This is equal to 0·4536 kg, and multiplying by 9·81 one obtains 4·450 N as the force exerted by 1 lb. Sometimes moments are expressed in inch-pounds, or foot-pounds. With 1 inch = 0·025 m and 1 foot = 0·30 m, one obtains the equivalence: 1 inch-pound = 0·1112 Nm and 1 foot-pound = 1·3349 Nm. One can also encounter kg-cm or kg-inch but, luckily, SI units are increasingly being used.

Prediction of strength and task feasibility using moments

Many experimental studies have been carried out to measure the maximum voluntary static strength of men and women. This can be overall body strength or else the strength of individual joints. The first category includes data on lifting strength in various postures. This type of information is very useful when it is directly applicable to a task; however maximum strength can be very different if the posture adopted is slightly different from the one tested, with joints not at their best angles and body weight contributing a different moment due to the different posture. In spite of these reservations, maximum voluntary static strength tests are frequently used for comparison with task loads.

Another approach to task assessment is to evaluate the moments created at each joint by a task and compare them with joint strength data from the population. If the moment evaluated for a given joint is higher than the estimated maximum capabilities within the population, one can predict that the task will not be feasible for most people.

A practical way of expressing the feasibility of a task is to state what percentage of the male or female working population is likely to be capable of it. Chaffin (1988a) gives examples of this method. One of these applies to pulling carts for moving stock, and it is shown how the percentage of women capable of this drops as the force required increases. In another example, it is shown that most of the working population is capable of performing a particular lifting task (however evaluation of loads on the spine shows the task actually is hazardous).

The method of comparing evaluated moments with population maxima is sometimes used to predict the maximum strength possible in a particular posture: one raises the hand-force in the calculations until the maximum capability of one joint is reached. That joint is the limiting factor, the weak link or bottleneck. If this method of raising the hand load is used to predict maximum strength, results may be inaccurate because as the hand-force is increased, a subject's real posture is likely to change to ensure body balance is maintained. Subjects may also change their posture to use other muscles and avoid the restrictions arising from weaker muscle groups.

Reliability of maximum joint strength data

Another problem is the reliability of the data on a particular joint's maximum strength. A very wide range of results can be found in the literature, and this may be due to different testing methods as well as to human variability. For instance, elbow strength results are sometimes given as the force subjects are able to exert using their hand. This method is not satisfactory as the wrist may be the weak link in this task. To avoid this problem there have been experiments in which the subjects exert a force on a device attached proximal to the wrist. These results are incomplete if the distance between this point and the elbow is not quoted. The most useful data are those where elbow strength is expressed directly as a moment and the method for its determination is noted. Low-back strength is particularly difficult to define, as there is no obvious point from which to define moments when testing for strength.

Dependence of joint strength with joint angles

The next limitation concerns joint angles. The force a muscle can exert depends on its length, so moment capabilities depend on joint angles. For example elbow strength depends not only on the elbow angle but also on the shoulder angle, as muscles span across both joints. Yet many results on elbow moments do not report the shoulder angle.

For a compilation from the literature of moment-angle curves of major joints, the reader is referred to Svensson (1987). Figure 24.7 has been adapted from one of these results and shows the dependence of trunk extensor moment and trunk angle. The hatched area includes curves from four different studies.

A word of caution is in order when referring to published results concerning joint angles: the field of biomechanics does not appear to have any angle conventions and the position of the zero angle varies across studies. One useful standard may be that set by the British Orthopaedic Association (1966) in their booklet describing terminologies used in joint motion. Their method is the 'Zero Starting Position': to accept the 'anatomical position' of a limb as zero degrees. For instance (see Figure 24.8) the elbow angle is zero for the extended straight arm, and its range of movement is from about 150° flexion to 10° hyperextension.

Interpretation of results

Readers can refer to prediction equations in Chaffin and Andersson (1984, p. 224) for maximum moments as a function of joint angles. However, as a wide range of values can be found from other sources, a rough compilation of these ranges is presented in Table 24.1. It is only intended to give the reader an order of magnitude with which to compare evaluated moments, and further work is needed in this area. The ranges given in Table 24.1 include all those found in some of the literature, and they all refer to the so-called 'fit and healthy' volunteer. The angle notation in Table 24.1 follows the British Orthopaedic Association convention described earlier.

Results refer to moments along a single axis, for instance pure flexion or

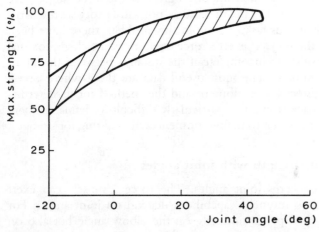

Figure 24.7. Back extensor strength as a function of trunk angle. Results from 4 studies, normalized by denoting the top value of each curve as 100%. (From Svensson, 1987)

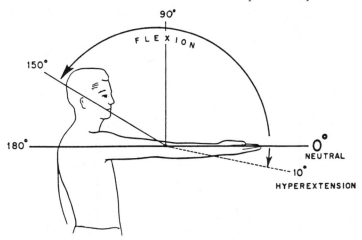

Figure 24.8. The elbow—flexion and hyperextension (from British Orthopaedic Association, 1966)

pure abduction, so caution must be exercised in using them for a task in which moments in several directions are combined.

An additional word of caution is given by Grieve (1987) who showed that at some joints, antagonistic muscles (i.e., muscles that have opposite effects) contract simultaneously. For example at some elbow angles, voluntary elbow flexion recruits not only the biceps (flexor), but also to some extent the triceps (extensor). This co-contraction, found also at the knee and shoulder, is believed to have a joint stabilizing role. Therefore, it is possible that prediction models overestimate a subject's strength, because co-contraction at a joint reduces the net moment provided by it.

At this stage there are very few data on maximum moments exerted dynamically, probably because of the problem in recording moments that cover a wide enough maximum dynamic range of angles, velocities and accelerations.

In conclusion, comparing the moments required by a task with data on maximum capabilities provides some useful guidance in terms of orders of magnitude, but the method should be used with caution and complemented with other methods, especially for repetitive tasks which may cause fatigue.

Forces on the low back

A study by Chaffin (1988a) was mentioned earlier, in which a particular lifting task was analyzed. From the moments at the joints it was estimated that most of the working population would be capable of performing the task; however evaluation of loads on the spine indicated that the task could put the back at risk. This section discusses what criteria are available for assessing such risk. A simple example of a 2D model to calculate the com-

Table 24.1. Maximum voluntary joint strengths (Nm) from some of the literature. The ranges presented include the ranges from these studies

Joint strength	Joint angle (degrees)	Range of moments (Nm) of subjects from several studies		Variation with joint angle
		Men	Women	
Elbow flexor	90	50–120	15–85	Peak at about 90°
Elbow extensor	90	25–100	15–60	Peak between 50° and 100°
Shoulder flexor	90	60–100	25–65	Weaker at flexed angles
Shoulder extensor	90	40–150	10–60	Decreases rapidly at angles less than 30°
Shoulder adductor	60	104	47	As angle decreases, strength increases then levels at 30° to −30°
Trunk flexor	0	145–515	85–320	Patterns differ among authors
Trunk extensor	0	143	78	Increases with trunk flexion
Trunk lateral flexor	0	150–290	80–170	Decreases with joint flexion
Hip extensor	0	110–505	60–130	Increases with joint flexion
Hip abductor	0	65–230	40–170	Increases as angle decreases
Knee flexor	90	50–130	35–115	In general, decreases with knee flexion but some disagreement with this, depending on hip angle
Knee extensor	90	100–260	70–150	Minima at full flexion and extension
Ankle plantarflexor	0	75–230	35–130	Increases with dorsiflexion
Ankle dorsiflexor	0	35–70	25–45	Decreases from maximum plantar flexion to maximum dorsiflexion

pression force on the spine was given earlier in the chapter (Figure 24.3). This section will describe other models, including 3D ones for asymmetrical postures.

A relatively direct method to evaluate forces on the spine is to insert a pressure-measuring needle into the intervertebral disc (Nachemson and Morris, 1964). Results have been used to evaluate various seated or lifting tasks (e.g., Nachemson and Elfström, 1970; Andersson and Örtengren, 1974). Following this, intradiscal pressure measurements have been used mostly in conjunction with electromyography (EMG) and intra-abdominal pressure measurements to verify the validity of low-back models (Schultz *et al.*, 1982). It is presumably more pleasant for subjects to have their spine compression calculated than their disc pressure measured. Nachemson and Morris (1964) and Aspden (1989) discuss how the compression force between vertebrae can be evaluated from intradiscal pressures.

Guidelines from low-back forces

The most commonly used guideline for task assessment is the value of the compression force between vertebrae. Experiments on cadaveric spines have shown that fractures appear above certain levels of compression. The level is lowest for older people; female spines are, as a general rule, weaker than male spines (Hutton and Adams, 1982). The National Institute for Occupational Safety and Health (NIOSH, 1981) reviewed the evidence in the literature and agreed on the following guidelines: tasks causing a compression on the lumbo-sacral joint greater than 6400 N are above the 'maximum permissible limit': they are unacceptable and engineering controls are required. On the other hand compressions under 3400 N can be tolerated by most young, healthy workers (over 75% of women and over 99% of men). It must be noted that these guidelines relate to lifting in the sagittal plane, and the spine may be much more vulnerable under axial rotation or hyperflexion (Adams and Hutton, 1981). The majority of compression tests have sought the ultimate compression strength, but work by Brinckmann *et al.* (1987) is promising, for data relating to repetitive tasks and the strength of intervertebral joints under cyclic loading. For instance they have shown that for a cyclic load of about half the ultimate compression strength, the probability of a fatigue fracture after 100 cycles is nearly 50%.

The value of the compression force may not be the most relevant parameter related to back injury, and guidelines such as the two NIOSH limits should be used with their limitations in mind, especially at extremes of trunk motion and in dynamic tasks. Additional indications of the severity of a task are given by the forces evaluated for the muscles of the low back. As a general rule, the maximum strength of a muscle is proportional to its largest cross-sectional area, and is approximately $50-100 \text{N/cm}^2$ (Schultz and Andersson, 1981). (Some cross-sectional areas can be found later in Table 24.5).

Applications

The value of lumbar spine compression is the most frequently used criterion in the evaluation of tasks that may put the back at risk. It has been used extensively in the analysis of lifting tasks, for instance to determine a good lifting method (Bejjani *et al.*, 1984) or to determine maximum acceptable weights (Hutton and Adams, 1982; Jäger and Luttmann, 1986).

Gagnon *et al.* (1986) used low-back estimates to compare three methods to lift a patient out of a wheelchair. Their paper points out the low-back model's limitations and how this puts uncertainty on absolute values for forces; however modelling did allow comparison between the three methods. Energy expended to lift patients was also taken into account in the choice of a lifting method.

More examples of the use of low-back modelling in the analysis of industrial tasks can be found in Jäger and Luttmann (1986).

Models of the low back

In the simple example of Figure 24.3, one set of back muscles, situated posterior to the spine, was used to resist a trunk flexion moment. This created a compression force on the spine. This is only one model of the low back: many others, varying in complexity, have been devloped. Some, used in the orthopaedics field, take into account a large number of muscles and ligaments attached to several points of each vertebra. For the ergonomics field, the main guideline is obtained from the value of the compression force between two lumbar vertebrae, and a less detailed model is usually considered adequate.

The rest of this section describes how to evaluate muscle and compression forces with 2D and 3D models. It has been written for readers who wish to carry out calculations themselves.

A simple 2D model

We will now complete the simple 2D model used at the beginning of the chapter (Figure 24.3). It allowed back muscles (representing the erector spinae group) to resist a trunk flexion moment. If trunk extension is to be resisted, these muscles can relax and abdominal muscles (rectus abdominis) take over. Figure 24.9 summarizes the results of this model. The lumbar spine compression is shown as a function of the low-back moment. The vertical force on the hands also compresses the spine, as shown from the parallel lines for 0 N, 1000 N upwards and 1000 N downwards.

The equations describing this 2D model, and from which Figure 24.9 was obtained are now described.

A flexion moment is provided by the rectus abdominis (R), and an extension moment is provided by the erector spinae (E) (see Figure 24.10). E acts at a distance y_E from the centre of the spine, and the lever arm for R is y_R.

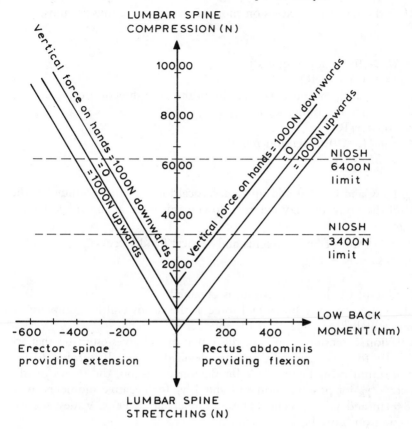

Figure 24.9. Lumbar spine compression as a function of low-back moment, with a simple 2D model

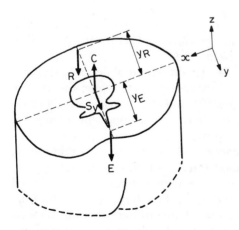

Figure 24.10. A simple 2D model, represented on a section of the low-back (R = rectus abdominis, E = erector spinae, C = compression, y_R, y_E = lever arms)

If M_x is the flexion or extension moment which these muscles must provide,

> if $M_x > 0$ (flexion required)
> $M_x = (-y_R) \times (-R)$
>> (both the y and z axes are in an opposite direction to y_R and R, hence the minus signs)
> so $M_x = y_R R$
> else if $M_x < 0$ (extension required)
>> $M_x = -y_E E$

Finally, E, R and C must add up to a vertical force F_z that counteracts the weight of the body and any external downwards force acting at the hands:

> $F_z = -$ (sum of body weights and vertical hand force)
> $F_z = C - R - E$

so the value of C, the compression, is obtained.

In the same way, any horizontal force at the hands will be compensated by a horizontal shear force S_y at the intervertebral joint.

For flexion/extension moments, this crude model gives estimates that are very similar to those given by a more detailed model. Returning to the graphical summary in Figure 24.9 for flexion/extension, the slopes on the graph are $1/y_R$ for positive moments and $1/y_E$ for negative moments, with $y_R = 8 \cdot 0$ cm and $y_E = 5 \cdot 8$ cm. (These numbers are average values: see the later section on 'Low-back geometry'.)

Simple 3D models

The previous model did not have any muscles accounting for lateral flexion or axial rotation of the trunk. A model described by Chaffin and Andersson (1984, p. 206) will be briefly summarized here.

It consists of six muscles (Figure 24.11): the rectus abdominis (R) and erector spinae (E), provide flexion and extension, respectively; the vertical components of the left and right obliques (VL and VR) provide lateral flexion to the left and to the right. Finally, the horizontal components of the left and right obliques (HL and HR), provide respectively positive (anti-clockwise) and negative axial moments round the z axis. Other low-back forces in the model are the compression force C on the intervertebral joint (if $C < 0$, the force is on the contrary an extension force), and the lateral and antero-posterior shear forces, S_x and S_y on the intervertebral joint. One more force not mentioned so far is the force P due to intra-abdominal pressure: it is believed that the rise in pressure in the abdominal cavity, that occurs during heavy manual handling, supports the trunk and effectively produces

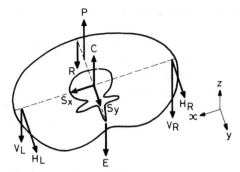

Figure 24.11. A schematic diagram of a simple 3D model, adapted from Chaffin and Andersson, 1984. (R, E, C = rectus abdominis, erector spinae and compression; P = force due to intra-abdominal pressure; VL, HL are the vertical and horizontal components of the obliques on the left side of the body; VR, HR on the right side)

an extensor moment. This moment is equivalent to a force P acting on the centre of the diaphragm. This topic will be discussed in more detail later.

The equations for this model are presented in Table 24.2. Antero-posterior lever arms are denoted by y, lateral lever arms by x, so y_E is the distance of the erector spinae force E behind the spine, for instance, x_O is the lateral lever arm of the obliques. M_x, M_y, M_z are the low-back moments to be provided by the muscles. F_x, F_y, F_z are the forces provided by the low back to counteract body weight and hand-forces.

One limitation of this model is that the vertical and horizontal components of the obliques are made to act independently, whereas oblique muscles in reality always pull simultaneously in the vertical and horizontal directions. As this model allows an oblique to provide purely a horizontal force, the compression on the spine may be underestimated for tasks involving axial rotation. We will discuss later a 10-muscle model by Schultz and Andersson (1981), which models the obliques in a more realistic way, with internal

Table 24.2. Equations for the simple 3D model (adapted from Chaffin and Andersson, 1984)

IF	M_x	$\geqslant$	0	THEN E	=	0	(flexion required)
	M_x	$\leqslant$	0	THEN R	=	0	(extension required)
IF	M_y	$\geqslant$	0	THEN VR	=	0	(flexion to left required)
	M_y	$\leqslant$	0	THEN VL	=	0	(flexion to right required)
IF	M_z	$\geqslant$	0	THEN HR	=	0	(anti-clockwise rotation required)
	M_z	$\leqslant$	0	THEN HL	=	0	(clockwise rotation required)
F_z		=	S_x				
F_y		=	S_y + HL + HR				
F_z		=	C + P − R − E − VL − VR				
M_x		=	$-y_R P + y_R R - y_E E$				
M_y		=	x_o (VL − VR)				
M_z		=	x_o (HL − HR)				

obliques acting posteriorly downwards and external obliques acting anteriorly downwards.

This 10-muscle model, and others involving more muscles require a computer to carry out a particular mathematical procedure (linear optimization). This is a handicap to some users wanting a simple program quickly written on a microcomputer, or even just worked out on a calculator. Accordingly a 'micro-model' is proposed (Tracy, 1988), to model the obliques more realistically and produce results that are closer to those of models requiring linear optimization.

The rule for oblique action is as follows:

Internal and external obliques pull respectively posteriorly and anteriorly downwards, as in Figure 24.12. Suppose a clockwise axial rotation moment must be provided ($M_z < 0$): this can be done by the external on the left (XL) and by the internal on the right (IR). If lateral flexion to the left is also required ($M_y > 0$), XL and IL will be in action. With this model, XL acts strongly to provide both M_z and M_y, and either IR or IL act, depending on which moment is the largest. So if axial rotation is more important than lateral flexion, XL and IR are active.

The equations for this model are given in Table 24.3. Its predictions come close to those of the 10-muscle model which is to follow, but a computer is not needed. Its main limitations are that the erector spinae and the rectus abdominis (E and R) are placed in the mid-sagittal spine, whereas in reality they are groups of muscles situated to the left and the right. They contribute to lateral flexion moments, whereas the simple models seen here only allow obliques to do this.

3D models requiring optimization

If the erector spinae and the rectus abdominis are placed as separate forces on either side of the sagittal plane, and if any other trunk muscles are also represented, special techniques are required to decide how several muscles

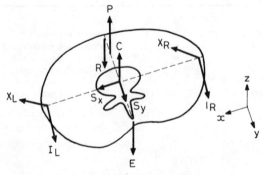

Figure 24.12. Schematic diagram of the micro-model—a simple 3D model with more realistic representation of the obliques. (Symbols described in Table 24.3)

Table 24.3. Equations for the micro-model shown in Figure 24.12

Conventions

R	rectus abdominis
E	erector spinae
XL, XR	left and right external obliques, acting in the (y, z) plane, downwards and towards the ventral part of the trunk, at 45° to the transverse plane
IL, IR	left and right internal obliques, acting in the (y, z) plane, downward and towards the dorsal part of the trunk, at 45° to the transverse plane
P	force due to intra-abdominal pressure
C, S_x, S_y	compression and shear forces on the intervertebral joint
Axes (x, y, z)	centered on the spine, (x, y) in the transverse plane with x to the left of the body, y directed posteriorly, z directed upwards
F_x, F_y, F_z	resultant reaction forces at the level of the section
M_x, M_y, M_z	resultant reaction moments at the level of the section
ABS (x)	absolute (positive) value of x
SUM =	$\dfrac{M_y + M_x}{2\,(x_o \cos 45)}$
DIFF =	$\dfrac{M_y - M_z}{2\,(x_o \cos 45)}$

Equations

$F_z = C + P - E - R - (IL + IR) \cos 45 - (XL + XR) \cos 45$

$F_y = (IL + IR) \sin 45 - (XL + XR) \sin 45 + S_y$

$F_x = S_x$

$M_x = -y_E\,E + y_R R - y_P\,P$

$M_y = x_O\,(IL - IR) \cos 45 + x_o\,(XL - XR) \cos 45$

$M_z = x_O\,(IL - IR) \sin 45 + x_O\,(XR - XL) \sin 45$

If $M_x \geq 0$ then $E = 0$

If $M_x < 0$ then $R = 0$

---

that can all do the same job are going to share it out. Lateral flexion to the right, for instance, can now be provided by the right obliques, the right erector spinae, or the right rectus abdominis. Mathematically, there are more variables (forces) than equations. The indeterminacy must be solved by making assumptions, and giving rules as in Tables 24.2 or 24.3 is not possible or practical when many muscles are involved.

A technique is then to use a computer library for linear programming: this optimizes one variable while making sure a number of equations are satisfied. Schultz and Andersson (1981) and Schultz *et al.* (1983) have established models with 10 to 22 muscles, based on the assumption that the spinal compression C is to be at a minimum. A linear programming routine will ensure that all muscle forces provide the required moments and do not exceed a maximum capability of 100 N/cm² and that at the same time the forces have been distributed so that C is as low as possible. Although the problem may seem complicated, linear programming routines are in principle straightforward to use. For asymmetrical tasks, models such as the ones developed by Schultz *et al.* (1983) are far superior to simple calculator-based models, as shown by Schultz *et al.* in validation experiments. The reader is referred to Schultz and Andersson (1981) and Schultz *et al.* (1983) for details of these models.

Table 24.3. Contd

___

If $M_y \geqslant 0$ and $M_z \geqslant 0$
  and ABS $(M_y) \geqslant$ ABS (M_z) then

    IR = 0
    XR = 0
    IL = SUM
    XL = DIFF

  and ABS $(M_y) <$ ABS (M_z) then

    IR = 0
    XL = 0
    IL = SUM
    XR = −DIFF

If $M_y \geqslant 0$ and $M_z < 0$
  and ABS $(M_y) \geqslant$ ABS (M_z) then

    IR = 0
    XR = 0
    IL = SUM
    XL = DIFF

  and ABS $(M_y) <$ ABS (M_z) then

    IL = 0
    XR = 0
    IR = −SUM
    XL = DIFF

If $M_y < 0$ and $M_z < 0$
  and ABS $(M_y) \geqslant$ ABS (M_z) then

    IL = 0
    XL = 0
    IR = −SUM
    XR = −DIFF

  and ABS $(M_y) <$ ABS (M_z) then

    IL = 0
    XR = 0
    IR = -SUM
    XL = DIFF

If $M_y < 0$ and $M_z \geqslant 0$
  and ABS $(M_y) \geqslant$ ABS (M_z) then

    IL = 0
    XL = 0
    IR = −SUM
    XR = −DIFF

  and ABS $(M_y) <$ ABS (M_z) then

    IR = 0
    XL = 0
    IL = SUM
    XR = −DIFF

___

The condition to minimize the compression force will have the effect that a muscle with a larger lever arm will provide a force in preference to one with a smaller lever arm. Muscles close to the spine act only if other muscles have reached their maximum capability.

This assumption produces good predictions but it must be pointed out that validation experiments have only been carried out on small loads. Bean *et al.* (1988) have proposed a more sophisticted assumption: first use linear programming to minimize not C but muscle intensities (force per cross-sectional area): the largest of all the muscle intensities, $I^\star$ must be as small as possible. This ensures that no muscle is giving its maximum while other muscles which could also contribute are inactive. The next step is to solve the problem all over again, minimizing C, but with the condition that no muscle intensity exceeds $I^\star$. This recent method seems to provide more realistic modelling than when to minimize C is the only criterion, and validation with experimental data is expected.

Minimizing C also implies that antagonistic muscles do not co-contract, for instance that if trunk extension is required the rectus abdominis are inactive. This is a simplification of reality, as co-contraction can be important in some tasks. Nachemson *et al.* (1986) for instance observed co-contraction when a Valsalva manœuvre (voluntary raising of intra-abdominal pressure) was performed. This implies a higher value of spine compression than predicted by the previous models.

Inputs to biomechanical calculations

Posture input

Biomechanics is sometimes used as a predictive tool, on a posture that has not been observed but has been estimated as a likely posture for a task. However, the posture may not be realistic and it may be worth ensuring that the body is in balance by checking that the resultant of all external forces lies in the area between the two feet. If the posture is asymmetric, each leg can take a different proportion of the resultant force at the feet, and this needs to be measured with a force plate, which brings us back to the laboratory. Therefore, biomechanics must be used with caution when used as a predictive tool.

The reader is referred back to the beginning of the chapter for an overview of methods to record posture.

Body segment weights

Table 24.4 summarizes masses and the locations of the centre of gravity of body segments, compiled by Pheasant (1986). Other anthropometric data used in modelling, such as link lengths, can be found in the same reference. Segment mass data originate from very small, poorly representative samples, so may be a source of error in biochemical calculations. This and other sources of error are discussed in the last section of this chapter.

Low-back geometry

There have been recent improvements in the data available for low-back models. Both *computed axial tomography* (CAT) and *magnetic resonance imaging* (MRI) have been used to measure muscle lever arms and cross-sectional areas, whereas previously values came from a limited sample of cadaveric data. Some of the results needed for the models described in this chapter can be found in Table 24.5. These are lever arms for 96 females (Chaffin *et al.*, 1990) and at the L3–L4 level for 26 males (Tracy *et al.*, 1989). Data for more

Table 24.4. Segment masses and locations of centre of gravity, from Pheasant (1986)

Segment	Mass (percentage body mass)	Location of centre of gravity
1. Head and neck	8·4	57% of distance from C7 to vertex
1a. Head	6·2	20 mm above tragion
2. Head and neck and trunk	58·4	40% of distance from hip to vertex
2a. Trunk	50·0	46% of distance from hip to C7
2b. Trunk above lumbo-sacral joint	36·6	63% of distance from hip to C7
2c. Trunk below lumbo-sacral joint	13·4	Approximately at the hip joint
3. Upper arm	2·8	48% of distance from shoulder to elbow joints
4. Forearm	1·7	41% of distance from elbow to wrist joints
5. Hand	0·6	40% of hand length from wrist joint (at centre of an object gripped)
6. Thigh	10·0	41% of distance from hip to knee joints
7. Lower leg	4·3	44% of distance from knee to ankle joints
8. Foot	1·4	47% foot length forward from the heel (half height of ankle joint above the ground)— mid-way between ankle and ball of foot at the head of metatarsal III

Total body mass* (kg)

	Men				Women			
Percentiles	5th	50th	95th	S.D.	5th	50th	95th	S.D.
British (19–65 years)	55.3	74.5	93.7	11.7	44.1	62.5	80.9	11.2

* Masses (in kg) to be multiplied by 9·81 to obtain weights (or forces in N) for the calculation of moments.

muscles and at other lumbar sections can be found in these same studies. Lever arms for both sexes have also been measured by Nemeth and Ohlsen (1986) at the lumbo–sacral joint, and by Kumar (1988) at L3 and L4 (as well as T7 and T12).

Table 24.5. Lever arms of some muscles at L3–L4 level, from a CT study of 96 women (Chaffin *et al.*, 1990) and an MRI study of 26 males (Tracy *et al.*, 1989). Standard deviations in parentheses)

	Erector spinae	Rectus abdominis	Oblique
Females			
Antero-posterior lever arm (mm)	52 (4)	70 (19)	20 (10)
Lateral lever arm (mm)	34 (4)	43 (11)	113 (16)
Males			
Antero-posterior lever arm (mm)	58 (5)	80 (18)	17 (12)
Lateral lever arm (mm)	38 (3)	34 (10)	122 (11)

At the moment it is common to use muscle lever arms observed at a particular cross-section, but ideally the values used should take into account the line of action of the muscles and their points of attachment. More information will no doubt become available with the recent developments in CAT and MRI scanning.

Role of intra-abdominal pressure

The most widespread theory of the role of intra-abdominal pressure (IAP) in low-back force production is that the pressure supports the trunk, and its action on the diaphragm and pelvic floor is equivalent to a force for trunk extension. According to this model, the force produced by IAP is calculated by multiplying the pressure by the area of the diaphragm. The extensor moment created by this force is the product of the force with the lever arm of the centroid of the area on which IAP acts. Using this model, IAP reduces lumbar compression by 4–30% according to Schultz *et al.* (1982), 2–8% according to Leskinen and Troup (1984). This range is large because the percentage reduction depends on the value of IAP and on the value of compression. The reduction in spinal compression due to IAP may be underestimated, because calculations of moments and lumbar loads, ignoring IAP, sometimes still result in excessive compression values although no structural failure is observed (Jones, 1983; Chaffin and Andersson, 1984). On the other hand, both Krag *et al.* (1985) and Nachemson *et al.* (1986) have argued from experimental evidence that IAP does not reduce lumbar compression; EMG readings showed that trunk extensor muscle action was not reduced when the abdominal cavity was voluntarily pressurized.

A number of theories for the role of IAP have been put forward, and the reader is referred to other texts (e.g., Aspden, 1987) for a review of these various theories. Until the controversies on the role of IAP are resolved, the most common approach is to represent it as an extensor force, as in Figure 24.11. Shown in Table 24.6 are some values found in the literature for various tasks, but many authors choose to ignore IAP and set the force to zero. It is possible to measure IAP with a swallowed radio-pill (Davis and Stubbs, 1977), but some experience is required to use this technique. Chaffin and Andersson (1984, p. 192) have published a prediction equation for IAP, using the hip moment and angle, for lifting in the sagittal plane.

It must be added that the measure of IAP is sometimes used directly to evaluate manual handling tasks. Davis and Stubbs (1977) conducted some studies in which they found IAP increases that were proportional to weights lifted. They found higher incidences of back pain among males undertaking heavy physical work, during which peak abdominal pressures frequently exceeded 13 kPa (100 mmHg). Accordingly they developed a set of recommendations for safe limits of force from values of forces which produce 12 kPa (90 mmHg) in a 5th percentile man.

However, there is some disagreement on the validity of the relationship

Table 24.6. Intra-abdominal pressure for various tasks (1 kPa = 7·6 mmHg)

Task	IAP (kPA)	Force (N) developed by IAP over 299 cm²
Schultz *et al.* (1982)		
Relaxed standing	1·0	30
Uprights, arms in, holding 8 kg in both hands	1·5	45
Flexed 30°, arms out	4·2	125
Flexed 30°, arms out, holding 8 kg in both hands	4·4	130
Davis and Stubbs (1977)		
Breathing	1	30
90 mmHg 'safe limit'	12	560
Grieve and Pheasant (1982)		
Competitive weight lifting	40	1196
Nachemson *et al.* (1986)		
Valsalva manoeuvre	4	120

Estimate for diaphragm area: 299 cm (Leskinen and Troup, 1984)
Estimate for IAP lever arm: 48 mm (Schultz *et al.*, 1982)

between IAP and spinal load. For instance Andersson *et al.* (1977) and Ört-engren *et al.* (1981) found a linear relationship between intradiscal and intra-abdominal pressures, whereas Schultz *et al.* (1982) and Nachemson *et al.* (1986) did not. Schultz *et al.* (1982) found a poor correlation between esti-mated spine compression and IAP values. Chaffin and Andersson (1984) reported a relationship between IAP and hip moment which is not linear but nearly quadratic, while Mairiaux *et al.* (1984) found it to be linear.

Angle of discs

So far we have referred to forces on 'the lumbar spine' without specifying a particular vertebra or disc. The low-back models discussed are too crude to differentiate between different vertebrae, and the main difference between calculations at L3 or at L5–S1 is the weight of the trunk above it. There is one other difference, though, and that is the angle of the intervertebral discs. The force referred to as the compression force earlier on is actually partly compression and partly shear if the intervertebral joint is not perpendicular to the line of action of the erector spinae or rectus abdominis. Unfortunately there is very little information on disc angles for various postures, and as there are large variations in the degree of lordosis in the population, predictions on disc angles are associated with a large uncertainty. Chaffin and Andersson (1984, p. 192) obtain the L5–S1 angle from hip and thigh angles, for postures in the sagittal plane. Another approach is to infer the shape of the spine from the shape of the surface of the back (Stokes and Moreland, 1987; Tracy *et al.*, 1989) but this work is still somewhat inconclusive. Until more information is available, one solution is to make the approximation that L3 remains perpendicular to the line of action of the erector spinae and rectus abdominis, whatever the posture (Schultz *et al.*, 1983, for instance), or to use Chaffin and Andersson's (1984) relationship for the L5–S1 angle. Either way the uncertainty will mean that some of the compression force evaluated may in fact be a shear force, and vice versa.

Uses and limitations of biomechanics

This chapter has surveyed the application of biomechanics in the ergonomics field. Calculating moments at joints provides an estimate of the severity of the task. Moments calculated in 3D will also highlight possible twisting efforts which could be eliminated. Biomechanical calculations allow applied forces or posture to be varied, so that problems and solutions can be identified.

Repetitive work and fatigue

Unfortunately, biomechanics cannot on its own answer questions of the type: 'What force can be applied safely and without fatigue x times a minute for y hours, given n rest pauses of m minutes are provided?'

Biomechanics can only give some indications of the effect of fatigue if the moment required of a joint is close to its maximum capability, for the onset of fatigue is near. Rohmert (1973) provides information for rest allowances in static work as a function of the percentage of maximum strength a task requires. However Rohmert *et al.* (1986) question the universality of these relationships; it appears that postures where passive structures (skeleton and ligament) are playing the key role can be held longer than those requiring mainly active muscular force.

For intermittent static work a rough guideline is to keep the force exerted under 15–30% of the maximum capability (Bjorksten and Jonsson, 1977; Pheasant and Harris, 1982) to avoid fatigue. Even less can be said at the moment about dynamic work, presumably because of the large number of variables in the problem. In general, whether the work is static, intermittent or dynamic, biomechanics cannot on its own give reliable answers, except in extreme cases where the task can be shown to be so strenuous it can only be performed occasionally.

Safe limits

Human variability is the main problem when determining acceptable limits, especially where back pain is concerned. Limits on lumbar spine compression, such as those used by NIOSH (1981), are based on results of ultimate compression strength tests performed on cadaveric spines. However, as discussed in this chapter, these do not usually produce the effects observed in real back injuries; also it is not possible in many back-pain patients to define the precise source of back pain (Wells, 1985).

There may be some confusion on how often a task can be repeated if it creates a spine compression of the order of magnitude of the ultimate compression of strength of cadaveric specimens. These tests imply that one single exertion would damage the spine, yet the NIOSH guidelines (Waters *et al.*, 1993) allow this type of task to be done quite frequently and these tasks are indeed regularly performed in industry (NIOSH guidelines are relevant to this chapter but are more appropriately discussed in the context of work systems and manual handling assessments—see chapter 30). Studies on fatigue fractures from repetitive loading should eventually bring some light on the matter. A further discussion of these problems can be read in Jones (1983).

Sources of inaccuracy

Results of low-back forces depend on the model used, so it is useful to bear in mind the assumptions and simplifications a model employs, and if several models are available to compare their predictions. Some inputs to low-back calculations are subject to uncertainty. These include the following parameters:

1. *Intra-abdominal pressure.* If IAP is not measured, an estimate needs to be made, and the contribution of IAP to forces in the low back depends on the model. Pressures range from about 1 kPa (relaxed standing) to about 40 kPa (competitive weight lifting) (Table 24.6). If the spine compression calculated without IAP is around 3400 N (first NIOSH limit), including IAP could reduce the estimate by up to 60%, depending on the value of IAP. At the second NIOSH limit (6400 N), the reduction is up to 30%.

2. *Angles of discs.* There are individual variations in spinal shape between subjects, and there is further uncertainty on disc angle changes with trunk motion. The largest compression estimates will be obtained with discs that are perpendicular to the line of action of the muscles.

3. *Geometry of the low back.* There are quite large inter-subject variations in some muscle lever arms—for instance in Table 24.5, the lever arm of the erector spinae has a standard deviation of nearly 10% of the mean value. Accordingly the uncertainty about the force provided by the erector spinae is also represented by a standard deviation of about 10% of the force calculated with the mean lever arm.

Other sources of uncertainty in biomechanical calculations have already been mentioned. How accurately were the posture and force recorded? Was an asymmetric task evaluated with a 2D model? Were accelerations ignored in a static model? There are also uncertainties on limb weights and their centres of mass. These have the greatest effect when the trunk and arms are held out at large lever arms. In this case the spine compression on a person with a 95th percentile body weight is nearly 30% greater than that for a male with a 50th percentile weight.

All these sources of uncertainty are not usually important when results are used to compare tasks, but are useful to bear in mind when results are used as absolute numbers, and perhaps evaluated against guidelines. There is a mathematical method of evaluating the effect of all the uncertainties on the final result (Barford, 1985), but another way is to experiment with different values of the input parameters.

Conclusion

Biomechanics is a useful tool to evaluate manual handling tasks, highlight problems, and test out possible improvements. Models used can be more or less sophisticated; some require computers while a lot can be achieved just with a calculator. Results should be interpreted with a basic knowledge of the simplifications and uncertainties that have been involved in the calculations. Other methods usefully drawn in to complement biomechanics include physiological measurements, EMG, injury statistics, discomfort charts and questionnaires.

Acknowledgement

The author is indebted to Diva Ferreira for her contribution to the development of this chapter.

References

Adams, M.A. and Hutton, W.C. (1981). The effect of posture on the strength of the lumbar spine. *Engineering in Medicine*, **10**, 199–202.

Andersson, B.J.G. and Örtengren, R. (1974). Lumbar disc pressure and myoelectric back muscle activity during sitting. II. Studies on an office chair. *Scandinavian Journal of Rehabilitation Medicine*, **3**, 115–121.

Andersson, G., Örtengren, R. and Nachemson, A. (1977). Intradiscal pressure, intra-abdominal pressure and myoelectric back muscle activity related to posture and loading. *Clinical Orthopaedics and Related Research*, **129**, 156–164.

Aspden, R.M. (1987). Intra-abdominal pressure and its role in spinal mechanics. *Clinical Biomechanics*, **2**, 168–174.

Aspden, R.M. (1989). The spine as an arch. A new mathematical model. *Spine*, **14**, 266–274.

Ayoub, M.M. and El Bassoussi, M.M. (1978). Dynamic biomechanical model for sagittal plane lifting activities. In *Safety in Manual Materials Handling*, edited by C.G. Drury (Cincinnati, OH: US Department of Health, Education and Welfare), pp. 88–95.

Barford, N.C. (1985). *Experimental Measurements: Precision, Error and Truth*, 2nd edition (New York: John Wiley).

Bean, J.C., Chaffin, D.B. and Schultz, A.B. (1988). Biomechanical model calculation of muscle contraction forces: a double linear programming method. *Journal of Biomechanics*, **21**, 59–66.

Bejjani, F.J., Gross, C.M. and Pugh, J.W. (1984). Model for static lifting: relationship of loads on the spine and the knee. *Journal of Biomechanics*, **17**, 281–286.

Björksten, M. and Jonsson, B. (1977). Endurance limit of force in long-term intermittent static contractions. *Scandinavian Journal of Work and Environmental Health*, **3**, 23–27.

Brinckmann, P. (1986). Injury of the annulus fibrosus and disc protrusions. An *in vitro* investigation on human lumbar discs. *Spine*, **11**, 149–153.

Brinckmann, P., Johannleweling, N., Hilweg, D. and Biggemann, M. (1987). Fatigue fracture of human lumbar vertebrae. *Clinical Biomechanics*, **2**, 94–96.

British Orthopaedic Association (1966). Joint motion. Method of measuring and recording. Published by the American Academy of Orthopedic Surgeons, reprinted by the British Orthopaedic Association, 1966.

Chaffin, D.B. (1988a). A biomechanical strength model for use in industry. *Applied Industrial Hygiene*, **3**, 79–86.

Chaffin, D.B. (1988b). Biomechanical modelling of the low back during load lifting. *Ergonomics*, **31**, 685–697.

Chaffin, D.B. and Andersson, G. (1984). *Occupational Biomechanics* (New York: Wiley-Interscience).

Chaffin, D.B., Redfern, M.S., Erig, M. and Goldstein, S.A. (1990). Lumbar muscle size and locations from CT scans of 96 women of age 40 to 63 years. *Clinical Biomechanics*, **5**, 9–16.

Corlett, E.N., Madeley, S.J. and Manenica, I. (1979). Posture targetting: a technique for recording working postures. *Ergonomics*, **22**, 357–366.

Davis, P.R. and Stubbs, D.A. (1977). Safe levels of manual forces for young males. *Applied Ergonomics*, **8**, 141–150; **8**, 219–228; **9**, 33–37.

Drury, C.G., Roberts, D.P., Hansgen, R. and Bayman, J.R. (1983). Evaluation of a palletising aid. *Applied Ergonomics*, **14**, 242–246.

Gagnon, M., Sicard, C. and Sirois, J.P. (1986). Evaluation of forces on the lumbo-sacral joint and assessment of work and energy transfers in nursing aides lifting patients. *Ergonomics*, **29**, 407–421.

Garg, A., Chaffin, D.B. and Freivalds, A. (1982). Biomechanical stresses from manual load lifting: a static vs dynamic evaluation. *IIE Transactions*, **14**, 272–281.

Grieve, D.W. (1987). Demands on the back during minimal exertion. *Clinical Biomechanics*, **2**, 34–42.

Grieve, D.W. and Pheasant, S.T. (1982). Biomechanics. In *The Body at Work – Biological Ergonomics*, edited by W.T. Singleton (Cambridge: Cambridge University Press), pp. 71–161.

Hutton, W.C. and Adams, M.A. (1982). Can the lumbar spine be crushed in heavy lifting? *Spine*, **7**, 586–590.

Ito, K., Minamizaki, Y. and Ito, M. (1980). *Computer-aided dynamic analysis of multi-link system for biomechanical applications*. Research Reports of

Automatic Control Laboratory, Faculty of Engineering, Nagoya University, Volume 27.

Jäger, M. and Luttmann, A. (1986). Biomechanical model calculations of spinal stress for different working postures in various workload situations. In *The Ergonomics of Working Postures*, edited by E.N. Corlett, J.R. Wilson and I. Manenica (London: Taylor and Francis), pp. 144–154.

Jones, D.F. (1983). Back injury research: have we overlooked something? *Journal of Safety Research*, **14**, 53–64.

Keyserling, W., Herrin, G., Chaffin, D.B., Armstrong, T. and Foss, T. (1980). Establishing an industrial strength testing program. *American Industrial Hygiene Association Journal*, **41**, 730–736.

Krag, M.H., Gilbertson, L. and Pope, M.H. (1985). Intra-abdominal and intra-thoracic pressure effects upon load bearing of the spine. *31st Annual Meeting Orthopedic Research Society*, Las Vegas, Nevada.

Kumar, S. (1988). Moment arms of spinal musculature determined from CT scans. *Clinical Biomechanics*, **3**, 137–144.

Leskinen, T.P.J. and Troup, J.D.G. (1984). The effect of intra-abdominal pressure on lumbosacral compression when lifting. *Computer-aided Biomedical Imaging and Graphics Physiological Measurement and Control: Proceedings*, Aberdeen, PMCS: 4.

MacDonald, E.B. (1984). Back pain, the risk factors, and its prediction in work people. Occupational aspects of low back disorders. *Society of Occupational Medicine, Symposium Proceedings*, pp. 1–17.

Mairiaux, P., Davis, P.R., Stubbs, D.A. and Baty, D. (1984). Relation between intra-abdominal pressure and lumbar moments when lifting weights in the erect posture. *Ergonomics*, **27**, 883–894.

McGill, S.M. and Norman, R.W. (1985). Dynamically and statically determined low-back moments during lifting. *Journal of Biomechanics*, **18**, 877–885.

Nachemson, A. and Elfström, G. (1970). Intravital dynamic pressure measurements in lumbar discs. *Scandinavian Journal of Rehabilitation Medicine* (Suppl. 1), 1–40.

Nachemson, A. and Morris, J. (1964). *In vivo* measurements of intradiscal pressure. *Journal of Bone and Joint Surgery*, **46A**, 1077–1092.

Nachemson, A.L., Andersson, G.B.J. and Schultz, A.B. (1986). Valsalva maneuver biomechanics: effects on lumbar trunk loads of elevated intra-abdominal pressures. *Spine*, **11**, 476–479.

Nemeth, G. and Ohlsen, H. (1986). Moment arm lengths of trunk muscles to the lumbosacral joint obtained *in vivo* with computed tomography. *Spine*, **11**, 158–160.

NIOSH (National Institute for Occupational Safety and Health) (1981). A work practices guide for manual lifting. Cincinnati: DHHS (NIOSH) publication no 81–122.

Örtengren, R., Andersson, G.B.J. and Nachemson, A.L. (1981). Studies of relationships between lumbar disc pressure, myoelectric back muscle activity, and intra-abdominal (intragastric) pressure. *Spine*, **6**, 98–103.

Pheasant, S. (1986). *Bodyspace. Anthropometry, Ergonomics and Design* (London: Taylor and Francis).

Pheasant, S. and Harris, C.M. (1982). Human strength in the operation of tractor pedals. *Ergonomics*, **25**, 53–63.

Rohmert, W. (1973). Problems in determining rest allowances. *Applied Ergonomics*, **4**, 91–5; **4**, 158–162.

Rohmert, W., Wangenheim, M., Mainzer, J., Zipp, P. and Lesser, W. (1986). A study stressing the need for a static postural force model for work analysis. *Ergonomics*, **29**, 1235–1249.

Samuelson, B., Wangenheim, M. and Wos, H. (1987). A device for three-dimensional registration of human movement. *Ergonomics*, **30**, 1655–1670.

Schultz, A.B. and Andersson, G.B.J. (1981). Analysis of loads on the lumbar spine. *Spine*, **6**, 76–82.

Schultz, A., Andersson, G., Örtengren, R., Haderspeck, K. and Nachemson, A. (1982). Loads on the lumbar spine. *Journal of Bone and Joint Surgery*, **64A**, 713–720.

Schultz, A., Haderspeck, K., Warwick, D. and Portillo, D. (1983). The use of lumbar trunk muscles in isometric performance of mechanically complex standing tasks. *Journal of Orthopaedic Research*, **1**, 77–91.

Stokes, I.A.F. and Moreland, M.S. (1987). Measurement of the shape of the surface of the back in patients with scoliosis. *Journal of Bone and Joint Surgery*, **69A**, 203–211.

Svensson, O.K. (1987). On quantification of muscular load during standing work. A biomechanical study. Dissertation from the Kinesiology Research Group, Department of Anatomy, Karolinska Institute, Stockholm, Sweden.

Tracy, M.F. (1988). Strength and posture guidelines: a biomechanical approach. Ph.D. Thesis, University of Nottingham.

Tracy, M., Haslegrave, C.M. and Corlett, E.N. (1987). Automating the measurement and biomechanical analysis of posture. In *New Methods in Applied Ergonomics*, edited by J.R. Wilson, E.N. Corlett and I. Manenica (London: Taylor and Francis), pp. 267–272.

Tracy, M.F., Gibson, M.J., Szypryt, E.P., Rutherford, A. and Corlett, E.N. (1989). The geometry of the lumbar spine determined by magnetic resonance imaging. *Spine*, **14**, 186–193.

Troup, J.D.G., Foreman, T.K., Baxter, C.E. and Brown, D. (1987). The perception of back pain and the role of psychophysical tests of lifting capacity. *Spine*, **12**, 645–657.

Tsuang, Y.H., Schipplein, O.D., Trafimow, J.H., Andersson, G.B.J. (1992). Influence of body segment dynamics on loads at the lumbar spine during lifting. *Ergonomics*, **35**, 437–444.

Waters, T.R., Putz-Anderson, V., Garg, A., and Fine, L.J. (1993). Revised NIOSH equation for the design and evaluation of manual lifting tasks. *Ergonomics*, **36**, 749–776.

Wells, N. (1985). *Back Pain* (Office of Health Economics).

Winter, D.A. (1979). *Biomechanics of Human Movement*. (New York: Wiley Interscience).

Chapter 25

Techniques in mental workload assessment

Najmedin Meshkati, Peter A. Hancock, Mansour Rahimi and
Suzanne M. Dawes

Introduction

The question of mental workload (MWL) assessment is relatively new and
important; new in comparison to companion techniques for the assessment
of physical load, whose origins are the contemporary of the Industrial Revolution, and important in that an increasing proportion of work taxes the information processing capabilities of operators, rather than their physical capacity.
It is the load placed upon such cognitive capabilities that mental workload
assessment is designed to measure. The techniques used to measure this load
are the primary focus of this chapter. Four contemporary groups of methods,
each comprising several techniques, are evaluated below from the view point
of their practicality and utility for the working ergonomist:

1. *Primary task measures*. These are probably the most obvious method of
mental workload assessment. For example, if we want to know how driving
is affected by differing task demands, e.g., traffic conditions, fatigue or lane
width, we should be able to utilize the driving performance itself as a criterion
(Hicks and Wierwille, 1979).

2. *Secondary task measures*. A secondary task is a task which the operator
is asked to do in addition to his or her primary task. If he or she is able to
perform well on the secondary task, this is taken to indicate that the primary
task is relatively easy; if he or she is unable to perform the secondary task
and at the same time maintain the primary task performance, this is taken
to indicate that the primary task is more demanding (Knowles, 1963). The
difference between the performances obtained under the two conditions,
with and without inclusion of the primary task, is then taken as a measure,
or index, of the workload imposed by the primary task.

3. *Subjective rating measures*. These include direct or indirect queries of the
individual for their opinion of the workload involved in a task. The easiest

way to estimate the mental workload of a person who performs a certain task is to ask him or her what he or she feels about the mental load level of the task.

4. *Physiological (or psychophysiological) measures.* Individuals who are subjected to some degree of mental workload commonly exhibit changes in a variety of physiological functions. As a result, several researchers have advocated the measurement of these changes to provide an estimate of the level of workload experienced.

Primary task measurement

There are several methodological approaches to the measurement of performance, or system output measures (Chiles and Alluisi, 1979). From the practical standpoint the *analytical approach* appears most appropriate. Welford's (1978) concept of the analytical approach looks in detail at the actual performance of the task to be assessed, examining not only overall achievement, but also the way in which it is attained. The advantage of this method is that the various decisions and other processes that make up performance are considered in the context in which they normally occur, so that the full complexities of any interaction between different elements in the task can be observed. For instance, it has been shown that the presence in a cycle of operations of one element which has to be carried out more deliberately than the rest slows the performance of all the elements in the cycle, so that a prediction of the time taken made on the basis of the time required to carry out each element in isolation would be too low.

The analytical approach requires that several scores be taken of any one performance. For example, the component parts of a complex cycle of operations should be measured separately, errors may be recorded as well as time taken, and different types of errors need to be distinguished. Welford (1978) has argued that the greatest value of this approach is probably that it enables the more subtle effects of workload to be examined by showing the strategies in use, such as maintaining a balance between speed and accuracy or between errors of omission and commission, and methods of operation which in various ways seek to increase efficiency and to reduce excessive load. This approach has two difficulties. First, the detailed scores required may be difficult to obtain for tasks such as process monitoring in which most of the decisions made do not result in any overt action, and second that even where there is sufficient observable action, recording may have to be elaborate and analysis of results laborious.

Synthetic methods comprise another major approach to performance measurement as a MWL assessment technique. According to Chiles and Alluisi (1979) these methods start with a task analysis of the system, in which the proposed operating profile is broken down into segments or phases that are relatively homogeneous with respect to the way the system is expected to operate. For each phase, the specific performance demands placed on the operator are then identified through task-analytic procedures. Performance times and operator reliabilities are assigned to the individual tasks and sub-tasks on the basis of available or derived data. The information on perform-ance time is then accumulated for a given phase and the total is compared with the predicted duration of the phase. This comparison of required time with available time can be employed as an index of workload.

Welford's (1978) approach to this method is relatively similar. He con-sidered loads imposed by task demands (e.g., data gathering, choices, actions) which are separate, in terms of time taken or other measures, either in the laboratory or in artificially simplified work conditions. The total load is then assessed by adding the components together. This is the approach on which standard time analysis of manual work is based. For the assessment of mental workload based on this approach, the work of Kitchin and Graham (1961) is considered of seminal importance. They questioned the applicability of conventional work measurement techniques to mental workload and were able to show that those techniques could give a satisfactory quantitative expression to three types of mental activity on a time basis. The first type of mental activity involves the direction and co-ordination of muscular activity during all defined physical movements of the body, including highly manipulative work. The second type of mental activity is that of perception (taken to mean the actual receipt of information by any of the senses), while the final activity is that of the senses searching for a random (but likely) signal demanding instant action. In summary, work measurement, although predominantly a technique for measuring work which can be observed as being carried out physically, satisfactorily recognizes for practical purposes those mental activities which are an integral part of some physical activity and which are defined by the physical activity which they accompany in time.

The third major approach to mental workload assessment via performance analysis is the *multiple measurement of primary task performance*, which is a com-posite technique. These techniques might be considered useful for workload assessment when individual measures of primary task performance do not exhibit adequate sensitivity to operator workload because of operational adaptivity due to perceptual style and strategy. According to Williges and Wierwille (1979), using multiple measures in a combined analysis has the beneficial effect of reducing the likelihood that important strategy changes will go unnoticed. There are numerous studies that report both positive and negative results for the application of multiple measures which can be found in this chapter.

Although use of the multiple measures approach is potentially advan-tageous, it may have a detrimental effect similar to noise amplification. By measuring a large number of variables, it becomes more likely that some will not change reliability as a function of workload. Another area of concern is the differential sensitivities of the individual members of the multiple meas-ures to the different aspects of the task. For instance, there might be two or more different measures which appear almost equal in ability to discriminate change in operator workload, but which may in fact have large differences in sensitivity. Consequently, this might cause disarray in scaling of the different workloads.

The lack of sensitivity of performance measures to changes in mental workload levels is one of the major problems of these methods. This issue has been addressed by many authors such as Gaume and White (1975) and Gartner and Murphy (1975). Gaume and White argue that the level of mental workload may increase while performance is unchanged so that performance may not be a valid measure of workload. Gartner and Murphy propose that an operator may show equal performance for two different configurations, but in reality effort on one system may greatly exceed the effort on the other. Generalization to different task situations poses an additional problem in the application of performance measures, since for each experimental situation a unique measure must be developed (Hicks and Wierwille, 1979). Williges and Wierwille (1979) also refer to this point and argue that the measures of performance of the primary task are task-specific. Each time a new situation is examined, new measures must be developed and tested. In several other techniques of workload assessment, the same measures can be used regardless of the application.

Williges and Wierwille (1979) cite numerous studies which support the same concept; namely, that no substantial change occurs in the primary task as a function of workload. In general, the studies cited appear to have been performed at workload levels where the operator had sufficient reserve capacity to adapt to the increased load. Rouse (1979) also refers to long-term performance measures as the indicator of relative workload, although this would seem to provide at best an ordinal scale of workload measurement. Further, unless one is willing to assume that humans always operate to capacity and that all humans have the same capacity, the performance-based workload scales would only reflect the states of particular individuals for which the data were collected. In other words, inter-individual comparisons may not be valid for this measure.

Secondary task measures

The concept of using a secondary task as a measure of mental workload is grounded on the assumption of the limited channel capacity of the human information processing system (Welford, 1959; Kalsbeek, 1968, 1973). This approach assumes that an upper limit exists on the ability of a human operator to gather and process information. The secondary task is a task which the operator is asked to do in addition to the primary task. There are two types of secondary tasks; 'loading' and 'subsidiary' (or 'non-loading'). If the subject is instructed to aim for error-free performance on the secondary task at the expense of the primary task, the secondary task is called a loading task (Rolfe,

1976). In this case the operator must always attend to the secondary task which may cause performance degradation on the primary task (Sheridan and Johannsen, 1976). If the subject is instructed to avoid making errors on the primary task, the secondary task is called a non-loading or subsidiary task. In this case the operator attends to the secondary task when time is available.

According to Knowles (1963), one of the best ways of measuring operator load is to have the operator perform an auxiliary or secondary task at the same time as performing the primary task under evaluation. If the operator is able to perform well on the secondary task, this is taken to indicate that the primary task is relatively easy. If he or she is unable to perform the secondary task, and at the same time maintain primary task performance, this is taken to indicate that the primary task is more demanding. The difference between the performance obtained under the two conditions is taken as a measure or index of the workload imposed by the primary task. There are two related, but different, reasons for using the secondary or loading task. The first one, according to Knowles (1963), is to compensate for any deficiency in the loading of the primary task and to stimulate aspects of the total job that may be missing. Therefore, the secondary task is used simply to bring pressure on the primary task with the idea that as the operator becomes more heavily stressed, performance on difficult tasks will deteriorate more than performance on easy tasks. In the first application of the secondary task, the emphasis is upon stressing the primary task. Differences in operator workload are indicated by differences in the primary task performance measures taken under the stress induced by the auxiliary task. In the second application of the secondary task as a subsidiary task, the auxiliary task is used not so much with the intention of stressing the primary task as with intention of finding out how much additional work the operator can undertake while still performing the task to meet satisfactorily the system criteria.

Some examples of secondary tasks are arithmetic addition, repetitive tapping, choice reaction time, critical tracking tasks and cross-coupled dual tasks. Varying combinations of these tasks have been employed by many investigators in their measurement of operator workload. Ogden *et al.* (1978) reviewed 144 experimental studies which used secondary task techniques to measure, describe or characterize operator workload. In addition to the secondary task techniques that have been reported by Ogden, his colleagues and Williges and Wierwille (1979), there are two other related (secondary task) techniques. The first of these is occlusion and the second is handwriting analysis.

Occlusion is actually a time-sharing technique. However, in occlusion, the time-sharing is forced rather than voluntary. The operator is given samples over time of the visual information required to perform the primary task; the time-sharing is thus accomplished by suppressing the information input (Hicks and Wierwille, 1979). The usual method is to block the operator's visual input from the display. Senders *et al.* (1967) and Farber and Gallagher (1972) used this technique in measuring the attentional demand of driving an automobile and reported some positive results. However, according to Hicks and Wierwille (1979) the occlusion method is not particularly sensitive and is more intrusive compared with other techniques in a simulated automobile driving situation.

Handwriting analysis is another potential measure of mental workload because of its deterioration due to distraction of the individual by other tasks. Kalsbeek and Sykes (1967) utilized handwriting as a secondary task while the primary task was to respond, via pedal depression, to a random series of binary choice signals which were presented to either the auditory or visual senses. They were able to show a step-by-step disintegration of writing performance provoked by the increasing number of binary choices.

The secondary task as a mental measurement technique has many shortcomings. Perhaps the most difficult aspect of secondary task methodology for assessing workload is intrusion. When the secondary task is introduced, performance on the primary task is known to be modified and usually degraded (Williges and Wierwille, 1979). This problem has been addressed by other authors such as Welford (1978), who regarded the extra load imposed by the secondary tasks as a factor that might produce a change of strategy in dealing with the primary task and consequently distort any assessment of the load imposed by the primary task alone. Brown (1978) argued that since the dual task method is essentially a resource-limiting device (human-processing resources being limited), interference should occur within the processing mechanisms, rather than at sensory input or motor output. He claimed there is empirical support for the idea that interference is maximal at the level of response selection. Brown (1978) indicated that the dual task interference is greater when the tasks share the same response modality than where responses occupy different modalities.

The nature of the primary task and its informational load and structural characteristics can cause problems of efficiency and a reduction in the utility of the secondary task. Workload may be largely a function of the structural characteristics of a task rather than of the informational load imposed by its component parts (Brown, 1978). Therefore, the more interesting tasks (i.e., those which most closely assimilate real-life situations) may be relatively inaccessible to study by the dual task methodology. This fact and other expected and unexpected interactions between certain tasks, drove Ogden *et al.* (1978) to point out that the choice of the secondary task is problematic (see also McCormick and Sanders, 1982).

The question of individual differences in secondary task performance has been addressed by some investigators through association with personality constructs. The introduction of an additional task may increase arousal which has been shown to affect differing personality types in contrasting ways (Gibson and Curran, 1974; Huddleston, 1974). Motivation is another related factor which may play an important role in secondary task performance (Kalsbeek and Sykes, 1967). Knowles (1963) attempted to provide a set of criteria against which to judge the desirability of a secondary task. These criteria included unobtrusiveness with respect to the primary task, ease of learning, self-placing (in order for the secondary task to be neglected in maintenance of primary task performance) and compatibility with the primary task. Ogden *et al.* (1978) added sensitivity and representativeness to the above set in order to address a wide range of human abilities and functions. Kalsbeek (1971) suggested the dual task method for use in two ways. First, in the traditional way, measuring the so-called spare mental capacity, and second, in experiments where the main task, to which preference has to be given, is a simple or repetitive one. For instance, a binary choice task can be regarded as a stress condition in the performance of a secondary task.

The various problems outlined above point to Brown's (1978) conclusion that:

> The dual task method should be used only for the study of individual difference in processing resources available to handle workload . . . . If it is so used, it should probably be in the form of an additional, secondary task, presenting discrete stimuli of constant load, on a forced paced schedule, and competing for processing resources only.

Subjective rating measures

As mentioned in the introduction, the easiest way to estimate the mental workload of a person who performs a certain task is to ask what he or she subjectively feels about the load of the task. Sometimes a list of key words or definitions describing different levels of load can be given. The subject then has to rate the load with reference to these levels (Sheridan and Stassen, 1979). Sheridan (1980) has also argued that mental workload should be defined in terms of subjective experience, and he continued: 'subjective scaling is the most direct measure of such subjective experience'. (For a general discussion of subjective assessment methods see chapter 3 of this book.)

Subjective estimates of load have often been obtained through either the use of *rating scales* or *interviews/questionnaires*. A widely used rating scale in systems evaluation is the Cooper and Harper (1969) scale, originally developed to measure the handling characteristics of aircraft by using the subjective reports of test pilots. The Cooper-Harper scale was designed primarily to assess flight characteristics and the descriptors of this scale pertain to 'flyability' of an aircraft. Therefore, it can be applied to manual control tasks (Moray, 1982). However, according to Williges and Wierwille (1979), if this scale was used for workload assessment the assumption must be made that handling difficulty and workload are directly related. The assumption is that if a pilot states that an aircraft is difficult (or impossible) to fly, this is equivalent to saying that the task of flying imposes a very heavy or unsupportable load (Moray, 1982). The modified Cooper-Harper scale (Rahimi and Wierwille, 1982; Wierwille and Casali, 1983a; Wierwille *et al.*, 1985b), which is a modified version of the original Cooper-Harper, is considered as further development for the subjective measurement of mental workload. This scale is applicable to a wider variety of task workloads, especially for systems which load perceptual, mediational and communication activities (Wierwille *et al.*, 1985a).

Wewerinke (1974) has confirmed the validity of the Cooper-Harper scale as an indicator of MWL. He was able to report an extremely high correlation coefficient (0·8) between subjective difficulty rating and objective workload level in his study. Gartner and Murphy (1975) echoed the above idea and referred to the positive qualities of the Cooper-Harper scale as operational relevance, convenience and unobtrusiveness. Some of the other examples of application of rating scales can be found in the work of Williges and Wierwille (1979).

Another widely used subjective rating technique is SWAT (Subjective Workload Assessment Technique), which was designed specifically to measure operator workload in a variety of systems for a number of tasks. It uses the conjoint measurement technique (Nygren, 1982) to combine ratings on three different dimensions of workload: time load, mental effort load, and stress load. It should be noted that SWAT, like the Cooper-Harper scale, can be applied to workload in a number of different settings, although cockpit evaluation has been most common. An extended discussion of SWAT, in the context of subjective assessment methods generally, appears in an appendix at the end of this chapter.

Application of subjective-rating technique in strictly cognitive tasks has been addressed by Borg *et al.* (1971) who were able to achieve a high correlation between subjective and objective measures of difficulty. The strictly cognitive tasks, unlike the manual control tasks and time-stressed tasks (e.g., some signals must be processed before the processing of predecessors is completed), are single-trial tasks. The subject is not under pressure associated with a continuous or arbitrary stream of signals that may arrive before he or she has finished dealing with an earlier signal. However, Phillip *et al.* (1971) argued that in time-stress tasks it is not possible to differentiate unambiguously between the criteria of stress time and difficulty of the control task, based upon subjective rating methods. Therefore, the subjective feeling of difficulty in work processing seems to be essentially dependent on the time-stress involved in performing the task. Gaume and White (1975), in their study of mental workload evaluated the subjective estimates of stress levels obtained during the experiment, and they considered them as potentially valid and reliable indicators of mental workload.

The second approach to subjective rating is through the application of interviews/questionnaires. The procedures used in this approach are not as structured as rating scales. They range from completely open-ended debriefing sessions to self-reporting logs of stressful activities, from carefully chosen questionnaire items (Williges and Wierwille, 1979). Usually, interviews and questionnaires have been used primarily as supplementary measures to other techniques. For example, Sherman (1973) demonstrated a high correlation between subjective measures of workload and various physiological measures. The advantages of the approach stem from its unobtrusiveness and extreme ease of application.

Since the subjective rating of the difficulty of a task is primarily a function of the raters' perception, the concept of perceived difficulty has to be given importance and analyzed directly. Audley *et al.* (1979) proposed that the perceived difficulty of scaling the task demands, should be the primary consideration and attempts should be made to dissociate this from other facets of subjective aspects of workload. The perceived difficulty of a task might alter the human operator's attitude to it. This, in turn, could affect the time operators would be prepared to spend and the level of confidence in their decisions (Moray, 1982). The perceived difficulty for the individual is influenced by at least three groups of factors. The first group deals with the content of long-term memory including both general experience and memories of similar tasks. The second group is of background factors such as personality traits, habits and general attitudes including likes and dislikes, aspiration and expectation levels. The third group of factors represents momentary conditions, e.g., one's emotional state, general fatigue, motivation and the importance ascribed to the task, as well as actual anticipated success or failure (Borg *et al.*, 1971).

The subjective rating of task difficulty could also be affected by the situation and job as a whole rather than by only the task–induced or individual rater's factors. Borg (1978) proposed that it is necessary to point out the set of factors which seem to cause the experience of the difficulty in one job which may be different from those in another job. Finally, it should be noted that general individual differences and differences in the adaptivity of the operator to the system, the task and the resultant impression of the task as viewed by the operator, causes higher than normal rating (Williges and Wierwille, 1979).

Physiological measures

Individuals engaged in cognitive activities provide indirect indices of their level of effort through changes in the status of a number of physiological systems. Ursin and Ursin (1979) recognized that these physiological methods do not measure the imposed load but rather they give information concerning how the individuals themselves respond to the load and, in particular, whether they are able to cope with it. In what follows we have attempted to differentiate the main current physiological approaches on the basis of their validity as a measure of workload and their applicational utility. From this analysis, the most practical method to emerge is heart rate or one of its derivatives (e.g., Kalsbeek, 1968) and the most valid measures are changes in central nervous system (CNS) activity (e.g., Wickens, 1979). We suggest that a tympanic temperature measure, taken in the ear canal, can provide a potentially useful compromise in the trade-off between the concern for validity and practicality (see Hancock and Brainard, 1981; Hancock and Dirkin, 1982; Hancock, 1983, 1984). The interested reader can find specific details of differing physiological measures in the review articles of Ursin and Ursin (1979). Williges and Wierwille (1979), and Hancock *et al.* (1985).

Use of physiological measures as indicators of mental workload is influenced by a combination of several factors, such as the cost of both hardware and software to operate the equipment, the training level of the personnel who administer the physiological tests, environmental conditions of the workplace and the willingness of the employees to be connected to a physiological recording mechanism. Due to problems caused by various combinations of these factors at the present time, physiological measures are among the least practical methods of mental workload assessment for use in complex machine-person systems. We should note, however, that technological innovations such as telemetric monitoring, have reduced some of the problems associated with these measures and use of these techniques in the working environment may be realized in the very near future. Two frequently used physiological methods of mental workload measurement are *event-related potentials* and *heart rate variability*.

Event-related potentials

Event-related potentials (ERPs) are fluctuations in the activity of the nervous system recorded in response to environmental stimulation in association with psychological processes, or in preparation for motor activity (Martin and Venables, 1980). Repeated presentations of a physical stimulus elicit waves which are subsequently signal-averaged to reduce or eliminate random variation and to yield a stimulus locked wave or ERP. Various ERP components have been taken to reflect information processing activity, and change in mental workload. Workload inferences are based upon the amplitude and latency elements of the elicited wave. For example, it has been observed that the amplitude of the wave at a latency of 200 ms following stimulus onset (P2) and the overall maximum power in the evoked response provided a metric of subjective difficulty of task performance (Spyker *et al.*, 1971).

The major advantage of ERPs is their representation as direct reflections of the information processing activity of the operator. In addition, ERPs are multivariate measures characterized by differing latency peaks which provide a considerable amount of information per observation. Also, as ERPs are elicited by discrete events in the environment there is more specificity between stimulus and response compared with other more global methods. Wierwille (1979) acknowledged that the dependence of the ERP upon the operator's perceived utilities, attitude and understanding, in addition to the imposed load, represents both an advantage and a limitation with respect to its power as a workload assessor.

The major disadvantages of this approach include: first, the single recorded trial response contains a high noise-to-signal ratio (Wickens, 1979). Consequently, either multiple trial recording and subsequent signal averaging, or filtering and application of analysis such as template matching, is required to extract meaningful information from single observations (Squires and Donchin, 1976). Second, the response may be contaminated by motor artefacts. Also, because of individual differences, 'calibrations' must be undertaken for each different operator. In practical terms, the technique requires considerable supporting instrumentation (e.g., computer facilities) and trained personnel for operation and interpretation. These limitations, however, represent technical barriers which are subject to constant change. Therefore, ERPs potentially represent the most promising physiological measure of mental workload for future exploitation.

Heart rate variability

If ERPs represent the most valid physiological measure, then measures pertaining to heart rate and its derivatives are currently the most practical physiological method of assessing imposed mental workload. Among such measures perhaps none is more thoroughly investigated than that of heart rate variability (HRV).

In research involving this measure, Kalsbeek (1971) noted a gradual suppression of the heart rate irregularity due to increases in the difficulty of a task. In consequence, it was posited that such a measure could be used to reflect mental workload. Several empirical investigations attest to the strength of this assertion (Kalsbeek, 1968, 1973). Since such observations were first made, there have been many experimental studies in which the connection between HRV and mental workload has been observed. Detailed reviews of these efforts are available (Wierwille, 1979; Meshkati, 1983).

Measures of HRV have been assessed through the use of three major calculational approaches (see Meshkati, 1988a): (1) scoring of the heart rate data or some derivative (e.g., standard deviation of the R-R interval); (2) through the use of spectral analysis of the heart rate signal; and (3) through some combination of the first two methods. There are two advantages to such a measure—a relative and an absolute advantage. First, the absolute advantage refers to the sensitivity of the measure to change in mental workload as demonstrated in the previously cited investigations. The second advantage is its practical utility and relative simplicity in both administration and subsequent interpretation, when compared with alternative physiological techniques. However, it should be acknowledged that since HRV is a particularly sensitive physiological function, it is vulnerable to potential contamination from the influence of both stress and the ambient environment (Kalsbeek, 1971). For the interested reader, further information on the details of each of these measures can be found in reports in Meshkati *et al.* (1984), and in Moray (1979).

Some other physiological means that may be of interest to ergonomics practitioners are pupil diameter, body chemical analysis and auditory canal temperature (ACT). For instance, Wierwille and Connor (1983) examined five different physiological measures (mean pulse rate, pulse rate variability, respiration rate, pupil diameter, and voice pattern) elicited by digit shadowing and mental arithmetic tasks. According to their results, only the mean pulse rate demonstrated some limited 'sensitivity' to some of the differences in the psychomotor load conditions. In a related study, Casali and Wierwille (1983) monitored respiration rate, heart rate mean, heart rate standard deviation, pupil diameter and eye blinks, and concluded that the sole physiological measure to display sensitivity to changes in communications load is the pupil diameter measure.

There are studies which utilized relatively unconventional and novel physiological approaches to assess human mental workload. Hyyppa *et al.* (1983) investigated psychoneuroendocrine responses to mental workload. They were able to find a significant decline of the cortisol and prolactin levels of subjects undergoing psychologically demanding achievement-oriented tasks. Loewenthal (1983) proposed alveolar gas concentration level could be a 'cleaner' physiological measure than the others (e.g., respiratory arrhythmia). In his extensive study, Loewenthal cited several studies that tried to demonstrate a relationship between alveolar gas pressures and mental workload.

Hancock (1983) and Hancock *et al.* (1985) considered tympanic temperature or, more correctly, deep auditory canal temperature (ACT) as an alternative measure which circumvents certain problems associated with other physiological measures. It has been observed that subjects beginning work on a simple mental task, after a period of quiescence, exhibit small but constant increases in ACT (Hancock, 1983). Also, subjects encountering a number of different computational problems embedded in a series of simple mathematical additions show an increase in ACT (Hancock and Brainard, 1981).

Relevance and coping with individual differences

A common and important aspect of all mental workload assessment methods is their relative sensitivity to individual differences. Moray (1984) asserted that:

> Individual differences in workload research is far more important than has hitherto been acknowledged. Without taking this into account we are seriously delaying the development of a useful measure.

Many investigators who failed to obtain significant results in applying mental workload measurement techniques have suggested that either the sample population must be homogenized or, alternatively, personality traits, individual differences, and other related factors should be incorporated into the experimental design. The following is a summary of such expert recommendations (for further discussion see Meshkati and Loewenthal, 1988a).

Kitchin and Graham (1961) refer to the character of the human operator as 'a very important area of concentration', without which the assessment of operator's physical and mental abilities are of little value to industry. Mulder and Mulder-Hajonides Van Der Meulen (1973) acknowledged the large differences among subjects and, therefore, recommended single subject analysis, and Leplat (1978) stated that the characteristics of personality could intervene in a far from negligible manner in regard to workload. Hamilton *et al.* (1979) tried to analyze the activation responses as a function of the task characteristics. Furthermore, they acknowledged that the subject's active information processing involves personality traits. Hopkin (1979) considered personality variables as potentially relevant to mental workload.

According to Firth (1973), in the real-life working environment individual differences in operators' characteristics very much influence information processing of the individuals. These differences arise from a combination of past experience, skill, emotional state, motivation and the estimation of risk and cost in a task. The influence of these individual differences is important, since many of these factors have been shown to influence cardiac responses directly. There are some other indications of the relationship between personality traits and physiological reaction parameters, e.g., Rotter's (1966) internal-external locus of control and heart rate control. Ray and Lamb (1974) and Gatchel (1975) found that internal locus of control subjects were better able to increase their heart rate as compared with their external counterparts.

Duffy (1962) reported that individual differences in responsiveness have been observed in many forms, in the frequency and amplitude of rhythms in the EEG, in the occurrence of 'spontaneous' changes in skin resistance, peripheral blood flow, heart rate, muscle tension and other functions. The author referred to the work of Armstrong (1938), who detected correlation between cardiovascular reactivity and emotional stability in 700 Army Corps candidates. Offerhaus (1980), based upon his study of hospital staff (normal subjects) and psychiatric patients, concluded that by employing the concept of heart rate variability, it is possible to differentiate between two pairs of groups of subjects: first, the high anxiety group from the anxiety one (i.e., psychotic patients from non-patients), and second, the stress reactor group from the non-stress reactor group (i.e., acute patients and neurotic staff from chronic patients and stable staff).

The issue of individual differences and psychological variables and their substantial effects on automatic responses has been addressed by Cleary (1974), Van Egeren *et al.* (1972) and Sutton and Tueting (1975). The concept was experimentally evaluated and confirmed by Bryson and Driver (1969).

They found that 'cognitively complex' subjects manifest higher GSRs in attending to stimuli. Lykken (1968), in the same regard, referred to two additional areas of consideration of individual differences, as the tonic psycho-physiological level and phasic response to specific stimuli. The effect of personality traits and individual differences on the performance of a mental task bears a great amount of significance. Hopkin (1979) stated that on many occasions, individual differences have precluded general judgements on whether the task-induced workload is excessive as distinct from high. He considered this typical inability to generalize the findings as mostly due to the fact that any given individual characteristic becomes a pertinent factor in workload only insofar as the task being performed brings that characteristic into play. Schroder *et al.* (1967) also reiterated this fact by arguing that if the task requires the processing of large amounts of descriptive information, and if this information must be integrated into a flexible, comprehensive system, then it can be expected that the 'integratively complex' person would perform better than integratively simple persons. They also postulated and later demonstrated that superior performance may be expected of a simple person, in an open situation, if the environment is complex and the criterion is simple.

Thackray *et al.* (1973) studied the role of personality in performance decrement and attention. Their results indicated that individuals scoring high on a distractibility scale (i.e., extrovert) found it difficult to maintain a uniform mode of performance. This group of subjects exhibited increasing lapses in attention, while introverted ones failed to show any evidence of a decline in attention.

According to an experimental study of mental workload by Meshkati and Loewenthal (1988b), operators' individual information processing behaviour affects their sinus arrythmia and subjective rating of task difficulty.

Wickens (1979) also regarded the relatively large differences among subjects in time-sharing abilities as the cause of substantial variance in dual–task performance. His proposal to tackle this problem was to 'calibrate' particular workload measurement techniques for different operators. Furthermore, with reference to Pew (1970), who recommended that at the same time these individual differences might actually be exploited to enhance system performance by employing them to provide guidelines for merging operators to specific systems, or by modifying systems to the limitations and strengths of individual operators, Kahneman (1973) rates a system possessing these qualities as 'perfect'.

Guidelines for use

In measuring human mental workload it is extremely important to define the measure and the nature of the loading task as accurately as possible. Suppose, for example, that an investigator uses the secondary task of time estimation as a means of assessing mental workload. Time estimation is a second-ary task technique for measuring the spare mental capacity of the human operator while performing a primary task. Estimating a 10-s time interval is a popular approach for this measure. First, the investigator has to define the actual measure. Next, the procedure and equipment by which this measure is to be taken should be precisely defined. Otherwise, another investigator may obtain different results because of a difference in the type of measure and/or procedure.

In time estimation, to continue the example, the procedure includes many aspects, such as instructed interval, method of prompting, method of responding, amount of training, instructions regarding counting, and instruc-tions regarding relative importance of the estimation task compared with the primary task. Measure definition includes type of estimates to be included (e.g., only those actively initiated and actively completed; or perhaps those actively initiated whether completed or not, with all incompleted estimates scored as the maximum interval between prompts). Measure definition also includes the types of computation, such as mean, median, standard deviation, or root mean square value of the estimated intervals. For example, our experi-ence with these experiments has shown that standard deviation is by far the most sensitive type of computation for time estimation. Also, there are no guarantees that the most sensitive computation remains viable for all task-loading situations. The following is a short description of guidelines and pit-falls in application of some mental workload measures.

Primary task measures

One difficulty in application of these measures is determining which compo-nent of the task is paramount to the overall task performance. In particular, tasks which are complex and multidimensional may need multiple primary task measures. It may even be necessary to divide the task components into different classes of performance (e.g., Berliner *et al.*, 1964) and assign primary measures to each category. That is, to break the global operator tasks into individual task elements and to determine what subset of those elements drives the workload metric.

A potential problem with application of primary task measures is intrus-iveness, particularly in field evaluations. Another problem is in the interpret-ation of the results. Primary task data may reflect a wide variety of influences such as motivation and learning effects. The effects of training and the ability of the operators to muster more effort may reduce the sensitivity of this measure to the real changes in mental workload.

Secondary task measures

In general, secondary task measures are relatively more expensive and difficult to administer. One reason is that both primary and secondary task performances need to be monitored and measured at the same time. One study even found a significant effect of secondary task measurement on the primary task performance of pilots while flying a simulator (Wierwille *et al.*, 1985b). This may indicate a certain degree of danger in measuring mental workload for real-world tasks that require critical operator attention. Also, the secondary task measures should not be combined with subjective measures, because operators may inadvertently include loading due to the secondary task in their ratings.

Subjective rating measures

There has been significant attention given to the use of subjective measures, mostly in the form of opinion scales. Subjective measures are used to assess the mental difficulty of the assigned task. One of the most useful opinion scales, the modified Cooper-Harper scale, shows a consistent level of sensitivity in tasks which have perceptual and cognitive components. Yet it requires after-the-fact evaluation of the task difficulty, using a guide scale. It is applicable to a wide variety of critical tasks (Wierwille and Casali, 1983a). Most opinion scales are easy and inexpensive to administer. However, care must be given to the interpretation of these scales since they contain words that may be interpreted differently under different task situations.

Physiological measures

In using these measures, care must be exercised in their selection and application. In general, the sensitivity of these measures is task dependent (Wierwille *et al.*, 1985b). For example, heart rate is one of the simplest physiological measures used in a number of early studies, yet the results indicate lack of generalizability, sensitivity and reliability across some mentally demanding tasks. Another difficulty with this measure is the nonmonotonicity of some physiological measures. That is, the measure may indicate increase in mental activity when task difficulty increases from low to medium, but it may show decrease after the workload continues to increase (Noel, 1974).

It should be mentioned, however, that physiological measures might be more useful in assessing the effects of time on task and task 'strain'. Also, in regard to intrusion, physiological measures have been fairly successful indicators without influencing the primary task performance of the operators.

Evaluation of techniques

The following criteria are useful in evaluating which mental workload assessment technique is appropriate for which particular setting. The listing below provides general guidelines which are elaborated in the sections on specific techniques (Meshkati, 1988b, and see also chapter 5).

1. *Validity*. The chosen mental workload measure should satisfy three validity constraints, those of content, predictability and construct.
2. *Reliability*. The measure should provide stable and repeatable results across repeated administrations.
3. *Sensitivity*. Sensitivity refers to the capability of a technique to discriminate significant variations in the workload levels imposed by a task or group of tasks (Eggemeier, 1985).
4. *Diagnosticity*. Diagnosticity refers to the capability of a technique to discriminate the amount of workload imposed on different resources or capabilities of the human operator (Eggemeier and O'Donnell, 1982; Wickens, 1984).
5. *Intrusion*. The mental workload measurement technique should not interfere with and/or cause degradations in the task being performed.
6. *Focus*. The measurement technique should be focused only on the changes in mental workload levels and should not reflect changes in status due to artefacts from variation in environmental conditions.
7. *Ease of field utilization*. The chosen technique should be robust and easy to administer within the constraints of the work environment under consideration. Factors which contribute to making a technique cumbersome include the instrumentation, analyst and operator training and data recording and analysis.
8. *Operator acceptance*. The success of a mental workload measurement technique is largely dependent upon operator acceptance and co-operation. This implies the necessity to understand the psychological profile of the typical end-user population prior to the application of any measurement technique.

By referring to the above criteria, some investigators have advocated a multiple battery approach, using representative elements from each of the four major groups of workload assessment methods. In such an approach, however, the advantage of non-interference is sacrificed for the greater specificity of workload information obtained. In the operation of dynamic person-machine systems, the limits of human mental capabilities need to be considered. The limit of such abilities and the operators' approaching of their own individual load tolerances are of particular importance with respect to safety and efficiency of overall action.

In this brief review, four major groups of mental workload assessment methods, their applications, advantages, and disadvantages and feasibility, have been presented. Subjective rating measures, due to their apparent simplicity, ease of administration and interpretation, seem to be the most widely used by practitioners for the measurement of non-physical workload. However, as mentioned previously, the problem of perceived task difficulty and the resultant artefacts inherent in operators' responses are as yet unresolved issues (cf. Meshkati and Driver, 1984). Until these questions are answered, subjective ratings of task difficulty should be normalized for individual differ-

ences and analyzed with extreme caution. Also, the validity and reliability
of this method could be enhanced by the use of multiple scaling techniques
which are validated through cross-checking procedures.

References

Acton, W.H. and Rokicki, S.M. (1986). Survey of SWAT use in operational
    test and evaluation. *Proceedings of the Human Factors Society 30th Annual
    Meeting*, pp. 221–224.
American National Standard. *Guide to Human Performance Measurements*
    (1992). BSR/AIAA, G-035-1992.
Armstrong, H.G. (1938). The blood pressure and pulse rate as an index of
    emotional stability. *American Journal of Medical Science*, **195**, 211–220.
Audley, R.J., Rouse, W., Senders, J. and Sheridan, T. (1979). Final report
    of mathematical modeling group. In *Mental Workload: Its Theory and
    Measurement*, edited by N. Moray (New York: Plenum Press).
Berliner, C., Angell, D., and Shearer, D.J. (1964). Behaviors, measures, and
    instruments for performance evaluation in simulated environments.
    Paper presented at the *Symposium and Workshop on the Quantification of
    Human Performance*, Albuquerque, NM.
Biers, D.W. and McInerney, P. (1988). An alternative to measuring subjec-
    tive workload: use of SWAT without the card sort. *Proceedings of the
    Human Factors Society, 32nd Annual Meeting*, pp. 1136–1139.
Borg, G. (1978). Subjective aspect of physical and mental load. *Ergonomics*,
    21, 215–220.
Borg, G., Bratfisch, O. and Dorinc, S. (1971). On the problem of perceived
    difficulty. *Scandinavian Journal of Psychology*, **12**, 249–260.
Boyd, S.P. (1983). Assessing the validity of SWAT as a workload measure-
    ment instrument. *Proceedings of the Human Factors Society, 27th Annual
    Meeting*, pp. 124–128.
Brown, I.D. (1978). Dual task methods of assessing workload. *Ergonomics*,
    21, 221–224.
Bryson, J.B. and Driver, M.J. (1969). Conceptual complexity and internal
    arousal. *Psychonomic Science*, **17**, 71–72.
Casali, J.G. and Wierwille, W.W. (1983). A comparison of rating scale, sec-
    ondary task, physiological and primary task workload estimation tech-
    niques in a simulated flight task emphasizing communications load.
    Human Factors, **25**, 623–641.
Chiles, W.D. and Alluisi, E.A. (1979). On the specification of operator or
    occupational workload with performance-measurement methods.
    Human Factors, **21**, 515–528.
Cleary, P.J. (1974). Description of individual differences in autonomic reac-
    tions. *Psychological Bulletin*, **81**, 934–944.
Cooper, G.E. and Harper, R.P. (1969). *The Use of Pilot Rating in the Evalu-
    ation of Aircraft Handling Qualities*. Report No. ASD-TR-76-19.
    (Moffett Field, CA: National Aeronautics and Space Administration).
Courtright, J.F. and Kuperman, G. (1984). Use of SWAT in USAF System
    T & E. *Proceedings of the Human Factors Society*, pp. 700–703.

Damos, D.L. (1988). Individual differences in subjective estimates of workload. In *Human Mental Workload*, edited by P.A. Hancock and N. Meshkati (Amsterdam: North-Holland).

Duffy, E. (1962). *Activation and Behavior* (New York: John Wiley).

Eggemeier, F.T. (1985). Workload measurement in system design and evaluation. In *Proceedings of the 29th Annual Meeting, the Human Factors Society* (Santa Monica, CA: Human Factors Society).

Eggemeier, F.T. and O'Donnell, R.O. (1982). A conceptual framework for development of a workload assessment methodology. In *Text of the Remarks made at the 1982 American Psychological Association Annual Meeting* (Washington, DC: American Psychological Association).

Ericsson, K. and Simon, H. (1980). Verbal reports as data. *Psychological Review*, **87**, 215–251.

Farber, E. and Gallagher, V. (1972). Attentional demands as a measure of the influence of visibility conditions on driving task difficulty. *Highway Research Record*, **414**, 1–5.

Firth, P.A. (1973). Psychological factors influencing the relationship between cardiac arrhythmia and mental load. *Ergonomics*, **16**, 5–16.

Gartner, W.B. and Murphy, M.R. (1975). *Pilot Workload and Fatigue: A Critical Survey of Concepts and Assessment Techniques*. Report No. ASD-TR-76-19 (Washington, DC: National Aeronautics and Space Administration).

Gatchel, R.J. (1975). Change over training sessions of relationships between locus of control and voluntary heart rate control. *Perceptual and Motor Skills*, **40**, 424–426.

Gaume, J.G. and White, R.T. (1975). *Mental Workload Assessment. III. Laboratory Evaluation of One Subjective and Two Physiological Measures of Mental Workload*. Report MDC-J7024/01. (Long Beach, CA: McDonnell-Douglas Corp.).

Gibson, H.B. and Curran, J.B. (1974). The effect of distraction on a psychomotor task studied with reference to personality. *Irish Journal of Psychology*, **2**, 148–158.

Hamilton, P., Mulder, G., Strasser, H. and Ursin, H. (1979). Final report of the physiological psychology group. In *Mental Workload: Its Theory and Measurement*, edited by N. Moray (New York: Plenum Press), pp. 367–385.

Hancock, P.A. (1983). The effect of an induced selected increase in head temperature upon performance of a simple mental task. *Human Factors*, **25**, 441–448.

Hancock, P.A. (1984). An endogenous metric for the control of perception of brief temporal intervals. *Annals of the New York Academy of Sciences*, **423**, 594–596.

Hancock, P.A. and Brainard, D.M. (1981). *Tympanic Temperature: A Noninvasive Physiological Measure of Workload*. Technical Report (MA: Endeco).

Hancock, P.A. and Dirkin, G.R. (1982). Central and peripheral visual choice reaction time under conditions of induced cortical hyperthermia. *Perceptual and Motor Skills*, **54**, 395–402.

Hancock, P.A., Meshkati, N. and Robertson, M.M. (1985). Physiological reflections of mental workload. *Aviation, Space, and Environmental Medicine*, **56**, 1110–1114.

Hart, S.G. and Staveland, L.E. (1988). Development of NASA-TLX (Task Load Index): results of empirical and theoretical research. In *Human Mental Workload*, edited by P.A. Hancock and N. Meshkati (Amsterdam: North-Holland).

Hicks, T.G. and Wierwille, W.W. (1979). Comparison of five mental workload assessment procedures in a moving-base driving simulator. *Human Factors*, **21**, 129–143.

Hill, S.G., Zaklad, A.L., Bittner, Jr., A.C., Byers, J.C. and Christ, R.E. (1988). Workload assessment of a Mobile Air Defense Missile System. *Proceedings of the Human Factors Society, 32nd Annual Meeting*, pp. 1068-1072.

Hopkin, V.D. (1979). General discussion based upon interactive group sessions. In *Mental Workload: Its Theory and Measurement*, edited by N. Moray (New York: Plenum Press), pp. 484–487.

Huddleston, H.F. (1974). Personality and apparent operator capacity. *Perceptual and Motor Skills*, **38**, 1189–1190.

Hyyppa, M.T., Aungola, S., Lahtela, K., Lahti, R. and Marniemi, J. (1983). Psychoneuroendocrine responses to mental load in an achievement-oriented task. *Ergonomics*, **26**, 1155–1162.

Kahneman, D. (1973). *Attention and Effort* (Englewood Cliffs, NJ: Prentice Hall).

Kalsbeek, J.W.H. (1968). Measurement of mental workload and of acceptable load: possible applications in industry. *International Journal of Production Research*, **7**, 33–45.

Kalsbeek, J.W.H. (1971). Standards of acceptable load in ATC tasks. *Ergonomics*, **14**, 641–650.

Kalsbeek, J.W.H. (1973). Do you believe in sinus arrhythmia? *Ergonomics*, **16**, 99–104.

Kalsbeek, J.W.H. and Sykes, R.N. (1967). Objective measurement of mental load. *Acta Psychologica*, **27**, 253–261.

Kitchin, J.B. and Graham, A. (1961). Mental loading of process operations: an attempt to devise a method of analysis and assessment. *Ergonomics*, **4**, 1–15.

Knowles, W.B. (1963). Operator loading tasks. *Human Factors*, **5**, 155–161.

Leplat, J. (1978). Factors determining workload. *Ergonomics*, **21**, 143–149.

Loewenthal, A. (1983). *Alveolar Gas Concentration and Mental Workload*. Technical Report 83–2 (Department of Industrial and Systems Engineering, University of Southern California).

Lykken, D.T. (1968). Neuropsychology and psychophysiology in personality research. In *Handbook of Personality Theory and Research*, edited by E.F. Borgatta and W.W. Lambert (Chicago: Rand McNally), pp. 413–509.

Martin, I. and Venables, P.H. (1980). *Techniques in Psychophysiology* (New York: John Wiley).

McCormick, E.J. and Sanders, M.S. (1982). *Human Factors in Engineering and Design* (New York: McGraw-Hill).

Meshkati, N. (1983). A conceptual model for the assessment of mental workload based upon individual decision styles. Dissertation, University of Southern California.

Meshkati, N. (1988a). Heart rate variability and mental workload assessment. In *Human Mental Workload*, edited by P.A. Hancock and N. Meshkati (Amsterdam: North Holland).

Meshkati, N. (1988b). Toward development of comprehensive theories of mental workload. In *Human Mental Workload*, edited by P.A. Hancock and N. Meshkati (Amsterdam: North Holland).

Meshkati, N. and Driver, M.H. (1984). Individual information processing behavior in perceived job difficulties: A decision style and job design approach to coping with human mental workload. In *Human Factors in Organizational Design and Management*, edited by H.W. Hendrick and O. Brown, Jr. (Amsterdam: North Holland).

Meshkati, N., Hancock, P.A. and Robertson, M.M. (1984). The measurement of human mental workload in dynamic organizational systems: an effective guide for job design. In *Human Factors in Organizational Design and Management*, edited by H.W. Hendrick and O. Brown, Jr. (Amsterdam: North Holland).

Meshkati, N. and Loewenthal, A. (1988a). An eclectic and critical review of four primary mental workload assessment methods: a guide for developing a comprehensive conceptual model. In *Human Mental Workload*, edited by P.A. Hancock and N. Meshkati (Amsterdam: North Holland).

Meshkati, N. and Loewenthal, A. (1988b). The effects of individual differences in information processing behavior on experiencing mental workload and perceived task difficulty: an experimental approach. In *Human Mental Workload*, edited by P.A. Hancock and N. Meshkati (Amsterdam: North Holland).

Moray, N. (1979). *Mental Workload: Its Theory and Measurement* (New York: Plenum Press).

Moray, N. (1982). Subjective mental workload. *Human Factors*, **24**, 24–45.

Moray, N. (1984). Mental workload. *Proceedings of the 1984 International Conference on Occupational Ergonomics*, Toronto, Canada, pp. 41–46.

Mulder, G. and Mulder-Hajonides Van Der Meulen, W.R.E.H. (1973). Mental load and the measurement of heart rate variability. *Ergonomics*, **16**, 69–83.

Noel, C.E. (1974). Pupil diameter versus task layout. Master's Thesis. Monterey, California, Naval Postgraduate School.

Nygren, T.E. (1982). *Conjoint Measurement and Conjoint Scaling: a User's Guide (AFAMRL-TR-82-22)*. DTIC No. ADA 122579 (Wright-Patterson Air Force Base, OH: Air Force Aerospace Medical Research Laboratory).

Offerhaus, R.E. (1980). Heart rate variability in psychiatry. In *The Study of Heart Rate Variability*, edited by R.I. Kitney and O. Rompelman (Oxford: Clarendon Press), pp. 225–238.

Ogden, G.D., Levine, J.M. and Eisner, E.J. (1978). *Measurement of Workload by Secondary Tasks: A Review and Annotated Bibliography*. Prepared under contract no. NAS2-9637 to National Aeronautical and Space Administration. (Ames Research Center, Washington, DC: Advance Research Resources Organization).

Papa, R.M. and Stoliker, T.E. (1987). Pilot workload assessment—a flight test approach. *AGARD Conference Proceedings No. 425, The Man Machine Interface in Tactical Aircraft Design and Combat Automation*, pp. 8–1–8–12.

Pew, R.W. (1970). Comments on promotion of man: challenges in sociotechnical systems: design for the individual operator. *Proceedings of the Global Systems Dynamics International Symposium*, Charlottesville, NJ, pp. 59–65.

Phillip, V., Reiche, D. and Kirchner, J. (1971). The use of subjective ratings. *Ergonomics*, **14**, 611–616.

Potter, S.S. and Bressler, J.R. (1989). Subjective Workload Assessment Technique (SWAT): A user's guide. Wright Patterson Air Force Base, Armstrong Aerospace Medical Research Laboratory, OH.

Ray, W.J. and Lamb, S.B. (1974). Locus of control and the voluntary control of heart rate. *Psychosomatic Medicine*, **36**, 180–182.

Rahimi, M. and Wierwille, W.W. (1982). Evaluation of the sensitivity and intrusion of workload estimation techniques in piloting tasks emphasizing mediational activity. *Proceedings of the IEEE International Conference on Cybernetics and Society*, pp. 593–597.

Reid, G. and Colle, H.A. (1988). Critical SWAT values for predicting operator overload. *Proceedings of the 32nd Annual Meeting of the Human Factors Society*.

Reid, G.B. and Nygren, T.E. (1988). The subjective workload assessment technique: a scaling procedure for measuring mental workload. In *Human Mental Workload*, edited by P.A. Hancock and N. Meshkati (Amsterdam: North-Holland).

Reid, G.B., Singledecker, C.A., Nygren, T.E. and Eggemeier, F.T. (1982). Development of multidimensional subjective measures of workload. *Proceedings of the Human Factors Society*, pp. 403–406.

Rolfe, J.M. (1976). The measurement of human response in man vehicle control situations. In *Monitoring Behavior and Supervisory Control*, edited by T.B. Sheridan and G. Johannsen (New York: Plenum Press).

Rotter, J.B. (1966). Generalized expectancies for internal versus external control of reinforcement. *Psychological Monographs*, **80**, 1–22.

Rouse, W.B. (1979). Approaches to mental workload. In *Mental Workload: Its Theory and Measurement*, edited by N. Moray (New York: Plenum Press).

Schlegel, B. (1986). Evidence for a subjective measure of excessive mental workload. Unpublished master's thesis, University of Oklahoma, Norman.

Schroder, H., Driver, M. and Streufert, S. (1967). *Human Information Processing* (New York: Holt Rinehart and Winston).

Senders, J.W., Kristofferson, A.B., Levision, W.H., Dietrich, C.W. and Ward, J.L. (1967). The attentional demand of automobile driving. *Highway Research Record*, **195**, 15–33.

Sheridan, T.B. (1980). Mental workload, what is it? why bother with it? *Human Factors Society Bulletin*, **23**, 1–2.

Sheridan, T.B. and Johannsen, G. (1976). *Monitoring Behavior and Supervisory Control* (New York: Plenum Press).

Sheridan, T.B. and Stassen, H.G. (1979). Definitions, models and resources of human workload. In *Mental Workload: Its Theory and Measurement*, edited by N. Moray (New York: Plenum Press).

Sherman, M.R. (1973). The relationship of eye behavior, cardiac behavior and electromyographic responses to subjective responses of mental fatigue and performance on a doppler identification task. Masters Thesis, Naval Postgraduate School, Monterey, California.

Spyker, D.A., Stackhouse, S.P., Khalafalla, A.S. and McLace, R.C. (1971). *Development Techniques for Measuring Pilot Workload*. Technical Report, NASA CR-1888.

Squires, K.C. and Donchin, E. (1976). Beyond averaging: the use of discriminant functions to recognize event related potentials elicited by single auditory stimuli. *EEG Clinical Neurophysiology*, **41**, 449–459.

Sutton, S. and Tueting, P. (1975). The sensitivity of the evoked potential to psychological variables. In *Research in Psychophysiology*, edited by P.H. Venables and M.J. Christie (New York: John Wiley).

Thackray, R.I., Jones, K.N. and Touchstone, R.M. (1973). *Personality and Physiological Correlates of Performance Decrement on a Monotonous Task Requiring Sustained Attention*. Report No. AM-73-14 (Washington, DC: FAA Office of Aviation Medicine).

Ursin, H. and Ursin, R. (1979). Physiological indicators of mental workload. In *Mental Workload: Its Theory and Measurement*, edited by N. Moray (New York: Plenum Press).

Van Egeren, L.F., Headrick, M.W. and Hein, P.L. (1972). Individual differences on autonomic responses: illustration of a possible solution. *Psychophysiology*, **9**, 626–633.

Welford, A.T. (1959). Evidence of a single-channel decision mechanism limiting performance in a serial reaction task. *Quarterly Journal of Experimental Psychology*, **11**, 59–66.

Welford, A.T. (1978). Mental workload as a function of demand, capacity, strategy, and skill. *Ergonomics*, **21**, 157–167.

Wewerinke, P.H. (1974). Human operator workload for various control conditions. *Proceedings of the 10th Annual NASA Conference on Manual Control*, Wright-Patterson AFB, Ohio, pp. 167–192.

Wickens, C.D. (1979). Measures of workload, stress and secondary tasks. In *Mental Workload: Its Theory and Measurement*, edited by N. Moray (New York: Plenum Press).

Wickens, C.D. (1984). *Engineering Psychology and Human Performance* (Columbus, OH: Charles E. Merril Publishing Co.).

Wierwille, W.W. (1979). Physiological measures of air crew mental workload. *Human Factors*, **21**, 575–593.

Wierwille, W.W. and Casali, J.G. (1983a). A validated rating scale for global mental workload measurement applications. In *Proceedings of the 27th Annual Meeting of the Human Factors Society* (Santa Monica, CA: Human Factors Society), pp. 129–133.

Wierwille, W.W. and Casali, J.G. (1983b). Mental workload estimation—an IE problem. *IE Ergonomics News*, **XVII** (3), 1–4.

Wierwille, W.W. and Connor, S.A. (1983). Evaluation of 20 workload measures using a psychomotor task in a moving-base aircraft simulator. *Human Factors*, **25**, 1–16.

Wierwille, W.W., Casali, J.G., Connor, S.A. and Rahimi, M. (1985a). Evaluation of the sensitivity and intrusion of mental workload estimation techniques. *Advances in Man Machine Systems Research*, **2**, 51–57.

Wierwille, W.W., Rahimi, M. and Casali, J.G. (1985b). Evaluation of sixteen measures of mental workload using a simulated flight task emphasizing mediational activity. *Human Factors*, **27**, 499–502.

Williges, R.C. and Wierwille, W.W. (1979). Behavioral measures of air crew mental workload. *Human Factors*, **21**, 549–574.

Winter, D.H. (1988). Validation of the Subjective Workload Assessment Technique in a simulated flight task. *AGARD Methods to Assess Workload, Proceedings No. 216*.

Appendix

Practical Mental Workload Assessment: Subjective Workload Assessment Technique (SWAT)

Subjective techniques are probably the most widely used methods for assessing mental workload in a practical context. As well as the Subjective Workload Assessment Technique (SWAT), these techniques include the NASA Task Load Index (TLX) (Hart and Staveland, 1988), Bedford Workload Scale, Cooper-Harper Scale, McDonnell Rating Scale and the NASA Bipolar Rating Scale (ANSI, BSR/AIAA, 1992). However, SWAT has seen more widespread use and acceptance as compared with other available techniques. Developed for the aerospace industry, SWAT is well-researched and replicated. It is difficult to find reports of general industrial use of SWAT, but there would be no reason why it could not be adapted for such purposes.

Problems with mental workload have been occurring routinely in aerospace over the last twenty years, and the aerospace industry has begun to take a careful look at the mental workload imposed on crew and has attempted to avoid high workload situations by task allocation, redesign of tasks and equipment and by automation. However, automation may not represent an optimum remedy because monitoring is still required, and conventional or manual backup must usually be provided (Wierwille and Casali, 1983b).

As military systems performance has improved and such systems have become more complex, the human operator has increasingly been identified as a potential limiting factor in system capability (Courtright and Kuperman, 1984). The role of the human operator in modern high-technology systems is, increasingly, that of a systems monitor, systems manager and decision-maker. Computer technology has advanced to the point at which enormous amounts of information can be displayed to the operator. In fact, the sources and quantity of information available frequently exceed people's capacity to process them.

Individual operator-related variables such as fatigue, stress, motivation and a perception of success or failure may contribute to workload for some, but may be irrelevant to others. These different perceptions of what contributes to workload might account for the inconsistencies in ratings found among individuals who experience apparently identical situations. People may not be aware of the influence that different factors have on the workload they experience, and it may be difficult for them to verbalize their biases abstractly or as related to a specific experience. Multidimensional approaches to workload assume:

1. Workload is a complex phenomenon derived from other independent or related factors.
2. Variation in the component dimensions can be defined and evaluated more precisely than can workload itself.
3. Different individuals place more emphasis on some dimensions than others.
4. People can verbalize the relative importance of different factors to their workload experience.

The development of SWAT

SWAT was developed by the Workload and Ergonomics Branch of the Air Force Armstrong Aerospace Medical Research Laboratory (AAMRL) (Reid *et al.*, 1982), as part of a research programme to develop a comprehensive battery of tests for assessing pilot workload. Mental workload has loosely been defined as how hard one has to work in order to accomplish a certain job. Many previously established methods have been designed for specific applications. The other categories of readily available measures of subjective workload include the Cooper–Harper rating, which assesses handling qualities, and the Systems Operability Measurement Algorithm (SOMA), which evaluates the system's operability. These measures are focused on systems evaluation with workload as a component. SWAT was designed specifically to assess human mental workload. The major goals of this measure are to be precise and sensitive enough to quantify the existence of high workload and to be simple to administer so that the applications in such as operational test and flight test are possible.

In order to develop SWAT, the researchers defined workload as primarily comprising three dimensions: 1. time load, 2. mental effort load, and 3. psychological stress load. Each of the three dimensions has three levels corresponding roughly to high, medium and low loading (Table 25.1).

SWAT is a technique whereby mental workload is measured subjectively, e.g., simply asking operators how hard they were working. This rating of workload is based on subjects' direct estimate or comparison estimate of the workload experienced at a particular point in time. Advantages of this method are that it is direct, quite easy, timely, no bulky measuring equipment is required and it is relatively non-intrusive.

Table 25.1. SWAT scales

Time Load
1. Often have spare time. Interruptions or overlap among activities occur infrequently or not at all.
2. Occasionally have spare time. Interruptions or overlap among activities occur frequently.
3. Almost never have spare time. Interruptions or overlap among activities are frequent or occur all the time.

Mental Effort Load
1. Very little conscious mental effort or concentration required. Activity is almost automatic, requiring little or no attention.
2. Moderate conscious mental effort or concentration required. Complexity of activity is moderately high due to uncertainty, unpredictability, or unfamiliarity. Considerable attention required.
3. Extensive mental effort and concentration are necessary. Very complex activity requiring total attention.

Psychological Stress Load
1. Little confusion, risk, frustration, or anxiety exists and can be easily accommodated.
2. Moderate stress due to confusion, frustration, or anxiety noticeably adds to workload. Significant compensation is required to maintain adequate performance.
3. High to very intense stress due to confusion, frustration, or anxiety. High to extreme determination and self-control required (Potter and Bressler, 1989, pp. 12–14).

Source: American National Standard Guide to Human Performance Measurements, 1992, page B-86)

SWAT is a two-phase process in which each subject or group of subjects completes a scale development phase and an event scoring phase. During the scale development phase, data necessary to develop a workload scale are obtained from a group of subjects. During the event scoring phase, subjects rate the workload associated with a particular task and/or mission segment. These are two distinct events and occur at different times.

The entire SWAT procedure is based on the application of a psychometric technique known as conjoint measurement. In conjoint measurement, the joint effects of several factors are investigated and the rule of composition principle that relates the factors to one another is extracted from the data. One major advantage of conjoint measurement is that only the ordinal aspects of the data are required for the production of an interval level scale, which represents the joint effects of the factors. Because only ordinal properties of the data are needed to generate a scale, the first step in the procedure is to order the dependent variables as a function of the joint effect of the factors being investigated. Next, the composition rule of the ordered data is defined through a series of axiom tests. Several combinatory results are possible including an additive rule, distributive rule and joint distributive rule. Finally, when the appropriate rule has been identified, the scaling transformation is computed.

In summary, the application of conjoint scaling to the measurement of workload is based on the assumption that workload is a multidimensional construct. SWAT is the application of conjoint scaling that comprises three levels for each of three dimensions—time, mental effort and psychological stress.

SWAT scale development

In the portion of the SWAT procedure named Scale Development, information is obtained from subjects regarding the way the three dimensions theoretically combine to produce workload. This information is collected by having subjects order all possible combinations of the definitions. Therefore, the ordering of all possible combinations of the three levels of the three dimensions, or 27, are conducted prior to starting an experiment. The data are then tested for agreement among subjects. If sufficient agreement exists (a Kendall's Coefficient of Concordance greater than 0·75) then the remainder of the procedure is based on group data. However, if the Kendall's Coefficient of Concordance is less than 0·75, an independent scale for each subject can be generated.

Once the experimenter has decided whether to obtain one scale to represent the group or a scale for each subject, the data are tested by a procedure called conjoint measurement. This procedure tests the ordering data to determine the combination rule used by the subjects.

Once the mode is defined, the data are transformed via an iterative procedure known as conjoint scaling. This transformation fits the data to the defined model and maintains the order inherent in the original data, which results in a unidimensional interval level scale. These values are then utilized during the event scoring phase.

Considerations in the use of subjective methods

Generally, subjective measures present special problems to researchers and designers. One of the primary considerations of methods within the applied world is the cost required to implement that methodology. SWAT, for example, is limited in that it requires a large investment of time by both the experimenter and the subjects. Another major factor to be considered when selecting a particular workload measure is the ease of implementation applicable to system evaluation techniques. Considerations include instrumentation requirements and operator training, which affect the ease with which a technique can be utilized. These criteria are of critical importance when applications to high-fidelity simulation or operational test and evaluation environments are considered. Operator acceptance is a third relevant criterion, relating to the degree of operator acceptance enjoyed by a particular assessment technique. In general, assessment techniques perceived by system operators as artificial or bothersome tend to be ignored and, therefore compromise the effectiveness of the resulting measure.

One of the primary limitations found when conducting workload studies in operational and test environments, such as flight test programmes, is the often limited pool of subjects available. For example, in recent studies of advanced aircraft, one aerospace contractor could only commit six crews (twelve crew members) to be available for the duration of the test, presenting a potential problem of having one individual overly influencing the workload ratings; omitting their data creates an even smaller sample size.

While a subjective rating such as SWAT may indicate a potential 'choke point', the experimenter or tester must still analyze that task in order to ascertain the primary factors contributing to that high workload rating. Two tools that are extremely useful in determining these key factors are question- naires and videotapes. Questionnaires have been used extensively to solicit information from crew members as part of their debriefing, but are limited in their ability to provide the context in which the high workload rating was made. In this regard, videotapes are one of the most useful tools for understanding the primary factors contributing to workload. The videotape can answer questions such as, was it a training issue or does this problem require a design modification? Considerations, such as the cost and time required to instrument the aircraft or system of interest and to review the videotapes must be addressed prior to beginning the study.

Training requirements

One of the critical components required for the correct application of SWAT is sufficient training of the crew members. Sufficient training is required so that meaningful event-related scores can be obtained with a minimum of intrusion. To date, no established criteria exist for the level of training required to familiarize subjects with the SWAT rating scales; it is often left up to the mental workload practitioner to make the appropriate determi- nation. A word of caution: training time provided is often dictated as a result of schedule and costs irrespective of the subjects' training needs. A minimum of 3–4 practice sorties or flights in flight test simulation has proved to be a reasonable number of practice sessions. In addition, a review of the three categories of SWAT in the pre-flight briefing is suggested. An additional consideration is the use of an abbreviated description for each of the three levels of workload components on either the test cards or map routes. Com- plete recall of the verbal content of the dimensional levels or memorization of these levels is complicated by the tests, the environment and the individual. In the flight environment, every second is critical, especially when flying close to the ground. In this instance, requiring the crew member to expend any time or energy recalling the verbal content of the dimensional levels is inappropriate. However, adequately and appropriately assessing workload relative to these 'points in the sky' is critical. One solution has been the development of mnemonic devices that facilitate relatively quick and easy SWAT level designations (Papa and Stoliker, 1987). In this instance, the mnemonic was a highly succinct variation on the level wording. Therefore, for the time dimension the level choices were: 1—spare, 2—some spare, and 3—no spare. This abbreviation can be practised prior to the flight and, if feasible, can be included on the test cards or map route. While including this technique may not be appropriate in some flight test scenarios, it has proved valuable during flight test simulation, serving as a useful reminder for the crews.

Crew member input should be solicited during the selection of SWAT data collection points. This input will serve to select SWAT points that encompass all tasks in that segment of flight, such as refuelling, and contribute to the refinement of points that minimize intrusiveness and interference with the task. One of the major drawbacks of SWAT, like many other subjective techniques, is that it is generally limited to discrete events and does not provide a continuous measure of workload. Therefore, the mental workload practitioner must be able to establish and clearly define for the subject the boundaries of the task of interest, to ensure that the SWAT score basis is comparable within and between subjects. Field results have demonstrated that some subjects tend to provide SWAT ratings that are inconsistent relative to the aspect of the task being investigated (Acton and Rokicki, 1986).

Some subjects have a tendency to report scores that are equivalent to an overall mission rating as opposed to a task-specific rating, whereas other subjects may be inclined to report a score that is indicative of 'peak' experience that occurred during the test of interest, as opposed to the overall task rating. This variability in response has to be anticipated and carefully controlled with explicit instructions as to what it is about the task that the subjects are to actually evaluate (i.e., peak load, overall tasks, mission load, or task/mission element-specific load). If there is considerable variability in the phase of the tasks that the subjects have evaluated, then the SWAT results are confounded so that the workload assessment based on this measurement technique alone is inappropriate. It is especially critical to define adequately the primary focus of continuous tasks that are to be evaluated by the subject so as to minimize this type of confusion. With discrete tasks, it is important that extraneous variables are minimized or controlled so that the workload comparisons are appropriate. While additional training may reduce or minimize errors and perceptual awareness is important, subject acceptance of the SWAT approach is also critical.

Subject acceptance of any mental workload measurement technique is not to be assumed a priori. Acceptance has to be developed and nurtured carefully. If the subjects are familiar with another scale (for instance, the Cooper-Harper Handling qualities rating scale, Cooper and Harper, 1969), getting them to accept SWAT may be most challenging, especially since it appears to present them with unfamiliar and sometimes difficult tasking (in terms of the card sort). SWAT training is critical in overcoming and minimizing the effect of these problems. If workload studies are to be completed in an integrated fashion over the course of a large-scale development programme, it is important that the same measures of mental workload be employed from programme inception to completion of testing. Far too often, personnel and organizational changes result in different measurement techniques being used over the course of the programme. Using a common set of techniques will facilitate user acceptance and familiarity with the techniques selected. This may also allow the researcher to compare results found in the initial design to those found during operational tests.

The initial card sort task requires a considerable amount of cognitive ability (Boyd, 1983) on the part of the subjects as they make relative comparisons between the phrases on the cards and their recalled representative mental workload situations. Normally, in an applied setting, if the practitioner has adequately briefed the subjects on SWAT theory and method the subjects take approximately 45 minutes to complete the card sort procedure. As stated earlier, one of the shortcomings of many workload studies in the field is the limited sample size, which can lead to additional problems if the group sort does not result in a Kendall's Coefficient of Concordance greater than 0·75. If the statistic is greater than 0·75, then the group data are said to be sufficiently homogenous and a combined interval scale can be used during subsequent SWAT event score analysis. If the resultant Kendall's is not at or above the 0·75 value, then it is suggested that the data grouping needs to be further defined since an excessive amount of intersubject variance exists. This partitioning of subjects into smaller and smaller subgroups, in order to maximize the Kendall's coefficient reduces the ability of the mental workload practitioner to make global comparisons of mental workload across subjects and task events.

SWAT has been tested extensively and used in a multitude of applied settings (Reid and Nygren, 1988). Most of these applications have been within aerospace environments such as development test and evaluation and operational test and evaluation. SWAT has also been translated into German and is used to assess aircrew workload in that country (Winter, 1988). It can be assumed that SWAT has been employed in these settings for a variety of reasons. Ease of administration, face validity of opinion data, the rigorousness of SWAT development and its purported reliability and sensitivity (construct validity) are suggested to be among the key reasons.

Subject characteristics

The scale development phase occurs when a subject takes the 27 card combinations of dimensions and levels and orders them in ascending order from lowest to highest workload. This particular task is not easy and to be effective often requires considerable explanation on the part of the workload practitioner. Some researchers (Hill *et al.*, 1988; Schlegel, 1986) have found that, regardless of the amount and type of explanation, some individuals do not seem capable of providing an acceptable SWAT card sort.

Another problem occasionally encountered appears to be that subjects are unable to recall some perceived workload event from their past that would suggest a certain ordering of the cards. When this occurs they have been encouraged to place the cards in an order to resemble their predicted workload. Therefore, the assumption is that people can accurately predict the amount of mental workload they would experience under various levels of the three dimensions. This assumption requires a considerable leap of faith.

It is also conceivable that some subjects may not be capable of performing the card sort because of verbal or cognitive limitations. According to Ericcson and Simon (1980), subjective ratings, as a form of verbal data, are limited to the extent that subjects can recall and report information that is either in short-term memory at the time of recall or retrievable from long-term memory. It follows then that unattended information or information based on cognitive processes that appears to bypass short-term memory (i.e., automatic processes) cannot be rated accurately (Damos, 1988).

Another problem is that some mental workload researchers may not be effective in allaying the concerns or difficulties that some subjects have with this particular procedure (Acton and Rokicki, 1986). Acton and Rokicki also found that some subjects approach this procedure in a very casual manner, as if lacking motivation to perform the task appropriately. They believe this to be the case especially for subjects they suspect are mentally capable of performing the card sort solution.

Finally, as stated earlier, the card sort procedure is fairly time-consuming and can be quite tedious. To find a solution to this problem, Biers and McInerney (1988) investigated the possibility of using SWAT to measure workload effectively without the requirement of having the subjects perform the card sort. A SWAT interval score could be obtained by forming a composite workload measure of the three rating scales based upon either simple sum or multivariate statistics. They admitted that this adaptation is not as psychometrically sound.

As a result of these known problems the scale development phase, with concomitant card sort, would have to be very carefully briefed and managed in order to maximize the opportunities for successful accomplishment. The workload practitioner is advised to monitor closely the performance of the subjects during the card sort. In addition they must encourage questions to facilitate the subjects' performance of this task. Even so, the literature and experience of applied human factors specialists suggest that not all people will be capable of performing this task adequately.

Missed data points

One of the major issues when deciding to utilize SWAT or any subjective rating measure is to decide what to do with SWAT points missed during the evaluation study. Two major approaches have been suggested: either to assume the workload was too great to provide a SWAT rating and, therefore, assign the maximum score; or to provide the score for the segment after the study. In general, differences are noted between subjects in delayed versus immediate rating. Recent flight test and simulator study data indicate a lowering of rated workload with delay. Delays may be of unknown length when operational priorities dictate ratings being part of a standard post flight questionnaire. Another problem is often that each crew member does not experience the same scenarios, and, in many cases, certain segments of flight are only seen by certain crews. It is not unrealistic to expect that out of an eight-hour sortie only 3–4 segments of flight may be comprehensive and similar enough to be considered adequate SWAT points.

Critical SWAT values

One of the questions applied researchers have often asked of subjective measures in general, and SWAT values in particular is, 'What are the critical SWAT values?' Reid and Colle (1988) illustrate two major types of evaluation, system comparison and system acceptability. In absolute evaluations, a system must be tested to determine whether it will adequately perform a particular design goal when employed in normal operational environments and operated by people who are representative of the normal population of operators. The question is whether the workload associated with the system operation under normal and emergency conditions is acceptable. We need a specific workload value to use as a design target or at least a critical range of values that should not be exceeded. This area has not been well-addressed by research in mental workload.

Reid and Colle (1988) discuss the concept of 'redline' and state that this 'redline' is not a single point but rather is a range of values. One of the important factors influencing workload ratings is dwell time. Dwell time is the amount of time the operator or subject must maintain a certain level of workload. The redline concept proposed by Reid and Colle (1988) has a cautionary zone, which could be considered an average indication. However, the concept should not be used to evaluate individuals, but rather should be used as a general guideline not a fixed criterion and subjects should be carefully briefed regarding the frame of reference to be used when making their SWAT ratings.

In the case of SWAT, a critical value of 40, plus or minus 10, does a reasonable job of predicting where performance difficulties might be encountered. It is a guideline to direct further evaluations. The 30–50 range of SWAT values reveal conditions that warrant closer examination. Decisions may need to be made to redesign equipment or modify operational procedures. However, modifications such as these should be based on a more extensive evaluation of existing data (such as questionnaires, performance data and videotape content analysis). These data can be used to direct the formulation of questions for additional research before a final decision is made. For the redline concept proposed by Reid and Colle (1988), the basic philosophy is that if the redline is exceeded for a short period of time, there will probably not be any consequences. However, if the subjects operate within this range for an extended period of time the workload may be considered excessive. There is a range of scales associated with an increased probability and possible performance failure; thus, if the system tests demonstrate that operator workload was consistently or frequently within the redline, the system could be re-evaluated for design changes, procedure changes, crew composition modifications or additional operator training.

Another consideration within flight test programmes is the fact that crew members are continually entering or leaving the programme. Thus, for any particular evaluation sortie, crew members will vary in the amount of experience they will have with a particular technique. One of the most useful approaches is that during the kickoff meeting of an evaluation a refresher SWAT briefing is provided for all crew members. This briefing will refresh those crew members who have completed their base sort previously and provide necessary training for newer crew members.

When selecting data points for an evaluation, using the initial training sorties in a simulator is an ideal way to flush out the appropriate points. Initial data points can be selected and crew members can then aid in determining the appropriate point at which the data should be collected. In one training session, crew members were instructed to provide SWAT ratings verbally over the radio. They were also instructed during the initial training that providing the SWAT rating was not to interfere with performance of their tasks; if it did the subjects were to provide the SWAT ratings at their first opportunity, in order not to increase the crew members' workload and to provide the experimenters with information so they could best determine points during a mission when the intrusiveness of SWAT points would be minimized. Crew members were also advised to note any additional points during the mission when they felt a SWAT point would be warranted.

Until recently, limited practical information was available to the mental workload research practitioner but the *American National Standard Guide to Human Performance Measurements* (1992) is an excellent guide. An example of the information provided on workload is found in Table 25.2. This table describes questions that should be asked when workload is based on the effort the human operator must expend to complete the required tasks. Based on the task characteristics, representative questions and candidate measures are included.

In addition to each of the candidate workload measures, a general description, strengths and limitations of the measures, data requirements, thresholds and reference materials are provided for each human performance measure.

As many complex systems have evolved, the role of the operator has become that of a systems manager, thus, the mental workload researcher must remember that any one of the current measures of workload may not be sufficient to address concerns, such as information management and crew coordination, as evidenced by problems in many of the systems in operation today.

Table 25.2. Factors in workload assessment

Workload aspect	Representative questions	Candidate measures	Comments
Sustained workload	• What was the average overall workload? • How will various intensities of sustained workload affect performance	• Subjective rating scales • Overall performance on salient tasks	
Momentary workload	• What was the magnitude of workload during peak periods? • How was human performance affected during periods of high demand?	• Subjective rating scales • Secondary task performance	• Global measures of task performance are inappropriate
Reserve capacity	• What margin of full performance did this task require? • How much more can the operator handle effectively?	• Secondary tasks • Subjective rating scales	• Primary task measures are inappropriate

(Source: American National Standard Guide to Human Error Performance Measurements (1992, p. 35))

Acknowledgments

The authors would like to thank Dr Valerie J. Gawron, Principal Human Factors Engineer, Calspan Corporation, New York, and Ms. Regina M. Papa, Acting Chief of Human Systems Integration Flight, Edwards Air Force Base, California, for their suggestions and comments concerning subjective measures.

Chapter 26

The nature and measurement of work stress: theory and practice

Tom Cox and Amanda Griffiths

Introduction

Over the past two decades there has been an increasing belief in all sectors of employment that the experience of stress at work has undesirable consequences for individual health and safety. Such a belief has been reflected in public and media interest and in concern voiced by the trades unions, professional bodies and organizations such as the International Labour Office and the World Health Organization, and national bodies such as the UK Health and Safety Executive. Stress has now become a major issue in occupational health and safety.

This chapter answers a series of questions about the nature, recognition and measurement of stress at work. Together these questions represent the first steps in a systematic problem-solving approach to the management of stress at work (see Cox and Cox, 1993). The chapter begins by considering different approaches to the definition of stress and explores the adequacy of those approaches and their implications for recognition and measurement. It then offers a framework for measurement, based on the principle of triangulation and the concept of a stress process defined in terms of antecedent factors, the experience of stress and the correlates of that experience.

The definition of stress

The definition of stress is not simply a question of semantics: it is important that there is broad agreement on its nature. A lack of such agreement would seriously hamper the development of effective measurement systems and subsequent research. Given this, it is an unfortunate but popular misconception that there is little consensus on the definition of stress as a scientific concept or, worse, that stress is in some way undefinable and unmeasurable. Such belief belies a lack of knowledge of the relevant scientific literature.

It has been concluded in several different reviews of the relevant literature that there are essentially three different, but overlapping, approaches to the definition and study of stress (Appley and Trumbull, 1967; Cox, 1978; Cox and Mackay, 1981; Fletcher, 1988; Lazarus, 1966). Each has its own associated baggage of concepts, methods, theories and prejudices and can be seen as originating from: (1) engineering and ergonomics, (2) medicine and physiology, or (3) psychology, respectively.

The engineering approach conceptualizes occupational stress as an aversive or noxious characteristic of the work environment and as an environmental *cause* of strain. The physiological approach, on the other hand, defines stress in terms of the common physiological effects of a wide range of aversive or noxious stimuli and as a *response* to a threatening or damaging environment. The third approach—the psychological approach—conceptualizes work stress in terms of the dynamic interaction between individuals and their work environment. When studied, stress is either inferred from the existence of problematic person–environment interactions or measured in terms of the cognitive processes and emotional reactions which underpin those interactions. The engineering and physiological models are obvious among the earlier theories of stress, while the more psychological models characterize contemporary stress theory. These three approaches are described in more detail below.

Engineering approach

The engineering approach has treated stress as a *stimulus characteristic* of the person's environment and as an independent variable, usually conceived in terms of the load or level of demand placed on the individual, or some aversive (threatening) or noxious element of that environment (Cox, 1978; Cox and Mackay, 1981; Fletcher, 1988). Occupational stress is thus treated as a property of the work environment, and usually as an objectively measurable aspect of that environment. In 1947, Symonds wrote, in relation to psychological disorders in RAF flying personnel, that '. . . stress is that which happens *to* the man, not that which happens *in* him; it is a set of *causes* not a set of *symptoms*'. Somewhat later, Spielberger (1976) argued, in the same vein, that the term stress should refer to the objective characteristics of situations. According to this approach, stress was said to produce a *strain* reaction which although often reversible could, on occasions, prove to be irreversible and damaging (Cox and Mackay, 1981; Sutherland and Cooper, 1990). The concept of a stress threshold grew out of this way of thinking and individual differences in this threshold have been used, within this approach, to account for differences in stress resistance and vulnerability.

Physiological approach

The physiological approach to the definition and study of stress received its initial impetus from the work of Selye (1950). He defined stress as 'a state manifested by a specific syndrome which consists of all the non-specific changes within the biologic system' that occur when it is challenged by aversive or noxious stimuli. Stress is treated as a generalized and non-specific physiological response syndrome and as a dependent variable. For many years, the stress response was largely conceived in terms of the activation of two neuroendocrine systems, the anterior pituitary-adrenal cortical system and the sympathetic-adrenal medullary system (Cox and Cox, 1985; Cox *et al.*, 1983).

Selye (1950, 1956) argued that the physiological response was triphasic in nature involving an initial alarm stage (sympathetic–adrenal medullary activation) followed by a stage of resistance (adrenal cortical activation) giving way, under some circumstances, to a final stage of exhaustion (terminal reactivation of the sympathetic-adrenal medullary system and death). Repeated, intense or prolonged elicitation of this physiological response, it was suggested, increases the wear and tear on the body, and contributes to what Selye (1956) described as the 'diseases of adaptation'. This apparently paradoxical term arises from the contrast between the immediate and short–term advantages bestowed by physiological responses to stress (energy mobilization for an active behavioural response) and their long–term disadvantages (increased risk of certain 'stress-related' diseases).

Criticisms of engineering and physiological approaches

Three sets of criticisms of these approaches have been offered.

First, engineering and physiological models do not adequately account for the existing data. Consider, for example, the effects of noise on performance and comfort. The engineering model would seem to suggest that exposure to loud noise as stress would cause, in a direct and fairly simple manner, impairment of performance, discomfort and possibly ill health. However, the effects of noise on task performance, for example, are not a simple function of its loudness or frequency but are subject to its nature (type of noise) and to individual differences and situational effects (Cox, 1978; Melamed *et al.*, 1993). Noise levels which are normally disruptive may help maintain task performance when subjects are tired or fatigued (Broadbent, 1971) while even higher levels of music may be freely chosen in social and leisure situations (see chapter 17).

In the same vein, Scott and Howard (1970) have written '. . . certain stimuli, by virtue of their unique meaning to particular individuals, may prove problems only to them; other stimuli, by virtue of their commonly shared meaning, are likely to prove problems to a larger number of persons'. This statement implies the mediation of strong cognitive as well as situational (context) factors in the overall stress process (see below). This point has been forcefully made by Douglas (1992) with respect to the perception of risk. Such perceptions and related behaviour, she has argued, are not adequately explained by the natural science of objective risk and are strongly determined by group and cultural biases.

The simple equating of demands—such as loud noise—with stress has been associated with the belief that a certain amount of 'stress' is associated with maximal performance (Welford, 1973) and possibly good health. This erroneous belief in optimal levels of stress has been used on occasions to justify poor management practices.

The physiological model is equally open to criticism. Both the non-specificity and the time course of the physiological response to aversive and noxious stimuli have been shown to be different from those described by Selye (1950, 1956) and required by the model (see Mason, 1968, 1971). Mason (1971), for example, has shown that some noxious physical stimuli do not produce the stress response in its entirety. In particular, he has cited the effects of heat. Furthermore, Lacey (1967) has argued that the low correlations observed among different physiological components of the stress response are not consistent with the notion of an identifiable response syndrome. There is also a difficulty in distinguishing between those physiological changes which represent stress and those which do not, particularly as the former may be dissociated in time from the stressor.

There is now much research which suggests that if the stress response syndrome exists it is not non-specific. There are subtle but important differences in the overall pattern of response. There is evidence, for example, of differentiation in the response of the catecholamines (reflecting sympathetic-adrenal medullary activation) to stressful situations (Cox and Cox, 1985). Several dimensions have been suggested as a basis of this differentiation but most relate to the expenditure of different types of effort, for example, physical versus psychological (Dimsdale and Moss, 1980a, 1980b; Cox *et al.*, 1985). Dimsdale and Moss (1980b) studied plasma catecholamine levels using a non-obtrusive blood withdraw pump and radioenzymatic assay. They examined 10 young physicians engaged in public speaking, and found that although levels of both adrenaline and noradrenaline increased under this set of demands, the levels of adrenaline were far more sensitive. This sensitivity was associated with feelings of emotional arousal which accompanied the public speaking. Cox and her colleagues (1985) examined the physiological response to three different types of task associated with short cycle repetitive work: urinary catecholamine excretion rates were measured using an adaptation of Diament and Byers (1975) assay technique. She found that both adrenaline and noradrenaline were sensitive to work characteristics, such as pay scheme and pacing, but differentially so. It was suggested that noradrenaline activation was related to the physical activity inherent in the various tasks, and to the constraints and frustrations present, while adrenaline activation was more related to feelings of effort and being stressed.

The second criticism of the engineering and physiological models of stress is that they are conceptually dated, being set within a relatively simple stimulus-response paradigm. As a result, they largely ignore individual differences of a psychological nature together with the perceptual and cognitive processes which might underpin them (Sutherland and Cooper, 1990). These models treat the person as a passive vehicle for translating the stimulus characteristics of the environment into psychological and physiological responses. They largely ignore the interactions between individuals and their environments which are an essential part of systems-based approaches to biology, behaviour and psychology. Further, they ignore the psychosocial and organizational contexts to work stress (see Cox and Cox, 1993).

The third criticism is focused on the physiological model. The ideas promoted by Selye and his more recent advocates have imposed false expectations on the nature of the relationships between different markers of stress, and very much dictated a 'medicalization' of the subject area. This has been associated with an attendant narrowing of focus within the practical field of stress management. This approach encourages strategies which concentrate on individuals and their responses to stress independent of the organizational context within which the problem occurs. Partly as a result of this, we have witnessed the development of 'band aid' solutions (relaxation, for example) to stress problems in the workplace. Such views also tend to encourage the attribution of responsibility for 'breakdown' to the individual. Attempts to rework the physiological approach have not resulted in a more coherent or clearer account of stress research or stress management. The development of notions of 'good' stress (eustress) and 'bad' stress (distress) have not resolved any of the existing problems, but have added to the semantic confusion. They have further encouraged the misguided view that 'some stress is good for you'.

Psychological approach

The third approach to the question of definition is a psychological one, and offers what is essentially a cognitive model of stress. In doing so, it attempts to overcome some of the criticisms that the other two models have attracted, and there is now some consensus developing around the adequacy of this psychological approach to stress.

Variants of this psychological approach dominate contemporary stress theory, and among them two distinct types can be identified: the interactional and the transactional. The former focuses on the structural features of individuals' interactions with their work environment, while the latter is more concerned with the psychological processes underpinning those interactions. Transactional models are primarily concerned with cognitive appraisal and coping. In a sense they represent a development of the interactional models, and are largely consistent with them. This chapter adopts a transactional approach but integrates the most useful aspects of more interactional theories. It proposes the unifying concept of a *stress process*, beginning with (1) its antecedent factors and (2) the cognitive perceptual process which gives rise to the emotional experience of stress; it then considers (3) the correlates of that experience.

Stress as a psychological phenomenon

Stress has been defined as a cognitive state (Cox, 1985a) which is part of a wider process reflecting the person's perception of and adaptation to the demands of the (work) environment. This approach emphasizes the person's cognitive appraisal (Lazarus, 1966) of the situation, and treats the whole process of perceiving and reacting to stressful situations within a problem solving context (Cox, 1987). Stress is not treated as a dimension of the physical or psychosocial environment: it cannot be defined simply in terms of workload or the occurrence of events determined by consensus to be stressful. Equally, it cannot be defined in terms of responses that are sometimes correlates of stress, such as physiological mobilization or performance dysfunction. Framed in this way, the study of stress is about normal people coping with (or failing to cope with) the problems that face them.

A central feature of such approaches to stress is the process of cognitive appraisal (Lazarus, 1966), which Holroyd and Lazarus (1982) have defined in terms of being 'the evaluative process that imbues a situational encounter with meaning'. Later refinements of the theory (Folkman *et al.*, 1980) suggest both primary and secondary components to the appraisal process. Primary appraisal involves a continual monitoring of the person's transactions with his or her environment, focusing on the question, 'Do I have a problem?'. Secondary appraisal is contingent upon the recognition that a problem exists and involves a more detailed analysis and the generation of possible coping strategies (see below). It has been suggested (Cox, 1987) that this process of appraisal involves a continual monitoring of at least four aspects of individuals' transactions with their environment, and a continual evaluation of the balance between them (Cox, 1987). Cognitive appraisal appears to take account of individuals' perceptions of:

1. the demands on them;
2. the constraints under which they have to cope;
3. the support they receive from others in coping; and
4. their personal characteristics and coping resources—their knowledge, attitudes, behavioural and cognitive skills, and behavioural style.

Demands

Demands are requests for action or adjustment, whether cognitive, emotional, behavioural or physiological in nature. They require some degree of decision making and the exercise of skill. They may be imposed by the external environment, say as a function of work or the work–home interface, or may be internal, reflecting the person's needs: material, social or psychological. There may be several important dimensions describing external (job) demands, for example, 'pleasantness/ unpleasantness' and 'ease/difficulty' (see Cox, 1985b). Demands usually have a time base, the effects of which may be amplified by an acute sense of time urgency (type A behaviour: Zyzanski and Jenkins, 1970).

The absolute level of demand would not appear to be the important factor in determining the experience of stress. More important is any discrepancy which exists between the perceived level of demand and individuals' perceptions of their ability to meet that demand. The size of this discrepancy appears to be an important determining factor in the stress process. However, the relationship between the discrepancy and the intensity of the stress experience may be curvilinear rather than linear (Cox, 1978). Within reasonable limits stress can arise through either overload (demand > abilities) or through underload (demand < abilities), or through some combination of the two. A person's ability to cope with such an imbalance may be constrained or supported in different ways and to different extents.

Constraints

Constraints operate as restrictions or limitations on free action or thought, reflecting a loss or lack of discretion and control over actions. These may be imposed externally. For example, constraints may be imposed by the requirements of specific jobs or by the rules of the organization. They may also be role related or reflect the beliefs and values of the individual.

There is some discussion in the literature as to whether the effects of job demands and constraints (job discretion) are additive or subject to a true interaction. Karasek (for example, Karasek, 1979; Karasek et al., 1981) has argued that a true interaction exists, while others, for example Warr (1990), have found evidence only for a simple additive effect. Despite this, the notion that high job demands and constraints create a disproportionately risky condition for worker health remains.

Support

Support can be made available in different ways, most essentially through 'social interaction': through advice and information, through practical assistance or by providing understanding and declaring empathy. It is possible that women need and are more sensitive to social support than men (Cox et al., 1983, 1984).

Coping resources and personal characteristics

There are many different ways in which a person's coping resources might be conceptualized; however, it could be useful to think of them in terms of energy, knowledge, attitudes, behavioural style (or personality) and skills. The idea of skill has to be extended beyond traditional conceptualizations in terms of psychomotor and technical skills, to include social and cognitive skills.

The attitudes and behavioural style as well as personal knowledge and skills can be developed both through formal education and training and more informally through untutored experience. Furthermore, several elements of this 'package' of resources are subject to change, influenced by factors such as time of day, fatigue, and state of health. In addition to any consideration of its cognitive and perceptual elements, the state of stress is often defined by the person's experience of negative emotion, unpleasantness and general discomfort.

Antecedent conditions: work hazards

According to contemporary stress theory, as described above, situations which are experienced as stressful may have at least three characteristics:

1. Individuals are faced with demands and pressures which are not matched to personal characteristics and resources and therefore have difficulty in coping.
2. Individuals are constrained in the way they carry out their work and cope with its demands: they have little control over their work or how they cope with it.
3. Individuals are relatively isolated and receive little support from colleagues, supervisors, friends or family.

These three characteristics of stressful situations may be expressed and combined in different ways in the workplace and may reflect exposure to various hazards (Cox and Cox, 1993). Work hazards, aspects of work which have been shown to carry the potential for harm, can be broadly divided into the *physical*, which include the biological, biomechanical, chemical and radiological, and the *psychosocial*. The latter, although perhaps initially a difficult concept to grasp, concern aspects of work such as organizational function and culture, career development, decision latitude/control, role in the organization, interpersonal relationships, work–home interface, task design, workload, workpace and work schedule. Such hazards might be usefully conceived of as relating to the *context* or the *content* of work (Hacker, 1991). Thus, stress may be experienced as a result of exposure to a wide range of work hazards and, in turn, contribute to an equally wide range of health outcomes (Cox, 1993): it is one link (see Figure 26.1) between hazards and health (see also the section on 'Stress and health' below).

Coping

An awareness that an unmanageable problem exists together with an associated negative emotional experience (see the Stress Arousal Checklist below), normally initiate a cycle of changes in individuals' perceptions and cognitions (secondary appraisal), and in their behavioural and physiological function.

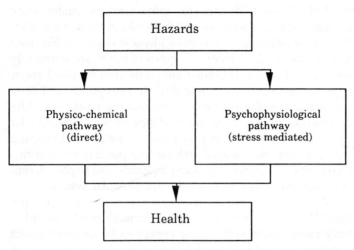

Figure 26.1. Pathways from hazards to health

Some of these changes are attempts at mastering the problem or attenuating the experience of stress, and have been termed 'coping' by Lazarus (1966). Coping usually represents either an adjustment *to* the situation or an adjustment *of* the situation. Elsewhere, the process of coping with stress has been described within a 'problem solving' framework (Cox, 1987).

Physiological changes

In addition to these psychological responses to stress, there may be significant changes in physiological function, some of which might facilitate coping, at least in the short term, but in the longer term may threaten physical health.

These physiological correlates of stress have been studied largely in terms of the activities of two neuroendocrine systems: the sympathetic-adrenal medullary system, and the pituitary–adrenal cortical system (see above). More recently, interest has also focused on the immune system, and its sensitivity to the experience of stress. For many people, and on many occasions, such changes in the function of these systems do not have implications for pathology. However, for some that experience can be of aetiological or prognostic significance.

Stress and health

Many of the psychological and physiological changes associated with the experience of stress may represent a modest dysfunction and possibly some associated discomfort. Many are easily reversible although still damaging to the quality of life at the time. However, for some workers, under some circumstances, they might translate into poor performance at work, into other psychological and social problems and into poor physical health. It has been suggested that under certain circumstances, *all* aspects of health are potentially susceptible to stress effects. If this is true then questions must be asked about which are the more susceptible or the most directly susceptible, and how that susceptibility is affected by the nature of work and the workplace. The more susceptible conditions appear to be those relating to the cardiovascular and respiratory systems (for example, coronary heart disease and asthma), the immune system (for example, rheumatoid arthritis and possibly some forms of cancer), the gastro-intestinal system (for example, gastric and peptic ulcers), and those relating to the endocrine, autonomic and muscular systems.

If significant numbers of workers are experiencing and expressing the effects of stress, then the organization itself can be deemed to be unhealthy. The most frequently cited effects which may appear to be of more direct concern to organizations are reduced availability for work, involving high turnover, absenteeism and poor time keeping, impaired work performance and productivity, an increase in client complaints and an increase in employee compensation claims.

The measurement of stress

It has been suggested that the available evidence supports a psychological approach to the definition of stress, and that transactional models are among the most adequate and useful of those currently available. Within this framework, stress is defined as a psychological state which is part of and reflects a wider process of interaction between individuals and their (work) environment. Part of that process is the sequence of relationships between the objective work environment and the worker's perceptions, between those perceptions and the experience of stress, and between that experience and changes in behaviour, physiological function and health. This sequence provides a basis for measurement in which individual perceptions and cognitions are critical. The measurement of the stress state should be based primarily on self–report measures which focus on the appraisal process and on the emotional experience of stress (Cox, 1985a). Measures relating to appraisal need to consider workers' perceptions of the demands on them, their ability to cope with those demands, their needs and the extent to which they are fulfilled by work, the control they have over work and the support they receive in relation to work. Dewe (1991) has argued that it is necessary to go beyond simply asking workers whether particular demands are present (or absent) in their work environments and measure various dimensions of demand such as frequency, duration and level. Furthermore, such measures need to be used in a way which allows for the possibility of interactions between perceptions, such as demand with control (Karasek, 1979; Warr, 1990) or demand

and·control with support (Cox, 1985a; Karasek and Theorell, 1990; Payne and Fletcher, 1983). The importance to the worker of coping with particular combinations and expressions of these work characteristics needs also to be taken into account (Sells, 1970; Cox, 1978).

Triangulation

Despite their obvious centrality and importance, self-report measures of appraisal and the emotional experience of stress are, on their own, insufficient. While their reliability can be established in terms of their internal structure or performance over time without reference to other data, their validity cannot. Therefore data from other domains are required, and triangulation* is a recommended process for both reliability testing and/or validation in many different disciplines (we see the same need in the assessment of mental work load—see chapter 25). Applying this principle would require data to be collected from, at least, three different domains. This can be achieved by considering evidence relating to (1) the objective and subjective antecedents of the stress experience, (2) self-report on the stress state, and (3) the various changes in behaviour, physiology and health status which may be correlated with the antecedents. The influence of moderating factors, such as individual and group differences, may also be assessed.

Bailey and Bhagat (1987) have recommended a multi-method approach to the measurement of stress which is consistent with the concept of triangulation. They have argued in favour of balancing the evidence from self-report, physiological and unobtrusive measures. Their unobtrusive measures relate to what Folger and Belew (1985) and Webb *et al.* (1966) have called non-reactive measures, and include: physical traces (such as poor housekeeping), archival data (such as that on absenteeism), private records (such as diaries), and nonintrusive observation and recordings. Bailey and Bhagat (1987) have also pointed out that obtrusive measures often change the very nature of the behaviour or other response being assessed (see chapter 2).

Triangulation would require evidence drawn from the three domains with (1) an audit of the work environment (including both its physical and its psychosocial aspects), (2) a survey of workers' perceptions of and reactions to work, and (3) the measurement of workers' behaviour in respect to work, and their physiological and health status. It is not possible here to offer a comprehensive review of the plethora of measures which might be used in such audits and surveys. However, some discussion is devoted to two self-report instruments, developed at Nottingham, which have proved useful in the study of occupational stress—the Stress Arousal Checklist and the General Well-being Questionnaire (see below).

---

*The concept of triangulation in measurement relates to the strategy of fixing a particular position or finding by examining it from, at least three, different points of view. The degree of agreement between those different points of view gives some indication of the reliability of the finding, and/or, depending on the measures used, its concurrent validity (see chapter 1).

The use of any measure must be supported by data relating to its reliability and validity, and its appropriateness and fairness in the situation in which it is being used (Dane, 1990). The provision of such data would conform to good practice in both occupational psychology and psychometrics but may also be required if any subsequent decisions are challenged in law. Preferably such data collection would take the form of continuous monitoring and thus be capable of mapping work-related changes in all three domains. Ideally, the principle of triangulation should be applied both within and between domains. This should help overcome the problem of missing data and help resolve inconsistencies in the data given that these are not extreme. Its use between domains has been briefly discussed above. Within domains, several different measures should be taken and preferably across different measurement modalities to avoid problems of common method variance. This may be most relevant and easiest to achieve in relation to the measurement of changes in the third domain: behaviour, physiology and health status. There are no available studies to suggest that the various measures from the different domains can be statistically combined into a single and defensible 'stress index'.

What is being measured is a process: 'antecedents—perceptions and experience (and moderating factors)—correlates and effects'. This can be simplified conceptually to 'work hazards—stress—harm'. This approach underlines both the complexity of measurement, when approached scientifically, and the inadequacy of asking for or using single one-off measures of *stress* (however defined).

The Stress Arousal Checklist (SACL)

The measurement of mood may offer one direct method of tapping the individual's experience of stress, and there was a resurgence of interest in this issue in the 1980s as witnessed by a series of articles in the *British Journal of Psychology* (Cox and Mackay, 1985; Cruickshank, 1984; King *et al.*, 1983) and elsewhere (Burrows *et al.*, 1977; Ray and Fitzgibbon, 1981; Russell, 1979, 1980; Watts *et al.*, 1983). Most of these studies have employed the 'Stress Arousal Checklist' (SACL) developed by Cox and Mackay and originally published in the *British Journal of Social and Clinical Psychology* (Mackay *et al.*, 1978).

This adjective checklist (SACL) was developed at Nottingham, using factor analytical techniques (Cox and Mackay, 1985; Gotts and Cox, 1990; Mackay *et al.*, 1978) for the measurement of self-reported mood. It presents the respondent with 30 relatively common mood describing adjectives, and asks to what extent they describe their current feelings. The model of mood which underpins the checklist is two dimensional. One dimension appears to relate to feelings of unpleasantness/pleasantness or hedonic tone (stress) and the other to wakefulness/drowsiness or vigour (arousal). Such a model is well represented in the relevant psychological and psychophysiological literature (see, for example, Mackay, 1980; Russell, 1980).

It was suggested by Mackay *et al.* (1978) that the stress dimension may reflect the perceived favourability of the external environment, and thus have a strong cognitive component in its determination. This view is consistent with the psychological model of stress. Arousal, it was suggested, might relate to ongoing autonomic and somatic activity, and be essentially psychophysiological in nature. It has now become obvious that stress may partly reflect how appropriate the level of arousal is for a given situation, and the effort of compensating for inappropriate levels (Cox *et al.*, 1982).

The split half reliability coefficients for the two scales which tap into these dimensions have proved acceptable: arousal 0·82 and stress 0·80 (Watts *et al.*, 1983; Gotts and Cox, 1990). Both were conceived of and developed as state measures, and are thus seen as transient in nature. Together the two dimensions can be used to describe a four quadrant model of mood within which characteristic emotions and related states may be identified: high arousal and high stress (anxiety), high arousal and low stress (pleasant excitement), low arousal and high stress (boredom), and finally, low arousal and low stress (relaxed drowsiness). There are now many reported studies using the SACL, and reporting data from its two scales (Burrows *et al.*, 1977; Cox *et al.*, 1982; King *et al.*, 1983; Ray and Fitzgibbon, 1981; Watts *et al.*, 1983). There have also been a number of studies which have used modified versions of the checklist (Cruickshank, 1982, 1984), although locally inspired changes in the instrument cannot always be defended (Cox and Mackay, 1985).

A third scale has been suggested based on the use of a '?' category on the response scale associated with the different mood adjectives. This category signifies, in part, uncertainty about whether the adjective given currently describes the respondent's mood. A score based on the frequency of '?' responses might reflect an inability to report feelings, and this may be symptomatic of a disordered psychophysiological state. Such a scale has an acceptable split half reliability coefficient: 0·89 (Cox and Mackay, 1985).

A recent compilation of the available British and Australian data has allowed the publication of mean levels for different groups, broken down by country of origin, age, sex and occupation (Gotts and Cox, 1990). Some of these 'normative' data are presented in Table 26.1.

The nature and measurement of health

The definition of health has been no less a subject for debate than that of stress, although the broad view espoused by the World Health Organisation (WHO, 1946) is often cited as a starting point for further discussion and development. In its constitution, the WHO offered a dynamic and positive definition of health in terms of psychological and social as well as physical well-being, and emphasized that it is both dynamic and changeable (like the

Table 26.1. Some normative data for the SACL (derived from Gotts and Cox, 1990*)

Sample	Dichotomized Scores						Q		
	Stress			Arousal					
	x	SD	n	x	SD	n	x	SD	n
Mixed population	6·0	4·6	1027	6·4	3·2	1040	4·2	4·2	1079
Males: mixed sample	6·0	4·7	296	6·6	3·2	297	4·9	4·6	266
Females: mixed sample	6·0	4·6	731	6·3	3·3	743	3·9	4·1	584
Students	6·3	4·9	515	5·7	3·6	518	4·7	4·1	535
Ages 16 to 30	6·2	4·6	466	6·0	3·2	469	5·0	4·3	379
Ages 31 to 45	5·9	4·9	344	7·2	3·3	353	3·5	4·2	334
Age more than 45	5·1	4·2	122	6·4	3·3	123	3·7	4·1	1132

* More complete normative data have been published as part of a manual for the SACL (Gotts and Cox, 1990). Further information can be obtained from the authors.

weather). Somewhat later, Rogers (1960) proposed that the health state was a function both of the individual's heredity, and the accumulated and current effects of the person's environment as 'they act upon the psyche and body'. This offered an alternative approach to the traditional medical emphasis on constitutional and genetic factors. Rogers (1960) also suggested that health might be usefully viewed as a continuum, the opposing poles of which are 'complete well-being' and 'death', with a significant watershed existing at the point where the person is recognized as being obviously ill or injured. Accepting this simple model immediately highlights the area between complete well-being and obvious illness; an area which Rogers (1960) referred to as sub-optimum health, and the advertising world of the early 1960s discussed in terms of being 'one degree under'.

For most people, the day to day variation in their state of health occurs within this 'grey' area of sub-optimum health, and therefore it may be changes in sub-optimum health which reflect similar variation in the impact of the environment on the person, and any subsequent experience of stress.

The General Well-being Questionnaire

There are several different questionnaire instruments which have been used to tap into subjects' sub-optimum health, and which by the nature of their scales and internal structure offer some description of that area of health (Crown and Crisp, 1966; Derogatis *et al.*, 1974; Goldberg, 1972; Gurin *et al.*, 1960).

Studies at Nottingham have also attempted to map sub-optimum health using self-reported symptoms of general malaise (Cox and Brockley, 1984; Cox *et al.*, 1983, 1984). Initially a compilation of general non-specific symptoms of ill health was produced from existing health questionnaires (see above) and from diagnostic texts. These symptoms included reportable aspects of cognitive, emotional, behavioural and physiological function, none of which were clinically significant in themselves. From this compilation, a prototype checklist was designed with each symptom being associated with a five point frequency scale ('never' through to 'always') which referred to a six month response window. In a series of classical factor analytical studies, on British subjects, now variously reported (Cox *et al.*, 1983, 1984), two clusters of symptoms or factors were identified (see Table 26.2). These factors were derived orthogonal. The first factor (GWF1) was defined by symptoms relating to tiredness, emotional lability, and cognitive confusion; it was colloquially termed 'worn out'. The more cognitive items would appear to imply difficulties in decision making (in the specific context of feeling 'worn out'): (a) Has your thinking got mixed up when you have had to do things quickly? (b) Has it been hard for you to make up your mind? and (c) Have you been forgetful? These may have implications for personal problem solving and coping (see later; also Cox, 1987). The second factor (GWF2) was defined by symptoms relating to worry and fear, tension and physical signs of anxiety; it was colloquially termed 'up tight and tense'. This model of sub-optimum health appeared to have some face validity in that it was accept-

Table 26.2. Items defining the GWBQ Scales (international version)

GWF1

    Have your feelings been hurt easily?
    Have you got tired easily?
    Have you become annoyed and irritated easily?
    Has your thinking got mixed up when you have had to do things quickly?
    Have you done things on impulse?
    Have things tended to get on your nerves and wear you out?
    Has it been hard for you to make up your mind?
    Have you got bored easily?
    Have you been forgetful?
    Have you had to clear your throat?
    Has your face got flushed?
    Have you had difficulty in falling or staying asleep?

GWF2

    Have you worn yourself out worrying about your health?
    Have you been tense and jittery?
    Have you been troubled by stammering?
    Have you had pains in the heart or chest?
    Have unfamiliar people or places made you afraid?
    Have you been scared when alone?
    Have you been bothered by thumping of the heart?
    Have people considered you to be a nervous person?
    When you have been upset or excited has your skin broken out in a rash?
    Have you shaken or trembled?
    Have you experienced loss of sexual interest or pleasure?
    Have you had numbness or tingling in your arms or legs?

---

able to a conference audience of British general practitioners and medical and psychological researchers (see Cox *et al.*, 1983).

The General Well-being Questionnaire (GWBQ) was derived from this factor model, and has now been used in a number of studies conducted at Nottingham, and elsewhere by other researchers.

Recently, new data have been collected by Cox and Gotts through a series of linked studies in Britain and Australia. These data have recently been re-analyzed and the model and its associated scales have been slightly amended to increase their robustness in relation to this international sample (and also to diverse homogenous samples). A number of symptoms have now been deleted from the two original scales, although no new symptoms have been added. The two new 'international' scales are now defined by twelve symptoms (see Table 26.2), but retain their essential nature: 'worn out' and 'up tight and tense'. The deleted symptoms were among the weaker ones in terms of scale definition (and item loadings). New norms have been computed for the 'international' scales; the conversion weights for transforming the old scores into new international scale scores are also to be published. Table 26.3 presents some of these new data. It is thus suggested that sub–optimum health, the 'grey area' between complete well-being and obvious illness, is made up of two states, one related to being 'worn out', and the other related to being 'up tight and tense'. The former has an interesting cognitive component,

Table 26.3. Some normative data for the GWBQ (derived from unpublished data of Cox and Gotts*) for mixed populations

	INTERNATIONAL VERSION (1987)					
	'Worn out' (12 Items)			'Up tight' (12 Items)		
Sample	x	S.D.	n	x	S.D.	n
All	16·7	8·3	2300	10·7	7·4	2312
Males	15·9	7·8	1031	8·2	6·5	1042
Females	17·4	8·6	1262	12·8	7·4	1262
British sample by age (years)						
16–20	16·5	8·7	141	11·5	7·9	141
21–25	16·9	9·2	147	11·3	7·6	147
26–30	15·6	8·4	236	10·2	7·5	236
31–35	17·2	8·6	239	9·0	6·5	239
36–40	16·1	8·1	201	9·2	7·5	201
41–45	15·5	8·6	199	10·4	7·7	199
46–50	16·0	8·3	175	9·7	7·7	175
51–55	14·5	8·0	174	9·1	7·4	174
56–60	13·7	8·0	127	7·7	6·6	127
> 60	13·5	6·4	26	4·8	5·8	26

* Further information can be obtained from the authors.

possibly related to decision making and coping, while the latter is partly defined by physical symptoms of anxiety and tension. It has been shown that people vary in the extent to which they report these feelings, both between individuals and across time, and it has been suggested that this variation may not only (a) reflect the experience of stress, but also (b) affect other responses to stress, such as self-reported mood (see Mackay *et al.*, 1978; Cox and Mackay, 1985).

Conclusion

Work stress is an important contemporary issue in occupational health, and there is a growing literature concerned with its nature and measurement. The current consensus defines stress as a psychological state with cognitive and emotional components, and acknowledges its effects on the health of both individual employees and their organizations. Furthermore, there are now theories which can be used to relate the experience and effects of work stress to exposure to work hazards and to the harmful effects on health that such exposure might cause. Applying such theories to the understanding of stress at work facilitates the measurement and management of work stress.

Despite such advances, more research and development is required in relation to the measurement of the experience of stress and the overall stress process. The inadequacy of single one-off measures is widely recognized in the literature but despite this they continue to be used, and across studies focused on different aspects of the stress process. This diversity may account for much of the disagreement within stress research. Part of the solution to this problem lies with agreeing the theoretical framework within which measurement is made, but part lies with the development of a more adequate technology of measurement based in 'good practice' in a number of areas including psychometrics, knowledge elicitation and knowledge modelling. A forced standardization of measurement is not being argued for here and should be resisted for its effects on scientific progress. What is being argued for throughout is better measurement processes, conforming to recognized good practice in relevant areas, and applied within a declared theoretical context.

References

Appley, M.H. and Trumbull, R. (1967). *Psychological Stress* (New York: Appleton-Century-Crofts).

Bailey, J.M. and Bhagat, R.S. (1987). Meaning and measurement of stressors in the work environment. In *Stress and Health: Issues in Research Methodology*, edited by S.V. Kasl and C.L. Cooper (Chichester: Wiley and Sons).

Broadbent, D.E. (1971). *Decision and Stress* (New York: Academic Press).

Burrows, G.C., Cox, T. and Simpson, G.C. (1977). The measurement of stress in a sales training situation. *Journal of Occupational Psychology*, **50**, 4–51.

Cox, S., Cox, T., Thirlaway, M. and Mackay, C.J. (1985). Effects of simulated repetitive work on urinary catecholamine excretion. *Ergonomics*, **25**, 1129–1141.

Cox, T. (1978). *Stress* (London: Macmillan).

Cox, T. (1985a). The nature and measurement of stress. *Ergonomics*, **28**, 1155–1163.

Cox, T. (1985b). Repetitive work: occupational stress and health. In *Job Stress and Blue Collar Work*, edited by C.L. Cooper and M. Smith (Chichester: Wiley and Sons).

Cox, T. (1987). Stress, coping and problem solving. *Work and Stress*, **1**, 5–14.

Cox, T. (1993). *Stress Research and Stress Management: Putting Theory to Work* (London: HMSO).

Cox, T. and Brockley, T. (1984). The experience and effects of stress in teachers. *British Educational Research Journal*, **10**, 83–87.

Cox, T. and Cox, S. (1985). The role of the adrenals in the psychophysiology of stress. In *Current Issues in Clinical Psychology*, edited by E. Karas (London: Plenum Press).

Cox, T. and Cox, S. (1993). Psychosocial and Organizational Hazards: Monitoring and Control. Occasional Series in Occupational Health, No. 5. World Health Organization (Europe), Copenhagen, Denmark.

Cox, T., Davis, A. and Beale, D. (1989). Repetitive VDU-based Work: Duration of Optimal Exposure and Organization of Rest breaks. Final Report to the Health & Safety Executive (Executive Summary). Centre for Organizational Health, Department of Psychology, University of Nottingham.

Cox, T. and Mackay, C.J. (1981). A transactional approach to occupational stress. In *Stress, Work Design and Productivity*, edited by N. Corlett and J. Richardson (Chichester: Wiley and Sons).

Cox, T. and Mackay, C.J. (1985). The measurement of self-reported 'stress and arousal. *British Journal of Psychology*, **76**, 183–186.

Cox, T., Thirlaway, M. and Cox, S. (1982). Repetitive work, well-being and arousal. In *Biological and Psychological Basis of Psychosomatic Disease*, edited by H. Ursin and R. Murison, *Advances in the Biosciences*, **42**, 115–135 (Oxford: Pergamon Press).

Cox, T., Thirlaway, M. and Cox, S. (1984). Occupational well-being: sex differences at work. *Ergonomics*, **27**, 499–510.

Cox, T., Thirlaway, M., Gotts, G. and Cox, S. (1983). The nature and assessment of general well-being. *Journal of Psychosomatic Research*, **27**, 353–359.

Crown, S. and Crisp, A.H. (1966). A short clinical diagnostic self-rating scale for psychoneurotic patients. The Middlesex Hospital Questionnaire (MHQ). *British Journal of Psychiatry*, **112**, 917–923.

Cruickshank, P.J. (1982). Patient stress and the computer in the waiting room. *Social Science and Medicine*, **16**, 1371–1376.

Cruickshank, P.J. (1984). A stress and arousal mood scale for low vocabulary subjects. *British Journal of Psychology*, **75**, 89–94.

Dane, F.C. (1990). *Research Methods* (Pacific Grove, CA: Brooks/Cole).

Derogatis, L.R., Lipman, R.S., Rickels, K., Uhlenhuth, E.H. and Convi, L. (1974). The Hopkins Symptom Checklist (HSCL). In *Modern Problems in Pharmacopsychiatry*, volume 7, edited by P. Pichot (Basel: Karger).

Dewe, P. (1991). Measuring work stressors: the role of frequency, duration and demand. *Work and Stress*, **5**, 77–91.

Diament, J. and Byers, S.O. (1975). A precise catecholamine assay for small samples. *Journal of Laboratory and Clinical Medicine*, **85**, 679–693.

Dimsdale, J.E. and Moss, J. (1980a). Plasma catecholamines in stress and exercise. *Journal of the American Medical Association*, **243**, 340–342.

Dimsdale, J.E. and Moss, J. (1980b). Short-term catecholamine response to psychological stress. *Psychosomatic Medicine*, **42**, 493–497.

Douglas, M. (1992). *Risk and Blame* (London: Routledge).

Fletcher, B.C. (1988). The epidemiology of occupational stress. In *Causes, Coping and Consequences of Stress at Work*, edited by C.L. Cooper and R. Payne (Chichester: Wiley and Sons).

Folger, R. and Belew, J. (1985). Non-reactive measurement: a focus for research on absenteeism and occupational stress. In *Organizational Behaviour*, edited by L.L. Cummings and B.M. Straw (Greenwich, CT: JAI Press).

Folkman, S., Schaefer, C. and Lazarus, R.S. (1980). Cognitive processes as mediators of stress and coping. In *Human Stress and Cognition*, edited by V. Hamilton and D.M. Warburton (Chichester: Wiley and Sons).

Goldberg, D.P. (1972). The Detection of Psychiatric Illness by Questionnaire. Maudsely Monograph No. 21, Oxford University Press, London.

Gotts, G. and Cox, T. (1990). *Stress and Arousal Checklist: A Manual for Its Administration, Scoring and Interpretation* (Melbourne, Australia: Swinburne Press).

Gurin, G., Veroff, J. and Feld, S. (1960). *Americans' View of Their Mental Health* (New York: Edinburgh).

Hacker, W. (1991). Objective work environment: analysis and evaluation of objective work characteristics. Paper presented to: A Healthier Work Environment: Basic Concepts & Methods of Measurement. Hogberga, Lidingo, Stockholm.

Holroyd, K.A. and Lazarus, R.S. (1982). Stress, coping and somatic adaptation. In *Handbook of Stress*, edited by L. Goldberger and S. Breznitz (New York: Free Press).

Karasek, R.A. (1979). Job demands, job decision latitude and mental strain: implications for job redesign. *Administrative Science Quarterly*, **24**, 285–308.

Karasek, R.A., Baker, D., Marxer, F., Ahlbom, A. and Theorell, T. (1981). Job decision latitude, job demands and cardiovascular disease. *American Journal of Public Health*, **71**, 694–705.

Karasek, R.A. and Theorell, T. (1990). *Healthy Work: Stress, Productivity and the Reconstruction of Working Life* (New York: Basic Books).

King, M.G., Burrows, G.D. and Stanley, G.V. (1983). Measurement of stress and arousal: validation of the stress arousal checklist. *British Journal of Psychology*, **74**, 473–479.

Lacey, J.I. (1967). Somatic response patterning and stress: some revisions of activation theory. In *Psychological Stress*, edited by M.H. Appley and R. Trumbull (New York: Appleton-Century-Crofts).

Lazarus, R.S. (1966). *Psychological Stress and the Coping Process.* (New York: McGraw-Hill).

Mackay, C.J. (1980). The measurement of mood and psychophysiological activity using self-report techniques. In *Techniques in Psychophysiology*, edited by I. Martin and P. Venables (Chichester: Wiley and Sons).

Mackay, C.J., Cox, T., Burrows, G.C. and Lazzerini, A.J. (1978). An inventory for the measurement of self-reported stress and arousal. *British Journal of Social and Clinical Psychology*, **17**, 283–284.

Mason, J.W. (1968). A review of psychoendocrine research on the pituitary-adrenal cortical system. *Psychosomatic Medicine*, **30**, 576–607.

Mason, J.W. (1971). A re-evaluation of the concept of non-specificity in stress theory. *Journal of Psychiatric Research*, **8**, 323–333.

Melamed, S., Harari, G. and Green, M. (1993). Type A behaviour, tension, and ambulatory cardiovascular reactivity in workers exposed to noise stress. *Psychosomatic Medicine*, **55**, 185–192.

Payne, R. and Fletcher, B. (1983). Job demands, supports and constraints as predictors of psychological strain among school teachers. *Journal of Vocational Behaviour*, **22**, 136–147.

Ray, C. and Fitzgibbon, G. (1981). Stress, arousal and coping with surgery. *Psychological Medicine*, **11**, 741–746.

Rogers, E.H. (1960). *The Ecology of Health* (New York: Macmillan).

Russell, J.A. (1979). Affective space is bipolar. *Journal of Personality and Social Psychology*, **37**, 345–346.

Russell, J.A. (1980). A circumplex model of affect. *Journal of Personality and Social Psychology*, **39**, 1161–1178.

Scott, R. and Howard, A. (1970). Models of stress. In *Social Stress*, edited by S. Levine and N. Scotch (Chicago: Aldine).

Sells, S.B. (1970). On the nature of stress. In *Social and Psychological Factors in Stress*, edited by J. McGrath (New York: Holt, Rinehart and Winston).

Selye, H. (1950). *Stress* (Montreal: Acta Incorporated).

Selye, H. (1956). *Stress of Life* (New York: McGraw-Hill).

Spielberger, C.D. (1976). The nature and measurement of anxiety. In *Cross-Cultural Anxiety*, edited by C.D. Spielberger and R. Diaz-Guerrero (Washington DC: Hemisphere).

Sutherland, V.J. and Cooper, C.L. (1990). *Understanding Stress: Psychological Perspective for Health Professionals. Psychology and Health, Series: 5* (London: Chapman and Hall).

Symonds, C.P. (1947). Use and abuse of the term flying stress. In *Air Ministry, Psychological Disorders in Flying Personnel of the Royal Air Force, Investigated during the War, 1939–1945* (London: HMSO).

Warr, P.B. (1990). Decision latitude, job demands and employee well-being. *Work and Stress*, **4**, 285–294.

Watts, C., Cox, T. and Robson, J. (1983). Morningness-eveningness and diurnal variations in self-reported mood. *Journal of Psychology*, **113**, 251–256.

Webb, E.J., Campbell, D.T., Schwartz, R.D. and Sechrest, L. (1966). *Unobtrusive Measures: Non-reactive Research in the Social Sciences* (Chicago: Rand McNally).

Welford, A.T. (1973). Stress and performance. *Ergonomics*, **16**, 567–580.

World Health Organisation (1946). *Constitution of the World Health Organisation (3)* (Geneva: WHO).

Zyzanski, S.J. and Jenkins, C.D. (1970). Basic dimensions within coronary-prone behaviour pattern. *Journal of Chronic Diseases*, **22**, 781–795.

Chapter 27

Assessment of visual performance

Mark A. Bullimore, Peter A. Howarth and E. Jane Fulton

Introduction

This chapter is concerned with the assessment of visual performance. Let us
begin by considering what we mean by visual performance and why we
might want to assess it. By visual performance we mean how well people
can deal with visual information. This can involve a range of levels of complexity from simply detecting a light to integrating complex qualitative and
quantitative information from a display or visual scene. Visual performance
is a function of:

(a) the abilities of the observer—the inherent capacities and limitations
    of the human visual system, individual idiosyncracies and special
    characteristics like levels of arousal and fatigue;
(b) the characteristics of the observed objects—'displays'—how bright,
    how much contrast, how big and for how long viewed; and
(c) the characteristics of the visual environment in which viewing takes
    place.

We will deal principally with the first two aspects: characteristics of the
observer and of the viewed objects. For each we will consider the important
aspects which affect visual performance and describe ways of assessing it.
Environmental factors affecting visual performance are dealt with in chapter
15. We recommend strongly that these two chapters be read together. It is
rarely possible to deal with a practical issue of visual performance without
some concern also for environmental characteristics. Also, we should consider
the consequences of performing a visual task and amongst these may be visual
fatigue, discussed in chapter 28.

Except for a few special circumstances, such as camouflage, the ultimate
purpose of assessment will most often be to optimize a visual task. In an
ergonomics framework we are concerned with issues such as:

Design: How should this display be designed so that, say, bleary-eyed night-shift nurses will be able to read, from the other side of the bed, how much of the drug has been infused into a patient's arm?

Trouble shooting: Why do quality controllers/inspectors continually miss flaws in the seal of this milk powder packaging? How can we improve their performance?

Evaluation: It may be important to know about the visual function of specific people in consideration of the needs of a particular population or to check an experimental sample, e.g., what proportion of a group show colour vision deficiencies and of what type?

These scenarios might lead an ergonomist to ask questions about visual performance, such as:

How accurately can people absorb this kind of visual information? How much information do they miss?

Would performance improve if we changed the way information was displayed?

What kinds of error are common?

Does this performance deteriorate over time or improve with practice or experience?

In this chapter we will cover basic information about the human visual system and the important characteristics of objects/displays which affect visual performance. This will provide the reader with information about what might be important in particular contexts and help in deciding what to measure and how. We aim to clarify what is feasible for the general ergonomist to attempt in the way of measurement; some kinds of visual performance assessment can be carried out fairly readily using easily obtainable equipment but other kinds are complex and are more appropriately performed by vision experts. The chapter will also refer to other sources which give more detail about particular issues raised here.

The human visual system

In this section we will give an overview of the human visual system and discuss how its physiology can influence visual performance. The basic structure of the human eye is shown in Figure 27.1. Light enters the eye through the transparent cornea, where the majority of the refraction, or focusing of the light, occurs. It then passes through the pupil, the circular aperture in the iris which regulates the amount of light reaching the retina. The pupil normally looks black because little or no light comes back from inside the eye (however in a flash photograph it can look red because the inside of the eye is lit up by the light). The light is then refracted further by the lens, to form an inverted image on the retina. (For a detailed treatise of retinal image formation see Charman, 1983). Constriction of a muscle within the eye, the ciliary muscle, modifies the shape of the lens, and hence its power, so that

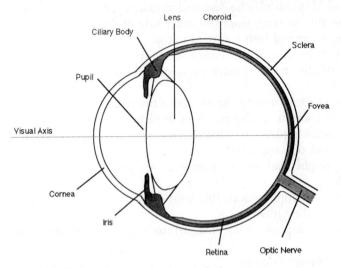

Figure 27.1. Horizontal cross-section through the human eye

objects at various distances from the eye can each be brought into focus on the retina, a process termed *accommodation*. Children and young adults possess large amounts of accommodation, and hence have no difficulty in focusing on an object as close as 10 cm although sustained viewing at this distance may cause fatigue. An observer's ability to accommodate will, however, decrease with age (see Figure 27.2). This is termed *presbyopia* and most people over the age of 40–45 years will require spectacles for near vision.

Sensitivity to light

Light falling on the retina stimulates light sensitive cells called *photoreceptors*. These convert the light energy into electrical signals which are transmitted to the visual cortex, in the rear portion of the brain. The complex processing which occurs in the retina, visual pathways and the cortex is discussed elsewhere (e.g., De Valois and De Valois, 1988; Zeki, 1993). The photoreceptors are divided into two kinds, *rods* and *cones*, which have different characteristics and properties. The distribution of rods and cones across the retina is shown in Figure 27.3. The cones, which are responsible for vision at higher light levels, discrimination of fine detail and the perception of colour, are most abundant in the central or foveal region. When we 'look at' something we turn the eye so that the image of what we are interested in falls on the fovea—an eccentricity of 'zero' in Figure 27.3. The rods, which are responsible for vision at low levels of illumination, are found in greater numbers in the peripheral retina. The implications of these relative distributions will be considered later. The visual system is unable to detect light at levels below 10^{-6} cd m^{-2} and in the *scotopic* range, between 10^{-6} and 10^{-3} cd m^{-2}, only

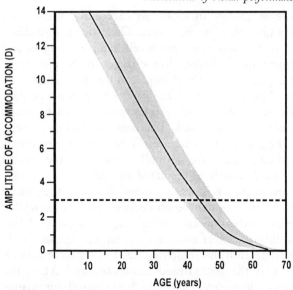

Figure 27.2. Variations in accommodation with age. The shaded area represents the differences between people, the dashed line represents a distance of ⅓ metre

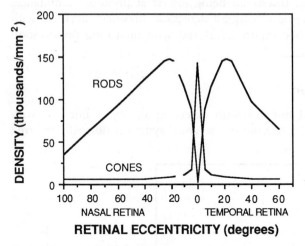

Figure 27.3. Density of rods and cones across the human retina in the horizontal meridian. The gaps in the functions are due to the optic nerve

rods are functioning. At light levels above 3 cd m^{-2}, the *photopic* range, cones play the major role in vision. The area between 10^{-3} and 3 cd m^{-2} is called the *mesopic* range wherein both rods and cones are operating.

In the dynamic visual environment the eye has to adapt to changing light levels and does so in three different ways. First, the pupil can change size; however the maximum area change is less than 100 times—far too small to account for the eye's immense change in sensitivity. Secondly, small rapid changes in neural sensitivity take place in the retina. These occur in milliseconds and compensate for small changes in light levels, e.g., walking in and out of shade. The third mechanism involves slow changes in the photopigments in the rods and cones and is seen, for example, in the slow adaptation after entering a cinema or a photographic darkroom. The process of dark adaptation may be observed by measuring the eye's increasing ability to detect a dim light over a period of time in darkness. The dark adaptation curve is a bi-phasic function (see Figure 27.4) where the first portion represents changes in the sensitivity of cones and takes around 10 min, and the second portion shows changes in visual sensitivity mediated by rods.

The processes above relate to the eye's *absolute* sensitivity. It may be more relevant, in a practical context, to consider the luminance of the target *relative* to the background. Extensive experimental studies (e.g., Blackwell, 1946) have examined the eye's ability to detect small circular targets against a uniform background, with the detection threshold described in terms of *contrast*, defined as $\Delta L/L$—where L is the background luminance and ΔL is the difference between the target luminance and the background luminance (although alternative definitions of contrast may be employed in other circumstances). Threshold contrast was found to be dependent on the adaptation level of the retina, thresholds being lowest at photopic luminances (see Figure 27.5). Furthermore, contrast detection thresholds decrease with increasing stimulus size (see Figure 27.5) and with increasing presentation time (see Figure 27.6).

Spatial aspects of vision

We are often concerned not just with detecting an object but also with discriminating detail. This attribute of the visual system is normally referred

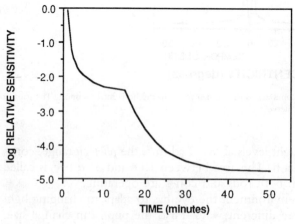

Figure 27.4. The dark adaptation curve

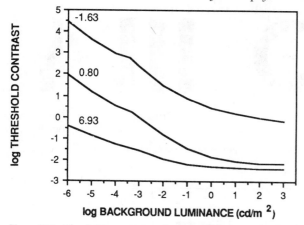

Figure 27.5. Threshold contrast as a function of luminance for three target sizes (after Blackwell, 1946). Target sizes are displayed in log mrad²

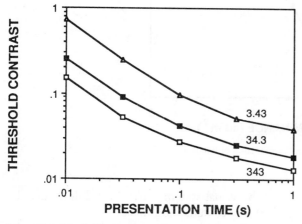

Figure 27.6. Threshold contrast as a function of presentation time for a 4 min arc target and for three background luminances (after Blackwell, 1959)

to as *visual resolution* or *visual acuity* (VA). Visual acuity is usually defined as the minimum angular separation between two lines which is necessary to perceive two lines rather than one. A variety of targets can be employed in its measurement, such as letters, Landolt Cs or gratings (see Figure 27.7). Visual acuity values may be expressed in terms of minutes of arc (min arc) or as a Snellen fraction, e.g., 6/6 (or 20/20 in the USA). The fraction is more commonly used by clinicians where the numerator refers to the test distance in metres (or feet) and the denominator signifies the distance at which the limbs of the letter would subtend 1 min arc. Under optimal conditions the range of normal visual acuity is 6/4 to 6/6 (0·67 to 1·00 min arc). Not surprisingly, a reduction in luminance or contrast will result in a decrease in visual acuity (see Figure 27.8).

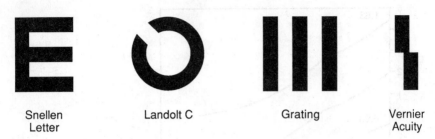

| Snellen Letter | Landolt C | Grating | Vernier Acuity |

Figure 27.7. Various targets which may be used in the measurement of visual acuity

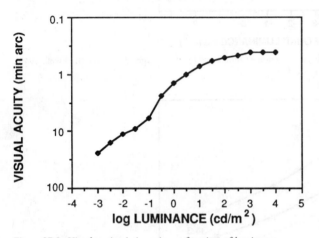

Figure 27.8. Visual acuity (min arc) as a function of luminance

A more complete picture of the visual system's spatial capabilities may be determined by testing people's ability to detect luminance sine wave gratings. The grating is defined in terms of its contrast and its spatial frequency—the number of cycles (a light and dark bar) per degree. Threshold contrast is determined as a function of the spatial frequency to yield the *contrast sensitivity function* (CSF)—a graph of (the reciprocal of) threshold contrast as a function of the spatial frequency. Although such a function may appear of little practical interest, any object or scene can be represented as a series of sine waves of different contrast and spatial frequency and hence its visibility can be predicted from known contrast sensitivity values. (For further reading see Cornsweet, 1970.)

A further important aspect of the eye's spatial sense is its extraordinary ability to detect the misalignment of two lines. Berry (1948), Westheimer (1979a) and others have shown that observers can identify misalignment to an accuracy of 5 seconds of arc. This threshold is referred to as *vernier acuity* and is relevant to the reading of micrometers, slide rules and other tasks where the precise judgement of alignment is required.

Temporal aspects of vision

The human visual system is fairly good at detecting a target that is changing with respect to time. Observers can detect that an object is moving for velocities as low as 7 min arc sec^{-1} with no frame of reference (Boyce, 1965) or 1 min arc sec^{-1} with a reference frame (Salaman, 1929). Further work has shown that target velocities of up to 5 deg sec^{-1} have little influence on visual acuity or vernier acuity (Westheimer and McKee, 1975).

The visual system is also very sensitive to detection of flicker and two thresholds are important. The *critical fusion frequency* (CFF) is the maximum temporal frequency (in Hz) at which flicker can be detected. Under photopic conditions the CFF is around 60 Hz and hence the typical 100 Hz flicker of fluorescent lights is undetectable (see chapter 15). Like visual acuity, the CFF declines with luminance and at low light levels, such as in the cinema, flicker of a given frequency is less detectable. The second threshold which may be of interest is the minimum modulation (or change) in luminance required for the detection of flicker. This is termed *temporal contrast sensitivity* and has been shown to be a function of both temporal frequency and luminance (de Lange, 1958).

Colour vision

An important feature of our visual system is the ability to discriminate colour. Colour can be a powerful tool in the design of visual displays and it may be defined in terms of hue and saturation. Hue describes the perceived colour, e.g., red or blue, and saturation describes how pale or how dark the colour is.

The nature of human colour vision is discussed more fully in specialized articles and texts (e.g., Hurvich, 1981; Adams and Haegerstrom-Portnoy, 1987) but it suffices to say that there are three types of cone in the retina with peak sensitivity to short (S-cones), medium (M-cones) and long wavelengths (L-cones). These receptors are sometimes called blue, green and red cones respectively, based on the peak sensitivity to colours but these terms are misleading because the cones themselves are not coloured, nor do they actually signal these colours. Our ability to discriminate between different colours arises from the fact that a given wavelength of light will stimulate each cone type to a different extent, in the same way that colour televisions produce a range of colours by varying the luminance ratio of the blue, green and red pixels. Optimally, we are able to discriminate between colours as close as 2 nm in wavelength (see Figure 27.9) although our ability to discriminate between desaturated colours is poorer. Colour vision is, however, defective in some individuals. This will be discussed later.

The optical power of the eye is dependent upon the wavelength of light and objects of different wavelengths are focused at different points within the eye. This phenomenon is called 'chromatic aberration', and while its practical consequences are generally not severe, focusing difficulties can occur

Analysis of work activities

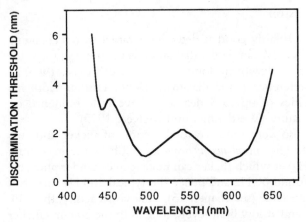

Figure 27.9. The variation in wavelength discrimination with test wavelength

when wavelengths from extremes of the spectrum are viewed together. To generalize, this means that if a red and a blue object are adjacent, one may seem to be blurred compared with the other. Also, they may appear to be at different depths, a phenomenon known as chromeostereopsis, because of the eyes' chromatic aberration (Thibos *et al.*, 1990).

The visual field and visual search

The majority of the aspects of visual performance discussed previously have concerned optimal or foveal viewing. Each eye has, however, a wide field of vision extending 100 degrees temporally, 50 degrees nasally, 60 degrees superiorly and 90 degrees inferiorly from the visual axis. Our visual capabilities vary across the visual field; for example, visual acuity is best at the fovea (see Figure 27.10, and compare this with the cone distribution shown in Figure 27.3).

Our ability to detect a static object is in part a function of its position within the visual field. The probability of detection, within a single fixation pause, may be plotted against eccentricity to yield the characteristic *visual detection lobe* (see Figure 27.11). The visual lobe varies with exposure time and target size, hence the detectibility of a peripheral target can be improved by increasing its size (see Figure 27.11). The visual lobe is an important concept in visual search and inspection tasks since most detection takes place away from the visual axis.

Unlike static visual performance, detection of a dynamic target can actually be better in the periphery. CFF and temporal contrast sensitivity do not decline rapidly with eccentricity. On the contrary, CFF is actually higher in the peripheral visual field than the central field. This can be demonstrated easily with a conventional TV or VDU. If you look directly at the screen

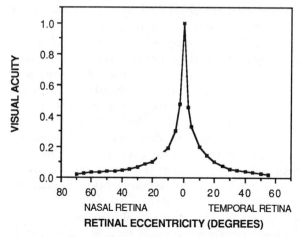

Figure 27.10. The variation in visual acuity (normalized) with retinal eccentricity (after Wertheim, 1891)

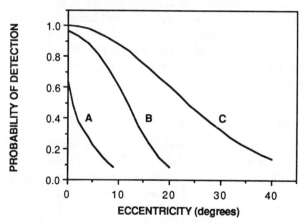

Figure 27.11. The visual detection lobe, the probability of detection of a target within a single fixation pause as a function of eccentricity, for three targets (C is the largest and A is the smallest)

you can probably not perceive the flicker whereas if you shift your gaze to the left or right of the screen it will appear to be shimmering.

Stereopsis and eye movements

One of the most important attributes of the human visual system is that we have two eyes which can move together. Possessing two eyes enables us to perceive the world in three dimensions—an ability termed stereopsis—with which we can make extremely accurate judgements of the relative distance of objects from ourselves. Stereopsis is usually described in seconds of arc, and thresholds of 10 sec arc or less are possible: that is, an object 1 m away can be perceived to be closer than another which is 1·00075 metres away.

The muscles which move the eyes are controlled by visual feedback so that with both eyes open they remain pointed towards the object of interest. Covering one eye will break the feedback loop—there is no visual information as to where that eye is directed—and the eye may take up a different position. This change in eye position is termed *heterophoria* and while it is quite normal for someone to have a small degree of heterophoria, for some individuals this can lead to discomfort and symptoms such as headaches.

The eyes can make rapid movements—saccades—to enable the object of interest to be imaged on the fovea. This is particularly important, for example, in the context of visual search. Alternatively the eyes can track a moving object—a pursuit movement—in order to keep the image on the fovea. Furthermore, the eyes can move rapidly in order to compensate for voluntary and involuntary movements of the head. For all of these 'version' movements the eyes move left or right together. The eyes can also make 'vergence' movements, which alter the angle between the visual axes, in order to look at objects at different distances. The eyes converge in order to view a near object and diverge to view a more distant object. An important characteristic of the vergence system is that it fatigues relatively easily and hence sustained convergence or frequent changes in vergence may produce discomfort. Most people will be able to converge closer than 10 cm: hold a pencil in front of your nose and bring it towards you. When it appears double, you've passed your *near point of convergence*.

Inter-subject variations in visual performance

We have now considered the major characteristics of the normal human visual system. We must also consider factors which will decrease the visual capabilities of the observer. Disease and poor health, for example, may influence visual performance, as will the intake of tobacco, prescribed (and non-prescribed) drugs and alcohol (see Adams *et al.*, 1978; Gilmartin, 1987).

In many individuals the optical components of the eye do not form a clear image on the retina due to a *refractive error*. There are three types of refractive error, the most commonly considered being *myopia* or near-sightedness. Myopia affects 20–25% of the working population and, because the cornea and lens are too powerful or the eye is too long, the image of a distant object is brought to focus in front of the retina. Because of this, distant objects appear blurred. Myopes can, however, see near objects clearly and hence uncorrected myopia may not decrease visual performance in the near environment. Myopia is corrected with concave spectacle or contact lenses. In 10–15% of people the eye is too short, or the optical components are not powerful enough, and the image will be focused behind the retina, a condition termed *hyperopia* (hypermetropia) or far-sightedness. Unlike myopia, the visual effects of hyperopia are often not obvious. This is because many hyperopes can exert their accommodation in order to bring distant objects and, if the hyperopia is not too severe, near objects into focus. Convex spectacle or contact lenses will be required for clear and comfortable vision in the older hyperope and the younger hyperope performing sustained visual tasks. The third class of refractive error, which affects the majority of the population, is *astigmatism*. Like myopia, this produces a decrease in visual performance which cannot be compensated for by accommodation, but unlike either myopia or hyperopia it is equally detrimental to distance and near vision. In the astigmatic eye, lines of different orientations are focused at different positions relative to the retina. For example, an astigmat may see the horizontal poles of a scaffold clearly while the vertical poles appear blurred. Virtually everyone has *some* astigmatism, and it is only when the amount is large that visual performance is affected. As for myopia and hyperopia, astigmatism may be corrected with spectacles or contact lenses.

The visual performance of an observer may change considerably with age (Figure 27.2). Not only does the over 45-year-old have to come to terms with their decreased ability to accommodate (presbyopia) but also with a reduction in pupil size and changes in the crystalline lens which will result in less light reaching the retina. Hence people over 50 may require higher levels of illumination in order to perform as well as their younger colleagues and take longer to adapt to changes in illumination. Furthermore, changes in the crystalline lens may increase their susceptibility to disability glare (see chapter 15). Finally, the correction of presbyopia with bifocals or trifocals may cause focusing problems if people are looking through the wrong part of the spectacle lens.

Earlier we discussed the characteristics of normal colour vision. It should be acknowledged, however, that some 8% of males and 0·5% of females have defective colour vision. The relative frequencies and characteristics of the various types of defect are given in Table 27.1. All congenital colour deficiencies are due to anomalies in the retina, the most common type being

Table 27.1. Prevalence and properties of colour vision defectives in the male population. Although the prevalence of each type is much less in the female population, the *relative* proportions are similar

Type of defect		Spectral colour discrimination	Prevalence (%)
Anomomalous Trichromacy			
	Protanomalous }	Reduced for green, yellow,	1·0
	Deuteranomalous }	orange and red	5·0
	Tritanomalous	Reduced for blue-green cyan, and blue	0·001 (?)
Dichromacy			
	Protanope }	Absent for green, yellow,	1·0
	Deuteranope }	orange and red	1·0
	Tritanope	Absent for blue-green cyan, and blue	0·001 (?)
Rod Monochromacy		Little or no discrimination	0·003 (?)

anomalous trichromacy where one of the three cone types in the retina is abnormal. A more severe defect is dichromacy where one of the cone types is absent. The most dramatic defect occurs in the rod monochromat who has no colour discrimination, a scotopic spectral sensitivity function and reduced visual acuity, but these people make up a minute proportion of the population.

Characteristics of tasks and viewed objects which affect visual performance

In the visual working environment we are generally concerned with more than the detection of simple spots of light, distinguishing single characters, or discriminating two colours. We are concerned with the acquisition of visual information from various sources. These are usually complex rather than simple stimuli and are most often well above threshold levels for visual detection.

This section describes briefly characteristics of visual tasks and viewed objects which affect how well they can convey information to people through the visual system. Here 'viewed objects' means all those things from which people receive visual information. These may be 'displays' in the traditional sense, they may be the focus of an inspection task, or they may be any other kind of material such as printed documents or vehicles on the road.

Types of visual task

What constitutes good visual performance depends on the requirements of the task. In thinking about assessment of visual performance it is important to appreciate the nature of the tasks being carried out—this may affect the type of assessment which is appropriate. Three examples will illustrate task differences:

1. *Detection.* Some visual tasks require simply that an observer detects the presence or absence of something or finds out where something is. Examples are detecting that a warning light has come on, checking a manufactured unit for breaks in a seal, or finding the cursor on a computer screen. Here good visual performance requires only that the observer see the object against its background—no other discrimination is needed.

2. *Recognition.* Most often a visual task will require that an observer detect *and* recognize what something is—this demands a higher level of discernment because there has to be discrimination between stimuli. This is the case, for example, in obtaining information from graphic displays and text or carrying out complex inspection tasks.* Here good performance involves being cor-

---

* *Visual inspection.* Visual inspection, usually associated with monitoring product quality, represents a specific kind of visual task notable for its sustained and invariable nature. The same general principles affecting visual performance and assessment of other tasks also apply to visual inspection. Visual inspection

rect in the judgement of what it is that has been detected. In counting out change, for example, it is important to distinguish between different coins.

3. *Interpretation.* Most tasks also require observers to interpret what they have seen in terms of what it means for their subsequent actions. Examples include establishing which of a row of warning lights has come on, the significance of a blemish on a photograph, what a dial is indicating and what a text message means.

Here we are not considering this cognitive level of extracting meaning from visual information, but rather the sensory capacity to obtain information such that cognitive factors can begin to play. We should not, however, lose sight of the fact that both sensory and cognitive factors are important in the design and evaluation of visual material; no matter how lucid and interesting the prose, it will be without value if it is written in tiny grey characters on a grey page; conversely, no matter how legible the message it is useless if it makes no sense. Beware!—cognitive problems can be mistaken for problems of visual performance.

Types of visual display

Visual information comes to us in the form of light either reflected by objects or emitted by them. Nowadays more and more visual information, specifically from displays in the working environment, comes to us via sophisticated emissive technologies in the form of cathode-ray tubes (CRTs), light emitting diodes (LEDs), backlit liquid crystal displays (LCDs), plasma displays, etc. Before new technologies were widely available, visual information was displayed principally by the traditional reflective technologies of inks and paper, printed labels for electro-mechanical dials and some simple emissive displays like warning lights. There is a rather large body of knowledge associated with these traditional media in terms of guidelines for good design for performance (e.g., McCormick and Sanders, 1987; Helander, 1987; Boff and Lincoln, 1988).

The important special characteristics of the new emissive displays with respect to visual performance are those concerned with the stability of images and their resolution. These include raster screen refresh rates, pixels density, phosphor persistence, dot sizes, brightness and spacing on matrix displays. There are reference books (e.g., ANSI, 1988; Berlinguet and Berthelette, 1990; BS 7179, 1990; Luczak *et al.*, 1993; National Research Council, 1983; Snyder, 1988; Travis *et al.*, 1992) which give guidelines for visual information presented on these kinds of display, but these need to be interpreted circumspectly because the new technologies are themselves developing very rapidly in terms of their ability to support good quality images.

---

is, however, an area of industrial ergonomics which, because of the direct impact of its performance on profitability, has received special attention over the last 20 years. While we make reference to inspection in a general way, for more detailed discussion we refer the reader to writings dedicated to the subject (Smith and Lucaccini, 1977; Drury, 1973; Drury and Addison, 1973; Megaw, 1979).

It is more difficult to produce robust standards and guidelines about this class of display. This makes *assessment of visual performance* more important since it is often possible to refer with confidence to guidelines about the physical characteristics of images on the displays; the chances are that there are none researched sufficiently for the particular quality of display with which you are concerned. This fact is reflected in the current emphasis of the International Standards Organization's (ISO) efforts to produce ergonomics standards for visual displays which attempt to define standard test procedures for visual performance rather than physical characteristics (e.g. the ISO Standard 9241/3). Nevertheless, although existing guidelines may be inappropriate for state-of-the-art displays there will always be work environments where people are exposed to examples of outdated technology.

Basic principles of appropriate design for visual performance

It is the intention in this section to explain underlying principles which govern the suitability of visual information for the human visual system, rather than to present exhaustive guidelines for design or evaluation. The section is divided into two. In the first part, simple physical aspects of the display are considered, while the second part considers aspects which involve some cognitive component.

Physical aspects

As a general rule, visual performance is better the brighter the ambient lighting, the greater the contrast between object and background, the larger the object and the longer the viewing time. The influence of these physical parameters on visual performance can be seen in a general model presented in Figure 27.12. If the parameter value is too low (e.g., the size is too small)

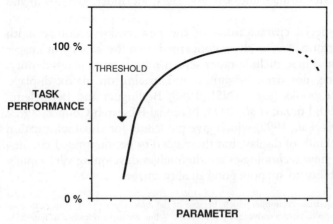

Figure 27.12. The influence of parameter on visual performance. The 'parameter' may be size, contrast, luminance or time. The broken portion of the line signifies a potential decrease in performance

then the task will be below threshold. As the parameter increases threshold is reached and subsequent increases will improve performance until the optimal level is reached. In some cases though, if the parameter continues to increase performance will eventually decline. We shall now consider the relevant parameters.

ILLUMINATION

Within the normal range of illumination levels which we encounter naturally, or produce artificially, visual performance is improved by increasing illumination. This is because the eye is relatively more sensitive to change at higher illumination levels. Chapter 15 discusses the importance of maintaining a relatively constant illumination level within the visual field so that visual performance is not affected by adaptation to either of the extreme levels. There are some rare occasions when illumination levels can become too high, for example, in a visual environment combining bright sunshine and wide expanses of snow, visual performance deteriorates.

CONTRAST

While it is generally true that the greater the luminance contrast the better the task performance, it is important to qualify this statement. For a moment consider, as an example, driving at night. Performance in the task of detecting oncoming vehicles is enhanced by their displaying bright headlights, but the contrast between these and the rest of the scene commonly causes discomfort and disability and hence a reduction in performance of the visual driving task as a whole. This example demonstrates the need to think about the whole task context rather than simply parts of it. It is also important to think of contrast both in terms of bright objects against dark backgrounds and of dark objects against light backgrounds. For some tasks there are advantages in illuminating the background to improve the observer's ability to see a stimulus—for example, in checking for flawed items using backlighting or shadows. Conversely, for tasks involving written character recognition it seems that *in general* dark text and symbols on a light background ('positive contrast') is preferable to light text on a dark background ('negative contrast') (Gould *et al.*, 1987a).

SIZE

Generally the larger the object (the greater the angular subtense at the eye) the more easily it will be seen and discriminated from other objects. For a resolution task this generalization holds for sizes greater than 1 min arc (the normal resolution threshold) up to the point where optimum performance is reached—the value for which depends upon other factors discussed here. As demonstrated by the model in Figure 27.12, above a certain point increasing the size will not improve performance and might in fact degrade it. Imagine standing directly in front of a large advertising hoarding and trying to read it! Applied to alphanumeric characters, size recommendations are that character heights should be large enough to subtend an angle of around 20 min arc at the observer's eye, i.e., about 4 mm height at a viewing distance of 600 mm.

EXPOSURE TIME

Visual performance is better the longer an observer gets to look at, or look for, something. This has implications both for the design of tasks—it is important to ensure that presentations are for an adequate length of time—and for the measurement of performance. Indeed, 'time to detect' can be used to assess the visibility of a stimulus.

Aspects which include a cognitive component

All the above physical parameters apply to the basic sensory processes of the human visual system. The parameters considered now all have, in addition, some cognitive aspects to them, and require a more sophisticated appreciation of their interactions. Consider Figure 27.13 where the physical aspects of the stimulus do not easily reveal the way the visual system responds to the pattern elements in this illustration.

DYNAMIC ASPECTS

Most often when we consider criteria for the design of visual material we are concerned with stable and static images—for most tasks this is an optimum condition. Images might, however, be unstable due to vibration of the observer or the viewed object or characteristics of the display itself. Furthermore, dynamic displays are becoming more widespread. Care must be taken to compensate for the effects of these movements by ensuring that speed of movement is controlled and that illumination, contrast, image size and viewing time are increased above that which permits adequate performance with static and stable images. The conspicuity of an object, i.e., its capacity to attract our visual attention, is in part a function of its dynamic

Figure 27.13. The text is more conspicuous because of the pattern of lines, an effect which is not predictable from simple parameters of luminance, size or contrast

characteristics—movement or intermittency. A moving or flashing stimulus is more conspicuous than a static one.

CHANGE AND COMPARISON OF VISUAL STIMULI

Relative judgements are easier to make than absolute judgements. For example, you can easily detect that a vehicle brake light has come on if you notice the increase in intensity; however, if you miss the brightness change then because they are the same colour it is difficult to know whether you are looking at brake lights or rear lights. In this example there is no external brightness reference to compare the lights with, and the task is extremely difficult. A reference item for comparison will improve performance in many circumstances, such as inspection tasks and monitoring tasks, and the use of reference lines and markers are of significant value in assisting visual search and judgements.

PATTERN RECOGNITION AND CODING

We are adept at pattern recognition and tend to group visual information in terms of similarities of its physical appearance; colour, brightness, shape, size and orientation. These factors can be used to enhance visual performance by helping the viewer organize visual information. Designing a bank of dials so that the pointers all line up in the same direction (particularly either horizontally or vertically) when status is normal makes it easier to detect when one of them is registering an abnormal condition. We are better able to see and distinguish objects if they have unbroken lines and boundaries and whole regular shapes. These factors are exploited in the design of camouflage, where a major principle is to break up boundaries and outlines with colour or shading. As another example, in the design of alphanumeric characters the implication, and empirically supported wisdom, is that it is important to use clear, non-slanting and simple fonts without serifs.

REDUNDANCY

Visual performance can sometimes be enhanced by providing observers with redundant information. The detection and discrimination of warning lights for different functions, for example, can be improved by making them a different colour *and* a different shape. Similarly, by being colour- and size-coded, British paper money scores over its US counterpart for visual discriminability. It is sometimes appropriate to use other senses as a redundant cue to aid visual performance, for example auditory cues will improve the detection of visual warnings.

USE OF COLOUR

Colour has an important role as a coding device in separating and grouping elements in a design. When colour is used for coding, or for grouping information, it is important not to use too many colours, although authorities differ in the maximum number advisable (e.g., Grether and Baker (1972) recommend using no more than 10 colours but preferably 3!). Of course, many more colours and shades can be used to render form and depth in visual displays. The saturation level of colour can be important, and desaturated (pastel like) colours should be avoided. In using colour for coding the colour discrimination abilities of the user population must be considered. An example of failure to do so is the use of self-administered glucose tests for diabetics, where the test involves colour matching even though diabetes is known to cause colour vision defects.

Colour is also important in providing contrast and it can be used to make objects conspicuous. The success of a colour used as a highlight depends upon the visual context in which it is used; all else being equal, orange has better contrast with green, for example, than it does with red. The concept of generic 'high visibility' colours can be misleading. Those colours which we generally refer to as 'high visibility', such as bright and fluorescent yellows, oranges and yellow-greens are effective in many environments because on average they contrast well with their backgrounds. In other specific background circumstances these colours can also be 'low visibility'. For example, red flags by the roadside will be highly conspicuous as few natural scenes are bright red whereas soccer linesmen no longer use red flags because they merge in with red garments in the crowd and with red seats in some stadiums.

DISPLAY LOCATION, SIZE AND AREA

The best location for display of visual material is roughly perpendicular to the observer's line of sight, unobstructed and preferably requiring a minimum of eye movement. The most frequently accessed information is, therefore, best placed centrally. Standardizing location of specific types of information is useful in reducing search time. The appropriate size to make a display depends on characteristics of the task, mainly the amount of information which must be displayed and its relative importance. If the display area is too small and too dense this will increase search time and decrease legibility. Clutter and complexity in layout reduce performance, and issues such as these have received attention in the user-interface design literature (e.g., Shneiderman, 1987).

Conclusion on task and object characteristics

This section has reviewed basic general principles of the design of visual displays and tasks which affect people's visual performance. It is worth bearing in mind, in the following section about methods, that a widely used and economical method of assessing visual performance is to compare the particular circumstances of interest with standards and guidelines of good practice. Some sources for these guidelines have been referenced here. Applying these principles, with discretion, can often save a great deal of time and energy by preventing assessments of performance which effectively repeat other people's work.

Methods for assessment of visual performance

Assessment of visual performance may be important in a number of different circumstances. Just what it is appropriate to measure, and how, will depend on these circumstances. Assessments are generally necessary either to trouble-shoot unsatisfactory conditions or as a tool in design and research.

When something is wrong there is often objective and quantifiable evidence of poor performance. It may be evident from people making errors (e.g., confusing alphanumeric characters or failing to detect flaws) or performing more slowly than anticipated (e.g., taking longer to do specific tasks or spending more time idle). Unsatisfactory conditions can also become evident as the result of subjective complaints from people about fatigue,* discomfort (e.g., glare, 'eyestrain'), or general matters (e.g., ill-health). (See chapter 28 in this book for some clarification.) In all cases the assessor's job is to try to identify the causes of poor performance and usually to devise and evaluate ways of improving it. This might involve assessment of:

1. *Individual's visual functional abilities.* For example, are specific individuals displaying decrements in particular aspects of vision related to their work? Is their visual system deficient in any way?

2. *Characteristics of the viewed objects and visual task.* For example, are the contrast and luminance values too close to threshold levels for optimum performance? Does spatially reorienting the task improve the situation?

3. *Visual environmental factors.* For example, is the spectral output of over-head lighting adversely affecting colour discrimination?

4. *Non-visual factors.* For example, is the general health of employees good? How satisfactory are social and organizational factors within the workplace? Are other environmental factors, heat and humidity or vibration for example, having a detrimental influence on visual performance?

In design and research it is sometimes important to assess what level of visual performance can be expected from a specific group of people or from a specific design of task or display. In these circumstances the most appropriate assessment will involve:

(a) *measuring abilities of people* doing tasks typical of those they might be required to do,

(b) *reference to guidelines* for comparison of the characteristics of a task or display,

---

Visual Fatigue. One of the problems in considering visual fatigue is that the term itself is used in a variety of contexts. Some authors use visual fatigue to describe subjective complaints of discomforts while others apply the term to changes in visual function. This led the National Research Council's Committee on Vision (1983) to conclude that:

'The terms *visual fatigue* and *eyestrain* are frequently used in ill-defined and differing ways. These terms do not correspond to known physiological or clinical conditions. We suggest instead that researchers and others use terms that specifically describe the phenomena discussed, such as *ocular discomfort, changes in visual performance* and *changes in oculomotor functions*.'

(c) *evaluation tests* of the display/task with a sample of people typical of the likely user population.

In sampling from populations, *assessment of visual function* is sometimes necessary to describe or screen a group of people and decide whether they represent the abilities of a specific population. This is important in selecting people to take part in empirical assessments of visual stimuli. (See also chapter 10 for a discussion of user trial sample selection.) Whenever assessment is necessary the common elements are the person, the task and the environment. The assessment of the visual environments and of non-visual factors is dealt with in other chapters of this book. The other two elements, namely assessment of personal visual function and assessment of tasks and displays, are discussed separately here, although some overlap will be evident.

Assessment of task/display

Performance based measures

Performance-based measures involve the observation and measurement of people's performance on visual tasks, either in their natural environment (e.g., factory, driving cab or office) or in laboratory settings where selected attributes of the task can be simulated and examined under more controlled circumstances. These measures can and have been employed extensively in the evaluation of lighting conditions, display quality and the effects of prolonged visual performance. However, there are a number of problems in designing and interpreting these measures, and a variety of approaches have been taken to account for these problems.

First, performance usually involves both speed and errors and people often make complex trade-offs between them (see chapter 5). For tests that allow subjects to establish their own criteria for time and errors, a slight change in error rate could be reflected in a relatively large change in speed. This can make the use of performance-based measures extremely difficult unless an underlying model of the trade-off is available. The problem can sometimes be overcome in the design of the tests themselves. For example, people can be allowed to take as long as they want, and the performance measure will then be accuracy alone; this approach is seen in the reading of the optometrist's letter chart. As an alternative, the performance measure could be the time taken to achieve a certain level of accuracy: an example of where the criterion is 100% accuracy is how long it takes someone to locate a particular town on a map. Similarly, other accuracy levels could be fixed by rejecting trials where the subject's error rate is greater or less than a predetermined level, and then using speed alone as the performance measure.

A second problem with performance-based measures is that, in short-term studies, visual performance may differ in unpredictable ways from when the task is performed on a prolonged or permanent basis. This is of particular concern in long-term inspection tasks when vigilance and tiredness may be involved. While this is a general problem in ergonomics, particular difficulties can come about in visual tasks because of the long-term demands on accommodation and convergence.

Finally, there is a complex relationship between the visual demands imposed by the task, the amount of effort and attention allocated and the resultant performance levels. Again, it may be possible for the design of tests to control subjects' arousal and attention to some extent, for example, by rewarding good and penalizing poor performance or by employing secondary tasks.

Analytical and empirical approaches

Two types of performance method can be identified, the 'analytical' approach and the 'empirical' approach (Hopkinson and Collins, 1970; Boyce, 1981). In the analytical approach the performance of simple contrived tasks is observed so that a quantitative model may be developed to relate visual performance to visual conditions. For example, Blackwell (1946) describes a method which involves detecting a spot of light against a darker or lighter background. In this way the relationship between contrast, luminance and visual performance can be modelled, and this model can subsequently be applied to more complex tasks such as the legibility of characters on a given background. This approach has a great deal of merit when comparing between tasks or displays which differ only with respect to one or two variables.

In the empirical approach, the speed and accuracy with which a task is performed is measured under real or simulation conditions. Weston (1945), like Blackwell (1946), investigated the relationship between task contrast and visual performance. He used a large number of Landolt Cs ('C's oriented in various directions with the subject's task being to identify the location of the gap; see Figure 27.7) to test people's speed and accuracy under a variety of contrast and light levels. Various other types of tasks have been employed such as reading text (Carmichael, 1948; Kruk and Muter, 1984; Nordqvist *et al.*, 1986; Gould *et al.*, 1987b), simulated inspection (Brozek *et al.*, 1950; Murch, 1983), and search (Bodmann, 1962; Neisser, 1964). Modifications of these tasks have been used to examine the effects of contrast, luminance and size (Khek and Krivohlavy, 1966; Boyce, 1974; Stone *et al.*, 1980).

In many real-life situations the empirical approach has great practical value because it allows for comparison between a number of options where multiple variables are involved and where there are not resources to develop a complex model to help predict performance. For example, suppose that a choice must be made between three different liquid crystal displays for use on a chemical analysis machine. If these displays vary in a single dimension, say the sizes of character that they can support, then a good decision can be made confidently on the basis of an analytical approach. Knowing the range of distance from which chemists will need to read results it is possible to select the most appropriate display sizes. However, if the choice had to be made between three different displays, one of which was liquid crystal, one was a vacuum fluorescent display and the other an LED display, there are many more variables differentiating them: e.g., display colour, luminance, contrast, character form, size and effective viewing angles. There is no ready model to help make the decision about which would be best. Who knows what the appropriate weightings are for each variable? This choice can be made empirically in a user test by comparing the legibility of characters on each of the displays in ambient lighting conditions and from angles and distances to the display which cover the range expected in the machine's use. The advantage of a performance based experiment like this is that it provides useful and sound predictive information for the specific application. The disadvantage of the approach is that it adds little to the body of theoretical knowledge about visual performance, since it has compared the performance of discrete complex objects but revealed nothing quantitative about the interactions between the many variables which were involved.

The empirical approach is also particularly useful when visual performance is being affected by higher level factors beyond the simple physical attributes of the task (such as size and contrast) or the physiological attributes of the visual system (e.g., Figure 27.13). These factors range from the legibility of characters to the organization of visual information in certain ways to capitalize on our visual pattern recognition abilities. For example, an analytical approach could help in improving inspection performance in the detection of stitching irregularities in the seams of jeans, by increasing ambient illumination levels and changing the lights' spectral characteristics to enhance colour contrast between stitching and cloth and/or allowing inspectors longer to look at each pair of jeans. On the other hand, there might be vast scope for improvement in the visual performance of railway timetable-enquiry clerks, even if they are using full-colour high-resolution visual displays. For example, it might be helpful to organize the listings graphically and to introduce different grouping and colour coding conventions on to the screens. Here an analytical approach would not be a suitable way to assess different ways of organizing the visual layout. An empirical approach, simulating the clerks' search tasks and measuring visual performance with each of several layout options, would be a more appropriate way to assess potential improvements.

Task evaluation

How can we assess the effects of changing the physical attributes of a task? Suppose we know from the analytical approach that an increase in illumination level might be expected to improve the performance of someone reading documents. How can we assess first, whether there is any need for improvement, and second whether the strategy we have adopted has been successful?

In assessing visual tasks and performance the approach taken by the Commission Internationale de l'Eclairage (CIE, 1972, 1981) was to use the parameter of contrast to define a measure they termed 'visibility level'. By determining what contrast reduction is necessary to reduce the task to threshold you effectively determine how far above threshold the task was in the first place. The higher above threshold, the more 'visible' the task. The approach has been successfully applied to a variety of lighting situations and to paper-based tasks (Boyce, 1981). To use this approach you need to have some means of reducing contrast without affecting overall luminance, and this is the function of a 'visibility meter'. In assessing a visual task, a vision or a lighting specialist would probably either use this approach or would measure the physical attributes of the task and then apply the values obtained to an existing visual performance model.

But what if you haven't got a visibility meter or a sophisticated photometer? As discussed earlier in this chapter there is a general relationship between task performance, and each of the physical attributes of size, contrast, illumination and viewing time. The relationship between performance and any of these parameters is shown graphically in Figure 27.12, and knowledge of this function provides us with a simple, yet elegant, means of assessing whether a visual task is adequate. If we knew for a given parameter how far above threshold the performance is optimal, then we could devise a strategy to reduce the parameter by that amount. If the visual task was originally above the optimum level, then this reduction would still leave the task above threshold. On the other hand, if the task was sub-optimal, then this reduction would leave the task below threshold!

Size is an appropriate candidate for this approach, and as a good rule of thumb, if the task is reduced in size by a factor of three and can still be performed, then the initial conditions were acceptable for adequate visual performance. The integrative aspect of this simple approach can be seen by considering that by reducing *any* parameter such as contrast, task luminance or illumination, the whole curve relating performance to size (size being the 'parameter' in Figure 27.12) will be altered. With this alteration the threshold size will increase. The strategy of reducing the task size could then take the task below threshold, and it could not be seen. This size reduction can be achieved easily by increasing the distance from the eye to the task by a factor of three (Bailey, 1987).

A major advantage of this simple strategy is that the match between the task and the individual performing it can be assessed by using the person themself as the observer. Alternatively, the task alone can be assessed by using an observer with good eyesight. A word of caution is in order! It is important that size is the only parameter that is being varied and that the measurement procedure itself should not affect the task. Since, in this example, the task is moved further away, the focusing demands on the observer are less. However, supposing the person who normally performs the task wears spectacles designed to focus at the task distance and not at further distances. The task itself could be quite acceptable but when the viewing distance is increased detail could become unclear due to focusing rather than image size reasons. This could lead to an incorrect conclusion that at the normal working distance the task was inadequate.

Subjective reports

Another class of methods involves the use of subjective measures based on questionnaires, interviews or informal discussion. Typically, this approach has been used both to assess visual performance and to investigate complaints of visual discomfort. The advantage of these methods for environments and tasks outside the laboratory (where testing procedures can be strictly controlled) is that the effect of complex variables can often be rapidly assessed. These are relatively easy and economical methods but are not without their drawbacks. Subjective reports should not always be taken at face value—people are often mistaken in their assessment of their own visual system and its performance. (Chapter 3 provides a review of subjective assessment and advantages and disadvantages.)

Subjective reports are also likely to be biased by popular beliefs and topical misconceptions. A complaint about glare on screens, for example, could be prompted by a belief that VDUs would be better if provided with a special anti-glare screen rather than because there is a real performance problem. The investigator needs to develop methods to avoid being misled by the subjects' analysis; a brief investigation with placebo treatments or use of subjective reports from people other than those who were party to the original analysis would be useful techniques to adopt as controls.

Subjective reports are of different kinds; they can be more or less structured and are often most reliable when they are most structured and specific. For example the choice between specific options—which of these fonts is more legible—is more likely to yield useful results than asking an open-ended question. On the other hand, open-ended questions can often reveal unforeseen problems which might affect performance, for example, with equipment cleaning and maintenance practices or seasonal variations in light levels.

A good example of the use (and misuse) of subjective reports is the literature concerning reports of ocular discomfort and visual display units. The increasing use of VDUs in the early 1970s brought with it a plethora of studies reporting a high incidence of complaints of visual discomfort. However, reviews of these early studies are invariably critical: Helander *et al.* (1984), for example, stated that 21 of the 28 studies they surveyed had serious design faults. These flaws included a lack of control groups and biased samples (there is anecdotal evidence that, in at least one early study, subjects were encouraged to over-report difficulties by one of the participants because this would bring problems to the attention of the management). Subsequently,

Howarth and Istance (1986) suggested that in many of these studies of visual discomfort the use of one-off questionnaires was inappropriate. If groups are well-matched and appropriate measures used (such as *change* in discomfort over the day, rather than simply discomfort at the end of the day), then no difference is found between VDU users and non-users (Howarth and Istance, 1985). This finding does not negate results of studies which show significant differences between VDU users and non-users (e.g., Knave *et al.*, 1985), but rather indicates that these reports are demonstrating problems other than the use of VDUs *per se*.

Assessment of personal visual function

In certain circumstances we may wish to assess a person's visual capabilities. This could be because we suspect that an individual's poor visual performance has a physiological basis. Alternatively, we may wish to evaluate a task or display and need to ensure that the group of observers to be used are 'normal'. In the same way, assignment of subjects to experimental groups may be on the basis of their visual capabilities. Many people will be able to tell you something about visual problems they have; however, they may be totally unaware of visual disabilities like colour vision defects. Finally, a change in visual function could itself be the metric of interest. This section is divided into two parts. In the first, basic tests of visual function are described which a competent ergonomist should be able to perform. In the second, visual functions needing elaborate (and expensive) equipment not generally available to the non-specialist are reviewed.

Basic tests of visual function

Test charts are available that allow visual acuity measurements to be made easily. Distance visual acuity charts contain rows of letters which decrease in size down the chart. These letter sizes are labelled by the distance at which they would subtend 5 mm arc (and the limbs, 1 mm arc) at the eye. Hence, an '18m' letter on the chart would be three times as large as a '6m' letter. Most charts are designed for use at 6 m and this distance should be adhered to wherever possible in order to avoid confusion. To use the standard Snellen chart, the observer is instructed to read as far down it as possible and the lowest line in which most of the letters are read may be taken as the threshold (the visual acuity values are marked clearly on most charts). Visual acuity may also be measured for near vision with appropriate charts, which usually employ lower case Times Roman print. An observer with normal visual acuity should have no difficulty in reading 5 point (N5) print at 40 cm. This raises an important point, which is that 6/6 distance visual acuity does not itself guarantee good intermediate or near vision, particularly in observers over 40 years of age. Hence visual acuity should always be assessed at a distance relevant to the task or display that the observer is or will be using. Also, as well as measuring both eyes together visual acuity should be measured for each eye separately since an imbalance may be contributing to any reported symptoms. Careful attention should also be paid to the luminance of the test chart (see Figure 27.8): there is a variety of international standards for chart luminance and as a guideline we recommend a value of between 80 and 300 cd m^{-2}.

The standard Snellen chart described earlier has been used for many years with little change. Over the last 15 or so years a number of new charts have been developed. These range from charts consisting of luminance sine waves at various orientations to letters embedded in random-dot noise. An interesting recent development has been the introduction of low-contrast test charts (see e.g., Bailey and Bullimore, 1991; Reeves *et al.*, 1993; Regan and Neima, 1983). This appears to be very much the future for visual performance assessment.

The type of high contrast test-chart we recommend was introduced by Bailey and Lovie (1976)[*] and consists of rows of five black letters on a white background. The size of the letters on each row is related logarithmically to the rows above and below, and with this chart we record the *logarithm of the minimum angle of resolution* (logMAR). The work of Westheimer (1979b) and Hallden (1972) suggests that a logMAR scale is a perceptually equal-interval scale. Being logarithmic, the scale does not have a 'true' zero, although 6/6 is recorded as a logMAR of zero, and so the scale can be taken to be at an interval level of measurement, but not at a ratio level. The chart has five letters on each line, and each letter correctly read increases the person's score by 0·02 log units. The person reads as much of the chart as they can, and their vision is then scored according to the number of lines and letters they correctly identified. Despite the scientific advantages of the logMAR charts, the standard Snellen chart is more likely to be encountered. This latter chart is quite adequate for most purposes, however keep in mind that we *cannot* assume that measurements using this chart are at a measurement level higher than ordinal (or possibly ordered metric). The practice of averaging vision scores is, therefore, incorrect.

The ergonomist should be aware of the contribution of accommodation and vergence problems to visual discomfort. A subject's near point of accommodation (NPA) can be measured with a near vision chart (or even a newspaper!). The print is moved slowly towards the observer until they report it beginning to blur—this point is the near point of accommodation. As mentioned earlier the person's accommodative ability declines with age, and so while a NPA of < 10 cm might be normal for a teenager, a 35-year-old might not be able to focus much closer than 20 cm from their eyes. It is desirable that an individual should have a near point of accommodation significantly closer than their required viewing distances for a display. Reading

---

[*]Available from Ian Bailey, School of Optometry, University of California, Berkley, CA 94720.

glasses and bifocals alter a subject's near point, and it will be more appropriate to take this measurement with the subject wearing their spectacles. The near point of convergence can be measured in a similar fashion using no more than a pencil. This is held vertically and moved towards the observer until it first appears 'double'—this represents the near point of convergence. Most people, irrespective of their age, should have near points of convergence no further than 8–10 cm and any value much beyond this range may give rise to symptoms. Do not confuse the near points of accommodation and convergence: in the former you are looking for *blurring* of the target, while in the latter you are looking for the target to appear *double*.

Colour vision may also be assessed relatively simply by the ergonomist. The simplest and most common type of test uses pseudo-isochromatic plates. These are book tests of numbers, letters or symbols in which the background camouflages the task for the colour defective. The Ishihara Plates are an example of this type of test. These tests are fairly efficient at detecting colour defectives and will often differentiate between protan- and deutan-type defects. It is important, if the correct standard illuminant (Illuminant C) for which the tests were designed is not available, that daylight is used to illuminate such tests. If daylight is not available either, then cool fluorescent tubes can be used. Other light sources, e.g., incandescent lights, will unacceptably alter the apparent colour of the plates (see chapter 15), possibly producing incorrect results.

Stereopsis is the final visual function that one can realistically assess without a large amount of equipment. Inexpensive tests are readily available, such as the Titmus Fly Test and the TNO Test. In these tests, the two eyes are dissociated with either crossed polarizing filters or red and green filters, and a composite picture or pattern (e.g., of random dots) is placed in front of the person. Because of the filters employed the two eyes will see different images, in the same way that the two eyes see slightly different views of a 3D object. If the person has stereopsis, a 3D pattern will be seen coming out from, or going into, the page.

Instead of using the above techniques, the ergonomist may use a 'vision screener' in order to evaluate the vision of an observer. These instruments are based on the principle of the Wheatstone stereoscope, and eyes are tested either singly or together. In most instruments the following aspects of vision are assessed:

1. Distance and near visual acuity.
2. Colour vision.
3. Heterophoria.
4. Stereopsis.

Several vision screeners are commercially available including the Bausch and Lomb Ortho-Rater and the Mavis Vision Screener. Although instrument norms are available, the quantitative results obtained from these machines should be treated with caution since their false alarm rate is generally high. As a screening instrument they are, however, generally excellent and will usually detect people who should be referred for expert evaluation. Individuals should normally be referred to an optometrist for a visual examination, who, on request, will provide a written report. Although there may be a charge for this service, it may be the most economical way to solve problems.

Specialized tests

A variety of visual functions have been assessed in the evaluation of visual workload. We shall examine briefly some of the techniques described in the literature although the practising ergonomist may not have the resources to perform most of them. It is important that when measuring these functions we understand the relevance of any recorded changes. Indeed, there is clearly a need to distinguish 'fatigue', as described in the literature, from an adaptation process.

Malmstrom *et al.* (1981) employed an objective optometer to measure the accommodative response to a far and near sinusoidally moving target. They found that the accommodative response diminished significantly over a 6·5-min period and propose that this is due to fatigue of the accommodation system. There is an abundance of literature demonstrating that a period of sustained near vision can induce changes in the accommodation and vergence systems (e.g., Fisher *et al.*, 1987; Gilmartin and Bullimore, 1987; Jaschinski-Kruza and Schubert-Alshuth, 1992; Ostberg, 1980; Owens and Wolf-Kelly, 1987; Pigion and Miller, 1985). Ostberg (1980) demonstrated that 2 h of close work induced a proximal shift in both the resting focus and the far-point of accommodation, although Murch (1983) could not replicate these findings. There is, however, no evidence that such changes represent fatigue rather than simply the adaptability of the human visual system. Fisher *et al.* (1987) found that although symptomatic and asymptomatic individuals showed accommodative adaptation of similar magnitudes, there were significant differences in the baseline measures and the temporal characteristics of the adaptation.

Haider *et al.* (1980) demonstrated that distance visual acuity decreased from 0·93 to 1·22 min arc following 3 h of near work whereas Dainoff *et al.* (1981) found no change in distance visual acuity for 23 subjects who undertook near work. Jaschinski-Kruza (1984) demonstrated that contrast sensitivity for high spatial frequency gratings presented at 5 m was significantly reduced after 3 h of near work and that the results of these studies were due to optical effects. It should be noted that these changes are for distance visual acuity and may not imply any change in visual function for near work nor explain any associated discomfort. Conversely, Lunn and Banks (1986) showed that after reading text presented on a VDU, contrast sensitivity was reduced for a limited range of spatial frequencies. In this instance the reduction was neural rather than optical in origin, hence it can be seen that a change in contrast sensitivity does not itself tell us anything about causal factors.

It has been suggested that visual fatigue can result in a change in eye movement behaviour. Megaw (1986) and Megaw and Sen (1984) were able to demonstrate effects of continuous VDU viewing on some eye movement parameters, although the effects also showed significant intersubject variations. Wilkins (1986) has shown that the presence of 50 Hz flicker causes the eye to overshoot its target and therefore increases the frequency of corrective saccades. Wilkins was unable, however, to explain any relationship between these changes and reports of visual discomfort. Leermakers and Boschman (1984) have shown that fixation times and the length of primary saccades are determined by the contrast of the text and that these effects are correlated with subjective reports of comfort. It is unlikely, therefore, that changes in saccadic behaviour reflect anything other than the difficulty in extracting information from the display.

A variety of other methods has been used in an attempt to evaluate visual performance. These include measuring changes in critical fusion frequency, pupil size and blink rate. The plethora of discrepant tests suggested for assessing visual workload indicates that no single visual function adequately reflects visual work. (Megaw provides some comparative assessment of different tests in the context of visual fatigue in chapter 28 of this book).

Summary — an example

Along with assessment methods, this chapter has discussed aspects of the human visual system and of viewed objects and tasks which are relevant to visual performance. By way of conclusion we present an example which illustrates the variety of factors which may need to be considered.

A client informs you that someone in their office keeps complaining of difficulty seeing information on their VDU and is definitely getting through work more slowly than expected. The office has 10 other people, performing similar work, who seem to be symptom-free. How should you approach this problem?

The person and task are apparently not well-matched; the question is whether you can assess how to improve the situation. You visit the office on a Monday morning and interview the person about matters such as the nature of the work, how it varies through the day, whether any other visual problems are experienced at home or when driving. From this discussion you establish that the difficulty actually occurs only at the end of the day. Given that no-one else in the room is having problems, even though they are all using similar equipment, the most likely explanation is that the complication is related to this person's visual system. The onset of the difficulty, towards the end of the day, suggests that there may be muscular problems either of co-ordinating the two eyes (vergence problems) or focusing (accommodation problems). To investigate these physiological aspects you could measure them yourself using a vision screener or, preferably, refer the person to their optometrist. This person is in their mid-thirties and if they were far-sighted (hyperopic) they could well be approaching presbyopia. You try out this possibility by having them move a sheet of printed text towards them until the characters begin to blur. You discover that their near point of accommodation appears to be around 15 cm which is adequate for VDU use. At this point you decide to refer them for an eye examination; as well as their vision and muscular co-ordination, the health of their eyes will be evaluated.

In the meantime, you decide to examine whether the display is adequate by using two strategies. First you suggest that the person having problems changes places temporarily with someone else in the office and that you will check back in a few days to see whether either of them are experiencing any difficulty; this will act as a check on both the VDU and the individual. Second, you assess the VDU and its image quality yourself. You note that the characters are stable, they are dark on a light background (positive contrast), glare does not appear to be a problem; the VDU is facing away from windows and even moving the screen around on its adjustable base does not lead to the reflection of luminaires in the office. Also, with your good eyesight you can read the screen at three times the normal viewing distance, not easily, but you can do so.

You re-visit the office a few days later. The optometrist's assessment was that the person's eyes were healthy, vision good, and muscular co-ordination excellent. Furthermore, neither the person themselves, nor the person now using their workplace is having difficulty seeing their VDU!

The person's vision seems normal, the physical task seems acceptable and has not altered since your first visit, and the person's job is the same as before. This seems to leave you with two possibilities: either there is a 'Hawthorne effect' operating here, i.e., the mere fact that attention is given to a perceived problem is itself a temporary alleviation of it, or there is an environmental problem which you may not yet have uncovered. You are concerned that the first of these seems more likely and so test the idea by returning the operators to their original workstations. A day later the person reports difficulties seeing the screen just as before, and so you decide to visit the office that afternoon to find out what changes have been made in the environment, and this reveals the real problem. Late in the afternoon the temperature in the office becomes too high for comfort and the practice is to open the door to the corridor which is situated behind the problem workstation. The corridor is painted white and is brightly lit with fluorescent striplights—the open door causes a veiling glare to be cast over this one VDU, reducing its visibility. The small increase in luminance caused by the veiling glare is hardly noticeable on the light background display, and looking at the screen as a whole, the operator had not been aware of glare as a problem. However, the veiling effect on the dark characters is sufficient to reduce the contrast of its dark characters significantly and hence affect the operator's performance. The other operator had been intolerant of noise from the corridor while seated by the door and had kept it closed.

Having discovered that the screen contrast has been reduced by veiling reflections successful remedies you can recommend include fitting an anti-reflection cover over the screen, moving the screen on the desk, or fixing the office heating system!

This example illustrates the range of aspects which need to be considered in the assessment of visual performance. Visual performance encompasses a variety of person-, task- and environment-related issues, each of which the ergonomist must be aware of, and any of which can be detrimental to a person's health, well-being and performance.

References

Adams, A. J., Brown, B., Flom, M.C., Jampolsky, A. and Jones, R. (1978). Influence of socially used drugs on vision and vision performance. *AGARD Conference Proceedings*, No. 218, C5. 1–11.

Adams, A.J. and Haegerstrom-Portnoy, G. (1987). Color deficiency. In *Diagnosis and Management in Vision Care*, edited by J.F. Amos (Boston: Butterworths), pp. 671–713.

ANSI (1988). *American National Standard for Human Factors of Visual Display Terminal Workstations*. ANSI/HFS 100–1988 (Santa Monica, CA: Human Factors Society).

Bailey, I.L. (1987). Mobility and visual performance under dim illumination. In *Night Vision: Current Research and Future Directions*, National Research Council Committee on Vision (Washington, DC: National Academy Press) pp. 220–230.

Bailey, I.L. and Bullimore, M.A. (1991). A new list for the evaluation of disability glare. *Optometry and Visual Science*, **68**, 911–917.

Bailey, I.L. and Lovie, J.E. (1976). New design principles for visual acuity letter charts. *American Journal of Optometry and Physiological Optics*, **53**, 740–745.

Berlinguet, L. and Berthelette, D. (Eds) (1990). *Work with Display Units 89* (Amsterdam: North-Holland).

Berry, R.N. (1948). Quantitative relations between vernier, real depth, and stereoscopic depth acuity. *Journal of Experimental Physiology*, **38**, 708–715.

Blackwell, H.R. (1946). Contrast thresholds of the human eye. *Journal of the Optical Society of America*, **36**, 624–643.

Blackwell, H.R. (1959). Specification of interior illumination levels. *Illumination Engineering*, **54**, 317–353.

Bodmann, H.W. (1962). Illumination levels and visual performance. *International Lighting Review*, **13**, 41–47.

Boff, K.R. and Lincoln, J.E. (1988). *Engineering Data Compendium: Human*

Perception and Performance, Volumes I, II and III (New York: John Wiley).

Boyce, P.R. (1965). The visual perception of movement in the absence of a frame of reference. *Optica Acta*, **12**, 47–52.

Boyce, P.R. (1974). Illumination, difficulty, complexity and visual performance. *Lighting Research and Technology*, **6**, 222–226.

Boyce, P.R. (1981). *Human Factors in Lighting* (New York: Macmillan).

Brozek, J., Simonson, E. and Keys, A. (1950). Changes in performance and in ocular functions resulting from strenuous visual inspection. *American Journal of Psychology*, **63**, 51–66.

Campbell, F.W. and Durden, K. (1983). The visual display terminal issue: a consideration of its physiological, psychological and clinical background. *Ophthalmic and Physiological Optics*, **3**, 175–192.

Carmichael, L. (1948). Reading and visual fatigue. *Proceedings of the American Philosophical Society*, **92**, 41–42.

Charman, W.N. (1983). The retinal image in the human eye. In *Progresses in Retinal Research*, Volume 2, edited by N. Osborne and G. Chader (Oxford: Pergamon), pp. 1–50.

Commission Internationale de L'Eclairage (CIE) (1972). *A Unified Framework of Methods for Evaluating Visual Performance Aspects of Lighting*. Publication CIE 19 (TC 3.1) (Paris: International Commission on Illumination).

Commission Internationale de L'Eclairage (CIE) (1981). *An Analytic Model for Describing the Influence of Lighting Parameters Upon Visual Performance*. Publication CIE 19/2.1 (TC 3.1) (Paris: International Commission on Illumination).

Cornsweet, T.N. (1970). *Visual Perception* (London: Academic Press).

Dain, S.J., McCarthy, A.K. and Chan-Ling, T. (1988). Symptoms in VDU operators. *American Journal of Optometry and Physiological Optics*, **65**, 162–167.

Dainoff, M.J., Happ, A. and Crane, P. (1981). Visual fatigue and occupational stress in VDU operators. *Human Factors*, **23**, 421–428.

de Lange, H. (1958). Research into the dynamic nature of the human fovea–cortex systems with intermittent and modulated light: I. Attenuation characteristics with white and coloured light. *Journal of the Optical Society of America*, **48**, 777–784.

De Valois, R.L. and De Valois, K.K. (1988). *Spatial Vision* (New York: Oxford University Press).

Drury, C.G. (1973). The effect of speed working on industrial inspection accuracy. *Applied Ergonomics*, **4**, 2–7.

Drury, C.G. and Addison, J.L. (1973). An industrial study of the effects of feedback and fault density in inspection performance. *Ergonomics*, **16**, 159–169.

Fisher, S.K., Ciuffreda, K.J., Levine, S. and Wolf-Kelly, K.S. (1987). Tonic adaptation in symptomatic and asymptomatic subjects. *American Journal of Optometry and Physiological Optics*, **64**, 333–343.

Gilmartin, B. (1987). The Marton Lecture: ocular manifestations of systemic medication. *Ophthalmic and Physiological Optics*, **7**, 449–459.

Gilmartin, B. and Bullimore, M.A. (1987). Sustained near-vision augments inhibitory sympathetic innervation of the ciliary muscle. *Clinical Vision Sciences*, **1**, 197–208.

Gould, J.D., Alfaro, L., Finn, R., Haupt, B. and Minuto, A. (1987a). Reading from CRT displays can be as fast as reading from paper. *Human Factors*, **29**, 497–517.

Gould, J.D., Alfaro, L., Barnes, V., Finn, R., Grischkowsky, N. and Minuto, A. (1987b). Reading is slower from CRT displays than from paper: attempts to isolate a single-variable explanation. *Human Factors*, **29**, 269–299.

Grether, W.F. and Baker, C.A. (1972). Visual presentation of information. In *Ergonomic Aspects of Visual Display Terminals*, edited by H.P. Van Cott and R.G. Kincade (Washington, DC: American Institutes for Research), pp. 41–121.

Haider, M., Kundi, M. and Weisenbock, M. (1980). Worker strain related to VDUs with differently coloured characters. In *Ergonomic Aspects of Visual Display Terminals*, edited by E. Grandjean and E. Vigliani (London: Taylor and Francis), pp. 53–64.

Hallden, U. (1972). Notes on the statistical treatment of the visual resolution. *Acta Ophthalmologica*, **50**, 47–57.

Helander, M.G. (1987). Design of visual displays. In *The Handbook of Human Factors*, edited by G. Salvendy (New York: John Wiley), pp. 507–549.

Helander, M.G., Billingsley, P.A. and Schurick, J.M. (1984). An evaluation of human factors research on visual display terminals in the workplace. In *Human Factors Review: 1984*, edited by F.A. Muckler (Santa Monica: The Human Factors Society), pp. 55–129.

Hopkinson, R.G. and Collins, J.B. (1970). *The Ergonomics of Lighting* (London: MacDonald).

Howarth, P.A. and Istance, H.O. (1985). The association between visual discomfort and the use of visual display units. *Behaviour and Information Technology*, **4**, 131–149.

Howarth, P.A. and Istance, H.O. (1986). The validity of subjective reports of visual discomfort. *Human Factors*, **28**, 347–351.

Hurvich, L.M. (1981). *Color Vision* (Sunderland, MA: Sinauer Associates).

Jaschinski-Kruza, W. (1984). Transient myopia after visual work. *Ergonomics*, **27**, 1181–1189.

Jaschinski-Kruza, W. and Schubert-Alshuth, E. (1992). Variability of fixation disparity and accommodation when viewing a CRT visual display unit. *Ophthalmic and Physiological Optics*, **12** (4), 411–419.

Khek, J. and Krivohlavy, K. (1966). Variation of incidence of error with visual task difficulty. *Light and Lighting*, **59**, 143–145.

Knave, B.G. (1983). The visual display unit. In *Ergonomic Principles in Office Automation* (Stockholm: Ericsson Information Systems), pp. 11–41.

Knave, B.G., Wiborn, R.I., Voss, M., Hedstrom, L.D. and Berqvist, U.O. (1985). Work with video display terminals among office employees: 1. Subjective symptoms and discomfort. *Scandinavian Journal of Environmental Health*, **11**, 457–466.

Kruk, R.S. and Muter, P. (1984). Reading of continuous text on video screens. *Human Factors*, **26**, 339–345.

Leermakers, M.A.M. and Boschman, M.C. (1984). Eye movements, performance and visual comfort using VDTs. *IPO Annual Progress Report*, **19**, 70–75.

Luczak, H., Cakir, A. and Cakir, G. (Eds) (1993). *Work with Display Units 92* (Amsterdam: North-Holland).

Lunn, R. and Banks, W.P. (1986). Visual fatigue and spatial frequency adaptation to video display of text. *Human Factors*, **28**, 457–464.

Malmstrom, F.V., Randle, R.J., Murphy, M.R., Reed, L.E. and Weber, R.J. (1981). Visual fatigue: the need for an integrated model. *Bulletin of the Psychonomic Society*, **17**, 183–186.

McCormick, E.J. and Sanders, M.S. (1987). *Human Factors in Engineering and Design*, 6th edition (New York: McGraw-Hill).

Megaw, E.D. (1979). Factors affecting inspection accuracy. *Applied Ergonomics*, **10**, 27–32.

Megaw, E.D. (1986). VDUs and visual fatigue. In *Contemporary Ergonomics 1986*, edited by D.J. Oborne (London: Taylor and Francis), pp. 254–258.

Megaw, E.D. and Sen, T. (1984). Changes in saccadic eye movement parameters following prolonged VDU viewing. In *Ergonomics and Health in Modern Offices*, edited by E. Grandjean (London: Taylor and Francis), pp. 352–357.

Murch, G.M. (1983). Visual fatigue and operator performance with DVST and raster displays. *Proceedings of the Society for Information Display*, **14**, 53–61.

Muter, P., Latremouille, S.A., Treurniet, W.C. and Beam, P. (1982). Extended reading of continuous text on television screens. *Human Factors*, **24**, 501–508.

National Research Council Committee on Vision (1983). *Video Displays, Work and Vision* (Washington: National Academy Press).

Neisser, U. (1964). Visual search. *Scientific American*, **210**, 94–100.

Nordqvist, T., Ohlsson, K. and Nilsson, L. (1986). Fatigue and reading text on videotext. *Human Factors*, **28**, 353–363.

Oborne D.J. (1982). *Ergonomics at Work* (New York: John Wiley).

Ostberg, O. (1980). Accommodation and visual fatigue in display work. In *Ergonomic Aspects of Visual Display Terminals*, edited by E. Grandjean and E. Vigliani (London: Taylor and Francis), pp. 41–52.

Owens, D.A. and Wolf-Kelly, K. (1987). Near work, visual fatigue and variations in oculomotor tonus. *Investigative Ophthalmology and Visual Science*, **28**, 743–749.

Pigion, R.G. and Miller, R.J. (1985). Fatigue of accommodation: changes in accommodation after visual work. *American Journal of Optometry and Physiological Optics*, **62**, 853–863.

Reeves, B.C., Wood, J.M. and Hill, A.R. (1993). The reliability of high and low contrast letter charts. *Ophthalmic and Physiological Optics*, **13** (1) 17–26.

Regan, P. and Neima, D. (1983). Low-contrast letter charts as a test of visual function. *Ophthalmology*, **90**, 1192–1200.

Salaman, M. (1929). *Some Experiments on Peripheral Vision*. MRC Special Report 136 (London: HMSO).

Shneiderman, B. (1987). *Designing the User Interface: Strategies for Effective Human–Computer Interaction* (Reading, MA: Addison-Wesley).

Smith, R.L. and Lucaccini, L.F. (1977). Vigilance research: its application to industrial problems. In *Human Aspects of Man–Machine Systems*, edited by S.C. Brown and J.N.T. Martin (New York: Open University Press).

Snyder, H.L. (1988). Image quality. In *Handbook of Human–Computer Interaction*, edited by M. Helander (Amsterdam: North Holland), pp. 437–474.

Stone, P.T., Clarke, A.M. and Slater, A.I. (1980). The effect of task contrast on visual performance and visual fatigue at a constant luminance. *Lighting Research and Technology*, **12**, 144–159.

Thibos, L.N., Bradley, A., Still, D.L., Zhang, X. and Howarth, P.A. (1990). Theory and measurement of ocular chromatic aberration. *Vision Research*, **30**, 33–49.

Travis, D.S., Stewart, T.F.M. and Mackay, C. (1992). Evaluating image quality. *Displays*, **13**, 139–146.

Wertheim, T. (1891). Peripheral visual acuity, translated by I.L. Dunsky, 1980. *American Journal of Optometry and Physiological Optics*, **57**, 915–924.

Westheimer, G. (1979a). The spatial sense of the eye. *Investigative Ophthalmology and Visual Science*, **20**, 893–912.

Westheimer, G. (1979b). Scaling of visual acuity measurements. *Archives of Ophthalmology*, **97**, 327–330.

Westheimer, G. and McKee, S.P. (1975). Visual acuity in the presence of retinal image motion. *Journal of the Optical Society of America*, **65**, 847–850.

Weston, H.C. (1945). *The Relation between Illumination and Visual Performance*. Report No. 87 (London: Industrial Health Research Board).

Wilkins, A. (1986). Why are some things unpleasant to look at? In *Contemporary Ergonomics 1986*, edited by D.J. Oborne (London: Taylor and Francis), pp. 259–263.

Zeki, S. (1993). *A Vision of the Brain* (Oxford: Blackwell Scientific Publications Ltd).

Chapter 28

The definition and measurement of visual fatigue

E.D. Megaw

Definition of visual fatigue

The early literature is full of attempts to arrive at an acceptable definition of visual fatigue, beginning with the frequently referenced phrase of Plautus quoted by Ramazzini (1700)—'Sitting hurts your loin, staring your eyes'. This implies that the subjective feeling of pain is one of the essential features of visual fatigue. Weber (1950) claimed that ocular fatigue is a form of 'occupational neurosis' like writer's cramp. The term neurosis was taken more seriously by Adler (1962) who believed that many ocular complaints are really psychosomatic in origin, a view that would not be very popular with today's workforce.

The results of early attempts to define visual fatigue can be seen generally to have failed as reflected in the following two quotes. Simmerman (1950) concluded that 'The subject of fatigue has been investigated from many angles by many research workers. However, the results of all investigations have failed to give us a clear and concise definition of the meaning of fatigue'; Collins (1959) found that 'In speaking about visual fatigue, we are speaking about something which has no precise and unique definition. Visual fatigue means different things to those concerned with different aspects of vision'. There is, in fact, a strong temptation to take the view expressed by Muscio who in 1921, after considering whether any kind of general fatigue test was possible, concluded that the term fatigue should be totally banished from precise scientific discussion and that attempts to develop a fatigue test be abandoned. Rather, he suggested that one looked for direct correlations between task conditions and various measures of human functioning so that so-called tests of fatigue could be used to determine the effects of the task factors rather than to determine the presence or absence of fatigue itself.

Many early studies have shown that several measures one might have suspected as being sensitive to various visual task factors are not so. In an early study of Carmichael (1948), subjects read one of two books for 6 continuous hours while their eye movements were recorded. One of the books was Adam Smith's *Wealth of Nations* which was considered by many of the subjects to be dull while the other, Blackmore's *Lorna Doone*, was often considered interesting! The text was presented either as hard-copy or projected on a microfilm device. The lighting levels in both cases were around 170 lux. Results showed no effects of continuous reading on any of the eye movement measures and on comprehension scores. There was no change in visual acuity measures taken before and after the reading task. However, some subjects did report an increase in reported feelings of fatigue as the experiment progressed including 'wishing to stop'. All these results were independent of the reading material and the method of presentation. While the low accuracy and reliability of early methods of recording and analysis should not be forgotten, this typical lack of significant results on objective measures of fatigue explains why so much importance is attributed to subjective measures or, to give them their proper title, symptoms of asthenopia. These symptoms can be grouped into ocular symptoms such as discomfort of the eyes and feelings that the eyes are tired, itchy, dry and burning, visual symptoms such as difficulty in focusing and blurred vision and, finally systemic symptoms reflected in reports of headaches, postural fatigue and general tiredness. It is these frequently reported symptoms of asthenopia associated with the introduction of visual display units (VDUs) over the last 10 years that have rekindled the interest in the subject (Matula, 1981; Stellman *et al.*, 1987). Moreover, as a result of improved recording and analysis methods and a closer understanding of the visual mechanisms involved, we are now in a better position to look at the possible relationships between task factors and human functioning along the lines suggested by Muscio (1921).

Not wishing totally to avoid giving a definition of visual fatigue, it is proposed that the following points should be included if and when a more formal definition is proposed:

(a) Visual fatigue does not occur instantaneously but involves being subjected to the same visual task conditions over a period of time during the course of which the fatigue tends to build up. If a person suffers from some visual defect such as uncorrected refractory errors, the time course may be comparatively rapid.

(b) Visual fatigue should be distinguished from mental workload which is associated with the information and cognitive demands of the task.

(c) Visual fatigue can be overcome, often very rapidly, either by rest or by changing the task conditions. The presence of fatigue does not have any long–term harmful effects except possibly if an individual is prone to some related disease.

(d) Visual fatigue should be distinguished from an adaptive response of the visual system. That is to say, like any other biological system, the functioning of the visual system adapts to the particular conditions it meets. Sometimes this adaptation process takes a comparatively long time as reflected, for example, by the typical dark adaptation curve. On other occasions, it is very fast. The neural adaptation in the retina to small changes in luminances takes only a matter of milliseconds (Rushton and Westheimer, 1962).

(e) Symptoms of asthenopia are the main reasons for assuming the existence of visual fatigue.

(f) Individual factors, as well as task and environmental ones, can contribute to fatigue. If a person has uncorrected refractive errors, particularly say if they are presbyopic as well as already being hyperopic, they are bound to experience feelings of discomfort when performing a reasonably demanding visual task. The same applies to those suffering from uncorrected abnormal muscle balance.

(g) Symptoms of asthenopia can result from conditions which are not visual in origin. To take a very obvious example, a dry environment containing noxious particles can lead to complaints of visual discomfort (Messite and Baker, 1984).

Origins of visual fatigue

The origins of visual fatigue are probably threefold. Most likely it is fatigue of one or more of the oculomotor control systems. The other two relate, firstly, to fatigue of the neural processes involved in visual processing ranging from effects at the retina, the optic nerve, the lateral geniculate body through to the striate cortex and, secondly, to relatively non-specific effects such as arousal levels and the amount of so-called effort exerted by subjects. There are obvious problems in disassociating the latter from effects related to mental workload (see chapter 25). Moreover, it is obvious that these three sources of fatigue are not independent. A lowering of arousal may affect any of the oculomotor control systems, as exemplified by the effects of alcohol and various drugs on the execution of rapid eye movements (Wilkinson et al., 1974; Gentles and Llewellyn Thomas, 1971). Similarly, a lowering of the resolution powers of the retina will have effects on the accommodation demands as blur is one of the main stimuli for accommodation (Phillips and Stark, 1977). A further problem is that the presence of other factors can obscure the origins of fatigue. For example, the pupil response is sensitive to short term memory load and arousal level (Kahneman and Beatty, 1966) as well as to visual factors such as levels of illumination and depth of focus requirements. It is not difficult to appreciate that in these circumstances the use of well controlled experimental procedures is essential.

The main oculomotor control systems in regard of each eye can be listed as follows:

– Accommodation control achieved by the ciliary muscle which is intermediate between striated and non-striated muscles.

- Vergence control and muscle balance achieved by six extraocular eye muscles, the four recti and two oblique muscles.
- Version control in relation to both saccadic and pursuit eye movements achieved by the same muscles responsible for vergence.
- Pupil control achieved by two muscles which are non-striated, the sphincter pupillae causing constriction of the pupil and the dilator pupillae causing dilation.
- Blink control achieved mainly by a sphincter muscle, the orbicularis palpebrarum, although the upper eyelid can be raised by the action of the levator palpebrae superioris to expose the eye ball.

The extent of the interaction between these control systems means that it is often very difficult to ascertain whether any changes in the functioning of one of these reflects fatigue or whether it is an adaptive change whereby the emphasis on the control of the visual system is shifted from one oculomotor system to another. The most often quoted example of such an interactive adaptive response is reflected by the close dependence of the accommodation, vergence and pupillary mechanisms. In particular, pupil constriction, which increases the depth of focus, can reduce the demands on the accommodation system.

Measurement of visual fatigue

Methods for measuring visual fatigue can be divided into four categories. Firstly, there are those methods which attempt to measure more or less directly changes in functioning of the various oculomotor systems that have been described earlier. Then there are methods based on measures of visual acuity which may reflect changes in functioning of the oculomotor systems but also may reflect changes in visual processing ability. Thirdly, there are methods which look more generally at visual task performance. Naturally, any changes in performance may reflect effects on both visual acuity and oculomotor functioning. Finally, there are methods based upon reported symptoms of asthenopia.

Measures of oculomotor functioning

Eye movements

Many different types of eye movement are executed by the extraocular muscles. They range in amplitude from less than one minute of arc to 40 degrees or more and in acceleration from a very few deg s^{-2} to 40,000 deg s^{-2}. They are often accompanied by head movements. The essential eye movement types are listed in Table 28.1.

There is a variety of methods to record eye movements and these have

Table 28.1. Types of eye movement

Saccadic eye movements	rapid voluntary conjugate movements resulting in changes of fixation observed during visual search and reading executed in order to bring the retinal image of the object being viewed onto the fovea where there is high spatial resolution
Corrective saccades	conjugate saccadic movements to compensate for target undershoot or overshoot by the main saccadic eye movements
Tremor, slow drift and microsaccades	a variety of very small amplitude movements, less than one degree, associated with visual fixation of a target
Pursuit or slow tracking movements	slow smooth conjugate movements executed to stabilize the retinal image as a moving target is tracked, normally not under voluntary control
Compensatory eye movements	smooth movements, closely related to pursuit movements and the slow phase of vestibular nystagmus which compensate for active or passive movements of the head or trunk
Vergence eye movements	movements of the two eyes in opposite directions to facilitate fusion of the retinal images of the two eyes when objects are viewed at different distances
Optokinetic and vestibular nystagmi	a mixture of smooth image stabilizing movements to either a continuously moving visual field or stimulation of the semicircular canals and rapid 'return' saccadic eye movements
Dynamic overshoot	small amplitude saccadic-like eye movements which sometimes occur in the opposite direction to the saccadic eye movements immediately preceding them, usually monocular
Glissades	slow movements which sometimes occur in either the same or opposite directions as the saccadic eye movements immediately preceding them, often monocular

been reviewed very thoroughly by Young and Sheena (1975). Apart from direct observation, the following methods are commonly used—

Electro-oculography. By placing surface electrodes around the eye, changes in the corneoretinal potential can be recorded as the eye rotates in respect of the head.

Corneal-reflection. Because of the geometry of the cornea, the angle of reflection of a light source directed onto the surface of the cornea changes as the eye rotates in relation to the head. The reflected light source is often recorded using a video system. The most well known commercial system based on this method is the NAC Eye Mark Recorder which also registers pupil size (Figure 28.1a illustrates version V).

Limbus tracking. A light source, usually an infrared one, is directed at the boundary between the iris and the sclera (the limbus). As the eye rotates, there is a change in the amount of reflected light due to the differences in the reflection characteristics of the iris and the sclera which can be detected using photosensitive diodes or some other method of recording the amount of light (see Figure 28.1b).

Contact lens methods. There are several different contact lens methods. With the original method, a mirror was attached to the cornea so that a beam of

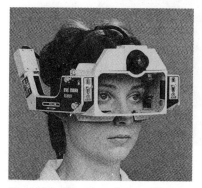

Figure 28.1. Illustration of some of the common methods used to register eye movements and point of eye fixation (a) A corneal-reflection system manufactured by NAC which permits free head movements and is suitable for registering point of regard

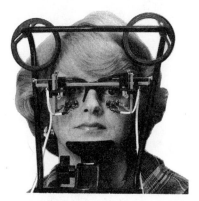

Figure 28.1. (b) A limbus tracking system manufactured by Applied Science Laboratories which requires the head to be fixed

light could be directed onto the mirror and reflected from it to the recording device, the angle of the reflected beam changing as the eye rotates. A more recent development involves attaching a small induction coil onto the limbus. Eye movements are detected by changes in magnetic induction as the coil moves in relation to a uniform magnetic field generated by a series of coils arranged around the head. Such a system is manufactured by Skalar and is illustrated in Figures 28.1c and 28.1d.

Pupil-centre corneal reflection. If one looks at a remote source of light reflected in the pupil, its position in relation to the centre of the pupil will vary as a function of head or eye movements. By video recording the pupil, it is possible to identify where somebody is looking by measuring the relative position in the pupil of the reflected source. This is the basis of the remote oculometers manufactured by Applied Science Laboratories, ISCAN and by

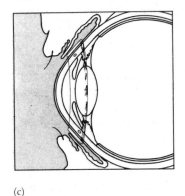

(c)

(d)

Figure 28.1 (c) and 28.1 (d) A magnetic induction system manufactured by Skalar which requires the head to be fixed (c illustrates the attached scleral induction coil)

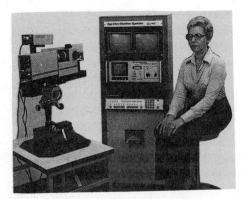

Figure 28.1 (e) A pupil–centre corneal–reflection system manufactured by Applied Science Laboratories which permits head movements

Demel (see Figure 28.1e). These systems also permit pupil size to be monitored.

Double Purkinje image method. This method is employed by the Stanford Research Institute system and takes advantage of the relative changes in angle of the first (front of the cornea) and fourth (rear of the lens) reflections from the eye as it rotates.

The choice of methods is influenced by many factors, the most important of which may often be cost, which can vary from a few hundred pounds to over £80,000. The questions to consider are:

– are you mainly interested in where somebody is looking (point of regard) or in the characteristics of the eye movements themselves, or both?

- is it necessary to permit head movements or is a fixed-head system acceptable?
- do you wish to record movements other than horizontal ones?
- what are the prevailing lighting conditions?
- do you wish to record eye movements from people wearing glasses or contact lenses?
- should the system be unobtrusive?
- what is the smallest amplitude of eye movement you wish to register?
- what accuracy of recording is required?
- do you require measurements of pupil size?

Since the review of Young and Sheena (1975), developments in computerized recording and analysis have had an enormous effect on the quality of the data that can be obtained. Not only has accurate data calibration become easier to achieve but, by using high sampling rates and digital filtering, the undesirable effects of system noise are greatly reduced (McConkie, 1981; Harris *et al.*, 1984).

That eye movement parameters can be affected by fatigue to the saccadic eye movement system was suggested by the results of Bahill and Stark (1975). They showed that the longer subjects performed a step-tracking task, the lower was the peak velocity of their saccades, the more likely it was for saccades to overlap and the greater was the probability of glissades being executed. Using a standard infrared limbus tracking technique, Megaw and Sen (1984), Sen and Megaw (1984) and Megaw (1986) were unable to demonstrate similar effects as a result of subjects performing tasks requiring continuous VDU viewing. On the other hand, they did observe changes in the probabilities of execution of glissades and dynamic overshoot in some subjects but not always in the direction predicted by the results of Bahill and Stark (1975). It should be said that the laboratory task demands imposed by Bahill and Stark are unlikely to be met in the course of day-to-day visual behaviour.

One factor that has been shown to disturb the control of saccadic eye movements is flicker or intermittent task illumination. Whether originating from a CRT display screen or from conventional fluorescent lighting, Wilkins (1986) and Kennedy and Murray (1991) have reported that the presence of flicker caused alterations to primary saccade amplitudes and an increase in the frequency of corrective saccades when performing reading tasks. Wilkins (1991) has speculated that the flicker results in the short presentation of a stimulus during the course of the primary saccade, so that the normal process of saccadic suppression resulting from masking (Martin *et al.*, 1972) does not occur, and the stimulus can, therefore, be seen during the course of the primary saccade. Such an explanation is supported by the results of Neary and Wilkins (1989) demonstrating the occurrence of a higher frequency of corrective saccades with CRT displays having a short rather than long persistence phosphor. What remains uncertain is whether these disturbances to the control of saccadic movements do contribute to visual fatigue, particularly as they are not normally supported by reported symptoms of visual discomfort, although Wilkins et al. (1989) have reported a decrease in the frequency of reported headaches when room lighting is achieved by using a relatively steady form of fluorescent lighting. Even if flicker is thought to be potentially fatiguing, there is the opposing evidence that watching television is rarely accompanied by reports of visual discomfort to the same extent as viewing computer screens. For this reason, Wilkins (1991) has suggested that there must be additional factors which cause fatigue when viewing CRT display screens, for example, the spatial ambiguity of text in resembling striped pattern with relatively small horizontal and vertical spacings. This last point is taken up later in this chapter (see chapter 29 for discussion of flicker perception—CFF—as a measure of workload).

In addition to the effects of flicker, Leermakers and Boschman (1984) have shown that fixation times and the length of primary saccades are determined by the contrast of reading material and that these effects are correlated with reports of visual discomfort. However, it is extremely unlikely that the changes in saccadic behaviour reflect anything other than the increased difficulty in extracting information from the display as contrast is reduced and does not reflect fatigue to the saccadic system.

If subjects continuously track a sinusoidally moving target, the amplitude of the accompanying pursuit movements decreases (Malmstrom et al., 1981). However, when it came to comparing pursuit eye movement performance before and after reading aloud from either a normal or high resolution display terminal, Miyao et al. (1988) were unable to demonstrate any effect on root mean square tracking error as a result of performing the intervening reading task. Nor were they able to demonstrate any differences between the two displays. On the other hand, there was an increase in reported symptoms of visual fatigue following the reading task.

Studies on vergence eye movements and blink rate are described under subsequent headings.

Accommodation

Because of the limitation in recording methods, most of the early studies relating to accommodation concentrated on measures of accommodation time or of the nearest point of accommodation, rather than on measurements of the accommodation power of the lens while viewing a particular stimulus. Results from the early studies are inconclusive. Both Collins and Pruen (1962) and Krivohlavy et al. (1969) have reported an increase in accommodation time following performance on interpolated tasks while Brozek et al. (1950) found no increase in the near point of accommodation following strenuous inspection work. Similar techniques have been used more recently by Stone et al. (1980), Mourant et al. (1981) and Gunnarsson and Soderberg (1983). Stone et al. (1980) failed to find any consistent differences in accommodation time before and after performing one hour of various visual tasks with different levels of contrast. On the other hand, Mourant *et al.* (1981) reported that accommodation time significantly increased after performing a visual search task displayed on a video display but no such increase when the task was in the form of hard copy. These results were supported by the results from Gunnarsson and Soderberg (1983) who reported a decrease in the near point of accommodation following strenuous VDU work accompanied by increases in subjective symptoms of visual discomfort.

There are some doubts about the validity of all these results, mainly because of the relative inaccuracy of the recording methods. On more specific grounds, the results of Mourant *et al.* (1981) have been queried because the near and far targets they used to measure speed of accommodation were sufficiently close together so as not to require any accommodative response as a result of the available depth of perception. It should also be remembered that measures of speed of accommodation and near point of accommodation typically involve a perceptual element, in that subjects have to perform some kind of target detection task such as detecting the gap in a Landolt ring; it may be, therefore, that the reported findings reflect sensory rather than oculomotor fatigue.

As a result of recent technical developments, it has been possible to monitor more or less directly the accommodative power of the eyes. Three main techniques have been used: laser optometry (Hennessey and Leibowitz, 1972), infrared optometry (Cornsweet and Crane, 1970) and polarized vernier optometry (Simonelli, 1980). With some optometers, subjects have to make subjective estimates in order that the accommodative state can be assessed while in others the state is assessed directly and it has been claimed that, for example, the necessity to make judgements about the direction of speckle movement when using traditional laser optometers may influence the recorded states of accommodation (Post *et al.*, 1984).

Using a tracking task where subjects had to alter their accommodation between targets moving sinusoidally between 0.0D (infinity) and 0.4D (25 cms), Malmstrom *et al.* (1981) were able to show a decrease in the amplitude of accommodation over time. Iwasaki and Kurimoto (1987) proposed the low frequency component of accommodative oscillations as a possible indicator of fatigue. They demonstrated that these oscillations significantly increased after performing a VDU-based task but not if the task was presented in traditional hard-copy form. The results were supported by subjective reports of visual discomfort. In a later study, Iwasaki and Kurimoto (1988) have reported an increase in the accommodation time when changing focus from a near to a far target from subjects who performed one hour of VDU-based work. They claimed that this increase was accompanied by some changes in the latencies and amplitudes of visually evoked potentials (VEPs). Further studies on accommodative state are considered under the heading of viewing distance.

Convergence and phorias

Convergence can be measured by some of the same methods used to measure accommodation. Traditional methods usually involve measuring the near point of convergence or the amplitude of convergence. For example, Luckiesh and Moss (1935a) developed a method whereby a pair of prisms were automatically rotated in front of the subject's right eye in order gradually to bring a stimulus closer to a subject until the prism motion was stopped by the subject when he or she reported seeing the stimulus as double. After reading for one hour under a low level of illumination (11 lux), results showed that the amplitude of convergence as measured by prism strength was reduced from around 16 to 13 dioptres. Gunnarsson and Soderberg (1983) measured the near point of convergence of VDU operators by moving a pointer along a ruler towards the base of the nose until they reported seeing a double image. An increase in the near point of convergence was observed (i.e., a reduced ability to converge) as the working day progressed. There are, however, doubts over the reliability of these results suggesting a fatigue to the convergence system. Stone *et al.* (1980), based on similar methods of measurement to Gunnarsson and Soderberg, reported that subjects showed a greater ability to converge after performing visually demanding tasks, irrespective of the contrast of the visual material. In addition, using laser optometry to measure the point of task convergence, Murch (1983) was not able to find any effects following 2.5 hours of intense display work.

Rather than using measures of convergence, some authors have looked for changes in muscle balance following visually demanding work. Muscle balance can be crudely measured by industrial visual screening equipment such as the Bausch and Lomb Vision Tester or the Keystone Ophthalmic Telebinocular. To measure the balance, separate stimuli are displayed to the two eyes and because there is no meaningful way that the two images can be merged, it is possible to measure the tendency of the eyes to deviate inwards (esophoria) or outwards (exophoria) when binocular fusion is impossible. Some studies have found no changes in phoria following strenuous visual work (e.g., Collins, 1959) while others (Stone *et al.*, 1980) have shown a trend to esophoria. What seems likely is that whether or not changes occur in near point convergence or phoria will depend upon the actual task viewing distance. Recently, Marek *et al.* (1988) have reported an increase in the instability of phorias as VDU work progressed.

Viewing distance

A commonly expressed view on visual fatigue is reflected in the quote from Tyrrell and Leibowitz (1990) where they state 'Although scientists and grandmothers seem to agree that visual fatigue is related to near work, there is no consensus as to why the relationship exists'. In this context, much of the recent interest in the use of accommodation and convergence measures has been concerned with the relationship between task viewing distance and visual fatigue. The starting point for this approach is that when convergence and accommodation are measured in darkness or in a homogeneous light field (Ganzfeld), the resting point of the relevant eye muscles is not at a position equivalent to infinity but a position intermediate between infinity and the near point of vision. These values are usually referred to as the dark focus or dark vergence points. There is a considerable amount of individual variation in these values. For example, in the study of Owens and Leibowitz (1980), the recorded mean dark focus of accommodation for 220 college students was 1·52D (66 cms) and ranged from around 3·0D (33 cms) to near 0·0D (infinity). With respect to dark vergence, Tyrrell and Leibowitz (1990) reported a mean dark convergence point of 62 cms with values ranging from 30 cms to 120 cms for a group of 104 persons.

A number of points should be made before discussing the implications of dark focus and vergence for visual fatigue. Firstly, it is not uncommon for different studies to report quite disparate values of both accommodation and dark vergence. This is partly a result of the different recording techniques employed (mentioned previously in relation to the measurement of accommodation, Post *et al.*, 1984) as well as reflecting individual differences particularly when small sample sizes are used. Additionally, recorded values can be affected by the conditions under which the experiments are conducted. For some individuals, the dark vergence points are influenced by the vertical angle of gaze such that it is greater when subjects elevate their gaze compared with lowering their gaze (Heuer and Owens, 1989), a phenomenon (now referred to as the Heuer effect) that may have important consequences for understanding visual fatigue arising from VDU-based work. Moreover, there is only a weak relationship between a person's dark focus and dark vergence points so that for some individuals their dark focus point is longer than their dark vergence points and vice versa (Owens and Leibowitz, 1980). This may reflect a tendency of the two oculomotor control systems to become decoupled in the absence of visual cues in the environment.

Despite these reservations, there is considerable support for the hypothesis that visual fatigue might originate from a mismatch between a person's task viewing distance and his or her dark focus or vergence points. Tyrrell and Leibowitz (1990), for example, reported a small but significant correlation between reported symptoms of eye strain and dark vergence point, suggesting that those subjects who had a longer dark vergence point would require greater vergence effort when viewing a VDU screen. Similarly, they found that for those subjects who demonstrated the Heuer effect, greater visual discomfort was experienced when viewing a VDU with gaze elevation 20 deg above the base line elevation than 20 deg below it. Jaschinski-Kruza (1988) divided subjects into those with near or far dark focus points. Subjects then performed a search and comparison task while viewing the display screen at distances of 50 or 100 cms. For the 'far' group, higher levels of reported fatigue and slower task performance were associated with a screen distance of 50 cms than with one of 100 cms. For the 'near' group no such differences were found. In general both groups of subjects preferred the 100 cm to the 50 cm condition. In a later study, Jaschinski-Kruza (1991) reported similar results based on dark convergence point. To some extent these results are reflected in the data collected by Schleifer *et al.* (1990). Rather than using measures of dark focus, they obtained estimates of preferred comfortable reading distance and found that ocular discomfort was increased the greater the mismatch between a person's preferred reading distance and his or her actual VDU viewing distance.

If accommodation is measured while subjects perform a visual task, the accommodative state of the eye tends to be some way intermediate between the task viewing distance and the subject's dark focus point (Ostberg, 1980; Murch, 1983). In addition, Ostberg (1980) and Ostberg and Smith (1987) have found that after performing a demanding VDU-based task, there is a shift both in the dark focus point of individuals and in their accommodation to targets presented at distances greater than their dark focus point, in the direction of greater myopia. Because this increase in myopia is accompanied by increased reports of visual discomfort, the authors suggest that accommodative state should be considered seriously as an objective measure of visual fatigue.

There is, however, a danger in placing too much emphasis on the results concerning accommodative state and viewing distance. Most importantly, it should be remembered that people frequently perform different kinds of visual task such as reading from hard copy, fine assembly work and driving, at distances which do not coincide with their dark focus and vergence points, and without experiencing visual discomfort. Ideal viewing distance is just as likely to be influenced by task features such as image quality and display format, and by the amount of variation in viewing distance available during the course of performing the task.

Pupil size

Pupil size has been monitored by Zwahlen *et al.* (1984) and by Geacintov and Peavler (1974). The former study used the pupil-centre corneal-reflection technique and failed to find any effects on pupil diameter as a result of performing a lengthy VDU-based task. This result confirms Campbell and Whiteside's (1950) conclusion that it is probably difficult, if not impossible, to fatigue the sphincter muscle of the iris. On the other hand, the study of Geacintov and Peavler based on infrared photographic records of the pupil did show a gradual constriction of the pupil with continuous work when using material either printed on paper or displayed on a microfilm reader. It was difficult to conclude if the extent of the constriction varied between the two presentation methods. There is the possibility that the increased constriction reflected either a lowering of arousal or an adaptive response to increase the depth of field and thus reduce the demands on accommodation control.

Blinking

Blinking can be measured comparatively easily by either direct filming techniques or by using electromyographic (EMG) recording techniques. Bitterman and Soloway (1946) compared both methods and failed to find any effects of glare on blink rate. Results of Brozek *et al.* (1950) failed to show any consistent reduction in blink rate as a result of performing a strenuous visual task. On the other hand, Poulton (1958), on re-analyzing the results of Carmichael and Dearborn (1947), found evidence of increased blink rate as a function of the time spent on reading. These results were confirmed by Krivohlavy *et al.* (1969) who were concerned with close machine work. It seems likely that blink rate is affected by mental workload and this is supported by the results of Stern and Skelly (1984) which showed that fewer blinks were executed by pilots when controlling an aircraft compared with when they were not controlling the aircraft. In this context, Stark (National Research Council, 1983) has suggested that blinking be considered as a 'cybernetic windshield wiper' for the retina. Because a reduction in blink rate leads to drier eyes, it is not unexpected to find that more demanding tasks should yield complaints of irritation to the eyes. In the same way, a dry atmosphere is likely to cause similar complaints or an increase in blink rate.

Rather than using blink rate as an index of fatigue, Berg *et al.* (1988) have used the electrical activity originating in the muscles surrounding the eye socket—paraorbital electromyography. These muscles, in addition to controlling blinking, control the muscles which determine the facial expressions, including squinting, that are often associated with eyestrain, and particularly those said to result from specular glare. Berg *et al.* (1988) found significant changes in the spectral power distribution of the recorded muscle activity as a result of prolonged VDU work and they also showed that the distribution was affected by whether subjects performed the proofreading task under what could be described as either good or poor VDU viewing conditions.

Visual acuity measures

Using the so-called 'li' test developed by Ferree and Rand (1927), Tinker (1939) reported decreases in clear seeing after reading text under fairly low levels of illumination (around 10 lux). With improved recording techniques, Haider *et al.* (1980) have proposed transient myopia to far targets as a possible indicator of fatigue, having observed significant decreases in acuity following 3 hours of continuous VDU work. Apart from the fact that other investigators have failed to replicate these results (e.g., Dainoff *et al.*, 1981), it is difficult to decide whether to attribute reduced acuity to fatigue of one of the oculomotor control systems such as the vergence or accommodation systems, to a reduced sensitivity of the visual processing processes or to a general reduction in arousal.

As well as the reported transient myopia following prolonged VDU viewing (Haider *et al.*, 1980; Ostberg and Smith, 1987), a number of studies have reported a decrease in spatial sensitivity (Jaschinski-Kruza, 1984; Lunn and Banks, 1986: Watten *et al.*, 1992). There is some argument as to whether this results from fatigue to oculomotor function or from contrast adaptation to the dominant spatial frequencies present in the display, thus reflecting adaptation within the visual processing system. Lunn and Banks' results supported the contrast adaptation hypothesis by demonstrating that the maximum increase in threshold occurs at the dominant display frequencies (i.e., single spaced text viewed at approximately 75 cms equivalent to a spatial frequency of 2·6 cycles per deg.). On the basis of this, Lunn and Banks suggested that, because this is in the range of frequencies important for accommodation control, this could account for the disturbed accommodation response observed during VDU viewing as well as the subjective reports of visual fatigue. On the other hand, in the study of Watten *et al.* (1992) a significant decrease in contrast sensitivity was obtained across all spatial frequencies (from 1·5 to 18·0 cycles per deg.). The task involved computer programming and it is not clear what were the dominant frequencies in the display. Additionally Watten *et al.* reported no difference in the decrease in sensitivity following 4 hours of programming compared with 2 hours. In fact Magnussen *et al.* (1992) have reported decreases in contrast sensitivity after only 10 minutes of performing a reading task and that the effect was very much larger with bright characters on a dark screen than with the other polarity in respect of a spatial frequency equivalent to the line spacing of 1·4 cycles per deg. Complete recovery to pre-work contrast sensitivity took a time approximately equivalent to the duration of the work period, but the relationship was not linear and 50% recovery had taken place after only one or two minutes.

In contrast to the above findings, Woo *et al.* (1987) were unable to demonstrate any consistent changes in contrast sensitivity at the end of a day of VDU-based work. Additionally, there remains the question whether this possible explanation can account for the lack of reported symptoms of asthenopia accompanying the reading of hard copy.

As well as spatial acuity, there have been studies which have used temporal measures. Aoki *et al.* (1984) failed to find any effect on critical fusion frequency (CFF) of continuous VDU work but Osaka (1985) reported a reduction in CFF for both foveal and peripheral tests particularly if the VDU task was performed with either blue or red characters. As with measures of spatial acuity, it is difficult to establish what is responsible for this decrease in CFF. That general arousal levels are affected by continuous performance of a visual task was demonstrated as early as 1935 by Luckiesh and Moss (1935b). They reported that heart rate level progressively decreased as the subject read text under an illuminance of 11 lux. Since then, CFF has frequently been interpreted as a measure of general arousal (e.g., Weber *et al.*, 1980).

Visual performance measures

Numerous studies have reported results on performance measures in relation to the main fatiguing task. Apart from reading tasks (Carmichael, 1948; Nordqvist *et al.*, 1986) and simulated inspection tasks (Brozek *et al.*, 1950; Murch, 1983), a number of specially designed laboratory tasks have been used. These include Weston's original Landolt rings test (Weston, 1953), Bodmann's search tasks (Bodmann, 1962) and Neisser's search tasks (Neisser, 1964). These tasks were originally used to compare the effects of various task factors including target illuminance, contrast and size. Modifications to these tasks have been used more recently to examine the same factors (Khek and Krivohlavy, 1966; Boyce, 1974; Stone *et al.*, 1980). Occasionally, these specially designed tasks are given in addition to the main fatiguing tasks. Thus Nordqvist *et al.* (1986) presented subjects with a Neisser-type search task after completing every 15 mins of the main reading task. The results in relation to scanning speeds and errors on the Neisser-type search task were inconclusive and showed little evidence of a decline in performance as a result of the intervening demanding tasks presented either as hard copy or displayed on Videotex.

There are three major limitations to performance measures. Firstly, any decline in performance over time may reflect factors unrelated to visual fatigue—factors such as boredom, low arousal and possibly mental fatigue. Secondly, it is possible for people to prevent performance from falling off either by supplying increased 'effort' or as a result of learning. Finally, it is often difficult to obtain valid measures, particularly as most tasks involve a speed/error trade-off (see chapter 5). For example, Wilkinson and Robinshaw (1987) reported that, over time, more errors were committed but reading speed increased as subjects performed a 50-min proofreading task. Thus there was the possibility that subjects changed their speed/accuracy trade-off as time proceeded rather than that there was a decline in performance. Using a somewhat ad hoc index of change in performance, the authors were able to conclude that the increase in errors could not be totally attributed to an increase in reading speed. Comparing hard copy with VDU presentation of the stimulus material, they also demonstrated that the decline in performance over time, indicative of fatigue they claim, was much greater for the VDU presentation.

Symptoms of asthenopia

As mentioned before, it is the high frequency of reported symptoms of asthenopia accompanying the introduction of VDU-based work into the workplace that has led to the revival of interest in the topic of visual fatigue. For this reason, the possible reasons for these complaints should be explored. The introduction of VDUs obviously alters the physical characteristics of the displayed information. Amongst the obvious differences from printed material are the size of the characters, their shape, their contrast and colour and their luminance profile, the presence of flicker, the viewing angles and distances, and the general lighting conditions under which the material is viewed. In addition, the introduction of VDUs alters the job characteristics. Not only are working methods and workload levels changed but the amounts of pacing and autonomy are altered. The effects of all these factors are reflected in measures of job attitudes.

When it comes to considering the characteristics of the display, Campbell and Durden (1983) have implied that there is often little a priori evidence why several of the characteristics of VDU-based work should lead to visual fatigue, such as the adaptability of the visual system. In fact, if one looks at those studies which have used well designed control groups, there is not as much evidence as commonly thought to suggest that working at a VDU itself does lead to more complaints of visual fatigue. Muter *et al.* (1982) who made a direct comparison between continuous reading of material presented as hard copy and on a VDU found no significant differences in subjective measures of discomfort although subjects did read more slowly in the case of the VDU group. In the study of Nordqvist *et al.* (1986), there were again no significant differences in reported visual fatigue between those subjects who read from hard copy and those that read from a VDU and, unlike the previous study, there were no differences in the number of pages read under the two conditions.

Turning to less controlled field studies where, nevertheless, care was taken to balance control and experimental tasks in respect of job characteristics, results often fail to show any differences in the frequency and severity of reported symptoms of visual fatigue (Starr *et al.*, 1982; Howarth and Istance, 1985). However, against this, Knave *et al.* (1985) have reported that groups of subjects exposed to over 5 hours of VDU work per day recorded more eye discomfort symptoms than the control groups. Unfortunately, the authors do not tell us the extent to which the groups were matched. More significant, and less controversial, are their results concerning gender effects. Generally, the women tended to report a higher frequency and severity of symptoms of eye discomfort than the men, particularly under the VDU condition, a result that has been confirmed in several other studies (e.g., Levy and Ramberg, 1987). There is no obvious reason why there should be such gender effects solely arising from the way visual information is presented and these results strongly suggest that the tendency to report symptoms of visual discomfort must, therefore, often be related to factors which are not visual in origin.

The importance of individual disposition to registering complaints of asthenopia is supported by the results of Lindner and Kropf (1993). In assessing the prevalence of visual complaints about fluorescent lighting, the authors confirmed that women have a greater tendency to complain of asthenopia than men, especially those aged between 20 and 30 years. Not surprisingly, it was found that those people who had relatively poor binocular and stereoscopic vision and those who were relatively sensitive to peripheral flicker showed a greater disposition to registering complaints. In addition, the results revealed that a significantly greater proportion of those who reported visual complaints had 'neurotic symptoms' compared with those who did not report such symptoms.

Summing up the results based on subjective reports, Helander *et al.* (1984) have gone so far as to conclude that of the 28 studies they surveyed, 21 of which were field studies, the majority had serious design faults to the extent that the results were often useless. Further discussion on the validity of subjective reports of visual discomfort is provided by Howarth and Istance (1986). Some faith in their validity is restored by the recent results of Bergqvist *et al.* (1993). They extended the earlier study of Knave *et al.* (1985) to provide longitudinal data by administering the same questionnaire to the same group of operators in 1987 as in 1981. Most significantly, those operators who originally worked within a computerized work environment but who were not involved in a high proportion of screen–based work exhibited a significant increase in prevalence of symptoms of eye discomfort as a result of transferring in the interim period to more extensively screen-based work.

Conclusions

Just because some doubts have been cast over the validity of reported symptoms of fatigue and because of the absence of any coherent evidence of more objective indications, the existence of visual fatigue should not be denied. Laboratory studies have demonstrated that the vergence, version and accommodation mechanisms can be fatigued, but that usually this requires subjects to perform tasks that can be considered 'unnatural'; that is to say, tasks that are unlikely to be performed during the course of normal visual behaviour. Moreover, recovery from fatigue induced by such laboratory methods is extremely rapid, less than minutes and often in the order of a few seconds. At the same time, there is some evidence to suggest that certain display characteristics such as stimulus contrast and screen flicker can modify oculomotor performance and that these modifications are accompanied by subjective reports of visual discomfort. However, this does not necessarily mean that these changes in oculomotor performance reflect fatigue of the oculomotor control systems. One cannot over-emphasize the close interplay that takes place between the accommodation, convergence and pupil control systems, and that, therefore, any changes in oculomotor performance may only reflect an adaptive response to the prevailing visual conditions where the burden of control is redistributed amongst the different control systems.

To establish the presence of fatigue more unequivocally, it will be necessary to examine functioning of the control systems in relation to factors which would on a priori grounds be expected to lead to fatigue. For example, there is the possibility that both the accommodation and vergence systems might become fatigued as a result of a 'hunting' action when viewing displays that are untextured and which, therefore, provide few distance cues or that have images reflected in them. Similarly, the idiosyncratic effects that VDU viewing may have on certain eye movement parameters might reflect a disturbance to the vergence and version control mechanisms. Until our knowledge of the various oculomotor systems and their complex interactions is extended, it will remain difficult to decide whether or not the reported symptoms of visual discomfort do reflect fatigue of the oculomotor mechanisms or whether they are more a consequence of non-visual factors related to job design and job satisfaction.

References

Adler, F.H. (1962). *Textbook of Ophthalmology* (Philadelphia: W.B. Saunders).

Aoki, K., Yamonoi, N., Aoki, M. and Horie, Y. (1984). A study of the change of visual function in CRT display tasks. In *Human–Computer Interaction*, edited by G. Salvendy (Amsterdam: Elsevier), pp. 465–468.

Bahill, A.T. and Stark, L. (1975). Overlapping saccades and glissades are produced by fatigue in the saccadic eye movement system. *Experimental Neurology*, **48**, 95–106.

Berg, W.K., Krantz, J.H., McGovern, J.B., Donohue, R.L. and Silverstein, L.D. (1988). The development of an objective electromyographic measure of VDT-induced visual stress. In *Proceedings of the 1988 Society for Information Display International Symposium Digest of Technical Papers, Volume XIX*, edited by J. Morreale (Playa del Rey, CA: Society for Information Display), pp. 348–351.

Bergqvist, U., Knave, B., Voss, M. and Wibom, R. (1993). A longitudinal study of VDT work and health. *International Journal of Human–Computer Interaction*, **4**, 197–219.

Bitterman, M.E. and Soloway, E. (1946). Frequency of blinking as a measure of visual efficiency: some methodological considerations. *American Journal of Psychology*, **59**, 676–681.

Bodmann, H.W. (1962). Illumination levels and visual performance. *International Lighting Review*, **13**, 41–47.

Boyce, P.R. (1974). Illuminance, difficulty, complexity and visual performance. *Lighting Research and Technology*, **6**, 222–226.

Brozek, J., Simonson, E. and Keys, A. (1950). Changes in performance and in ocular functions resulting from strenuous visual inspection. *American Journal of Psychology*, **63**, 51–66.

Campbell, F.W. and Durden, K. (1983). The visual display terminal issue: a consideration of its physical, psychological and clinical background. *Ophthalmic & Physiological Optics*, **3**, 175–192.

Campbell, F.W. and Whiteside, T.C.D. (1950). Induced pupillary oscillations. *British Journal of Ophthalmology*, **34**, 180–189.

Carmichael, L. (1948). Reading and visual fatigue. *Proceedings of the American Philosophical Society*, **92**, 41–42.

Carmichael, L. and Dearborn, W.F. (1947). *Reading and Visual Fatigue* (Boston: Houghton Mufflin).

Collins, J.B. (1959). Visual fatigue and its measurement. *Annals of Occupational Hygiene*, **1**, 228–236.

Collins, J.B. and Pruen, B. (1962). Perception time and visual fatigue. *Ergonomics*, **5**, 533–538.

Cornsweet, T.N. and Crane, H.D. (1970). Servo-controlled infrared optometer. *Journal of the Optical Society of America*, **60**, 548–554.

Dainoff, M.J., Happ, A. and Crane, P. (1981). Visual fatigue and occupational stress in VDU operators. *Human Factors*, **23**, 421–438.

Ferree, C.E. and Rand, G. (1927). An investigation of the reliability of the 'li' test. *Transactions of the Illuminating Engineering Society*, **22**, 52–75.

Geacintov, T. and Peavler, W.S. (1974). Pupillography in industrial fatigue assessment. *Journal of Applied Psychology*, **59**, 213–216.

Gentles, W. and Llewellyn Thomas, E. (1971). Effect of benzodiazepines upon saccadic eye movements in man. *Clinical Pharmacology and Therapeutics*, **12**, 563–574.

Gunnarsson, E. and Soderberg, I. (1983). Eye strain resulting from VDT work at the Swedish telecommunications administration. *Applied Ergonomics*, **14**, 61–69.

Haider, M., Kundi, M. and Weisenbock, M. (1980). Worker strain related to VDUs with differently coloured characters. In *Ergonomic Aspects of Visual Display Terminals*, edited by E. Grandjean and E. Vigliani (London: Taylor and Francis), pp. 53–64.

Harris, C.M., Abramov, I. and Hainline, L. (1984). Instrument considerations in measuring fast eye movements. *Behavior Research Methods & Instrumentation*, **16**, 341–350.

Helander, M.G., Billingsley, P.A. and Schurick, J.M. (1984). An evaluation of human factors research on visual display terminals in the workplace. In *Human Factors Review: 1984*, edited by F.A. Muckler (Santa Monica, CA: The Human Factors Society), pp. 55–129.

Hennessey, R.T. and Leibowitz, H.W. (1972). Laser optometer incorporating the Badal principle. *Behavior Research Methods & Instrumentation*, **4**, 237–239.

Heuer, H. and Owens, D.A. (1989). Vertical gaze direction and the resting posture of the eyes. *Perception*, **18**, 363–377.

Howarth, P.A. and Istance, H.O. (1985). The association between visual discomfort and the use of visual display units. *Behaviour & Information Technology*, **4**, 131–149.

Howarth, P.A. and Istance, H.O. (1986). The validity of subjective reports of visual discomfort. *Human Factors*, **28**, 347–351.

Iwasaki, T. and Kurimoto, S. (1987). Objective evaluation of eye strain using measurements of accommodative oscillation. *Ergonomics*, **30**, 581–587.

Iwasaki, T. and Kurimoto, S. (1988). Eye-strain and changes in accommodation of the eye and in visual evoked potential following quantified visual load. *Ergonomics*, **31**, 1743–1751.

Jaschinski-Kruza, W. (1984). Transient myopia after visual work. *Ergonomics*, **27**, 1181–1189.

Jaschinski-Kruza, W. (1988). Visual strain during VDU work: the effect of viewing distance and dark focus. *Ergonomics*, **31**, 1449–1465.

Jaschinski-Kruza, W. (1991). Eyestrain in VDU users: viewing distance and the resting position of ocular muscles. *Human Factors*, **33**, 69–83.

Kahneman, D. and Beatty, J. (1966). Pupil diameter and memory load. *Science*, **154**, 1583–1585.

Kennedy, A. and Murray, W.S. (1991). The effects of flicker on eye movement control. *Quarterly Journal of Experimental Psychology*, **43A**, 79–99.

Khek, J. and Krivohlavy, J. (1966). Variation of incidence of error with visual task difficulty. *Light and Lighting*, **59**, 143–145.

Knave, B.G., Wiborn, R.I., Voss, M., Hedstrom, L.D. and Berqvist, U.O. (1985). Work with video display terminals among office employees: 1. Subjective symptoms and discomfort. *Scandinavian Journal of Work, Environment & Health*, **11**, 457–466.

Krivohlavy, J., Kodat, V. and Cizek, P. (1969). Visual efficiency and fatigue during the afternoon shift. *Ergonomics*, **12**, 735–740.

Leermakers, M.A.M. and Boschman, M.C. (1984). Eye movements, performance and visual comfort using VDTs. *IPO Annual Progress Report*, **19**, 70–75.

Levy, F. and Ramberg, I.G. (1987). Eye fatigue among VDU users and non-VDU-users. In *Work with Display Units 86*, edited by B. Knave and P.G. Wideback (Amsterdam: North-Holland), pp. 42–52.

Lindner, H. and Kropf, S. (1993). Asthenopic complaints associated with fluorescent lamp illumination (FLI): the role of individual disposition. *Lighting Research and Technology*, **25**, 59–69.

Luckiesh, M. and Moss, F.K. (1935a). Fatigue of convergence induced by reading as a function of illumination intensity. *American Journal of Ophthalmology*, **18**, 319–323.

Luckiesh, M. and Moss, F.K. (1935b). The effect of visual effort upon the heart-rate. *Journal of General Psychology*, **13**, 131–138.

Lunn, R. and Banks, W.P. (1986). Visual fatigue and spatial frequency adaptation to video display of text. *Human Factors*, **28**, 457–464.

Magnussen, S., Dyrnes, S., Greenlee, M.W., Nordby, K. and Watten, R. (1992). Time course of contrast adaptation to VDU-displayed text. *Behaviour & Information Technology*, **11**, 334–337.

Malmstrom, F.V., Randle, R.J., Murphy, M.R., Reed, L.E. and Weber, R.J. (1981). Visual fatigue: the need for an integrated model. *Bulletin of the Psychonomic Society*, **17**, 183–186.

Marek, T., Noworol, C., Pieczonka-Osikowska, W., Przetacznik, J. and Karwowski, W. (1988). Changes in temporal instability of lateral and vertical phorias of the VDT operators. In *Trends in Ergonomics/Human Factors V*, edited by F. Aghazadeh (Amsterdam: North-Holland), pp. 283–289.

Martin, E., Clymer, A. and Martin, L. (1972). Metacontrast and saccadic suppression. *Science*, **178**, 179–182.

Matula, R.A. (1981). Effects of visual display units on the eyes: a bibliography (1972–1980). *Human Factors*, **23**, 581–586.

McConkie, G.W. (1981). Evaluating and reporting data quality in eye movement research. *Behavior Research Methods & Instrumentation*, **13**, 97–106.

Megaw, E.D. (1986). VDUs and visual fatigue. In *Contemporary Ergonomics*

1986, edited by D.J. Oborne (London: Taylor and Francis), pp. 254–258.

Megaw, E.D. and Sen, T. (1984). Changes in saccadic eye movement parameters following prolonged VDU viewing. In *Ergonomics and Health in Modern Offices*, edited by E. Grandjean (London: Taylor and Francis), pp. 352–357.

Messite, J. and Baker, D.B. (1984). Occupational health problems in offices: a mixed bag. In *Human Aspects in Office Automation*, edited by B.G.F. Cohen (Amsterdam: Elsevier), pp. 7–14.

Miyao, M., Allen, J.S., Hacisalihzade, S.S., Cronin, S.A. and Stark, L.W. (1988). The effect of CRT quality on visual fatigue. In *Trends in Ergonomics/Human Factors V*, edited by F. Aghazadeh (Amsterdam: North-Holland), pp. 297–304.

Mourant, R.R., Lakshmanan, R. and Chantadisal, R. (1981). Visual fatigue and cathode ray tube display terminals. *Human Factors*, **23**, 529–540.

Murch, G.M. (1983). Visual fatigue and operator performance with DVST and raster displays. *Proceedings of the Society for Information Display*, **14**, 53–61.

Muscio, B. (1921). Is a fatigue test possible? *British Journal of Psychology*, **12**, 31–46.

Muter, P., Latremouille, S.A., Treurniet, W.C. and Beam, P. (1982). Extended reading of continuous text on television screens. *Human Factors*, **24**, 501–508.

National Research Council Committee on Vision (1983). *Video Displays, Work and Vision* (Washington, DC: National Academy Press).

Neary, C. and Wilkins, A.J. (1989). Effects of phosphor persistence on perception and the control of eye movements. *Perception*, **18**, 257–264.

Neisser, U. (1964). Visual search. *Scientific American*, June, 94–100.

Nordqvist, T., Ohlsson, K. and Nilsson, L. (1986). Fatigue and reading text on videotex. *Human Factors*, **28**, 353–363.

Osaka, N. (1985). The effect of VDU colour on visual fatigue in the fovea and periphery of the visual field. *Displays*, **6**, 138–140.

Ostberg, O. (1980). Accommodation and visual fatigue in display work. In *Ergonomic Aspects of Visual Display Terminals*, edited by E. Grandjean and E. Vigliani (London: Taylor and Francis), pp. 41–52.

Ostberg, O. and Smith, M.J. (1987). Effects of visual accommodation and subjective discomfort from VDT work intensified through split screen technique. In *Work with Display Units 86*, edited by B. Knave and P.G. Wideback (Amsterdam: North-Holland), pp. 512–521.

Owens, D.A. and Leibowitz (1980). Accommodation, convergence, and distance perception in low illumination. *American Journal of Optometry & Physiological Optics*, **57**, 540–550.

Phillips, S. and Stark, L. (1977). Blur: a sufficient accommodative stimulus. *Documenta Ophthalmologica*, **3**, 65–89.

Post, R.B., Johnson, C.A. and Tsuetaki, T.K. (1984). Comparison of laser and infrared techniques for measurement of the resting focus of accommodation: mean differences and long-term variability. *Ophthalmic & Physiological Optics*, **4**, 327–332.

Poulton, E.C. (1958). On reading and visual fatigue. *American Journal of Psychology*, **71**, 609–611.

Ramazzini, B. (1700). *De Morbis Artificum—Diseases of Workers* (English edition, New York: Hafner Publishing, 1964).

Rushton, W.A.H. and Westheimer, G. (1962). The effect upon the rod threshold of bleaching neighbouring rods. *Journal of Physiology*, **164**, 318–329.

Schleiffer, L.M., Sauter, S.L., Smith, R.J. and Knutson, R.J. (1990). Ergonomic predictors of visual system complaints in VDT data entry work. *Behaviour & Information Technology*, **9**, 273–282.

Sen, T. and Megaw, E.D. (1984). The effects of task variables and prolonged performance on saccadic eye movement parameters. In *Theoretical and Applied Aspects of Eye Movement Research*, edited by A.G. Gale and F. Johnson (Amsterdam: North-Holland), pp. 103–111.

Simmerman, H. (1950). Visual fatigue. *American Journal of Optometry and Physiological Optics*, **27**, 554–561.

Simonelli, N.M. (1980). Polarized vernier optometer. *Behavior Research Methods & Instrumentation*, **12**, 293–296.

Starr, S.J., Thompson, C.R. and Shute, S.J. (1982). Effects of video display terminals on telephone operators. *Human Factors*, **24**, 699–711.

Stellman, J.M., Klitzman, S., Gordon, G.C. and Snow, B.R. (1987). Work environment and the well-being of clerical and VDT workers. *Journal of Occupational Behaviour*, **8**, 95–114.

Stern, J.A. and Skelly, J.J. (1984). The eye blink and workload considerations. In *Proceedings of the Human Factors Society 28th Annual Meeting*, edited by M.A. Alluisi, S. de Groot and E.A. Alluisi (Santa Monica, CA: Human Factors Society), pp. 942–944.

Stone, P.T., Clarke, A.M. and Slater, A.I. (1980). The effect of task contrast on visual performance and visual fatigue at a constant illuminance. *Lighting Research and Technology*, **12**, 144–159.

Tinker, M.A. (1939). The effect of illumination intensities upon speed of perception and upon fatigue in reading. *Journal of Educational Psychology*, **30**, 561–571.

Tyrrell, R.A. and Leibowitz, H.W. (1990). The relation of vergence effort to reports of visual fatigue following prolonged near work. *Human Factors*, **32**, 341–357.

Watten, R.G., Lie, I. and Magnussen, S. (1992). VDU work, contrasts adaptation, and visual fatigue. *Behaviour & Information Technology*, **11**, 262–267.

Weber, A., Fussler, C., O'Hanlon, J.F., Gierer, R. and Grandjean, E. (1980). Psychophysiological effects of repetitive tasks. *Ergonomics*, **23**, 1033–1046.

Weber, R.A. (1950). Ocular fatigue. *Archives of Ophthalmology*, **43**, 2.

Weston, H.C. (1953). The relation between illumination and visual performance. *MRC Industrial Health Research Board Report* (London: HMSO).

Wilkins, A. (1986). Intermittent illumination from visual display units and fluorescent lighting affects movements of the eyes across text. *Human Factors*, **28**, 75–81.

Wilkins, A.J. (1991). Visual display units versus visual computation. *Behaviour & Information Technology*, **10**, 515–523.

Wilkins, A.J., Nimmo-Smith, I., Slater, A.I. and Bedocs, L. (1989). Fluorescent lighting, headaches and eye-strain. *Lighting Research and Technology*, **22**, 103–109.

Wilkinson, I.M.S., Kime, R. and Purnell, M. (1974). Alcohol and human eye movement. *Brain*, **97**, 785–792.

Wilkinson, R.T. and Robinshaw, H.M. (1987). Proof-reading: VDU and paper text compared for speed, accuracy and fatigue. *Behaviour & Information Technology*, **6**, 125–133.

Woo, G.C., Strong, G., Irving, E. and Ing, B. (1987). Are there subtle changes in vision after use of VDTs? In *Work with Display Units 86*, edited by B. Knave and P.G. Wideback (Amsterdam: North-Holland), pp. 490–503.

Young, L.R. and Sheena, D. (1975). Survey of eye movement recording methods. *Behavior Research Methods & Instrumentation*, **7**, 397–429.

Zwahlen, G.T., Hartmann, A.L. and Rangarajulu, S.L. (1984). Video display work with a hard copy—screen and a split screen data presentation. *Final Report, Department of Industrial and Systems Engineering, Ohio University*.

Chapter 29

Practical measurement of psychophysiological functions for determining workloads

Masaharu Kumashiro

Importance of workload

The loads induced by work are psychosomatic responses appearing as a result of factors associated with accomplishing one's duties and an imbalance between work conditions and the work environment and the adaptive capacity of the people who work there. When the psychosomatic response induced by work exceeds a person's range of tolerance, a special effort is required for that person to continue accomplishing his or her daily work. Typical related phenomena include, for example, the need for rests. The state where one has lapsed into this condition is referred to as workload. Consequently, workload expresses the degree of qualitative and quantitative load induced by work and is closely related to stress and industrial fatigue. (See, for example, chapter 26 for a comparison of a number of views and definitions of stress.)

The objective of workload-related research is to seek optimization of the work task. Specifically, this will contribute to the design of work conditions and work environments that impose minimal stress and industrial fatigue due to work. However, review of those workload-related studies that have been conducted in a great many past ergonomics research leads to a certain level of apprehension. Past research has pursued the relationship between external tasks and internal responses and much of this research has sought to minimize the external task. Such an approach would best be described as the intervention of risky ergonomics. When ergonomists intervene in workload research, it must always be from a position of consistent consideration of creating a balance between the external task and internal response. Optimization of the work task does not mean minimizing the workload. Taking this concept one step further, in fact it means providing flexibility for the balance between task and load. Viewing work among certain coherent units as a total, we find the relationship between task and load to be balanced, that is, ± 0. When we view changes over time, we find the relationship sometimes favours the plus side while at other times it favours the minus side; that is to say, the relationship requires flexibility. The premise for this concept is that one must investigate the workloads. In general, we may separate the types of practical technology for studying workload into the following:

- Subjective aspect;
  Symptoms of self-conscious fatigue and other psychosomatically related complaint symptoms and physical conditions.
- Objective aspect;
  Operational behaviour such as subsidiary behaviour, performance and human error.
- Psychophysiological function aspect;
  Maladjustment of psychophysiological functions such as the lowering of cerebration.
- Others;
  Transitions in fatigue phenomena in the workplace, such as work history, labour conditions, adaptation symptoms, patient's history and sequela, proclivity to contract disease, physical strength, physical constitution, age and sex.

Types of industrial fatigue due to evolving workplaces

When selecting workload measurement tools, we must focus on the relationship between the workers and the work, in other words the work mode. Work modes change along with production equipment and production technology advances, and may bring about many changes in the fatigue phenomena of the workers subjected to such changes. As might be expected, work mode will play a large part in selection of measurement tools. When we survey production systems at manufacturing sites we find a transition from work in a human-tool system to work in a human-machine system. Today, there is a continuing shift towards human-computer systems, and changes in fatigue phenomena will result.

Work and fatigue phenomena under the human-tool system

In the human-tool system, indeed in the classic human-machine system, the production system itself is manual, relying on physical operations performed by workers. In terms of the sources of fatigue, the work could generally be classified as full-body, physical work. Such work requires considerable consumption of calories by workers, because of a high relative metabolic rate. The fatigue phenomenon is thus physical fatigue primarily related to muscular fatigue. The mechanism for the onset of this type of physical fatigue is relatively simple and its evaluation is not overly difficult. In other words, such work causes the immediate appearance of fatigue conditions. Its occurrence is easily sensed in the form of worker exhaustion or hunger and the intake of high calorie foods. A short rest or brief sleep in response to physiological demands serves as an effective means of relieving the fatigue.

This is a somewhat bold explanation, but the acute fatigue resulting from rigorous physical labour, can be readily comprehended when considered as a result of the balance between the amount of physical energy consumption and the prescribed calories. Putting it another way, the fatigue conditions brought about by muscle-related work in the human-tool system have a very simple make-up compared with those which are encountered today.

This type of work is categorized as muscular work. Primarily classical physiological techniques, such as relative metabolic rate (RMR), oxygen consumption, heart rate (HR) and electromyograms are used as effective work strength indices. Hence fatigue is pursued from the perspective of human body physiology.

Work and fatigue phenomena under the human-machine system

Machines or automatic machines have replaced tools as the primary means of production and a semi-automatic method of production has come into use. Along with this, people's work has become more sensori-motor in nature. Fatigue phenomena occurring in this work system has shifted to localized muscular fatigue and psychological fatigue.

The work of human-machine systems has led to paced work methods that inevitably compel the worker to cope with increasingly controlled and high-speed work. Under such paced work methods, worker freedom has been reduced, causing tension under higher speed conditions. Moreover, the work itself has also been configured in a fixed operational structure, where the same work is performed repetitively, thus increasing the worker's workload further. In other words, the mechanical repetition of motions causes workers to lose zest in their consciousness periphery. This increases a sense of monotony and weariness, eventually leading to abnormalities in the central nervous system, and, as a result, gives rise to psychological nervous fatigue.

From these developments it has become necessary to understand fully the fluctuations in psychophysiological functions, thereby giving greater consideration to the psychological aspects of work. At the same time, the observation of subsidiary behaviour has become an effective evaluation tool..

Work and fatigue phenomena in the human-computer system

The byword in today's manufacturing scene is factory automation. Here the focus is increasingly placed on automation of work through computer control, thus increasing work involving information input and processing systems requiring human recognition—judgement—control. Decision making and information processing capabilities involving such characteristics as accuracy, speed and immediate actions are required more than ever before. Having arrived at this stage, the relationship between task and load changes from [A] to [B] as shown in Figure 29.1, creating a need for cerebral-physiological studies. In particular, there is a need to focus on fluctuations in the levels of cerebral cortex activity, involving techniques to ascertain physiological phenomena using multiple channel recorders and there is a need to find the optimum approach to determining psychophysiological functions. Visual display terminal work in particular, in process control and other operational systems, is a classic example of this.

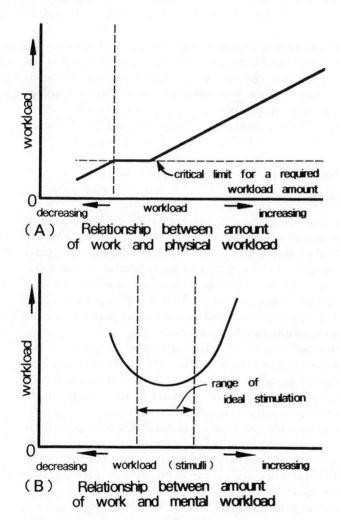

Figure 29.1. A model of task-workload relations

A number of studies have been undertaken since the beginning of the 1980s in relation to the impact of such human-computer system work on the human body. Many of these studies have pointed to the occurrence of visual fatigue (see chapter 28) and an increase in the static muscular load (see chapters 23 and 30). This sort of work has the potential to affect visual function, ultimately leading to damage to the visual information processing ability that plays a role in the vision—cerebrum—peripheral function system (although see chapter 28). This mode of fatigue (which is also increased in human-machine systems work), thus induces increased central nervous system stress and increases the nervous sensation type of fatigue.

The occurrence of a stress phenomenon that triggers these sorts of computer work–related fatigue and mental fatigue problems was referred to by Brod (1984) as 'technostress', and it now has considerable impact on today's industrial society.

In the light of such changes in industrial technology, work systems and jobs, with consequent changes in fatigue phenomena, ergonomics assessment needs techniques from sensory physiology, psychophysics, and anatomic nerve function measurement. This is in addition to conventional physiology techniques, subjective assessment, psychophysiology, and performance measures. We also need to consider personality and behavioural type effects.

Methods for investigating psychophysiological functions

Outline of investigative methods

Various methods are used to investigate psychophysiological functions in order to evaluate the degree of fatigue at the workplace. Table 29.1 lists several of these. Sensibility function tests are techniques to produce qualitative and quantitative responses by loading a certain fixed performance requirement onto the subjects under investigation. One of the methods classified under muscle function investigations in Table 29.1 is the knee jerk percussion threshold reflex test, but this is not categorized as a sensory performance test because it measures an involuntary muscle response.

Physiological function techniques can only be used after having previously investigated the conditions that are likely to affect the results obtained from the test technique adopted. In other words, through the test results it must be possible to attach a psychophysiological meaning to the changes in the psychosomatic function of the subject observed.

Muscle function tests are different from continuous observation methods such as recording the heart rate, brain waves, and electromyograms. The work must be interrupted to set a time for the test, which should take place at least before and after the work and, as far as possible, at intervals while the work is being performed.

For respiratory function tests, the condition of the test equipment must be calibrated constantly using a fixed standard. With autonomic nerve tests,

Table 29.1. List of the main methods for examining psychophysiological functions

1. Sensory function test	Vision function	Accommodation test (far point and near point distance, contraction and relaxation time)	Measurement of recognized percussion value
	Auditory function	Minimum audible pure tone percussion value test	Measurement of recognized percussion value
	Skin sensibility	Two point touch discrimination percussion value method	Measurement of discriminated percussion value
	Parallel functions	The body's centre of gravity test	Measurement of automatic oscillation phenomenon in erect position maintenance condition
2. Psychological function test (including information processing function test)		CFF test	Measurement of discriminated percussion value
		Reaction time test (simple and choice types)	Measurement of information processing capability
		Blocking test	Measurement of performance under uncontrolled conditions
		TAF test	Measurement of overall functions as one of the performance tests
		Dual task method	Measurement of information processing capability
3. Muscle function test	Muscular strength	Gripping power test Back muscle strength test Leg strength test	Measurement of body motion functions using a dynamometer
	Coordinated motion	Finger tapping ability test Two-handed co-ordination test	Measurement of motion control function
	Knee jerk	Test for hyper- or hypoactive	Measurement of patellar tendon reflex (PTR)
4. Respiratory function test		Lung capacity test Maximum air ventilation test	Measurement of body motion function
5. Autonomic nerve testing			Tests such as cold pressure rise test are available, but inappropriate for industrial sites.

the subject(s) must be proficient with respect to the test equipment and methods.

As a rule,when conducting psychophysiological function tests at the work-place, a great effort is made promptly to observe and measure any physiologically and psychologically abnormal phenomena of the subjects. Usually more attention is directed to the results obtained by the various test equipment used rather than to how the interactions and combinations of the various psychophysiological functions in response to the work are deduced. For this reason, the true character of the workload and of the industrial fatigue arising as a result of it are not investigated in depth, and the background factors that induce and amplify the task and its cause and effect relationship are neglected. Instead, credence is given only to the phenomena as depicted by the test equipment and, as a result, the task and fatigue at the workplace are not accurately pin-pointed and a deceptive image may be conveyed to those responsible for job design and health management. Consequently, in order also to make an accurate evaluation of the effects of the task on people who are working, the true character of fatigue and the background factors must be considered from the perspective of a worker group and its working conditions, rather than just considering individual fluctuations. Moreover, the correlation between fatigue and conditions must be studied in detail at the time of analysis.

In the remainder of this chapter, three of the many methods available are discussed. The critical flicker frequency (CFF) test is a representative example of threshold value measurement. The concentration maintenance function (CMF) test represents performance measurement. Finally, heart rate *variability* is considered as perhaps the way forward in workload measurement.

Critical flicker frequency (CFF) test

Light and dark phases of light are alternated rapidly by continuously shining or flickering an impulse light using a sector (shield) or neon tube or similar device. When the flashing frequency per unit of time is low flicker is sensed, but when the frequency becomes high the flickering fuses and appears as a continuous light. This flashing effect is referred to as flicker and the flashing frequency per unit time is referred to as the flicker value (in Hertz—Hz). The frequency at the boundary between flicker and fusion is called the frequency of fusion and the test to measure this value is known as a critical flicker frequency (CFF) test. This critical fusion frequency becomes the index for visual power (resolution power) that changes over time under various conditions. Generally, the value tends to be low when a fatigue phenomenon arising from the central nervous system appears (advocated by Simonson and Enzer, 1941). Thereafter Walker *et al.* (1943) and Kogi and Kawamura (1960) performed encephological investigations using animals such as monkeys and cats. They revealed that a remarkable correlation could be perceived between the activity levels of the cerebral cortex and the CFF values. Based on this explanation of central fusion, the CFF test became a significant factor in physiological studies and today is frequently used in research on industrial fatigue. Specifically, a drop in the CFF value reflects a drop in the sensory perception function that is attributable to a decrease in the level of alertness. This is explained by a decline in the visual information processing ability in the cortex associated with sensory perception that includes the vision system (Hashimoto, 1963). A general discussion of the perception of flicker and its effects on visual fatigue appears in chapter 28.

In general, two methods of measurement are used in the CFF test. One is the sector method in which the light from a light source is continuously covered and uncovered using a sector (or shield). The other is the light source flashing method in which a sawtooth voltage from an electron tube oscillator is obtained to cause a neon lamp or LED to flash. The Industrial Fatigue Research Committee of the Japan Association of Industrial Health (1953) indicated a draft standard for sector type CFF measuring instruments in Japan. The basic functions of the many CFF test devices commercially sold today generally conform to this standard. Both methods should progress from a high to a low frequency of flashing and measurement should be made using a descending (psychophysical) procedure whereby the subject responds at the point where he or she senses the flicker.

When performing the test, every effort should be made to keep the impulse conditions for the retina as fixed as possible so that the values measured for the same subject under the same conditions will closely agree; experience in CFF testing is required to do this. At the time of the CFF test, the individual's work is interrupted to perform the test, but any additional behaviour such as movement of the subject is undesirable since changes in the conditions should be avoided as far as possible. Representative values should be measured about five to ten times on each occasion and the average of those values used.

Since the time required for the CFF test is short and a relatively simple measuring instrument is used, measurements can be made frequently and therefore the fluctuation over time of the obtained CFF values can be indexed. Figure 29.2 shows variation in CFF values over a 24 h period, plotted together with proportions of time spent in sitting and standing postures.

The fluctuation in the CFF value over time showed a significant drop in the values at 03:30 and 05:30 compared with the values before the day shift (08:30) and before work for the night shift (20:30) ($p < 0.05$). The phenomenon of fluctuation in the balance of sitting/standing posture is thought to be explained by the fact that the person exhibits behaviour referred to as spontaneous position change in response to changes in the excitability of the cerebrum, e.g. during night turning into morning. The increase in sitting position accompanied by a drop in the CFF values during the night is thought to indicate the phenomenon of exhaustion that occurs because the psychosomatic activity of having a load of night-time work imposed on the body runs counter to the characteristic physiological functions of humans. In con-

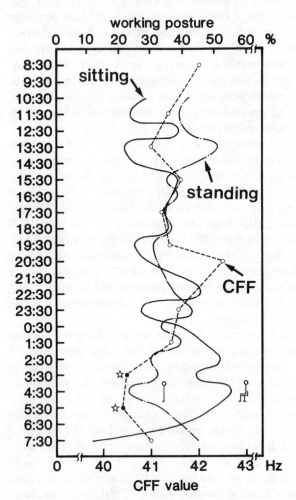

Figure 29.2. Circadian rhythm of worker's CFF and time spent in standing posture and sitting posture during day and night shift work (Kumashiro, 1986)

trast, an increase in the time spent in a sitting position was not observed with daytime work despite a drop in the CFF values. This is thought to indicate social restraints on behaviour, as the worker was conscious of supervisors in the workplace. This phenomenon represents attitude behaviour by the individual worker directed toward the management control structure in relation to factors such as employment, wages and promotion. From the workload perspective, it must be explained as a counter-physiological phenomenon.

As seen in the above example, when persons become fatigued, CFF values exhibit low levels. For this reason, at production sites in Japan, CFF values are still widely used by specialized personnel as a simple fatigue index for study and research. For example, despite the fact that ergonomics researchers are abandoning research related to the physiological significance of CFF values, occupational nurses, occupational physicians and labour hygiene managers are adopting the CFF test, making use of abundant results from past on-site studies as references. In the majority of cases, however, they are adopting it as one of *many* tools being used to understand workload. For their evaluations, these professionals are conducting comparative studies of the results of work observation and fluctuations in CFF values, so as to deduce the information processing capability in the cerebral cortex and to evaluate the degree of workload.

The first organ to perceive flicker is the eye, and, for this reason, perception is greatly affected by the level of visual function. The threshold value of the CFF, for example, is measured by the activity of vision cells and biporal cells in the retina (Kelly, 1978). Iwasaki and Akiya (1990) describe a large drop in the CFF value for one loaded eye in comparison with the control eye that was not assigned a visual load. The drop in the CFF value is regarded as reflecting an adaptation to light and a drop in the signal transmission function of the optic nerve. Figure 29.3 shows a comparison of the CFF value levels for six elderly men ranging in age from 60 to 64 years and six young men aged 20 to 25 years. The average eye adjustment power of the persons in the target elderly group was 3·5 diopters. For persons in their sixties, the adjustment power was comparatively good, but they exhibited a drop of

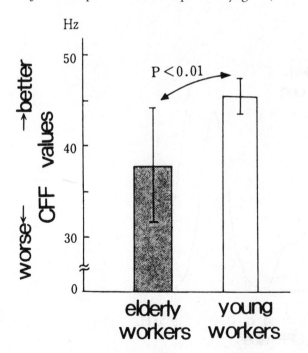

Figure 29.3. Comparison of CFF values for elderly and young men (Kumashiro, 1986)

more than twice that of the young persons in their twenties. One hundred measurements were taken using the descending procedure for the CFF test whereby they responded at the point of sensing the flicker. Of course, the CFF test was implemented at intervals so that the subjects were not fatigued by the test. As a result, the mean CFF level of the elderly group was found to be 37·9 Hz, while that for the young group was 45·4 Hz. Figure 29.4 shows the distribution of the CFF values. It was found that the CFF value for the elderly group was spread over a broad range of 23·5 Hz ~ 52·0 Hz. It also exhibited a two-peak characteristic with peaks of 31·1 Hz ~ 33·0 Hz and 43·1 Hz ~ 45·0 Hz. Conversely, the flicker value of the young group was in the range 40·0 Hz ~ 52·0 Hz and exhibited a regular distribution.

Figure 29.5 shows a comparison of the CFF values for three groups. One is a group of 22 young men (average age: 24 years) and another a group of 22 middle aged men with an average age of 52 years, both of whom worked in a machine factory. The third is a group of more elderly men with an average age of 62 years who had retired from the machine factory one to two years earlier. We can note a phenomenon of declining CFF value accompanying an increase in age. Consequently, the CFF values were measured to compare the degrees of workload from a psychophysiological aspect for the young group and the older worker group to whom the same work task had been assigned at the production site. The results are shown in Figure 29.6.

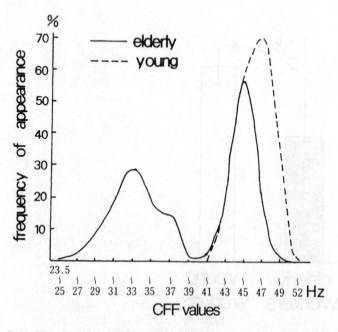

Figure 29.4. Comparison of the distribution of CFF values for elderly and young men (Kumashiro, 1986)

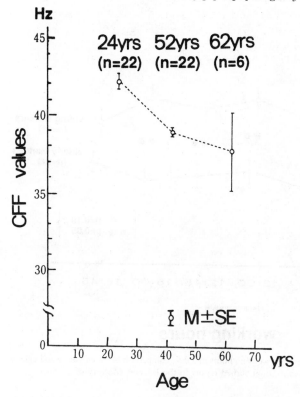

Figure 29.5. Relationship between the increase in age and CFF values (Kumashiro, 1986)

Vision function and visual information processing capacity expressed in CFF values decline with advancing age. At the same time, however, individual differences with respect to this function appear to increase. It is generally recognized that the vision function deteriorates with age. At the same time, however, it appears certain that there is also a phenomenon of decline in information processing capacity.

Examination of the concentration maintenance function (TAF)

Seeking an objective evaluation of industrial fatigue, in 1960, Takakuwa proposed a concept he referred to as concentration maintenance function, in which the state of concentration in continuous target aiming is described as a continuous curve TAF—target aiming function; he developed a TAF measuring instrument (Takakuwa, 1962). TAF is based on the concept of making an index for the 'concentration of attention' representing the psychological function, and for the 'maintenance of concentration' representing the physiological function.

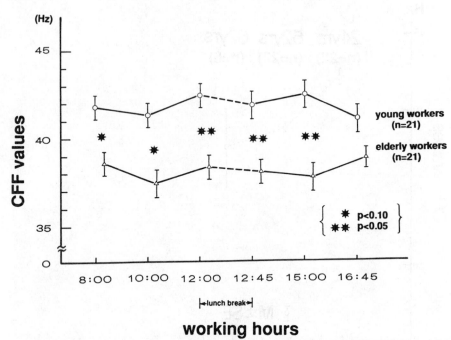

Figure 29.6. Changes in CFF values with working hours in a day; comparison between young and elderly workers in a machine shop (Kumashiro, unpublished report to the Japanese Ministry of Labour)

TAF measurement is performed as follows. The subjects sit down in front of a light source (24 V, 40 W) target with a diameter of 2·5 cm at a distance of 2 m. They focus on the target through focusing glasses with a one point support on which a photo-transistor is mounted (aiming). The subjects focus for 1 min then rests for 10 s. This operation is repeated three times for a total of 3 min. During the 10 s resting period subjects take their hand off the glasses' handle and also stop focusing on the target. The accuracy of the focus aim is marked on a recorder as a continuous curved line (Takakuwa *et al.*, 1963). Figure 29.7 shows an example of results from one of the three concentrated target aiming periods. Since subjects will reach 'true' levels of concentration about 5–10 s after the beginning of the test, the curves showing performance during the initial 5–10 s should be discarded and the plotting of curves should begin after this period. The recording paper revolves at the rate of 1 cm/5 s, and the plotting is done every 2·5 s. In this manner, 22 points are plotted for a 1 min curve (the first and last points being discarded). Sometimes the number of points can be less than 22 when the subject is slow in reaching his or her normal level. The plotting points are scored in the following manner. The uppermost line is counted as 0·0, and the lowest line as 10·0, and the points are scored to one decimal place. Calculations are made from the scores thus obtained. In a 3 min test, the average value of

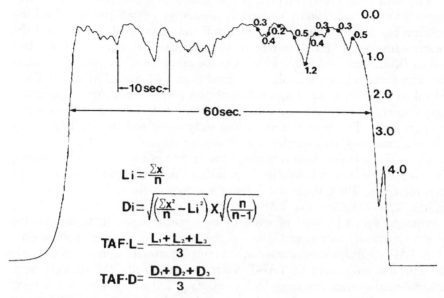

Figure 29.7. Plots on a TAF curve (one minute)

three L*i*s is called TAF-L, and that of three D*i*s TAF-D. The former indicates the level of concentration, while the latter shows the degree of fluctuation in maintaining that concentration. Usually the pre- and post-work values of TAF-L and TAF-D are compared statistically, as well as values of different groups under various working conditions.

TAF-L and TAF-D values determined in this manner are now widely applied to judge fatigue and in particular to evaluate psychological stress in places prone to industrial fatigue. For example, it has been found to be effective for objectively evaluating the task load due to factors such as end of year operations of banks (Takakuwa *et al.*, 1964a), the alternating shift work of nurses (Yamamoto *et al.*, 1966), inspection work (Kumashiro *et al.*, 1973) and repetitive work with conveyor-paced systems (Kumashiro *et al.*, 1976). TAF is said to be closely related to IQ, which is based on the examination of intelligence, and is recognized to be correlated with the results of the Uchida–Kreplin psychological work test (Koizumi *et al.*, 1964). Moreover, it has been shown to be effective for objective evaluation of intoxication with alcohol (Takakuwa *et al.*, 1964b) and for accurately determining the effects of changing cacophony and white noise that cause symptoms, such as psychological indecisiveness and loudness (Takakuwa *et al.*, 1966). For physiological studies, TAF is effective as a means of inferring the excitability of the central nervous system (Nakadaira, 1969). It has also been demonstrated that there is a relationship between TAF and the degree of control of the brain waves (Takakuwa *et al.*, 1967). .

The basic 3 min TAF method is recognized as appropriate for measurement (Matsui *et al.*, 1965). Moreover, studies have been conducted on the relationship between the TAF values and the amount of practice and the practice time for TAF tests in relation to quantitative studies of TAF standard values (Kumashiro and Saito, 1978). On the other hand, the following points must be noted when using values obtained from TAF tests. TAF-L and TAF-D values fluctuate depending on fatigue and psychological stress. However, upon observing the TAF curve in the morning before going to work when there is little effect from factors such as daily living and labour, TAF values show a pattern of individual variation among subjects.

Generally, a comparison is made of the degree of fluctuation after loading as compared with that before loading, taking the labour load conditions into consideration. When there is a difference among individuals of 0·5 or more in the TAF-L values, the TAF is regarded as having gone down. When comparing values by type of work or by target group, it is recommended that a statistically meaningful study of the group concerned be performed.

A TAF test that was conducted at a certain department store, as an example of statistical judgement of TAF-L values, is described. The subjects were middle-aged women averaging 50·7 years of age and young women averaging 30·3 years of age who performed sales work in the store. Figure 29.8 shows a comparison based on the computation of the average TAF-L values taken before and after work for a five-day period. In comparing the pre- and post-work values of the two groups a declining trend in the TAF-L values of the younger group (0·91) in comparison with those of the middle-aged group (0·76) was observed (TAF expresses an increasing decline of a function as its

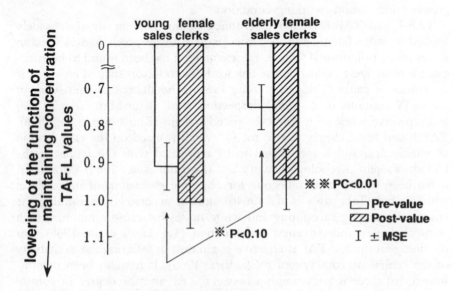

Figure 29.8. Comparison of TAF-L values for elderly and young female sales clerks (Kumashiro, 1988)

numerical value increases). Conversely, when the values before work and after work of the two groups were compared, the post-work value (0·95) compared with the pre-work value showed a significant decline at the risk factor level of 1% for the middle-aged group only.

The way forward? Heart rate variability

Measurement of heart rate (HR) fluctuation has long been used as an index for comprehending workloads, and observation of HR is a traditional means for measuring the functions of living bodies. HR sharply and faithfully expresses the reaction of the functions of the body to stimulation and has thus remained until today as an effective index that is difficult to discard. When used as a means of evaluating workload, however, unless we can express quantitatively the type and amount of load which is being imposed on workers, it is difficult to evaluate the reaction of the living body as indicated by the HR. As demonstrated by a great deal of past data, HR is sensitive to both physical and psychological loads and shows the same reaction to both. In order to eliminate such ambiguity, it is usually used in conjunction with observation of work behaviour. Figure 29.9 shows a superimposition of HR fluctuation and activity records taken for a single day's work of one type at a certain worksite, thereby allowing us to draw a correspondence between work behaviour and HR.

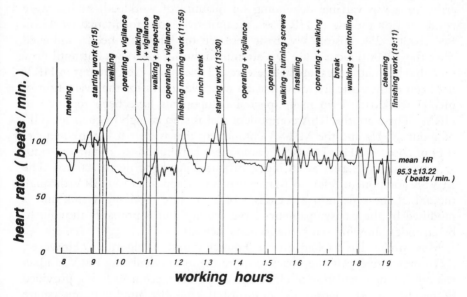

Figure 29.9. Fluctuation of heart rate with working hours in a day of a control valve operator at a plant (Kumashiro, unpublished, 1988 company report)

Although these sorts of data are convenient for observing the general workload situation, they are not adequate for determining workload quantitatively. In the latter half of the 1960s, a new approach recognized the importance of HR but took into consideration also its inadequacies for this type of research. This approach focused attention on the R-R (interbeat) interval even during HR observation. For example, a great deal of research was reported on converting the amount of cardiac output into an index as a means of evaluating the psychological stress reaction that paced work systems had on the human body. Hokanson *et al.* (1971) pointed to a rise in the blood pressure value for conveyor-paced work as compared with self-paced work systems. In the same manner, Salvendy and Knight (1979) also indicated an increase in the work task load under conveyor-paced work systems using fluctuations in blood pressure values and arrhythmia as indices.

Manenica (1977) reported that even if the average prescribed time value per unit obtained from a (pacing) free work system were established as the conveyor feed speed, the average value and the fluctuation of the R-R interval of HR during conveyor work still increased in comparison with that for self-paced work systems. More recently, it has been recognized that the phenomenon of increases in the cardiac response under paced work systems is present regardless of differences in the characteristics of the work itself (Sharit *et al.*, 1982).

Subsequently, from about 1980, serious studies began on the practical use of HRV (heart rate variability) which took the analyses of this R-R interval one step further.

Generally, since Kalsbeek (1971) had pointed out that respiratory sinus arrhythmia was variable depending on quantity of workload, studies were conducted on the use of HRV as a quantitative index for mental workload (see chapter 25). Historically, basic research on HRV has been conducted since the 1960s. Practical use was attempted during the 1980s but until now there has been little clear applied research. Today, however, interest in HRV has been enthusiastically revived as an effective means for the evaluation of mental workload. Two main spectral components have been reported in HRVs. These are the RSA component that is seen at high frequencies (HF) of about 2·5 Hz and the MWSA component that is seen at low frequencies (LF) of about 0·1 Hz. The RSA component corresponds to respiratory sinus arrhytyhmia and serves as an index of parasympathetic nerve activity. The MWSA component refers to Mayer waves that appear in the HRV through the aid of the reflection mechanism of a pressure receiving vessel, and is modified by the parasympathetic nerve activity, but is primarily thought to be an index showing sympathetic nerve activity.

Most studies have relied on the Fourier transform algorithm. However, an autoregressive (AR) approach is more useful in calculating HRV, and an example of the calculation of HRV by AR power spectral analysis is provided in this chapter. AR algorithms can furnish the number, amplitude, and centre frequency of the oscillatory components automatically, without requiring a priori decisions. Because short segments of data are more likely to be station-ary, the AR algorithms, which are capable of operating efficiently even on shorter series of events, appear to provide an additional advantage (Malliani *et al.*, 1991). Data obtained through ECG (CM5 induction) measurement are A/D converted at a sampling frequency of 100 Hz and stored on a floppy disk or similar device. The data are saved for the purpose of detailed analysis. Thereafter, the first and second steps are to produce a time series and trend graph for the R-R interval over 256 heartbeats, not including artefacts, using data such as premature ventricular contractions (Figure 29.10a and 29.10b). The third step is to determine the AR power spectrum of the trend graph (Figure 29.10c). The fourth step is to compute the middle frequency of each spectral component (Figure 29.10d). The power of each spectral component is standardized by dividing it by the total power minus the DC component (if there is one). The optimum AR model was determined by minimizing the value of final prediction error (Akaike, 1970). Stationarity was confirmed by the pole diagram analysis (Baselli *et al.*, 1987; Pagani *et al.*, 1986). These procedures permit the spectral components to be seen.

Figure 29.11 shows an example of each spectral component of normal healthy males at rest, lying face up. Two main spectral components are recog-nized, the LF component (middle frequency: 0.09 Hz eq, 0.08 cycles/beat, power: 1,094 msec^2, standardized power: 61.4%) and the HF component (middle frequency: 0.27 Hz eq, 0.24 cycles/beat, power: 616 msec^2, stan-dardized power: 34.6%). These components can be studied by normalizing them with such expressions as %LF, %HF, and LF/HF.

As demonstrated in the example it is possible to deduce the LF and HF components of HRV. Thus, research using this approach to evaluate the mental workload has gradually been promoted in the domain of ergonomics. Considering the number of years that have passed since HRV began to draw attention, however, the number of research reports is quite small. Up to the present time, there have been very few reports of studies where HRV has been applied to field studies, while conversely, the number of laboratory reports is strikingly high. The reason for this is attributed to the difficulty of quantifying such factors as respiratory control or the complex body motions of workers when targetting actual work. Nevertheless, if one quantitatively observes motion and behaviour during work and painstakingly removes the respiratory influence during EGG analysis, this problem can be solved to some extent. HRV is an index of workload that can be useful, with this sort of effort. There are many reports from laboratory research where mental arithmetic has been used in tasks to study the HRV response (for example, Cerutti *et al.*, 1988; Miyake *et al.*, 1990; Sloan *et al.*, 1991; Zwiener *et al.*, 1982). In addition to this, there are reports on memory research (Aasman *et al.*, 1987), tracking (Vicente *et al.*, 1987; Backs *et al.*, 1991), target detection (Pagani *et al.*, 1991) and VDT data entry work (Itoh, 1988). The results in all these cases appear to suggest that HRV can be an effective index for evaluating mental workload.

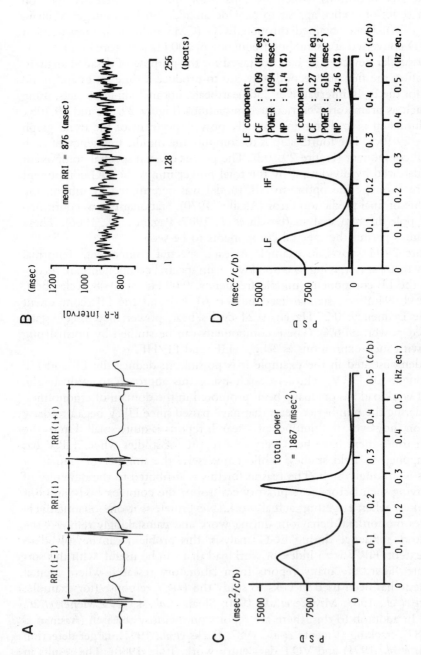

Figure 29.10. Schematic outline of autoregressive spectral analysis of R-R interval (RRI) variability. A: from surface electrocardiogram a time series of RRIs was calculated as function of beat number. B: RRIs trendgram. C: autoregressive power spectrum from B. D: individual spectral components. Note: PSD, power spectral density; c/b, cycles/beat; Hz eq, Hertz equivalent; LF, low frequency; HF, high frequency; CF, centre frequency; NP, normalized power. (Inoue, *et al.*, 1990)

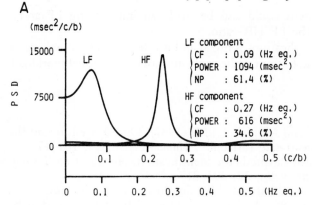

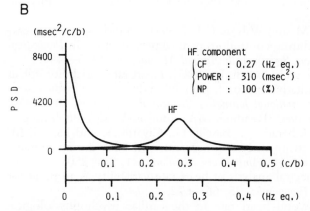

Figure 29.11. Individual spectral components of subjects at rest in supine position. In healthy male (A) there were 2 major spectral components, i.e., LF component and HF component. On the contrary, in neurological complete quadriplegic male (B), HF component was observed, whereas LF component was not observed. See caption of Fig. 29.10 for definitions of abbreviations. (Inoue, *et al.*, 1990)

Cases such as the following are available as examples of the use of HRV in industrial field studies or related domains: highway driving (Egelund, 1982), punch card operation work (Kamphuis and Frowein, 1985), piloting the work of aircraft pilots using flight simulators (Itoh *et al.*, 1990). These various reports suggest that low cycle components of 0·1 Hz serve as useful indices of mental workload.

In general, it is felt that the mental workload can be investigated from fluctuations in sympathetic activities (Pomeranz *et al.*, 1985; Pagani *et al.*, 1986) where this low cycle region primarily is significant, by focusing on the 0·1 Hz components referred to as the Mayer waves.

On the other hand, due to the fact that the HF component corresponds to the respiratory sinus arrhythmia and reflects only parasympathetic activities (Akselrod *et al.*, 1985), it is thought that the degree of imbalance in the autonomic nervous system could also be made an index using LF and a comparison with LF—the LF/HR ratio.

The range of HRV applications is broad and it is felt that HRV will come into increasing use in the future for the evaluation of occupational workload.

References

Aasman, J., Mulder, G. and Mulder, L.J.M. (1987). Operator effort and the measurement of heart-rate variability. *Human Factors,* **29**, 161–170.

Akaike, H. (1970). Statistical predictor identification. *Annals of the Institute of Statistics and Mathematics,* **22**, 203–217.

Akselrod, S., Gordon, D., Madwed, J.B., Snidman, N.C., Shannon, D.C. and Cohen, R.J. (1985). Hemodynamics regulation: investigation by spectral analysis. *American Journal of Physiology,* **249** (Heart Circ. Physiol. 18), H867–H875.

Backs, R.W., Ryan, A.M. and Wilson, G.F. (1991). Cardiorespiratory measures of workload during continuous manual performance. In *Proceedings of the Human Factors Society 35th Annual Meeting,* pp. 1495–1499.

Baselli, G.S., Cerutti, S. and Civardi, F. (1987). Heart rate variability signals processing: a quantitative approach as an aid to diagnosis in cardiovascular pathologies. *International Journal of Biomedical Computing,* **20**, 51–70.

Brod, C. (1984). *Technostress* (Reading, MA: Addison-Wesley Publishing).

Cerutti, S., Fortis, G., Liberati, D., Baselli, G., Civardi, S. and Pagani, M. (1988). Power spectrum analysis of heart rate variability during a mental arithmetic task. *Journal of Ambulatory Monitoring,* **1**, 241–250.

Egelund, N. (1982). Spectral analysis of heart rate variability as an indicator of driver fatigue. *Ergonomics,* **25**, 663–672.

Furedy, J.J. (1987). Beyond heart rate in the cardiac psychophysiological assessment of mental effort: the T-wave amplitude component of the electrocardiogram. *Human Factors,* **29**, 183–194.

Hashimoto, K. (1963). Physiological meaning of the Critical Flicker Frequency (CFF) and some problems in their measurement—theory and practice of the flicker test. *Japanese Journal of Industrial Health,* **5**, 563–578.

Harbin, T.J. (1989). The relationship between the Type A behavior pattern and physiological responsivity: a quantitative review. *Psychophysiology,* **26**, 110–119.

Hokanson, J.E., Degood, D.E., Forrest, M.S. and Brittain, T.M. (1971). Availability of avoidance behaviours in modulating vascular-stress responses. *Journal of Personality and Social Psychology,* **10**, 60–68.

Houston, B.K. (1988). Cardiovascular and neuroendocrine reactivity, global Type A, and components of Type A behavior. In *Type A Behavior Pattern: Research, Theory, Intervention,* edited by B.K. Houston and C.R. Shyder (New York: Wiley Interscience), pp. 212–253.

Industrial Fatigue Research Committee of the Japan Association of Industrial Health (1953). A draft standard for sector type CFF measuring instruments in Japan (translated into English) *The Journal of Science of Labour,* **29**, 305–306.

Inoue, K., Miyake, S., Kumashiro, M., Ogata, H. and Yoshimura, O. (1990). Power spectral analysis of heart rate variability in traumatic quadriplegic humans. *American Journal of Physiology*, **258**, H1722–H1726.

Itoh, Y. (1988). The relation between the changes of biophysiological reactions and subjective mental workload in the typing tasks under time pressures. *The Japanese Journal of Ergonomics*, **24**, 253–260.

Itoh, Y., Hayashi, Y., Tsukui, I. and Saito, S. (1990). The ergonomic evaluation of eye movement and mental workload in aircraft pilots. *Ergonomics*, **33**, 719–733.

Iwasaki, T. and Akiya, S. (1990). Changes in CFF values and that physiological meaning during the experimental visual task with CRT display screen. *The Japanese Journal of Ergonomics*, **26**, 181–184.

Kalsbeek, J.W.H. (1971). *Measurement of Man at Work* (New York: Van Nostrand Reinhold).

Kamphuis, A. and Frowein, H.W. (1985). Assessment of mental effort by means of heart rate spectral analysis. In *Psychophysiology of Cardiovascular Control: Models, Methods, and Data*, edited by J.F. Orlebeke, G. Mulder and L.J.P. van Doornen (New York: Plenum), pp. 841–853.

Kelly, D.H. (1978). Human flicker sensitivity: two stages of retinal diffusion. *Science*, **202**, 896–899.

Kogi, K. and Kawamura, H. (1960). On the variation of flicker fusion frequencies of visual pathway with special reference to activating system of the brain. *The Journal of Science of Labour*, **36**, 459–473.

Koizumi, K., Sukegawa, H. and Takakuwa, E. (1964). Studies on the function of concentration maintenance (TAF) (part 3): comparison of TAF with I.Q. and kraepelin character types. *Japanese Journal of Hygiene*, **19**, 8–11.

Kumashiro, M. (1986). Psychological and physiological functions in the middle to elderly workers. In *Job Re-design and Its Theory and Practice* (translated into English), edited by M. Nagamachi (Tokyo: The Association of Employment for Senior Citizens).

Kumashiro, M. (1988). The measuring for psychophysiological functions. In *The Handbook for Industrial Fatigue* (translated into English), edited by The Industrial Fatigue Research Committee and the Japan Association of Industrial Health, Rodokijyunchosakai, Tokyo.

Kumashiro, M. and Saito, K. (1978). Studies on the learning effects of target-aiming performance. *Japanese Journal of Industrial Health*, **20**, 212–217.

Kumashiro, M., Saito, K. and Takakuwa, E. (1973). Optimal performance in conveyor system from a viewpoint of the physiological functions. *The Japanese Journal of Ergonomics*, **9**, 207–214.

Kumashiro, M., Saito, K. and Takakuwa, E. (1976). Work load due to stamping task as a simple repetitive hand work. *Japanese Journal of Industrial Health*, **18**, 117–122.

Kumashiro, M., Mikami, K. and Saito, K. (1982). Psychophysiological effects of shift work in a small-medium sized factory. In *Proceedings of the International Ergonomics Association Congress* (London: Taylor and Francis), pp. 198–199.

Kumashiro, M., Hasegawa, T., Mikami, K. and Saito, K. (1984a). Stress and aging effects on female workers. In *Proceedings of the Human Factors Society 28th Annual Meeting*, **2**, pp. 751–755.

Kumashiro, M., Mikami, K. and Hasegawa, T. (1984b). Effects of visual and mental strain on VDT performance. *Japanese Journal of Industrial Health*, **26**, 105–111.

Malliani, A., Pagani, M., Lombardi, F. and Cerutti, S. (1991). Cardiovascular neural regulation explored in the frequency domain. *Circulation*, **84**, 482–492.

Manenica, I. (1977). Comparison of some physiological indices during paced and unpaced work. *International Journal of Production Research*, **15**, 261–275.

Matsui, K., Sakamoto, H., Kojima, T., Takakuwa, E. and Ikeda, H. (1965). Studies on the function of concentration maintenance (TAF)—discussion on three-minute method of the TAF-test. *Japanese Journal of Hygiene*, **20**, 26–33.

Miyake, S., Inoue, K., Kamada, T. and Kumashiro, M. (1990). Cardiovascular response in long term mental arithmetic. *The Japanese Journal of Ergonomics*, **26**, 142–143.

Nakadaira, S. (1969). Studies on the function on concentration maintenance (TAF)—with special reference to autonomic nervous function. *The Hokkaido Journal of Medical Science*, **44**, 177–189.

Pagani, M. et al. (1986). Power spectral analysis of heart rate and arterial pressure variabilities as a marker of sympatho-vagal interaction in man and conscious dog. *Circulation Research*, **59**, 178–193.

Pagani, M., Mazzuero, G., Ferrari, A., Liberati, D., Cerutti, S., Vaitl, D., Tavazzi, L. and Malliani, A. (1991). Sympathovagal interaction during mental stress: a study using spectral analysis of heart rate variability in healthy control subjects and patients with a prior myocardial infarction. *Circulation*, **83** [supp II], II-43-II-51.

Pomeranz, B. et al. (1985). Assessment of autonomic function in humans by heart rate spectral analysis. *American Journal of Physiology*, **248**, H151–153.

Salvendy, G. and Knight, J.L. (1979). Physiological basis of machine-paced and self-paced work. In *Proceedings of the Human Factors Society 23rd Annual meeting*, pp. 158–162.

Sharit, J., Salvendy, G. and Deisenroth, M.P. (1982). External and internal attentional environments: I. The utilization of cardiac deceleratory and acceleratory response data for evaluating differences in mental work, load between machine-paced and self-paced work. *Ergonomics*, **25**, 107–120.

Simonson, E. and Enzer, N. (1941). Measurements of fusion frequency of flicker as a test fatigue of the central nervous system. *Journal of Industrial Hygiene and Toxicology*, **23**, 83–89.

Sloan, R.P., Korten, J.B. and Myers, M.M. (1991). Components of heart rate reactivity during mental arithmetic with and without speaking. *Physiology & Behavior*, **50**, 1039–1045.

Takakuwa, E. (1962). The function of concentration maintenance (TAF)—as an evaluation of fatigue. *Ergonomics*, **5**, 37–42.

Takakuwa, E., Takahashi, M., Koizumi, K. and Sukegawa, H. (1963). A method for describing the function of concentration maintenance (TAF). *Japanese Journal of Hygiene*, **18**, 241–246.

Takakuwa, E., Sukegawa, H. and Koizumi, K. (1964a). Studies on the function of concentration maintenance (TAF) Part 2: Evaluation of fatigue in overtime clerical work at a bank. *Japanese Journal of Industrial Health*, **6**, 257–262.

Takakuwa, E., Iida, N., Koizumi, K., Nakadaira, S., Ikeda, H. and Sukegawa, H. (1964b). Studies on the function of concentration maintenance (TAF) Part 4: Influences of alcohol on TAF. *Japanese Journal of Hygiene*, **19**, 25–32.

Takakuwa, E., Ikeda, H., Nakadaira, S., Domon, H. and Masukawa, T. (1966). Studies on the function of concentration maintenance (TAF) Part 6: Evaluation of the qualitative differences in noise with TAF-test. *Japanese Journal of Industrial Health*, **8**, 11–14.

Takakuwa, E., Domon, H., Saito, K., Ikeda, H., Ohnaka, Y. and Sukegawa, H. (1967). Studies on the function of concentration maintenance (TAF) Part 9: In relation to brain wave (alpha wave). *Japanese Journal of Hygiene*, **21**, 397–402.

Vicente, K.J., Thornton, D.C. and Moray, N. (1987). Spectral analysis of sinus arrhythmia: a measure of mental effort. *Human Factors* **29**, 171–182.

Walker, A.E., Woolf, J.I., Halstead, W.C. and Case, T.J. (1943). Mechanism of temporal fusion effect of photic stimulation on electrical activity of visual structures. *Journal of Neurophysiology*, **6**, 213–219.

Yamamoto, S., Shimoda, M., Kanemoto, T., Nakadaira, S., Takakuwa, E., Ikeda, H., Domon, H., Onaka, Y. and Anei, T. (1966). Fatigue in nursing work—with special reference to the function of concentration maintenance (TAF). *Iryo*, **20**, 15–21.

Zwiener, U., Bauer, R. and Scholle, H. Ch. (1982). The influence of mental arithmetic on autospectra, coherence, and phase spectra of autonomic rhythms in man. *Automedica*, **4**, 113–121.

Part VI

Analysis and evaluation of work systems

One of the most obvious changes in ergonomics practice during the past two decades or so has been the expansion in its areas of application. Most in the field would now accept that the 'system' in the human-machine system embraces more than the interface controls and displays, the workplace and environment. We must be interested in the wider context of work, the psychosocial environment and the impact of organizational factors such as production technology, organization structure and finance. A very good example of this is the field of human-computer interaction, and specifically the introduction and use of Visual Display Terminals (VDTs). In the context of workers' health, comfort and performance, as much is written about good job design, work organization, support and training and technology implementation as about the purely physical factors of lighting, seating, keyboards etc.

This part of the book then, and Part VII following, takes a fairly broad view of work systems. The first three chapters by Haslegrave and Corlett (30), Kirwan (31) and Brown (32) concentrate upon assessments which can be made in order to evaluate the 'quality' or success of a work system—performance measurement—and also to identify areas for system improvement through redesign. They are most of all three different approaches to risk assessment. Recent changes in legislation and in the underlying attitudes within industrial society have given impetus to the search for ergonomics risk assessment methodologies, especially in the context of manual handling and exposure to risk of musculoskeletal disorders. Haslegrave and Corlett review the background, need, basis and contemporary approaches to such risk assessment.

Along with recognition of the vital role of the human-computer interface in determining the success of computer systems, a major boost to the perceived importance of ergonomics has been the Three Mile Island incident in 1979, and subsequent concern for nuclear power plant safety. This concern has continued through the occurrence of a number of well publicized disasters or near disasters involving complex systems in power plants, chemical processes, transport systems and so on. Fundamental to many of these incidents and to safety in such systems is the potential for human error. Therefore there is great interest in developing methods for the analysis and measurement, and subsequent enhancement, of human reliability. Kirwan discusses this as a component of a ten part generic methodology, and provides very much the perspective of an active practitioner.

There is a strong link between the consideration of human reliability and the use of accident reporting and analysis techniques. In chapter 32, Brown shows how theories of human error may well be very productive in terms of ergonomics measures for safety improvements. He distinguishes human error theories from theories of accident causation; in very much a conceptual review, he shows how vital it is to understand such theories in order best to develop an accident reporting system and, most importantly, use its data in accident reduction. Many of the points he makes and the cautions he gives about reporting and analysis could be read in the context of event observation generally (see chapters 2 and 3).

If we are to take a true systems approach in occupational ergonomics, and if we are to have any meaningful impact in our work, then we must understand much about the organization which is the site of our investigations. We must be able to assess, for instance, how the way in which an organization is structured, its management philosophy and its willingness to change may influence behaviour or may affect any developments we initiate. Shipley in chapter 33 provides a view on such issues which will be food for thought for any ergonomics practitioner, and within it presents a powerful argument about *how* ergonomists should tackle investigations into people's work, whatever the *what* of the particular study's focus.

In chapter 1 it was noted that ergonomics can have aims which relate to organizational as well as individual well-being, although the strong linkage between these was stressed. Whilst many of us would, in an ideal world, put individuals' interests as our major priority, we must work in the real world of industry, at least as far as occupational ergonomics is concerned. Here, Simpson and Mason argue (in chapter 34), an economic case generally must be made before an organization will fund, or even play host to, an ergonomics investigation. This consideration can be extended; a powerful motivation for ergonomics input in consumer product design for instance is the economic threat of strict product liability provisions. Thus, we do require some means of showing the economic returns on our efforts, and Simpson and Mason present some relevant techniques.

Chapter 30

Evaluating work conditions and risk of injury—techniques for field surveys

Christine M. Haslegrave and E. Nigel Corlett

Introduction

Industry has become more generally aware of the important effects of work and workplace design on health, safety and quality performance. This has created a corresponding demand for measures of ergonomics quality of workplace design and of working conditions, both to demonstrate compliance with legislation and, where problems are found to exist, to set priorities for making changes. Moreover, such measurements, where they are related to health and safety, incorporate the concept of risk assessment in ways very analogous to the fields of occupational health and safety and human reliability (see chapters 31 and 32).

Assessments need to be made at two different levels: surveys at company or plant level to identify jobs which may present risks; and in-depth investigations of individual jobs or workplaces. In undertaking these, techniques are needed both for evaluating the demands of work tasks and for assessing whether these present risks of traumatic injury or of long-term cumulative damage. Auditing may be carried out by ergonomics professionals but many of the personnel undertaking such assessments within companies will have had little training in ergonomics. They require fairly simple and well-defined procedures and ones which can easily be applied in the field. It is also useful if the procedures can incorporate help with devising efficient solutions and guidance on the circumstances in which specialist ergonomics expertise should be called in to assess the more complicated situations.

These needs involve a change in emphasis for ergonomics methodology, which has heretofore been used largely by trained ergonomics professionals, and techniques are being developed to meet them. This chapter reviews some of the techniques currently available and provides guidance on the situations in which they are most applicable. Since this non-specialist methodology is currently in the early stages of evolution, techniques are not yet available for answering all the questions posed by industry, particularly those concerned with determining actual levels of risk or for making quantitative statements about the factors presenting hazards. (The reader is referred also to chapter 35 for discussion of how to run an ergonomics study in the field.)

One concern when developing these methods is to prevent their use without an understanding of their wider implications, which could in fact create more problems, which a better understanding of ergonomics would have avoided. The provision of such techniques may also encourage users to apply their conclusions as if they were absolute values. To reduce this possibility, many ergonomists advocate the introduction of participative methods of making changes to work or work organization, involving people throughout the work force in designing their own workstations and working systems (Corlett, 1991; Noro and Imada, 1991; Wilson, 1991, 1994). When implementing changes, a broadly-based team which can bring a breadth of viewpoints and experiences to bear, armed with techniques which point up key weaknesses in the situation, can protect the operators at the workstations from many mistakes. If basic ergonomics knowledge is disseminated amongst the operators themselves, they are enabled to improve even further the designs of their own workstations.

Recognizing the presence of a hazard

The evaluation of risk at a work situation requires, first, the recognition of the presence of a hazard and then judgement of its 'strength'. Neither step is easy, but the second is much more difficult than the first. Chapter 31 by Kirwan on Human Reliability Assessment, which is an application of risk assessment trying to quantify the likelihood that people will make errors in predicted situations, will give some idea of the complexity of the problem.

Let us start with the less difficult part, recognizing the presence of a hazard. As far as field studies are concerned, it is rarely possible to undertake experiments, so we are confined to studying the existing situation and to examining past history. Past history can be a rich source of information, particularly accident books, medical records and personnel records. Confidentiality must not be breached in many of these, so data compression and analysis must take this into account. Chapter 4 on the use of archival data gives information on the identification and use of such data; Brown's chapter 32 on Accident Reporting and Analysis is essential reading. In that chapter, he discusses some of the major accident models and theories of human error, which are necessary information to recognize some of the dimensions of potential mismatch, and he also discusses the analysis of accident data which gives direction to the investigator who is seeking to identify hazards.

Assessing for hazards is to assess for workplace inadequacy. The inadequate match between the person and the total working environment introduces the potential for risk of injury as well as for restrictions on people's abilities to perform well. Workplace assessment, by its nature, is a risk assessment although it may need forms of analysis which vary depending on the anticipated use of the results.

Assessment of the risk

Study of the current situation is clearly a necessity, but also clearly needs a sound ergonomics knowledge. What is being sought is evidence of mismatches between people and their environment, under all circumstances where they are likely to interact, and then some evidence of the severity of the consequences of each mismatch.

A rapid overview of a particular office or factory area can be gained by the use of a Loads and Causes Survey (Åberg, 1981), and one version of the recording form is shown in Figure 30.1. Here the degree of mismatch is assumed to be represented by the magnitude (or severity) judged during the assessment of each load factor. Such a survey, which can be modified to suit the area of concern, can be used both to make an initial survey for hazardous features and to identify priorities for starting an ergonomics programme of change.

A Loads and Causes Survey form would typically be used in an assessment by a group of people concerned with the area under investigation (for example, safety officer, foreman, some operators and plant engineer). Initially, of course, they must agree on their recording criteria and procedures, and then after making recordings get together to examine their results and discuss and resolve any differences.

In some circumstances, direct observation of operators and their work tasks is possible. Branton (1970) watched four operators of capstan lathes over a period of 16 weeks to identify potential causes of accidents. He created simple forms to assist in recording the movements of operators, the purposes of their movements and their endpoints. From his data, he was able to identify how injuries might occur, to get an estimate of frequency of exposure, and to make proposals for hazard- and injury-reduction. Branton's recognition of hazard came not just from his observations, which would not have required much ergonomics, but also from his recognition of the information processing—or motor control—difficulties in locating the components and in locating and operating certain of the controls. Such recognitions permitted him to make more valid proposals for improvement, due to his deeper understanding of the causes of the errors he saw.

This would be true also for the use of techniques such as OWAS or RULA (described in chapter 23). The purpose of such techniques is to identify postures, combined with forces and repetitions, which may indicate a risk of low back injury or of upper limb injury. However, the posture or muscle

LOAD FACTOR		CAUSE FACTOR								
MAGN-ITUDE	TYPE	TECHNOLOGY/ PROCESS	ROOM, BUILDINGS	MATERIAL FLOW	LAY OUT	MACHINES	TOOLS, AIDS	WORK ORGANIZATION	WORK METHOD	WORK PLACE DESIGN
	NOISE									
	DUST, SMOKE									
	CLIMATE									
	BODY									
	SPACE									
	CIRCULATORY LOAD									
	INFORMATION LOAD									
	CONSTRAINT MONOTONY									
	CONTACT POSSIBILITIES									
	COOPERATION POSSIBILITIES									
	DISTURBANCE OF OTHERS									
	OTHERS DISTURBING									
	SUM									

Figure 30.1. Åberg's Loads and Causes survey form

loading as recorded via these two techniques is but one factor in the context of hazard. A lack of co-worker or management support, as for example in the case of a nurse handling a patient or when pressure for output encourages the reduction of rest breaks, could be key factors tipping the situation from the undesirable but feasible to the impossible. This would not be recognized simply from a study of record charts, but could be identified by a skilled and informed observer.

In any study, observation is often the best start and gives a much better indication of where mismatches lie than is possible from simply hypothesizing about the problem. Seeing what people really do, and talking to them about what the work feels like, and what they are really trying to accomplish, leads to a clearer knowledge of the links between people and their work. The work, indeed, may be only a minor factor in risk: the organization, workplace structure or other factors may well be of greater importance.

Checklists and questionnaires

Given that a good understanding has been developed, questionnaires can be useful for the more detailed investigation. Chapter 3 gives basic advice on questionnaire design and reference to the index of this book and Table 1.3 will indicate sources for questionnaires for particular purposes. It will be evident that questionnaires, and their cousin the checklist, can be of value in the context of hazard investigation, although their use on their own is usually insufficient.

Checklists which require only a 'yes' or 'no' response are most widely used as memory aids, being good for making sure that all aspects of a situation have been covered. In the context of hazard analysis they are also valuable to check whether all practical and known measures have been taken.

Figure 30.2 gives an extract from a broadly based set of questions in a checklist suggested by the UK Health and Safety Executive in a guide to preventing work-related upper limb disorders (HSE, 1990). The extract deals with environmental factors, and it will be noted that the questions as given can be answered by 'yes' or 'no'. The questions draw attention to a range of environmental factors seen to be relevant, but do little more than this. It will be clear, however, that most questions could be broken down to give the components of the factors concerned, which could then be given a response scale crossing the 'adequate' point from one extreme of inadequacy

---

Environmental factors

(a)   Are levels of noise sufficient to cause mental stress or interfere with communications or safety?
(b)   Does music impose rhythmic patterns that are inappropriate for the task?
(c)   Are lighting levels causing operators to adopt awkward postures to avoid shadows or to see properly?
(d)   Are flickering lights causing stress to operators?
(e)   Is the air temperature avoidably low at any time of the year?
(f)   Is protective clothing issued because of the environment constraining posture or do gloves affect grip?
(g)   Is poor environment a source of discontent among operators?
(h)   Are there chemicals in the air that might be affecting the operators' coordination or muscular system?

Figure 30.2. Extract from checklist for the identification and reduction of work-related upper limb disorders (HSE, 1990). Crown copyright is reproduced with the permission of the Controller of HMSO

to the other. The resulting answers would reveal, with very little more effort, the extent of the major mismatches in the environment. Rather than ask questions about what is wrong with a situation, many checklists present what a good situation should consist of, leaving the user to check the departure from this ideal in the case under investigation. Sometimes the wording is imprecise, indicating goodwill but little guidance. It is of little use to say that loads should be minimized or trunk deflection kept low. The set of guidelines (checklist) on manual handling (presented later in this chapter as Figure 30.6) has its share of these, but also gives specific statements against which situations can be tested, such as

"Handling loads at heights below the knees or above the shoulder should be eliminated wherever possible."

or

"Use grasps which are two-handed, with the load or force evenly distributed between both hands/arms."

These examples give specific requirements, so that a hazard assessment could document when these were breached and investigate the reasons. Moreover, a study of operator behaviours 'before' and 'after' intervention could then demonstrate a reduction (or otherwise!) in the hazard.

A similar checklist, for workplace design, is given in Table 19.2, where each statement has endeavoured to be specific. Again, a workplace could be assessed against such specifics, even in some cases with numbers attached, and changes made and evaluated.

Other forms of checklist may be developed to incorporate scaled answers. A good example of this is the Chair Feature Checklist given in chapter 21. Here each feature of the seat relevant to its use has been set down and provided with a scale running from one extent of inadequacy to the other, e.g., 'too high' to 'too low'. Each question can be answered by introspection whilst seated on the chair, and a profile of the seat's adequacy drawn. This will specify what needs to be done to improve the seat, and in this case a modest knowledge of the ergonomics of seating would probably enable the sitters to make their own improvements. The application of such a feature checklist can be considered in other areas. Figure 30.3 shows a proposal to explore the match between an office job and the individuals' needs. From a knowledge of the jobs surveyed, and the mismatch on the cognitive dimensions, sources of error could be surmised, which could be compared with recorded errors. The risk of error is relative to the severity of the mismatch, even though the exact relationship is not known. Nevertheless, analysis of such a checklist, in the light of concerns regarding the severity of the effects of various types of error, would present a starting point for further investigation.

For the most difficult parts of your job, answer these questions:

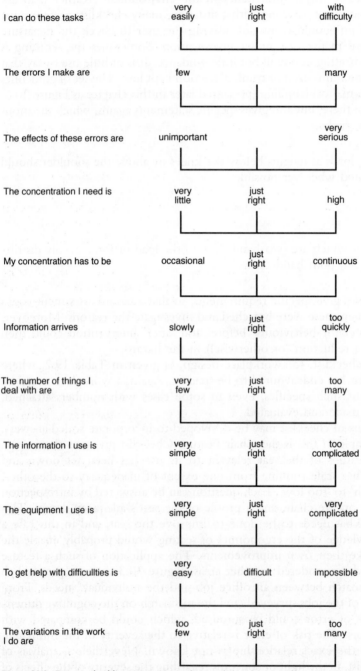

Figure 30.3. Feature checklist to investigate the match between an office job and the individual's needs

Psychosocial aspects

As mentioned earlier, psychosocial factors can influence performance and work demands, thus affecting risks of industrial injuries. Examples of psychological work demands, which can be associated with physiological indicators of strain, are the levels of monotony, boredom, job satisfaction, autonomy or social support.

The general approach for assessing risk must be to understand, by investigation, the holistic nature of the work. It is therefore consonant with our approach to investigating work conditions and risks to look at psychosocial work aspects from the perspective of a mismatch between what exists and what people expect. It is also necessary to note that what exists may well be 'what is perceived to exist', in the sense that people's perceptions of a situation are what is relevant if we are speaking of a mismatch. This point has been amply illustrated in an early study by Kjellstrand (1974) which showed that, as workplace constraints such as pacing increased, perceptions of environmental stress increased, although the actual environmental situations were substantially the same.

The psychosocial factors which are usually of interest to the ergonomist are work stress (a term needing definition before it is measured—see chapter 26), fatigue and attention, work attitudes, and social support or cohesion. Apart from identifying the presence and strength of these factors for research purposes, there is interest in their contributions to general stress levels and, of course, to the potential mismatches mentioned above.

Questionnaires may be suitable for investigating some of these aspects. The Modified Work APGAR questionnaire (Bigos *et al.*, 1991) is a type of factor checklist, presenting a number of factors relevant to personal associations at work. Karasek's (1979) demand control questionnaire provides measures of psychological demands in a job, intellectual discretion and authority over decisions together with indicators of mental strain symptoms (exhaustion, depression, job satisfaction and life satisfaction). Where work pressures exist, job stress is suspected or alienation is evident (which are all factors liable to lead to personal risk), such questionnaires can give indications of mismatch.

In general, questionnaires will be suitable where features relevant to the work (and where mismatch could be deemed to exist) can be answered by reference to things as they are at the time of the enquiry or in the recent past. These aspects are then separated into their component features, the features making part of the 'model' or concept of the work and its hazards, which are under investigation. It is important that the investigator has such a model, for single issue studies are unlikely to be very useful since most psychosocial responses are multifactorial. Investigations which ignore this will be sub-optimal.

Moreover, the simple use of psychosocial data collected by questionnaires can be unwise. Answers can only be interpreted where the questionnaire is based on an appropriate model of the effects of psychosocial factors, and where this model is recognized together with some understanding of the culture of the site of the study, and of its organizational and social structures. Many investigators stress the desirability of complementing questionnaires and other instruments with interviews, so that a clearer knowledge of the meaning of the responses is then available.

The reader will gather that there are difficulties in compressing methods for investigating psychosocial factors into part of a single book chapter! There is no substitute for deeper study in this area. Nevertheless there are some well developed instruments, with normative values, which can be introduced. One widely used technique is the Stress Arousal Check List (SACL) of Cox and McKay (1985) described in chapter 26 in this book. The underlying model and robustness of this technique is discussed there and in the quoted references.

Fatigue is both a physical and a psychological state, and earlier chapters have discussed the former in terms of muscular and cardiovascular symptoms. Several fatigue inventories exist, to explore people's feelings of fatigue, and one which is widely used is that of the Industrial Fatigue Research Committee of the Japan Association of Industrial Health (1970). It consists of thirty-one items covering feelings of physical states, such as stiffness, drowsiness, or difficulties in concentration. This inventory is often used by Japanese researchers within a battery of tests and measures. In a study of checkout operators in supermarkets, Kishida (1991) looked at job structures, using observation and questionnaire survey, and responses on the Industrial Fatigue Inventory. He analyzed the results to identify correlations between fatigue responses and work times and activities, highlighting aspects of the work which gave rise to the peak responses.

Job attitude questionnaires have a long history, as have studies of the effects of social support in the workplace and outside it. Even more than for the previous two areas, the use of the available instruments, and more especially their interpretation, is strongly linked to the models on which they are based. It is deemed inappropriate, therefore, to expand on this, beyond making the cautionary point that, when studying the literature, the reader should keep in mind that there is a strong cultural dimension which may make the transfer of methods gleaned from the literature unsuitable for transfer directly to a local problem.

Quantifying the hazards

The purpose of hazard investigation is to define and, as far as possible, quantify the hazards, and then to present what should be done to reduce or eliminate them. Clear presentation of the analyses is vital, but the analyses themselves must reveal the picture. So having used workplace survey, questionnaire, checklist or interview, the investigator then needs to reduce the data.

Moreover, where hazards have been identified, some indication of their importance is needed and decisions have to be taken on how to eliminate or control these. If the hazard can be eliminated with no significant effect on the rest of the system, then that is the decision to be taken. If this is not the case, then a risk assessment is needed, so that undue effort is not put into highly unlikely events with minimal human risk or commercial impact.

We present here the outline of a procedure which can be used for hazard assessment and for deciding on subsequent actions. Its purpose is to provide a framework around which any company can build its own assessment procedure to meet its particular requirements (see Figure 30.4).

The approach is to assemble a group of relevant people within the company and ask them to assess each hazard which has been identified. How do

Hazard severity	Probability of occurrence		
	3	2	1
3	stop job and eliminate	eliminate as high priority	eliminate as priority
2	eliminate as high priority	eliminate; guard	eliminate; guard
1	eliminate; guard; protect against	eliminate; guard; protect against	eliminate when routine allows; guard; protect against

Note: Levels of hazard, probability of occurrence and action should be decided beforehand in relation to experience with the plant.

Hazard Severity: Could be defined in terms of consequences.

Thus 3 =   Death or serious injury; high cost; major damage
      2 =   Minor injury; moderate cost; moderate damage
      1 =   No injury; minor cost

Probability of occurrence:

e.g. 3 =   frequent
      2 =   infrequent, e.g. once or twice a year
      1 =   rare, unlikely but possible

Actions, in the body of the matrix:

The sequence should be: (i) eliminate, (ii) guard, (iii) protect against, e.g. special clothing, (iv) train for special work behaviours. In all cases knowledge of the hazard and warning notices should be provided. Actions may, of course, be combined as needed. They are not to be seen as independent.

Figure 30.4. Initial assessment of levels of hazard—an action matrix

we find out where the hazards lie? Past records have already been mentioned, which reveal what has already happened, probably a random selection from what could happen. When looking at work situations, a listed sequence of work activities (as for instance from a method study) should be augmented with all the associated or nearby equipment, such as the presence of forklift truck movements or high pressure lines. The group studies each of these for potential hazards, possibly using a checklist compiled with the help of an expert in the relevant part of the system, which lists all possibilities for malfunction. Possible hazards from both common and exceptional situations should be considered (e.g., from poor access for maintenance to flooding).

When looking at this list, the task demands for physical or mental effort can have important influences, and required activities should be assessed with 'worst case' situations in mind. As the group pursues the assessment, a note is made of each identified hazard, and possible ways for overcoming it. However, all known hazards are not equal, and some priority has to be determined to see that an effective programme is established. This is done by assessing both the likelihood of occurrence and the consequences for the company and its people if the hazard caused an accident.

The consequences can be presented as a sequenced list, in order of increasing severity perhaps starting with 'no delay or cost' and extending through damage and product delay to injuries of different degrees of severity. For example, the severity of injury might be classified as *high* (involving fatality, permanent disablement or long-term health effects), *moderate* (where a fatality is unlikely but long–term disability or long periods of lost-time would probably arise from the event), and *low* (where minor injuries might occur or the task demand could prove difficult for some operators).

Some estimate also has to be made of the likelihood or chance of each hazard causing harm in order to judge the potential 'cost' of the hazard (which can be expressed in terms of the product of severity and probability). This enables hazards to be prioritized for decision purposes. The probability of occurrence might quite simply be defined by the group using the categories: for example, frequent = 'once per shift', moderately often = 'once per week', and occasional = 'once in period longer than a week'. The actual choice of rankings, though, would need to be appropriate to the types of work and hazard involved in a particular company. It might also need to take account of the number of people exposed.

If the chosen levels of consequences and probabilities are then set out across the top and down the side of a matrix, as in Figure 30.4, the levels of priority for remedial action can be inserted by the group, so that the relative priorities for actions appropriate to each hazard situation and its likelihood can be specified. More important than the exact rankings or codings is the discussion and agreement among the group in deciding upon appropriate strategies for action. Moreover, a written record should be made for reference and for updating as experience develops within the company or industry concerned. It is also important that the agreed procedure should be pilot tested and reviewed before full scale implementation.

One further point is important to assessing both the likelihood and severity of hazards. Earlier, reference has been made to psychosocial factors as contributory to work demands and thus to accidents and injuries. In chapter 31 on human reliability assessment, is a discussion of 'performance shaping factors', those many influences which modify performance. The entire chapter is relevant to the analysis of hazards in general, but attention is drawn here to performance shaping factors because they will affect the importance of any objectively identified hazard. For example, oil on the floor is a hazard, but it is an increased hazard where operators are under time pressure and have to run. Hence, having assessed as objectively as possible the inadequacies and mismatches in any situation, these should be considered in conjunction with a table of performance shaping factors. For example, one feature which gives managers and safety officers much concern is risk-taking behaviour. Risk-taking can often be generated by a company climate, familiarity with the situation, a personality trait or other reasons. The influence of such behaviour on a particular hazard can be to increase its severity, and a company's hazard analysis report should take these factors into account.

Techniques for assessing the effect of work activities on musculoskeletal problems

Having discussed the needs for assessment of work activities and risk of injury, we can turn to a more specific application, that of assessing the hazards of musculoskeletal injuries in industry. Such assessments are largely concerned with physical work activities, although environmental, organizational and psychosocial aspects of the work may well have an impact on these and also deserve investigation.

The present discussion draws on many techniques already widely used by ergonomists for in-depth studies and seeks to show how these can be applied in audits and field assessment surveys. Some of the methods are described in more detail than others but this is not intended to reflect their relative importance. Where methods have already been presented in other chapters of this book they are simply mentioned here and any features specific to field assessments are noted. A few methods are introduced which are not included elsewhere in the book, and these are discussed at greater length.

Musculoskeletal problems arise from the responses in the human body to the physiological and biomechanical demands of physical activity (see chapters 22, 23 and 24). The nature of those demands (and the responses) changes with the type of task performed, across the spectrum of jobs from those which require very static postures to be held for long periods (by keyboard operators, for example) to the heavy dynamic work of brewery draymen or forestry workers. However, simple observation of work in industry will very quickly show that most tasks involve both static and dynamic elements. The data entry keyboard operator or sewing machinist works in the same static posture all day long but performs highly intricate and dynamic tasks with hand and wrist. The brewery drayman performs dynamic work, lifting and moving heavy weights, but this places heavy and sustained loads on the muscles of the shoulders and arms in supporting the barrels or crates which are carried. Any tasks performed standing at a fixed workplace impose static loads on back and leg muscles throughout the working period.

Ergonomics audits of physical activities at work and assessment of consequential risks have to take account of the mixed elements of static and dynamic muscular work. The techniques for measurement of these demands and evaluation of the effects will draw on knowledge of responses to static and dynamic work as discussed in chapters 22, 23 and 24. Risk assessment requires tools that can be used in the field rather than the laboratory, and methods for surveillance of large populations. It may not be necessary to provide measures of absolute risk but it is important to have comparative measures which assist in establishing priorities for making changes and redesigning the work.

The ergonomics factors contributing to musculoskeletal disorders are now well-recognized, but dose-response relationships have yet to be accurately defined for most musculoskeletal problems and it has to be acknowledged that our current understanding of the risks borne by individual workers is still limited. The methods presented here cannot therefore, as yet, answer all the questions posed during workplace risk assessments.

Choice of assessment techniques

Having recognized that work tasks may have both dynamic and static elements, it is obvious that any tasks which are predominantly dynamic should be assessed in terms of the physiological demands of the work, primarily in terms of energy expenditure (following the guidance in chapter 22). These jobs now largely occur in traditional and specialized industries. The techniques for assessing jobs which require the holding of very static postures are discussed in chapter 23. Most jobs, however, fall somewhere between these categories. Figure 30.5 shows the appropriate techniques which might be used for different types of work.

In many situations, the effects of musculoskeletal loadings may well be cumulative and long-term. The demands and consequences of this type of work are more difficult to assess than the risks inherent in single overexertions, but a variety of techniques for assessing these is presented in the following sections of this chapter.

Careful thought is needed in deciding which techniques are most appropriate in a particular work situation. It is important first to consider the nature of the work, in order to ascertain where the demands arise (and specifically whether the muscle activity comes from dynamic activities or from sustained

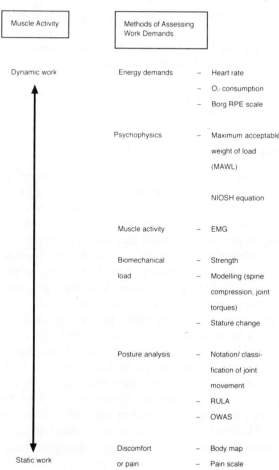

Figure 30.5. Methods available for assessing physical workload

and relatively static loading on the muscles). This should be firmly based on observation and analysis of the work activities to identify the tasks performed, although this is often far from easy in real work situations. The mixture of tasks and their timing and duration can be highly variable in many jobs, and the range of this variation over the day should be established as far as possible.

Once the principal demands of the work have been identified, the next step is to assess whether these pose risks for any of the workers involved. The major risks of musculoskeletal injuries would seem to fall into three main categories (not by any means mutually exclusive in any one job):

(i) long-term effects of static postures or of continuous loading of particular muscle(s)
(ii) biomechanical stresses involved with manual handling tasks and over-

loading of musculoskeletal structures, possibly presenting a risk in even a single exertion

(iii) cumulative effects of repeated loading on the body's structures resulting in work-related musculoskeletal disorders.

The possibility of each of these types of injury needs to be considered when assessing the risks of physical work activities.

Survey techniques to assess the scale of the problem

The body's responses to strain and injury manifest themselves in discomfort and pain, and survey techniques have been developed to use these signs to identify problems occurring even in the earliest stages. Many musculoskeletal injuries will also have an effect on movement capability and performance, and these effects are obviously important in clinical diagnosis and treatment, but such symptoms are not particularly useful in large scale surveys because of the wide variation in population characteristics, the unreliability of the symptoms in the early stages of developing musculoskeletal disorders, and the practical fact that the diagnostic tests are time-consuming.

Where musculoskeletal injuries are of major concern, people's perceptions of fatigue, discomfort or pain can be used—even prior to reported injuries— to identify possible hazards. It has been shown that people can reliably report the local body regions in which such problems occur (see chapter 23). A knowledge of the site of discomfort or pain can frequently point to the actions or tasks which may be causing the stresses. However, it has to be remembered that in some cases pain can be referred to sites other than the body tissues under stress, particularly when it arises from entrapment or compression of nerves, which presents the same problems for identifying the cause as it does in clinical diagnosis.

Various checklists have been published as tools for investigating musculoskeletal disorders and identifying related factors in the workplace (as, for example, Putz-Anderson (1988), HSE (1990) and Keyserling *et al.* (1993)), but there are fewer tools for making assessments of the risks of injury. Many tools can help in more detailed investigations of layout aspects and biomechanical loading, and these are covered elsewhere in this book. Wells and Moore (1992) provide a review of some of the computer aided analysis techniques and models which are available to assist investigations concerning musculoskeletal disorders, dividing them for convenience into techniques for assessing workplace layout (CAD), evaluating instantaneous strength demands or musculoskeletal loading, monitoring postures and events, and measuring muscle activation (through EMG).

Two questionnaires which are helpful in collecting workforce information on musculoskeletal problems are the Nordic Questionnaire (Kuorinka *et al.*, 1987) and Kemmlert and Kilbom's questionnaire for assessing musculoskeletal stress factors (Kemmlert *et al.*, 1987). These questionnaires were both designed and widely tested in the Nordic countries, and have also been tested in the UK (Dickinson *et al.*, 1992).

The English language version of the Nordic Questionnaire of Dickinson *et al.* (1992) is included as an appendix to chapter 23. The Nordic Questionnaire is being used in Scandinavia for epidemiological analysis, with the potential for identifying problems by industry, work practices or types of equipment. It is designed to identify sites of pain or discomfort and then to collect further information on duration of the problem (over the course of the previous year) and the extent to which this has affected work activities, both of which are important in assessing the severity of the problem.

Kemmlert and Kilbom's (1986) Questionnaire for Assessing Musculoskeletal Stress Factors was developed at the National Board for Occupational Safety and Health, Solna, Sweden, to provide a simple instrument for widespread use by Labour Inspectors for assessing workplaces. The form is illustrated in Figure 23.21; chapter 23. It complements the personal reports of musculoskeletal problems collected in the Nordic Questionnaire by linking such reports to factors in the workplace which may be causing the musculoskeletal stress, the checklist of 'stress factors' being developed from knowledge of common causes of injury resulting from poor workplace design. In addition to the checklist, a list is included of other important factors related to the organization of the work and the psychosocial environment.

Investigating work-related musculoskeletal disorders

Musculoskeletal disorders arise in many forms and the symptoms are frequently non-specific. They have been given many names: WRULD—work related upper limb disorder; CTD—cumulative trauma disorder; MSD—musculoskeletal disorder; RSI—repetitive strain injury; and OOS—occupational overuse syndrome, being some of the more common, reflecting the various emphases placed on potential causes. From an ergonomic point of view, the main concern is that work activities impose stresses which cause long–term damage to body tissues and structures (and particularly to muscles, tendons, ligaments and nerves), which result in both personal consequences (discomfort, pain and injury) and reduced performance. Research has shown that the causation mechanisms of such disorders are highly complex and may involve many different factors (psychosocial and organizational as well as physical). Nevertheless, it is recognized that the three main factors implicated are *force exerted* (in relation to the strength capacity of the muscles used), *posture of body segments* involved, and *repetitive nature* of the actions. This latter factor is most important in relation to the ability of the body's structures to recover from exertion, so that the fourth factor in any investigation is the *rest breaks* provided within the task cycle and between working periods.

The assessment of the physical risk factors therefore requires investigation of both layout (for instance, how the arrangement matches the anthropometry of the operators working there and affects the working postures which can be adopted—reach, clearance and access issues) and the musculo-skeletal loading. However, many other organizational, environmental, and psychosocial factors also need to be considered and evaluated in a full assessment of any work where musculoskeletal problems are of concern.

This serves to emphasize the difficulty of investigating such problems. Certainly both static and dynamic aspects of the tasks need to be identified; however, in most cases, the dynamic aspects of the tasks usually only involve hands or arms (or occasionally legs) and so the techniques given in chapter 22 for assessing the physiological responses to whole-body exertion are less applicable. Techniques which may be used specifically for assessing hand, wrist and arm motions are discussed below, while this section addresses the more general assessment of the risk factors in the tasks.

RULA (Rapid Upper Limb Assessment) is a survey method specifically designed for use in investigations of workplaces where upper limb disorders are suspected (McAtamney and Corlett, 1993 and described in detail in chapter 23). It provides a method of quickly assessing posture (whether seated or standing), paying particular attention to the neck, trunk and upper limb segments. Moreover, it assesses the contribution of the muscular effort, whether arising from exerting external force, from postural effort, or from muscle loading in the task activities such as the holding of tools. RULA is intended to be used by non-ergonomists as well as by ergonomists, but requires some training or previous skills in posture observation. It can be used as a tool for surveying large numbers of workplaces within a company in order to identify tasks where significant risks of upper limb disorders might exist and further to be able to prioritize the changes and redesigns which might be required. Practical experience gained in its use has shown that it is equally valuable in investigating individual workstations, where it highlights major postural problems. This concentrates attention on aspects of the workplace layout or task factors which constrain the operator's posture and thus identifies aspects of the work which need modifying.

Investigating hand, wrist and arm motions

Video records are valuable for investigating hand and wrist motions adopted while performing work tasks, unless of course goniometers are used to measure the motions of specific joints. Detailed analysis of the task requires consideration of typical work cycles, taking measures of time (or repetition rate) and the forces involved in addition to the joint postures.

Force criteria have proved very difficult to establish because the stress caused by the loading depends on the muscles which are being used to perform the task. Moreover, although accurate transducers are available for measuring the forces involved in direct pushing and pulling (as, for example, in depressing a control knob), as yet there are no suitable transducers which can be used in field studies to measure contact pressures, force distribution over different surfaces of the fingers or palm of the hand, or grip forces commonly required in industrial tasks. Any such transducers would be liable to interfere with the task itself and so alter the work method which they were attempting to measure. However, in a field survey it is rarely important to measure the actual magnitude of the force—the presence of a high force exertion or of exertion during a significant proportion of a repetitive work cycle is usually sufficient to warrant changes to the task itself. Any force exertion is a problem if applied in postures with joints towards the extreme ranges of their motion.

It is therefore possible to identify stressful postures and to determine the duration of the exertions involved, but there are as yet no criteria which enable us to set specific limits on the duration or speed of work which might be regarded as tolerable for different types of work. In investigations in this area, information concerning accidents, injuries or absence related to the work conditions provide the evidence for unsatisfactory conditions.

Repetition in tasks

There is indeed little guidance on the definition of the 'repetitive' nature of a task, in terms either of cycle time or of repetition rate. In the context of muscle activity, Silverstein *et al.* (1986) have suggested that the definition of a task with a 'high' repetition rate might be taken as one in which the cycle time is 30 seconds or less, or where repeating sub-cycles occupy more than 50% of the fundamental cycle.

Moore and Wells (1992) have pointed out that the repetitiveness of a job has several components: the amount of tissue movement, the number of repetitions, the cycle time, and an estimate of 'sameness'. Sameness of actions or of muscle involvement is an important element both for the closely defined repetitive jobs typical of work on many production lines and also when considering the possibility of job rotation. It is important also to distinguish repetitive sub-cycles in jobs where the basic cycle time is quite long. Kivi (1984) suggested that repetitive movements could be considered monotonous (lacking in variation) if a movement took less than a minute to perform, and very monotonous if less than 30 seconds.

Moore and Wells (1992) have proposed an analytical technique to calculate the autocorrelation function of a posture record as a means of assessing the 'sameness' or 'monotony' of posture variables. The autocorrelation function is calculated by extracting one cycle of movement and cross-correlating step by step against the whole work record (perhaps 20–40 minutes of recording). The technique gives two pieces of information, the duration (or frequency) of the movement and an indication of its monotony judged from the relative value of the autocorrelation function.

Armstrong et al.*'s (1982) posture analysis technique*

The notation developed by Armstrong *et al.* (1982), which has already been shown in Figure 23.7 in chapter 23 is very useful for recording the postures adopted during performance of a task. The technique allows the posture of the arm and force exertion to be recorded concurrently and, if desired, plotted on a time base. This helps to give an understanding of the durations of task elements and rest pauses.

Joint angles are estimated and classified according to a coding system which defines the zones of angular deviation for the shoulder (in three axes), elbow (in two axes) and wrist (in two axes). The hand posture is also coded according to six different categories and there is provision for noting where any contact pressures in the task are exerted on the fingers or palm of the hand. These can all be noted rapidly on a recording form and thus provide a record which can then be used to determine the percentage of the job cycle spent with the hand or arm in different posture ranges.

Assessing manual handling tasks

Investigation of manual handling tasks is to some extent a special case in the general consideration of risk of work-related musculoskeletal disorders. There is no clear dividing line between work causing insidious cumulative damage and work requiring high force exertions which may risk overloads on body structures. One of the major concerns for people engaged in manual handling activities is the high incidence of back pain: this may be due to a single overexertion or to the effects of cumulative microtrauma, and it is sometimes difficult to determine which.

Nevertheless, a practical distinction is usually made, so that manual handling assessments take account of jobs which involve the movement of 'heavier' loads (mainly lifting and carrying activities although they may include tasks such as movement of loaded trolleys or exertion of high forces when using tools). Some jobs have elements both of manual handling of heavy products and of repetitive work or long-held static postures: the work of supermarket checkout operators or assembly line operators mounting heavy components are examples. In these cases, an ergonomics audit will need to use techniques appropriate for both aspects of the work, referring both to this section and to the previous two sections.

A variety of factors affects the risk of injury in manual handling tasks, so that different types of assessment may be necessary. As a general guideline, biomechanical and muscle strength criteria, together with assessment of posture, are most appropriate for determining limits for infrequent lifts of large or heavy objects. For objects of more moderate weight when lifted frequently but for periods of, say, less than 1 hour, limits can be based on psychophysical measures of acceptable loads. Physiological measurements are more appropriate when identifying limits for tasks involving frequent lifting over periods of several hours, where physical fatigue could be relevant.

In jobs where manual handling is involved, the loads and efforts can often vary over the working day and, before the risks are analyzed, some prior study is useful to identify the tasks where major loading is suspected. Moreover, the additional stresses which may occur at peak periods of activity should not be overlooked. As will be apparent in the following discussion, most techniques presuppose the selection of the tasks presenting the greatest risk, which are then subjected to in-depth analysis. However, tasks with lower loads which are repeated should also be analyzed, since a repeated load has a lower limit than a lift performed occasionally.

Identification of hazardous postures

Some ergonomics expertise may be necessary to ensure that all potentially hazardous postures are identified, but most can be selected by reference to the principal risk factors of

- working height (either for applying a force or throughout the lift or transfer of a load)
- reach distance
- trunk bending (fore/aft or lateral)
- twisted, awkward or constrained postures.

The OWAS posture classification system (described in chapter 23) is helpful in this context, although it was not designed specifically for identifying the biomechanical hazards in lifting loads. It is a good method of identifying postures which are unsuitable for general activities with or without high force exertion.

Checklist approach

Various checklists have been produced for assessing the risk of manual handling operations (see, for example, Department of Labour (1988) or HSE (1992)). These all draw attention to known sources of risk factors in the workplace, often grouping these under different aspects of the job: tasks performed, loads, environment, and individual capability. A summary of the essential points to consider when designing workplaces where manual handling takes place is given in Figure 30.6, and this can help in identifying posture and load situations for further analysis.

The checklist in Figure 30.7 could be used to provide guidance on routes to possible solutions, once a risk has been identified.

If a deeper analysis is deemed necessary, and this will be the case in many situations where load handling is a significant part of the job, there are several approaches. A glance down the items of Figure 30.6 will show that it is quite inadequate to specify a maximum load without making qualifications, since too many factors influence a person's ability in load handling to permit such

1. Avoid having to lift or transfer wherever possible.
2. The number of trunk flexion movements required to perform a task and the range of rotation and side bending should be kept to a minimum.
3. Any requirements for a load to be supported or force applied over a substantial part of the work cycle should be minimized.
4. Handling loads at heights below the knees or above the shoulder should be eliminated wherever possible.
5. Minimize congestion and confined work spaces to ensure that an adequate base can be used to perform the task.
6. Use grasps which are two-handed, with the load or force evenly distributed between both hands/arms. Adopt grasps with maximal body contact area and where the upper arms and elbows are by the side, elbow joints are in mid-range and wrists are straight.
7. When performing two-person lifts or transfers, wherever possible select a colleague of similar height. Ensure that your colleague is trained in the techniques you will use.
8. Avoid jolting or sudden movements, especially with a high load or force, or with trunk flexion or rotation.
9. Wherever possible, use the trunk and arm muscles to stabilize the load and posture, and use the leg muscles (especially the thigh muscles) to provide the force or movement required.
10. Do not attempt to lift or transfer a weight which *may* be beyond your capacity, or is likely to be unpredictable, without suitable mechanical or human assistance.
    This can arise in difficult workplace conditions (e.g., slippery floor) or, in health care, when handling an uncooperative human or animal patient. If in doubt, put the object/person in a safe position and fetch assistance.

Figure 30.6. Ten guidelines on manual handling

Is the task necessary?
Can the workstation be relocated to reduce carrying?
Can the work height/reach distance be improved?
Can the load be reduced, for example by using smaller units or counterbalancing the weight?
Would a handling aid help?
Can the task be mechanized?
Should job rotation be introduced to allow periods for the musculoskeletal system to rest and recover?

Figure 30.7. Checklist suggesting potential solutions

a simple solution. We will mention first the use of biomechanical modelling, which deals with the geometry of the posture and the loads exerted. Then other methods of assessment will be given, which incorporate biomechanics concepts but bring in, also, other factors.

Biomechanical analysis

Biomechanical analysis is important for assessing the effects on the musculo-skeletal system of lifting a known load or of any other force exertion. Few techniques available at present for assessing manual handling activities take full account of the dynamic nature of the work, and assessments of the inertial forces involved are particularly difficult with the current state of knowledge of their biomechanical effects. Although sophisticated measurement techniques have been developed for assessing dynamic biomechanical effects, no methods are yet suitable for general use in field surveys. In the field situation, the analysis has to be performed on a 'snapshot' of the task in which the forces on the body are estimated in a single posture taken at some point during the lift (usually the posture judged to be the most stressful). Methods suitable for recording the posture for analysis have already been described in chapter 23.

The biomechanics techniques used in analysis and estimation of musculo-skeletal loads are discussed in chapter 24. Simple and approximate two-dimensional calculations can be very helpful in assessing the effects of task layout and of particular loads and can give an insight into some of the important factors. However, three-dimensional analysis is necessary to analyze the effects of lifting in asymmetric postures outside the mid-sagittal plane. Biomechanical analysis packages are commercially available for detailed analy-sis and evaluation against population strength norms (Chaffin and Andersson, 1991; Wells and Moore, 1992).

However, it should be said that biomechanical analysis, whilst providing very useful quantitative measures of loading, is time consuming and probably not necessary in many manual handling assessments, unless there are difficult situations that are not easily addressed by other techniques. Once risk factors have been identified in the workplace, their presence would normally be sufficient to warrant changing the task or workplace arrangement. Practical experience has shown that biomechanical analysis is most useful in the follow-ing situations:

- comparative evaluation of alternative layouts
- evaluation of stresses caused by loads of moderate weight in very awk-ward postures
- assessment of designs for manual handling aids.

Although one of the major concerns in manual handling is injuries to the low back, and most assessment techniques focus on this, many injuries also occur to the shoulder and arm, and can occur elsewhere in the body. So, attention should be directed to stresses anywhere in the musculoskeletal sys-tem which may be affected through the posture or postures adopted while performing the task. However, the only criterion which has been set for a stress level likely to cause injury is for the lumbar spine (although the assess-ment of stresses at other joints is discussed in chapter 24). The evidence on biomechanical criteria was thoroughly reviewed during revision of the NIOSH guidelines (Waters *et al.*, 1993). They noted that the joints in the lumbar spine are particularly highly stressed during lifting (and many other materials handling activities) and confirmed the criterion set during develop-ment of the first NIOSH guidelines, deciding that a compressive force on the lumbar spine of 3·4 kN represented a risk of low back injury. Waters *et al.* (1993) recognized that shear and torsional forces can also be important, but there are as yet no widely accepted ways to combine these so that their contribution to low back injury can be specified.

Assessing the acceptability of a load in lifting

There is no such thing as a 'safe' load, since both the posture in which the lift or force exertion is carried out and the conditions in the workplace can profoundly affect the risk of injury. Various methods have been proposed for assessing whether loads are acceptable and three of these are discussed below. A limitation with many of the methods is that they can at present only be applied to two-handed lifts and, in some cases, only to those in symmetrical postures, essentially analyzing them as two-dimensional situations. In practice, of course, many handling tasks are carried out single-handed and most working postures are highly asymmetric.

NIOSH (1991) EQUATION
The NIOSH guidelines were developed in the USA to take account of more factors than just geometry and load in the evaluation of workplaces. They seek to integrate biomechanical, physiological and psychophysical criteria, as well as using epidemiological evidence of musculoskeletal injury rates of low back pain (Waters *et al.*, 1993). Each of the three criteria takes account of a different aspect of the stresses imposed on the body during lifting and the evidence suggests that using single criteria alone might lead to different limiting loads. Physiological data representing the energy demands of the work would suggest that it is more efficient to lift heavier weights less frequently, whereas biomechanical principles suggest that musculoskeletal stress is minimized by lifting lighter weights (which might lead to recommending more frequent lifts to perform a given task). In psychophysical studies, the limiting acceptable weights chosen by subjects lifting from the floor tend to be higher than would be recommended by either physiological or biomechanical criteria.

The NIOSH criterion is set in terms of a 'Recommended Weight Limit' which is expressed as a formula taking into account: height at which the lift commences, the vertical travel in the lift, the reach distance, and the frequency of lift. It was recognized that the risk factors involved are likely to interact and multiply the risk, and the formula was derived to take this into account. The NIOSH (1991) equation (given in Figure 30.8) is discussed in detail in Waters *et al.* (1993).

The equation can be used to calculate the 'Recommended Weight Limit' for given task conditions, assuming a baseline limit of 23 kg under the best conditions; these best conditions are a sagittal plane lift, occasional lifting, good coupling (handholds), less than 25 cm vertical displacement of load, and a situation in which the lift is made at a vertical height of 75 cm from the floor and a horizontal reach distance of no more than 25 cm from the mid-point between the ankles. This is believed to represent the weight that at least 90% of healthy US workers should be able to lift over the defined work period without an increased risk of developing lifting-related low back pain. It should however be noted that these values are likely to be high in relation to working populations in many other parts of the world.

Recommended Weight Limit (RWL) in kg =
23 kg × HM × VM × DM × AM × FM × CM

where

Horizontal multiplier HM is $\dfrac{25}{H}$

with H the horizontal distance (in cm) of hands from mid-point between the ankles. Measure at the origin and the destination of the lift.
Vertical multiplier VM is $1 - (0.003\,|V - 75|\,)$
with V the vertical distance (in cm) of the hands from the floor. Measure at the origin and destination of the lift.
Distance multiplier DM is $0.82 + \dfrac{4.5}{D}$
with D the vertical travel distance (in cm) between the origin and the destination of the lift.
Asymmetric multiplier AM is $1 - (0.0032\,A°)$
with A the angular displacement (in degrees) of the load from the sagittal plane. Measure at the origin and destination of the lift.
Frequency multiplier FM is input from Table A, based on the average frequency rate of lifting measured in lifts/min and duration (≤ 1 hour, ≤ 2 hours, ≤ 8 hours assuming appropriate recovery allowances).
Coupling multiplier CM is input from Table B on the basis of the degree of coupling between hand and load.

Figure 30.8. NIOSH (1991) equation (continued overleaf)

Limitations which are recognized in the application of the NIOSH (1991) equation (according to Waters *et al.*, 1993) are:

(a) that it applies to lifting or lowering tasks but not to carrying or other manual handling activities, nor to those which require significant energy expenditure (since the physiological criteria were based only on the need to restrict expenditure to avoid fatigue); not does it apply to the lifting of people, to shovelling, or to supporting handling aids such as barrows;

(b) that it only applies to standing tasks and not to tasks carried out sitting or kneeling;

(c) that it does not take account of sudden or unpredicted conditions such as shifts in load distribution, an unexpectedly heavy or light load, or foot slip;

(d) that it is not designed to assess lifting single-handed, constrained work-places, poor thermal environments, or unusual loads such as contami-nated objects.

We must point out that the NIOSH (1991) equation, although based upon the experience of using the earlier NIOSH (1981) equation and developed to provide the most conservative estimates of lifting limits, does not provide a proven result which can be used as a conclusive rule. Waters *et al.* (1993) emphasize that validation of the formula can only come from the long–term collection of evidence arising from its application. This by no means reduces

TABLE A   Frequency multiplier (FM)

Frequency lifts/min	Work duration					
	≤ 1 hour		≤ 2 hours		≤ 8 hours	
	V < 75 cm	V ≥ 75 cm	V < 75 cm	V ≥ 75 cm	V < 75 cm	V ≥ 75 cm
0·2	1·00	1·00	0·95	0·95	0·85	0·85
0·5	0·97	0·97	0·92	0·92	0·81	0·81
1	0·94	0·94	0·88	0·88	0·75	0·75
2	0·91	0·91	0·84	0·84	0·65	0·65
3	0·88	0·88	0·79	0·79	0·55	0·55
4	0·84	0·84	0·72	0·72	0·45	0·45
5	0·80	0·80	0·60	0·60	0·35	0·35
6	0·75	0·75	0·50	0·50	0·27	0·27
7	0·70	0·70	0·42	0·42	0·22	0·22
8	0·60	0·60	0·35	0·35	0·18	0·18
9	0·52	0·52	0·30	0·30	0·00	0·15
10	0·45	0·45	0·26	0·26	0·00	0·13
11	0·41	0·41	0·00	0·23	0·00	0·00
12	0·37	0·37	0·00	0·21	0·00	0·00
13	0·00	0·34	0·00	0·00	0·00	0·00
14	0·00	0·31	0·00	0·00	0·00	0·00
15	0·00	0·28	0·00	0·00	0·00	0·00
> 15	0·00	0·00	0·00	0·00	0·00	0·00

TABLE B   Coupling multiplier (CM)

Couplings	V < 75 cm	V ≥ 75 cm
	Coupling multipliers	
Good	1·00	1·00
Fair	0·95	1·00
Poor	0·90	0·90

Figure 30.8. Continued

the importance of what is probably the most reliable attempt so far to quantify the hazards of materials handling, but it does warn us that, in this complex area, we cannot avoid using our knowledge and judgement in deciding on the suitability of handling situations.

LIFTING INDICES FOR IDENTIFYING HAZARDOUS JOBS

A Lifting Index was also developed alongside the NIOSH (1991) equation as an index of relative physical stress in the task (Waters *et al.*, 1993). This is intended to assist in job assessment surveys to identify lifting jobs that pose a significant risk of low back injury. The Lifting Index is the ratio of the load to be lifted to the Recommended Weight Limit, as calculated by the equation above. Where this ratio is greater than 1, it is likely that an increased risk of low back injury exists for some fraction of the workforce. Where the index is 3 or greater, a serious level of risk exists even for fit and experienced workers.

The Lifting Index is similar in concept to two other indices, the Job Severity Index (Ayoub and Mital, 1989) and the Lifting Strength Rating (Chaffin, 1974). All three aim to assess the job demand against the capacity of the person or population working under the job conditions. The Job Severity Index is calculated from the job demands (identified through task analysis) and the capacity of the worker (which is either measured in psychophysical tests or estimated on the basis of published population data). The Lifting Strength Rating for a job is calculated from the weights handled in that job and the predicted strength of a strong person in the postures observed for handling the weights.

Both the Lifting Index and the Job Severity Index can be used to assess the lifting demands of tasks in which the load weights or the workplace factors vary. However, they do not quantify the precise degree of risk involved nor the risk for different groups of the worker population. They can be used to compare the relative severity of two jobs for the purpose of evaluating and modifying the conditions.

PSYCHOPHYSICAL MEASUREMENTS OF MAXIMUM ACCEPTABLE WEIGHT OF LOAD

Snook (1978) developed the psychophysical technique of asking operators themselves to judge what they consider to be a maximum acceptable load for a given work period (not to be equated with a 'safe' load). From a methodological point of view, this technique is essentially based on the psychophysical Method of Adjustment, but with the test standard being the subject's own concept of maximum load. The subjects are asked to adjust the weight of the load (or any other task variable such as frequency of lift) according to their perception of the strain and fatigue involved, the final workload being taken as the Maximum Acceptable Weight of Load (MAWL) under the working conditions.

The method has the advantage that the operator's judgement takes into account the whole job, integrating biomechanical and physiological factors. The disadvantage is that subjects have to extrapolate their judgements from a short period during the trial to the whole of a regular working day. A few validation studies have been performed (Legg and Myles, 1981; Ljungberg *et al.*, 1982; Mital, 1983; Karwowski and Yates, 1986), but the results are somewhat contradictory—many subjects overestimate their capacity but some underestimate it. The tests should be made with experienced workers and an initial training period is important to obtain reliable results. The tests should be repeated and the results of at least two trials averaged to ensure that repeated results lie within 15% of each other.

Snook has collected a large database of psychophysical judgements for lifting tasks (and for a few other forceful activities) under a variety of workplace arrangements (Snook and Ciriello, 1991). This is useful for assessing weights on the criterion of acceptability rather than that of safety.

Conclusions

A range of techniques has been discussed which can be used in ergonomics audits for workplaces, and in particular in those where musculoskeletal problems are suspected. The techniques should enable investigators to recognize poor features in the design of work and workplaces, to identify the aspects which need change, and to develop appropriate redesigns.

Although the chapter set out to review risk assessment techniques, it is apparent that the available techniques have not yet developed to provide quantitative levels of risk although some classify the levels of severity. However, we are able to recognize the risk factors associated with work-related injuries and have appropriate methods to assess the most important of them, but cannot yet set criterion levels for the risk of injury with any real confidence. Further research is needed to gain knowledge of dose-response relationships for upper limb disorders, for back injuries experienced in manual handling, and for musculoskeletal problems in general.

References

Åberg, U. (1981). Techniques in redesigning routine work. In *Stress, Work Design and Productivity*, edited by E.N. Corlett and J. Richardson (Chichester: John Wiley), pp. 157–163.

Armstrong, T.J., Foulke, J.A., Joseph, B.S. and Goldstein, S.A. (1982). Investigation of cumulative trauma disorders in a poultry processing plant. *American Industrial Hygiene Association Journal*, **43**, 103–116.

Ayoub, M.M. and Mital, A. (1989). *Manual Materials Handling* (London: Taylor and Francis), pp. 198–209.

Bigos, S.J., Battie, M.C., Spengler, D.M., Fisher, L.D., Fordyce, W.E., Hansson, T.H., Nachemson, A.L. and Wortley, M.D. (1991). A prospective study of work perceptions and psychosocial factors affecting the report of back injury. *Spine*, **16**, 1, 1–6.

Branton, P. (1970). A field study of repetitive manual work in relation to accidents at the workplace. *International Journal of Production Research*, **8**, 93–107.

Chaffin, D.B. (1974). Human strength capacity and low back pain. *Journal of Occupational Medicine*, **16**, 248–254.

Chaffin, D.B. and Andersson, G.B.J. (1991). *Occupational Biomechanics*, 2nd edition (New York: John Wiley and Sons), pp. 286–290.

Corlett, E.N. (1991). Ergonomics fieldwork: an action programme and some methods. In *Towards Human Work*, edited by M. Kumashiro and E.D. Megaw (London: Taylor and Francis), pp. 179–185.

Cox, T. and Mackay, C.J. (1985). The measurement of self reported stress and arousal. *British Journal of Psychology*, **76**, 183–186.

Department of Labour (1988). *Manual Handling. Regulations and Code of Practice* (Melbourne: Department of Labour).

Dickinson, C.E., Campion, K., Foster, A.F., Newman, S.J., O'Rourke, A.M.T., and Thomas, P.G. (1992). Questionnaire development: an examination of the Nordic Musculoskeletal Questionnaire. *Applied Ergonomics*, **23**, 197–201.

HSE (1990). *Work Related Upper Limb Disorders. A Guide to Prevention* (London: HMSO).

HSE (1992). *Manual Handling. Guidance on Regulations* (London: HMSO).

Industrial Fatigue Research Committee of the Japan Association of Industrial Health (1970). The inventory for subjective symptoms of fatigue (revised 1970). *Digest of Science of Labor*, **25**, 12–33.

Karasek, R.A. (1979). Job demands, job decision latitude, and mental strain: implications for job redesign. *Administrative Science Quarterly*, **24**, 285–307.

Karhu, O., Kansi, P. and Kuorinka, I. (1977). Correcting working postures in industry: a practical method for analysis. *Applied Ergonomics*, **8**, 199–201.

Karwowski, W. and Yates, J.W. (1986). Reliability of the psychophysical approach to manual lifting of liquids by females. *Ergonomics*, **29**, 237–248.

Kemmlert, K., Nilsson, Å., Andersson, B. and Bjurvald, M. (1987). Prevention of injuries related to physical stress through interventions by Labour Inspectors. In *Musculoskeletal Disorders at Work*, edited by P. Buckle (London: Taylor and Francis), pp. 146–152.

Keyserling, W.M., Stetson, D.S., Silverstein, B.A. and Brouwer, M.L. (1993). A checklist for evaluating ergonomic risk factors associated with upper extremity cumulative trauma disorders. *Ergonomics*, **36**, 807–831.

Kishida, K. (1991). Workload of workers in supermarkets. In *Towards Human Work*, edited by M. Kumashiro and E.D. Megaw (London: Taylor and Francis), pp. 269–279.

Kivi, P. (1984). Rheumatic disorders of the upper limbs associated with repetitive occupational tasks in Finland in 1975–1979. *Scandinavian Journal of Rheumatology*, **13**, 101–107.

Kjellstrand, L. (1974). Quality of life at the workplace. In *The Quality of Life at the Workplace, Proceedings of the Regional Trade Union Seminar*, OECD, Paris, pp. 33–42.

Kuorinka, I., Jonsson, B., Kilbom, Å, Vinterberg, H., Biering-Sørensen, F., Andersson, G. and Jørgensen, K. (1987). Standardised Nordic questionnaires for the analysis of musculoskeletal symptoms. *Applied Ergonomics*, **18**, 233–237.

Legg, S.J. and Myles, W.S. (1981). Maximum acceptable repetitive lifting workloads for an 8 hour day using psychophysical and subjective rating methods. *Ergonomics*, **24**, 907–916.

Ljungberg, A.S., Gamberale, F. and Kilbom, Å. (1982). Horizontal lifting—physiological and psychological responses. *Ergonomics*, **25**, 741–757.

McAtamney, L. and Corlett, E.N. (1993). RULA: a survey method for the investigation of work-related upper limb disorders. *Applied Ergonomics*, **24**, 91–99.

Mital, A. (1983). The psychophysical approach in manual lifting—a verification study. *Human Factors*, **25**, 485–491.

Moore, A.E. and Wells, R. (1992). Towards a definition of repetitiveness in manual tasks. In *Computer Applications in Ergonomics, Occupational Safety and Health*, edited by M. Mattila and W. Karwowski (Amsterdam: North-Holland), pp. 401–408.

Noro, K. and Imada, A. (Eds) (1991). *Participative Methods* (Chichester: Wiley).

Putz-Anderson, V. (Ed.) (1988). *Cumulative Trauma Disorders: A Manual for Musculoskeletal Diseases of the Upper Limbs* (London: Taylor and Francis).

Silverstein, B.A., Fine, L.J. and Armstrong, T.J. (1986). Hand, wrist and cumulative trauma disorders in industry. *British Journal of Industrial Medicine*, **43**, 779–784.

Snook, S.H. (1978). The design of manual handling tasks. *Ergonomics*, **21**, 963–985.

Snook, S.H. and Ciriello, V.M. (1991). The design of manual handling tasks: revised tables of maximum acceptable weights and forces. *Ergonomics*, **34**, 1197–1213.

Waters, T.R., Putz-Anderson, V., Garg, A. and Fine, L.J. (1993). Revised NIOSH equation for the design and evaluation of manual lifting tasks. *Ergonomics*, **36**, 749–776.

Wells, R. and Moore, A.E. (1992). A framework for computer assisted approaches to the prevention of work-related musculoskeletal disorders involving workplace design and modification. In *Computer Applications in Ergonomics, Occupational Safety and Health*, edited by M. Mattila and W. Karwowski (Amsterdam: North-Holland), pp. 55–62.

Wilson, J.R. (1991). Design decision groups: a participative process for developing workplaces. In *Participative Methods*, edited by K. Noro and A. Imada (Chichester: Wiley).

Wilson, J.R. (1994). Devolving ergonomics: the key to ergonomics management programmes. *Ergonomics*, **37, 579–594**.

Chapter 31

Human reliability assessment

Barry Kirwan

Introduction

Recent accidents such as at Chernobyl and Bhopal have demonstrated unequivocally the importance of considering human error in high risk systems (see USSR State Committee, 1986; Bellamy, 1986). For any existing plant, or new one being designed, it is important to try to assess the likelihood of such accidents and prevent them from occurring. This requires the assessment of the impact of human errors on system safety and, if warranted, the specification of ways to reduce human error impact and/or frequency. These are the major goals of human reliability assessment and the primary domain for application and development of human reliability assessment methodologies has been high technology high risk plant.

Ironically, *human reliability assessment* (HRA) owes its current status to the very accidents it would wish to prevent. Whereas research and development in this field has been carried out since the early 1960s, the drive for development of sound and practicable methodologies received a hitherto unparalleled boost following the accident at Three Mile Island (USNRC, 1980). This accident in particular, which rocked the nuclear power world's foundations and beliefs that such accidents simply could not happen, brought home the realization that human error was of fundamental importance, and that the survival of the nuclear power industry (and other similar industries) would depend on the ability to prevent such accidents from recurring. Accidents since Three Mile Island, such as at Bhopal, the *Challenger* disaster (Rogers *et al.*, 1986) and Chernobyl, have entirely reinforced this view; accidents such as the Zeebrugge ferry disaster have similarly demonstrated the importance of human error in low-technology systems (Reason, 1988a). Human reliability assessment clearly has an important role to play, and this role is likely to extend to many industries, wherever human errors can propagate within systems to lead to unacceptable events.

HRA is a hybrid area, arising out of the disciplines of engineering and reliability on the one hand, and psychology and ergonomics on the other.

The former require human error probabilities to fit neatly into the logical mathematical framework of *probabilistic safety analysis* (PSA), and the latter urge more detailed and theoretically valid modelling of the complexity of the human operator (Wagenaar, 1986). PSA is the quantitive statement defining the expected frequencies of accidents, and hence it determines whether or not a plant's risk compares favourably or otherwise against pre-defined risk criteria (Green, 1983). Thus HRA must be incorporated into PSA if risk is to be properly estimated.

Significant attempts have been made to properly integrate HRA into PSA (Bellamy *et al.*, 1986), and to validate HRA approaches (Embrey and Kirwan, 1983; Comer *et al.*, 1984; Kirwan, 1988). It is likely that within a few years more cogent and valid methodologies will appear, and be adopted and prescribed by the bodies empowered to regulate the various industries at risk from human error. Until such a time it is only possible to describe the current general approach of HRA and the most prominent and promising of the current methods, and this is the intention of this chapter.

A generic approach or framework for HRA is presented. Although this framework has 10 steps within it (as will be described), there is a core of three goals, namely:

1. *Human error identification*. What can go wrong?
2. *Human error quantification*. How often will a human error occur?
3. *Human error reduction*. How can human error be prevented from occurring or its impact on the system reduced?

Most research has focused on (2), the development of human reliability quantification techniques, although logically (1) is at least (if not more) important, since unless all significant errors are identified a HRA will underestimate the impact of human error. Whilst a good deal of research has been carried out in the field of human error identification and particularly error classification, few practical techniques have been developed for use in risk assessments. It is likely that future research and development will focus on the development of such techniques.

Human error reduction has recently become more important since there is an obvious need for this capability once HRA is being applied in earnest, as there are bound to be identified some unacceptably probable human errors, which must in some way be reduced in frequency. This is where the role of the ergonomist in HRA is clearest and most useful, since the ergonomist can usually specify a number of ways to improve the reliabiilty of operators' performance.

This chapter concentrates on human error quantification, as this has been most heavily researched, but also discusses some of the most recent human error identification techniques, and elaborates on the human error reduction approaches currently available. It does not go into some of the more complex mathematical issues underlying the integration of HRA into PSA (see, for example, Apostolakis *et al.*, 1987; or Park, 1987), as these would require a chapter in themselves, and are not required for an appreciation of how HRA works and is applied. Furthermore, as quickly becomes apparent to the practitioner, human reliability analysis is far from being a precise science (and nor is probabilistic safety assessment itself; Nicks, 1981), but is a useful means of identifying and prioritizing plant safety vulnerabilities to human error, and thereby reducing the frequency of accidents.

The following section details the generic HRA methodology. Following this, there is a discussion of future areas of investigation and development needs within the field of HRA. For a fuller discussion of all issues raised see Kirwan (1994).

Human reliability assessment: a generic methodology

Once it is decided that a human error or human reliability problem requires analysis, a means of systematically solving this problem is required. The 'problem' may range from possible errors during a nuclear power plant emergency, to a desire for improved performance in an offshore maintenance task. Whatever the objective of the assessment, a systematic methodology of HRA will help ensure that the problem itself is dealt with reliably, minimizing biases or errors distorting the analysis.

Each of the 10 steps of the methodology defined below, from initial problem definition to final documentation of the results, is briefly explained and further references are given on available guidelines and current research in these areas.

The generic human reliability assessment methodology encompasses the following:

1. *Problem definition.* To define precisely the problem and its setting in terms of the system goals and the overall forms of human-caused deviations from those goals.
2. *Task analysis.* To define explicitly the data, equipment, behaviour, plans, and interfaces used by the operators to achieve system objectives, and to identify factors affecting human performance within these tasks.
3. *Human error analysis.* To identify all significant human errors affecting performance of the system, and ways in which human errors can be recovered.
4. *Representation.* To model the human errors and recovery paths in a logical manner such that their impact on the system can be quantitatively determined. This usually necessitates integrating human errors with hardware failures in a fault or event tree.
5. *Screening.* To define the level of detail and effort with which the quantification will be conducted, by defining all significant human errors and

interactions, and ruling out insignificant errors which can be effectively ignored by the study.

6. *Quantification.* To quantify human error probabilities and human error recovery probabilities, in order to define the likelihood of success in achieving the system goals.

7. *Impact assessment.* To determine the significance of human reliability with respect to the achievement of the system goals, to decide whether improvements in human reliability are required, and if so what are the primary errors and factors negatively affecting system reliability.

8. *Error reduction.* To identify error reduction mechanisms, means of supporting error recovery likelihood, and ways of improving human performance in achieving system goals, so that an acceptable level of system performance can be achieved.

9. *Quality assurance.* To ensure that the enhanced system satisfactorily meets system performance criteria, and will continue to do so in the future.

10. *Documentation.* To detail all information necessary to allow the assessment to be understandable, auditable, and reproducible.

Figure 31.1 shows the relationship between these various steps, as is discussed in detail later.

(1) Problem definition

There are two ways of defining the problem, dependent upon whether it is being considered as a problem in its own right, or as an integral part of a larger risk assessment. When a human reliability 'problem' has been identified (e.g., a desire to improve safety, or to assess risk or productivity), and defined in its system context (e.g., to assess the risks of offshore platform evacuation by lifeboat in severe weather), discussions should occur with system design and plant engineers, and if possible with operational and managerial personnel. Discussions around the problem will help define more precisely the scope of the project, and the range of conditions and scenarios which must be considered in order satisfactorily to resolve the problem, and/or fulfil the goals of a general safety assessment of a plant.

It is useful at this stage to define the system goals, at various levels, for which operator actions are required. This will in turn define higher level goals towards which the operators are aiming (e.g., maintain reactor core cooling, achieve highest output). It is also important to gain some understanding of these high level goals, and to determine in particular where and how the goal of safety fits in. Ideally there will be clear criteria via which the operators know when to 'drop' their production goals in favour of safety goals. If such criteria do not exist, then production goals may compromise safety goals and it is possible that the roots of a human reliability problem are already inherent in the system. Therefore, with existing systems it is well

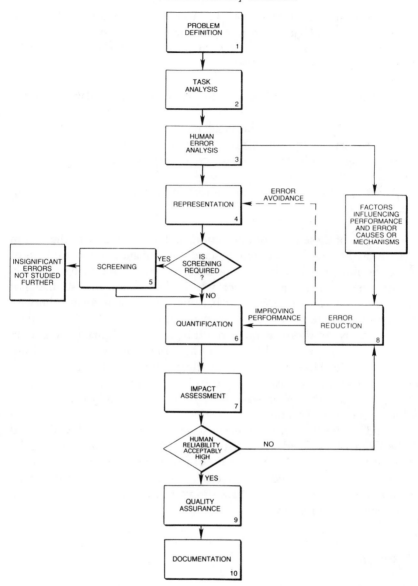

Figure 31.1. Steps in a human reliability assessment

worth investigating this 'safety culture' aspect of the plant, as it can influence human reliability predictions dramatically, and is an important aspect of the problem definition. If there are no clear criteria, it must not be assumed that operators will necessarily make the right decision in the heat of the moment.

If the HRA is being carried out as part of an overall risk assessment, the human reliability analyst will probably be given a set of scenarios which have been chosen for the risk analysis, and asked to consider human contributions to risk within these scenarios. In this case there are five types of human-system interaction which the analyst should consider with respect to an incident scenario (Spurgin *et al.*, 1987):

(a) maintenance/testing errors affecting safety system availability (latent errors),
(b) operator errors initiating the incident,
(c) recovery actions by which operators can terminate the incident,
(d) errors (e.g., misdiagnosis) by which operators can prolong or even aggravate the incident, and
(e) actions by which operators can restore initially unavailable equipment and systems.

Consideration of these types of interaction, and discussions with the system risk analysts at the problem definition stage will enhance the smooth integration of the human reliability analysis into the system risk analysis. It may also identify new important scenarios which the system analysts had not initially considered. This is important because the problem may otherwise be defined in too limited a scope. As an example, Kirwan's (1987) assessment of an emergency offshore depressurization system originally was only intended to look at operational failures during emergency scenarios and not maintenance aspects. However, when investigated further, a highly significant maintenance error was identified which had potentially dramatic effects on the whole platform; this error was probably of more significance than the entire set of operational failures put together.

At the end of the problem definition stage the problem to be addressed should be explicitly defined in its system context. A list of scenarios to be addressed and, within each scenario, a list of overall tasks required to achieve system and safety goals should also have been identified. This sets the scene for the task analysis phase. Figure 31.2 shows a brief example of the results of the problem definition phase.

(2) Task analysis

The object of task analysis is to provide a complete and comprehensive description of the tasks that have to be performed by the operator(s) to achieve the system goals. There are many forms of task analysis (Drury, 1983; Kirwan and Ainsworth, 1992), such as sequential task analysis which looks at operator actions as they occur in chronological order; hierarchical task analysis which considers tasks in terms of the hierarchy of goals the operator is trying to achieve (Shepherd, 1986); and tabular format decision task analysis (Pew *et al.*, 1987) which concentrates on cognitive decision-making aspects as a function of the information available, and operators' knowledge, expec-

Problem:	To effect emergency shutdown (ESD) of a chemical plant during a loss of power scenario.
Problem Setting:	A computer-controlled, operator-supervised plant suffers a sudden loss of main power. The VDU display system will also fail and so the operator, backed up by the supervisor, must initiate ESD manually using hardwired controls in the Central Control Room. However, due to valve failures on plant, these actions are only partially successful, and so the operator must send out another operator onto plant to determine which ESD valves have not closed. The CCR operator, via engineering drawings, can then determine which manual valves must be closed on plant. The outside operator must then go to close these valves, completing this action successfully within 2 hours from the onset of the scenario.
System Goals:	The overall system goals are safe shutdown of all feeds to the plant within 2 hours of loss of power. In this scenario there are no production goals once the event occurs since safety is clearly under threat. Prior to the event, the operator is concerned with achieving steady feed throughout via monitoring the top two levels of a VDU display hierarchy, and notifying the supervisor of any alarms higher than level 2. The outside operator (on plant) will have various duties associated with maintenance tasks.
Overall Human Error Considerations:	No operator initiating events were identified, and maintenance errors were not relevant except that identifying and moving the local manual valves could prove difficult. Recovery actions involve identifying the appropriate valves to close and closing them. Errors of failing to realise that ESD has not been 100% effective, and of mis-identifying the valves, appear most likely. Loss of power is so evident that misdiagnosis or failure to diagnose is not considered to be credible.

Figure 31.2. Example of problem definition

tations and beliefs about the situation. The latter type of analysis is useful for analyzing operator diagnosis scenarios (e.g., for nuclear power plant emergencies). Other forms of task analysis are discussed within an overview of its approach and techniques in chapter 6.

If a detailed HRA is being carried out, then a task analysis is essential since it provides a detailed description of the operators' tasks from which it will be possible to identify errors in the next phase. The basic methods of deriving information for the task analysis are: observation; structured and unstructured interviews with operators, maintenance personnel, supervisors, managers and system designers; analysis of procedures; incident analyses; structured walkthroughs of procedures (where the operator talks the analyst through the procedure); and examination of system documentation such as engineering/process flow diagrams. In practice it is important not to rely on procedures/operating instructions as the sole source for defining the task, since often actual operating practices differ somewhat from the formal written documents.

For a proceduralized task in which the operator (or supervisor, or maintenance person) is using familiar skills or written/remembered rules, a hierarchical task analysis is probably most appropriate. An example is shown in Figure 31.3, part of a task analysis for the task of starting up a plant, and considering the sub-task 'warm up furnace'. Sequential information is represented in this form of hierarchical task analysis, via the numbers on the boxes at each level which determine the order in which the tasks should occur. Plan 4 in the diagram shows a more complex plan the operator must use while monitoring temperature and pressure (Shepherd, 1986).

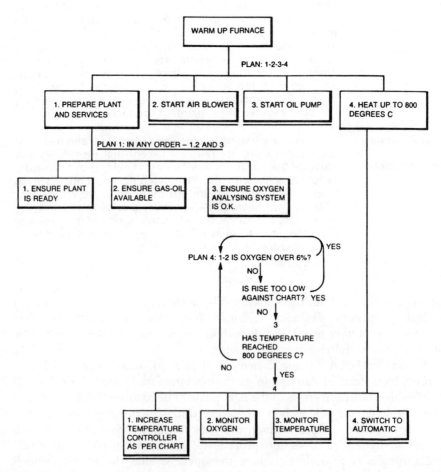

Figure 31.3. Example of hierarchical task analysis after Shepherd (1986)

Figure 31.4 illustrates another form of task analysis using a tabular format showing the type of information which can be recorded by the analyst during the task analysis. This type of task analysis may be more useful when the operator is in either a highly dynamic situation and/or one in which problem solving or diagnostic behaviour is required. The format focuses on the events as they occur in time, since the order in which events occur partly determines behaviour. However, the analyst must also be aware of the hierarchy of goals the operators are trying to achieve, which may shift during an emergency situation. In detailed analyses it may be necessary to utilize both hierarchical and tabular task analysis formats. Wherever possible the task analysis should be verified, if only by asking operations personnel to review it. Having defined the operators' tasks in detail, it is then appropriate to determine what can go wrong, in terms of what human errors can occur. Examples of task analyses carried out as part of real HRA's are given in Kirwan and Ainsworth (1992).

(3) Human error analysis

Human error analysis is arguably the most critical part of a human reliability analysis, since if a significant error is omitted at this stage then it will not appear subsequently in the analysis and hence the results may seriously underestimate the effects of human error on the system.

The simplest approach is to consider the following possible 'external error modes' (Swain and Guttman, 1983) at each step in the procedure defined in the task analysis:

Error of omission	– act omitted (not carried out)
Error of commission	– act carried out inadequately
	– act carried out in wrong sequence
	– act carried out too early/too late
	– error of quality (too little/too much)
Extraneous error	– wrong (unrequired) act performed

This approach is rudimentary but nevertheless can identify a high proportion of the potential human errors which can occur, as long as the assessor has a good knowledge of the task and a good task description of the operator–system interactions.

Another method for human error analysis is embedded within the *System-atic Human Error Reduction and Prediction Approach* (SHERPA: see Embrey, 1986a). This human error analysis method consists of a computerized question–answer routine which identifies likely errors for each step in the task analysis. The error modes identified are based on the 'skill rule and knowledge' model (Rasmussen *et al.*, 1981), and *Generic Error Modelling System* (GEMS: Reason and Embrey, 1986; GEMS: Reason, 1990).

T	Opr	System Status	Info Available	Operator Expectation	Procedure (written, memorised)	Decision/Communications Act	Equipment/location	Feedback	Secondary duties; Distractions; Penalties	Comments
0:30 mins	SS OO CRO	ESD; Cell 23 full of gas; mixture beyond explosion point at present; platform at muster status.	Gas cloud; Loud roaring noise;	Looking for leak in pipe union, flange, seal etc. Check near the gas detectors which were alarming.	1) Locate source and isolate if possible. 2) Maintain personal safety (use breathing apparatus) 3) Prevent ignition of gas	Ops search for leak in compressor module using sound and visual cues, as well as the gas detectors. Communicate to CCR.	Cell 23 gas detectors	Noise and smell of gas tactile cue if flesh exposed to gas jet path; if cold enough may see white gas plume from leak.	Maintain personal safety and avoid causing ignition. Extra delays if depressurise	Search will be more difficult in breathing apparatus. Deluge would make search safer, though less likely to succeed.
0:40	OFM CCRO SS	As at 0:30	Panel indications and from operators now outside of cell 23.	Gas leak now confirmed. Ignition possible.	Minimise chances of ignition.	CCR operator reviews vessel pressures on VDU system; gives OIM status report and confirms significant gas leak occurrence. CCR opts to consider whether to depressurise or not. Main			Many enquiries may block communication channels, and must be responded to by CCR operators.	Consider merits of removing compressor liquid – How long would the vessels withstand fire if ignited? Results of hazard analyses should be immediately transmitted to offshore personnel from

0:42	CCRO	As at 0:30	Deluge Available (from light on console)	Ignition still likely	concern is with the liquid in the vessels which can only be removed if gas pressure remains in the compressors. However, no remote controls exist and it is considered too hazardous to try the operation. Decide therefore to muster personnel at other end of the platform until deluge cools down compressors and pressure drops. Tells outside operators to clear cell 23. Muster points broadcast on public address system. Deluge activated.				onshore emergency centre to update their knowledge and enhance decision-making.

Figure 31.4. Example of tabular task analysis

Table 31.1 shows the psychological error mechanisms underlying the SHERPA system. An example of the tabular output from such an analysis is shown in Figure 31.5 (Kirwan and Rea, 1986). This 'human error analysis table' has similarities to certain reliability engineering approaches to identifying the failure modes of hardware components. One particularly useful aspect of this approach is the determination of whether errors can be recovered immediately, at a later stage in the task, or not at all, information useful if error reduction is required later in the analysis. This particular tabular approach also attempts to link error reduction measures to the causes of the human error, on the grounds that treating the 'root causes' of the errors will probably be the most effective way to reduce error frequency.

Another computerized system is the *Potential Human Error Cause Analysis* (PHECA) system (Whalley, 1988). Figure 31.6 shows the error causes and mechanisms inherent in the model, and Figure 31.7 shows the major performance shaping factors which interact with the error causes. This approach has also borrowed from the reliability world in the form of the well established HAZOP (Hazard and Operability Study) technique (Kletz, 1984), as all errors identified in PHECA can occur in only the following 'external error mode' forms:

Table 31.1. SHERPA: classification of psychological mechanisms (see Reason and Embrey, 1986)

1. *Failure to consider special circumstances.* A task is similar to other tasks but special circumstances prevail which are ignored, and the task is carried out inappropriately
2. *Short cut invoked.* A wrong intention is formed based on familiar cues which activate a short cut or inappropriate rule
3. *Stereotype takeover.* Owing to a strong habit, actions are diverted along some familiar but unintended pathway
4. *Need for information not prompted.* Failure of external or internal cues to prompt need to search for information
5. *Misinterpretation.* Response is based on wrong apprehension of information such as misreading of text or an instrument, or misunderstanding of a verbal message
6. *Assumption.* Response is inappropriately based on information supplied by the operator (by recall, guesses, etc.) which does not correspond with information available from outside
7. *Forget isolated act.* Operator forgets to perform an isolated item, act or function, i.e., an act or function which is not cued by the functional context, or which does not have an immediate effect upon the task sequence. Alternatively it may be an item which is not an integrated part of a memorized structure
8. *Mistake among alternatives.* A wrong intention causes the wrong object to be selected and acted on, or the object presents alternative modes of operation and the wrong one is chosen
9. *Place losing error.* The current position in the action sequence is misidentified as being later than the actual position
10. *Other slip of memory* (as can be identified by the analyst)
11. *Motor variability.* Lack of manual precision, too big/small force applied, inappropriate timing (including deviations from 'good craftsmanship')
12. *Topographic or spatial orientation inadequate.* In spite of the operator's correct intention and correct recall of identification marks, tagging, etc., he unwittingly performs a task/act in the wrong place or on the wrong object. This occurs because of following an immediate sense of locality where this is not applicable or not updated, perhaps due to surviving imprints of old habits, etc.

TASK 51: TERMINATE SUPPLY AND ISOLATE TANKER. (SEQUENCE OF REMOTELY – OPERATED VALVE OPERATIONS)

TASK STEP	ERROR TYPE	RECOVERY STEP	PSYCHOLOGICAL MECHANISM	CAUSES, CONSEQUENCES AND COMMENTS	RECOMMENDATIONS		
					PROCEDURES	TRAINING	EQUIPMENT
51.1	ACTION TOO LATE	NO RECOVERY	PLACE LOSING ERROR	OVERFILL OF TANKER RESULTING IN DANGEROUS CIRCUMSTANCE.	OPERATOR ESTIMATES TIME/RECORDS AMOUNT LOADED	EXPLAIN CONSEQUENCES OF OVERFILLING	FIT ALARM-TIMING/ VOLUME/TANKER LEVEL
51.2.1	ACTION OMITTED	5.2.4	SLIP OF MEMORY	FEEDBACK WHEN ATTEMPTING TO CLOSE CLOSED VALVE. OTHERWISE ALARM WHEN LIQUID VENTED TO VENT LINE			MIMIC OF VALVE CONFIGURATION
51.2.2	ACTION TOO EARLY	5.2.2	PLACE LOSING ERROR	ALARM WHEN LIQUID DRAINS TO VENT LINES	SPECIFY TIME FOR ACTIONS	OPERATOR TO COUNT TO DETERMINE TIME	
	ACTION OMITTED	5.2.2	SLIP OF MEMORY	AS ABOVE AND POSSIBLE OVER PRESSURE OF TANKER (SEE STEP 5 1.2.3)			MIMIC OF VALVE CONFIGURATION
51.2.3	ACTION TOO EARLY	NO RECOVERY	PLACE LOSING ERROR	IF VALVE CLOSED BEFORE TANKER SUPPLY VALVE OVERPRESSURE OF TANKER WILL OCCUR		STRESS IMPORTANCE OF SEQUENCE AND EXPLAIN CONSEQUENCES	INTERLOCK ON TANKER VENT VALVE
	ACTION OMITTED	5.2.6	SLIP OF MEMORY	AUTOMATIC CLOSURE ON LOSS OF INSTRUMENT AIR.			MIMIC OF VALVE CONFIGURATION
51.2.4	ACTION OMITTED	5.2.2	SLIP OF MEMORY	AUDIO FEEDBACK WHEN VENT LINE OPENED.		EXPLAIN MEANING OF AUDIO FEEDBACK	MIMIC OF VALVE CONFIGURATION
51.3	ACTION OMITTED	NO RECOVERY	SLIP OF MEMORY	LATENT ERROR.	ADD CHECK ON FINAL VALVE POSITIONS BEFORE PROCEEDING TO NEXT STEP		MIMIC OF VALVE CONFIGURATION

Figure 31.5. Example extract of human error analysis (Kirwan and Rea, 1986)

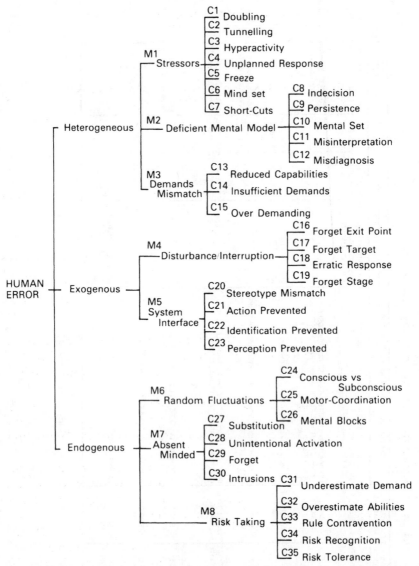

Figure 31.6. Error causes grouped by error mechanisms (Whalley, 1988)

Not done	Repeated
Less than	Sooner than
More than	Later than
As well as	Mis-ordered
Other than	Part of

Human error identification technique development has, as noted above, gen-

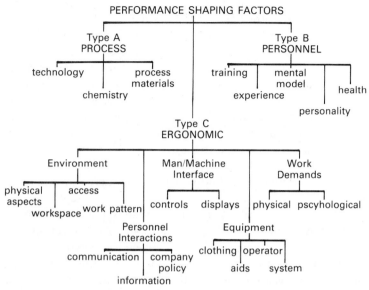

Figure 31.7. Major sections of the performance shaping factors classification structure (Whalley, 1988)

erally received less attention than human error quantification. However, a general pattern is already emerging in terms of the functions required of any human error identification technique. Useful techniques must address the following:

External error mode identification. All human errors can be categorized into these descriptions. They are the level of description which is put into the probabilistic safety assessment, e.g., operator fails to respond to alarm ('error of omission', or 'not done', etc.).

Error cause or mechanism identification. These descriptions define, in psychologically and/or ergonomically meaningful terms, how the error actually occurred (e.g., the above 'failure to respond' may be due to a 'misinterpretation' of the signal, or due to 'reduced capabilities', etc.).

Identification of performance shaping factors (PSF). Factors which affect performance can obviously be usefully considered during the human error identification phase, although frequently they are not identified until the quantification phase. PHECA is currently the only technique which links PSF to error causes, which can be a source of useful information if error reduction is required at a later stage.

The distinction between the external error mode and the error cause or mechanism is particularly important if error reduction is required. Knowing the error mechanism, effective error reduction mechanisms will be more readily specified. It is likely that future human error identification techniques will follow this basic pattern as PHECA, and to a lesser extent SHERPA already do, making the errors identified more psychologically/ergonomically meaningful, and more helpful in reducing accident potential.

There are very few reviews of human error identification techniques and approaches, although many paths have been explored (described in Kirwan, 1986; 1992a; 1992b). The one area in which all error identification methods have difficulty is in cognitive decision making/diagnostic tasks, e.g., when an operator is trying to diagnose a complex set of symptoms in an emergency situation. Recent history, e.g., the Three Mile Island and the Davis–Besse (USNRC, 1985) incidents, has shown that one of the most significant potential human errors is misdiagnosis during such an emergency, yet it is difficult to predict the form that a misdiagnosis may take. This form is important since a misdiagnosis may not only lead to an unsafe state or an accident, but may actually lead to a worse state of affairs than might have occurred had the operators done nothing. One method which has attempted to determine the nature of potential misdiagnoses is the confusion matrix approch (e.g., Potash *et al.*, 1981). This is basically a matrix of different possible scenarios, rated in similarly of symptoms (and hence confusibility) by operators and system dynamics experts. Other research has recently concentrated on expert-system based simulations of the operator (Woods *et al.*, 1987). However, usable tools are still not available, and so this is a primary target area for future research. The reader is also referred to Reason (1990), for a general and useful discussion of all error forms, including the more 'cognitive' error modes.

The identification of error recovery paths is largely carried out using the judgement of the analyst. The task analysis should also highlight points in the sequence at which discovery of an error will be possible, e.g., via indications (especially the occurrence of alarms) and checks or interventions by other personnel.

Analysts' judgement is a worthwhile resource to utilize in error identification. Many risk analysis practitioners build up experience of identifying errors in safety studies, either from their operational experience or from involvement in many different safety analyses. Whilst this is perhaps an 'art' rather than a science and as such is less accessible to the novice than a more formal method, its value must not be overlooked. Due to the specificity of every new safety analysis carried out, the human reliability practitioner 'standing on the outside' may often be more disadvantaged in terms of error identification than the hardware reliability analyst who knows the system details intimately. It is therefore worthwhile adopting a hybrid team approach to error identification as well as the use of formal systematic methods. The adage of 'two heads are better than one' is especially true in the area of human error identification.

The next step following identification of these error forms is to represent them in some logical format so that their effects on the system goals can be evaluated.

(4) Representation

A fault tree is a typical way of representing a set of human errors and their effects on the system goals. A fault tree is a logical structure which defines what events (human errors, hardware/software faults, environmental events) must occur in order for an undesirable event (e.g., an accident) to occur (Henley and Kumamoto, 1981). The undesirable event or outcome, usually placed at the top of the 'tree' and hence called the top event, may for example be 'failure to launch a lifeboat successfully at first attempt', or 'failure to achieve recirculation of primary coolant', etc. The tree is constructed primarily by using two types of 'gate' by which events at one level can proceed to the next level up until finally they reach the top event. The first type of gate is an 'OR' gate, and the event above this gate occurs if *any* one of the events joined below it by this gate occur. An event above an 'AND' gate only occurs if *all* the events joined below it by this gate occur. An example is shown in Figure 31.8.

A fault tree can be used to represent a simple or complex pattern of system failure paths, and may comprise human errors alone, or a mixture of human, hardware, and/or environmental events, depending upon the scenario. Once structured, the events (including human error probabilities) must be quantified to determine the overall top event frequency (or probability), and the relative contributions of each error to this undesirable event.

Another type of 'tree' is the operator action tree (OAT: see Figure 31.9). The OAT proceeds from an initiating event, usually placed at the left hand side of the tree (e.g., loss of power causes ESD demand), to consider a set of sequential events each of which may or may not occur, causing the tree to branch (usually) in a binary fashion at each event 'node'. The events and branching continue until an end state is reached for each path, which is either success in terms of achieving system safety, or else failure in terms of lost production, plant or equipment damage, injury, or fatality. OATs are especially useful when considering dynamic situations, and in general are preferable to fault trees when human performance is dependent upon previous actions/events in the scenario sequence (Hall *et al.*, 1982).

A recent variant on the OAT is to represent the emergency actions of operating personnel in the form of an event tree which uses three branches representing successful operation, failure to respond, and a mistake or misdiagnosis (USNRC, 1984). This type of OAT is able to represent important potential misdiagnoses that may have been identified in the human error analysis stage. If significant misdiagnoses are identified, the main event tree may branch off to another sub-event tree to consider the alternative consequences that may occur as a result of the mistake.

The above are formal methods used for representing moderately complex patterns and sequences of failures. Whilst the examples given here are simple, in practice such trees can become quite complex, with many 'nodes' and events, and in the case of event trees, a large number of possible final out-

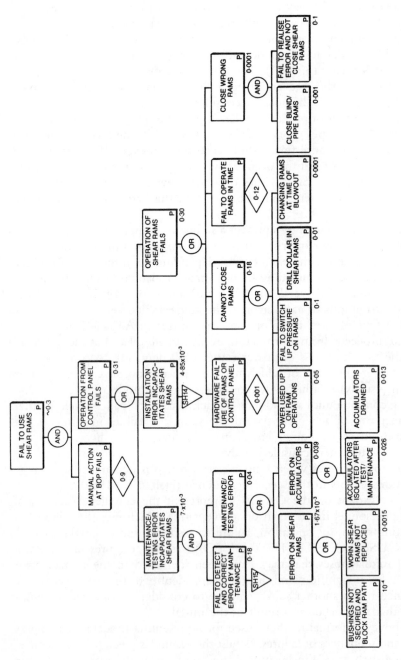

Figure 31.8. Offshore drilling blowout fault tree sub-tree: fail to use shear rams to prevent blowout

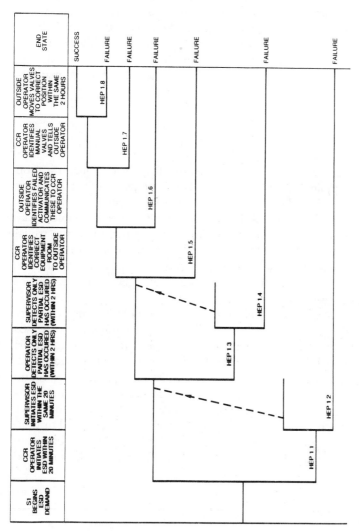

Figure 31.9. Operator action tree for ESD failure scenario (Kirwan, 1988)

comes. It is something of an art to develop such trees so that the human errors are adequately and accurately represented, without letting the trees become too complex and unwieldy.

If in a particular assessment the number of errors is small, and their effect on the system goals is very simple, such representation may be unnecessary. Furthermore, some analyses may stop at this point if their objective was merely qualitative in nature, i.e., simply to identify human error modes without quantifying their probability or consequential effect on the system. In all other cases however, quantification of the human error probabilities will be necessary. Due to resource limitations screening may be applied to limit the amount of quantification required. If screening is not utilized, then quantification is the next step.

(5) Screening

A screening analysis identifies where the major effort in the quantification analysis should be applied. There may for example be particular tasks which are theoretically related to the system goals being investigated, but which in fact make little contribution to risk if they fail (due to diverse reliable back-up systems which adequately compensate for human error, or to the trivial nature of the tasks themselves). It is efficient to expend little effort on such tasks and instead focus on those in which human reliability is critical. The identification of those errors which can be effectively ignored by the rest of the study is the purpose of a screening analysis.

The *systematic human action reliability procedure* (SHARP) methodology defines three methods of screening logically structured human errors (see Spurgin *et al.*, 1987). The first method 'screens out' those human errors which can only affect the system goals if they occur in conjunction with an extremely unlikely hardware failure or environmental event. The second method involves allocating each human error a probability of 1·0, and examining the effects of the various errors on the system goals. Those that have a negligible effect even with a probability of unity are not considered further. The third method assigns broad probabilities to the human errors based on a simple categorization (such as the one shown in Table 31.2). This method

Table 31.2. Generic human error probabilities

Category	Failure probability
Simple, frequently performed task, minimal stress	10^{-3}
More complex task, less time variable, some care necessary	10^{-2}
Complex, unfamiliar task, with little feedback and some distractions	10^{-1}
Highly complex task, considerable stress, little performance time	3×10^{-1}
Extreme stress, rarely performed task	10^{0}

works in the same way as the previous method but is a 'finer-grained' analytic method.

With most screening methods (particularly the third method above) there is a danger of ruling out of the study important errors and interactions, balanced against a need to reduce to a manageable level the complexity of, and resources required for, the analysis. As a general rule when applying any screening technique at any level in the study—if in doubt, leave the human error in the fault/event tree. The next step following screening, or following error analysis if screening was not applied, is quantification of the human errors in the fault or event trees.

(6) Quantification

Human reliability quantification techniques all quantify the human error probability (HEP), which is the metric of human reliability assessment. The HEP is defined as

$$HEP = \frac{\text{number of errors occurred}}{\text{number of opportunities for error to occur}}.$$

Thus, if when buying a cup of coffee from a vending machine on average one time in a hundred tea is accidentally purchased, the HEP is taken as 0·01 (it is somewhat educational to try and identify HEPs in everyday life with a value of less than once in a thousand opportunities, or even as low as once in ten thousand).

In an ideal world there would be many studies and experiments in which HEPs were recorded. In reality there are few such recorded data. The ideal source of human error 'data' would be from industrial studies of performance and accidents, but at least three reasons can be deduced for the lack of such data:

(a) difficulties in estimating the number of opportunities for error in realistically complex tasks (the so-called denominator problem);
(b) confidentiality and unwillingness to publish data on poor performance;
(c) lack of awareness of why it would be useful to collect data in the first place (and hence lack of financial incentive for such data collection).

There are other potential reasons (see Williams, 1983, for some of these) but the net result is a scarcity of HEP data (however, see Kirwan *et al.*, 1990, for a recent successful data collection exercise). Other sources are simulator data (e.g., exercises using high-fidelity simulators), and data derived from experimental laboratory-based studies, reported in the human performance literature. Two problems exist with respect to simulator studies, the first being that such simulators are used almost exclusively for training purposes (and in the USA nuclear power industry for operator re-certification), and hence personnel on the simulator are highly motivated and frequently know what is on the training curriculum (i.e., they know which scenarios to expect). Secondly, it is not clear how realistic it is facing an emergency in a simulator compared with the real thing (the 'cognitive fidelity' issue). (See Meister in chapter 8 of this book for a more complete review of simulation.)

The human performance literature has a similar problem in that studies in this vein are usually highly controlled, often looking at one or two independent variables (unlike industry where many PSFs vary and interact), and using reasonably motivated subjects for a short period of time. Generalizing from such studies to complex industrial multi-personnel situations is not easy, and is often a questionable exercise.

Overall, therefore, there is a 'data problem'. Furthermore, even if there is a sound datum for, e.g., a chemical plant operator failing to respond to an alarm in scenario X, how can this be generalized to scenario Y, or even to scenario X on a different chemical plant, with a different operating regime, or to a nuclear power plant? In other words, what defines the 'generalizability' of data? Such difficult and as yet unresolved issues as these have led to the development of non-data-dependent approaches, namely to the use of expert opinion. This is by no means necessarily a bad thing, and expert opinion has been used successfully in other areas (e.g., Murphy and Winkler, 1974; or Ludke *et al.*, 1977), and is in any case used at least occasionally in probabilistic safety assessments (Nicks, 1981) where similar problems often exist.

The human error quantification techniques described here all contain an element of expert judgement, even though some in particular give the appearance (not necessarily intended) of being based on 'hard' well-founded empirical data.

In a review by Kirwan *et al.* (1988), eight human reliability quantification techniques were qualitatively assessed. These were:

> *Absolute Probability Judgement* (APJ) (Seaver and Stillwell, 1983).
> *Paired Comparisons* (PC) (Hunns and Daniels, 1980),
> *TESEO* (Bello and Columbari, 1980),
> *Technique for Human Error Rate Prediction* (THERP), Swain and Guttmann, 1983),
> *Human Error Assessment and Reduction Technique* (HEART) (Williams, 1986),
> *Influence Diagrams Approach* (IDA) (Phillips *et al.*, 1983),
> *Success Likelihood Index Method* (SLIM) (Embrey *et al.*, 1984), and
> *Human Cognitive Reliability Model* (HCR) (Spurgin *et al.*, 1987).

Four of the techniques (APJ, PC, IDA, and SLIM) use a group of expert judges to evaluate HEPs. APJ and PC largely leave the judgemental task to the experts, with some help from the analyst or 'facilitator' who may point out inconsistent judgements or biases in the judgement-making process. SLIM and IDA also use expert judges, but the judges are asked to consider what factors affect performance, and from the assessment of these factors and modelling of their influence on performance, they then determine the human error probability. They are assisted by the analyst in creating a quantitative causal model of the influence of these factors on the HEP. Typical perform-ance shaping factors (PSF) utilized are stress, quality of interface design, degree of training and adequacy of procedures.

TESEO, THERP and HEART, in contrast, either include a database of HEPs or specify procedures for generating numerical HEPs directly. These techniques require only one analyst, rather than a group of experts. The data on which they rely are a mixture of field experience and judgement in the case of THERP, and a mixture of judgement together with data from ergo-nomics and psychological performance literature in the case of HEART and, to a lesser extent, TESEO.

Lastly the HCR model (also called the time reliability correlation approach) attempts to quantify diagnostic/cognitive errors as a function of time elapsed since the onset of the incident, and assumes that the likelihood of successful diagnosis and, consequently action, increases as the time available increases. This approach is currently a mixture of judgement and simulator data.

Within the scope of this chapter it is not possible to review all these tech-niques. Instead, therefore, three are reviewed, namely SLIM, HEART and THERP. The first two are probably of most interest to the ergonomist, and the latter exemplifies the reliability engineering oriented approach, and is also the technique which has been most widely used to date. For a review of the others, see Kirwan *et al.* (1988) or the indicated source references. The reader is also referred to Swain (1989) for an in-depth review of fourteen HRA approaches.

Success likelihood index method (SLIM)

SLIM can best be explained by means of an example human reliability assess-ment, in this case an operator decoupling a filling hose from a chemical road tanker. The operator may forget to close a valve upstream of the filling hose, which could lead to undesirable consequences, particularly for the operator. The human error of interest is 'failure to close V0204 prior to decoupling filling hose'. In this case the decoupling operation is simple and discrete, and hence failure occurs catastrophically rather than in a staged fashion.

PSF identification

The 'expert panel' would typically comprise, for example, two operators with 10 years experience, one human factors analyst, and a reliability analyst fam-iliar with the system who also has some operational experience.

The panel is initially asked to identify a set of *performance shaping factors* (PSFs), which are any factors relating to the individual(s), environment, or task, which affect performance positively or negatively. The expert panel could be asked to nominate the most important or significant PSFs for the scenario under investigation. In this example it is assumed the panel identify the following major PSFs as affecting human performance in this situation: training, procedures, feedback, perceived risk, and time pressure.

PSF rating

The panel are then asked to consider other human errors possible in this scenario (e.g., mis-setting or ignoring an alarm), and for each one, to decide to what extent each PSF is optimal or sub-optimal for that task in the situation being assessed. The 'rating' of whether a task is optimal or sub-optimal for a particular PSF is made on a scale of 1 to 9, in this case with 9 as optimal. For the three human errors under analysis, the ratings obtained are as follows:

		Performance Shaping Factors			
Errors	*Training*	*Procedures*	*Feedback*	*Perceived risk*	*Time*
V0204 Open	6	5	2	9	6
Alarm mis-					
set	5	3	2	7	4
Alarm					
ignored	4	5	7	7	2

PSF weighting

If each factor was equally important, one might simply add each row of ratings and conclude that the error with the lowest rating sum (alarm mis-set) was the most likely error. However, this expert panel, as with most panels, does not feel the PSFs are all equal. In this particular case (and with this particular panel of experts), the panel feels that perceived risk and feedback are most important, and are in fact twice as important as training and procedures, which are in turn one and a half times as important as time. (As it is a routine operation, time is not perceived by the panel to be particularly important). Weightings for the PSFs can be obtained directly from these considered opinions, as follows, normalized to sum to unity:

Perceived Risk	0·30
Feedback	0·30
Training	0·15
Procedures	0·15
Time	0·10
Sum =	1·00

SLIM, and the decision analysis technique it is based upon, called *simple multi-attribute rating technique* (Edwards, 1977) propose simply that preference can be derived as a function of the sum of the weightings multiplied by their ratings for each item (human error). SLIM does this and calls the resultant preference index a *success likelihood index* (SLI). This is illustrated using a table of weightings (W) × ratings (R): (SLI = WR) (see Table 31.3).

In this case, the lowest SLI is 4·3, suggesting that 'alarm mis-set' is still the most likely error. However, due to the weightings used, the likelihood ordering of the other two errors have now been reversed (close inspection of the figures reveals that this is because feedback is held to be important, and there is ample feedback for 'alarm ignored' but not for 'V0204 open'). Clearly at this point, a designer would realize that increased feedback about the position of V0204 to the operator might be desirable.

However, the SLIs are not yet probabilities. Rather, they are indications of the relative likelihoods of the different errors. Thus the SLIs show the ordering of likelihood of the different errors, but do not yet define the absolute probability values. In order to transform the SLIs into HEPs, it is necessary to 'calibrate' the SLI values. (Note: the paired comparisons technique also requires this calibration using the same basic formula.) Two earlier studies by Pontecorvo (1965) and Hunns (1982) have derived such a calibration relationship, both suggesting a logarithmic relationship of the form:

$$\text{Log}_{10} \, (\text{HEP}) = a \, \text{SLI} + b.$$

If two tasks for which the HEPs are known are included in the task/error set which are being quantified, then the parameters of the equation can be derived via simultaneous equations, and the other (unknown) HEPs can be quantified. If in the above example, two more tasks (*A* and *B*) were assessed which had HEPs of 0·5 and 10^{-4} respectively, and were given SLIs of 4·00 and 6·00, respectively, then the equation derived would be:

$$\text{Log} \, (\text{HEP}) = -1·85 \, \text{SLI} + 7·1.$$

The HEPs would then be: V0204 = 0·0007; alarm mis-set = 0·14; alarm ignored = 0·0003.

Table 31.3. SLI calculation

Weighting	PSF	Weighting × Rating	V0204	Alarm mis-set	Alarm ignored
0·30	Feedback	(0·3 × 2) =	0·6	0·6	2·1
0·30	Perceived Risk	etc.	2·7	2·1	2·1
0·15	Training		0·9	0·75	0·6
0·15	Procedures		0·75	0·45	0·75
0·10	Time		0·60	0·40	0·2
	SLI (Total)		5·55	4·30	5·75

This is the body of the rationale underlying SLIM, but in practice SLIM is more complex and is computerized to facilitate its ease of use and to prevent bias, often found in the elicitation of expert opinions. The computerized version, known as SLIM-MAUD (SLIM using *Multi Attribute Utility Decomposition*: Embrey *et al.*, 1984), due to the mathematics in the software which is present partly to avoid such bias, will produce slightly different values (HEPs) than the hand calculated method used above. In particular the simple summary of weightings and ratings is refined in several ways according to the more detailed mathematical requirements of multi-attribute utility theory. However, the above is the general rationale of SLIM, and enables the reader to understand more easily how SLIM works.

Human error assessment and reduction technique (HEART)

This technique is of particular interest to ergonomists as it is based on the human performance literature. It has been designed by its author as a relatively quick method for HRA, to be simple to use and easily understood. Its fundamental premise is that in reliability and risk equations one is interested in ergonomics factors which have a large effect on performance, e.g., causing a decrement in performance by a factor of three or more. Thus, whilst there are many well-studied ergonomics factors and consequent guidelines (e.g., lighting recommendations), many of these factors actually have (in reliability terms) a negligible effect on operator performance. HEART therefore concentrates on those factors which have a significant effect.

This point is important because in part it underlies something of a communications gap between engineers and ergonomists. Engineers and designers designing a plant cannot spend unlimited funds on the optimization of ergonomics aspects, and often ask how important (in quantitative terms) an ergonomics recommendation is. Frequently the ergonomist is unable to answer this question which to the engineer is fundamental. HEART in particular, and some of the other techniques (e.g., SLIM), allow the human reliability analyst to answer this question quantitatively.

The first part of the HEART assessment process is to refine the task in terms of its generic proposed nominal human unreliability, as shown in Table 31.4. Thus the task is first assigned a nominal human error probability by classifying it according to whether it is a complex task, a routine task, and so on. The next stage is to identify error producing conditions (EPCs) which are evident in the scenario and would negatively influence human performance. A table of the major EPCs in HEART is shown in Table 31.5.

Example

As a hypothetical example of how HEART is used to quantify a human error probability for a task, taken from Williams (1988), we assume that a safety, reliability, or operations engineer wishes to assess the nominal likelihood of an operative's failing to isolate a plant bypass route following strict procedures. The scenario necessitates a fairly inexperienced operator applying an opposite technique to that which he normally uses to carry out isolations and involves a piece of plant, the inherent major hazards of which he is only dimly aware. It is assumed that the man could be in the seventh hour of his shift, that there is talk of the plant's imminent closure, that his work may be checked and that the local management of the company is desperately trying to keep the plant operational despite the real need for maintenance because of its fear that partial shutdown could quickly lead to total permanent shutdown.

Using a simplified HEART, the safety reliability and operational engineer's assessment could look something like this:

Type of Task = F			*Nominal Human Unreliability* = 0·003
EPC - number and description	*Total HEART Affect*	*Engineer's Assessed Proportion of Affect* (from 0 to 1)	*Assessed Affect*
15. Inexperience	× 3	0·4	(3–1) × 0·4 + 1 = 1·8
9. Opposite Technique	× 6	1·0	(6–1) × 1·0 + 1 = 6·0
12. Risk Misperception	× 4	0·8	(4–1) × 0·8 + 1 = 3·4
18. Conflict of Objectives	× 2·5	0·8	(2·5–1) × 0·8 + 1 = 2·2
— Low morale	× 1·2	0·6	(1·2–1) × 0·6 + 1 = 1·12

Assessed nominal likelihood of failure
$$0·003 \times 1·8 \times 6·0 \times 3·4 \times 2·2 \times 1·12 = 0·27$$

Time-on-shift effects would be ignored as there is no indication of monotony.

Similar calculations may be performed if desired for the predicted 5th and 95th percentile bounds, which in this case would be $0·07 - 0·58$. As a total probability of failure can never exceed 1·00, if the multiplication of factors takes the value above 1·00 the probability of failure has to be assumed to be 1·00 and no more.

The relative contribution made by each of the error producing conditions to the amount of unreliability modification is as follows:

	% contribution made to unreliability modification
Technique unlearning	41
Misperception of risk	24
Conflict of objectives	15
Inexperience	12
Low morale	8

Table 31.4. Generic classifications (HEART, after Williams, 1986)

Generic task	Proposed nominal human unreliability (5th–95th percentile bounds)
(A) Totally unfamiliar, performed at speed with no real idea of likely consequences	0·55 (0·35–0·97)
(B) Shift or restore system to a new or original state on a single attempt without supervision or procedures	0·26 (0·14–0·42)
(C) Complex task requiring high level of comprehension and skill	0·16 (0·12–0·28)
(D) Fairly simple task performed rapidly or given scant attention	0·09 (0·06–0·13)
(E) Routine, highly-practised, rapid task involving relatively low level of skill	0·02 (0·007–0·045)
(F) Restore or shift a system to original or new state following procedures, with some checking	0·003 (0·0008–0·007)
(G) Completely familiar, well-designed, highly practised, routine task occurring several times per hour, performed to highest possible standards by highly-motivated, highly-trained and experienced person, totally aware of implications of failure, with time to correct potential error, but without the benefit of significant job aids	0·0004 (0·00008–0·009)
(H) Respond correctly to system command even when there is an augmented or automated supervisory system providing accurate interpretation of system stage	0·00002 (0·000006–0·0009)

Thus an HEP of 0·27 (just over one in four) is calculated, which is a very high predicted error probability, and unlikely to be acceptable. In this case technique unlearning is the major contributory factor to this poor performance, and so clearly either some form of retraining, or else redesign to make the isolation procedures consistent across plant, must be considered. However HEART goes further than other techniques in error reduction, as for each EPC it gives corresponding suggested error reduction approaches, e.g., for the task above, the remedial measures in Table 31.6 would be proposed.

Thus HEART offers a quick and simple human reliability calculation method which also gives the user (engineer or ergonomist) suggestions on error reduction. HEART has recently been reviewed qualitatively by its author and is under further development (Williams, 1992).

Technique for human error rate prediction (THERP)

THERP is in itself a total methodology for assessing human reliability. The quantification part of THERP comprises the following:

(a) a database of human errors which can be influenced by the assessor to reflect the impact of PSFs on the scenario;

Table 31.5. HEARTS EPCs (Williams, 1986)

Error producing condition	Maximum predicted nominal amount by which unreliability might change going from 'good' conditions to 'bad'
1. Unfamiliarity with a situation which is potentially important but which only occurs infrequently or which is novel	× 17
2. A shortage of time available for error detection and correction	× 11
3. A low signal-to-noise ratio	× 10
4. A means of suppressing or overriding information or features which is too easily accessible	× 9
5. No means of conveying spatial and functional information to operators in a form which they can readily assimilate	× 8
6. A mismatch between an operator's model of the world and that imagined by a designer	× 8
7. No obvious means of reversing an unintended action	× 8
8. A channel capacity overload, particularly one caused by simultaneous presentation of non-redundant information	× 6
9. A need to unlearn a technique and apply one which requires the application of an opposing philosophy	× 6
10. The need to transfer specific knowledge from task to task without loss	× 5.5
11. Ambiguity in the required performance standards	× 5
12. A mismatch between perceived and real risk	× 4
13. Poor, ambiguous or ill-matched system feedback	× 4
14. No clear direct and timely confirmation of an intended action from the portion of the system over which control is to be exerted	× 4
15. Operator inexperience (e.g., a newly-qualified tradesman, but not an 'expert')	× 3
16. An impoverished quality of information conveyed by procedures and person/person interaction	× 3
17. Little or no independent checking or testing of output	× 3
18.* A conflict between immediate and long-term objectives	× 2·5
19. No diversity of information input for veracity checks	× 2·5
20. A mismatch between the educational achievement level of an individual and the requirements of the task	× 2
21. An incentive to use other more dangerous procedures	× 2
22. Little opportunity to exercise mind and body outside the immediate confines of a job	× 1·8
23. Unreliable instrumentation (enough that it is noticed)	× 1·6
24. A need for absolute judgements which are beyond the capabilities or experience of an operator	× 1·6
25. Unclear allocation of function and responsibility	× 1·6
26. No obvious way to keep track of progress during an activity	× 1·4

* 18–26. These conditions are presented simply because they are frequently mentioned in the human factors literature as being of some importance in human reliability assessment. To a human factors engineer, who is sometimes concerned about performance differences of as little as 3%, all these factors are important, but to engineers who are usually concerned with differences of more than 300%, they are not very significant. The factors are identified so that engineers can decide whether or not to take account of them after initial screening.

Table 31.6. A subset of HEART remedial measures (Williams, 1986)

1.	Technique unlearning (× 6)	The greatest possible care should be exercised when new techniques are being considered to achieve the same outcome — they should not involve adoption of opposing philosophies
2.	Misperception of risk (× 4)	It must not be assumed that a user's perception of risk is the same as the actual level — if necessary a check should be made to ascertain where any mismatch might exist and what its extent is
3.	Objectives conflict (× 2·5)	Objectives should be tested by management for mutual compatibility, and where potential conflicts are identified these should either be resolved to make them harmonious or made prominent so that a comprehensive management control programme can be created to reconcile such conflicts as they arise, in a rational fashion
4.	Inexperience (× 3)	Personnel criteria should contain specified experience parameters thought relevant to the task — chances must not be taken for the sake of expediency
5.	Low morale (× 1·2)	Apart from the more obvious ways of attempting to secure high morale, by way of financial reward for example, other methods involving participation, trust and mutual respect, often hold out at least as much promise — building up morale is a painstaking process, which involves a little luck and great sensitivity — employees must be given reason to believe

(b) a dependency model which calculates the degree of dependence between two operator actions (e.g., if an operator fails to detect an alarm, then failure to carry out appropriate corrective actions reliably cannot be treated independent from this failure);

(c) an event tree modelling approach to combine HEPs for steps in a task into an overall task HEP;

(d) the assessment of error recovery paths.

The basic THERP approach is shown in Figure 31.10, from Bell (1984).

Probably because of its database and similarities to reliability engineering approaches, THERP has been used more than any other technique in industry applications. This small section can only cover the rudiments of the technique, and for more information the reader is referred to Swain and Guttmann (1983), Bell (1984) and Kirwan *et al.* (1988).

An example of the type of basic event data given in THERP is shown in Table 31.7.

A human reliability analysis event tree (HRAET, as shown in Figure 31.11) can be used to represent the operator's performance. Alternatively, as shown in Figure 31.12, an operator action event tree can be used (Whittingham, 1988). In each case the event tree represents the sequence of events and considers possible failures at each branch in the tree (omission, commission, and so on). These errors are quantified and error recovery paths are then added to the tree where appropriate. Figure 31.13 shows the PSFs which

Phase 1: Familiarisation

Plant Visit

Review Information From
System Analysts

Phase 2: Qualitative Assessment

Talk- or
Walk-Through

Task Analysis

Develop HRA Event Trees

Phase 3: Quantitative Assessment

Assign Nominal HEPs

Estimate the Relative
Effects of Performance
Shaping Factors

Assess Dependence

Determine Success and
Failure Probabilities

Determine the Effects
of Recovery Factors

Phase 4: Incorporation

Perform a Sensitivity
Analysis, if Warranted

Supply Information to
System Analysts

Figure 31.10. Outline of a THERP procedure for HRA (adapted from Bell, 1984)

can be used in a THERP study, although often in a THERP study only one
or at most a few of these are utilized quantitatively (e.g., stress).

It is important to model recovery, particularly when investigating highly
proceduralized sequences, since often an operator will be prompted by a later
step in the procedures to recover from an earlier error in a previous step.
For example, if an operator omits a step to turn the power on, and then a
second step is attended to which involves checking certain power-supplied
instruments, then the operator will rapidly recover the first error. If recoveries
in such highly proceduralized situations (for which THERP is typically used)
are not identified, then human error may be overestimated. THERP is in
fact the only technique which emphasizes error recovery in this way. Error
recovery paths are shown on the operator action tree in Figure 31.12 as
dashed lines, and in this particular tree these are largely recoveries of one
person's error by another person.

Table 31.7. Task analysis — initiation of flow via stand-by train (Webley and Ackroyd; in Kirwan *et al.*, 1988)

Task identifier*	Task of interest	Error identifier	Human errors	Human error probabilities (range of median values given in USSR State Committee, 1986)	
A	Identity loss of flow via duty train	A_1	Fail to identify loss of flow	$10^{-4} - 0 \cdot 25$	Alarm response model
B	Start correct procedure	B_1	Fail to start procedure	$10^{-3} - 10^{-2}$ $10^{-2} - 5 \times 10^{-2}$	Procedure used Procedure not used
C	Roving operator opens correct valve	C_1	Error of omission — verbal order	10^{-3}	
			Error of commission — select incorrect valve	$10^{-3} - 10^{-2}$	
D	1st operator start stand-by pump via remote control	D_1	Error of omission — written procedures available	$10^{-3} - 10^{-2}$ $10^{-2} - 5 \times 10^{-2}$	Procedure used Procedure not used
		D_2	Error of commission — select incorrect control	$10^{-3} - 10^{-2}$	
E	Supervisor checks 1st operator	E_1	Error of omission — written procedures available	$10^{-3} - 10^{-2}$ $10^{-2} - 5 \times 10^{-2}$	Procedure used Procedure not used
		E_2	Error of commission — select incorrect valve	$10^{-3} - 10^{-2}$	
G	Shift Manager checks activation	G_1	Fail to initiate checking function	$10^{-3} - 10^{-2}$	
			Fail to identify errors	$5 \times 10^{-2} - 0 \cdot 2$	

* Task F no errors identified for this step.

THERP also models dependency between human errors. If, for example, an operator is in a high stress situation trying to carry out a procedure quickly, or is demotivated or fatigued, etc., then a whole section of procedures and recovery steps within these procedures may be carried out wrongly. This is an example of a 'human dependent failure'. An important example of dependency modelling concerns one operator checking another operator. It is unlikely that the action by operator 1 and the check by operator 2 will be completely independent, since the first operator may assume that the second will detect any faults, and the second may assume the first did the job properly. THERP is one of the few techniques which actually quantitatively models this dependency between actions/errors. It uses a simple five-level model of dependency from zero dependence (i.e., total independence) through low, medium and high dependence levels to complete dependence. The effects of these levels of dependence are mathematically calculated in the HRAET/OAT.

As a more general point, human dependent failures can be extremely important in any human reliability assessment, and care should be taken to

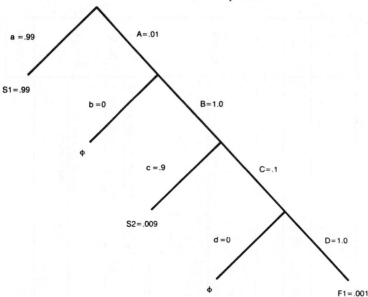

A = FAILURE TO SET UP TEST EQUIPMENT PROPERLY

B = FAILURE TO DETECT MISCALIBRATION FOR FIRST SETPOINT

C = FAILURE TO DETECT MISCALIBRATION FOR SECOND SETPOINT

D = FAILURE TO DETECT MISCALIBRATION FOR THIRD SETPOINT

φ = NULL PATH

Figure 31.11. HRA event tree of hypothetical calibration task (adapted from Swain and Guttmann, 1983)

identify these where possible (e.g., if an operator might become incapacitated, or a misdiagnosis might occur, or if production pressures might totally over-rule safety considerations).

Advantages and disadvantages of SLIM, HEART, and THERP

SLIM-MAUD is a highly structured approach to the use of expert opinion, and often has high credibility with the experts taking part. The computerized version enables the user to investigate how improvements, in particular PSFs (e.g., interface design) can affect the HEPs, allowing the cost effectiveness of error reduction strategies to be investigated (see Impact assessment). However, SLIM relies on 'experts' who may be difficult to find and verify as experts, and on calibration data (at least two known HEPs). SLIM is an exhaustive technique and hence can use up a relatively large amount of personnel resources and time.

HEART is one of the quickest techniques available, and does not require experts, the EPCs instead being based on extensive analysis of the human

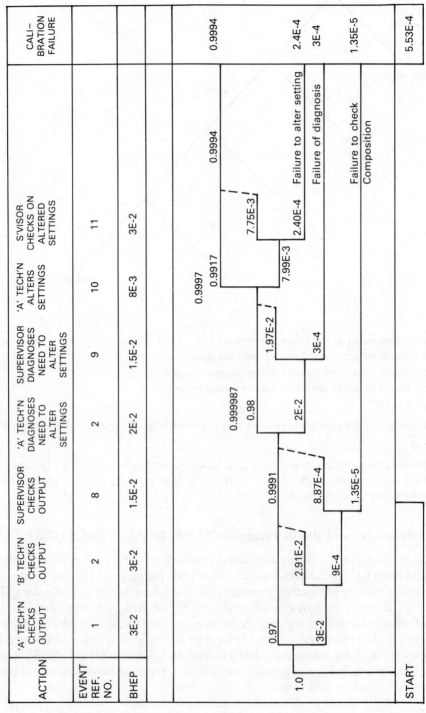

OPERATOR ACTION TREE

ACTION	'A' TECH'N CHECKS OUTPUT	'B' TECH'N CHECKS OUTPUT	SUPERVISOR CHECKS OUTPUT	'A' TECH'N DIAGNOSES NEED TO ALTER SETTINGS	SUPERVISOR DIAGNOSES NEED TO ALTER SETTINGS	'A' TECH'N ALTERS SETTINGS	S'VISOR CHECKS ON ALTERED SETTINGS	CALI-BRATION FAILURE
EVENT REF. NO.	1	2	8	2	9	10	11	
BHEP	3E-2	3E-2	1.5E-2	2E-2	1.5E-2	8E-3	3E-2	

Figure 31.12. Supervisory check on altered settings (Whittingham, 1988)

EXTERNAL PSFs		STRESSOR PSFs	INTERNAL PSFs
SITUATIONAL CHARACTERISTICS	TASK AND EQUIPMENT CHARACTERISTICS:	PSYCHOLOGICAL STRESSORS:	ORGANISMIC FACTORS:
THOSE PSFs GENERAL TO ONE OR MORE JOBS IN A WORK SITUATION	THOSE PSFs SPECIFIC TO TASKS IN A JOB	PSFs WHICH DIRECTLY AFFECT MENTAL STRESS	CHARACTERISTICS OF PEOPLE RESULTING FROM INTERNAL AND EXTERNAL INFLUENCES
ARCHITECTURAL FEATURES	PERCEPTUAL REQUIREMENTS	SUDDENNESS OF ONSET	PREVIOUS TRAINING/EXPERIENCE
QUALITY OF ENVIRONMENT: TEMPERATURE HUMIDITY. AIR QUALITY, AND RADIATION	MOTOR REQUIREMENTS (SPEED, STRENGTH, PRECISION)	DURATION OF STRESS	STATE OF CURRENT PRACTICE OR SKILL
LIGHTING	CONTROL-DISPLAY RELATIONSHIPS	TASK SPEED	PERSONALITY AND INTELLIGENCE VARIABLES
NOISE AND VIBRATION	ANTICIPATORY REQUIREMENTS	TASK LOAD	MOTIVATION AND ATTITUDES
DEGREE OF GENERAL CLEANLINESS	INTERPRETATION	HIGH JEOPARDY RISK	EMOTIONAL STATE
WORK HOURS/WORK BREAKS	DECISION MAKING	THREATS (OF FAILURE LOSS OF JOB)	STRESS (MENTAL OR BODILY TENSION)
SHIFT ROTATION	COMPLEXITY (INFORMATION LOAD)	MONOTONOUS, DEGRADING OR MEANINGLESS WORK	KNOWLEDGE OF REQUIRED PERFORMANCE STANDARDS
AVAILABILITY/ADEQUACY OF SPECIAL EQUIPMENT, TOOLS, AND SUPPLIES	NARROWNESS OF TASK	LONG, UNEVENTFUL VIGILANCE PERIODS	SEX DIFFERENCES
MANNING PARAMETERS	FREQUENCY AND REPETITIVENESS	CONFLICTS OF MOTIVES ABOUT JOB PERFORMANCE	PHYSICAL CONDITION
ORGANISATIONAL STRUCTURE (eg AUTHORITY, RESPONSIBILITY, COMMUNICATION CHANNELS)	TASK CRITICALITY	REINFORCEMENT ABSENT OR NEGATIVE	ATTITUDES BASED ON INFLUENCE OF FAMILY AND OTHER OUTSIDE PERSONS OR AGENCIES
ACTIONS BY SUPERVISORS CO-WORKERS, UNION REPRESENTATIVES, AND REGULATORY PERSONNEL	LONG AND SHORT TERM MEMORY	SENSORY DEPRIVATION	GROUP IDENTIFICATIONS
REWARDS, RECOGNITION, BENEFITS	CALCULATIONAL REQUIREMENTS	DISTRACTIONS (NOISE, GLARE, MOVEMENT FLICKER, COLOUR)	
	FEEDBACK (KNOWLEDGE OF RESULTS)	INCONSISTENT CUEING	
	DYNAMIC VS. STEP-BY-STEP ACTIVITIES		
	TEAM STRUCTURE AND COMMUNICATION	PHYSIOLOGICAL STRESSORS:	
	MAN-MACHINE INTERFACE FACTORS: DESIGN OF PRIME EQUIPMENT, TEST EQUIPMENT, MANUFACTURING EQUIPMENT, JOB AIDS, TOOLS FIXTURES	PSFs WHICH DIRECTLY AFFECT PHYSICAL STRESS	
JOB AND TASK INSTRUCTIONS:		DURATION OF STRESS	
SINGLE MOST IMPORTANT TOOL FOR MOST TASKS		FATIGUE	
PROCEDURE REQUIRED (WRITTEN OR NOT WRITTEN)		PAIN OR DISCOMFORT	
WRITTEN OR ORAL COMMUNICATIONS		HUNGER OR THIRST	
CAUTIONS AND WARNINGS		TEMPERATURE EXTREMES	
WORK METHODS		RADIATION	
PLANT POLICIES (SHOP PRACTICES)		G-FORCE EXTREMES	
		ATMOSPHERIC PRESSURE EXTREMES	
		OXYGEN INSUFFICIENCY	
		VIBRATION	
		MOVEMENT CONSTRICTION	
		LACK OF PHYSICAL EXERCISE	
		DISRUPTION OF CIRCADIAN RHYTHM	

performance literature. It also offers means of determining error reduction strategies and investigating the cost effectiveness of such strategies. However it does not consider possible interactions between the various EPCs, and the nominal HEPs have not yet been validated. Many of the EPCs can only be assessed for an existing plant, and not for a plant being proposed or developed (see Williams, 1992).

THERP is similar in appearance to conventional reliability assessment approaches, yet it can bring PSFs into the equation, explicitly models human errors and error recovery, and considers dependency between errors. However, THERP also requires a good deal of resources, albeit in terms of a single assessor rather than a panel of experts, and often two different assessors may carry out THERP in rather different ways. There is no substantiation of the THERP database, and detailed assessments, due to the fine level of description of the human errors in THERP analysis, can become quite complex.

The above are some of the basic pros and cons of the three techniques; for a more formal qualitative analysis of them, see Kirwan *et al.* (1988).

Selection of an appropriate quantification technique

Reviews by Kirwan *et al.* (1988) and Kirwan (1988) have attempted to aid the process of selecting which quantification techniques to use in a particular application. The first review qualitatively analyzed the eight techniques mentioned earlier, drawing conclusions from the published literature. It assessed the techniques against a set of criteria, namely:

Accuracy: numerical accuracy (in comparison with known HEPs); consistency between experts and assessors.

Validity: use of ergonomics factors/PSF to aid quantification; theoretical basis in ergonomics, psychology; empirical validity; validity as perceived by assessors, experts, etc.; comparative validity (comparing results of one technique with results of another for the same scenario).

Usefulness: qualitative usefulness in determining error reduction mechanisms; sensitivity analysis capability, allowing assessment of effects on HEPs of error reduction mechanisms.

Effective use of resources: equipment and personnel requirements; data requirements, e.g., SLIM and PC require calibration data (at least two known HEPs); training requirements of assessors and/or experts.

Acceptability: to regulatory bodies; to the scientific community; to assessors; auditability of the quantitative assessment.

Maturity: current maturity; development potential.

The eight techniques are assessed against these criteria in Table 31.8, and a second selection guideline table based on perceived user requirements is

Table 31.8. Summary of evaluation of techniques (adapted from Kirwan et al., 1988)

	APJ	PC	TESEO	THERP	HEART	IDA	SLIM	HCR
Accuracy	Moderate	Moderate	(Low)†	Moderate	Moderate	(Low)	Moderate	(Low)
Validity	Moderate/high	Moderate	Low	Moderate	Moderate	Moderate	Moderate	Low
Usefulness	Moderate/high	Low/moderate	Moderate/high	Moderate	High	Moderate/high	High	Low/moderate
Effective use of resources*	Moderate	Low/moderate	High	Low/moderate	High	Low/moderate	Low/moderate	Moderate
Acceptability	Moderate	Moderate/high	Low	High	Moderate/High	(Moderate)	Moderate/high	Low/moderate
Maturity	High	Moderate	Low	High	Moderate	Low/moderate	Moderate/high	Low

* A rating of high on this criterion means the technique is favourable with respect to effective use of resources (i.e. resource requirements are low), and vice versa.
† Ratings in parentheses were based on the subjective opinions of the authors and contributors to the document, since insufficient empirical evidence was available to justify a rating from applications alone.
Note: This table was constructed prior to the experimental study assessing accuracy (Kirwan, 1988).

presented in Table 31.9. With these two tables it is possible for the user to decide which technique or set of techniques to utilize.

A second, experimental review (Kirwan, 1988) looked particularly at the accuracy of five of the techniques (APJ, PC, SLIM, HEART, THERP), comparing their predictions against some known data points (e.g., see Figure 31.14), and comparing their predictions for a set of realistic PSA scenarios. The results showed APJ and THERP to be accurate techniques, with HEART exhibiting some degree of accuracy. The results for SLIM and PC were less clear since whilst they did not perform well in this experiment, this could have been due to the resources limitations of the experiment rather than the technique. It is likely that further validation studies will be carried out in the future, to narrow the most acceptable techniques down to two or three.

(7) Impact assessment

Once human error probabilities have been quantified, the system risk, or reliability, can be calculated and compared to an acceptable level to see if improvement is necessary. If so, it will first be necessary to determine whether human error was a major contributor to inadequate system performance. This involves carrying out an analysis of the importance of each 'event' in the fault tree, to see which events (quantitatively) most affect the predicted top event (accident) frequency. Many fault tree computer analysis packages will automatically determine the most important events.

If human errors do not significantly affect the predicted frequency of the event, then reduction of risk will be best effected by hardware/software improvements to the system. Otherwise, and particularly if human error

Table 31.9. Selection matrix (from Kirwan *et al.*, 1988)

	APJ	PC	TESEO	THERP	HEART	IDA	SLIM	HCR
Is the technique applicable for:								
Simple and proceduralized tasks?	Y	Y	Y	Y	Y	Y	Y	N
Knowledge-based, abnormal tasks?	Y	Y	Y	N	Y	Y	Y	Y
Misdiagnosis which makes a situation worse?	Y	Y	N	N	N	Y	Y	N
Are qualitative recommendations possible?	Y	N	Y	Y	Y	Y	Y	Y
Is sensitivity analysis possible?	N	N	Y	Y	Y	Y	Y	Y
Does the technique have:								
Requirements for calibration data?	N	Y	N	N	N	N	Y	N
Requirements for experts (judges)?	Y	Y	N	N	N	Y	Y	N

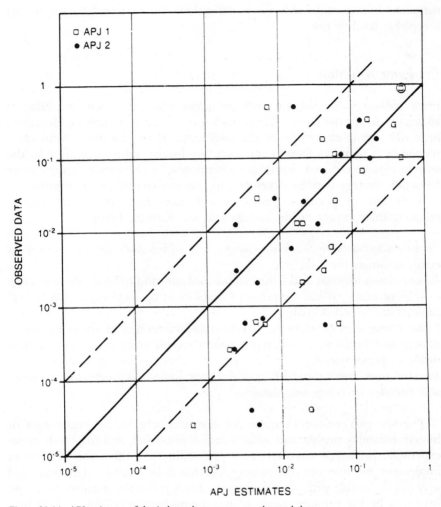

Figure 31.14. APJ estimates of the independent groups vs. observed data

'dominates' the undesired event frequency, error reduction mechanisms should be investigated. Identification of the major human errors will enable the most effective error reduction mechanisms to be specified in the next phase. The positive effects of these mechanisms can be calculated and factored back into the quantitative analysis until system performance reaches an acceptable level. Similar calculations and 'sensitivity analysis' can also be carried out for event-tree-based assessments.

If human error cannot be reduced to an acceptable level, even with additional hardware recommendations, then significant redesign of the system and/or its operation will be required. Usually however, an effective combination of human and hardware modifications can be found to achieve an acceptable level of risk.

(8) Error reduction

Error reduction mechanisms will be unnecesssary if human reliability is adequate, or not the most effective means of achieving system performance, or if not within the scope of the assessment. If on the other hand error reduction is required then either particular human errors identified in the analysis may be reduced in impact or frequency, or else a more general error reduction strategy may be developed to improve overall task performance.

In the case of specific identified critical errors, there are several ways of reducing their impact on the system (see also Kirwan, 1990):

Prevention by hardware or software changes: Use of interlock devices to prevent error; automate the task, etc.
Increase system tolerance: make the system hardware and software more flexible or self-correcting to allow a greater variability in operator inputs which will achieve the intended goal.
Enhance error recovery: enhance detection and correction of errors by means of increased feedback, checking procedures, supervision and automatic monitoring of performance.
Error reduction at source: reduction of errors by improved procedures, training, and interface or equipment design.

The first two measures require collaboration between the ergonomist or human reliability analyst and system design personnel, and may well prove expensive. Improved error recovery probabilities are often the simplest to implement, but may not reduce error likelihoods by a sufficient amount, and may not be feasible with all critical errors. Error reduction at source therefore may well be the primary means of improving human reliability. This may require consultation with a human factors specialist, although some quantification techniques (e.g., HEART) do prescribe specific error reduction mechanisms for identified errors. Error reduction measures can also be identified by considering the error causes and the PSFs identified in the human error analysis phase (e.g., via PHECA or a comparable system). Such measures should be effective as they are aimed at the root causes of the human error.

A second additional analysis of the results will be possible only with quantification methods which use a structured performance shaping factor (PSF) approach (e.g., SLIM, IDA, HEART, THERP, TESEO). With these approaches it is possible to determine the contributions of individual PSFs to human error goals (see the earlier HEART worked example). For example, the most significant PSF in a particular scenario may be 'quality of procedures' and, therefore, error reduction measures aimed at improving the quality of procedures will be most effective at reducing error likelihood. The potential reduction however, can actually be calculated since with these methods error probability is a direct function of the PSFs. Thus it may be that improving the quality of procedures for a particular error makes the error ten times less likely in a particular assessment, which may render the error probability 'acceptable'. If however, the error probability is still not low enough, then clearly other additional PSFs should be investigated. Furthermore, if for example quality of procedures is the most important PSF for a number of human errors, this then suggests that a single global error reduction strategy generally to enhance performance can be specified. This type of investigation of the results will enable the cost effectiveness of potential error reduction strategies to be assed. One recent technique which allows all the above types of error reduction to be carried out is the Human Reliability Management System (HRMS: described in Kirwan and James, 1989; Kirwan, 1990).

If an overall improvement in performance is required, then the principal PSFs identified in the task analysis, human error analysis, and quantification phases should be addressed. It may be, for example, that training was continually cited as a prominent factor affecting performance in a particular process plant emergency scenario, and in fact was rated the most important factor when quantification took place. Clearly, focusing on training in this particular scenario should have an overall positive effect on performance, and may be the best single way of reducing the impact of human error on the system goals. Some case studies of PSAs, including the derivation of error reduction strategies, are given in Kirwan *et al.* (1988).

Where possible, the positive effects of the error reduction strategy adopted should be factored back into the quantitative analysis and it should be checked that the HEPs and overall system risk calculated become acceptable. This requires not only the precise operational definition of each aspect of the error reduction strategy, but also a method of ensuring that the strategy is properly implemented and maintained throughout the remaining life of the plant. This is part of the quality assurance phase.

(9) Quality assurance

If the assessment has generated certain error reduction mechanisms the implementation and effectiveness of these should be ensured. For example, if as a result of an analysis equipment design recommendations are accepted, the successful implementation of the design changes should be monitored, and if possible the effects on performance verified at a later stage.

A continual performance monitoring system represents a powerful quality assurance system. A risk assessment generally predicts reliability for many years ahead, and much can change in this time: a gradual degradation in performance standards; an increasing maintenance loading; the loss of personnel involved in the design; impromptu changes during commissioning and start-up of the plant; and increasing 'retrofit' changes to the plant. Due to such factors the plant risk level may slowly and almost imperceptibly rise above the initial predicted and acceptable level to one which is unacceptable. Long-term performance monitoring (i.e., recording human errors, near-misses, incidents, and accidents) will detect this. A performance monitoring system can identify the point in time at which the assumptions and results of human reliability analysis are no longer applicable, and signify the need for a further analysis to justify the acceptability of the risk of the plant. This type of quality assurance system can avoid the gradual erosion of the safety barriers (the Bhopal syndrome).

(10) Documentation

It is important for the auditability and justifiability of the results of the study that it is fully documented. This will enable personnel unconnected with the assessment to understand how it was carried out, including all assumptions and judgements made during the study, and allow the study to be independently examined, updated, or even reproduced if necessary. It will also provide a potentially useful database from which to monitor future progress of the system's performance in comparison to the predictions made in the analysis. It is also entirely possible that at some stage in the future an accident investigation may take recourse to reviewing the predictions of an earlier safety assessment, to assess why the PSA failed to predict the occurrence of the accident which has now occurred. This perhaps would at least allow HRA to learn from its own mistakes, should they occur.

Future directions in HRA

Three areas are noted below in which human unreliability can affect system risk, but which are not currently adequately addressed by HRA techniques. These are therefore predicted to be the most likely future directions for human reliability developments.

Low technology risk

The HRA field has mainly concerned itself with the high risk, high technology industry sector, including nuclear power plants and chemical plants. There exist however, a large number of other lower technology sectors, e.g., mining, which often incur a high risk via a large number of 'small' accidents (say one or two fatalities), rather than (high risk technology) industries where the high risk is caused by a very small probability of an accident with many and serious consequences. This is clearly an area where applied human reliability should be able to help reduce risk. Some workers have already begun to use HRA approaches in such areas (Collier and Graves, 1986).

Cognitive errors and misdiagnosis

If the operating crew in a plant misdiagnose a situation, they may be slow to realize they have made a 'mistake' or incorrect diagnosis. They may reinterpret system feedback within their mental view of what they think the problem is, rather than what it actually is (also called 'mind set'). This is in fact what occurred during the Three Mile Island accident in 1979, and can result in the human operators actually defeating safety systems and making matters worse than if they did nothing at all. Also there is evidence to suggest that there is a low probability of operators recovering such mistakes once made (Woods, 1984).

The 'modelling' and analysis of cognitive errors is a difficult problem area, and one that has not yet been resolved (Woods and Roth, 1986). Some work has been carried out on the psychology underlying cognitive errors in an attempt to predict the forms cognitive errors are likely to take (Reason and Embrey, 1986). The Influence Modelling and Assessment System (IMAS) can be used to elicit mental maps of operators' perceptions of causal relationships in nuclear power plant operation (see Embrey, 1986a). These maps could be used to consider what misdiagnoses may occur. Also, in the Artificial Intelligence field, an attempt is being made to simulate the cognitive operator with an expert system, and estimate what errors will occur (Woods *et al.*, 1987). These and other developments may eventually solve the difficult problem of cognitive errors.

Management, organizational and sociotechnical contributions to risk

Three of the most salient accidents emphasizing the impact of human error are the Bhopal (1984), Chernobyl (1986), and Challenger Space Shuttle (1986) disasters. Each of these contained a significant human error contribution, without which the disasters would not have occurred. However, it is arguable that current HRA techniques would have had difficulty in predicting them, because the fundamental types of error leading to the accidents were neither procedural nor diagnostic in the conventional sense (Watson and Oakes, 1988). Rather, all three involved a management and organizational error component. The Bhopal plant was possibly subject to economic pressures, resulting in management decisions which may have been more in favour of production than safety (Bellamy, 1986). The Challenger Space Shuttle disaster also appeared to contain a significant element of management-type decision error, also influenced by economic considerations (Roger *et al.*, 1986). Lastly, the Chernobyl disaster (USSR State Committee, 1986) contained such a bizarre series of events affected by management that, had an analysis predicted this scenario prior to the accident, it possibly would have been treated with derision. Nevertheless, even such accidents as Chernobyl can be rationalized after the event, and indeed similar sequences can and have occurred (Reason, 1988b).

Errors may also occur when dealing with small groups, at a 'lower' management and organizational level and caused by individual, organizational, and sociotechnical factors (e.g., social pressures, personality conflicts; Bellamy, 1983). These factors and the errors they cause are rarely assessed in HRAs, yet clearly can influence risks; this may happen via risk-taking: decisions to ignore procedures; failure to communicate information due to personality conflicts, etc. Techniques could therefore be developed in the future which identify, quantify and specify how to reduce these types of error.

It is clear that HRA must eventually consider not only the operators in the control room, but also the management running the plant, since the latter can have a profound effect on safety.

In broad terms HRA is becoming an integral part of risk assesments and its usage is steadily increasing. In the future it would seem likely that this will continue and the 'science' of HRA will become an accepted tool for use in many areas, not only those dominated by the need to assess high risk systems.

Summary

This chapter has overviewed the field of human reliability assessment as a tool for reducing the risk of large scale accidents, although it has other uses (e.g., in improving productivity). A broad generic methodology has been outlined, detailing the various steps in HRA and its interactions with PSA. The more prominent methodologies and issues in HRA have been outlined. From the discussions in this chapter it is apparent that HRA is itself not yet a mature science, and has some way to go in the development of sound and theoretically and empirically valid methodologies. Nevertheless, a great deal can still be achieved with the existing approaches, and current research is progressing possibly faster than at any other time in HRA's history. The usefulness and veracity of HRA will, however, ultimately be proven one way or the other by the incidence of human-error induced large scale accidents in future years.

References

Apostolakis, G.B., Beer, V.M. and Mosleh, A. (1987). A critique of recent models for human error rate assessment. Paper presented at the *International POST-SMIRT 9th Seminar on Accident Sequence Modelling*, Munich, Federal Republic of Germany. Reprinted in *Reliability Engineering and System Safety* (1988), **22**, 201–217.

Bell, B.J. (1984). Human reliability analysis for probabilistic risk assessment. *Proceedings of the 1984 International Conference on Occupational Ergonomics*, pp. 35–40.

Bellamy, L.J. (1983). Neglected individual, social and organisational factors in human reliability assessment. *Proceedings of the 4th National Reliability Conference*, Reliability 85, NEC, Birmingham.

Bellamy, L.J. (1986). The safety management factor: an analysis of human aspects of the Bhopal disaster. Paper presented at the *1986 Safety and Reliability Symposium*, Southport.

Bellamy, L.J., Kirwan, B. and Cox, R.A. (1986). Incorporating human reliability into probabilistic risk assessment. Paper presented at *5th International Symposium in Loss Prevention and Safety Promotion in the Process Industries*, Société de Chimie Industriale.

Bello, G.C. and Columbari, V. (1980). The human factors in risk analysis of process plants: the control room operator model, TESOC. *Reliability Engineering*, **1**, 3–14.

Collier, S.G. and Graves, R.J. (1986). Improving human reliability: practical ergonomics for design engineers. In *Proceedings of the 9th Advances in Reliability Technology Symposium*, University of Bradford.

Comer, M.K., Seaver, D.A., Stillwell, W.G. and Gaddy, C.D. (1984). *Generating Human Reliability Estimates Using Expert Judgement*. USNRC Report Nureg/CR-3688 (Washington, DC: USNRC).

Drury, C.G. (1983). Task analysis methods in industry. *Applied Ergonomics*, **14**, 19–28.

Edwards, W. (1977). How to use multi-attribute utility measurement for social decision-making. *IEEE Transactions on Systems, Man, and Cybernetics*, **SMC-7-5**.

Embrey, D.E. (1986a). Approaches to aiding and training operator's diagnosis in abnormal situations. *Chemistry and Industry*, **7**, 454–459.

Embrey, D.E. (1986b). SHERPA: a systematic human error reduction and prediction approach. Paper presented at the *International Topical Meeting on Advances in Human Factors in Nuclear Power Systems*, Knoxville, Tennessee.

Embrey, D.E. (1987). Error Analysis in Proceduralised Tasks. Material presented in the *Human Reliability Analysis Course*, SRD, UKAEA, Culcheth, Cheshire.

Embrey, D.E. and Kirwan, B. (1983). A comparative evaluation study of three subjective human reliability qualification techniques. In *Proceedings of the Ergonomics Society's Conference 1983*, edited by K. Coombes (London: Taylor and Francis), pp. 137–141.

Embrey, D.E., Humphreys, P., Rosa, E.A., Kirwan, B. and Rea, K. (1984). *SLIM-MAUD: An Approach to Assessing Human Error Probabilities Using Structured Expert Judgement*. USNRC Report Nureg/CR-3518 (Washington, DC: USNRC).

Green, A.C. (1983). *Safety Systems Reliability* (Chichester: John Wiley).

Hall, R., Fragola, J. and Wreathall, J. (1982). *Post Event Human Decision Errors: Operator Action Tree/Time Reliability Correlation*. USNRC Report Nureg/CR-3010 (Washington, DC: USNRC).

Henley, E.J. and Kumamoto, H. (1981). *Reliability Engineering and Risk Assessment* (New Jersey: Prentice Hall).

Hunns, D.M. (1982). The method of paired comparisons. In *High Risk Safety Technology*, edited by A.E. Green (Chichester: John Wiley).

Hunns, D.M. and Daniels, B.K. (1980). The method of paired comparisons. In *Proceedings of the 6th Symposium on Advances in Reliability Technology*, Report NCSR R23 and R24. NCSR, UKAEA, Culcheth, Cheshire.

Kirwan, B. (1986). *Techniques to Aid Human Error Identification in Human Reliability Assessment, Human Reliability Course Manual* (London: IBC Technical Services Limited).

Kirwan, B. (1987). Human reliability analysis of offshore emergency blow-down system. *Applied Ergonomics*, **18**, 23–34.

Kirwan, B. (1988). A comparative evaluation of five human reliability assess-ment techniques. In *Human Factors and Decision Making*, edited by B.A. Sayers (Oxford: Elsevier), pp. 87–104.

Kirwan, B. (1990). A resources flexible approach to human reliability assess-ment for PRA, Safety and Reliability Symposium, Altrincham, Sep-tember. London: Elsevier Applied Sciences, pp. 114–135.

Kirwan, B. (1992a). Human error identification in human reliability assess-ment. Part 1: overview of approaches. *Applied Ergonomics*, **23**, 5, 299–318.

Kirwan, B. (1992b). Human error identification in HRA. Part 2: Detailed comparison of techniques. *Applied Ergonomics*, **23**, 6, 371–381.

Kirwan, B. (1994). *A Guide to Practical Human Reliability Assessment* (London: Taylor and Francis).

Kirwan, B. and Ainsworth, L.A. (Eds) (1992). *A Guide to Task Analysis* (London: Taylor and Francis).

Kirwan, B. and James, N.J. (1989). The development of a human reliability assessment system for the management of human error in complex sys-tems, in *Reliability 89*, Brighton Metropole, June, pp 5/A/2.1– 5/A/2.10.

Kirwan, B. and Rea, K. (1986). Assessing the human contribution to risk in hazardous materials handling operations. Paper presented at the *First International Conference in Risk Assessment of Chemicals and Nuclear Materials*, Surrey University.

Kirwan, B., Embrey, D.E. and Rea, K. (1988). *The Human Reliability Assessor's Guide*. Report RTS 88/95Q. NCSR, UKAEA, Culcheth, Cheshire.

Kirwan, B., Martin, B.R., Rycraft, H. and Smith, A. (1990). Human error data collection and data generation. *International Journal of Quality and Reliability Management*, **7**, 4, pp. 34–66.

Kletz, T. (1984). *HAZOP and HAZAN—Notes on the Identification and Assess-ment of Hazards* (Rugby: Institute of Chemical Engineers).

Ludke, R.L., Strauss, F.F. and Gustafson, D.H. (1977). Comparison of five methods for estimating subjective probability distributions. *Organis-ational Behaviour and Human Performance*, **19**, 162–179.

Murphy, A.H. and Winkler, R.L. (1974). Credible interval temperature fore-casting: some experimental results. *Monthly Weather Review*, **102**, 784–794.

Nicks, R. (1981). Probabilistic approach to problems in reactor safety risk assessment. Paper presented at *Convegno Internazionale sul Fondamenti della Probabilita e della Statistica*, Luino, Italy.

Park, K.S. (1987). *Human Reliability* (Oxford: Elsevier).

Pew, R.W., Miller, D.C. and Feehrer, C.G. (1987). *Evaluation of Proposed Control Room Improvements Through Analysis of Critical Operator Decisions* (Palo Alto, CA: Electric Power Research Institute).

Phillips, L.D., Humphreys, P. and Embrey, D.E. (1983). *A Socio-technical Approach to Assessing Human Reliability.* Technical Report 83–4. London School of Economics, Decision Analysis Unit.

Pontecorvo, A.B. (1965). A method of predicting human reliability. *Annals of Reliability and Maintenance*, **4**, 337–342.

Potash, L., Stewart, M., Dietz, P.E., Lewis, C.M. and Dougherty, G. (1981). Experience in integrating the operator contribution in the PRA of actions operating plants. In *Proceedings of the ANS/ENS Topical Meeting on PRA*, New York, American Nuclear Society.

Rasmussen, J., Pederson, O.M., Carnino, A., Griffon, M., Mancini, C. and Gagnolet, P. (1981). *Classification System for Reporting Events Involving Human Malfunctions.* Report Riso-M-2240, DK-4000 (Roskilde, Denmark: Riso National Laboratories).

Reason, J. (1988a). Errors and violations: the lessons of Chernobyl. Paper presented at the *IEEE Conference on Human Factors in Nuclear Power*, Monterey, CA.

Reason, J. (1988b). Human fallibility. *Consortium of Local Authorities Proof of Evidence in the UK Hinkley 'C' Power Station Enquiry*, COLA 22.

Reason, J.T. (1990). *Human Error* (Cambridge: Cambridge University Press.).

Reason, J.T. and Embrey, D.G. (1986). *Human Factors Principles Relevant to the Modelling of Human Errors in Abnormal Conditions of Nuclear and Major Hazardous Installations.* Report for the European Atomic Energy Community (Lancs: Human Reliability Associates Ltd).

Rogers, W.P. et al. (1986). *Report of the Presidential Commission on the Space Shuttle Challenger Accident*, June 6.

Seaver, D.A. and Stillwell, W.G. (1983). *Procedures for Using Expert Judgement to Estimate Human Error Probabilities in Nuclear Power Plant Operations.* Nureg/CR-2743. (Washington, DC: USNRC).

Shepherd, A. (1986). Issues in the training of process operators. *International Journal of Industrial Ergonomics*, **1**, 49–64.

Spurgin, A.J., Lydell, B.O., Hannaman, G.W. and Lukic, Y. (1987). Human reliability assessment—a systematic approach. In *Reliability 87*, NEC, Birmingham.

Swain, A.D. (1989). Comparative evaluation of methods for human reliability analysis, GRS-71, Gesellschaft für Reaktor Sicherheit, GRSmbH, Köln, Germany.

Swain, A.D. and Guttmann, H.E. (1983). *A Handbook of Human Reliability Analysis with Emphasis on Nuclear Power Plant Applications.* Nureg/CR-1278 (Washington, DC: USNRC).

USNRC (1980). *Three Mile Island: A Report to the Commissioners and to the Public (The Rogovin Report)*, USNRC Report Nureg/CR-1250-V. (Washington, DC: USNRC).

USNRC (1984). *Seabrook Station Probabilistic Safety Analysis. Section 10: Human Actions Analysis.* (Washington, DC: USNRC).

USNRC (1985). *Loss of Main and Auxiliary Feedwater at the Davis–Beese Plant on June 9, 1985.* Nureg-1154. (Springfield, VA: National Technical Information Service).

USSR State Committee (1986). *The Utilisation of Atomic Energy: The Accident at the Chernobyl Nuclear Power Plant and Its Consequences.* Information compiled for the IAEA Experts' meeting, 19–25 August, Vienna (Part 1).

Wagenaar, W. (1986). *The Cause of Impossible Accidents.* Speech delivered on the occasion of the 6th Dujiker Lecture, 18 March, Netherlands Institute of Psychologists, University of Amsterdam.

Watson, I. and Oakes, F. (1988). *Human Reliability Factors in Technology Management.* SRD Report, UKAEA, Culcheth, Cheshire.

Whalley, S.P. (1988). Minimising the cause of human error. In *10th Advances in Reliability Technology Symposium,* edited by G.P. Libberton (London: Elsevier).

Whittingham, R.B. (1988). The design of operating procedures to meet targets for probabilistic risk criteria using HRA methodology. Paper presented at *IEEE Conference on Human Factors in Nuclear Power,* Monterey, CA, pp. 303–310.

Williams, J.C. (1983). Validation of human reliability assessment techniques. Paper presented at the *4th National Reliability Conference,* NEC, Birmingham.

Williams, J.C. (1986). HEART—a proposed method for assessing and reducing human error. In *Proceedings of the 9th Advances in Reliability Technology Symposium,* University of Bradford.

Williams, J.C. (1988). A data-based method for assessing and reducing human error to improve operational performance. Paper presented at the *IEEE Conference on Human Factors in Nuclear Power,* Monterey, CA, pp. 436–450.

Williams, J.C. (1992). Towards an improved evaluation analysis tool for users of HEART. In *Proceedings of the International Conference on Hazard Identification and Risk Analysis, Human Factors and Human Reliability in Process Safety (Centre for Chemical Process Safety, AIChemE),* January 15–17, Orlando, Florida.

Woods, D.D. (1984). Some results on Operator Performance in Emergency Events. In *Proceedings of the Institute of Chemical Engineers Symposium on Ergonomics Problems in Process Operations,* Symposium Series No. 90 (Rugby: Institute of Chemical Engineers), pp. 21–32.

Woods, D.D. and Roth, E.M. (1986). *Models of Cognitive Behaviour in Nuclear Power Plant Personnel: A Feasibility Study,* Nureg/CR-4532 (Washington, DC: USNRC).

Woods, D.D., Roth, G. and Pople, H. (1987). *An Artificial Intelligence Based Cognitive Model for Human Performance Assessment.* Nureg/CR-4862 (Washington, DC: USNRC).

Chapter 32

Accident reporting and analysis

Ivan D. Brown

Introduction

The nature and frequency of malfunctioning have important implications for the safety, productivity and efficiency of all human work and technological systems. This importance is recognized by the variety of accident reporting procedures which has developed as a means of work system evaluation. Indeed, in certain instances such procedures are a statutory requirement of system operation.

This widespread reporting of accidents is clearly in the interests of both society and the individual worker, given satisfactory interpretation of the data and translation of relevant findings into viable accident countermeasures. For a variety of reasons, including interindividual differences, intra-individual variability, fatigue and stress, it is virtually impossible to design and operate a perfectly safe human-technological system. Accident reporting thus represents the only practical way of evaluating system safety under real operating conditions and of identifying factors which may be contributing to accident causation.

The ubiquitous nature of accident reporting procedures should not, however, be allowed to obscure the essential differences in the way they are used to evaluate different areas of human activity. Good accident reporting systems are purpose designed. The information collected and the way it is sought will reflect an organization's need for that information and its expectancies about the various causes of malfunctioning. As Hale and Hale (1972) point out:

> The task of driving a car safely is very different from most industrial tasks in that it involves different skills, different motivation and a different degree of interaction with other people. Hence factors which could be expected to influence road accidents would not be expected to affect industrial accidents and vice versa.

Such differences between socio-technological systems may appear to be self-evident, yet all too often it is assumed that 'an accident is an accident is an accident'. This narrow view of accident causation has sometimes inhibited progress in accident prevention, at the level of both theory and practice. Hale and Hale (1972) go so far as to 'regard the influence that road accident research has had on industrial accident theories and conclusions as unfortunate in many instances'. This does not mean that it is impossible to develop a general model of human error and accident causation, but it may reasonably be concluded that no single, useful, accident reporting system will have universal applicability across the entire range of human activity.

In practice, the design of accident reporting systems will, in many instances, be compromised by the resources available within an organization to collect data, analyze and interpret them, and translate any important findings into viable accident countermeasures. Such compromises may be presented as an inescapable fact of working life. However, they must never be allowed to deflect attention from the fact that accident reporting is a means to an end, not an end in itself. Ease of data collection is of little importance if it compromises validity and renders data uninterpretable. It may appear more acceptable to compromise on the type of accident data collected in order to maximize reliability, but even this could have serious implications for the analysis of the data in question and its application to countermeasure design.

The rest of this chapter will discuss these and other aspects of accident reporting in more detail and consider ways in which it may be used more effectively to evaluate and improve human work and technological systems.

Definition of an 'accident'

Defining the term 'accident' would appear to be a prerequisite for the design and use of any accident reporting procedure. However, this is often not done explicitly, but is implied by the criteria used for categorizing those events which are to be reported as 'accidents'. Even where an explicit definition is offered, it may simply describe the subset of behavioural *outcomes* which the reporting procedure can record with an acceptable level of confidence, rather than describing the nature of system *malfunctioning* itself. For example, the British road accident recording procedure (Stats 19), like many other such procedures, defines an 'accident' as:

> One involving personal injury occurring on the public highway (including footpaths) in which a road vehicle is involved and which becomes known to the police within 30 days of its occurrence. The vehicle need not be moving and it need not be in collision with anything. (Department of Transport, 1991, p. 3)

Thus behavioural outcomes which result only in vehicle or other property damage are excluded from this 'accident' reporting system largely to maximize the reliability of the available data and ease the burden of collecting and analyzing them. However, such restricted definitions can have serious implications for the analysis and interpretation of accident *causation* and thus for the design of appropriate accident countermeasures, as will be discussed later.

In addition to these adverse effects of expediency, definitions of an 'accident' are biased by the specific interests of professional groups working in the fields of accident causation, prevention and treatment. The vast majority of accidents are probably multicausal, yet engineers and physicists will be interested mainly in technological malfunctioning and its rectification; behavioural scientists will be interested mainly in causes of human error and its prevention or reduction; physicians will be interested in injury patterns and ways of preventing or treating injury. Thus Shannon and Manning (1980), for example, classify accidents in terms of uncontrolled transfers of energy that result in injury, whilst Farmer and Chambers (1926) asserted that, 'from a psychological point of view an accident is merely a failure to act correctly in a given situation'. In their terms, an 'accident' is simply equated with erroneous behaviour. By contrast, medical usage has equated 'accident' with 'injury', which may confuse the interpretation of road accident statistics, for example, where one 'accident' can give rise to several 'casualties' (see Department of Transport, 1991, p. 3).

Because accident research and safety efforts cross a number of disciplinary boundaries, there is a need for an agreed definition of the term 'accident', which minimizes such confusion and maximizes communication among the many professional groups contributing to research and practice in this field. Unfortunately, lay language and dictionary definitions tend to equate 'accident' with 'chance' or 'luck', which appears to deny any attempt to understand or deal with the causes of accidents. For example, dictionary definitions include: 'arising from unknown or remote causes'; 'unforeseen, unplanned, or unpredictable event'; and 'any fortuitous or non-essential property, fact, or circumstance'. 'Suddenness' is often seen to be an essential characteristic of accidents, although it is clear that many accidents (e.g., those associated with operator fatigue) can result from relatively prolonged insidious developments of discrepancy between actual and perceived circumstances. Other definitions come closer to capturing the essential features which accident reporting procedures aim to identify, such as: 'combination of causes producing an unfortunate result'; or 'injurious consequences'; and, in particular, 'lack of intention'. However, none of these definitions reliably aids the design and use of accident reporting procedures. For this reason it has been argued, e.g., by the police, in relation to road traffic 'accidents', that such behavioural consequences are not random, chance events, but are usually predictable, or at least explainable. Therefore, they suggest, the term 'incident' should replace 'accident'. Whilst this alternative recognizes the small part played by chance or luck in accident *causation*, it does little to produce an informative, useful definition of the antecedent behavioural process which accident reporting procedures aim to record.

It seems preferable to base a definition of the term 'accident' on the fact that most human work involves one or more persons, using 'equipment' of greater or lesser complexity, and an 'environment' (physical or social) within which the work is performed. Each of these three 'main factors' may be solely responsible for malfunctioning of the work system. Alternatively, malfunctioning may be attributable to any of the three first-order interactions or the single second-order interaction between the three main system factors. This ergonomics approach (e.g., see Edwards, 1981) recognizes the multicausal nature of 'accidents', thus providing an opportunity to identify the behavioural antecedents of accidental consequences whilst retaining flexibility in the design and implementation of accident countermeasures. That is to say, an accident which is attributable to first- or second-order interactions between people, equipment and environmental factors may be remediable by treating any one or more of the relevant interface characteristics.

Using this approach, Brown (1976) defined an 'accident' as the 'unplanned outcome of inappropriate behaviour'. He defended this definition by pointing out that:

(a) It draws a clear distinction between antecedent *behaviour* and the *consequence* of that behaviour. By contrast with Farmer and Chambers' (1926) definition, only the *outcome* of the 'erroneous' behaviour is termed 'accidental'.

(b) This accidental outcome is characterized by its *unplanned* nature, rather than by the unpredictability which lay language and certain dictionary definitions have associated with accidents. Many accidental outcomes may in fact be quite predictable in probabilistic terms, even by the actor in question, but they are often assigned a negligible probability of occurrence and then ignored for all practical purposes.

(c) The antecedent behaviour is termed *inappropriate* because it is mismatched to the actual demands of the task or the environment. (Although it may be perceived as appropriate where display faults misrepresent actual task demands.) It may be intentional, but unfortunately unwise behaviour, for example, when a car driver brakes sharply on an icy road. It may be intentional but misdirected or misplaced, for example, when a pilot attempts to land on the wrong runway. Alternatively, it may be unintentional, for example, when the captain of a cross-channel ferry puts to sea unaware that its bow loading doors are still open.

(d) The association between 'accidents' and 'chance', favoured by certain dictionary definitions, has been avoided because it has often blurred the distinction drawn above between antecedent behaviour and accidental outcome. Since identical inappropriate behaviour will not inevitably result in an accident each time it is repeated, it might be claimed that there is a chance relationship between such behaviour and the *occurrence* of accidents. However, there are clearly identifiable categories of accidents resulting from given combinations of inappropriate behaviour when performing specific tasks in known environmental conditions. The actual *nature* of the probable accidental outcome is therefore not simply a chance occurrence and it is the nature, rather than the frequency of occurrence of accidental consequences, which is more informative of the causes of inappropriate behaviours and which it is therefore important to capture in any accident reporting system.

This definition of an 'accident' can now be set within the general theoretical context of accident causation and human error, before considering in more detail the desirable characteristics of an accident reporting procedure and how its data analysis may be translated into effective accident countermeasures.

Theories of accident causation

A critical review of the main theories of accident causation has been provided by Hale and Hale (1972). These theories will be discussed here only from the standpoint of their possible implications for the recording and analysis of accident data.

The 'pure chance' theory

This theory holds that everyone exposed to the same objective risk has an equal liability to accidents, which are thus entirely chance determined. It is virtually impossible to test this theory, empirically, because of the difficulty of finding sufficiently large samples of working populations among which objective risk is actually equated. The theory also appears to beg the question of what constitutes 'objective risk'. Is it simply the risk associated with a particular task performed under given conditions? Or is it the risk associated with a particular level of operator skill when performing the task under given conditions? Clearly, novice and experienced workers would not be expected to have an equal liability to accidents, therefore the theory appears only to discount other individual differences which might contribute to accident causation, such as age, sex, intelligence, personality, temperament and motivation. But does it also discount, say, effects of impairment? As a null hypothesis, the 'pure chance' explanation may be a way of directing attention to accidents as an index of risk associated with particular tasks, working situations, or environmental conditions. But it would be a very uninformative accident reporting system which concentrated on these sources of causation to the exclusion of individual differences in accident liability.

The 'biased liability' theory

This theory holds that an individual's accident involvement either increases or decreases their liability to subsequent involvement. This appears eminently reasonable, since involvement in an accident may well increase apprehension, with its tendency to impair performance, when the circumstances surrounding that accident are perceived to recur. Alternatively, accident involvement may produce a tendency to avoid its attendant circumstances in future, or encourage the victim to improve the skills and knowledge required to perform more safely under those conditions. This theory may serve to explain the unreliability of accident data in predicting an individual's liability to future involvement. However, the theory is unhelpful in specifying the duration of biased liability, or the extent to which this bias will generalize to accidental circumstances similar to and different from those of the initial involvement. It therefore provides little guidance on the type of data which should be collected in any accident reporting system.

The 'unequal initial liability' theory

This theory is probably better known as the theory of 'accident proneness'. For a conceptual review of the theory, see McKenna (1983). There are two versions of the basic theory. One holds that certain individuals are prone to accidents because of their innate personal characteristics. Thus their accident liability is assumed essentially to be a stable feature of their performance, irrespective of task, working conditions, time, or other non-personal factors. The other version, which owes much to the work of Cresswell and Froggatt (1963), holds that accident proneness is a variable factor, being associated with 'critical events' in the life of an individual rather than with situational risks. Both versions are inherently attractive, because they fit the common experience that some people have more accidents than others and that this pattern tends to vary with, for example, age and experience. The fallacies in this line of theoretical argument have been exposed by McKenna (1983), who suggests that it would be advantageous to abandon the concept of 'proneness' and employ a new term such as 'differential accident liability'. This would certainly permit a more flexible approach to the interpretation of accident data, by diverting attention away from the search for 'scapegoats' and towards the identification of contributory situational factors.

The 'stress' theory

This theory holds that accidents happen when a task, environmental, or individual stressor reduces the capacity of an individual to meet task demands, or when the demands of a task increase beyond the normal capacity of an individual to meet them. Example of the former type of stressor would be fatigue, illness, or environmental heat, cold, noise, and so on. Examples of the latter would be increased informational load or work-rate requirements, or demands falling outside an operator's repertoire of skills. Whilst limited in scope, this theory does at least permit accident causation to be explained in terms of intra- and inter-individual differences and also in terms of job, task and situational factors. (See chapter 26 in this book for a review of theories of stress.)

The 'arousal/alertness' theory

This has been developed from research on physiological activation and human behaviour (e.g., see Duffy, 1962). The central hypothesis here is that a relationship exists between an individual's level of arousal/alertness and their performance on any task, the efficiency of which rises to a peak as arousal increases, but then declines as arousal becomes inappropriately high. This became known as the 'inverted-U hypothesis'. It predicts that accidents are more likely to occur both when arousal is low (e.g., when the person is underloaded, bored or drowsy) and also when arousal is high (e.g., when the person is anxious, or excessively motivated). This arousal theory is sometimes confused with stress theory because overarousal, like stress, tends to degrade performance. However, the concepts should be kept distinct because effects of stress are, by definition, harmful, since they represent a reduction of coping ability. Effects of arousal may, or may not be, harmful, depending upon the optimal level of arousal for performance of the task in hand. Thus the theories provide different guidance on the interpretation of accident data and on the design of accident countermeasures. In particular, it is important to avoid conceptualizing 'stress' as a factor which can be employed as a countermeasure against accidents attributed to boredom, monotony, or other sources of underarousal among workers.

'Psychoanalytic' theories

These attribute accidents to subconscious processes with self-punitive aims, initiated by feelings of guilt, anxiety or motivational conflict. Although it is difficult to incorporate such theories into the design of accident reporting systems, clearly they have some guidance to offer when data analysis points to an individual's unexpectedly high level of differential accident liability.

'Epidemiological/ergonomics/situational' theories

These appear to have so much in common that, for all practical purposes, they are indistinguishable. Epidemiology, as its name implies, developed as an approach to the study of epidemics. It holds that causation is essentially a conjunction of a 'host' (the victim of the epidemic), an 'agent' (which transmits the disease) and an 'environment' within which 'host' and 'agent' interact. This seems perfectly analogous to the 'ergonomics', or 'situational' approach to explanations of accidents, in which a person (the 'host') interacts with a tool, or technological system (the 'agent') in a working environment (either physical or social). Both the medical and the ergonomics theories hold that it is the conjunction of all three factors which causes the problem (accident or epidemic) and that a solution can be found only by altering this conjunction. Furthermore, both approaches maintain a flexible view of accident/epidemic reduction, in that countermeasures are seen to result from manipulations of the person (host) and/or the technology (agent) and/or the environment.

This theoretical view of accident causation thus offers the greatest potential for improving safety, by directing attention to possible malfunctioning among all three components in human-technological systems and by keeping open all options for improving their independent and interacting performance characteristics.

The 'domino' theory

This theory has been developed in both industrial and road safety fields in order to explain the sequential, multicausal, nature of accident causation. Hale and Hale (1972) list five stages in this sequential process, attributed to Heinrich (1950). These are:

1. Ancestry and social environment.
2. Individual fault.
3. Unsafe act and/or mechanical hazard.
4. Accident.
5. Injury.

These 'stages', as postulated, tend to confound the sequential with the situational factors contributing to accident production. They encompass the epidemiological/ergonomics view of accident causation, in that they high-light the involvement of people and objects within an environment, but they give more weight to the temporal nature of the accident process. Thus the value of this theory for accident reporting and analysis is that, apart from directing attention to the fact that accidents seldom have a single cause, it alerts us to the importance of, perhaps temporally distant, antecedent behaviour and events (see Leplat, 1990; Malaterre, 1990; Wagenaar and Reason, 1990). In particular, it draws attention to the possibility that certain antecedent behaviours, representing attempts to recover from initial errors, may actually be counterproductive in terms of accident avoidance, rather than simply 'inappropriate', as described in the earlier definition of an accident.

Theories of human error

Accident theories originated in attempts to describe statistical distributions of accidents using 'curve-fitting' exercises which often made unjustified assumptions about the homogeneity and stability of operator characteristics among the working populations in question. The specific nature of the accidents in question and the antecedent behaviour of victims and other involved persons were given little or no consideration. This lack of a conceptual basis for accident theorizing led to years of academic controversy and little of practical value for safety workers. By contrast, approaches to accident reduction via theorizing about human error, whilst not entirely uncontroversial, seem likely to be highly productive of ergonomics measures for improving safety; especially where the approach to explanation of error is made via theory of normal, error-free behaviour.

Reason (1988) has provided a good, succinct, overview of such current models of human performance and error. His account of these models will briefly be discussed here only in relation to the guidance they provide for accident reporting and its analysis.

Shiffrin and Schneider's (1977) theory

This theory of controlled and automatic information processing developed from laboratory studies of search, detection, memory and retrieval. It distinguishes two types of non-sensory memory: a long-term store, the function of which is to retain information permanently but passively, and a short-term store which has a dual function; (a) to provide an activated subset of relevant information from the long-term store and (b) to provide a 'workspace' for conscious decision making and control processes.

This theory highlights the explanatory value of information obtained by introspection among accident-involved individuals, preferably shortly after the event. Given access to such information, the theory directs attention to two aspects of human functioning which are important for accident prevention and/or reduction.

1. Identification of operators' general knowledge of the task in question and the particular subset of that knowledge which was perceived as specifically relevant to the task during the period prior to the accident. Information of this kind has obvious practical applications to operator training and the design of informational displays.

2. Identification of those components of task performance which were under the operator's conscious control during the run-up to the accident and those which were sufficiently automated that they could be run off subconsciously. Again, this type of information could provide guidance on the need for attention to training and/or display design as an approach to accident reduction and prevention. It could also highlight certain aspects of inefficient

work organization which are imposing an unnecessarily high workload on conscious control processes, thus resulting in accidents via stress or distraction.

Broadbent's (1984) 'Maltese Cross' model of memory

This model was developed from studies of interference among concurrent cognitive activities. It distinguishes 'representation', in the form of persisting memory records, from 'processes' of translations between these memory records. Representations, in the form of sensory store, abstract working memory, motor output store and long-term associative store, communicate via a central processing system.

As with Shiffrin and Schneider's theory, the 'representation' component of this model highlights the need to obtain information on accident-involved individuals' task knowledge and the intentions associated with their antecedent behaviour. But, in addition, the 'process' component of the model emphasizes the importance of the sequential nature of operator behaviour in providing an understanding of the error(s) which contributed to the accident. Thus the practical applications of this theory to accident reduction and prevention will be via operator education and training and via work design.

Norman and Shallice's (1980) 'attention to action' model

This is one of several models which have developed from observation of, usually inconsequential, cognitive failures, often in domestic or clinical contexts (e.g., see Reason, 1979; Norman, 1981). The control structure envisaged in this model consists of horizontal 'threads' which govern habitual activities, not under continuous conscious control, and vertical 'threads' representing the attentional processes which govern behaviour in novel or emergency conditions. The model also incorporates motivational factors, which are assumed to act via the vertical 'threads', with longer time constants than the attentional effects.

Thus this model, like the previous two discussed, emphasizes the need for accident reports to include information on the relative extent to which antecedent behaviour was under the operator's conscious control, rather than being automated. It also emphasizes the importance of knowing whether automated antecedent behaviour may have been rendered 'inappropriate' by the persistent activity of some specialized processing structure (or 'schema') beyond the point at which it was no longer required to be active by the task in hand. The inclusion of motivational factors in the model highlights the need for accident reports to include information on intentions of and reasons for antecedent behaviour, so that the 'why', as well as the 'what' and 'how' aspects of accidents can be examined, understood, and dealt with by remedial measures (see Duncan, 1990).

Rasmussen's (1982) 'skills, rules and knowledge' model

This model of behaviour was developed to account for the rather more serious errors and accidents that can occur in complex, industrial, process control operations, particularly the emergency situations that result from breakdown of hazardous nuclear and chemical processes. This three-level model of performance and error developed from investigations of trouble-shooting behaviour using verbal protocols. The information these provided on covert behaviour enabled the model to describe two types of error: 'slips' (unintended actions) and 'mistakes' (inappropriate plans or intentions). Skill-based behaviour is governed by stored patterns of preprogrammed instructions. Rule-based behaviour enables familiar situations to be dealt with by learned production rules (e.g., if 'this' state occurs then implement 'that' remedial action). Knowledge-based behaviour is used to deal with completely novel situations to which no actions have been pre-planned. Skill-based errors will be associated with inappropriate behavioural combinations of space, time and effort. Rule-based errors will be associated either with a misperception of situational demands, or the incorrect recall of appropriate procedures. Knowledge-based errors will result from limitations in operator resources, or from incomplete or incorrect knowledge.

This model again emphasizes the need to obtain detailed information on antecedent behaviour from accident-involved personnel, if the precise nature of contributions from human error are to be understood. The model also associates particular categories of error with different levels of skill among trained personnel and thus emphasizes the benefits that a well-designed accident reporting system can have for identifying training, or retraining, needs as part of any required countermeasure programme. More subtly, the model identifies the need to examine possible sources of interference between these different levels of skill-, rule- and knowledge-based behaviour, as potential contributors to accidents. For example, did the concurrent requirement to deal with a novel situation disrupt a skilled action sequence, or lead to the implementation of an inappropriate rule, thus provoking an accident? The model therefore has implications for the understanding of work design as a possible contributory factor in accidents, as well as for the place of operator training in accident reduction and prevention.

Baars' (1983) 'global workspace' model

This model was developed from the view of human cognition as a parallel distributed processing system, in which specialized processors cover all aspects of mental functioning without the need for central executive *control*, although co-ordinated by a central information *exchange*. This takes place within a 'working memory' to which the specialized processors compete for access as a function of their current level of activation. Once within this 'global' workspace' they can recruit and control other processors. 'Consciousness' is identified with the current content of this workspace.

The model thus has similarities with Shiffrin and Schneider's model, in that it identifies a subset of cognitive operations on which attention is concentrated at any one time. It, too, therefore emphasizes the need to understand the 'how' and 'why' of behaviour antecedent to accidents, with its attendant implications for introspection and verbal reporting by accident-involved personnel as a method of revealing the relevant conscious contents of the operator's 'workspace'. Again, it identifies the possibility of reducing or preventing accidents by highlighting specific needs for training and/or work design.

Card *et al.*'s (1983) 'model human processor'

This theory is an attempt to construct a limited set of 'working approximations' to a broad range of cognitive activity. It has similarities with Broadbent's 'Maltese Cross' model, in that it consists of two parts: representations in memory, and principles of operation. These are represented at each of three interacting sub-system levels; perceptual, motor and cognitive. These three systems can operate in series or in parallel. Certain basic principles of operation are specified in the model; such as 'recognize–act', 'discrimination', 'uncertainty' and 'rationality'. Implications for the understanding of human error follow from the model's incorporation of constraints on goal attainment imposed by task structure, information input, limited resources and incomplete knowledge. These have obvious applications to the design of accident-reporting systems and implications for accident countermeasures based on (re)training and task design.

Anderson's (1983) ACT theory of cognitive architecture

This is a development of the 'Adaptive Control of Thought' model. It distinguishes three memory systems: working, declarative and production. Working memory interacts with five processes: encoding, performance, storage, retrieval and execution. Knowledge can be represented as: temporal strings, spatial images, or abstract propositions. Productions by the system can be matched in five ways in order to resolve conflicts: degree of match; production strength; data refractoriness; specificity; and goal dominance. 'Activation' controls the rate of information processing for production and is a function of environmental stimulation, production execution, and goal influences in working memory.

Immediate applications of this highly structured model to taxonomies of human error used in accident reporting systems are probably more apparent than real. However the model does, once again, emphasize the need to distinguish between representation and process in the knowledge that guided the antecedent behaviour of accident-involved individuals. It also highlights the need to examine and understand the operators' goals and the reasons for their behaviour (the 'why' of accident causation), as well as the 'what' and 'how' of accident production.

Implications for accident reporting

From the foregoing discussion it is possible to identify the following implications for the reporting of accident occurrence.

Purposeful reporting

There is little point in spending time and effort on the collection of comprehensive data with little idea of the use to which its analysis will be directed, or in the absence of resources required to implement any necessary accident countermeasures. The latter will essentially be of two types:

1. *Primary safety measures.* The reduction or prevention of accident occurrence.

2. *Secondary safety measures.* The prevention, or reduction in severity, of injury associated with accidents that do occur.

The principal aim of any accident reporting system will be to highlight the need for primary safety improvements and identify the particular types of countermeasure which seem likely to be most efficient in preventing or reducing accident occurrence. The emphasis here will be on the collection of comprehensive data relevant to the antecedent behaviour of accident-involved personnel and its relation to concurrent task demands. This procedure has been carried to a high level of efficiency in, for example, civil aircraft operations, where the 'black box' flight recorder stores valuable information on technological functioning and ongoing activity at various behavioural interfaces for a prescribed period prior to the occurrence of any accident. It should also be possible to report antecedent behaviour comprehensively in accidents occurring within, say, the process industries, where tasks and operations are highly structured and sequences of operations are logged. It will be somewhat more difficult, even in the process industries, to report and reconstruct antecedent behaviour patterns and discrepancies at congruences between different tasks, e.g., process operation and equipment maintenance. Furthermore, it may be extremely difficult to obtain accurate behavioural reports from industries where work patterns are relatively unstructured, such as on building construction sites, or in small engineering workshops. Nevertheless, if inappropriate behaviour is to be avoided and accidents reduced or prevented, then the detailed behavioural antecedents of accidents must be captured in as much detail as possible, even in these difficult situations.

The purposeful needs of secondary safety are often thought to be met by reporting the nature, severity and causes of accidental injury, so that wounding agents can be removed, resited, or redesigned, and/or so that their potential to inflict injury can be minimized by the introduction of protective clothing or equipment. However, it will be clear that an understanding of inappropritae antecedent behaviour patterns can also contribute substantially to secondary safety. It will achieve this by identifying undesirable and avoidable associations between specific operator activities, transfer of energy, and injury accident occurrence. This knowledge can then be used to design and introduce safety devices, such as interlock systems and machine guards, which prevent such injurious associations from occurring. The type of behavioural data of importance here will not consist simply of physical movement patterns; although these will obviously contribute to the design of certain secondary safety measures, such as guard rails and protective clothing. Even where the purpose is secondary safety, advantage will be gained from reports on covert activity, such as intentions, reasons and judgement, since these too are susceptible to change via the introduction of secondary safety measures.

Adequate reporting

As mentioned earlier, 'accident' reporting is often a misnomer since many systems report only that subset of accidents which result in injury. Such types of truncated accident reporting may appear to meet the principal needs of secondary safety and they are usually justified by their concentration on the 'more important' accidents and on 'cleaner' data. These claims have a certain face validity; however, they ignore the potential value of 'control' data on the behavioural antecedents of damage–only accidents, or near accidents which resulted in neither injury nor damage. Injury, damage and 'near misses' may all result from apparently identical human–technology interactions. The clue to the different consequences may be discernible only by detailed comparisons of the antecedent behaviours in question. If damage–only accidents and near accidents are not recorded, such comparisons are impossible and accident investigators can only speculate on the reasons why particular behaviour patterns sometimes result in accidental injury and sometimes do not. A concentration on injury accidents also severely limits the total number of data available for safety work and makes it difficult, if not impossible, to estimate the objective risk associated with specific behaviour patterns (see Brown, 1991). A useful accident reporting system will therefore be one which collects data on *all* accidents, to the same level of detail, regardless of their consequences.

Factual reporting

Accident reports are frequently completed with the sole aim of attributing blame for human error. The result is usually a highly subjective, excessively brief account of system failure, which identifies a scapegoat but provides little or no information on the need for specific accident countermeasures to protect all operators in the work situation in question and certainly adds nothing to the general pool of knowledge on human error and accident causation.

In order to meet even the first of these objectives, the 'first line' of any accident reporting system should avoid subjectivity and particularly the apportioning of 'blame' and, instead, concentrate on the factual reporting of task demands and operator behaviour in the relevant period prior to and surrounding the accident. As Rasmussen (1987) has pointed out: '. . . instead of focusing on human errors, data should be collected to represent situations of man–task mismatch and characterized accordingly'. In other words, a comprehensive factual description of technological system demands and human antecedent behaviour must be attempted initially, before there is any attempt to classify technical faults or human errors. These faults and/or errors should be identifiable from study of any obvious mismatches between task demand and operator response. But their nature and their 'causes' will become clear, if that stage is reached at all, only by further detailed analysis of the mismatches and the context in which they occurred.

Task-specific reporting

It follows from the above considerations that accident reporting systems which aggregate data across different tasks and thus profess to represent organizational safety via global statistics will contribute little to the understanding of human error, or to the design and development of accident countermeasures. They will merely index organizational safety, in a far from meaningful sense, since risk will largely be unquantifiable, although they may provide management with a crude overall measure of the cost benefits of accident prevention and reduction. If accident data are to be more meaningful than this and contribute usefully to safety improvements, they must be recorded in a task-specific form and not aggregated across dissimilar tasks until an initial data analysis has been completed on the different behavioural contributions to accidents involving those tasks.

Verbal reporting

An important aspect of task-specific reporting is that it retains and enhances the value of introspective verbal reports provided by accident-involved personnel (see Patrick, 1987, for an overview of methodological issues associated with verbal reporting; see also chapter 7 on verbal protocol analysis and chapter 3 on subjective assessment). Verbal reports will often be the only method of reconstructing the behavioural sequences antecedent to an accident and of supporting these reconstructions with information on participants' covert reasons, intentions, perceptions, decisions, judgements, etc. Within tasks, such reconstructed sequences may be aggregated or compared in order to highlight any mismatches between actual system demands, perceived system demands and operator behaviour which may have contributed to accident causation.

Analysis of accident data

The first general point to be stressed is that any accident analysis should aim principally to identify the potential for improving system safety, rather than simply identifying the 'causes' of past events. The latter objective may be considered essential to meeting the former, but this is true in principle rather than in practice, because of differences in the definition of human error as a 'cause' of accidents.

Where tasks are highly structured and operator behaviour can be clearly specified in a predetermined manner, error may be defined as any departure from operating instructions. In such work situations, mismatches between operator behaviour and task demands may be relatively easy to identify (depending upon the method used for concurrent automatic storing of antecedent behaviour in accidents and the truthfulness of subsequent verbal reporting). This will enable 'causal' or 'contributory' factors in operator performance to be identified with apparent ease and hence aid the production of global statistics on an organization's accident patterns. But it may be unhelpful in identifying acceptable options for accident countermeasures, because it will bias attention towards the elimination of any diversity in operator performance. Such diversity will be associated with differences between the various goals within system operation and it may not constitute 'erroneous' behaviour which needs to be eliminated if the system is to function safely.

A more productive view of human error, from the standpoint of safety improvement, is that it represents 'the effect of human variability in an unfriendly environment' (e.g., see Rasmussen, 1987; Kjellén, 1984a, b). On this view, accident analyses should focus not simply on the frequency of discrepancy between prescribed task performance and actual operator behaviour, it should examine the precise nature of such discrepancies and the behaviour range they represent. Instead of reflecting error to be eliminated in the name of accident prevention, antecedent behaviour analyzed in this manner should suggest remedial measures which will result in a more 'error'-tolerant system, capable of accepting safely a much wider range of behavioural diversity.

A second, related, general point is that accident analyses should include not only an examination of the range of antecedent behaviours actually exhibited by accident involved personnel, they should include consideration of alternative behaviours which would have met the task and system goals in question, but which were not exhibited (e.g., see Leplat's (1987) description of 'fault tree analysis'). The results of such broader analyses are then capable of being utilized beyond the immediate aims of causal attribution, apportioning of 'blame' and error-tolerant system redesign; they may be suggestive of ways in which 'error-recovery' procedures can be developed and taught to operators, or incorporated into system software for automatic implementation. In other words, the analysis of accident data should aim not simply to make technological systems less 'unkind'; it should actively seek to identify any potential for improving operator support and overall user-friendliness of the system.

A third and final general point to make represents a criticism of the heavy reliance on operator biographical data by many accident reporting systems. This seems based largely on convenience in data collection, rather than usefulness in understanding accident causation or relevance to the design and implementation of acceptable remedial measures. Certainly it is simple and usually convenient to collect factual information on sex, age, education, socio-economic class, etc. However, these variables usually contribute less than informatively to data analysis and the identification of countermeasures. Sex and age information may be statutory requirements of certain accident data collection systems and these variables have some importance in certain work situations (e.g., the identification of radiation risks for pregnant women, or the possibility of 'thrill-seeking' as a risk factor among young male drivers). However, biographical data may be misleading in accident analyses, because many of them are confounded with experience factors (e.g., see Brown and Groeger, 1988). In addition, when biographical factors appear to be playing a significant role in the accident process, they usually point to selection and training as appropriate remedial measures. Training may be acceptable, if potentially expensive, but considerable further analysis of the accident data is usually required in order to identify the nature and extent of any training requirements. Selection may be unacceptable, in social if not in supply-and-demand terms, if it conflicts with 'equal opportunities' legislation.

Where one aim of analysis is the classification of accident consequences, in order to identify methods for treatment and prevention of injury, it will be convenient to employ international injury coding systems (e.g., see Baker, 1982; Langley, 1982; Somers, 1983a, b). This will simplify the aggregation of data and their comparison across national or international organizations. But injury prevention will, primarily, require analysis of those specific operator activities which are associated with injury production.

The considerations and arguments advanced in foregoing sections have highlighted the need to analyze antecedent behaviour in some detail, if the contributory factors in accidents are to be revealed and attendant options for remedial measures presented. How is this to proceed, given that the 'first line' of accident reporting has produced an adequate description of the accidental circumstances and consequences, of the operator's antecedent behaviour and of the related sequence of task demands?

The prime concern will be with the analysis of 'human error', in its broadest sense and specifically in relation to relevant task demands. What form is this analysis of error to take? Given that one main aim is to identify mismatches between task demands and operator behaviour, it would seem appropriate to base the investigation of human error on a detailed hierarchical task analysis of the work in question (see Patrick *et al.*, 1986; and chapter 6 in this book). Such an analysis will break down a job into a hierarchical set of tasks, sub-tasks and 'plans' for describing when and in which sequence sub-tasks are performed. Task objectives form the basis of description. The structure of antecedent behaviour, composed from verbal reports and other more objective sources of data, can then be contrasted with this task analysis in order to reveal mismatches. However, it should be remembered that a task analysis may often represent the system designer's, or manager's view of operational requirements (although not if carried out correctly and fully). Mismatches will thus reflect 'errors' in the form of departures from these predetermined requirements, but they may not represent 'mistakes' in the sense of divergence from system goals (Reason *et al.*, 1990). With this reservation, task analysis appears to provide an appropriate framework for the investigation of behaviour antecedent to accidents: it should readily permit the location and identification of 'error' in the task sequence, in a form which allows either operator behaviour or system demands to be modified and/or error-recovery mechanisms introduced, in order to eliminate future errors of that kind or increase system tolerance to them.

Using task analysis as the structure for investigation of error will, naturally, tend to produce a classification of antecedent behaviour based largely on met and unmet system demands. By contrast, it may be argued (see Rasmussen *et al.*, 1987) that human error must be classified in terms of *human* characteristics if the results of such analyses are to be applied to new as well as to existing systems. This seems eminently reasonable, given that human behaviour is the common element in any system requiring manual control or monitoring, and provided that any system for classifying human error can be identified with the system demands to which it referred.

A number of systems, or frameworks, have been developed for the classification of human error, based on the theories of error briefly described earlier. For example, Reason (1987a) presents a *Generic Error Modelling System* (GEMS) for locating common human error forms. It identifies three basic error types: skill-based slips, rule-based mistakes, and knowledge-based mistakes. Reason (1987b) also presents a classification based on the identification of eight 'primary error groupings' and eight 'information-processing domains', the interactions of which define five 'basic error tendencies' (see Table 32.1). The model also allows 'predictable error forms' to be associated with 'situational factors' known to promote that form of occurrence. It therefore appears to meet the joint criteria of classifying human error, relatively simply, in terms of normal human characteristics, without losing sight of system demands and goals. As Reason (1987a) himself points out:

> To be of value in either a theoretical or a practical context, a classificatory framework must both simplify the available error data and acknowledge the multiplicity of possible causal interactions. Most importantly, however, it must recognise that predictable error and correct performance are two sides of the same coin, and hence demand common explanatory principles.

Table 32.1. Matrix for classifying primary error groupings (reproduced from Reason (1987a) with permission of John Wiley and Sons Ltd)

	Ecological constraints	Change enhancement	Resource limitations	Scheme properties	Strategies heuristics
Sensory registration	False sensations* X	X			
Input selection			Attentional failures* X	X	O
Volatile memory			Memory lapses* X	X	O
Long-term memory			Inaccurate recall* O	X	X
Recognition processes	O	O	Misperceptions* X	X	X
Judgemental processes		X	Errors of judgement* X	X	X
Inferential processes			Reasoning errors* X	X	X
Action control			Unintended words/actions* X	X	O

X, Primary node; O, secondary node; *, primary error groupings.

Rasmussen (1982) has presented a taxonomy of human error which appears less well-founded in the cognitive theories reviewed earlier, but which develops clearly from the concept of accidents as the consequence of 'inappropriate behaviour'. It describes features of the human–technological system mismatch in terms of inappropriate task performance, that is, omissions of activity in procedural sequences, action involving wrong components, reversals in an action sequence, inappropriate timing of action, and so on. The complete multifaceted taxonomy is illustrated in Figure 32.1, where Rasmussen uses the terms 'external mode of human malfunction' to describe inappropriate task performance, in order to avoid the term 'human error', with its connotations of 'guilt' and 'blame'.

In order to characterize the covert mental activities involved in the mismatches examined by this taxonomy, Rasmussen's approach requires an analysis of the relevant decision processes in order to reveal the underlying cognitive errors (or 'internal mode of malfunction'). Given sufficient objective, factual information on the accidental circumstances and reliable verbal reports, an experienced accident investigator might pursue the analysis of cognitive error using Rasmussen's (1976) 'step-ladder' model of decisoin making (see Figure 32.2).

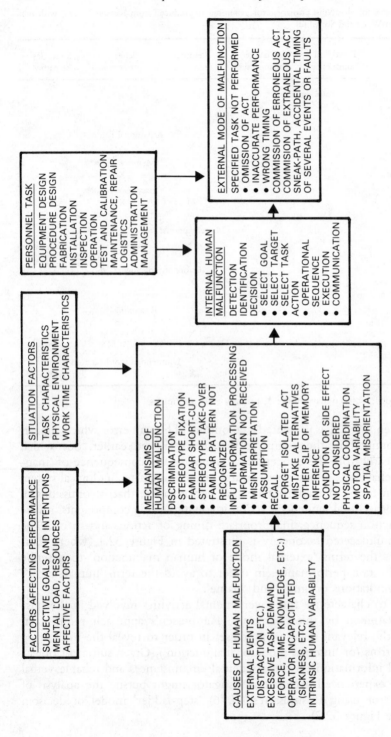

Figure 32.1. Multifacet taxonomy for description and analysis of events involving human malfunction (reproduced from Rasmussen (1982) with permission of Elsevier Science Publishers)

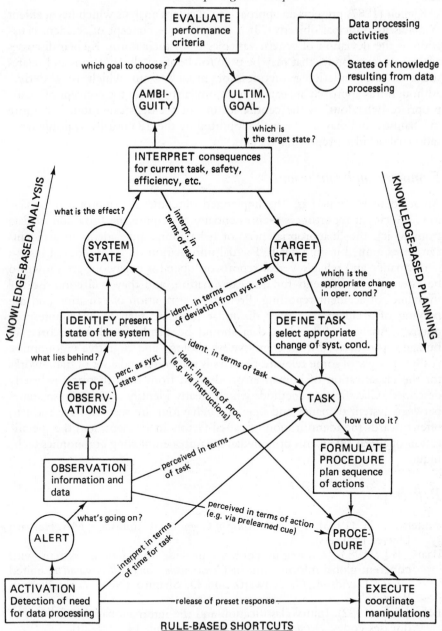

Figure 32.2. The 'step-ladder' model of decision making. Rectangles represent data processing activities, circles are states of knowledge from data processing. Reproduced from Rasmussen (1976) with permission of Plenum Press

Kjellén (1987) presents an approach to accident analysis which has accident *control* as its principal objective. It is based on the concept of accident causation as the deviation of system variables from their norm. Kjellén discusses a range of taxonomies that may be used to classify such deviations and relates them to the underlying theories, or models from which they derive. Although this approach appears to have similarities with the concept of 'inappropriate behaviour' as the central feature of accident causation, it is more mechanistic and may not be well supported by the data usually available from industrial accident reporting systems.

Summary and conclusions

An accident is defined as: 'the unplanned outcome of inappropriate behaviour'. Accident reporting is seen essentially to require the provision of data from which the 'inappropriateness' of behavioural antecedents to accidents can be examined in fine detail. Factual information on technological system performance and reliable verbal reports on operator behaviour are shown to be essential requirements for this examination, and they will comprise the 'first line' of accident reporting. Premature attribution of causation and the pooling of accident data over dissimilar tasks are seen to be counterproductive. Analysis of accident data should be firmly based on theories of human error, derived from cognitive psychology and cognitive ergonomics. A short review of some relevant theories is presented and a set of frameworks for the classification of human error, derived from such theories, is briefly described. Classification methods which clearly identify errors as mismatches between system demands and operator behaviour are seen to have most to offer towards the identification of causal factors in accidents and the specification of alternative forms of remedial measures employing ergonomics techniques.

References

Anderson, J.A. (1983). *The Architecture of Cognition* (Cambridge, MA: Harvard University Press).

Baars, B.J. (1983). Conscious contents provide the nervous system with coherent global information. In *Consciousness and Self-Regulation*, edited by R. Davidson, G. Schwartz and D. Shapiro (New York: Plenum Press).

Baker, S.P. (1982). Injury classification and the international classification of diseases codes. *Accident Analysis and Prevention*, **14**, 199–201.

Broadbent, D.E. (1984). The Maltese Cross: a new simplistic model for memory. *The Behavioural and Brain Sciences*, **7**, 55–94.

Brown, I.D. (1976). Psychological aspects of accident causation: theories, methodology and proposals for future research. Unpublished report prepared for the Medical Research Council, Environmental Medicine Committee's Working Party on Specific Aspects of Accident Research.

Brown, I.D. (1991). Prospects for improving road safety during the 1990s. In *Ergonomics, Safety and Health*, edited by W. Singleton, R. Bouwen, J. Dirkx, A. Meers and B. Overlaet (Leuven, Belgium: Leuven University Press), pp. 315–334.

Brown, I.D. and Groeger, J.A. (1988). Risk perception and decision taking during the transition between novice and experienced driver status. *Ergonomics*, **31**, 585–597.

Card, S.K., Moran, T.P. and Newell, A. (1983). *The Psychology of Human–Computer Interaction* (Hillsdale, NJ: Lawrence Erlbaum).

Cresswell, W.L. and Froggatt, P. (1963). *The Causation of Bus Driver Accidents: An Epidemiological Study* (London: Oxford University Press).

Department of Transport (1991). *Road Accidents Great Britain 1990: The Casualty Report* (London: HMSO).

Duffy, E. (1962). *Activation and Behaviour* (New York: John Wiley).

Duncan, J. (1990). Goal weighting and the choice of behaviour in a complex world. *Ergonomics*, **33**, 1265–1279.

Edwards, M. (1981). The design of an accident investigation procedure. *Applied Ergonomics*, **12**, 111–115.

Farmer, E. and Chambers, E.G. (1926). *A Psychological Study of Individual Differences in Accident Rate*. Industrial Health Research Board, Report No. 30 (London: HMSO).

Hale, A.R. and Hale, M. (1972). *A Review of the Industrial Accident Research Literature* (London: HMSO).

Heinrich, H.W. (1950). *Industrial Accident Prevention*, 3rd edition (New York: McGraw-Hill).

Kjellén, U. (1984a). The deviation concept in occupational accident control—I. Definition and classification. *Accident Analysis and Prevention*, **16**, 289–306.

Kjellén, U. (1984b). The deviation concept in occupational accident control—II. Data collection and assessment of significance. *Accident Analysis and Prevention*, **16**, 307–323.

Kjellén, U. (1987). Deviations and the feedback control of accidents. In *New Technology and Human Error*, edited by J. Rasmussen, K. Duncan and J. Leplat (Chichester: John Wiley), pp. 143–156.

Langley, J. (1982). The international classification of diseases codes for describing injuries and the circumstances surrounding injuries: a critical comment and suggestions for improvement. *Accident Analysis and Prevention*, **14**, 195–197.

Leplat, J. (1987). Accidents and injury production: methods of analysis. In *New Technology and Human Error*, edited by J. Rasmussen, K. Duncan and J. Leplat (Chichester: John Wiley), pp. 133–142.

Leplat, J. (1990). Relations between task and activity: elements for elaborating a framework for error analysis. *Ergonomics*, **33**, 1389–1402.

McKenna, F.P. (1983). Accident proneness: a conceptual analysis; *Accident Analysis and Prevention*, **15**, 65–71.

Malaterre, G. (1990). Error analysis and in-depth accident studies. *Ergonomics*, **33**, 1403–1421.

Norman, D.A. (1981). Categorization of action slips. *Psychological Review*, **88**, 1–15.

Norman, D.A. and Shallice, T. (1980). *Attention to Action: Willed and Auto-mated Control of Behavior.* Centre for Human Information Processing, Report 99 (La Jolla, CA: University of California).

Patrick, J. (1987). Methodological issues. In *New Technology and Human Error*, edited by J. Rasmussen, K. Duncan and J. Leplat (Chichester: John Wiley), pp. 327–336.

Patrick, J., Spurgeon, P. and Shepherd, A. (1986). *A Guide to Task Analysis: Applications of Hierarchical Methods* (Birmingham: Occupational Services).

Rasmussen, J. (1976). Outlines of a hybrid model of the process plant operator. In *Monitoring Behaviour and Supervisory Control*, edited by T.B. Sheridan and G. Johannsen (New York: Plenum Press), pp. 371–384.

Rasmussen, J. (1982). Human errors: a taxonomy for describing human malfunction in industrial installations. *Journal of Occupational Accidents*, **4**, 311–335.

Rasmussen, J. (1987). The definition of human error and taxonomy for technical system design. In *New Technology and Human Error*, edited by J. Rasmussen, K. Duncan and J. Leplat (Chichester: John Wiley), pp. 23–30.

Rasmussen, J., Duncan, K. and Leplat, J. (Eds) (1987). *New Technology and Human Error* (Chichester: John Wiley).

Reason, J.T. (1979). Actions not as planned: the price of automation. In *Aspects of Consciousness*, Volume 1, *Psychological Issues*, edited by G. Underwood and R. Stevens (London: John Wiley).

Reason, J.T. (1987a). Generic error-modelling system (GEMS): a cognitive framework for locating common human error forms. In *New Technology and Human Error*, edited by J. Rasmussen, K. Duncan and J. Leplat (Chichester: John Wiley), pp. 63–83.

Reason, J.T. (1987b). A framework for classifying errors. In *New Technology and Human Error*, edited by J. Rasmussen, K. Duncan and J. Leplat (Chichester: John Wiley), pp. 5–14.

Reason, J.T. (1988). Framework models of human performance and error: a consumer guide. In *Tasks, Errors and Mental Models*, edited by L.P. Goodstein, H.B. Andersen and S.E. Olsen (London: Taylor and Francis), pp. 35–49.

Reason, J., Manstead, A., Stradling, S., Baxter, J. and Campbell, K. (1990). Errors and violations on the roads: a real distinction? *Ergonomics*, **33**, 1315–1332.

Shannon, H. and Manning, D. (1980). The use of a model to record and store data on industrial accidents resulting in injury. *Journal of Occupational Accidents*, **3**, 57–65.

Shiffrin, R.M. and Schneider, W. (1977). Controlled and automatic human information processing: II. Perceptual learning, automatic attending and a general theory. *Psychological Review*, **84**, 155–171.

Somers, R.L. (1983a). The probability of death score: an improvement of the injury severity score. *Accident Analysis and Prevention*, **15**, 247–257.

Somers, R.L. (1983b). The probability of death score: a measure of injury severity for use in planning and evaluating accident prevention. *Accident Analysis and Prevention*, **15**, 259–266.

Wagenaar, W.A. and Reason, J.T. (1990). Types and tokens in accident causation. *Ergonomics*, **33**, 1365–1375.

Chapter 33

The analysis of organizations as an aid for ergonomics practice

Pat Shipley

Introduction

It is proposed in this chapter that the analysis of organizations is a useful conceptual tool for practitioners of ergonomic science. Whether acting in the role of practising ergonomist in industry, or researcher in the laboratory, the ergonomist works as part of an organization of some kind. Even the freelance consultant encounters organizations which affect the practice of consultancy; government bodies that regulate practice and organizations that consume products and services. Organization is the endemic phenomenon of modern life in the developed world. To have some insight into how an organization functions, and why it functions in the way it does, is to appreciate how it influences you, the ergonomist, or aspirant practitioner, as one of its constituent parts. In turn you will learn how you can, and do, influence the organization as an agent of change (or indeed as an agent of the status quo).

There is a philosophy underlying the chapter, that social relations are mutually constitutive, which is to say that both of the two theoretically extreme positions are avoided, of complete determinism on the one hand or pure voluntarism on the other. Ergonomists are not omnipotent, but if our powers of action and influence within organizations are quite definitely limited we are never, in principle, mindless and powerless dupes, wholly incapable of changing anything. In the last resort we are invariably free, at least in our democratic society, to leave the organization altogether. In the course of the following argument, therefore, the ergonomist's power base as an organizational member will be examined. Also, it is taken as given that the ergonomist's intentions include the promotion in an ethically acceptable way of ergonomics principles and practices in the production of goods and services, and in the quality characteristics of the goods and services themselves. It is also taken as given that those intentions include the promotion of their own and others' job satisfaction and well-being.

The concept of 'organization' is considered briefly, and a particular view of the development of organization studies, including the classical functionalist way of theorizing about organizations, is sketched. Ergonomics science is positioned in this framework. A stark comparison of the different world-views and paradigms broadly associated with the natural and social sciences is made, and their significance for understanding organizational life considered. The implications of these differences for the effectiveness of ergonomics interventions in organizations will also be considered. The collaborative mode of intervention is discussed as a possible way out of dilemmas posed by traditional ergonomics practice. The argument concludes with notes on assessing organizational culture, tempered with caveats about the unintended consequences of research and intervention practices, caveats which also point forward to other issues discussed in this book, and in particular to participation (see chapter 37 as well as later in this chapter) and ethics (see chapter 38).

Originating, we are told, in World War II in the UK, and located then in the research squadrons of the defence services, ergonomics science has since considerably broadened in practice beyond the science of aircraft seats and the design of the knobs and dials of military equipment to embrace activities and applications falling roughly within the general ideology of the 'quality of life', the 'quality of working life' especially. The prioritizing of safe and healthy practices at the workplace within resources limitations, the design of work activities to include an individual's complete job, even the job of a whole work team, as well as the traditional concern for the ergonomic design of engineered products, could now form part of this wider brief. Service functions such as communication networks and information providers, at railway stations, airports and town halls, for example, may in theory be subject to ergonomics scrutiny and design. These terms of reference cover the needs of consumers outside the workplace as well as of worker producers within. Such a wide and open frame of reference presupposes that to be more effective ergonomists need, *inter alia*, an organizational survival kit, and the capacity to win others round to their point of view.

The study of organizations

There are those who would argue that the study of organizations as an 'organizational science' is an independent and discrete discipline. A contrary view is that the subject matter has no valid claim to special status but should be viewed as a branch of sociology, or social theory and philosophy, as applied to organizations. With the exception of some notable sociological studies, as an object of formal study organizations have come only quite recently under the spotlight. Given their relative youth, and (some would argue) given the social nature of their subject matter, the attainment of the goal of theoretical coherence is far removed from organization studies. In view of this, to propound a 'one best way' of analyzing and modelling organizations would be folly. Any theory of organization is partial, at best. This partiality applies too, to the methodology of organizational analysis. Although a wide range of small and medium sized organizations exist, many of them family businesses, our understanding about organizations has in the main derived from study of large private enterprises and public utilities.

Students of organizations have identified and described key organizational dimensions of structure, function, goals, decisions and environment, and more recently, values, power, conflict and culture. The definition of 'organization' has attracted debate. To Buchanan and Huczynski (1985) organizations are distinguishable from other social arrangements because their leaders are pre-occupied with a need for control. The reasons for organizing are of equal interest. Given that the human race is a remarkably adaptable species our ability to organize into effective working groups, where the needs and wishes of individuals are in principle subordinated to group goals, can be seen as a crucial strategy in this adaptability. In erstwhile imperial quests resources would have been located and efficiently utilized to explore and colonize successfully; similarly to build and manage public institutions, to conquer space, to run effective mountain rescue teams, to develop and market a product, the 'bottom line' is control, for non-profit and profit-making organizations alike. To enrol in an organization is to trade some of our freedom as individuals in exchange for benefits bestowed by its membership.

The allocation and distribution of resources and their accountability require skilled and committed management. Control and resource utilization are commonly achieved through the deployment of classic organizational tools of hierarchical structuring and the division of labour, enshrined in the typical organizational chart with the chief executive at the top. Established hierarchical organizations such as the Church of Rome have grown to be big, powerful and complex. More recent examples are giant multinational profit-making companies. Institutions have proliferated and some appear to have lost sight of the goals they were originally set up to attain. The classic organizational structure, bureaucracy, projects a depersonalized image to the lowly servant toiling within its depths, or to the person in the street who relies on its services. To the consumer it may seem to be a law unto itself.

The place of ergonomic science and the classical view of organizations

Ergonomics can be seen as squarely within the 'managerialist' tradition of organizational theorizing. In practice ergonomics has usually operated in the role of management technology, except where consumer interests have been directly promoted by ergonomists on behalf of those consumers. The managerialist view is governed by a particular approach and a number of assumptions. Management *practice* is nothing if not pragmatic; its bias is in favour of what seems to work out well in practice. Much of actual management is intuitive rather than grounded in explicit theory. Management *science*, on the other hand, presupposes that it is helpful to know something in advance of trying to change it; a kind of 'cause-effect' knowledge, in particular about those things that inhibit the attainment of management's goals, such as restriction of output, wastage, absenteeism and so on, which it is hoped will be more effectively controlled as a result. Management practitioners vary in how far they regard management science as valuable, either in itself or as an attractive way of selling ideas and changes to the workforce.

A popular line of theorizing about organizations adopted by management scientists, often unconsciously, is the classical functional one. The functional view highlights formalities and structures and classical principles of organizational operation loosely defined as 'rationalist', rather than the social, informal, cultural and value-orientation side of organizations. For the functionalists, and the classicists, the organization is theoretically governed by rational principles; the rule of law, order, fact and logic, in the interests of efficiency. It is expected that management has an inalienable right to manage at all times in all conditions. The view of Burrell and Morgan (1979) is that functionalists stress the regulative and control features of organization: order and consensus, integration, and the status quo. It follows that questions of conflict, disharmony, imbalance and change will not be emphasized by functionalists, or are regarded as aberrant if they are discussed by them.

Early management and organization theorists of the classical school included Frederick Winslow Taylor of '*Scientific Management*' fame, Henry Fayol and Mary Parker Follett. They chose to specialize in one area that was extracted from the rich and seminal work of Max Weber at the turn of the century. Weber saw bureaucracies as a rational solution to the problems of efficiency posed by social arrangements and institutions which permitted non-rational influences, such as personal favours and feelings, to interfere with goal attainment. The classicists promoted simple formal principles which produced the dominant hierarchical organizational form as most of us know it today and which embodies the 'logic of efficiency'. The unforeseen negative and 'pathological' consequences of this over-used prescription, where classically designed organizations have generated practices defeating the institution's original objectives, has since led to some attempts to dismantle common-mode structures and to the substitution of alternatives, such as matrix organizations.

Challenges to classic rationalism have come from quarters other than those primarily concerned with efficiency problems. Less well known among these challenges is a feminist one. In feminist psychoanalysis, powerful and heavily structured organizations are seen to be projections of the male ego. Rationalist values are dismissed as male values, as are the values of achievement, competitiveness and efficiency as priorities. Underlying rationalism, it is argued, is a deeper value—that of the active male principle of environmental mastery, in which man presumably copes with his own insecurities about life and nature by over-exploiting these for his own purposes. The feminist counterpoint to this asserts the value of living more in harmony with one's surroundings whilst prioritizing caring for others in authentic relationships, rather than relating to others in an exclusively instrumental way. (In a text which may be the first thoroughgoing analysis of Weber's ideas in feminist terms Bologh (1990) challenges Weber's patriarchal belief that *love* has to be contained in the *private* sphere in order that the *greatness* of the 'masculine' ideals of rational action and achievement can be played out in the *public* sphere.) Lest the reader should think the author, as female, has an axe to grind, no evaluation of either position is being offered. The feminist challenge is merely put forward out of interest, not as prescription nor as panacea, but as different.

The natural and social sciences contrasted

Ergonomics shares the epistemological values of the natural sciences. This epistemology is in the rationalist tradition, and a central feature of it is rigorous scientific method as a powerful form of enquiry. Through the vigorous application of this method nature has yielded up some of her secrets to man the inquisitor. In this tradition complex phenomena are reduced to analyzable and controllable fragments by scientific experts (see Shipley and Harrison, 1974). This reductionism has its parallel in the division of labour principle and work specialization carried to extremes in the Taylorized factory and office. The rationalist western intellectual tradition acknowledges more readily the logical, cognitive, and goal-oriented sides of human functioning than it does the affective, intuitive and trans-rational. The asocial human is the ergonomist's chief focus as labourer or consumer.

Weber was a social theorist, not a management consultant. His was a wide frame of reference sweeping outside and beyond the narrow confines of particular organizations, though these forms of social arrangement are microcosms of the society in which they are embedded. Statistical procedures invented by the State to serve its own control function were copied by corporations within that State. The functionalist bias is toward so-called hard, objective data. Functionalists, if Burrell and Morgan (1979) are to be believed, have a preference for explaining social functioning by appealing to the presumed underlying cohesion and unity of organizations and societies.

Illuminating metaphors are those which depict harmony and smooth functioning, such as machine metaphors, and organicist metaphors in which constituent parts of systems are taken to be dedicated to the viability of the system as a whole. 'Man-machine systems' and 'socio-technical systems' are the figures of speech of ergonomics. The epistemology, concepts and language of the natural sciences are consonant with these metaphors.

If the functionalist paradigm can be carved out of Weberian scripts, so also can the interpretive paradigm. In 'verstehen' an understanding of social activity is sought through the analysis of subjective experience as recounted by the social actors themselves. The natural sciences' predilection is to focus on behaviour; for the social scientist (or more precisely, the student of society) accounts of personal experience are of greater interest. The epistemology of the social sciences, its concepts, languages, methods and assumptions, belongs to a wholly different paradigm from that of the natural sciences, of which the interpretive, 'verstehen' tradition is a good example. A radically different way of understanding organizations is presented, and ergonomics science may benefit from studying this alternative tradition along with the natural sciences tradition. Its subject matter is people and their relationships; its metaphors are not those borrowed from biology and engineers. Dimensions of organizational structure and function, beloved of classicists, lose some of their salience and others come to the fore, such as the anthropological notion of culture. Discussion and dialogue are about issues normally suppressed in the classical view; about relationships between people, about co-operation and conflict, power, values, mythology and beliefs. The rational mind of the individual worker, seemingly operating in social isolation, dissolves as a figment in the minds of its inventors, scientists reared in the tradition of methodological individualism. Individual minds, like holograms, mirror the minds of those around them, just as organizations contain and institutionalize the beliefs of the wider society of which they form a part. It follows that an understanding of the organization as a whole can be achieved through an understanding of a single member, an organizational 'gate-keeper' perhaps.

Hard objective data and fundamental truths waiting to be discovered elusively slip the grasp of the organizational scientist. The data are the shifting sands of shared and negotiated meaning rather than measurable, logically-generated information. Buildings, documents and furniture physically exist, it is true, but the meaning they embody is symbolic, ambiguous and open to interpretation. No wholly automated system completely manned by robots as yet exists to my knowledge. It could be argued that in a society which has mastered the natural elements the important dimension of the environment is now the social, and this social environment is in our heads and hearts.

About 20 years ago the classical myth of how managers spent their time was exploded as a result of empirical studies by Rosemary Stewart in this country and Henry Mintzberg in the States (see Mintzberg, 1973; Stewart, 1979). Face to face talk, telephoning, frequent absences, and crisis management were much more common than the supposed rational and reflective management activities of planning, co-ordinating and organizing. In reality, managers seem little able to exercise control over their own work activities, reacting often to other people's behaviours rather than to their own priorities.

That senior management rationally uses carefully accumulated information provided by a formal information system was found to be folklore.

Furthermore, managers actually much prefer to operate informally and verbally, and need the skills to do so; skills which are typically ignored in the textbooks. They need skills for making unprogrammed decisions, coping with ambiguity, managing conflicts and developing informal interpersonal information networks. The 'symbolic interactionist' school of the social sciences dwells on human relationships, and their symbolic expression, primarily through language as a symbolic medium. For the American social interactionist, George Herbert Mead (1863-1931), the concept of selfhood is generated through continuous social interaction, in which meaning and reality are mutually constituted by both parties to the interaction.

Such analyses are typically 'gender-blind', however. Women are not normally a dominant part of the power culture of the typical organization (except perhaps when they enact the role of the 'honorary male'), and are usually barred from powerful male networks and clubs by virtue of their gender. In their recent paper about major stumbling blocks to women's equality in the workplace Cassell and Walsh (1993) discuss ways in which gender determines work organization, and describe the linkages between power and culture and gender.

Culture and organization

Greenfield sites apart, every organization has its history, and culture is a reflection of that history. It would be easy to dismiss the concept of culture applied to organizations as a fashionable fad, or as a ragbag into which we dump all the unexplained and poorly understood loose ends of organizational life. But that would be throwing away a set of ideas which convey the realities of the organization that have stood the test of time. Culture is shared meaning. Currently, many social theories and management scientists increasingly view culture as central to understanding control and resistance to change in organizations and in society.

Schein (1985) suggests that there are three levels of culture; *artefacts* such as buildings, documents and policies; *values* such as what people agree should be the case, e.g., that safety and welfare should take precedence over profit and efficiency; and *basic assumptions*, the often unquestioned guesses and hunches about how things work and how problems should be dealt with. Artefacts are surface symbols, while assumptions and values lie at a deeper level.

The British management theorist, Charles Handy, offers us an organization typology in which organizational types are differentiated as distinct cultures. The structural features of organizations are embodiments of the culture (see Handy, 1985). We are introduced by Handy to four organizational cultures: power, role, task and person. The metaphor for the power culture is the spider's web with power at the centre. The role culture is the bureaucratic norm, and the metaphor is the Greek Temple with structural columns holding up the pediment or strategic unit. In the task culture the prime value is to get the job done and organizational form will be used as a means to this end. Its metaphor is the matrix. The remaining stereotype is the person culture of the alliance of consultants or craftsmen sharing common facilities and working as partners in mutual consent. The metaphor of the person culture is of a cluster or galaxy of stars.

When an ergonomist enters an organization, absorption into its culture is hard to resist. The older the institution the stronger and more entrenched the culture and consequently its resistance to change. We follow the dictates of cultures often intuitively and unconsciously. The ergonomist's background, a disciplinary and educational background which is usually that of the methodological individualism of the natural scientist, may represent an alternative culture that clashes with the host culture. The process of socialization or social conditioning begins once the portals of the organization have been passed through. Handy (1985) defines socialization as a process which is designed to encourage the individual to adapt to the organization's values and customs. This includes the accepted norms of behaviour (how the individual is expected to behave), and the accepted mode of organizational operation ('how things get done around here'). Contact with people outside the organization, a professional reference group like the Ergonomics Society, for example, can be a solace and a support to the single ergonomics practitioner coping with organizational reality. The chances are that the ergonomist will blend quite well into the host culture where classical organizational principles are in operation on the surface. Ergonomics science has, after all, perpetrated its own myths about rational man and rational practice. The clashes then might arise at the level of the informal subculture, if they arise at all.

As a consultant the ergonomist is probably first introduced to the organization by a client from a subculture within that organization; someone, say, from production, marketing, design or occupational health. To be effective, ergonomists may have to work across subcultural boundaries, as practitioners who apply a comparatively holistic perspective to their task, although they will be located physically in a single unit within a particular subculture. A worse fate may be a base in a corporate department at HQ, where trying to work with antipathetic people in the field from that base proves difficult.

These occupational or functional subcultures and identities cut across and through the organization, just as ethnic and gender identities do so, serving as a basis for stereotyping. We hear of 'hard-nosed engineers', 'software types', and so on. Each functional area has its own vested interests, and competing cultures have communication problems which frustrate attempts at co-ordination. Each is socialized and rewarded differently. Identities are shared with reference groups outside the organization, such as professional bodies. The production man is rewarded by getting the product out, the research scientist by producing data for the next scientific paper.

Clegg and Wall (1984) report case studies from the Sheffield (UK) Applied Psychology Unit demonstrating communication problems arising across functional divisions. They warn us that 'specialisms will attract and recruit experts in their own areas, people with specific professional training and standards who may well have quite dissimilar perspectives as well as a language of their own' (p. 436), and that 'the very factors which promote the need and opportunity for participation, at the same time encourage organizations to develop in ways which undermine its efficacy' (p. 437). "Put bluntly, the non-expert should not 'interfere'". (p. 438).

Ergonomists typically work in multidisciplinary teams, and the paper by Clegg and Wall (1984) is a reminder that what they refer to as 'lateral participation' can be as important for organizational and task efficiency as 'vertical participation', and is indeed much less controversial. These authors point out that the participation literature is focused almost exclusively on the latter and 'how to transcend the boundaries between the "managers" and "managed"', (p. 429) to the neglect of the 'integration of traditionally separate functional hierarchies.' (p. 430). They see this as especially problematic in a contemporary era of rapid and continuous change when organizations have to develop quick-acting and transient organic structures, such as temporary project teams, under conditions of chronic uncertainty in order to remain viable.

Culture enables an organization and individuals to cope with uncertainty, about the future, the present, the meaning and purpose of life and so on. Because the culture and the organization have survived this long there will be widespread dependency on it; it is consolidated by powerful executives and reinforced by the unquestioning habitual practice of countless peons who are 'just following the rules'. It conveys a comfortable feeling of security and outsiders who appear to be rocking the boat may soon find themselves rapidly ejected. Violent and powerful change forced from outside however, such as from a take-over or merger, may reveal the falseness of that security, and under such circumstances tensions held just below the surface can erupt into conflict. Many a nominal take-over and merger has failed to work out in practice, because of the failure of two cultures in the forced marriage to hit it off. Organizational change can be extremely painful under such conditions. Where a powerful status quo is resistant, to technological change, for example, the ergonomist expecting a rational response to a rational suggestion may be the only one to suffer when that suggestion is rebuffed.

There are two sides to an organization: its task side and its relationships side. Changes intended to improve or change the task requirements may disrupt relationships. Human beings have needs from peers for support and friendship. Some form of informal dominance hierarchy may exist too. This informal side to organizations, not visible in the organizational chart, office layout and job specifications, can be equally powerful as a block to change or as a facilitative channel. A new technology may be seen to threaten a way of working and of relating which has built up around an existing technology. A good example of this is the study in the 1950s by the Tavistock researchers of the short-wall form of mining in North East England (Trist and Bamforth, 1951). A common set of beliefs, languages and practices evolve to enable a group to adapt and survive and cannot be demolished overnight.

Power in organizations

A bland symbolic interactionist view of people negotiating meaning, and sharing a definition of social reality, is consistent to some critics with a *unitarist* model of organizational life. Organization members are united and grouped under a common banner in this account; they are rather like a well-conducted orchestra. For some theorists, this view omits an important variable; it largely overlooks the role of power (the capacity to control and influence others) in organizations. People may be unequal parties to a negotiating process. Indeed, they may have no bargaining rights whatsoever. The *pluralist* view on the other hand emphasizes the diverse vested interests and potential conflicts being contained and managed within organizational boundaries; the organization is a loose coalition or set of coalitions pursuing different interests. The introduction of information technology will be supported by a coalition, or alliance, of like-minded people whose status and power is bound up with the promotion of such technology. Dominant coalitions change as conditions change. A powerful technology one day may become obsolete the next. For pluralists the existence of power is a fact of organizational life; manipulation and conflict is as commonplace as co-operation. Conflicts are resolved through the manipulation of power. The 'political' facts of organizational life present problems of access for ergonomics consultants, either access to people or information, whether operating from outside or inside the organization.

The social commentator, Alan Fox, contrasts the pluralist with the *radical* position (Fox, 1985). For Fox even the pluralist view is too benign. Agreements not entered into freely, but which are the outcome of coercive or manipulative power rather than of equitable bargaining between parties of roughly comparable strength, cannot be morally binding. The unitarist/pluralist views both unwittingly legitimize much current amoral and immoral practice, whereas the radical perspective draws out the power inequalities more sharply. The trades unions, for example, do not challenge management on fundamental social issues about the public implications of the corporation's goals, the hierarchical structure of the organization, the massive inequalities in pay differentials, and so on. The reasons why this challenge is not taken up include the industrial indoctrination of people through the mass media and training and educational programmes which are closely controlled by the power elite. Socially-dominant ideas and rhetoric are simply taken for granted by the masses. All management strategies designed to secure compliance and commitment are dismissed as manipulative in Fox's radical critique. They rob people of dignity and self-respect. Even taken for granted routines and practices may be responsible for the recreation of injustices. Unlike mindless indoctrination and socialization, commitment entails conscious choice.

In hierarchically structured organizations specialization and task hierarchy reduce power for the many at the base of the hierarchical pyramid; power or lack of it is structured into the organization. The management prerogative, the right to manage at all times, and to make all the important decisions, is usually unquestioned, taken for granted and assumed to be rational. This shared ideology, so the argument goes, leads to habitual practices which legitimate the ideology. People can be trusted to follow instructions, after socialization into the norms of the workplace. Socialization, therefore, transforms power into authority. 'Technical rationality', of procedures, rules, norms and techniques, is legitimated and authenticated as rational and reasonable. In contrast, with 'substantive rationality' the values underlying a practice are fully understood and espoused, and practice is based on genuine commitment to that rationality.

Organizations are stratified into power positions and ergonomist newcomers learn their position in the organization geology. Almost certainly, as scientific experts, their *position* power will be quite strictly limited, relegated to an administrative and executive support role. The scientist is often a member of service management, a back-up to front-line management. This position power is supplemented by *expert* power, depending on the attitude other organization members hold towards experts. Handy (1985) describes the various forms of individual power that may be available to the organization member. *Coercive* power usually springs to people's minds whereas *reward* power, *legitimate* power, *expert* and *resource* power, and *charismatic* power perhaps do not do so readily. Expert power can be derived from valued contributions made on the job; it also derives from the status of science and the professions in society at large. The professions, like the sciences, have their pecking order, and ergonomics science is far from being near the top.

In their paper outlining alternative models for introducing new information technology into organizations, Blackler and Brown (1986) criticize traditional ergonomics and Tavistock socio-technical theory and contrast them with their scenario for more participative systems design. The former they argue are examples of a misleading and oversimplified linear and rational 'task and technology approach' to introducing new systems, in comparison with the more organic and cyclical 'organization and end-user approach' of participative method. Although they readily acknowledge that ergonomics has had some success in persuading designers to make their software and hardware more 'user-friendly' they believe there is considerable room for improvement; that ergonomics' boundaries are too limited and its political position too weak to have much real effect. Ergonomics apparently fails to take into account the powerful social and structural factors which influence strategic decision making.

Illuminated by ideas from Perrow (1983) on American human factors engineers, Blacker and Brown (1986) suggest that traditional ergonomists may unwittingly contribute to mechanistic social systems (where machines are installed to prop up existing social structures, such as authoritarian political arrangements), because of their predominantly narrow physiological perspectives and individualistic levels of analysis. These authors' views are not entirely theoretical. They are both empirical researchers and their paper was inspired also by an empirical study assessing British industrial practices in the utilization of microelectronics. They found the standards of implementation to be generally poor.

Because of the restraints on the ergonomist's position power, and because the expert power of traditional ergonomics is largely limited by training and indoctrination in the natural sciences, the organizational power base of ergonomics needs to be extended to improve the chances for effective ergonomic intervention in organizations. In theory there are means of doing this through a better understanding of organizational culture and function, and an awareness of the power that trans-rational factors possess to resist or facilitate changes. Hanging on to outmoded models, such as machine and systems models, retards this development. The acquisition of social and interpersonal practical skills to enhance information assimilation and persuasive powers, in the way that successful managers themselves have done, as a supplement to rational skills, is another valuable addition (see Shipley and Harrison, 1974). Consorting with top management, as a management agent, is often not enough. In the organizational underground information flows up from the bottom and across peer groups, and is often blocked at the interface with management. Informal social activity is spontaneous and can be emotionally-charged, even if at their higher levels management would prefer such communication to be overt, and that people's feelings were left at home and not brought into the workplace. To get safety standards improved usually requires more than a change to explicit company policy, although that helps. The value of it has to permeate through to the grassroots. People have to believe in it and want to change.

My colleague at Birkbeck College, Jean Hartley, reminds me of the usefulness of the distinction between direct and indirect participation, in her paper on industrial relations (Hartley, 1984). Direct participation involves employees in decision making and control at the level of their jobs. Indirect participation is involvement through channels of representation at various levels, plant or company. This could stretch as far as worker directors on company boards. Hartley maintains that both forms should be seen as complementing rather than conflicting with each other. No legislation exists in the UK or America, in contrast with Scandinavia and the European Continent, to promote worker participation and democracy. The Swedish Work Environment Law of 1977 is explicit in requiring employers and employees to collaborate in various ways, including the provision of opportunities in the design of work, to enable workers to influence their local work situations themselves. The Co-determination Act is a legal shrine for a principle, that of collaboration between labour and employer, which was already a relatively long-established norm in Sweden. Swedish labour already had considerable power compared with Britain and elsewhere, although Swedish labour critics of the Act claim that the balance of power still rests firmly in the hands of the employer.

As a stimulus to member states to improve workplace health and safety the European Commission adopted a 'Framework Directive' on Health and Safety at Work in 1989, from which a number of 'daughter' Directives will emanate, six of which having already become enshrined in national law in early 1993. An underlying objective is the increase in worker participation through worker rights for information, consultation, training and representation in health and safety matters in those member states which do not have such formal provisions. The UK Health and Safety at Work Act of 1974 requires employers to exercise a general duty of care only 'as far as is reasonably practicable', which often can seem to boil down to economic considerations in practice; i.e., what can be afforded. This is backed up by voluntary use of the few codes of practice that are available. The EC Directive is more explicit in the setting of standards for health and safety at work than the resulting UK Law, whose underlying principle is one of *self-regulation*, and urges the provision of structures for worker consultation and representation as a means by which such standards can be achieved. In contrast to the Continental Model the British way is more voluntaristic, which is perhaps a reflection of a wider cultural ethic that stresses individual freedom over more prescriptive external regulation. The British government apparently succeeded in getting some of the EC draft requirements watered down and, as a result, it was agreed that member states could interpret and implement the Directives in accordance with their own national laws and practices. The UK has continually resisted the Continental Model of Worker Democracy, with its formal rights of worker representation on Works Councils, to be found in Germany and The Netherlands, for example, preferring to leave such matters to local negotiations instead. The EC view, however, does indicate a shift of opinion in Europe away from the British and more in the direction of the Scandinavian Model.

There may be no shortcut, therefore, to the laborious groundwork involved in identifying and developing the appropriate coalitions and alliances. This groundwork may have to be done at all levels from the level of company policy, through group levels, and at the level of the individual opinion leader. Even senior management is restricted in how far it can get things changed, hence the notorious use of subversive and secretive tactics at the top. Sufficient numbers of the right people have to recognize they have an ergonomics problem to accept an ergonomics intervention. Having recognized the problem is not enough, because there has to be a willingness to accept the change required, as opposed to the alternative of continuing to live with the problem. Then the practical implementation of the agreed change has to be achieved.

As in the master–slave paradox power is a relational process; power has to be 'accepted' by both parties in the relationship. The balance of power can alter with the conditions. The most powerful will be those seen to have the key to the organization's viability; those who safeguard the status quo in a stable period as in the public or professional bureaucracy, and those who have the vision to lead it through turbulence in an unstable period. Unlike a machine bureaucracy, an adhocracy is a flexible organization and its power distribution is dispersed to enable it to produce rapid responses as the changing situation requires (see Mintzberg, 1979). The unwieldy bureaucracy, however, is cumbersome in the face of rapidly shifting demands. Role structure and rule-following behaviour are meant to ensure continuity, reliability and predictability; a strong culture which can act as a curb on divergent and aberrant behaviour. It does not cope well in emergencies when non-programmed decisions are called for.

The contingency view of organizations presupposes that, like adaptable species in nature, organizations take on the shape most suitable to their habitat and conditions. Some would be tempted to believe, no doubt, that certain ossified institutional forms are rather like organizational dinosaurs in the contemporary world where change rather than stability appears to be the reality, as pressure to compete and to innovate spirals upward. But, however well planned in advance these changes are by ergonomists and their collaborators, the consequences are rarely wholly predictable, although a valuable ergonomic contribution can be made through systemic and disciplined enquiry to reduce some of that ambiguity. It seems that options and choices may be open to us all the while we have a future, but complete power and control is never one of those options. To think otherwise is to be less effective in one's interventions than might have been possible.

The unintended consequences of the attempts by ergonomists at applying expert power single-mindedly in organizations is a case in point (Shipley and Harrison, 1974). The collaborative or participative mode of intervention, where power is equalized between consultant and client, is one possible way forward for ergonomic practice as an alternative to the expert mode.

Shipley and Harrison (1974) attempted to analyze why methods based on the politics of expertise can fail in a culture like our own. It was proposed, for example, that the expert too readily assumes solutions to the client's problem, recommending prescriptions which the system either rejects or may implement only in amended form, and that a psychological reason for this rigid expert behaviour is a love of status and the other symbols of expertise, which may bring a warm glow to the ego. The problem of limited ergonomics influence may however have other causes, such as limited imagination and competence.

Changes to ergonomics training to include the study of the theory and practice of client-consulting relationships in a collaborative, power sharing way were also proposed in Shipley and Harrison (1974). This participative approach is often referred to as 'interventionist research', or 'action research' and a useful text by a British exponent is that by Clark (1972). Shipley and

Harrison (1974) also advocated an expansion of ergonomics training to incorporate knowledge of organizations, the diagnosis of complex systems, and the management of change to enhance competence. As an example of resistance to ergonomics recommendations, we pointed to the management tactic of searching for ways to avoid or cast doubt on the consultant's findings; questioning the sampling and statistical analysis, for example.

It may be easier to indulge in scape-goating and denial rather than deal with the rational improvements called for. A further difficulty is that it takes time to achieve results and few of us, consultants or management, are willing these days to find that time. The diagnosis and problem solving phases in action research must proceed at a pace the client feels happy with. A continuous and unhurried organizational educational process may be required which is designed to give ergonomics gradually away to the client system. Then the consultant has to retain a hold in the organization to be able to help that process along. Retaining this hold can be difficult.

If it is seen to be of value, then practical problems must be the main barrier to the implementation of action research in ergonomics, coupled with perhaps a too naive view taken by some theorists of the personal implications for practitioners as members of organizations in trying to be participative. The ideology of participation should perhaps be tempered by a careful consideration of the practical realities of organizational life.

An extreme view of the more traditional approach to ergonomics is that the expert is reluctant to give away control, the client is relegated to the status of object, and the detached observer stance creates psychological distance and a failure to generate trust and client commitment.

An ergonomics project using an action research approach was carried out by a team from an ergonomics laboratory in Paris. In their presentation to a symposium in Zadar (Croatia), Teiger and Laville (1987) described how the methods of ergonomics can be used by non-ergonomists, in their report of shiftworking changes involving 2000 workers in a chemical plant. The company management gave the 'solution' to the ergonomics team initially along with the problem; a 'scientific' solution using biological criteria. Instead, the ergonomists were instrumental in setting up a task force which included some of the shift operators, with themselves acting as facilitators. The result was an outcome that was acted upon; a revised practice based on a number of compromises between ergonomic, social, economic, technical and organizational criteria. The final choice was made by the operators themselves, and the company retained the ergonomics team for other projects. The ergonomists here lent their research skills and knowledge expertise as members of a team operating on lateral participative lines in which the potential users, as the experts on the job, had considerable say and influence about how the project was managed.

Learning about what goes on in organizations: beyond dualisms

The ergonomist is well-versed in the principles of the scientific method, and these have their place in the testing of hypotheses in controlled settings, such as that of the laboratory, or its near-equivalent 'in the field', the industrial simulation. The physical dimensions of the environment and the cognitive aspects of people's minds lend themselves better to analysis by such principles than do many social and organizational variables, where social context is of great importance.

Some theorists may argue that the organization has a life of its own, independently of its constituent human parts. This is a reification of the organization, and the implications of this reification for the measurement of organizational dimensions remains to be worked out. The ontological status of organizations is a controversial issue and a contrasting view is that organizations are no more than the people who make them up. The Gestalt idea that the organization is more than the sum of its parts does not appear to be inconsistent with either view, but it does have specific implications for organization assessment and data collection. Organizational members do not work at their jobs in a social vacuum. Work is done in constant interaction with other people, and even in the most closely-prescribed work tasks there is always room for a little discretion, cutting corners or filling in gaps, supplementing or bending the rules, as opposed to 'working to rule'. Without the exercise of discretion by proactive, creative members organizations would grind to a halt. Because the future is always open, because it is never wholly predictable, then this must be so. Collecting data in organizations, therefore, must capture the processes and products of those interactions. Collecting the contents of individuals' minds and observing routine behaviours involves learning about the organizational culture and its body of shared knowledge and beliefs, in a way that transcends the dualism inherent in the semantic opposition of 'individual versus environment'.

A generalist or nomothetic view is that it is possible to discover universals about organizations and therefore to generalize across organizations. The alternative position is that it is never possible to generalize. The midway position is that organizations do share similarities but at a very general level— the preoccupation of management with control, the development of a set of operating rules and so on—and that the forms these generalities take, such as culture, vary in particular ways within each organization.

Behaviour can be monitored and documents and other records can be scrutinized, but these do not tell us anything about why people have or have not taken various actions. Social accountability theory (see Shotter, 1984) seeks to explain people's behaviour as illuminated by their own accounts of the situation. However, attributional error alerts us to the biases in our adjudication of our own or another's behaviour; the observer is prone to attribute causality to the actor, the actor is more likely to draw attention to constraining situations that influenced what was done or not done. To attribute accidents so frequently to human error is a case in point.

To scapegoat a relatively powerless person may be a good way for management to get out of a sticky situation. Whereas crises and emergencies may be reacted to negatively, and covered up, they could be opportunities for a courageous management to understand better, re-evaluate and revise their organization's culture. Attribution theory warns us to expect attributional bias in evaluating other people's actions even if our intentions are just and fair, and that to solicit the actor's account is to solicit a more balanced picture. After all, in the courtroom those charged are given an opportunity for self-defence when they are called to account.

When organizational members act out their roles there is tacit knowledge or agreement about how things should be done and what situations mean; a kind of 'negotiated order'. The environment is 'internalized', taken inside one's thoughts and worked on by the exercise of imagination. These mental models will be more or less accurate versions of organizational reality, approximations which guide individual behaviour and actions. To learn about these models, to get inside people's heads, we can talk to people, and we can explore with them the meanings of particular situations, how an individual's views compare with others, and how far behaviour is related to intentions or how far it appears to be outside individual control. This is the mutual exploration of organizational members' views, but largely on their terms. Interviews with individuals and group discussions can be recorded systematically and then analyzed *post hoc* using analysis protocols, such as content analysis of interviews and discourse, or dialogue analysis of interactions (see chapter 7). Pauses and silences can be analyzed as well as verbal content, and the 'music' or tone of the utterances as well as the semantics. Analyses by different observers can be compared for their degree of consensus and treated statistically. The analyst may wish to use some underlying perspective, such as psychodynamic theory, in the analysis. Contradictions can occur between what is said and what is done, or between two or more verbal statements. Emotionally neutral language used to describe emotionally laden behaviour may, for example, lend clues about the organization's culture.

Ethnography is the name for a body of anthropological techniques, whose object of study is culture. Culture can reside in physical embodiments, such as buildings, or in the fluidities of behaviours and languages. Oral language is a primary transmitter of all cultures, whereas written, formal language, such as the academic and scientific, and the bureaucratic, is not common to all. Literacy is a central feature of rationalist culture.

Language is also a vehicle for getting things done. Organizational vocabularies express communal values, reinforcing and legitimizing the status quo. They express shared and collective beliefs about reality. The organizational ideology conveys a set of beliefs about what the organization is supposed to be about; how to get things done. What is *not* said can be as illuminating as what *is* said. 'Morality' may generally be a taboo word; 'stress' may be regular verbal currency in non-macho organizations; members may be enjoined not to bring their feelings into the workplace. Ethics as concept and practice is now penetrating the world of organizations (e.g. conferences on business ethics and growth in ethical stock market funds), partly perhaps because of growing public concern (at least in the west) about environment, conservation and public health issues. Symbols such as 'old school' ties, and rituals such as annual factory outings and retirement parties, serve their own cultural purposes—all these cultural processes can provoke emotion and action and a sense of shared identity.

Language can be examined through dialogue and documentation. The analyst may or may not be part of the dialogue. There is no one best way of finding things out; no rigorous methodology in uncontrollable naturalistic settings. The change agent who ignores culture prevalent in the organization is not going to be effective always. Sometimes culture may be ignored with impunity, sometimes change can be managed around it. But sometimes attempts may need to be made to change the culture. Anyway, an effective choice of strategy depends on a good prior sense of what is acceptable, and what is changeable.

Often the 'old hands' can be the least aware of the basic assumptions underlying the culture. To ask them is to attract a mixture of fact, fantasy and propaganda. But the 'native view' remains a good source of information to the wary, especially when supported with evidence from other sources, or even if it is contradicted by this other evidence. Contradictions are at least interesting, and sometimes illuminating. In the studies of management time a variety of methods have been used, including interviews, questionnaires, diary methods, and behavioural observations of meetings and telephone calls, which can produce differing records of time spent.

How we go about collecting this information has an important bearing on the quality of the data; the agent has to draw on more skill than the mere manipulation of numbers and binary logic. Perhaps we need organizational anthropologists in all our fieldwork teams to get behind the scenes. A preferred method of analysis for anthropologists is 'participant observation', where the culture temporarily absorbs and transforms its stranger researcher. But if the ergonomist is having access problems, is seen to be potentially disruptive, a threat even, then this cultural embrace is not readily forthcoming. However, the alternative status, non-participant observation, can raise its own brand of ethical problems, as discussed in chapter 2 on direct observation techniques.

The quality of such data, as for any source of data, must be considered carefully. How trustworthy, how valid are the data? It would be too easy to dismiss intrinsically qualitative data out of hand as too subjective. To do so may be to throw away the valuable. In my experience much qualitative data has been collected in an invalid way and this has contributed to the negative reputation acquired by this category of data. To use the jargon, the 'process' as well as the content is important. The conditions under which such data collection takes place affect the quality of those data in an important way (see Shipley, 1987). Under collaborative conditions the chances of acquiring valid data are much greater.

Intentions and effects: ethical dilemmas in ergonomics

How far ergonomists are prepared to offer their services depends on who is paying and how much is offered, and on the ergonomists' own value orientations. It may not be clear whose interests are being met, your own or your client's, or both. Sometimes it may not be clear who your client is, whether it really is the user of the equipment, for example, or the supplier, the workforce's representatives or the firm's management. To practise according to our professional ethical code, it is encumbent upon us to be constantly on our guard. Yet our training as scientists may not have included these ethical and value considerations. If so, we could be acting unconsciously in a complicit and collusive manner which may shock us if it were explicitly pointed out to us (see Shipley, 1982).

We would have to be open to such criticism, although the chances are that over the years we have defended ourselves well against challenges to our practices and the effects such practices have. In other words, we may block our minds and our ears. To hide behind science as an objective value-neutral enterprise may be a common defence and be responsible for our own resistance to change.

A father figure of industrial psychology claimed that the business of the applied psychologist was wholly instrumental, to supply the means for the fulfilment of another's (usually the business owner's) aims. The job of the new discipline was to 'produce most completely the influences on human minds which are desired in the interest of business' (Munsterberg, 1913, p. 24). He earlier stated: 'But no technical science can decide within its limits whether the end itself is really a desirable one' (Munsterberg, 1913, p. 17 ff.). In his day such beliefs were commonplace but I doubt whether any contemporary psychologist would get away so easily with such a bold prescription. In a special issue of the *BPS Occupational Psychology Newsletter* dedicated to the debate about values and ethics, a Birkbeck colleague introduced an added slant to the debate (Hollway, 1986). For her, good intentions are simply insufficient, and applied scientists should reclaim responsibility as human beings for the effects their practices have.

One way of doing this is by enlarging our own awareness to take account of social factors, such as cultural blocks and power dynamics at work in those organizations in which ergonomics functions. Another way is to extend our power base, as discussed previously, to increase the prospects of the desired effects following from our good intentions. The possibilities afforded by the collaborative approach for dealing with ethical dilemmas has been discussed above.

The Ergonomics Society in Britain expects its registered practitioners to practise in accordance with its code of ethical professional practice. This code enjoins us to safeguard the interests of our clients, and the participants in our research programmes, particularly their safety and welfare interests. It enjoins us not to allow our standards of practice to be compromised by politics. Ethical practice is not the central subject of this chapter on organizational analysis (but see chapter 38), but the role, values, skills, imagination and power base of the ergonomics practitioner as a proactive agent for change are (see also Buchanan and Boddy, 1992).

What kind of change are ergonomists involved in, and whose agents are they, and why are they? All organizations have to be managed, even worker co-operatives. They are not easily managed effectively and fairly. How can ergonomists extend their powers to help the good intentions of management decisions and practices to be translated into good effects? And what can they do to protect themselves and others in their care should management expect them to collude in practices which are inconsistent with their personal values and their professional code of practice? My observation is that the ergonomist is first of all a person, a fellow or sister human being, and the expert role comes second to that.

Consultants, I suspect, do not often collude knowingly, but is that sufficient to absolve us of responsibility? A comfortable presumption is that there is no underlying conflict of interests. But there is now no excuse for holding onto obsolete models and presumptions. A body of literature on social and organizational studies, and a relevant set of practical skills for learning about organizations, many of them quite everyday, exist, from which the practitioners of ergonomics science and their clients stand to benefit.

References

Blackler, F. and Brown, C. (1986). Alternative models to guide the design and introduction of the new information technologies into work organisations. *Journal of Occupational Psychology*, **59**, 287–313.

Bologh, R.W. (1990). *Love or Greatness: Max Weber and Masculine Thinking— A Feminist Inquiry* (London: Unwin Hyman).

Buchanan, D.A. and Boddy, D. (1992). *The Expertise of the Change Agent.* (London: Prentice Hall).

Buchanan, D.A. and Huczynski, A.A. (1985). *Organizational Behaviour: An Introductory Text* (London: Prentice Hall).

Burrell, G. and Morgan, G. (1979). *Sociological Paradigms and Organizational Analysis* (Aldershot: Gower).

Cassell, C. and Walsh, S. (1993). Being seen but not heard: barriers to women's equality in the workplace. *The Psychologist: Bulletin of the British Psychological Society*, **6**, 110–114.

Clark, P. (1972). *Action Research and Organisational Change* (London: Harper and Row).

Clegg, C.W. and Wall, T.D. (1984). The lateral dimension to employee participation. *Journal of Management Studies*, **21**, 430–442.

Fayol, H. (1916). *General and Industrial Management* (London: Pitman). (Translated into English by C. Storrs, 1949).

Follett, M.P. (1918). *The New State* (London: Longman).

Fox, A. (1985). *Man Mismanagement,* 2nd edition (London: Hutchinson).

Handy, C.B. (1985). *Understanding Organisations*, 3rd edition (Harmondsworth: Penguin).

Hartley, J. (1984). Industrial relations psychology. In *Social Psychology and Organizational Behaviour*, edited by M. Gruneberg and T. Wall (Chichester: John Wiley), pp. 149–183.

Hollway, W. (1986). Effects not intentions: the ethical criterion for occupational psychology. *Occupational Psychology Newsletter, Special Issue on Values and Ethics in Occupational Psychology*, **23**, pp. 5–8.

Mintzberg, H. (1973). *The Nature of Managerial Work* (New York: Harper and Row).

Mintzberg, H. (1979). *The Structuring of Organisations* (London: Prentice Hall).

Munsterberg, H. (1913). *Psychology and Industrial Efficiency* (Boston: Houghton Mifflin).

Perrow, C. (1983). The organisational context of human factors engineering. *Administrative Science Quarterly*, **28**, 521–524.

Schein, E. (1985). *Organisational Culture and Leadership* (London: Jossey-Bass).

Shipley, P. (1982). Psychology and work: the growth of a discipline. In *Psychology in Practice*, edited by S. Canter and D. Canter (Chichester: John Wiley), pp. 165–176.

Shipley, P. (1987). The methodology of applied ergonomics: validity and value. In *New Methods in Applied Ergonomics*, edited by J. R. Wilson, E. N. Corlett and I. Manenica (London: Taylor and Francis).

Shipley, P. and Harrison, R.G. (1974). The Ergonomics Practitioner as Change Agent. Unpublished paper to Ergonomics Society Annual Conference, St. John's College, Cambridge (Available from Birkbeck College).

Shotter, J. (1984). *Social Accountability and Selfhood* (Oxford: Blackwell).

Stewart, R. (1979). *The Reality of Management* (London: Pan Books).

Taylor, F.W. (1911). *Principles of Scientific Management* (New York: Harper and Row).

Teiger, C. and Laville, A. (1987). How ergonomic methods can be used by non-ergonomists. Paper to the *International Occupational Ergonomics Symposium: Applied Methods in Ergonomics*, Zadar, Yugoslavia.

Trist, E.L. and Bamforth, K.W. (1951). Some social and psychological consequences of the longwall method of coal-getting. *Human Relations*, **4**, 3–38.

Weber, M. (1947). *The Theory of Social and Economic Organisation* (New York: Free Press).

Edited readings

There are some British (Penguin) paperbacks which are edited readings, manageable in size and good prices. Readers may like to have copies of the following:

Pugh, D.S. (Ed.) (1984). *Organisation Theory*, 2nd edition (Harmondsworth: Penguin).

A mixture of classic articles by famous people such as F.W. Taylor and Max Weber through to modern theorists like Fred Fiedler and Henry Mintzberg. Organizational level of analysis bias.

Warr, P. (Ed.) (1987). *Psychology at Work*, 3rd edition (Harmondsworth: Penguin).

A successful seller and until recently virtually the only text available for use in occupational psychology courses which was not written by Americans. Bias is less towards individual level of analysis and the classical theory prominent in the earlier editions. The third edition covers an even wider range of topics moving from emphasis on individuals, through groups, to the study of organizations.

Annotated bibliography

Beetham, D. (1987). *Bureaucracy* (Milton Keynes: Open University Press).

An excellent review of theory on bureaucracy, eclectic, and concluding with a critique which incorporates a personal perspective emphasizing the democratic objection to bureaucracy . . . a specialist text.

Buchanan, D.A. and Huczynski, A.A. (1985). *Organisational Behaviour* (London: Prentice Hall).

A clearly written, good value, comprehensive paperback by University-based theorists with consultancy experience. Not noticeably extremist and controversial—more an unbiased review but does not go deeply into important areas. A worthwhile first level course text which is somewhat like (but not much) 'distance learning' in style: i.e., it might just have come out of the Open University. Solid and basic.

Burrell, G. and Morgan, G. (1979). *Sociological Paradigms and Organisational Analysis* (Aldershot: Gower).

Another good-value paperback by University theorists: well-written, and a stimulating but more controversial critique and historical analysis of organizational sociology, organized around four paradigms—i.e., functionalist, interpretive, humanist and structuralist. More appropriate for sociologists than practitioners perhaps. Should be guided reading and informed by practical experience.

Fox, A. (1985). *Man Mismanagement*, 2nd edition (London: Hutchinson).

A modest-sized well-written, forceful and stimulating paperback by this Ruskin fellow deservedly running into a 2nd edition (first published in 1974) to accommodate Britain's much altered industrial relations climate under 'Thatcherism'. Out of the eight chapters, three are devoted to the subject of participation. Fox's position is a radical one and his treatise is rhetorical in flavour. It may be too polemical for some people's taste, however. Every industrial relations specialist should be familiar with it, even if in disagreement with it.

Handy, C.B. (1985). *Understanding Organisations*, 3rd edition (Harmondsworth: Penguin).

A popular good-value 'business school' introductory paperback in the managerialist mould clearly written by a popular energetic management theorist with substantial industrial experience. Of very 'handy' proportions, too! Uncritical and relatively superficial but not a bad start to the field. Need more than this for a solid course textbook, however.

Katz, D. and Kahn, R.L. (1978). *The Social Psychology of Organisations*, 2nd edition (New York: John Wiley).

An American text, unrivalled in its field for a long time, until the stranglehold over the field by Americans was broken by some British competitors in recent years. Replete with 'evidence' and carefully put together as an argument with the theoretical stamp on it of the Institute for Social Research, Michigan. Solid. A good price.

Klein, L. and Enson, K. (1991). *Putting Social Science to Work: the ground between theory and use explored through case studies in organisations* (Cambridge: Cambridge University Press).

The *structural* is one level of explanation (the rational and objective) of organizational life in the modern world; another important perspective, and more subjective, is the *psychosocial*, and what Klein and Eason refer to as the 'dynamics of action' within those structures. The authors are particularly interested in the dynamics of professional practice in their concerns to persuade organizations to utilize expertise from the social sciences and ergonomics. They illustrate their ideas and experiences of organizational complexities from detailed case studies when working in or for large UK and (previously)

West German service and manufacturing industries in the areas of 'work humanization' and job design, and management development. The *manifest* or intended effects of decisions made in organizations are distinguished from the *latent* and unintended effects. The job of these practitioners was to help the organizations' decision-makers to develop a clearer understanding of the effects of their decisions, so as to be more able to make informed and rational choices. This required them, (Klein and Eason) as potential agents of change and 'action researchers', to be highly flexible and to tolerate high levels of ambiguity. Their interventions were in reality only partially successful which is no cause for optimism about the potential for organizational change, at least in large and complex organizations. Organizational structures were often powerful blocks to change, but sometimes useful facilitative channels for it. Klein uses psychoanalytic concepts in her theorizing about her practice; the latent side of the organization is like a subconscious mind, for example.

Ottaway, R.N. (Ed.) (1979). *Change Agents at Work* (London: Associated Business Press).

This small volume consists of contributions from eight 'change agents' or practitioners; as trainers and as consultants. Their problems and achievements in changing organizations in actual practice is the volume's focus, not organizational change in abstract. A number of the contributors are women. Mary Weir of the Manchester Business School, for example, shares with us her experiences and reflections of an action research project commissioned by the Work Research Unit, which involved her in improving work life in an American factory based in Scotland, the goal of which was the encouragement of 'human development according to a set of criteria which would be defined by the people themselves' (p. 86). What follows is an account of her largely successful intervention to build a 'self-maintaining learning process' (p. 91) in that organization.

This was done through assuming the role of learner herself as well as educator and facilitator, and by helping promote appropriate structures, such as work groups and a work improvement committee; formal but not rigid entities capable of working independently after her exit, continuing the process of organizational learning and change. Traditional measures, such as absentee rates and turnover levels, she reminds us, help us very little, and the generation of baseline data is often important but by methods more valid than the traditional unilateral questionnaire method. Talking to people, once you have established trust and credibility, is far more effective, i.e., the 'tried and true' method of everyday exchange in folk communities.

The gains were lasting: 'everyone cared enough to want to make it work and not let the Project fade through apathy or disillusion' (p. 99); the costs were 'far more time and energy than the casual observer would imagine' (p. 101). Both action researcher and collaborative participants have to be ready to give real commitment to it.

Chapter 34

Economic analysis in ergonomics

Geoff Simpson and Steve Mason

Introduction

Several authors (e.g., Alexander, 1985; Galloway, 1985; Schneider, 1985; Simpson, 1985a, 1988) have argued strongly that there is an increasing need to emphasize economic arguments in the promotion of ergonomic research and in the justification for ergonomics change. As Alexander (1985) states: 'A manager may allow ergonomics to be tried, but without bottom-line improvements (or other equally convincing measures of effectiveness) the use of ergonomics will soon diminish and then disappear'. A fundamental problem within this context is that health and safety research is still seen in many organizations as largely altruistic with no significant, or even tangible, returns on investment. This is remarkable when you consider the total cost of (un)safety to industry as given in, for example, a recent paper by Rimmington (1993). In an interesting and wide ranging paper Rimmington makes the following point:

> The studies we [the Health and Safety Executive] have done suggest that the non-injury costs of relevant incidents in work activity cost somewhere between £2 billion and £6 billion per annum, which when added to the injury costs adds up to between 5% and 10% of all UK companies gross trading profits or somewhere between 1% and 2% of gross domestic product.

Despite such evidence it often still remains true that, at local levels, management do not fully appreciate the economic implications of (un)safety and while ergonomics continues to be considered as primarily a health and safety discipline it will, inevitably, face the same problem.

Unfortunately, ergonomics, like its collaborators within health and safety, has tended to shy away from economic argument. In part this is due to genuine and justifiable reservations on the assumptions necessary to cost issues in health and safety. However, it is also in part due to a lack of realization of the extent to which equally 'debatable' assumptions are made in the 'legitimate' accountancy field and, in part, through a lack of awareness of the techniques and data which can be used to build an economic justification.

Ergonomics may have an additional problem in contrast to the 'traditional' health and safety disciplines, as Simpson (1993) has suggested:

> ... the very existence of (or the apparent need for) ergonomics is an implicit "threat" to other disciplines. For example if you need to invoke a "new" discipline to overcome the shortcomings of designers, managers, existing safety specialists, occupational physicians, etc. to improve safety/health/performance at work, then it can be inferred, implicitly at least, that such disciplines have failed.

If there is even a hint of truth in this, it becomes even more important that ergonomics encompasses all the arguments which can substantiate its contribution.

The purpose of this chapter, therefore, is to identify the kind of data which can be used by ergonomists to build an economic case and some of the procedures and calculations which can be used to present the case. By way of conclusion, a number of recent studies which have used economic analysis of ergonomics are quoted to show that not only is the objective advocated desirable, but it is also achievable.

The basic requirements for an economic analysis of ergonomics

Traditionally ergonomics has been justified on the basis of health and safety with occasional, though usually vague (and somewhat embarrassed), references to production improvements. The first step necessary in developing an economic basic for ergonomics is the resurrection of production improvements as a legitimate objective in ergonomics, for, as Campbell (1993) has pointed out 'safety does not pay, it is production that pays, but safe production pays better'. The second is the acceptance that health and safety issues can be legitimately considered as loss prevention topics. Most industrial organizations have, in recent years, undergone some form of rationalization in pursuit of a 'leaner and fitter' operation. This has almost inevitably involved staff reductions and often the breakdown of skill boundaries to promote an increasingly multi-skilled workforce. While the 'leaner and fitter' organization is undoubtedly more productive and profitable under normal circumstances, it is also less resistant to disruptions in its reduced workforce. If disruptions occur as a result of, for example, 'organizationally self-inflicted' sickness such as repetitive strain injury, then clearly it is in the organization's financial interest to remove that drain on resource—in other words to engage in a loss prevention exercise. Similarly the increasing use of human reliability approaches in ergonomics can be used in conjunction with the costing models used in engineering reliability studies. The third requirement, having defined some of the costing approaches relevant to ergonomics, is to identify the kind of data which can be used in costing calculations. Finally, the fourth stage is familiarization with some of the calculations and procedures appropriate to the analysis and promotion of ergonomics change in economic terms.

Developing economic arguments for ergonomics change

Whether or not ergonomists feel comfortable with the idea, industries are profit-making centres and every job has been created as a necessary part of a larger profit-making machine. An ergonomist, like any other person in the industry, will therefore be expected to help the organization achieve its goals. For example, if ergonomists conduct a study of a group of workers who are standing at a bench all day assembling small components, it is highly probable that they will conclude that the workstations should be redesigned so that the workforce can be seated. If they then approach the works manager and argue that money should be spent on modifying benches and buying seats to improve working postures, the chances are very slim that the changes will be authorized. The manager may say that there is no apparent problem, after all people have been working like that for as long as can be remembered with no complaints, that people are not paid to be comfortable, and that anyway the budget is not available.

The reality of the situation is that the ergonomists and managers are talking different languages. The management's remit is centred around profit-making, product quality, and meeting production time-scales. They are likely to see the benefits of ergonomics change as, at best, peripheral to their objectives. The ergonomists then get frustrated by the management's lack of immediate enthusiasm for their ideas and wander off wondering why the engineers/management cannot understand them. The simple fact is that if ergonomists want to communicate effectively with people in industry, then it is up to them to try and learn industry's language. This does not mean that the ergonomist's role has to concentrate exclusively on production-related issues. Getting recommendations implemented which improve the health and safety of the workforce will be much easier if the ergonomist talks the industry's language. To return to the previous example, the ergonomist discovered that the work study department had previously assessed the job and when working out the appropriate 'relaxation allowance' had given 3% to cover them standing all day. They then approached the manager and said 'Do you realise that you are paying that group of workers an extra 3% to stand up?'. The manager then told the ergonomist to change the workstations to allow operators to work seated. The end result was exactly what the ergonomist wanted but the route was novel and extremely effective.

Some of the many factors which can be used to develop such economic arguments are now discussed.

Available data

The data that an ergonomist may find useful are usually easy to obtain; they are probably being collected simply to run the business. The following six functions in a company are potential sources of information, issues or criteria that ergonomists are likely to find useful, either to provide an economic justification for new studies, or to show the benefits achieved as a result of ergonomics intervention. The first three functions relate to those data and issues which are normally considered to be 'close' to the ergonomics remit. The second group represent functions, issues and approaches which are perhaps less frequently considered by many ergonomists. (See chapter 4 on archival data.)

Personnel departments

    (1) Absenteeism records.
    (2) Turnover rates.
    (3) Training costs.
    (4) Compensation costs, e.g., for injury.

Safety departments

    (1) Accident black-spots (whether by site, job or equipment used can be particularly revealing).
    (2) Jobs needing special safety precautions.
    (3) Jobs needing unusual safety equipment.

These problems all cause delays either directly or through additional procedures needing to be followed to maintain safety standards.

Medical units

    (1) The nature, severity and length of absence from injury.
    (2) The nature and length of absence for health problems.
    (3) Type and frequency of minor injuries dealt with at the medical centre.
    (4) The type and frequency of symptoms reported (e.g., headaches, eyestrain).

Time off the job can easily be costed in terms of wage charges plus overheads, cost of temporary cover and lost production.

Work study/method study departments

    1. *Work study 'relaxation allowance' payments.* It is normal for work study engineers to agree to pay the workforce extra to 'compensate' for: standing, heat, physical effect, noise levels, thermal environment, and poor lighting.

These are usually termed relaxation allowances and are meant to compensate for reduced performances caused by the presence of the various influences. The accuracy of these allowances is debatable in many instances. However, since it is implicit that the management and workforce accept them, they are very useful to the ergonomist in building a proposal to improve all aspects of working conditions.

2. *High variation of individual performance.* The performance of people on incentive schemes will vary. However where this variation is larger than normal, this is likely to reveal elements of a task which demand extremes of physical effort or skills. An ergonomist could therefore be directed at reducing the needs for such excessive abilities through redesign.

3. *Inspection reject rates.* The costs of reject components are often much higher than generally appreciated. Of course some rejects will be passed and some acceptable parts rejected, and where this occurs an ergonomist could do well to study the actual inspection procedures. Where the inspection is generally accurate, high reject rates could be caused by a large number of factors many of which are within the remit of the ergonomist.

4. *High inspection costs.* A high investment in inspection will generally be justified where there are high costs and warranty claims associated with component failure. Ergonomics can make a considerable impact in controlling or maintaining the performance levels of inspectors and hence can be viewed as a loss prevention aid.

Plant engineers

1. *Excessive downtime.* Although it is traditionally considered that downtime is purely a function of poor engineering (and hence not concerned with ergonomics), recent studies suggest that operating mistakes, through poor ergonomic features, lead to some equipment breakdowns (see, for example, Williams, 1982). Likewise if routine or repair maintenance is not performed strictly in line with the procedures laid down in the manufacturer's handbooks, then the likelihood of subsequent breakdowns is increased. For example, if a fitter is under pressure to repair a machine quickly, or if the importance of the particular task is not perceived, then a hose union may not be cleaned sufficiently before disconnection. The subsequent ingress of even very small amounts of dirt can block filters very quickly and result in a rapid failure of an expensive hydraulic pump.

2. *Equipment difficult to maintain.* The costs of maintaining equipment can be up to 30% of the total operating costs of an organization and yet designers (and ergonomists) often neglect simple design solutions which can overcome or minimize basic faults. It has been estimated, for example, that repair times could be reduced by 30% in many industrial tasks by improving access alone (Seminara and Parsons, 1982).

3. *Excessive scrap wastage.* Many industries will not routinely collect this information but the costs of operating errors which result in scrapping a

component are surprisingly high, especially for those which have already undergone many machining operations.

Industrial relations departments

Problems of industrial relations could occur for a variety of reasons which may be considered as outside the scope of ergonomics (although see chapter 1 for a wider view of ergonomics). Nevertheless, poor working conditions may have been a contributor. The apparent 'cause' of the difficulty is often a symptom of a wider problem. For example, complaints of VDU operators' eyestrain through poor lighting, although often the case, may also arise from other sources, e.g., organizational changes. Solving industrial relations problems is difficult enough without diversions through irrelevant issues. The ergonomist often has information which can reduce the chances of the debate centring on symptoms rather than causes; solving industrial relations problems is difficult enough without being side-tracked into negotiating on the wrong issues.

Of course all these measures can be supplemented by data from an ergonomics survey. For example, near-miss accident data, attitudes to risk-taking and attitudes relating to job satisfaction can all combine to help construct valuable economic arguments for making ergonomics changes which can improve both the operating performances and the health and safety standards. However, moving immediately to a proposal to institute an ergonomics study to define/refine problems involves a rather strange argument in financial terms. In essence this states: 'Give me some money so that I can justify you giving me some more'! If at all possible, it is best to be armed with some information, from sources of the type described above, even if the study proposed is only an exploratory one.

Techniques and data which are useful for cost-justifying proposals

Making investment return predictions for ergonomics

The production of predictive costing figures is essentially part of the marketing exercise of presenting the proposal itself. Professionals in the marketing field emphasize the importance of presenting a product in the most favourable light without actually telling lies. The latter point is crucial, for if you exceed what is credible the proposal is likely to be lost and it is possible that all future proposals will be viewed very sceptically. Inevitably given a shortage of some data, assumptions have to be made. It is therefore important to specify the assumptions made and always to be conservative in terms of predicted achievements. It is better to offer little and deliver more, than to offer a great deal and risk disappointing people by delivering less than they expected. Anyone who has ever attempted a predictive cost-benefit analysis is fully aware of the dangers. For example, you predict a saving of £10 000, but in reality provide only £8000; you can almost guarantee the reaction which will not be (as you had hoped), 'Thanks, that £8000 is very useful', but rather 'What happened to the other £2000'!

High levels of accuracy are therefore not actually crucial. In fact it is probably advantageous if the calculations are seen to be only as reasonable estimates, as any subsequent debate with specialists will almost certainly improve the calculations. For example, if the manager does not agree with your approximation of, say, overheads on salary cost and proposes a different figure, when this is then incorporated there can be little further argument since it is after all the manager's own figure. Some examples of economic cost arguments are developed below for both production-related issues and for health and safety.

Productivity

Of the several accountancy approaches to predicting costs, two which are of particular value to ergonomics are shown below. One is to calculate the contribution of poor ergonomics to the total cost of ownership of a machine or group of machines using the methods of life-cost accounting (e.g., the accumulated costs throughout the life cycle of the machine). This framework can also be useful in 'loss prevention' arguments. The second approach is the prediction of increased revenue arising from the contribution of ergonomics to improved productivity.

An example of each is given in relation to a proposal to examine the ergonomics of a coal winning machine (shearer). Although the figures quoted are out of date, this is of little importance in the current context as the objective is simply to show the approach.

'Life-cost' calculation

An estimate of the total life-cost to the organization must be derived covering all of that type of machine in the industry. This is approximated by the equation:

Total life cost $= n (x + Lt)$

where, $L =$ the life expectancy of the machine, $n =$ the number of machines in use, $x =$ the capital cost per unit, $t =$ the operating cost of a machine per annum.

For shearers, the figures are: $L = 8$ years; $n = 540$; $x = £250 000$, $t = £500 000$. Using this equation, the total life cost of shearers in UK mining is £2·3 billion.

Predetermined time and motion systems (e.g., MTM 1; Maynard *et al.*, 1948) are very powerful tools for the ergonomist. Where these cannot be easily applied, even simple conservative estimates of improvements on a breakdown of the various operating costs may be all that are required to argue for change.

Ideally the performance implications of particular ergonomics limitations should be known, in this way the overall job implications can be built up from the component task limitations. Unfortunately, given the complexity of industrial jobs and the specific context in which they are carried out, this ideal approach is only possible retrospectively using current data. For predicting the influences of ergonomic design deficiencies on performance, a number of avenues can be used:

1. If possible an assessment of the machine (or sample of machines) should be made.
2. If relevant knowledge is not available and pilot studies are not possible, task synthesis techniques (e.g., Annett *et al.*, 1971; Maynard *et al.*, 1948) should be considered.
3. In addition, some aspects of the ergonomics literature can be used either by careful analogy with the machine under consideration, or by the use of task-oriented human error/reliability data (e.g., Swain and Guttmann, 1983).

This information on task-related ergonomics limitations can then be related to the operational cycle of the equipment involved and *conservative* estimates derived of the performance improvement likely to arise. In the shearer calculation, elements of (1), (2) and (3) were all used to derive the figures contained in Table 34.1. This example shows that if fully implemented, ergonomic improvements in shearer design are likely to save 6·4 min each shift for a typical shearer design. These machines are available for production for 330 min each shift and therefore the current ergonomic limitations lose 2% of the total potential operating time.

It should be noted that these estimates include a potential saving through improving the maintainability features of shearers. Maintenance aspects of machinery are often overlooked, however, as maintenance costs are typically 30% of the costs of ownership, substantial savings can usually be found simply through recommending improved access to those components requiring frequent attention during routine maintenance operations.

Life-cost accounting is being used increasingly in industry as an aid to purchasing decisions. For example, faced with a choice between two machines (A and B) both of which seem adequate for their purpose, but without any additional information most would normally choose the cheapest (say, A). However, a life-cost analysis may show, for example, that spares are more expensive for A, and/or that routine maintenance takes longer. When the cost of these issues are projected over the life expectancy of the machine, the capital cost differential may become irrelevant with B in fact proving to be cheaper over the long-term. While this may seem a long way from ergonomics, there are in fact many opportunities to incorporate ergonomic issues into life-cost analyses. Say, for example, that A is noisier

Table 34.1. Potential saving from improved ergonomics of coal winning machines

Shift breakdown	Average duration (min)	Percentage saving through application of ergonomics of coal winning machines	Potential saved time (min)
Men travelling	89	0	0
Preparation and meals	26	1% through better design, reducing preparation	0·26
Machine running	107	2% resulting from improved performance	2·14
Ancillary time	29	1% through better design	0·29
Operational time	63	0	0
Lost time			
Electrical	12	5% through diagnostics	0·6
Mechanical	31	10% through access and handling improvements	3·1
Mining/geological	77	0	0
		Total min/shift	6·39

than B. In this case it can be argued that the life–cost equation should include the cost of noise reduction screening or the provision of hearing defenders. Similarly it could also include the cost of routine audiometry, changes to warning signals, and even estimates of the possible cost of litigation for hearing loss. Even if these fail to negate the capital cost differential which favours purchase of the noisier machine, they will have sensitized the management to the financial benefit of an ergonomics investment in hearing protection.

Increased revenue calculation

The time penalty of poor ergonomics of shearers over a 1 year period = the number of machine shifts per week × number of working weeks × the cost of poor ergonomics per shift. This calculation shows the cost to be 32 431 h/year. An average high technology coal face produces 3 tonnes/min at a saleable price of £40/tonne, and cuts coal for an average of 107 min per shift. The revenue per shift is therefore £12 840 or £7200 per machine hour.

Assuming all the lost hours per year from poor ergonomics (32 431) could be converted into extra machine running time over all the shearers, then:

$$\text{increased revenue} = 32\ 431 \times 7200$$
$$= \text{£234 million for the 540 units,}$$
$$\text{i.e., £0·4 million per machine.}$$

The cost or ergonomic limitations per annum in this instance is equivalent to:

$$\frac{234 \times 10^6}{2{\cdot}3 \times 10^9} \times 100 = 10\% \text{ of the total life-cost of shearers in the UK.}$$

The increased revenue figures may look spectacular, however the additional production may not have a market, at least at the existing prices, and the cost of the retrofit design changes needed may be prohibitive.

This problem can be reduced or eliminated by providing designers with all the necessary ergonomic criteria in a form that can be used at the concept and drawing board stages of new designs. This may seem ambitious, however this is the approach adopted by the British mining industry. Ergonomists there have produced separate design handbooks for a number of different types of mining machine including both their operational and maintenance requirements. By producing separate design handbooks for different types of machine, the ergonomists were able to develop much more specific guidelines than would have been possible if a single handbook was provided for all mining machines (Simpson and Mason, 1983). Once produced, the same data can be used and re-used by the designers in all the supplier companies and hence appropriately derived and presented ergonomics information can in fact be used to influence all machines of a particular type. Thus the long-term benefits can be very considerable indeed.

One step short of the increased revenue calculation (which avoids some of the reservations such as, 'can the extra product be sold?') is to examine machine/system availability. Most managers would listen to any proposal to reduce downtime and reasonable, easy-to-use estimates have been proposed to calculate availability. For example:

$$\text{Availability} = \frac{\text{MTBF}}{\text{MTBF} + \text{MTTR} + \text{MTPM}}$$

where, MTBF = mean time between failure; MTTR = mean time to repair; and MTPM = mean time for preventive maintenance.

As several authors (e.g., Seminara and Parsons, 1982; Ferguson *et al.*, 1985) have emphasized the role of ergonomics in the design of machines for ease of maintenance, it is possible to see immediately two aspects of the equation (MTTR and MTPM) where an economic case for ergonomics can be made.

Health and safety

The costing of health and safety issues is often avoided because of the inability to 'cost a life' given the natural ethical and/or moral reservations surrounding such an exercise. However, if one accepts that no organization actively wants to kill its staff and thus work simply on the costs to the organization, then the moral and ethical reservations on costing health and safety disappear, especially if by doing so ergonomists increase the probability of obtaining funding to promote improvements in health and safety.

There are, of course, many cost implications to an organization arising from health and safety issues; however the 'core costs' are as follows:

1. Costs incurred by disruptions in manning from non-work related sickness absence, e.g., influenza.
2. Costs incurred by disruptions in manning from work-related, chronic health issues, e.g., back pain, respiratory disease, dermatitis.
3. Costs incurred by disruption in manning from lost time injuries.
4. Costs incurred from compensation payments against injuries, e.g., loss of limbs.
5. Costs incurred from compensation payments against health effects, e.g., back pain, hearing loss, tenosynovitis.
6. Direct cost in lost production arising from accidents, i.e., time lost in accident recovery, investigations.
7. Social costs incurred in sickness benefit payments.

Although generalized costs are usually relatively easy to obtain, it is much more difficult to find information on particular subgroups of the workforce, e.g., a particular job category. However, with some effort a reasonable approximation is often possible.

Using the shearer example, the potential health and safety costs could be developed as follows:

- Define the relevant costs from (1) to (7) above for the industry.
- Reduce these costs proportionately to the numbers of men associated with shearer driving.
- Multiply these costs by the life expectancy of the machines.

Note: The costs in (7) above can be ignored for this group as they are effectively off-set by payment in lieu of wages. The costs in (1) are irrelevant. The costs in (5) may be influenced but a significant improvement is unlikely. Ergonomic improvements should reduce the costs in (2) and (3) but the amount is difficult to estimate. The national costs of (4) and (6) are known and will be influenced by improvements in the ergonomics although the total cost is not recoverable.

For convenience, appropriate at this level, assume that the ergonomic savings in (2) and (3) are equivalent to the costs in (4) and (6) which cannot be related to ergonomics. The relevant national health and safety costs can therefore be approximated to the costs in categories (4) and (6) which, in this case, allowing for inflation since the last published data are £8 million

(Collinson, 1980). Assuming health and safety problems are equivalent across all production job categories, then:

$$\text{cost per underground worker} = \frac{\pounds 8 \times 10^6 \times (1 - s/u)}{u} = \pounds 50 \text{ p.a.}$$

where, s = surface workforce; and u = underground workforce.

Shearers are one- or two-man operated, working two or three shifts per day and therefore we can assume four operators for each of the shearers. The maintainers of the shearers will also benefit from improved design and therefore using the industry's comparison of production and maintenance man-shifts, it is reasonable to assume that the same number of maintenance men are also involved. Health and safety associated purely with the ergonomics of shearer design = cost per man ($\pounds 50$) × number of men (8) × number of shearers (540) = $\pounds 216\,000$ p.a.

Clearly, such a calculation can be made more accurate by knowing, for example, exactly the number of shearer drivers, by using a ratio which avoids the assumption of equivalent risk over all job categories, by a closer approximation to the actual costs in (2), (3), (4) and (6). However, the principle involved is the same and the degree of accuracy obtained will depend on the availability of information and the argument to be developed. For example, if it is simply to show that real costs are involved, a superficial approximation as above may suffice, whereas to predict the rate of return on a redesign investment would require a more careful analysis.

Examples of economic analysis in ergonomics studies

Previous sections have suggested the type of performance, health and safety issues which can be used to develop an economic case to support ergonomics studies. This section provides examples of a number of studies which have included economic analysis. Taken as a whole they cover each of the three main areas of ergonomics activity, showing clearly that ergonomics can, and often does, contribute not only to the health and safety of the workforce, but also to the financial health of the organization.

Problem of heat stress in the steel industry

Feinstein and Crawley (1968) described a study of the design and siting of a slab shear pulpit in a steelworks which proved to be of considerable economic benefit. The job was to remotely-operate a shear blade to remove the end defects in steel slabs prior to rolling. The defect of most importance was a 'pipe', a hollow indentation, which occurred at each end of the slab. The 'pipe' was an inevitable consequence of the cooling process and was therefore entirely predictable. Unfortunately however, the depth of the 'pipe' was not. The problem was exacerbated by the surface temperatures of the steel which created a considerable radiant heat problem (see chapter 16). In order to avoid the heat, the pulpit had been sited over 10 m away from the shear blades. This distance, together with line of sight problems (which meant that to check the front cut, the slab had to be reversed beyond the pulpit), forced the operator into one of two equally disadvantageous practices. In order to get an accurate cut, the operator had to loop the slab several times to and from the blades, thus creating a bottleneck. Alternatively, if he wished to avoid bottlenecks, he had to deliberately overcut which, of course, incurred the cost of 'wasting' good steel.

The benefits of a pulpit closer to the blades were of course obvious. However there remained the problem of how to protect the operator from the radiant heat. The ergonomists suggested that the pulpit should be glazed with gold laminate glass, which was known to significantly reduce heat flux, and that a number of other improvements should also be included at the same time. When the redesign was costed the management considered it to be far too expensive given that no estimates of the return on the cost had been presented. The authors went back to their laboratory and carried out a series of studies to approximate the performance improvements which could be expected from the changes proposed. When these were presented to the management, it was decided to build the new pulpit. It was also decided that the actual benefit should be examined. The plant immediately instigated a study to measure the delays caused by the bottlenecks at the shears and checked all the off-cuts to establish the amount of good steel being recycled due to overcutting. The same measurements were continued for a year after the pulpit had been installed. This study identified a saving of slightly over £120 000 in the first year. The new pulpit had cost £10 000. This represented a payback time on the capital invested of less than one month!

Back pain

This is in many senses the classic example of an ergonomics/health issue which can be treated as a loss prevention exercise. Details of the overall cost of the problem have been presented elsewhere (see, for example, Simpson, 1985b) so two particularly graphic statistics will suffice in this context. Manstead (1984) presented data which showed that in the UK during 1982 more than six times as many man–days were lost due to back pain than the total lost due to industrial disputes! Hyland (1992) has suggested that the total cost to the UK economy arising from back pain related sickness absence is in the order of £9 billion per annum. Moreover, it should be remembered that back pain is by no means restricted to the 'heavy' jobs or those involving manual handling. Lloyd *et al.* (1986), for example, have shown in a study which compared miners and office workers, that there was no significant difference in the incidence of back pain in the two groups under the age of

45 years. There can be no doubt that back pain is both widely prevalent and expensive to industry, however can it be shown that ergonomics intervention will reduce those costs to the benefit of both the individuals and their employers? Teniswood (1982) describes a study of back pain in an Australian mining company. The study covered both surface and underground operations and examined manual handling, workspace and vibration in the drivers' cabs of mobile plant and the introduction of a new training programme. The number of lost time back injuries were halved in two years. Although the author himself places no financial value on this achievement, a subsequent paper (Anon, 1983) states that the company's operations were \$A157 000 per annum more efficient.

A paper by colleagues of the authors (Chan *et al.*, 1987) describes an interesting use of the cost implications of back pain to promote ergonomics change. They had been asked to advise on the workstations for a new engine assembly line which, at that stage, was still on the drawing board. An initial examination of the plans for the new line and studies of a similar existing line suggested that four tasks on the line were likely to need improvement in ergonomics terms. Simulations of these tasks in the new configuration showed that the workplace design was likely to make the target cycle times set for the operations unachievable.

One of the major changes on the proposed new line was a significant reduction in the number of repair loops on the line in comparison with previous practice. The combination of less repair loops and failure to meet cycle times on the tasks studied suggested that bottlenecks would be inevitable. However there was also considerable concern that redesigning the four workplaces could incur unacceptable delays on the progress of the whole development which was on an exceptionally tight schedule.

It seemed possible that the ergonomics arguments would be lost. It was then decided to use the fact that on two of the tasks, back pain was likely to be a major problem in conjunction with the discovery that musculoskeletal problems (in particular, back pain) was the largest cause of sickness absence among the company's assembly line workers, to attempt an economic argument (see chapters 22, 23 and 30). Using information from the simulations and data from the company's medical service, it was possible to predict that the absenteeism from back pain on the new line (70 operators) would be of the order of 50 man-weeks/year. Covering this absence within the proposed manning levels was almost impossible and 'carrying' spare manning was also considered to be unacceptable by the company in terms of break-even costs. The predicted financial implications created a new basis for the ergonomics argument.

When the ergonomics changes were proposed on the basis of yielding improvements in cycle time of between 6 and 16% (dependent on task) and a reduction of approximately 20% in musculoskeletal absence for those tasks, they were accepted.

Although no information is yet available to show whether these predictions were achieved, this example shows how important an economic argument is, even when the potential of ergonomics was recognized by the company, as evidenced by the fact that they called in the advice at a relatively early stage.

Workplace design

The first study discussed in this section is another example of how a loss prevention argument can ensure that ergonomics considerations are taken seriously. Work carried out by the British mining industry on prototype mining machinery (Mason *et al.*, 1980) had shown that there were often serious ergonomics limitations in terms of both the control layout and the sightlines on a wide range of mobile plant. Unfortunately as new prototypes are hardly an everyday occurrence and have to be studied in surface simulations, it was difficult to obtain any feel for the real implications of such shortcomings. Moreover it became apparent that a fully developed working prototype was too late in the process to suggest major changes on ergonomics issues, especially as the penalties which would arise from ignoring the ergonomics improvements were unknown. A subsequent study (Chan *et al.*, 1985) was able to examine the operational implications of poor layout and restricted sightlines (as well as other ergonomics issues) in some detail for a particular class of mining equipment—underground development machines.

These machines are used to drive underground roadways. They have the facility to remove strata, either by using rotating picks to cut it down or by drilling for shot-blasting. They are also able to collect the debris and load it onto a conveyor system for removal, either to pack the roadway sides to improve stability, or to take it to the surface. Such machines tend to be rather large, yet by the nature of mining, they are expected to operate in relatively confined spaces. This creates particular problems in terms of both controls and sightlines. The position of the driver and the bulk of the machine often restrict vision. Additionally all the machines studied were tracked vehicles which, given their large size and the confined space, made them difficult to position accurately.

The fact that these reservations were not simply a failure to meet some form of academic/ergonomics ideal was shown by the field studies in that it was standard practice in 80% of the machines studied for an additional man from the development team to act as a 'spotter' for the driver. The spotter positioned himself at a point which gave unrestricted vision and signalled instructions to the driver using hand and caplamp signals. The use of the spotter immediately identified that there were economic implications behind the ergonomists' concerns. What was intended as a one man operation was in reality closer to a 1·5 man operation. The study was also able, using a variety of techniques (including some from work study), to estimate the contribution of poor sightlines and control layout to the overall cycle time for the operation. The studies suggested that the ergonomics limitations were adding approximately 5% to the cycle time. This, used with a knowledge of the cost of drivage operations, together with estimates of the salary costs incurred from the 'unnecessary' function of the spotter (multiplied across all similar machines in use in the industry) revealed a total cost of between £8 million and £18 million, depending on which figures were used for drivage and wage costs.

As a result of highlighting the loss prevention argument the initial reluctance to 'impose' ergonomics considerations on the designers, because of lack of tangible benefits, was reduced considerably. For example, all suppliers of such machinery are being issued with ergonomics design manuals which were also produced during the study. Achieving this endpoint would have been almost impossible without the economic argument.

The second example in this section concerns a Norwegian study (Spilling et al., 1986) of the influence of workplace design on musculoskeletal problems, and is one of the most thorough examples published so far showing the economic analysis of ergonomics change. The plant studied was primarily concerned with the assembly and wiring of telephone switching panels, and employed a predominantly female workforce. The workstations involved considerable muscular loading and awkward postures. This was reflected in high sickness absence records for the plant.

During 1975 the authors carried out an extensive ergonomics redesign. Particular emphasis was placed on the need to give each operator greater flexibility allowing, for example, for both seated and standing operation. Several other changes were also made, including improved seating, improved tools and major changes to both lighting and ventilation. A number of the factors mentioned earlier in this chapter, e.g., sickness absence and labour turnover, which could be expressed in financial terms were then monitored throughout the period to 1983.

Prior to the improvements, musculoskeletal sickness absence was running at 5·3% of the production time available; in the period from 1975–82 it had dropped to 3·1%, a difference which was significant. While this is obviously a major improvement, a closer look at the period around the change shows an even greater effect. Although the average prior to 1975 was 5·3% as stated, the trend was rising steeply—the figures for 1973 and 1974 being 6·8% and 10%, respectively. Between 1979 and 1982 however, the absence rate was almost static at just below 3%. The analysis of labour turnover was even more marked. Prior to 1975 turnover was running at about 30%, whereas in the period after the change (to 1982) it had been reduced to an average of slightly over 7·5%. Obviously the labour turnover could have been influenced by many factors other than the ergonomics improvements. However, in interviews with the staff the improved working conditions were the most frequently mentioned issue. These reductions in labour turnover also created additional financial benefits beyond the immediate production improvements, for example, there were attendant savings in terms of both training and recruitment costs, which at a turnover of 30% were considerable.

The authors report a long and detailed financial analysis including indirect savings such as the training cost reduction. All calculations were normalized and fully amortized (over a 12 year life). The savings totalled approximately 3·25 million NKr on an investment, covering both the study and the cost of implementation, of approxiamtely 0·5 million NKr.

While there are a number of other studies in the literature which include some form of economic analysis of ergonomics intervention (see, for example, Simpson, 1988) the above are sufficient in the present context to show that economics can be a powerful tool for the ergonomist. Unfortunately, however, despite these examples economic analysis remains rarely used in ergonomics, despite the fact that many studies collect the type of data which would make it extremely easy. A study reported by Ong (1984) is a good example. The study covered VDT operations in an airline computer centre in Singapore. Improvements were made to the workplace, the lighting and the work patterns and extremely detailed records taken covering reported fatigue and performance. Reported muscle fatigue dropped considerably, in some areas by a half, and visual fatigue reduced by about 30%. Performance in terms of keystrokes/hour improved by almost 25% simultaneously with an error rate reduction from around 1% to 0·1%. Obviously these results are impressive as they stand but with relatively little effort, they could have been turned into monetary figures, which would have had even more impact. Moreover, a financial statement is, in effect, context free—it is, if the pun can be forgiven, the 'universal currency' of management discussion.

Human error analysis as a basis for accident reduction

A recent study conducted in the mining industry (see, for example, Simpson, 1992; Simpson, 1993) used basic ergonomics principles supplemented by the human error classifications of Reason (e.g., Reason, 1990) and Rasmussen (e.g., Rasmussen, 1983) to predict the potential for human error in mining systems (see also chapters 31 and 32). New safety initiatives were then identified by using the potential human errors to target safety activity. Prior to the study the accident rate at the colliery was over 36 per 100,000 man shifts; one year after the study and the implementation of the initiatives, the rate had reduced to just over 8 per 100,000 man shifts, a reduction of 80%. The savings when measured against the cost of the study represent a return on investment of the order of 36:1. Even when an estimate of the cost of implementation of the initiatives is added, the return on investment was of the order of 15:1.

A recent book by Oxenburgh (1991) has presented a collection of short case studies each of which, in one way or another, addresses the question of the return on investment from ergonomics studies and action and which provides an excellent basis for developing a generic argument for the economic viability of ergonomics.

Conclusions

Arguably, if ergonomics is as important to industry as ergonomists believe, then the extent to which it is not routinely used must suggest some failing in the way it is presented to management. It would seem reasonable to presume that as most managers already have plenty of problems, they are unlikely to be enthusiastic when someone turns up telling them they have more, especially if they are of the previously unheard-of ergonomics variety! Given that they already know they have safety problems, health problems, productivity problems and cash flow problems amongst others, it can hardly be surprising if they 'turn a deaf ear' to someone telling them they also have ergonomics problems. Moreover, a manager's job is to solve problems not to go looking for them.

The acceptance of these simple points immediately identifies a new approach to the promotion of ergonomics. First, there are no such things as ergonomics problems—there are however a wide range of problems including safety problems, occupational health problems, productivity problems and labour turnover problems in which ergonomics knowledge may be of assistance. Second, ergonomics must be seen increasingly as the provider of solutions rather than as the identifier of problems. Once it is accepted that the role of ergonomics is to aid in the solution of 'accepted industrial problem areas', then the development of economic arguments to support and justify the expenditure on ergonomics becomes much easier as the majority of relevant data are already being collected somewhere in the organization. The presentation of ergonomics as relevant to accepted problem areas supported with an economic bias immediately gives the ergonomists the advantage of talking the same language as the manager they are trying to convince. The increasing use of economic benefit in support of ergonomics change will in no way undermine the basic objective of improving health and safety. In many circumstances, it may in fact be the only way to ensure that the improvements are implemented.

This chapter has shown the kind of data which can be used to develop a cost-benefit approach to ergonomics and some of the procedures available to utilize such data. It has also shown, from the literature, that such analyses are in fact feasible and that they often have a major influence on the acceptance of ergonomics proposals. All that remains is for the use of economics in ergonomics to become the rule rather than the rare exception. This will hopefully lead to more publications on the economic analysis of ergonomics benefits, which will in turn enable more ergonomists to present their proposals and results in a form of immediate interest to the industries who fund them.

References

Alexander, D.C. (1985). Making ergonomics pay: adopting a business approach. *Industrial Engineering*, July, 32–39.

Annett, J., Duncan, K.D., Stammers, R.B. and Gray, M.J. (1971). *Task Analysis*. Training Information Paper No. 6. (London: HMSO).

Anon (1983). The human factor. *Mining Magazine (MIMAG)*, December, 15–18.

Campbell, B. (1993). Towards safe production in the Ontario mining industry. In *Proceedings of MineSafe International 1993*. (Perth Western Australia: WA Chamber of Mines).

Chan, W.L., Pethick, A.J., Collier, S.G., Mason, S., Graveling, R.A., Rushworth, A.M. and Simpson, G.C. (1985). *Ergonomic Principles in the Design of Underground Development Machines*. Final Report on CEC Contract 7247/12/007. (Edinburgh: Institute of Occupational Medicine) (IOM Report TM/85/11).

Chan, W.L., Pethick, A.J. and Graves, R.J. (1987). Ergonomic implementation in the design of an engine assembly line. In *Contemporary Ergonomics*, edited by E.D. Megaw (London: Taylor and Francis), pp. 140–145.

Collinson, J.L. (1980). Safety—the cost of accidents and their prevention. *The Mining Engineer*, January, 561–571.

Feinstein, J. and Crawley, J.E. (1968). *The Ergonomic Design of a Slab Shear Pulpit at Colvilles Limited*. Report No. BISRA OR/HF/8/68. (London: British Steel Corporation).

Ferguson, C.A., Mason, S., Collier, S.G., Golding, D., Graveling, R.A., Morris, L.A., Pethick, A.J. and Simpson, G.C. (1985). Final Report on CEC Contract 7247/12/008. (Edinburgh: Institute of Occupational Medicine) (IOM Report TM/85/12).

Galloway, G.R. (1985). Marketing ergonomics: influencing developers, managers and customers. In *Ergonomics International 85*, edited by I.D. Brown, R. Goldsmith, K. Coombes and M. Sinclair (London: Taylor and Francis), pp. 991–993.

Hyland, F. (1992). The new legislative framework. In *Proceedings of the National Back Pain Association Seminar 'Back Pain at Work: Whose Responsibility?'* (Teddington: National Back Pain Association).

Lloyd, M.H., Gould, S.R. and Soutar, C.A. (1986). Epidemiologic study of backpain in miners and office workers. *Spine*, **11**, 136.

Mason, S., Simpson, G.C., Chan, W.L., Graves, R.J., Mabey, M.H., Rhodes, R.C. and Leamon, T.B. (1980). *Investigation of Face-end Equipment and Resultant Effects on Work Organisation*. Final Report on CEC Contract 6245-12/8/47. (Edinburgh: Institute of Occupational Medicine) (IOM Report TM/80/11).

Manstead, S.K. (1984). The work of the Backpain Association: past, present and future. In *Occupational Aspects of Back Disorders*, edited by J. Brothwood (London: Society of Occupational Medicine).

Maynard, H.B., Stegmarten, G.J. and Schwarb, J.L. (1948). *Methods—Time Measurement* (New York: McGraw-Hill).

Ong, C.N. (1984). VDT work place design and physical fatigue: a case study in Singapore. In *Ergonomics and Health in Modern Offices*, edited by E. Grandjean (London: Taylor and Francis), pp. 484–494.

Oxenburgh, M. (1991). *Increasing Productivity and Profit Through Health and Safety* (North Ryde, NSW: CCH Australia Ltd).

Rasmussen, J. (1983). Skills, rules, knowledge, signals, signs and symbols and other distinctions in human performance models. *IEEE Transactions on Systems, Man and Cybernetics*, **SMC-13** (3).

Reason, J.T. (1990). The contribution of latent failures in the breakdown of complex systems. *Philosophical Transactions of the Royal Society of London B*, **327**, 475–484.

Rimmington, J. (1993). Does health and safety at work pay? *Safety Management*, September, 59–63.

Schneider, M.F. (1985). Ergonomics and economics. *Office Ergonomics*, May/June, p. 8, 12, 30.

Seminara, J.L. and Parsons, S.O. (1982). Nuclear power plant availability. *Applied Ergonomics*, **13**, 177–189.

Simpson, G.C. (1985a). Some requirements for improving the industrial utility of ergonomics. In *Ergonomics International 85*, edited by I.D. Brown, R. Goldsmith, K. Coombes and M. Sinclair (London: Taylor and Francis), pp. 334–336.

Simpson, G.C. (1985b). Cost benefits of ergonomic action. In *Ergonomics in the ECSC industries 1980–1984*. Community Ergonomics Action Report 5, Series 3, Volume I (Luxembourg: European Coal and Steel Community).

Simpson, G.C. (1988). The economic justification of ergonomics. *International Journal of Industrial Ergonomics*, **2**, 157–163.

Simpson, G.C. (1992). Human error in accidents: lessons from recent disasters. In *Proceedings of the Institute of Mining Engineers International Symposium 'Safety, Hygiene and Health in Mining'*. (Doncaster: Institution of Mining Engineers).

Simpson, G.C. (1993a). Promoting safety improvements via potential human error audits. In *Proceedings of the 25th International Conference of Safety in Mines Research Institutes*. (Pretoria, RSA: Chamber of Mines of South Africa).

Simpson, G.C. (1993b). Applying ergonomics in industry: some lessons from the mining industry. Keynote Address to the Ergonomics Society Annual Conference. In *Contemporary Ergonomics 1993*, edited by E.J. Lovesey (London: Taylor and Francis).

Simpson, G.C. and Mason, S. (1983). Design aids for designers: an effective role for ergonomics. *Applied Ergonomics*, **14**, 117–183.

Spilling, S., Eitrheim, J. and Aaras, A. (1986). Cost benefit analysis of work environment investment at STK telephone plant at Kongsvinger. In *The Ergonomics of Working Postures*, edited by E.N. Corlett, J. Wilson, and I. Manenica (London: Taylor and Francis), pp. 380–397.

Swain, A.D. and Guttmann, H.E. (1983). *Handbook of Human Reliability Analysis with Emphasis on Nuclear Power Plant Applications*. Report NUREG/CR-1278 (Washington DC: US Nuclear Regulatory Commission).

Teniswood, C. (1982). Back injury prevention in metaliferous mining operations. In *Ergonomics and Occupational Health*, edited by P. Rawlings (Melbourne: Ergonomics Society of Australia and New Zealand).

Williams, J.C. (1982). Cost effectiveness of human factors recommendations in relationship to equipment design. *Proceedings of the Institution of Chemical Engineers Symposium*, Series No. 76 (London: Pergamon Press).

Part VII

Introduction and implementation of systems

How often, in our professional or personal lives, have we had what we believed to be an excellent idea, or proposal turned down, and how often has that happened not because of lack of intrinsic merits or due to flaws in the plan but because of the way we introduced it to our colleagues or friends? Likewise the success or failure of many new systems, or at least the degree of their successful utilization, will be determined by how they are implemented. Eason, in chapter 36, looks at this issue generally and with relevance to all the other content of this book. First however, we need to understand how best to manage any ergonomics intervention in industry. How do we get agreement for our investigation or development, how best do we manage the initiative, how can we communicate with all concerned and, again, how can we best aid successful implementation of our suggested changes? Haines and McAtamney cover this in chapter 35.

The type of implementation strategy adopted, as identified by Eason, will determine the degree of participation that is possible. What he sees as concepts of value—the user 'champion', user representation, and full user involvement—will be more feasible and relevant in a strategy of incremental implementation than in a 'Big Bang' scenario. Whilst there are some downsides—for instance the time and other resources required, participant coping problems and so on—Eason generally argues for user-orientation in systems implementation, which includes development and introduction. In doing so he provides apt commentary for the concerns of the whole of this book. Concern for and involvement of the people who work in offices, factories, transport and services etc., must underpin ergonomics methods and techniques.

With this in mind, many occupational ergonomists are currently very involved in the development and promotion of participative methods and processes. Although this philosophy and approach can have difficulties and unwanted side effects, generally it is *the* appropriate methodology for ergonomics *practice*; Wilson, in chapter 37, reviews types, processes and methods of participation.

Finally, Corlett in chapter 38 puts the content of the book into a context of both our role and duties as applied ergonomists.

Chapter 35

Undertaking an ergonomics study in industry

Helen Haines and Lynn McAtamney

Introduction

One of the greatest problems for ergonomists is gaining a clear understanding of how the available methods and techniques may be utilized in practical studies. Success is often as much a matter of understanding the organization involved and being able to structure an appropriate programme of work as it is of knowing about ergonomics methodology. This chapter will describe a step-by-step approach to undertaking an ergonomics study using the particular example of assessing physical work in industry (such as manual handling or static work—see chapters 23 and 30) and from the perspective of someone from outside the company. This particular focus derives in part from the authors' background. Nevertheless it is true that many of the stages described below will extend to other types of investigations. The approach outlined here is not meant to be read like a recipe. Depending on the particular nature of the workplace and the reasons for undertaking the study, certain sections or stages may prove much more relevant than others (see Figure 35.1).

The chapter begins with a brief discussion of study aims. From there we move on to the issue of commitment, explaining the importance of support from people throughout the organization and outlining strategies by which this might be achieved. The next section provides a number of techniques for identifying and prioritizing problem areas. Having done this, we then discuss the selection of appropriate field methods, data collection, and the analysis and interpretation of results. In the later sections we provide some basic guidelines concerning the writing-up and presentation of any findings along with advice on strategies for the effective implementation of changes.

In summary, our aim in this chapter is to provide an overview of project management for practical workplace ergonomics which may be read in conjunction with the more detailed examination of risk assessment in chapter 30, cost-benefit analysis in chapter 34, participative processes in chapter 37 and implementation in chapter 36. Other information on project management may be found in Dul and Weerdmeester (1993, chapter 6), papers in the journal *Applied Ergonomics*, contributors to three special issues (parts 3, 4 and 5) on 'Marketing Ergonomics' in the journal *Ergonomics* volume 33, and (about the change agent) in Buchanan and Boddy (1992). In addition some useful case study material can be found in Corlett (1991) and Alexander and Pulat (1985, chapters 9–16).

Study aims

Ergonomics studies are instigated for a variety of reasons and vary in their depth and range. However at the centre of any study is the issue of how well the demands of the task, work environment and work organization match with the skills and capacities of the workforce. In cases where a mismatch exists, the general aim of the study is to assess the problems which might result (for example, risks of injury; reduced work quality etc.) and to develop appropriate intervention strategies to counteract these problems.

In practice, therefore, the ergonomist is invariably engaged in addressing a set of much more specific concerns, such as to carry out an audit of manual handling operations within a plant, or to investigate the causes of musculoskeletal problems (e.g. neck and shoulder complaints amongst microscope operators). In certain cases the ergonomist may not be able to clarify the study aims until the problems have been better identified and prioritized (this is discussed later in this chapter). However it is important that the aims of the study are identified as soon as possible so that everyone involved understands the scope of the study and appreciates what can reasonably be achieved given the available resources (even if some of the details have to be subsequently modified as further insights are gained).

Securing commitment to the study

Commitment at all levels of the organization is one of the most important factors in undertaking a successful study. Without the support from those with the power to make decisions, control finances and provide the essential information about company culture and work practices, the whole project will be much more difficult and could even fail. Just because you have been called in to act as a troubleshooter does not always mean that everybody concerned agrees upon the nature or even the existence of a problem.

At times an ergonomist may be brought in to act as a kind of arbitrator between management and the workforce, to decide upon whether or not a genuine problem exists. For example, van drivers might complain of back problems due to inadequate seating. However management may suspect that this is just an excuse to obtain a new fleet of vehicles. In such cases securing everybody's commitment to the study will require careful negotiation beginning with some level of agreement about the overall aims of the study.

Clearly a number of the factors determining people's likely levels of involvement can be out of the ergonomist's hands. Nevertheless, through appropriate consultation and education the chances of securing an adequate level of commitment may be significantly enhanced.

A useful starting point is to organize a presentation, in the first instance for management and then, having gained their approval, to key members of the organization likely to be involved in the study. The presentation should aim to clarify the following points:

- the current situation and likely implications if no action is taken
- the aims of the study
- how the assessment will be undertaken
- what each department or level of the organization is expected to do during the course of the study
- an indication of the resources required in terms of time, money and personnel.

When undertaking an assessment, especially when covering large groups of people, it is often useful to arrange for a working group to be set up. Generally speaking this should include a senior manager within the company, or their appointed representative, who will have final responsibility for implementing the recommended changes. It is also important to include a number of representatives from other levels of the organization including operators, supervisors and line managers. These members of the working group will provide the kind of first hand knowledge of the work situation which is essential for a clear understanding of the problems that exist. Furthermore they will be a vital source of feedback later in the study as to whether or not the changes, when implemented, have proved effective. In addition, other members of the working group could be drawn from personnel, health and safety and industrial engineering departments as they can often provide important details concerning company culture and the way work is organized. They can also advise on any future plans for restructuring or investment, be a source of archive material (see chapter 4) and may be a useful source of support for the assessment process. Where they are not directly involved within the working group, the relevant trades union representatives should also be contacted before any decisions are made about planning or undertaking the study.

A more general awareness programme may then be used to disseminate information about the study throughout the organization. Depending on the type of company and number of people involved, this may range from an informal conversation to a formal presentation or written explanation. Issues that may be covered include: the purpose of the assessment, what it will

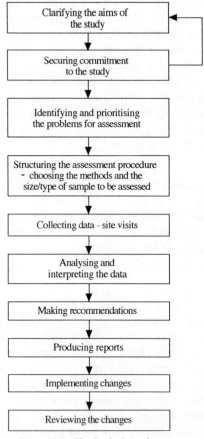

Figure 35.1. The Study Procedure

involve, who it will involve as well as giving the necessary reassurances about confidentiality.

Time taken at this stage can greatly influence the success of the investigation. First of all, communication as to proposed functions to give a broad range of people a greater sense of involvement in the study at the same time as allaying any fears that employees may have about the reasons for assessing their workplace. Indeed, taking a general approach of worker participation by actively encouraging involvement at this and subsequent stages of an ergonomics study has a wide range of potential benefits. For example, it can provide a broad base of experience from which the assessment and intervention is developed, ensure co-operation in fitting or mock-up trials, as well as facilitating a smooth transition to any modified work practices and equipment (see chapter 37 for a full presentation of participative practices in ergonomics).

Identifying and prioritizing the problems

In principle the first step that we need to take in any assessment is to identify the problem areas and then to decide, given the available resources, the order in which they should be tackled. In practice however, certain judgements may already have been made, especially as it is often the 'identification' of a problem which leads to the study being commissioned in the first place. Concern may have been expressed, for example, about injuries associated with particular work processes, high absenteeism or poor productivity in certain areas of the plant. In such cases the starting point for the study will be obvious. Where the picture is less clear the ergonomist will need to gather a wider range of information in order to help with setting priorities and structuring the assessment programme.

In any case, it is always worthwhile seeking as wide a range of perspectives on the problems as is practical (such as from management, trades union representatives, supervisors and other employees). Even in instances where a working group has already been set up, this may still be a valuable source of information.

In addition to this process of consultation there are a variety of other techniques available which can help with the identification and prioritization of problem areas. These are outlined below.

Analyzing medical/personnel records

A view of injury or absenteeism problems within the workplace may be obtained by analyzing existing company data sources (although of course the usefulness of this information is dependent upon the accuracy and reliability of the recording systems—see chapter 4). For example, in the case of the van drivers' alleged seating problems mentioned earlier, evidence came to light of complaints of back pain going back six years. This indicated that a genuine, long-term problem did, in fact, exist. Information about work related injuries will usually be found in medical or personnel records. However, if these are not sufficiently detailed, the accident reporting book may also be of use. Ideally, a whole year's data should be collected (i.e., for the immediately preceding twelve month period). Where these are not available, at least three months' worth of recent information should be gathered. The aim is to collect overall statistics and not details on individual conditions; under no circumstances should names or any other means of personal identification be recorded. After the data have been collected or collated, they should be analyzed systematically using appropriate groupings. For example, by analyzing the injuries of people performing similar jobs, it may be possible to identify links between particular pieces of equipment or work practices and specific problems. Similarly, an examination of the dates on which injuries occurred may indicate a seasonal link with certain lines or operations (for further information see McAtamney and Corlett, 1992).

From an analysis of accident/injury data, evidence may be gathered about the extent and severity of work related injuries and the jobs which put employees at highest risk. Not only will this help in prioritizing areas for assessment, but it may also provide the raw material for the kind of cost-benefit analyses used to advance the case for workplace modifications (see chapter 34).

Analyzing production records

In a similar way to medical and personnel records, analyzing production records may help identify problem areas. Lower than expected levels of production or problems with work quality might serve as useful signposts indicating, for example, inappropriate work organization or difficulties with poorly designed or maintained equipment. Comparison of productivity records from before and after the intervention programme might also provide some objective data which can be used in cost-benefit analyses (see chapter 34) and promotion of subsequent interventions.

Preliminary workplace observations

Another useful technique in identifying problem areas is the 'walkthrough' assessment. This involves the ergonomist conducting a brief tour of the workplace or plant, usually accompanied by a member of the company who is familiar with the area, in order to observe at first hand the layout of the workplace and normal working practices. The ergonomist can then begin to make judgements about where problems may lie and their likely severity (an important criterion in their prioritization).

A more structured approach is to focus upon particular departments or sections and to rank potential problems on some form of risk rating scale such as Åberg's matrix (Åberg, 1981—see chapter 30). The main advantage of this process is that it is very quick to carry out and yet still allows the investigator to construct a basic list of problems along with some idea of their magnitude. Furthermore, it also provides an opportunity for the investigator to become more familiar with the nature of both the tasks and the workplace.

Data on musculoskeletal problems

Alternatively, materials such as the Nordic Questionnaire (Kourinka et al, 1987) can be used to identify the locality and severity of musculoskeletal complaints (see chapter 23). Not only can this provide vital information about which areas of the body are under load, but if replicated some time after changes to the workplace or working practices have been made, can serve to indicate whether or not these actions have been effective.

Structuring the assessment procedure

By 'structuring the assessment' we mean outlining the steps by which it will progress and, in particular, deciding on both the most useful methods of data collection and the particular workstations/employees to be examined. By this stage the ergonomist should have an idea of where problems are occurring and be starting to get some idea about possible causes. However, it is important not to become too fixed in these initial ideas as this may lead to the assessment becoming unduly narrow, meaning that valuable information may be missed. This is sometimes more easily said than done particularly where, for example, the ergonomist has been called in to deal with an already specified problem. For in such cases, whilst the management might feel that, with the diagnosis in place, the job of the ergonomist is simply to prescribe the treatment, in practice the relationship between cause and effect may be altogether more complicated. For example, the management of a textiles factory were concerned about wrist problems amongst their sewing machinists. Consequently they brought in an ergonomics consultant to look at the layout of the machines. However the ergonomist discovered that not only were there many more finger-related problems than wrist problems, but also that the changes required were much more complex than the management had initially thought. (Haslegrave, 1990).

At the same time as thinking about the possible causes of problems it is also wise to get a clear idea about the resources available for data collection and analysis (e.g., time, money and expertise) as it is very easy to become overcommited in embarking upon too large or complicated an investigation. Similarly, where possible, you should avoid choosing methods and/or techniques that require specialist equipment or expertise that are difficult to obtain.

Sampling issues

Although some studies will involve examining all workstations or employees, in many cases this is neither practical nor necessary. Instead a sample from each task or work area is chosen. The type of sample required is dependent upon the nature of the study. For example, if we were interested in conducting a set of interviews designed to investigate problems associated with working in warehouses we might want a random sample of those working in the area. At other times it might be important to have more control over the make-up of the sample. For instance if we are looking at back problems associated with lifting loads from delivery lorries it may be better to go for a sample consisting of people at the extremes of anthropometric dimensions or the fastest and slowest workers and so on.

In general the issues to bear in mind are: firstly, can the sample size be smaller (i.e., quicker and easier to undertake) without sacrificing relevant information. Secondly, does the sample adequately represent all those who will be affected by the study including shift workers, maintenance and cleaning personnel. Finally does the sample accommodate the requirements of any workers with special needs? (For a discussion of the issues of sample selection and sample size see chapter 5.)

Choosing methods

It is not our intention in this chapter to describe in detail methods and techniques for use in the field. You will find these presented elsewhere in this book. However it is likely that most studies will draw from a range of fairly standard methods including direct observation, checklist procedures, interviews and questionnaires. When it comes to choosing the appropriate practical methods, a number of points should be kept in mind.

Firstly, because general study aims are often couched in the broadest of terms it may be necessary to formulate a new and more focused set of study questions in order to secure the relevant information. For example, if the general aim of the study is to examine the risk of upper limb disorders on a packing line we might need to know more about where high forces have to be applied or awkward postures held (see chapter 23). Clearly we should only begin thinking about methods of data collection once such decisions have been made.

As we have already mentioned, the resources available for data collection and analysis are likely to influence the choice of methods employed. Similarly, both the number and variety of tasks to be assessed and the level of access to both worksites and workforce will have a marked impact on which methods are most appropriate. Finally, a successful assessment must be sufficiently broad to cover the range of factors which may contribute to the situation under investigation. For example, if complaints are received from sedentary workers about neck and shoulder stiffness, we should investigate the posture, forces and loads exerted whilst working. However, it may also be necessary to assess room temperature and air flow as these can contribute to these same kinds of complaint (Sundelin, 1992). Ideally therefore, a combination of methods will be used to obtain as comprehensive view of the workplace and risk factors as possible (although see chapter 1 on the dangers of multimethod investigations).

Pilot studies

In a large scale study, it is often vital to undertake a pilot study, using a small sample of subjects to test the planned assessment procedure. This will provide an opportunity to check that the methods selected for data collection are appropriate and that any equipment functions satisfactorily. A pilot study may also give us important clues about how long the assessment proper will take. It also allows, where necessary, for a last minute streamlining of the procedure.

Collecting data—site visits

Before visiting the worksite there are a number of things that must be done which, at first sight, may appear trivial. For example, so as to avoid wasting time and effort it is always wise to check that key personnel will be available and that the particular areas due to be assessed will be running normally. Requests for any special facilities (e.g., a place to hold confidential interviews) should also be made around this time, as well as permission to use recording equipment such as cameras and videos. You will also need permission before speaking with employees. It is important that the people taking part in the assessment should not be penalized for the time they spend in interviews or being observed and therefore, where possible, encourage management to treat this time as 'productive hours' rather than as part of rest periods. Overall, try to be well organized. Remember, what might be a central concern to the ergonomist may represent just a minor detail in the busy working schedule of a manager or supervisor.

There is nothing more infuriating than arriving on the day of a carefully planned assessment to find that your tape recorder or video camera is malfunctioning. If possible take along back-up equipment as insurance against such an event. Give yourself plenty of time to conduct the investigations as there are often unforseen hitches, delays and complications, such as line stoppages, machine breakdowns or staff shortages causing a change in work routines.

Find out when things such as rest breaks and shift changes happen so that you can plan your assessment schedule more precisely. For example if you are planning to measure the dimensions of the workplace, it might be a good idea to do this when the workforce are elsewhere. Then when they return you can get on with conducting interviews and observing working practices.

During the site visit it is important to explain the rationale for the study to all those employees involved. In many instances, those taking part will need to be reassured that they will not be identified as the source of any given statement or opinion—although it must be acknowledged that sometimes, particularly where there are a small range of views being sought, this may not be easy. In such cases careful thought and planning is required before making any firm assurances.

Where it is possible, it might help the investigator to experience the work under assessment first hand; as this can sometimes provide extra insight into the nature of the problem. For example it would probably be far easier to visualize the component movements involved in a manual handling task if you have simulated the action yourself.

As a rule the key to successful data collection lies in being methodical and consistent in the way that you conduct your assessment. This ensures that you will obtain a reliable picture of the work situation on which to base any recommendations. When the site visits are completed, it may be useful to hold an informal debriefing session with appropriate company personnel.

This would provide an opportunity to give some initial feedback (although you should be careful to emphasize its provisional nature) as well as a further opportunity to gather potentially useful comments.

Analyzing and interpreting the data

Once the data have been gathered the next job is to try and make sense of them—in other words, we have to begin interpreting the data. Before starting it is worthwhile reminding yourself of the overall aims of the study; it is important to bear in mind that the analysis must provide specific information about the nature of the problems, where they are occurring, how severe they are and how widespread.

The type of analysis will partly be determined by the nature and volume of data as well as by the resources available to the investigator. Firstly a decision may have to be made about which data to analyze. For example, in cases where there are hundreds of postures to be analyzed it may be necessary to pick out either the most typical or important cases, rather than trying to examine them all. Analyzing data such as the measurement of workplace dimensions or environmental parameters can mean comparing them with ergonomics guidelines and standards. In certain circumstances, such as when analyzing a questionnaire survey, statistics may be used to summarise data and, where appropriate, to test for significant findings. However it is not always necessary or desirable to include statistical analyses especially if you are only looking at a small number of subjects or data points. Similarly, where qualitative information has been gathered such as through interviews with the workforce, the aim of the analysis will be to summarize and interpret the responses, for example, by looking for trends and points of consensus amongst the respondents.

The outcome of the analysis will be some overall findings or conclusions, the importance of which must now be considered. Firstly, do they suggest that the risks identified are high or low and to what extent, if at all, do they contravene ergonomics guidelines and/or legal requirements. Are the problems systemic, or do they place certain individuals at greater risk than others? Finally, can the problems or risks be grouped into different categories such as those related to engineering, training or work organizational issues?

Making recommendations

In cases where the existence of significant problems is confirmed by the analysis, decisions should be made about the kinds of remedial action that should be taken. It goes without saying that recommendations will vary as to how urgent they are and how easily they can be implemented. It is wise therefore to bear in mind the financial and technical feasibility of any recommendations made. In cases where solutions are bound to cause considerable inconvenience and/or expense, think where possible in terms of interim measures or short, medium and longer term aims. Be prepared to face the fact that there is rarely a simple and straightforward solution to any given problem. You may have to put forward a package of recommendations, each of which makes a contribution to improving the overall situation.

Sometimes a cost-benefit analysis, even if somewhat crude, can be helpful in demonstrating the value of your recommendations. (For examples of the application of cost-benefit analyses see chapter 34, with detailed discussion also in Oxenburgh, 1991. Similarly where a technique such as RULA has been used (McAtamney and Corlett 1993 and chapter 23) a scoring of the postures which would be adopted in the modified workstation could be used to indicate where risks will be reduced.

Producing reports

Most studies involve the production of a report to disseminate the findings. Although reports will vary in length and detail, they should take into account a number of basic points which are outlined below (for more detailed information about report writing see Turk and Kirkman, 1989 or Dane, 1990).

Perhaps the first consideration when preparing a report is to have a good idea of the intended readership. In particular a knowledge of their familiarity with ergonomics concepts is useful when making decisions about the tone and level of the report. In most cases it is good practice to avoid using technical jargon.

Be realistic about how the report is likely to be read. Typically recipients will be most interested in the conclusions and recommendations rather than in lengthy accounts of assessment procedures or tables of raw data. To this extent it may also be a good idea to begin the report with an executive summary which sets out the main findings of the study.

A draft version of the report could be produced for discussion with the working group. This will provide an opportunity to correct any errors or misunderstandings, clarify ambiguities and iron out any contentious issues before the final version is released to senior management and employee representatives.

A formal presentation of the final report may also be requested. As with the written version, the same considerations of level, tone and emphasis apply.

Implementing changes

For all the various problems involved in making an assessment and formulating recommendations, often the most difficult part of an ergonomics study comes when trying to get the recommended changes implemented (see chapter 36). As we have already mentioned, one way of encouraging the 'take-up' of recommendations is by creating and maintaining a wide base of commitment throughout the organization. Another is via the formulation of an action plan.

As the name suggests an action plan is a document outlining precisely how the recommended changes will be implemented. Typically written by management in consultation with members of the original working party (and also the ergonomist) its main purpose is to specify what are the main priorities, who will take responsibility for ensuring that particular changes are made and the timescales involved in each case.

In an ideal world everyone would welcome the recommended changes. In practice, however, things are often much more complicated. To begin with people may respond with varying levels of enthusiasm. For instance it may be hard for those people who have been relatively uninvolved in the assessment process to recognize the value of changes. Indeed they may be suspicious of the motives behind the study, fearing that it may lead to redundancies or deskilling (although the existence of a representative working party should help to minimize these concerns). Alternatively, they may appreciate the benefits of the changes but see them as small or insignificant compared with the problems associated with other work issues. However, it does also have to be recognized that while some changes may serve to benefit everyone in an organization, it is not always the case that everybody wins; indeed some people may definitely lose either in terms of status, role or even payment and other rewards. In addition, there will always be those who perceive themselves as 'losers' even when they are not.

Although many of the proposed changes to a workplace can be designed 'on the drawing board', it is worthwhile testing them before substantial amounts of money are invested. For example, where it is possible to get hold of sample equipment (such as seating, keyboards etc.) trials can be arranged to allow staff to evaluate their design. Alternatively proposed changes may be 'mocked-up' using inexpensive materials such as cardboard or plywood (see chapter 37). Fitting trials or user trials can then be carried out to evaluate the design (for further information on these see chapters 10 and 19). Where possible members of the workforce should also take part in this process in order to give them a greater level of involvement with the changes. What is more these kinds of trials can also be highly successful in illustrating the problems for management and helping to convince them of the need for change.

It is vitally important to maintain the impetus for change. In many organizations initial enthusiasm and commitment may disappear under ever-increasing workloads. To avoid this, it is important to review the progress of changes on a regular basis, and here the value of assigning names and dates to the action plan will become evident. The workplace should be re-evaluated several months after changes have been made to ensure that they have been successful and that no further modifications are necessary. This could involve the ergonomist returning for a follow-up session or alternatively the establishment of internal review mechanisms.

A legal postscript

When undertaking an assessment and preparing the report, consideration must be given to its legal implications. Sometime in the future the report may be used in litigation, so it is vitally important that analyses are accurate, and that all findings, interpretations, and recommendations are fully justified. Where possible the report should be supported by references to already published work. In addition, it is important to keep written records of any decisions or actions arising from the study. All ergonomists should carry full liability insurance and seek advice if unsure of any legal aspects of their work.

Conclusions

This chapter has presented an approach to undertaking ergonomics studies in industry. It has emphasized the need for careful planning, good communication and, most importantly, commitment both from top management and from those on the shop floor. Regardless of the size or type of study, the underlying aims of any ergonomics intervention remain the same: to discover improvements in job design and working conditions, such that people can operate productively and safely in jobs which suit their physical and psychological capacities.

Ideally an ergonomics study should be seen as part of a larger process within the company, designed to combine efficiency and profitability with the general well-being of the workforce (which takes us back to the twin aims of ergonomics outlined in chapter 1). As such, ergonomics studies should not be thought of as appropriate only in 'emergency' or 'fire fighting' situations. Rather they should be incorporated into general company policy on the development and review of working practices, processes and new operations. In this way, ergonomics will become both part of regular management and the property of everyone in the company.

References

Åberg, U. (1981). Techniques in redesigning routine work. In *Stress, Work Design, and Productivity*, edited by E.N. Corlett and J. Richardson (Chichester: John Wiley and Sons).

Alexander, D.C. and Pulat, B.M. (1985). *Industrial Ergonomics. A Practitioner's Guide*. (Atlanta, Georgia: Industrial Engineering and Management Press).

Buchanan, D. and Boddy, D. (1992). *The Expertise of the Change Agent.* (Hemel Hempstead: Prentice Hall).

Corlett, E.N. (1991). Ergonomics fieldwork: an action programme and some methods. In *Towards Human Work: Solutions to Problems in Occupational Health and Safety*, edited by M. Kumashiro and E.D. Megaw (London: Taylor and Francis).

Dane, F.C. (1990). *Research Methods.* (California: Brooks/Cole Publishing Company).

Dul, J. and Weerdmeester, B. (1993). *Ergonomics for Beginners. A Quick Reference Guide.* (London: Taylor and Francis).

Haslegrave, C.M. (1990). How well can ergonomists address problems identified in the workplace? In *Work Design in Practice*, edited by C.M. Haslegrave, J.R. Wilson, E.N. Corlett and I. Manenica (London: Taylor and Francis).

Kourinka, Il, Jonsson, B., Kilbom, A., Vinterberg, H., Biering-Sorenson, F., Andersson, G. and Jorgense, K. (1987). Standardised Nordic Questionnaires for the analysis of musculo-skeletal symptoms. *Applied Ergonomics*, **18**, 233–237.

McAtamney, L. and Corlett, E.N. (1992). Reducing the risks of work related upper limb disorders. A guide and methods. Nottingham: published by Institute for Occupational Ergonomics, University of Nottingham.

McAtamney, L. and Corlett, E.N. (1993). RULA: a survey method for the investigation of work-related upper limb disorders. *Applied Ergonomics*, **24**, 2, 91–99.

Oxenburgh, M. (1991). *Increasing productivity and profit through health and safety.* (Chicago: CCH International).

Sundelin, G. (1992) Electromyography of shoulder muscles—the effects of pauses, drafts and repetitive work cycles. *Arbete och Halsa* 1992:16. Stockhom: National Institute of Occupational Health.

Turk, C. and Kirkman, J. (1989). *Effective writing—Improving scientific, technical and business communication.* 2nd edition (London: Spon).

Chapter 36

New systems implementation

Ken Eason

Introduction

Technical change is a pervasive feature of organizational life. Enterprises devote a considerable proportion of their resources to the planning, purchasing and development of new technical systems to help them become more efficient and effective. At the end of each technical system development process there has to be an implementation phase when the ability of the new system is tested with real tasks and when the employees of the organization adapt to the changes in their working lives occasioned by the new system.

Implementation of a new technical system is a demanding time because different issues have to be addressed concurrently:

1. *Installing and testing the technical system.* It is one thing to have tested the system under experimental conditions, it is quite another to subject it to the full range and demands of everyday working conditions. There will therefore be many technical problems to overcome before the technical system operates as planned.
2. *Local workplace design.* If the system is to be used by many employees it will have to be installed in the workplaces of these employees. This may involve the creation of new workplaces as in a new factory, the redesign of existing workplaces as when the tills are modified in a bank to accommodate a terminal, or accommodating new equipment on an existing desk.
3. *Training and support.* People with new technical tools will need training to make good use of them. They may also need training in new ways of working. Training before implementation may have to be supplemented by support after implementation since it may take a considerable period of time to develop all of the skills necessary to cope effectively in the new situation.

4. *Organizational change.* Technical change of a substantive nature almost certainly goes hand in hand with organizational change which may require major adaptation from the employees in the organization.
5. *Acceptance of change.* All of the foregoing changes constitute a new and uncertain future for the employees in the organization and an important task in implementation is to develop strategies which will foster positive attitudes towards the change process by those to be affected by it.

It would be difficult to address all of these issues simultaneously under ideal conditions, but an additional factor often adds substantially to the problem. Most technical systems are introduced into organizations that are ongoing concerns and which have to maintain the volume and integrity of their activities throughout the process of change, such as serving customers and making products.

Given the range of issues to be addressed, it is perhaps not surprising that implementation of new systems is often problematic. It is quite common for systems never to be implemented, or for the implementation process to become a very painful and protracted period. The reasons may be many. If, for the purposes of analysis, we disregard the possibility that the technical system proves ineffective or inappropriate (a major cause of problems in fact) we can examine the problems of implementation itself.

The problem that has been widely recognized since the days of the Luddites is resistance to change; people acting against technical change in an attempt to preserve their jobs, their livelihoods and the ways of working that they have known. Change inevitably brings uncertainty and unfamiliarity and may well bring undesirable consequences, so it is hardly surprising that people who consider themselves to be losers in the change process mobilize their resources to protect their interests. Resistance to change is often presented as a blind, irrational attempt to hold back progress, but it can be a well developed strategy with clear and defendable goals. In reviewing the implementation of information technology, for example, Keen (1981) identifies a number of 'counter-implementation strategies' which user groups may employ to ensure the outcome is in their best interests. These include 'keeping the project complex', 'minimizing the implementer's legitimacy' and 'exploiting their lack of inside knowledge' which are all ways of deflecting or limiting the efforts of those attempting to introduce the system.

Technical systems design is usually portrayed as a rational, objective process but, at implementation, the presence of resistance to change may reveal that the different 'stakeholders' (groups of people affected by the proposed system) have different goals and the process has also to be seen as one in which conflicts have to be resolved or negotiated.

Even when resistance to change is not a major inhibitor of change, the human and organizational systems may take a considerable period of time to adjust to the new circumstances. Change is frequently followed by an 'initial dip' in performance and work output, and it may be weeks or months before the improved levels for which the change was introduced are achieved in practice. In this period individual users may be developing and consolidating the skills they need for the new form of working and new organizational structures and procedures may be evolving. It may be possible to switch from one technical system to another overnight but the human and social systems, depending as they do upon human motivation and learning, are better described as changing by evolution.

Implementation is therefore a complex socio-technical change process in which technical developments are accompanied by a wide range of human changes. The most common reason for failure in implementation is that the process is treated as a technical problem, and the human and organizational issues go unrecognized and untreated. In this chapter we will examine strategies and tactics for implementation which focus upon the process as socio-technical change. In the next section we will consider overall strategies to cope with the different circumstances of implementation with particular reference to the roles of potential users. Subsequently the function of user involvement is examined as the principal mechanism by which the human issues of implementation can be managed.

Implementation strategies

Figure 36.1 identifies six major strategies used for technical system implementation and is derived from Eason (1988). The strategies are loosely arranged on a dimension from the most revolutionary to the most evolutionary. At one extreme we have the major start up of a complete new system overnight

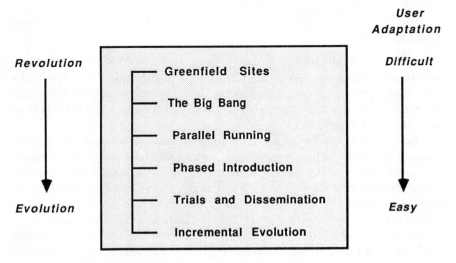

Figure 36.1. Implementation strategies

and, at the other, the steady introduction of technical facilities over an extended period of time.

The aim is to identify the main principles of each strategy and the circumstances in which it might be used. We will then subsequently consider the implications for the management of the human and organizational aspects of implementation.

The 'greenfield site' implementation

This is when the new system is introduced in an entirely new situation to any previous work system, i.e., a new factory is opened on a different site, or a new shop is opened in a town. Technically the system can be developed without too many constraints associated with existing forms of technology. Organizationally there is also the opportunity to break with tradition and to explore new working practices. The problems for implementation are often that planning has to be done without the help of the people who will operate the system because they have yet to be recruited, and there may be extra 'teething troubles' because the staff are initially strangers to one another. This kind of change process can be greatly facilitated by the recruitment of key staff sometime before the implementation date, so that they can play a major part in detailed planning. Systematic attention also has to be given to team building exercises because new staff will be entering new roles with no pre-existing knowledge of one another to help them. Team building exercises provide people with the opportunity to explore the nature of their role, its relationships with other roles and to get to know the people who will occupy these roles. If these explorations can be made before the new system is implemented, there will be a strong basis for co-operation and team action when it is demanded under operational conditions.

The big bang

One of the most difficult kinds of implementation to effect is when an existing system is being discontinued in its entirety at the end of one day, and a new system replaces it on the following day. The most publicized example in recent years was the overnight switch of the London Stock Market to electronic trading in 1986.

Obviously this can be high risk strategy because it attempts to maintain continuous work whilst, at the same time, making a complete technical change in a single phase. Every part of the new system, human and organizational, has to be functioning well if this strategy is to lead to effective operation immediately. It is almost inevitable that there will be an initial dip in performance before smooth, integrated operation can be achieved.

The chances of failure can be minimized by holding off-line trials before the launch, not only to test the technical system but also to allow staff to be trained and to rehearse new operating procedures. However, scaling up the operation to full line working will inevitably throw greater strain on both equipment and staff. Provision for manual back-up in the event of mishaps is a very wise precaution in this strategy. Before and after the launch date, there will be a need for extra resources. The need for specialist resources such as technical staff and trainers is expected, but what often comes as a surprise is the need for additional user resources. Since it is often the intention to run the new system with fewer staff, management may be surprised and resistant to the idea that extra staff are needed during the transition period. If normal work is to be sustained, staff are to be trained and be involved in off-line trials, overtime or additional resources will be required if the stress of overload is not to be added to the inevitable stress of change. Some large organizations, who accept that change is the norm rather than the exception, maintain a team of staff who can go into user departments and take over routine duties whilst the local staff prepare for change.

Given the problem of integrating the many faceted aspects of a complex process in a single implementation phase, there need to be good reasons for adopting this approach. One such reason is that the work system needs all the facets working together to be effective; a phased approach cannot be adopted. If, for example, all user sites need to be given simultaneous, real-time access to a new database, a 'big bang' strategy will be needed. Another reason is that the strategy ensures the change has a high profile and there can be no loss of momentum in the change process which is a characteristic of some of the other strategies. Finally, the opportunity of a complete change means that organizational changes can be made as well as technical change.

Parellel running

One popular way of minimizing the risks to the ongoing workload is to introduce the new system alongside the old one, and to run them in parallel until everybody is confident that the new system will be effective. When the maintenance of the quantity and quality of existing work is of the highest priority, this is the most popular strategy.

Whilst this strategy provides an insurance policy, it is not without costs. The common way of operating is to perform the work by the old system and then to repeat it with the new system. This will inevitably involve extra work and will constitute an additional strain on staff unless extra resources are provided. The process of using the new system for no real purpose can also produce negative attitudes amongst users. If the system keeps failing, they may not develop confidence in it and may cling to existing procedures. It it works well and saves a lot of time and energy, it can be very frustrating to continue using the old system. In the freightforwarding case study described in Klein and Eason (1991), the new system provided a way of capturing data once and using them in many ways whereas the old system required users to repeat the data capture for every use. After a short period of parallel running the staff pleaded to be allowed to use the new system for real tasks.

If parallel running is used as a strategy, it needs an agreed programme of tests so that everybody can see the progress that is being made towards the switch to the new system and can participate in the decision making.

Another problem with parallel running is how to make organizational changes. It is difficult enough for staff to operate two technical systems simultaneously; it is still more difficult for them to operate in two organizational structures simultaneously. There is a tendency to keep the existing organizational structures and procedures during parallel running which may inhibit the search for structures which can take best advantage of the new technical system.

Phased introduction

The problems of making massive changes can be eased by phasing in the changes over a period of time. There are two ways in which large scale changes can be subdivided to facilitate phased introduction. First, the functionality of the technical system can be introduced in phases so that the basic task processes can be supported in the early phases and subsequently facilities can be added which support, for example, the management and development tasks. Secondly, it may be possible to introduce the system in different parts of the organization at different times. A combination of these two approaches can also be used.

There are several advantages to introducing functionality in stages. It means that users do not have to master a very sophisticated system all at once but can engage in a progressive learning process as the system is gradually implemented. It also means that system implementation can start before the whole system has been designed. Indeed it may be possible for early user experiences with the system to shape the development of the facilities that are to be introduced later. This point will be elaborated further in the section on evolutionary design. A final advantage is that an explicit policy of phased introduction is in fact a recognition of an almost universal experience; even when a 'big bang' strategy is adopted, the system is never finished and revisions, elaborations and refinements are continually being added.

An important consequence of adopting this strategy is that decisions have to be made about what parts of the system to deliver first. There will be a tendency to deliver the foundation parts of the technical system in the first phase. If it is an information technology system, for example, the databases may be designed together with the data capture procedures and only a few of the major output facilities may be provided. The danger of this approach is that the first experience the users have of these systems is that they demand input whilst giving very little of value in return. Another possibility is that all the useful outputs go to one group of users whilst the input load falls on others. Neither of these strategies is likely to promote a good user reaction.

A good planning rule is that phasing should ensure that each group of users receives a service they value early in implementation to provide a positive experience of the system upon which later phases can build.

The alternative to introducing functionality in phases is to introduce the complete system in one site after the other. This is a popular strategy in organizations which have similar functional units spread over a geographical area, for example, a chain of shops or branches. In this strategy an implementation team can be developed which becomes very skilled at handling the implementation issues because they encounter them many times over. There is, however, the danger that they will forget how unfamiliar these issues are to the staff next on the list to make the change and they may expect implementation to be progressively faster with each new site they visit. The trainer for whom everything is 'obvious' and 'common sense' often serves only to increase the would-be users' sense of failure and frustration.

Trials and dissemination

This strategy explicitly recognizes that there will be 'teething troubles' when a new system is introduced, by holding a major trial before embarking upon full scale implementation. Unlike prototype testing which is often used to revise the specification for a system, the trial is usually undertaken with the technical system it is planned to implement. The purposes of the trial may be many; to test whether the technical system can do the job, whether it is robust and reliable and so on, to establish the training and support that is necessary, to examine the organization changes that are needed. To learn such lessons it is necessary to make the trial as realistic as possible. Ideally, a representative user community will use the new system for real work and thoroughly evaluate the implications, although off-line trials may be necessary when the maintenance of work standards is critical. Ways of undertaking full scale trials of systems are examined in chapter 12 on evaluating human computer interfaces.

This strategy provides a valuable opportunity for the user organization to prepare itself thoroughly for implementation before it is in the throes of full scale change. However, many organizations fail to make good use of the opportunities that the trial presents. They may, for example, choose as a trial site one in which the staff are positive in their attitudes and which is relatively isolated from other sites so that the change can be made without knock-on effects. These factors may make it relatively easy to implement the change but it should come as no surprise that it is much more difficult to make widespread changes subsequently. A related problem is the 'Hawthorne effect' (Rothlisberger and Dickson, 1939). This is the situation where change proves very effective in terms of productivity not because of the change itself but because the staff are 'in the spotlight' and respond to the extra attention they are receiving. This factor is often regarded as a confounding variable in experimental designs which compare the results of a change group with a control group where no change is introduced. In implementation it is a positive force and displays the motivational effects of user participation. The danger in a trial and dissemination strategy is that the staff in the trial get a lot of attention compared with the staff in the later dissemination. This factor, allied to the tendency of the implementation team to expect each successive implementation to go faster, can turn a very positive reaction in the trial into a very negative reaction to the wider dissemination.

Another typical failing is not to learn the lessons of the trials. Too often thorough evaluation is not undertaken and, as a result, detailed faults in the technical system are not identified and corrected, training regimes are not tailored to the needs of the users, and organizational changes are not planned. The trial simply becomes the first implementation and makes this strategy indistinguishable from a phased introduction across sites.

Incremental implementation

The logical alternative to a revolutionary change is gradual evolution. The advantage of such an approach is that users are never confronted by major change but have only to cope with small incremental steps. Ideally the learning they achieve from early steps can help them to select the next stages of the evolution. If this is achieved, the strategy may truly be called user led or user centred.

The growing sophistication and flexibility of technology is making the incremental implementation of systems an increasingly practical proposition. It is already a good characterization of the method of technical change used to implement technical tools for managers and professionals. When the user community is discretionary and powerful and their requirements are varied, not to say idiosyncratic, the service has to be tailored to individuals and incremental implementation may be the only option that is acceptable to users.

It is useful to distinguish between *ad hoc* and planned evolution. *Ad hoc* evolution is simply introducing technical change bit by bit with no overall objective to guide development. At its worst, it allows systems to 'grow like Topsy' so that the overall structure has incompatible parts, early stages are rendered redundant and later developments are forced into ineffective routes because of early decisions. To avoid this situation requires the identification of broad requirements and the planning of an overall system architecture and organizational structure which can accommodate many ways of meeting these requirements. Within such a structure, it should then be possible for individual users to follow an evolutionary route adding incremental parts to their own services to suit their requirements and their rate of learning.

A user's view of implementation strategies

The six strategies listed in Figure 36.1 form a range from the most revolutionary to the most evolutionary. Viewed from the perspective of the potential user, some are more daunting to cope with than others. The 'greenfield' and 'big bang' strategies are clean and comprehensive but mean the user has a lot of changes to cope with in a very short time. If the user has to cope with normal business at the same time, this kind of change can become a major strain even when the users are positively in favour of the change. 'Parallel running' and 'phased implementation' are safer and slower ways of introducing major changes in circumstances where normal work must be maintained. 'Trials' and 'incremental implementation' provide early opportunities for users to experience the technical change and can enable them to use this experience to guide the later stages of development. The more evolutionary forms of implementation therefore give users more opportunity to cope and to contribute to the form of development.

User involvement in system implementation

Whatever the strategy that is followed, a crucial factor affecting the success of implementation is the degree to which potential users are involved in the process. At worst, the users are confronted by a completely determined system which may have negative effects on their working lives and which they are expected to master very quickly whilst maintaining their normal work output. In these circumstances, which are not uncommon, even the most well designed system will run into implementation difficulties.

To avoid this situation, mechanisms for user involvement are necessary. Evolutionary developments give more opportunity for these mechanisms but they are possible and necessary in all implementation strategies. Figure 36.2 identifies the major mechanisms by which user involvement can be effected.

The first requirement is to establish a temporary organization structure which can carry the organization through the change. This structure will charge individuals with the variety of responsibilities that are necessary to implement the change. In may situations, these roles will be taken by professionals in the change process, i.e., technical designers, trainers and ergonomists. It is important, however, that many of the roles are taken by potential users.

It is, of course, necessary for users to participate in the design process from the outset to ensure, for example, that the specification meets the requirements of the users (see chapters 33 and 35 also). In a large scale development, it is likely that only a few users can actively participate in the design phases. When it comes to implementation, however, everybody will be affected and therefore a mechanism is needed to involve everybody. In a number of implementation strategies, we have found three concepts of value in creating temporary organizational structures; the user 'champion', user representatives and full user involvement.

Temporary Organizational Structures for Managing the Change Process	
Technical Roles	User Roles
Technical System Implementors	Senior User Champions
Trainers	User Representatives as Local User Support
Ergonomists	Local User Decision Making

Figure 36.2. Temporary design structures for system implementation

The user 'champion'

It is important to involve a senior user manager in the implementation process, if possible vesting responsibility for implementation in the manager. This will clearly identify the process as user driven and will announce the commitment of the user organization to the development. It will also recognize that, whilst there are technical issues to consider, the primary problems in implementation are human and organizational adaptation.

User representatives as local implementors

We will assume that, from the beginning of the system development, a small number of users have been involved in the process to represent user interests. If these people have been chosen to represent the major user groups they can play a significant role during implementation. By this time they are the users who are most knowledgeable about the system plans. They are in an excellent position therefore to help plan local implementation; the start-up procedures, training, workplace changes, etc. They may also come to play a significant role subsequent to implementation as the principal sources of support to users learning to cope with the new system. The combination of user experience and system knowledge can mean user representatives are much more able to understand and resolve the problems experienced by users than are technical experts.

Full user involvement in local implementations

If resistance to change is to be avoided it is necessary to involve all potential users in the change process, not merely a selected few. In a large implementation it is difficult to involve everybody in the strategic decisions but there are many local decisions in which everybody can participate; the timing and sequence of implementation, the extent of parallel running, allocations of duties to work roles are examples of such decisions. Working under the direction of local management, user representatives can play an important role in organizing working parties of local users to plan and undertake these activities. It is important to note that involvement of this kind gives people considerable influence over the decisions that affect them personally, and it is this kind of influence which most successfully combats feelings of external threat. For example, Wall and Lischeron (1977) demonstrated that whilst people liked to feel their interests were represented in strategic decision making, they wanted to be personally involved in the decisions that affected their daily work.

User-oriented implementation topics

The topics that the temporary organizational structure must address during implementation relate to both technical design and organizational change and some of the most important topics are listed in Figure 36.3. The most important feature of the technical design is that the earlier stages of development should have left considerable freedom during implementation for people to shape the system to meet local requirements. Involving people in a design process is not sufficient in itself to overcome resistance to change. It is important that their participation leads to decisions about significant issues and that can only be achieved if the technical system leaves options. In listing the principles of socio-technical design, Cherns (1976) cites 'minimum critical specification' as the vehicle which provides for discretion during implementation and beyond. This is a requirement laid on designers to specify only that which has to be specified and to leave open as much as possible for later decision making. Effective implementation can depend heavily upon the flexibility and adaptability built into the technical design so that it can be 'personalized' or 'customized' to local needs at the point of implementation. One encouraging feature of the increasing sophistication of the computer technology which underpins most new systems, is the degree of flexibility and adaptability that can be provided. Whereas a few years ago, for example, a user might be restricted to a standard printout provided for a

Technical Design	Organizational Change
Minimum Critical Specification	Training and Support
Customisation	Job Design
Participative Work Place Design	Team Building
Trials, Evaluations and Evolution	

Figure 36.3. Topics in user-oriented implementation

multitude of users, a system might now include a report generator, a tool which permits users to define their own reports. The provision of such tools within systems means that users, during implementation, may well find they have major areas of discretion and therefore important decision making tasks to undertake.

Another area which requires detailed local design constitutes one of the traditional areas of ergonomics. The integration of a new technical system within a workplace will involve issues of workstation layout, furniture selection and environmental design. In most large scale applications with a diverse user population there will be many different workplaces which require attention. There are, of course, a wide variety of standards, guidelines, data tables, techniques and tools available to support workplace design; see, for example, Grandjean (1986). For many ergonomists the principal contribution of the discipline to implementation is workplace and environmental design and it can be a major contribution to the subsequent health and efficiency of the workforce. (Many chapters in this book are indeed devoted to such contributions.) It is important to note, however, that if the practitioner approaches this role as the expert who designs all the workplaces for the implementation, the effect may be the same as the technical specialist who completely designs the technical service for the would-be users. The workforce may react negatively because, however well intentioned the specialist may be, decisions are being made *for* the users and not *by* them. Gower and Eason (in press) have provided a case study of office automation in which participative workplace design was employed. Secretaries were helped to select furniture, plan office layout and so on, with the ergonomists providing advice on relevant standards and guidelines whilst the secretaries provided information on local requirements, customs and culture.

Finally, with respect to technical design, in the implementation strategies that recognize the need for evolution there will be a need to conduct realistic trials. Users in trials need to be treated not as subjects in an experiment but as participants in the creation of their future work environment. This means that users not only get to use the future system but plan the trial, collect evidence about its value and its deficiences, and work with technical specialists on the future development of the system.

In addition to the decisions that relate to the technical design, there is in any socio-technical change process a wide range of human and organizational changes to be considered. Figure 36.3 also lists some of these topics. Whereas for technical design there are likely to be specialists available who will see it as their responsibility to be concerned with technical aspects of implementation, there are not likely to be specialists in organizational change available and these responsibilities will fall upon local management and workforce. One consequence of this is that these are topics which are frequently unrecognized and are dealt with in an *ad hoc* manner, as and when it becomes evident they must be given attention. The exception is training because it is widely recognized that potential users need to be able to understand and operate the new technical systems that are being provided. Too often, however, the training is limited to technical knowledge when much of the adaptation that is required relates to changing role responsibilities and relationships. There is also a tendency to attempt 'one shot' training before implementation in a single intensive burst, when many users can cope much better with a series of short training sessions spread over a period of time, combined with access to a range of 'point of need' support facilities (e.g., manuals, prompt cards and liaison staff).

Job design issues most clearly need attention in 'greenfield' and 'big bang' implementation strategies, when full scale change is being attempted in a single phase. There are a number of examples in the literature of procedures for job design in these circumstances, for example, Aberg (1981) in manufacturing industry, and Mumford (1983) in office automation. The most important requirement is that the technical system should have the flexibility to support a range of job design options so that participating users can choose a structure to which they can commit themselves and which will be effective. Eason and Sell (1981), for example, cite a case in which an order entry system was sufficiently flexible to permit branch offices around the country to allocate tasks to jobs in quite different ways and still make effective use of the technical system. One difficulty user groups often experience when planning for technical system implementation is that they cannot foresee the implications for job structures. Running a live trial offers a very effective way of revealing the implications, and it can be used to experiment with different job structures and test the flexibility of the technical system.

Job design issues take a different form in parallel running and evolutionary forms of implementation. Often there is no plan to change job structures but, as the technical system begins to be exploited, people begin to expand their job horizons and to encroach upon the work of others. The shop floor employees may, for example, begin to encroach on the role of the foreman or the inspector whilst the manager, by doing some of the typing, may encroach on a secretary's role. These factors can cause a gradual erosion of the existing job structures, often with poor relations developing between role holders. In the office automation case study described by Gower and Eason (in press) an evolutionary implementation strategy was followed. Regular monitoring audits were conducted to check for job design 'drift' in order to bring it to the attention of senior staff. By this mechanism emerging *ad hoc* practices were assessed and those job changes that were beneficial were formally recognized and implemented.

It will be apparent that, where major organizational changes are taking place, people will find themselves entering into new work relations with their colleagues. In the 'greenfield' situation they may be entirely new colleagues, whereas in the other strategies it may be the same people in different roles. In each case there is a need to help people appreciate not just the new technical system but also the new social system that is in operation. Team building in which people explore the new responsibilities and the relationships as well as getting to know how their encumbents may interpret them, is a very useful technique for preparing for implementation. It is particularly necessary in 'greenfield' and 'big bang' implementations because people will be expected to slot straight into the new job structure. Again the concept of running a trial can be used as a way of building a team understanding ahead of the time when it has to become an operational reality.

Conclusions: ergonomic contributions to content and process

Implementation procedures can make or break a technical innovation by the way potential users are introduced to the innovation and the extent to which they have an opportunity to shape the new system to meet their requirements. Different implementation strategies require different treatments. The evolutionary patterns offer more time for user learning and reflection and therefore give more opportunity to influence the subsequent course of events. The problem in these cases is to sustain the impetus of the implementation because it will take time and may become fragmented. In the strategies that have major implementation points, much more needs to be done before implementation if the users are not to be overwhelmed. In these cases there must be user involvement during planning and realistic trials are needed to minimize the shock of the implementation.

The need to involve users in planning and decision making for implementation raises an interesting question about the appropriate role for the ergonomist in this process. There are many areas in which the ergonomist has the methods to contribute directly by making 'design' decisions, for example in workstation layout, environmental design, and manuals and training. It is essential, however, to ensure that the ultimate users play their part in these processes which means there is also a much more challenging role the ergonomist can play. As one of the few disciplines that can take a socio-technical view of implementation the ergonomist can assist in the establish-ment of an implementation strategy which facilitates organizational change and human learning as well as technical change. The ergonomist can also assist in the creation of the temporary organizational institutions that are necessary to manage the transition from old to new.

In a survey of social science interventions in organizations Klein and Eason (1991) provide detailed accounts of a number of technical system implemen-tations. They note that in all successful cases the practitioners managed a complex interplay between two kinds of contribution. First they employed the many methods described in this book to generate and translate knowledge about people into useful design input (a direct contribution to the 'content' of the new system). Secondly, they contributed to the design and manage-ment of the 'process' by which other interested parties, especially the future users, were engaged in the development. Wilson (1994) also reviews many attempts to 'devolve ergonomics' in order that people at work can contribute directly to the development of their working conditions. In supporting such participatory practices, ergonomists are working as practitioners themselves and acting as change agents to help others make effective contributions. As noted elsewhere in this book, the combination of content expert and change agent creates many dilemmas for the ergonomist. Nevertheless every ergo-nomics practitioner has to make both kinds of contribution if he or she is to be successful.

The change process caused by the implementation of a new technical sys-tem leads inevitably to both technical change and human and organizational adaptation. Other professionals engaged in these changes tend to be con-cerned about technical issues. The human and organizational issues associated with each of the implementation strategies described in this chapter are often neglected. In this setting it is very important to the success of the change process, for the ergonomist to play a broad role and to contribute to both content and process.

References

Aberg, U. (1981). Techniques in redesigning routine work. In *Stress, Work Design and Productivity*, edited by E.N. Corlett and J. Richardson (Chichester: John Wiley), pp. 157–164.

Cherns, A.B. (1976). The principles of socio-technical design. *Human Relations*, **28**, 783–792.

Eason, K.D. (1988). *Information Technology and Organizational Change* (London: Taylor and Francis).

Eason, K.D. and Sell, R.J. (1981). Case studies in job design for information processing tasks. In *Stress, Work Design and Productivity*, edited by E.N. Corlett and J. Richardson (Chichester: John Wiley), pp. 195–208.

Gower, J.C. and Eason, K.D. (in press). The introduction of information technology in a city firm. In *The Application of Information Technology*, edited by S.D.P. Harker and K.D. Eason (London: Taylor and Francis).

Grandjean, E. (1986). *Ergonomics in Computerized Offices* (London: Taylor and Francis).

Keen, P. (1981). Information systems and organizational change. *Communications of the ACM*, **24**, 24–33.

Klein L. and Eason K.D. (1991). *Putting Social Science to Work.* (Cambridge: Cambridge University Press).

Mumford, E. (1983). *Designing Secretaries* (Manchester: Business School).

Rothlisberger, F.J. and Dickson, W.J. (1939). *Management and the Worker* (Cambridge, MA: Harvard University Press).

Wall, T.D. and Lischeron, J.A. (1977). *Worker Participation* (London: McGraw-Hill).

Wilson, J.R. (1994). Devolving ergonomics: The key to ergonomics management programmes. *Ergonomics*, **37**, 4, 579–594.

Chapter 37

Ergonomics and participation

John R. Wilson

Introduction

Throughout this book the issue of participation is raised, both implicitly and explicitly. In a sense it is impossible to carry out 'evaluation of human work' without at least some participation from those job holders actually doing the work. However, when we are talking about participation within systems design (chapter 1), investigations and intervention studies (chapter 35) or systems implementation (chapter 36), we need something more than user surveys or trials or the direct observation of people at work.

Participation within an ergonomics management programme at work will be taken in this chapter to be: *The involvement of people in planning and controlling a significant amount of their own work activities, with sufficient knowledge and power to influence both processes and outcomes in order to achieve desirable goals.*

It is not suggested that participants always have a 'blank cheque'; there will almost always be limits on what they can do and on the suggestions they would like to implement. However, limits on what is achievable should always be made clear at the outset—in terms of what is a 'significant amount' of the activities, what the 'desirable goals' are and thus what 'knowledge and power' are needed for participants to be sufficiently influential.

The needs for knowledge and power amongst participants are found in other views of participation also. Wall and Lischeron (1977) characterized participation in terms of the interactions between participants, the flow of information between all concerned, and the influence any one individual has over others with regard to decision making. Sen (1988) sees ergonomics participative change as embracing models of 'create, catalyze and care', 'design, develop and demonstrate', and 'implement, involve, and improve'. The first ensures the first optimum climate for change, the second provides a method to ensure feasibility and perception of quality for the new system, and the third optimizes the chances of successful change and subsequent enhancements.

Even if only the types of participation that are relevant and interesting to ergonomists and ergonomics are considered these are many and varied and range from true and complete participation through to manipulative and covert techniques. Fuchs-Kittowski and Wenzlaff (1987) divide participation into consultative (decisions rest with management, users express ideas), representative (elected staff members participate in development, management still make decisions), and consensus-oriented (all involved can put forward ideas and determine aims through discourse), and we can see examples of each type in the ergonomics literature. At a finer level of classification, again we may see different ergonomics initiatives at different points of Arnstein's Scale, quoted by Reuter (1987) as: user control; delegation; partnership; placating; consultation; information (one-way); user 'therapy'; user manipulation; public relations exercises.

It is apparent that a vast array of approaches, philosophies and techniques exists, which may come under the heading of participation; there will be dysfunctional consequences for any participative initiative which is in reality a sham. It is not the intention in this chapter to explore the philosophical or theoretical basis for participation at work. Nor will the vast amount of work on employee participation in the arena of industrial democracy, co-operatives or even in a wider political sense be touched on other than indirectly. The concentration is upon participative programmes and techniques which are of immediate relevance to mainstream ergonomics—that is in the analysis, design/implementation and evaluation required to improve an individual's tasks, equipment, interfaces, workplaces, jobs and work organization.

For a discussion of wider issues in participation the reader is referred to Doherty (1987), European Foundation for the Improvement of Living and Working Conditions (1985), Hornby and Clegg (1992), Knight (1989), Mumford (1991), and Wall and Lischeron (1977). The European Participation Monitor, published by the European Foundation for the Improvement of Living and Working Conditions, also has a valuable regular update on developments in participation.

Participation within ergonomics

Ergonomics participation can and should be at both organizational and individual levels. Effective ergonomics management programmes can best be sustained by building them within the organization's existing structures and processes, and within this programme using techniques and methods to encourage and support workforce participation in the analysis, redesign and evaluation of their own tasks, jobs and workplaces (Wilson, 1994).

A widespread general revival of interest in participation may be attributed to a number of circumstances, amongst them a growth in human centred manufacturing with emphasis upon the involved, responsible workforce, the recognition that motivation and performance are complex issues, increased

Trade Union acceptance of participation, more decentralized and flatter IT-based organizations, emphasis on quality and flexibility in manufacturing, and the continued introduction of new technical systems and methods of organizing them (Gaudier, 1988). These are all germane to ergonomics participation, as are recent developments in ergonomics and health and safety legislation. For instance, the European Framework Directive (89/391/EEC) 'on the introduction of measures to encourage improvements in the safety and health of workers at work' contains specific provisions about consultation and participation of workers (Article 11). Much of the remainder of the Directive can be shown also to have a strong relationship to, and implications for, participative processes. The Daughter Directives—e.g., on manual handling or display screen equipment—also make provision for a joint approach to health and safety by employers and workers. Whilst all the specific provisions have not yet all been translated into national legislation, nonetheless the changing legislative scene has already had considerable impact for ergonomics in Europe.

For ergonomists, participation—as a philosophy and a process—has always been of interest, partly due to the strong ergonomics emphasis on application. People-centred approaches are endemic, from user trials in consumer product testing (chapter 10) to user-centred design in human-computer interaction (chapter 12) to human-centred technology programmes in manufacturing industry. Most ergonomists also take a broad view of work redesign and of improving the work environment, arguing that the physical aspects of work neither can nor should be divorced from a consideration of psycho-social and organizational factors. Even when a work redesign programme establishes physical factors as being of greatest concern, to ignore other aspects of work may lead to only partial acceptance and success of any changes made, due to the central role of workforce attitudes (Wilson and Grey, 1990).

In order that workforce participation in the application of ergonomics methods be most successful, there must be a spreading of relevant skills and confidence throughout the company. There are growing numbers of efforts to train the workforce in ergonomics (see Hasle and Schmidt, 1991; Vasbinder and Maze, 1991). This training might be aimed at any or all of shopfloor workers, supervisors, industrial, production or design engineers, health and safety personnel, but the most complete programmes certainly include shopfloor staff in order to help them assess work and workplaces and propose solutions for ergonomics problems. As with ergonomics generally, the limited resources available to do this must be recognized, and as a consequence we might establish 'cascaded' training and awareness programmes, which are to some extent further extensions of a 'Train the Trainers' approach (see Nyran, 1991; Silverstein *et al.*, 1991, for examples). In essence, to train the trainers meets the need to spread ergonomics more widely, including to non-specialists. Ergonomists train a core team in ergonomics and in ergonomics education techniques, the training being largely practically-based and emphasizing 'learning by doing'. Initial training is passed on to others in the workforce by this core team.

An important series of studies by a group of French ergonomists have followed the line of providing tools, training, knowledge and confidence for workers to analyze their own work problems and to develop feasible alternatives. The focus of these studies varies from chemical plant control room design including display specifications (Boel *et al.*, 1985), to posture and health analyses (Montreuil and Laville, 1986), to introduction of new shift-work systems (Teiger and Laville, 1987). Their approach is to set up study groups comprising workers, ergonomists and technical staff; the worker representatives are equipped with the tools and skills needed to analyze and make recommendations about particular problems. In some of their work a task force was formed to oversee the whole project, keeping all levels in the company in touch with current developments. The researchers succeeded in achieving recognition amongst the workforce of the value of an ergonomics perspective and problem-solving process; through their involvement in the administration of questionnaires and direct observation, workforce groups managed to engage the interest of many colleagues in the issues and in finding solutions. Overall the researchers feel that the role of ergonomics has been widely accepted in these companies, not only in order to provide solutions but also to help the workers themselves iterate slowly towards solutions (see also Daniellou *et al.*, 1990).

Following a relationship built up with this French group at the Occupational Ergonomics Conferences held in Zadar, Croatia in the 1980s (Corlett *et al.*, 1986; Haslegrave *et al.*, 1990; Wilson *et al.*, 1987), the Institute for Occupational Ergonomics at the University of Nottingham has been instrumental in the formation of an 'Ergonomics Awareness and Action' programme. Investigators worked with a UK company of some 38,000 people based in over 100 plants, who were facing a growing number of cases of upper limb disorders (ULDs) in order to assist the company to promote the spread of ergonomics education and training. Ergonomics was 'sold' to senior management on a 1-day course; subsequently an Ergonomics Survey Team was formed and trained from amongst their industrial engineers. This team became responsible for training other industrial engineers, making presentations to the work force, and acting as a service operation to assist in ULD problem identification and elimination. Other industrial engineers in turn educated the work force so that within five months about 10,000 workers had some presentations and information (Corlett, 1991a).

Subsequently we have developed and run training and awareness programmes for companies in a number of industries including confectionery, automobiles, pharmaceuticals, household appliances, electronics and boilers. Training starts with a series of 2-day courses to mixed groups of company personnel. These are run on-site and, in addition to giving familiarization with practical and robust assessment tools, have a large shopfloor component with hands-on training in problem identification, prioritization, analysis, diagnosis and solution. Further 1-day training programmes emphasize how to set up an ergonomics management programme and how to devolve ergonomics to the shopfloor through group and participative activities. Parallel courses are run with marketing, sales and product development staff, to raise interest and awareness in addressing workplace ergonomics through product or package design, which neatly parallels current interest in design for manufacture. The final level of training is for experienced ergonomists to conduct workplace assessments and ergonomics interventions whilst being 'shadowed' by relevant company engineers or production staff.

In this way, by cascaded training which gives motivation, practical knowledge and some skills to spread the training further, we can ensure the spread of ergonomics within the ergonomics management programme, at the same time further strengthening its foundation (Wilson, 1994).

As well as an ergonomics management programme and training and awareness for as many as possible in the company, participative ergonomics also requires the active involvement of relevant job holders in task, job and workplace assessments and in the generation and testing of alternative solutions. Several examples of this can be found in Noro and Imada (1991), with other cases in Bengtsson (1990), Broms (1988), Forrester (1985; 1986), Hornby and Clegg (1992), Lehtelä and Kukkonen (1991), Leppänen (1991), Levary and Kalchik (1984) van der Schaaf and Kragt (1992) amongst many others.

Another example is Design Decision Groups, widely used by the author and adapted from O'Brien's (1981) 'Shared Experience Events'; these are discussed in much greater depth below.

For and against participation

Participation and participative techniques seem so attractive, and are embraced eagerly by so many ergonomists, that we must look at the case against them before we look at their advantages.

The negative aspects of participation cannot be ignored. Put bluntly, for whatever reason some people will just not want to participate (Neumann, 1989). Moreover participation, as philosophy or process, is neither an easy option nor, perhaps, always the most appropriate one (see Mossink, 1990), and half-hearted or cynical application will reap a harvest of problems, now or in the future. Use of participation within manufacturing industry has certainly often been viewed with suspicion by trade unions and work forces. Side–effects or systemic consequences may also give problems: for instance increasing autonomy of assembly work groups may affect the work of the purchasing or maintenance departments; enlarged job content for word processing may reduce requirements for personal assistants, clerks or even information specialists; improvements in one group's skill levels or learning may make workers in other offices or departments envious or dissatisfied. Related to this last, the very process of participation may have unexpected systemic effects, in that other groups may wish to be similarly involved but this may not always be feasible or desirable. Finally, a recent ergonomics participation case study has highlighted an age-old problem of work systems change, participative or not; at what point in the implementation should the 'change agent', or participation 'facilitator', leave the scene? If they leave too late, then true participation may be stifled; if they leave too early, a vacuum may be created in which nobody feels any 'ownership' for the process or change (Wilson, 1995).

The potential benefits of a participative approach are both direct and indirect. Use of participative techniques in change analysis and implementation may generate feelings of solution ownership—and therefore commitment to the change—amongst those affected. Participation may give better information in the design process and faster effective eventual use of a new system by the participants. The whole participation process can be a learning experience for 'designers' and 'users' alike. Importantly for this discussion, participative processes can aid the spread of interest and expertise in ergonomics and can help embed an ergonomics perspective within the company, if people really are allowed to make a genuine contribution and influence affairs (Wilson, 1991a, 1991b).

The relationship between participative processes and the development and gains from working in groups or teams is of great interest. Health and safety and workplace ergonomics investigations can be carried out through work groups. Self-directed team working anyway is often an appropriate mechanism to implement principles of enriched job design, and social support is seen as an effective reducer of strain resulting from stress at work. If work is to be organized on the basis of self-directed teams then it makes good sense if the groups that develop the ideas for such work organization and consequent physical workplace requirements will then subsequently become the work teams themselves. In this way the process of change enhances the content.

What are the requirements for a participative process to be successful generally, and for ergonomics change and job redesign in particular? Table 37.1 shows some of the most important requirements, classified by parts or aims of the participative process and largely relevant to experiences in physical work redesign and workplace layout. It can be seen that several of these requirements would be more easily met within investigation of concrete physical design issues than in redesigning people's jobs in terms of job characteristics like autonomy, responsibility, skill utilization and variety. Recent discussions between ergonomists interested in participation have identified the most important success factors to be: the nature of the facilitator; the ability to give the process time; the degree of competence and motivation of the participants (and as a requirement of these, their knowledge and power); and the 'visibility' of the problem.

Table 37.1. Requirements of participative processes in workplace ergonomics (adapted from Wilson (1991a) and Wilson (1991c))

Process elements	Requirements
Participative policy needs . . .	Support and resources, in terms of time, people and finance
Setting-up the process should be	Collaborative, stressing volunteers' participation, generating motivation and promoting likely benefits
Participative processes work best if they are . . .	Non-directive but at the same time have direction and purpose, iterative and allow compromises, promote creativity and learning
Participative techniques and methods must be . . .	Flexible and well-costed
Facilitators should be . . .	Respected as unbiased but knowledgeable, adaptable, sensitive and aware of when to 'leave' the process
Outcomes of the process should include . . .	Testable solutions, a spread of expertise in the organization, and the process becoming seen as internal and ordinary rather than external and extraordinary

Design decision groups and problem solving groups

Design decision group process

In the early 1980s Denis O'Brien of the UK Home Office proposed a number of methods for running creative meetings and for involving users in systems design (e.g., O'Brien, 1981). He adapted theories and techniques from market research and from the literature on creativity and innovation, in particular using non-directive and group discussions, developing 'thinking tools' and encouraging participants to talk about, write down or draw attitudes, experiences, visions or opportunities (see also Caplan, 1990). O'Brien described a number of techniques or tools within what he called Shared Experience Events (SEEs); amongst these are: the structuring of contributions from sub-groups whilst others listen in silence, only later criticizing; use of a dynamic agenda, formed at first from a 'word map' exercise; use of drawing as far as possible, which enhances concept development, communication of ideas, and also helps in structuring the process; and very careful pre-planning and preparation. His ideas were used in a study of fire brigade control room design (Langford, 1982).

Subsequently the present author has, with colleagues, adapted the method, called it 'Design Decision Groups' (DDG) and used it in a number of applications (Wilson, 1991c). The DDG process is summarized in Figure A.37.1 (p.1084), and further description and illustration of the stages appear in the appendix to this chapter.

Problem solving groups process

One criticism often made of DDG and similar techniques is that they are usually only employed with white collar or technical staff. An exception to this involved an adapted DDG, in what we termed a Problem Solving Group Process (Moreland and Wilson, 1992; Wilson, 1995). The workers concerned were crane drivers operating from a control room in an incinerator plant. Early observation identified that they had severe problems related to viewing the task, the design of joystick controls and a deadman's handle, seating and postural inadequacies, and general environmental impacts. A form of Design Decision Group technique was adapted for use with the drivers; critically this involved them in sourcing and costing their preferred design solutions within a budget set by management. The process is illustrated in Table 37.2. It will be seen that the final stage is that of continual improvement—the embedding, in this small company, of participation as *the* way to go about ergonomics problem assessment and solution.

Use of a participative process here probably did not lead to identification of any different problems and causes than would have been the case with a traditional ergonomics approach. However, the perspective on these problems and the priorities given them was different, and this guided subsequent improvements in directions which best met the needs and interests of the workers. The solutions which were eventually agreed were, at best, satisficing; that is, we could not argue that they were the best that could have been developed but they did entail a satisfactory answer to the problems. Since there is always the question of what is 'best' in ergonomics anyway, and since the process of continual change has led to further improvements, we are not at all unhappy with the outcomes. As change agents we withdrew from the process as soon as possible, with the consequence that the proposed redesigns were only partially implemented and therefore from our viewpoint there are still some ergonomics difficulties. However, this seems to be out-

Table 37.2. Stages of the problem solving group (PSG) procedure (from Wilson, 1995)

	Problem solving group procedure
Stage One:	Familiarization: of investigator with the work and workplace and of the participants with ergonomics and the aims of the PSG
Stage Two:	Field visits: by participants to 'similar' sites
Stage Three:	Design Decision Group (A): elements of session A of the DDG process
Stage Four:	Design Decision Group (B): elements of session C (+B) of the DDG process
Stage Five:	Lighting simulation: using screens, coverings etc. in the control room, to reduce reflection problems, etc.
Stage Six:	Sourcing and costing of solutions: crane drivers finding products or adaptations which best meet their preferred solutions, whilst staying within budget
Stage Seven:	Continual improvement: after the facilitators have left

weighed by finding not only that the changes made have been completely accepted, widely welcomed and have given rise to a dramatic fall in complaints and postural problems, but that the group has continued to discuss and implement further workplace and work improvements, demonstrating that an ergonomics perspective has been embedded for that group of workers to at least a limited extent.

Conclusions

Efforts to enable the widespread and successful implementation of ergonomics at work have multiplied in recent years; strenuous efforts are being made to take ergonomics out of the pages of text books and make it more applicable in practical use. We cannot just rely upon 'ergonomics experts'—there are not enough of these, this will not be cost effective for most smaller organizations and is anyway not the right way to approach the challenge. This challenge is to:

- motivate people throughout a company to become their own 'ergonomists', within a total ergonomics programme that has meaning and influence
- give them the tools and techniques and related training to make ergonomics interventions, ensuring that training comes at the right time for all relevant groups
- ensure they have enough understanding to know when and how to use these tools and techniques, *and when they cannot and must call in specialist assistance.*

[We must] give ergonomics away, ... transfer our knowledge and methods to others who are closer to the places where changes have to be made, so that they do much of the ergonomics for themselves. ... Of course there will be occasions when ergonomics is mis-used, or badly used, but with experience it will improve and the knowledge itself is a good protection from abuse. Until ergonomics is widely practised by other than professional ergonomists it is likely to remain something to be added on at the end. What is more, until it is more widely realised what ergonomics really is, we will not get it introduced into design specifications, where human performance and requirements become a normal part of the design criteria. Until this becomes the accepted practice, the implementation of ergonomics will always an uphill fight. (Corlett, 1991b, p. 418).

There is a strong argument for a participative approach to ergonomics; this is both participation at an individual level whereby workers are involved in workplace and work assessment and in providing solutions to any health and safety problems, and participation at an organizational level whereby

engineering, production, medical and other staff develop their own programmes for ergonomics awareness and action.

Of course there are limits to such participation; there are certain problems
and solutions which will require professional (outside) ergonomics expertise.
At least a part of the training we give should be aimed towards allowing
companies to understand not just what they can do internally but also what
they should not attempt. In this, an ergonomics programme should be no
different to many company activities involving a mixture of in-house and
outside expertise. In fact, at a more general level, there is little in an ergonomics and participative approach to health and safety which is not a part of what
should be general excellence within industrial and office management.

References

Bengtsson, P. (1990). Moving pictures as a method for dialogue in planning.
Paper presented at course on participative approaches to workplace
design. NIVA Nordisk, Åland, Finland.

Boel, M., Daniellou, F., Desmores, E. and Teiger, C. (1985). Real work
analysis and workers' involvement. In *Ergonomics International 85, Proceedings of the 9th Congress of the International Ergonomics Association*, Bournemouth, pp. 235–237.

Broms, G. (1988). Personal communication from the Institute for the Development of Production and Workplaces, Stockholm.

Caplan, S. (1990). Using focus group methodology for ergonomic design.
Ergonomics, **33**, 527–533.

Corlett, E.N. (1991a). Ergonomics fieldwork: an action programme and
some methods. In *Towards Human Work: Solutions to Problems in Occupational Health and Safety*, edited by M. Kumashiro and E.D. Megaw
(London: Taylor and Francis).

Corlett, E.N. (1991b). Some future directions for ergonomics. In *Towards
Human Work: Solutions to Problems in Occupational Health and Safety*
edited by M. Kumashiro and E.D. Megaw (London: Taylor and
Francis).

Corlett, E.N., Wilson, J.R. and Manenica, I. (Eds) (1986). *The Ergonomics
of Working Postures: Models, Methods and Cases* (London: Taylor and
Francis).

Daniellou, F., Kerguelen, A., Garrigou, A. and Laville, A. (1990). Taking
future activity into account at the design stage: participative design in
the printing industry. In *Work Design in Practice*, edited by C.M. Haslegrave, J.R. Wilson and E.N. Corlett (London: Taylor and Francis).

Doherty, P. (1987). Report from the working group on experience with
participation: application in administrative and health care. In *System
Design for Human Development and Productivity: Participation and Beyond*,
edited by P. Doherty *et al.* (Amsterdam: North-Holland), pp. 415–420.

European Foundation for the Improvement of Living and Working Conditions (1985). Participation in technological change. Consolidated
Report (Geneva: EFILWC).

Forrester, K. (1985). Adult education and workers investigation: an ergonomics survey by Leeds busworkers. In *Proceedings of the 9th Congress of the International Ergonomics Association*, Bournemouth, pp. 385–387.

Forrester, K. (1986). Involving workers: participatory ergonomics and the trade unions. In *Contemporary Ergonomics 1986*, edited by D.J. Oborne (London: Taylor and Francis).

Fuchs-Kittowski, K. and Wenzlaff, B. (1987). Integrative participation, a challenge to the development of informatics. In *System Design for Human Development and Productivity: Participation and Beyond*, edited by P. Doherty *et al.* (Amsterdam: North-Holland), pp. 3–18.

Gaudier, M. (1988). Workers' participation within the new industrial order: a review of literature. *Labour and Society*, **13**, 313–332.

Hasle, P. and Schmidt, E. (1991). Workplace improvements initiated by action-oriented training. In *Designing for Everyone, Proceedings of the Eleventh Congress of the International Ergonomics Association*, edited by Y. Quéinnec and F. Daniellou (London: Taylor and Francis).

Haslegrave, C.M., Wilson, J.R. and Corlett, E.N. (Eds) (1990). *Work Design in Practice: A Participative Approach* (London: Taylor and Francis).

Hornby, P. and Clegg, C. (1992). User participation in context: a case study in a UK bank. *Behaviour & Information Technology*, **11**, 293–307.

Knight, K. (Ed) (1989). *Participation in Systems Development* (London: Kogan Page).

Langford, J.B. (1982). West Sussex Fire Brigade: control room design. Scientific Research and Development Branch, Home Office.

Lehtelä, J. and Kukkonen, R. (1991). Participation in the purchase of a telephone exchange—a case study. In *Designing for Everyone, Proceedings of the Eleventh Congress of the International Ergonomics Association*, edited by Y. Quéinnec and F. Daniellou (London: Taylor and Francis).

Leppänen, A. (1991). Development of a complex productional system through worker participation. In *Designing for Everyone, Proceedings of the Eleventh Congress of the International Ergonomics Association*, edited by Y. Quéinnec and F. Daniellou (London: Taylor and Francis).

Levary, R.R. and Kalchik, S. (1984). A visual based group decision-making process for layout finalisation of large offices. *International Journal of Production Management*, **4** (2), 13–30.

Montreuil, S. and Laville, A. (1986). Cooperation between ergonomists and workers in the study of posture in order to modify work conditions. In *The Ergonomics of Working Postures*, edited by E.N. Corlett, J.R. Wilson, I. Manenica (London: Taylor and Francis), pp. 293–304.

Moreland, A.A. and Wilson, J.R. (1992). Energy efficient but not ergonomic: participative redesign in a crane control room. In *Contemporary Ergonomics '92*, edited by E.J. Lovesey (London: Taylor and Francis), pp. 469–473.

Mossink, J.C.M. (1990). Case history: design of a packaging workstation. *Ergonomics*, **33**, 399–406.

Mumford, E. (1991). Participation in systems design—what can it offer? In *Human Factors for Informatics Usability*, edited by B. Shackel and S.J. Richardson (Cambridge: Cambridge University Press), pp. 267–290.

Neumann, J. (1989). Why people don't participate when given the chance. *Industrial Participation*, No. 601, 6–8.

Noro, K. and Imada, A.S. (Eds) (1991). *Participatory Ergonomics* (London: Taylor and Francis).

Nyran, P.I. (1991). Cost effectiveness of core-group training. In *Advances in Industrial Ergonomics and Safety III*, edited by W. Karwowski and J.W. Yates (London: Taylor and Francis).

O'Brien, D.D. (1981). Designing systems for new users. *Design Studies*, **2** (3), 139–150.

Reuter, W. (1987). Procedures for participation in planning, developing and operating information systems. In *System Design for Human Development and Productivity: Participation and Beyond*, edited by P. Doherty *et al.* (Amsterdam: North-Holland), pp. 271–276.

Sen, T.K. (1988). Participative group techniques. In *Handbook of Human Factors*, edited by G. Salvendy (Chichester: Wiley), pp 453–469.

Silverstein, B.A., Richards, S.E., Alcser, K. and Schurman, S. (1991). Evaluation of in-plant ergonomics training. *International Journal of Industrial Ergonomics*, **8**, 179–193.

Tieger, C. and Laville, A. (1987). How ergonomic methods can be used by non-ergonomists. *Paper presented at 2nd International Occupational Ergonomics Symposium, Applied Methods in Ergonomics*, Zadar, Yugoslavia.

van der Schaaf, T.W. and Kragt, H. (1992). Redesigning a control room from an ergonomics point of view. In: *Enhancing Industrial Performance: Experiences of Integrating the Human Factor*, edited by H. Kragt (London: Taylor and Francis), pp. 165–178.

Vasbinder, D.M. and Maze, D.M. (1991). Ergonomics partners: the development of a participatory ergonomics program in the sewn products industry. In *Proceedings of the 24th Annual Conference of the Human Factors Association of Canada*.

Wall, T.D. and Lischeron, J.A. (1977). *Worker Participation: A Critique of the Literature and Some Fresh Evidence* (London: McGraw Hill).

Wilson, J.R. (1991a). A framework and a foundation for ergonomics? *Journal of Occupational Psychology*, **64**, 67–80.

Wilson, J.R. (1991b). Extending participative processes in ergonomics. In *Proceedings of the International Conference on Participatory Approaches to Improving Workplace Health*, Ann Arbor, Michigan, pp. 39–43.

Wilson, J.R. (1991c). Design Decision Groups: a participative process for developing workplaces. In *Participative Methods*, edited by K. Noro and A. Imada (Chichester: John Wiley), pp. 81–96.

Wilson, J.R. (1994). Devolving ergonomics: the key to ergonomics management programmes. *Ergonomics*, **37**, 579–594.

Wilson, J.R. (1995). Solution ownership in participative work redesign: the case of a crane control room. *International Journal of Industrial Ergonomics*, **15**, 329–344.

J.R., Corlett, E.N. and Mananica, I. (Eds) (1987). *New Methods in Applied Ergonomics* (London: Taylor and Francis).

Wilson, J.R. and Grey, S.M. (1990). But, what are the issues in work redesign? In *Work Design in Practice*, edited by C.M. Haslegrave, J.R. Wilson and E.N. Corlett (London: Taylor and Francis).

Appendix

Extract from: *Participatory Ergonomics*, Chapter 5, Design Decision Groups— A Participative Process for Developing Workplaces by John R. Wilson; edited by K. Noro and A. Imada (London: Taylor and Francis) 81–96.

Design decision groups procedure

Scene setting

In order to enable (not force) people to participate in meetings, exercises, investigations, etc., certain general principles can be observed. They should feel that there is some purpose to participating, that outcomes will be of benefit to themselves and/or to others. They should feel that they have enough information and knowledge to make a useful input, and should feel confident enough to ask for or to obtain such knowledge where they feel it is lacking. One important but tricky requirement is that the group must not be overtly directed or feel that this is the case; however, the group must have direction and be purposeful; disruptive or domineering members must be subtly reined in and not allowed to dominate meetings.

The above requirements for the DDG process can be achieved in several ways. We felt that holding meetings at an independent site (our laboratory) and emphasizing our independence and the confidentiality of the details of the process were important. Meetings were facilitated rather than run, generally with one investigator acting as introducer, prompter, interpreter and general resource, with a second investigator on hand as technical resource, including making and serving refreshments, distributing or showing graphical and visual aids, supplying materials and so on. The second investigator was also responsible for recording the proceedings, using video and photography (as unobtrusively as possible), and for note taking. All sessions were allowed to run in a relaxed manner, with everybody being gently encouraged to contribute, and with a very flexible agenda. At the same time, the investigators were well aware of certain tasks or exercises which had to be completed, and certain information which had to be derived, within particular time-scales. We have found that, although certain guidelines for doing this can be followed and can help, it is not surprising that it also takes a certain sort of person to facilitate the groups. Not everyone has the ability to communicate and subtly direct, without dictating or dominating.

All potential participants were contacted at their place of work and informed that they would take part in confidential discussions about the design of their workplace, how much they would be paid, and the dates and times when we would need them. Library staff largely came for evening sessions;

retail staff came on their afternoons off. All participants were collected and returned by taxi. Payment, which compared very favourably with their normal wages, was made in one lump sum at the end of all sessions.

In the remainder of this section the stages of the DDG process are described and illustrated. This is shown in figure A.37.1 as a three-session process; however, if time or other resource restrictions dictate, two, somewhat longer, sessions can be used, as in the retail case.

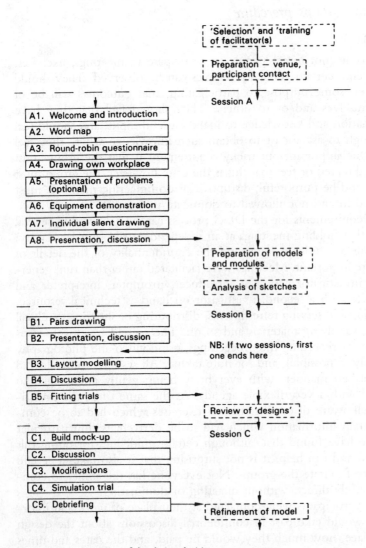

Figure A.37.1. Summary of the design decision group process

Session A

Introduction

This includes a welcome, introductions, an explanation of who the design meeting organizers are and a general overview of what is expected of the design sessions. Refreshments are served and every effort is made to establish a comfortable, relaxed environment.

Word map

'Word maps' or 'visual maps' are constructed by every participant, volunteering words, which they feel are connected with the particular design topic for a 5–10 min period. This technique defines the total problem space and helps produce the dynamic agenda. In our case we simply asked participants to make an inventory of objects and equipment required at their workplace. This acts as an 'ice-breaking' device and also participation is such an easy, non-stressful exercise it breaks down any inhibitions that quieter members of the group might have about expressing their ideas in front of their colleagues. Participation was encouraged by going 'round the table' several times (see figure A.37.2).

Round-robin questionnaire

Participants are presented with a series of very simple open-ended questions such as: 'A good work counter is ...?' or 'Problems at my workplace are ...?'. Each question is printed on a separate sheet, and these are passed around all participants with each attempting to complete the sentence with a different ending to that chosen by the others (figure A.37.3). This method is preferred to the alternative of each participant filling out an individual questionnaire for two reasons. Firstly, it makes the exercise a little more productive, in that each participants has to think of a different ending from those given by the others, thus providing more information by reducing duplication. Secondly, it can act as a stimulus. If, for example, participants had missed the point of the exercise, or were unable to think of any ideas for a certain sentence, then seeing what had been written by others may trigger a line of thought that could otherwise have lain dormant.

Drawing of own workplace

As a further ice-breaking exercise, as preparation for the more important drawing exercises to come, and to provide investigators with extra information, participants were asked to draw their existing workplace, indicating its good and bad points.

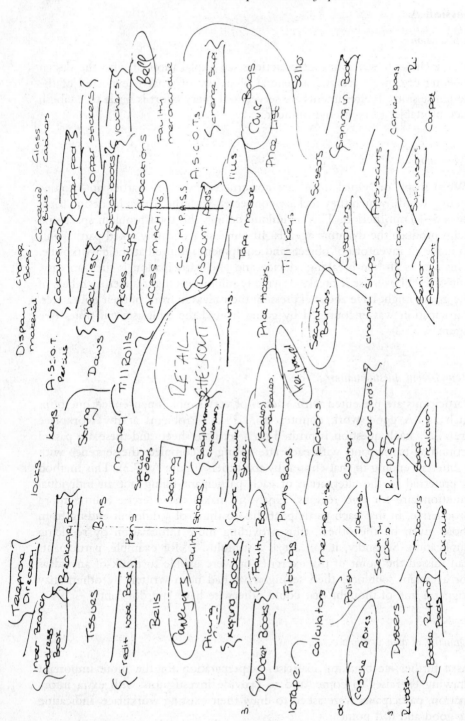

Figure A.37.2. Part of a simple word map

Seeing Rubbish and not being able to find anything
but piles of faulty goods stack up in one corner

When I'm working, the things I dislike in a checkout are. having

to stand and look down on customers,
not having enough room to write and
bag items.

not having the right equipment at hand to
deal with customers paying by ~~cheque~~ credit cards.

Having to stand, waiting for someone to answer
the bell.

To show a customer where something is. (point).

Having papers etc hanging around.

merchandise hanging around under your feet.

Having catalogues on the top near the tills.

Having no spare till Rolls.

People eg sales assistant not coming on Tills when small
Be etc and not having well

when you can't find price list or a price in book
when consultants a item checking over or looking at
prior buy in store

not having enough Time To Talk To Customers
when there are no customers to serve but you can't come
off till another person comes on
when you can't find anything when there are so many
draws.

Figure A.37.3. Example of a round-robin questionnaire

Slide presentation of ergonomics problems

Slides are selected from associated fieldwork, covering a range of topics: different types of layout, different types of equipment, etc., and, most importantly, a series showing how these pieces of equipment have been poorly installed in certain environments. The slides emphasize such features as poor seat height, lack of leg room, poor work–surface height, lack of work space and bad positioning of equipment and displays.

This stage is intended to provide a logical progression from the round-robin questionnaire. At that stage, the participants are asked to think about what makes good and bad designs, but only have their own experience to draw upon. The slide show is intended to make participants aware of systems and layouts that they may not have encountered before, and also to emphasize problems that they may not have considered whilst being influenced so heavily by their current system. Slides of a variety of different workplaces and systems can also emphasize the point that we are required to design something entirely new, and that suggested designs should be unimpeded by a narrow range of experience. We are conscious that such a procedure must not direct our participation in any particular direction, either of criticism of existing systems or of ideas for new ones. It may not be appropriate in all circumstances.

Equipment demonstration

When systems or equipment are to be installed which may be new to any or all participants it is advisable to demonstrate these and their operation. Actual examples should be used where possible; if not illustrations could be used.

Individual silent drawing

Each participant is asked to produce sketches on A4-sized paper of both plan and perspective views of their ideas for a good new workplace, incorporating any new systems. These should detail proposed layouts for people, work counters, modules and equipment (figure A.37.4). At this stage, only very general ideas need to be produced, and participants are discouraged from becoming too involved in detail (e.g., precise measurements of work surfaces, or exact location of documents). To call this stage a 'silent drawing exercise' is perhaps overstating the point, but it is beneficial to discourage communication in the hope that a variety of different ideas and designs will be produced for use in the later discussion stage. Each individual is encouraged to try out several rough ideas before producing a final sketch suitable for presentation. Different coloured pens are available for colour coding various items of equipment or furniture.

Individual presentation and discussion

Each participant is asked to present their final sketch to the rest of the group and to explain the reasoning behind the design. The designs are discussed and gently (the role of the facilitator) criticized.

After each of the sketches has been discussed many problems should be highlighted, and certain designs may be seen as 'superior'. The role of the facilitator is then to encourage a general discussion, concentrating on one or two particular designs and bringing up any points that may not have been previously considered, probably by referring back to issues arising from the word-map and round-robin stages.

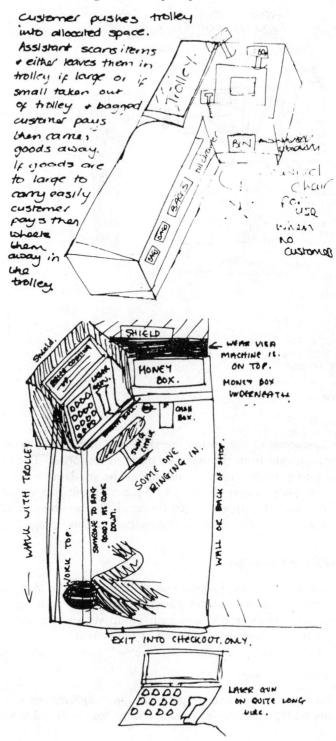

Customer pushes trolley into allocated space. Assistant scans items & either leaves them in trolley if large or if small taken out of trolley & bagged customer pays then carries goods away. If goods are to large to carry easily customer pays then wheels them away in the trolley

Trolley.

BOX

BIN

swivel chair for use when no customer

SHIELD

WEAR VISA MACHINE IS ON TOP.

MONEY BOX.

MONEY BOX UNDERNEATH

CASH BOX.

SOME ONE RINGING IN.

WALL OF BACK OF SHOP.

WALL WITH TROLLEY.

WORK TOP.

SOMEONE TO BAG GOODS AS CODES DOWN.

EXIT INTO CHECKOUT. ONLY.

LASER GUN ON QUITE LONG WIRE.

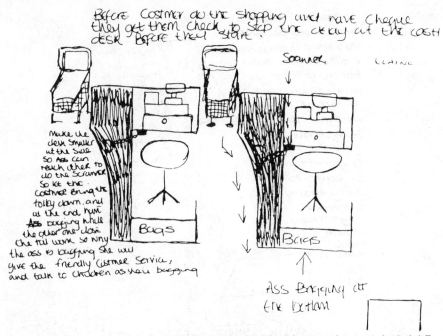

Figure A.37.4. Three examples of individually drawn proposed work spaces, showing individual differences in design ideas and how these can be illustrated and communicated. Considerable colour coding is used in the originals

Session B

Drawing in pairs

At the beginning of the session the facilitator briefly runs through some of the points arising from the sketches made in session A. The group is then divided into pairs or threes, according to the similarity of their ideas, and asked to produce a larger sketch, on A3 paper, of the plan for the workplace. This time more emphasis is placed on detail. Positioning of equipment is to be specified more precisely, as are seating and furniture design (figure A.37.5).

Presentation and discussion

Each drawing is presented to the rest of the group by its authors. Discussion at this point is geared toward choosing the 'best' drawing, but no firm decisions need to be reached as each group with get a further chance to demonstrate their ideas using models of the system.

Simple layout modelling

Participants are shown empty spaces of approximately the size available at the future workplace site. Very strong corrugated cardboard models of equip-

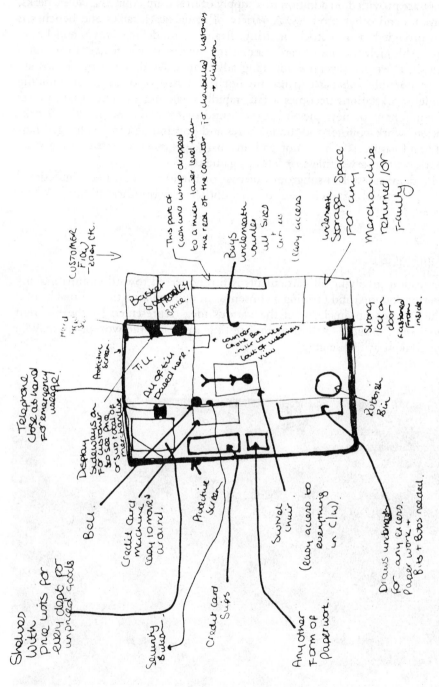

Figure A.37.5. Example of a workplace developed by a pair of participants, and used as the basis for a model-building exercise

ment are provided, in addition to a supply of mock-up counters, tables, desks, consoles and other modules. A variety of actual seats, tables and benches is also provided, as is a stock of string, flat card, modelling knives and heavy tape. Although the use of card models may seem, at first glance, to be amateurish, there are several compelling advantages. Models and mock-ups are more portable and participants are not frightened of breakage so building model workstations becomes a fast, rapidly changing exercise. Extra pieces of equipment or new pieces can be found or created very quickly. Workstation, work-counter and module sizes and positioning can be changed rapidly and easily. We have not had any participants who felt that they were taking part in a worthless or 'cheap' game.

Participants, still in subgroups of two or three, build a three-dimensional full-size model of their idea of the optimum workplace (figures A.37.6 and A.37.7).

Discussion

Discussion involving all participants is firmly geared toward eliminating any inferior design(s) and forming a consensus of opinion as to the optimal layout. This is often an amalgam of the various ideas represented by the different mock-ups. Limited 'walkthroughs', testing aspects of the workplace, may be undertaken at this stage.

Figure A.37.6. Initial planning for layout: two groups

Figure A.37.7. Discussion of the juxtaposition of equipment

Fitting trials

On reaching this stage, a general workplace layout and the positioning of all the equipment should largely have been agreed. The next crucial factor to be tackled is that of work-surface heights and depths. These are determined through fitting trials, with the aid of adjustable metal stands holding the cardboard models, using all group members as subjects. In addition, however, the facilitator must ensure, at the time or later, that anthropometric characteristics of other potential users are taken into account.

Session C

Building a mock-up

On returning for the third session, participants build a complete mock-up (or two if there is still internal disagreement) of the preferred workstation, integrating ideas on the placement and juxtaposition of equipment with work-surface height and depth data (figure A.37.8).

Discussion

The discussion entails detailed consideration of the operational requirements at each particular workstation; if two designs have been modelled, each sub-group must defend its ideas against those of the other group.

Figure A.37.8. Building a mock-up workplace

Modifications

As a result of the discussion, any modifications to the general structure of the workstations are made. If more than one design is still being considered, the best alternative is generally selected at this stage.

Simulation trial and assessment

The trials are aimed at simulating as many as possible of the scenarios, operations and tasks which are or might be undertaken in actual use. Participants take turns in playing the roles of staff and customers (figure A.37.9), including children (figure A.37.10). Final adjustments to the workplace model are made at this point.

Debriefing

Debriefing takes place after each session but this is the most important time of all. Participants are encouraged to discuss the value of the exercise and the perceived quality of their designs.

Refinements are made at the end of each stage, with the investigators accounting for constraints, equipment or tasks which may have been omitted (deliberately or accidentally) in the DDG sessions, or which may have come to attention later.

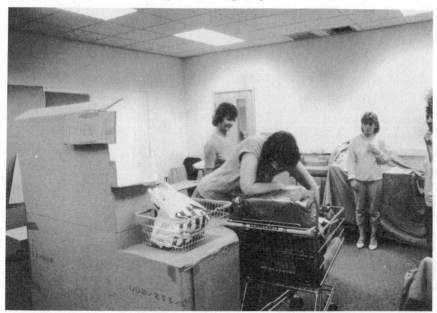

Figure A.37.9 A simulation trial

Design decision group outcomes

The outcomes of a DDG study, or other similar work, are varied. At the very least we have managed to find out, from the people who should know most about it, the problems or benefits associated with existing systems, the potential problems with the new system, and a number of concepts to be fed into the development process. At best, one or more feasible design alternatives for the whole workstation is provided; this is an invaluable outcome, regardless of whether these designs are 'new' to the investigators or are combinations of ideas already developed. The process and setting also allow flexible and iterative testing of design concepts, whether derived beforehand or during the process.

It has been noticed in all our work with DDGs that the creative, yet relaxed, atmosphere and the resultant concentration of minds seems to encourage the expression of opinion or facts about other aspects of work redesign. During the libraries study especially we learned much of interest about training and support, feedback and other communication, autonomy, role clarity and other job-content and organization issues. This allowed us to make a more informed analysis of current situations, and to make better suggested changes for improvement. Although not used by us in such a context, it is suggested that the spin-off benefits will be considerable when such an exercise is conducted at a local level with people whose own technology, workplaces and jobs are to be changed. Awareness of what should be done

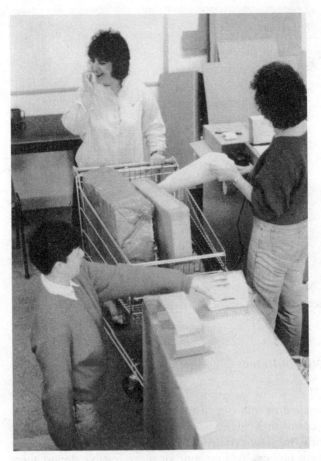

Figure A.37.10 A trial showing (initially sceptical!) participant simulating a child

to improve jobs and workplaces and of the criteria which should be applied, and the confidence to promote their own ideas can all help to create a more committed workforce, as well as healthier, safer, more effective and more satisfying jobs. The use of DDGs can thus give rise to direct gains through its outcomes in terms of design ideas, and can also give rise to systemic gains through the very application of the process.

Chapter 38

Methods in context: the future

E. Nigel Corlett

The method and the meaning

Which of us has not been aware, at some time in our professional lives, of a solution chasing a problem? The temptation to see problems from the standpoint only of certain techniques is one of which we must always remain aware. The measurements we make constrain our understanding of the problem and where we have a bias towards certain methods it is almost inevitable that we shall have limited the range of that understanding.

Since the above argument suggests that we have to understand the problem before we choose the methods it leaves us in something of a quandary. If we understand the problem we don't always need to do the study, but if we don't then how do we choose a method? The quandary is more a literary device than reality, however, since the argument just emphasizes that experimentation for problem solution is an iterative as well as an intellectual exercise. We choose methods according to our understanding but, as the results then give us additional information, we are in a position to re-assess the utility of our methods and re-work the investigation in other ways. We try, from which we learn, and then we try again, benefitting each time from a critical analysis of our results and their meanings.

Of course, as we gain experience we have a greater ability to see the likely shape of a problem, and become more adept at choosing methods which are appropriate for its investigation. We know the problem field better and we have a better understanding of the likely effects arising from the situations we are studying. However, with increased familiarity we can become biased. We use favourite methods or use favoured models of the situation, without sufficient initial thought.

People who move into ergonomics from another discipline can also fall into this trap at the start, seeing a problem in limited terms. If their work is presented only to a specialist group with similar backgrounds, the bias is reinforced and leads to a body of knowledge purporting to cover ergonomics issues but in reality providing a sub-optimal coverage of the problems. Terms can become misused and distorted, such as 'optimal' being equated with 'minimum', 'user friendly' representing childishly chatty dialogue, or 'efficient' being applied to an increase in short term output but not related to the human consequences of that output.

Contributing to knowledge

These are all very normal behaviours for people. We are not suggesting any evil intent or stupidity, but if we are aware that they are likely to occur, we must counteract their effects. Their likelihood gives good reason for the sorts of activity in which academics indulge but which commercial people sometimes feel are not productive. These are attendance at conferences, the widening of interests beyond the confines of one's own studies to see what others do, and the writing and publishing of papers.

Conferences present an opportunity, which we do not always take, to hear about a wider spectrum of work than just our own interests, as well as to hear of work in our own area done by people who have taken a different approach. Different approaches may lead to different conclusions, which is when we can gain increased insights into our own work.

To take an interest in other areas introduces us to different techniques, or methods used in different ways. Results, or their interpretation, in one area can illuminate our own work in another. Not least, studying people can be different from studying inanimate matter as the interactions within a person can be relatively unknown or only inferred. Thus considering other areas of work can help us to appreciate this complexity and, hence, to interpret better our own investigations.

The third leg in our support structure for better investigation is publication. Many people outside the academic world see this as an indulgence, the gathering of a list of publications for promotion purposes. Whilst this does happen it will be recognized that good publications demonstrate a person's contribution to the subject and, at the same time, their competence in the discipline. The reader will note the qualification 'good' in the previous sentence! All of us rely upon published contributions; we quote them not to demonstrate our own erudition, or even to convince others that we are right, but to outline the context of our own work and the foundations on which it is built. If we take our subject seriously, it is necessary that we describe to others the work we do, so that they may benefit from our experiences, use our knowledge and avoid repeating our mistakes. No matter how much the commercial world may financially reward an ergonomist, professional quality is established by the evidence of results publicly judged by our peers, as well as by the contributions we make to improving the capacity of our discipline to achieve its purposes. In particular, the commercial world which benefits from the discipline has a major responsibility to support its development, by allowing its professionals to take part in professional activities. This responsibility cannot be shirked by a company putting forward (sometimes spurious) claims of commercial confidentiality or by insistence on the need for a close attention to maximizing its own profitability.

Presentation and purposes

It could be argued with a good deal of justification that a book on methods should have had a section on presentation of results. Although the editors had some sympathy with this there is obviously a limit. A cogent argument for including presentation is in the area of interpretations. Statistical or other forms of analyses are only the first step, for it is what we make of the analyses which counts.

What we analyze depends to a great extent on what we have measured but many workers will have had the experience of the gradual illumination and understanding which comes from poring over the data; how new arrangements of the data demonstrate a better understanding of their meaning and how different ways of presenting the findings clarify the relationships. The recommendation to decide on the analysis and significance levels *before* starting the measurements is sound, particularly as the experimental design quite often defines the analysis method, and it should be decided beforehand if the method adopted is relevant to the effects being studied. However, this should not prevent workers from turning over their results in their minds and wondering, for example, why the dimensions they have measured are as they are, or whether there are yet other factors which will contribute to an understanding.

When an eminent scientist, many years ago, broadcast a talk on British radio on 'the fraud of the scientific paper', he pointed out that the standard sequence in a paper, of literature review, experimental design, experiment, analysis and conclusions, was not how science tended to be pursued. More usually, a question arises in the worker's mind, which is turned over and speculated upon until a means for trying it out is developed. It is then tried out, to 'see if it works', and then tested again and again, interspersed with thought and discussion, to understand its mechanisms and its boundaries. Good experimental and analytical controls are exercised throughout, but only at the writing-up stage is the conventional, logical sequence exposed which implies that the whole activity was the natural outcome from some previous stage.

The whole process, from idea to presentation, is one where a worker's curiosity must be alert. 'Why have these things happened?', What happens if . . .?', 'Is what has happened caused by what I think it is?' are the questions which should always be in a worker's mind. Although much work is inevitably narrowly focused, this should not necessarily restrain speculation about effects. For example, studying the likely problems occurring in routine assembly of electronic circuits on a 20 s cycle should also embrace the likely effects on the worker of the focused attention and suppression of internal or external mental activity needed to achieve consistent error-free production. Carpal tunnel syndrome and local muscular fatigue are not the only responses to this kind of repetitive work.

Cost-benefits?

The question of 'benefit' is a further matter for concern, since a narrow view of what it means can inhibit or distort the investigation of ergonomics problems. Quite often, in reported cost-benefit analysis studies, the benefits given are limited to the additional surplus on the direct activities accruing from the changes. Although this is undeniably a benefit, it is a limited interpretation of the word even for its everyday use. This is not to say that benefits should not be expressed in monetary terms, but that some effort should be made to express, on a common metric which is usually money, the many other gains from the ergonomics changes. As yet, our abilities in this respect are limited but a recent review (Corlett, 1988) has indicated that there are methods which can be pressed into service and developed in this area.

Developments in cost-benefit analysis are important because they will allow us to give a truer measure of the value of our work. 'Making workpeople comfortable', when expressed like that, can seem to have low priority in many studies; nice to do but hardly essential. To a great extent this is because we can put numbers on some things related to performance, but not on the effects of discomfort; we cannot quantify its costs. So it is not visible in the decision equation.

From a community health viewpoint, to take but one perspective, there is extensive evidence of the outcomes of situations which are recognized in their early stages through the presence of discomfort. These consequences may take a long time to develop, or be paid for by others than the employer, but these are not good reasons for seeing them as less important than performance in the short term. Their visibility, and consequent recognition and incorporation in the criteria for designing investigations will arise in great part from a quantification of the costs of their existence. A consequence of this development would be a change in the methodologies for many studies, which would incorporate techniques for cost-benefit analysis as a normal component in evaluation.

Who are we working for?

By looking at the wider effects of the interactions between people and their environment we are inevitably brought face to face with some other questions. These are not new questions; many people face them in the course of their professional lives, but they are quite fundamental to our ergonomics activities, to the interpretation of our results and to how we propose they should be implemented.

If our objectives include, for example, improving the effectiveness of human performance, do our measures of effectiveness stop at, say, speed and errors? Can we accept such a limited definition in every case? For whom are we working? Some employers would be happy if we accepted this limited definition, and in some cases of course it would be quite sufficient; but if there is anything in the claims by ergonomists to be a profession, then there are evidently many cases where a limited perspective on performance is quite insufficient.

To return to an earlier point, this is where the wider contacts and interactions within professional activities are important. The broadening of understanding about the whole person which comes from such widened contacts, together with discussions with one's peers, help to maintain a human-centred view of ergonomics work which maintains ethical boundaries appropriate for a profession which has people as its central interest. If misused, our profession has the means, within it, for the ruthless exploitation of others and unfortunately there are always a few people in the world who would wish to use this knowledge in such a way. It is, of course, the duty of every professional to block such a process and to see that there are true benefits for the subjects of our studies.

These are ethical matters, much influenced by the individual's personal beliefs and values. It has been found in many of the human sciences that it is not sufficient to leave these matters unspecified and rely on the assumption that all people will gravitate to the same point of view in caring for others. Professional associations, universities and others have drawn up codes of conduct to be observed when their members practise their profession. There are also codes of ethics for observance by society members in their professional activities which involve human subjects.

As examples of these, the code of practice presented to applicants to go on the Professional Register of the Ergonomics Society is given (Figure 38.1). This is similar to many such codes for professional bodies in the U.K. As our discipline deals also with people, there is need to consider in more detail the ethical principles to be observed when conducting research with them. The British Psychological Society (1993) requires all its members to observe the principles set out in Figure 38.2. These are extraced from a detailed statement which includes also the treatment of experimental animals and guidelines on professional relationships. The code of conduct and ethical principles given here will give guidance to readers when they have no sources of their own, and will enhance the practice of ergonomics in their environment. It is important for working groups to review, and formalize their activities, to

1. In pursuit of their profession, those on the Professional Register of the Ergonomics Society shall at all times value integrity, impartiality and respect for evidence, and shall sustain the highest ethical standards.
2. Within their obligations under the law, they shall hold the interest, safety and welfare of those in receipt of their services or affected by those services to be paramount at all times. Those carrying out research shall safeguard the interest of participants, and ensure that their work is in keeping with the highest standards of scientific integrity.
3. They shall endeavour to maintain and develop their professional competence, and to recognize and work within its limits, striving always to identify and overcome factors restricting this competence.
4. They shall not lay claim, directly or indirectly, to have competence in any area of ergonomics in which they are not competent, nor to have characteristics or capabilities which they do not possess.
5. They shall take all reasonable steps to ensure that their qualifications, capabilities or views are not misrepresented by others, and to correct any such misrepresentations of which they become aware.
6. If requested to provide services outside their personal competence, or if they consider that services of such nature are appropriate, they shall give every reasonable assistance towards obtaining such services from those qualified to provide them.
7. They shall take all reasonable steps to ensure that those working under their supervision act in concordance with this Code of Conduct.
8. They shall refrain from making misleading, exaggerated or unjustified claims for the effectiveness of their methods, and they shall not advertise services in a way likely to encourage unrealistic expectations about the effectiveness and results of those services.
9. They shall take all reasonable steps to preserve the confidentiality of information acquired through their professional practice or research, and to protect the privacy of individuals or organisations about whom information is collected or held. Subject to the requirements of the law they shall prevent the identity of individuals or organisations being revealed without their expressed permission. When working in a team or with collaborators, they should inform recipients of services or participants in research of the extent to which personally identifiable information may be shared between colleagues, and will ensure as far as lies within their powers that those with whom they are working will respect the confidentiality of the information.
10. With the exception of recordings of public behaviour, they shall only make sound or visual recordings of recipients of services or participants in research with the expressed agreement of the individuals or their representatives both to the recording being made and to the subsequent conditions of access to the recordings.
11. They shall conduct themselves in their professional activities in ways which do not damage the interests of the recipients of their services or participants in their research and which do not undermine public confidence in their ability to perform their professional duties.
12. They shall neither solicit nor accept from those receiving their services any significant financial or material benefit beyond that which has been contractually agreed, not shall they accept any benefits from more than one source for the same work without the consent of all the parties concerned.
13. They shall not allow their professional responsibilities or standards of practice to be diminished by considerations of religion, sex, race, age, nationality, class, politics or extraneous factors.
14. Where they become aware of professional misconduct by a professional colleague that is not resolved by discussion with the colleague concerned, they shall take steps to bring that misconduct to the attention of the General Secretary of the Society, doing so without malice.

Figure 38.1. Ergonomics Society current Code of Conduct for those admitted to the Professional Register

Ethical Principles for Conducting Research with Human Participants

1 Introduction

1.1 The principles given below are intended to apply to research with human participants. Principles of conduct in professional practice are to be found in the Society's Code of Conduct and in the advisory documents prepared by the Divisions, Sections and Special Groups of the Society.

1.2 Participants in psychological research should have confidence in the investigators. Good psychological research is possible only if there is mutual respect and confidence between investigators and participants. Psychological investigators are potentially interested in all aspects of human behaviour and conscious experience. However, for ethical reasons, some areas of human experience and behaviour may be beyond the reach of experiment, observation or other form of psychological investigation. Ethical guidelines are necessary to clarify the conditions under which psychological research is acceptable.

1.3 The principles given below supplement for reseachers with human participants the general ethical principles of members of the Society as stated in The British Psychological Society's Code of Conduct (q.v.). Members of The British Psychological Society are expected to abide by both the Code of Conduct and the fuller principles expressed here. Members should also draw the principles to the attention of research colleagues who are not members of the Society. Members should encourage colleagues to adopt them and ensure that they are followed by all researchers whom they supervise (e.g. research assistants, postgraduate, undergraduate, A-Level and GCSE students).

1.4 In recent years, there has been an increase in legal actions by members of the general public against professionals for alleged misconduct. Researchers must recognise the possibility of such legal action if they infringe the rights and dignity of participants in their research.

2 General

2.1 In all circumstances, investigators must consider the ethical implications and psychological consequences for the participants in their research. The essential principle is that the investigation should be considered from the standpoint of all participants; foreseeable threats to their psychological well-being, health, values or dignity should be eliminated. Investigators should recognise that, in our multi-cultural and multi-ethnic society and where investigations involve individuals or different ages, gender and social background, the investigators may not have sufficient knowledge of the implications of any investigation for the participants. It should be borne in mind that the best judge of whether an investigation will cause offence may be members of the population from which the participants in the research are to be drawn.

3 Consent

3.1 Whenever possible, the investigator should inform all participants of the objectives of the investigation. The investigator should inform the participants of all aspects of the research or intervention that might reasonably be expected to influence willingness to participate. The investigator should, normally, explain all other aspects of the research or intervention about which the participants enquire. Failure to make full disclosure prior to obtaining informed consent requires additional safeguards to protect the welfare and dignity of the participants (see Section 4).

3.2 Research with children or with participants who have impairments that will limit understanding and/or communication such that they are unable to give their real consent requires special safe-guarding procedures.

3.3 Where possible, the real consent of children and of adults with impairments in understanding or communication should be obtained. In addition, where research involves any persons under sixteen years of age, consent should be obtained from parents or from those "in loco parentis". If the nature of the research precludes consent being obtained from parents or permission being obtained from teachers, before proceeding with the research, the investigator must obtain approval from an Ethics Committee.

3.4 Where real consent cannot be obtained from adults with impairments in understanding or communication, wherever possible the investigator should consult a person well-placed to appreciate the participant's reaction, such as a member of the person's family, and must obtain the disinterested approval of the research from independent advisors.

3.5 When research is being conducted with detained persons, particular care should be taken over informed consent, paying attention to the special circumstances which may affect the person's ability to give free informed consent.

3.6 Investigators should realise that they are often in a position of authority or influence over participants who may be their students, employees or clients. This relationship must not be allowed to pressurise the participants to take part in, or remain in, an investigation.

Figure 38.2. British Psychological Society (1993). *Code of Conduct, Ethical Principles and Guidelines.* Published by the British Psychological Society, Leicester, UK

3.7 The payment of participants must not be used to induce them to risk harm beyond that which they risk without payment in their normal lifestyle.

3.8 If harm, unusual discomfort, or other negative consequences for the individual's future life might occur, the investigator must obtain the disinterested approval of independent advisors, inform the participants, and obtain informed, real consent from each of them.

3.9 In longitudinal research, consent may need to be obtained on more than one occasion.

4   Deception

4.1 The withholding of information or the misleading of participants is unacceptable if the participants are typically likely to object or show unease once debriefed. Where this is in any doubt, appropriate consultation must precede the investigation. Consultation is best carried out with individuals who share the social and cultural background of the participants in the research, but the advice of ethics committees or experienced and disinterested colleagues may be sufficient.

4.2 Intentional deception of the participants over the purpose and general nature of the investigation should be avoided whenever possible. Participants should never be deliberately misled without extremely strong scientific or medical justification. Even then there should be strict controls and the disinterested approval of independent advisors.

4.3 It may be impossible to study some psychological processes without withholding information about the true object of the study or deliberately misleading the participants. Before conducting such a study, the investigator has a special responsibility to (a) determine that alternative procedures avoiding concealment or deception are not available; (b) ensure that the participants are provided with sufficient information at the earliest stage; and (c) consult appropriately upon the way that the withholding of information or deliberate deception will be received.

5   Debriefing

5.1 In studies where the participants are aware that they have taken part in an investigation, when the data have been collected, the investigator should provide the participants with any necessary information to complete their understanding of the nature of the research. The investigator should discuss with the participants their experience of the research in order to monitor any unforeseen negative effects or misconceptions.

5.2 Debriefing does not provide a justification for unethical aspects of any investigation.

5.3 Some effects which may be produced by an experiment will not be negated by a verbal description following the research. Investigators have a responsibility to ensure that participants receive any necessary debriefing in the form of active intervention before they leave the research setting.

6   Withdrawal from the Investigation

6.1 At the onset of the investigation investigators should make plain to participants their right to withdraw from the research at any time, irrespective of whether or not payment or other inducement has been offered. It is recognised that this may be difficult in certain observational or organisational settings, but nevertheless the investigator must attempt to ensure that participants (including children) know of their right to withdraw. When testing children, avoidance of the testing situation may be taken as evidence of failure to consent to the procedure and should be acknowledged.

6.2 In the light of experience of the investigation, or as a result of debriefing, the participant has the right to withdraw retrospectively any consent given, and to require that their own data, including recordings, be destroyed.

7   Confidentiality

7.1 Subject to the requirements of legislation, including the Data Protection Act, information obtained about a participant during an investigation is confidential unless otherwise agreed in advance. Investigators who are put under pressure to disclose confidential information should draw this point to the attention of those exerting such pressure. Participants in psychological research have a right to expect that information they provide will be treated confidentially and, if published, will not be identifiable as theirs. In the event that confidentiality and/or anonymity cannot be guaranteed, the participant must be warned of this in advance of agreeing to participate.

Figure 38.2. Contd

8 Protection of Participants

8.1 Investigators have a primary responsibility to protect participants from physical and mental harm during the investigation. Normally, the risk of harm must be no greater than in ordinary life, i.e. participants should not be exposed to risks greater than or additional to those encountered in their normal lifestyles. Where the risk of harm is greater than in ordinary life the provisions of 3.8 should apply. Participants must be asked about any factors in the procedure that might create a risk, such as pre-existing medical conditions, and must be advised of any special action they should take to avoid risk.

8.2 Participants should be informed of procedures for contacting the investigator within a reasonable time period following participation should stress, potential harm, or related questions or concern arise despite the precautions required by the Principles. Where research procedures might result in undesirable consequences for participants, the investigator has the responsibility to detect and remove or correct these consequences.

8.3 Where research may involve behaviour or experiences that participants may regard as personal and private the participants must be protected from stress by all appropriate measures, including the assurance that answers to personal questions need not be given. There should be no concealment or deception when seeking information that might encroach on privacy.

8.4 In research involving children, great caution should be exercised when discussing the results with parents, teachers or others in loco parentis, since evaluative statements may carry unintended weight.

9 Observational Research

9.1 Studies based upon observation must respect the privacy and psychological well-being of the individuals studied. Unless those observed give their consent to being observed, observational research is only acceptable in situations where those observed would expect to be observed by strangers. Additionally, particular account should be taken of local cultural values and of the possibility of intruding upon the privacy of individuals who, even while in a normally public space, may believe they are unobserved.

10 Giving Advice

10.1 During research, an investigator may obtain evidence of psychological or physical problems of which a participant is, apparently, unaware. In such a case, the investigator has a responsibility to inform the participant if the investigator believes that by not doing so the participant's future well being may be endangered.

10.2 If, in the normal course of psychological research, or as a result of problems detected as in 10.1, a participant solicits advice concerning educational, personality, behavioural or health issues, caution should be exercised. If the issue is serious and the investigator is not qualified to offer assistance, the appropriate source of professional advice should be recommended. Further details on the giving of advice will be found in the Society's Code of Conduct.

10.3 In some kinds of investigation the giving of advice is appropriate if this forms an intrinsic part of the research and has been agreed in advance.

11 Colleagues

11.1 Investigators share responsibility for the ethical treatment of research participants with their collaborators, assistants, students and employees. A psychologist who believes that another psychologist or investigator may be conducting research that is not in accordance with the principles above should encourage that investigator to re-evaluate the research.

Figure 38.2. Contd

be sure that their working attitudes never drift into a perception of their work as dealing with people as objects.

A modest protection against this is to avoid pseudo-scientific terminology about people and implying a remote or impersonal approach. We are dealing with people, not biological (or cognitive) machines, and it is as well to say so in our reports. To speak of 'experimental subjects' is recognizably accurate and long established. But to refer to 'the worker', as if to an alien species (and the author is aware that he has done just that earlier in this chapter!) is to be avoided if possible. Even more to be avoided is reference to people as 'humans'—we do have animal ergonomics but are unlikely to confuse the two! We can refer to 'people', to 'the operator' or to them in their professional role of typist, manager or sky-diver. But we must remember, and emphasize, that they are people, with all the attributes, interests and complexities of people, and that if our studies and presentations reflect this, it will be easier to avoid the de-personalization of people and the risk of reaching and implementing conclusions inappropriate for the real problem.

Of course, it is not always simple to see the ethical problems. In a discussion dealing with moral questions which arise during the design and implementation of computer systems, Pullinger (1989) points to the difficulties in this respect of a programmer concerned with providing only one part of a complex body of software. The increased specialization reduces the opportunities for understanding and reflection on the functioning of the total system, and hence the opportunities for individuals to assess their contributions in terms of their own moral outlook.

Pullinger raises a number of points which are important for these workers developing computer interfaces, but he points out that the exercise of ethical judgement requires a wide understanding of the technological limits, the system's purposes and people's social and personal needs. It also requires a position to be taken on the forms of society which are acceptable, for example, on the question of security and confidentiality of information. He concludes that computer specialists are not in a position to exercise such judgements, primarily because their information base is deficient in much of the requisite knowledge. A group better able to consider this area are those who have moved into the human-computer interface area from ergonomics, occupational psychology or the like.

This recognition, that good work cannot contribute to a good society unless it is informed by more than technological competence, is one which we stress here. The presentation of information from an investigation which is clear as to its meaning and its limitations is as important as Pullinger's point that the designer of an expert system has a duty to make it clear that the system has limits, which will vary with the user's expertise, and that to transfer judgement from the individual to the system is not appropriate.

The position of the professional

Finally, there is a changing balance in ergonomics activities; an increase in the number of those engaged in ergonomics as professional practitioners and a reduction in the proportion of researchers. This change implies a change in the accepted criteria, from a primary interest in what are the causes and the effects, what can be relied upon to be a foundation for further understanding, to what is there in our discipline which can be used to improve a situation, and do it with the appropriate economy.

This is not a bad change, although it would be bad if research disappeared altogether, which is unlikely. After all, the practice of medicine is in this mode and is effective; the deficiences are small compared with the successes of its practice. But ergonomists will have to note some of the aspects which make medicine a successful profession in the context of its involvement with people. It is a profession which has a major effect on human welfare. Unlike others which also have a strong influence, e.g., politics or business management, it has set itself a moral standard of performance above an economic one. It is judged on its contribution to human welfare before it is judged on its costs. The latter are part of the equation, medicine must be efficient, but costs do not always predominate.

As mentioned earlier, ergonomics is a political activity, with both a direct and an indirect impact on human lives. But its primary purpose is a better match, for people, of all that is in their environment. 'People are central and the rest must adapt' is perhaps too extreme, but the adaptations required of people must be within boundaries which are set well within distortion or injury, be it physical, mental or social. Hence the measures of our effectiveness must include our successes in how far we are within these boundaries, as well as the costs and financial returns of our interventions.

These changes set the agenda for an extension of our methodology, not only in the cost-benefit area as called for earlier in this chapter and in chapter 34, but in applied methods which can put scientific knowledge to use without distortion. The recognition of the social dimension of our discipline also calls for its extension in bringing into our activities all those who are influenced by it. The growing recognition of the value of participatory practices in ergonomics applications is another area where methods will develop in the immediate future and where professional activities will change as this approach to our work grows.

All of this sets the agenda to require a code of conduct to be *not* a facade behind which we practice our profession, but a public statement of our position and actions in all that we do. The further we move into this future, of being primarily a body of professional practitioners, the more urgent will these matters become. There are boundaries to *The evaluation of human work*, and we should make it clear where they lie.

Conclusion

It will be evident that this book has but scratched the surface. It could have been more detailed on many methods which are mentioned, there could have been other areas introduced which are important for ergonomists, and there is much room for expansion in the ways that methods may be chosen, used and their results interpreted. What the Editors hope will be one result of this text is that there will be an increased awareness of the need to select methods in relation to the model of the system as seen by the investigator,

1108   *Introduction and implementation of systems*

and the careful selection of a minimum number of methods sufficient for the investigator's purposes. Of course, experimentation is expensive and if more data of relevance can be gathered without damaging a study, then they should be gathered. However we trust that the user of the book will be able to avoid a touching faith in the value of 'high tech' equipment rather than an intelligent use of the minimum necessary technology. We also hope that, no matter what the readers' opinion of the book may be, they will not throw it at the problem! The relevant analogy for experimental work is the precision of the scalpel rather than the universal attack of the shotgun. The latter is often combined with a blind faith in statistical packages to solve the problems of thoughtless data collection. If our text has contributed to the efficiency of experimentation, the clarity of understanding the results and not least a recognition of the important contributions which ergonomics investigations make to the well-being of individuals and society, we will be well content.

References

Corlett, E.N. (1988). Cost benefit analysis. *International Reviews of Ergonomics*, **2**, 85–104.
Pullinger, D.J. (1989). Moral judgement in designing better systems. *Interacting with Computers*, **1**, 93–104.

Contributors

Lisanne Bainbridge is former Reader in Psychology at University College London. Most of her research has been on industrial process operation, on the cognitive mechanisms of planning activities and using knowledge, including their implications for mental workload and advanced interface design. Much of this work has been based on analysis of verbal protocols. In 1976 she was awarded the Bartlett Medal of the Ergonomics Society.

Maurice C. Bonney is Professor of Production Management in the Department of Manufacturing Engineering and Operations Management at the University of Nottingham. His research areas have centred around the development of computer aids in the field of robot simulation, computer aided ergonomics design, computer aided work study, and computer aided production management. The work on the production management framework includes development of analysis and design methods for manufacturing systems and enterprise integration based on Petri-net methods with associated software. He is a director of BYG Systems Ltd and SAMMIE CAD Ltd. He is a recipient of the 1994 Otto Edholm Award presented by the Ergonomics Society 'in recognition of the development and application of SAMMIE as an ergonomics design tool'.

Rosemary A. Bonney graduated in Ergonomics from Loughborough University. She then undertook research at Nottingham University, specializing in the effects of vibration and other factors on spinal load which she extended when employed as an ergonomist at the Vermont Rehabilitation Engineering Centre, Burlington, Vermont, USA. She then retrained, and is now teaching in the UK.

Dr Ivan D. Brown OBE was Assistant Director of the Medical Research Council's Applied Psychology Unit in Cambridge, England, until his retirement in April 1993. He researched accident causation and prevention for some 35 years, specializing in behavioural aspects of road safety. He was also Extra-Mural Professor of Traffic Science at the University of Groningen, The Netherlands, from 1988 to 1991. He is currently Director of Ivan Brown Associates, providing advice internationally on human factors aspects of transport systems. He is a Fellow of the British Psychological Society, the Ergonomics Society, and the Human Factors & Ergonomics Society. He was honoured by the Human Factors Society in 1990 with their 'Distinguished Foreign Colleague' Award. In the 1991 Queen's Birthday Honours List he was appointed an Officer in the Order of the British Empire. In 1993 the Human Factors & Ergonomics Society presented him with the A.R. Lauer Award and in 1995 he was elected an Honorary Fellow of the Ergonomics Society.

Mark A. Bullimore is an Assistant Research Scientist and Assistant Clinical Professor at the University of California at Berkeley, School of Optometry. His research interests include basic and clinical applications of motion perception, the effect of cataract and refractive surgery on visual performance, the assessment of discomfort glare, and visual impairment. He received his bachelors in opthalmic optics (optometry) in 1983, and his PhD in 1987 from Aston University, Birmingham.

Mike Burton is Lecturer in Psychology at the University of Nottingham, where he also received his PhD. His principal research interests are Knowledge Acquisition and the Modelling of Human Cognitive Processes. He has published extensively in both these areas.

Dr Keith Case is Professor of Computer-Aided Engineering in the Department of Manufacturing Engineering at Loughborough University of Technology. His current research areas include computer aided ergonomics, applications of virtual reality techniques in design, integrated design and manufacture CAD systems and the CAD modelling of modular machines. He is a Fellow of the Ergonomics Society and a recipient of their 1994 Otto Edholm Award 'in recognition of the development and application of SAMMIE as an ergonomics design tool'.

Professor Bruce Christie is joint Course Organiser for the MSc in Multimedia Systems, the MSc in User–Interface Design and associated courses at London Guildhall University. Students include those working in the industry up to the level of senior management. He has worked in the field of Information Technology for over twenty years, including working as a consultant for a range of companies in the UK, USA, Germany, Brazil and elsewhere including major computer and telecommunications companies, user organizations, government departments and others. He has had management experience as Manager of Human Factors Technology for the ITT group of companies, research experience (in the UK and the USA) and research administration experience during a period working at the headquarters of the then Social Science Research Council. He has published several books in the field of human factors, user–interface design and related subjects, and has been responsible for patents (owned by ITT and Goldmark Communications Corporation) in the fields of computer-based behavioural analysis and audio conferencing. He is a Certified Management Consultant, a Fellow of the British Psychological Society and a Chartered Psychologist. He has a PhD in the field of communications psychology, an MSc by research in the field of display ergonomics and a First Class Honours degree in Psychology.

Jenny Collyer is joint Course Organiser for the MSc and associated post-graduate courses in Multimedia Systems at London Guildhall University. She was also a member of the team responsible for developing the MSc in User–Interface Design, and is currently a member of the course team. Her recent research has been in the field of multimedia systems, including managing the University's part in an international research project co-funded under the European TIDE Programme (now part of Telematics). She has a BA from the Open University.

Emeritus Professor E. Nigel Corlett worked as a mechanical and an industrial engineer in industry prior to moving into University education in 1957. He worked with aero engine, machine tool and domestic equipment companies, as designer and factory manager both in the UK and abroad. Before retirement he was Head of the Department of Production Engineering and Operations Management at the University of Nottingham. Prior to this he had been the Professor of Industrial Ergonomics at the University of Birmingham. His research focused on the effects of work on people's performance and its stressful consequences, as well as on ways to improve these in practice. He is currently Scientific Adviser to the Institute for Occupational Ergnomics at the University of Nottingham, as well as President of the Centre for the Registration of European Ergonomists. He is a Fellow of the Royal Academy of Engineering, an Hon. Fellow of the Ergonomics Society and holds the DSc from London and the PhD from Birmingham.

Tom Cox, BSc, PhD, CPsychol, FBPsS, FRSH, FRSA, is Professor of Organizational Psychology, Head of the Department of Psychology at the University of Nottingham, and Director of the Centre for Organizational Health & Development which is recognized by the World Health Organization as a Collaborating Centre for Occupational Health. His research interests focus on occupational health psychology and stress, and he has published extensively in this area for over 20 years. Professor Cox is an Expert Adviser on occupational health and stress to the European Commission and World Health Organization (European Region), and a member of the Occupational Health Advisory Committee for the UK Health & Safety Executive. He has previously been a member of several other UK HSE, Home Office and Ministry of Defence Working Groups (for example, on Stress in the Police Force) and is Editor of the international journal *Work & Stress* (Taylor and Francis).

Suzanne M. Dawes received her Bachelor Degree in Mathematics from Colby College, and a Master's Degree in Industrial Engineering and Operations Research from the University of Massachusetts at Amherst. She is currently pursuing a PhD in Industrial and Systems Engineering with a concentration in Human Factors Engineering at the University of Southern California. Over the past seven years, Ms Dawes has worked in a variety of positions relating to Human Factors. Prior to joining Northrop in 1990 where she is involved in workload studies, Ms Dawes worked in business product development at Xerox Corporation and was the Human Factors Engineer at Rancho Seco Nuclear Power Plant. Ms Dawes is a member of the Human Factors & Ergonomics Society and the Institute of Industrial Engineers.

Dr Colin G. Drury is Professor of Industrial Engineering at the State University of New York at Buffalo, USA. He has been active in human factors engineering for twenty-five years, both in industry as Manager of Ergonomics at Pilkington Glass, and in academia. His publications (over 200) concentrate on human factors in quality control, process control, maintenance and safety. In 1980, Colin Drury was awarded the Bartlett Medal of the Ergonomics Society for his work in industrial quality control, and in 1990 the State University of New York Excellence Award for implementing manufacturing change as Executive Director of the Center for Industrial Effectiveness (TCIE). Since 1990 Dr Drury has led a team applying human factors to aviation maintenance and inspection. He is a Fellow of the Human Factors and Ergonomics Society, the Institute of Industrial Engineers and the Ergonomics Society, and received the HFES' Paul M. Fitts award in 1992.

Professor Ken Eason is Director of the HUSAT Research Centre in Loughborough. Over a 20-year period he has researched the impact of computer systems upon their users. His doctoral studies investigated the impact of systems upon managers. Latterly, he has researched the system design process in order to identify methods by which the characteristics of users and their organizations can be better recognized by design teams. Professor Eason has published two books *Managing Computer Impact* and *Information Technology and Organisational Change*.

Martin Freer is a consultant ergonomist with SAMMIE CAD Ltd, who provide ergonomics consultancy services and market the SAMMIE system worldwide. He has made extensive use of the SAMMIE system in practical ergonomics evaluations across a wide range of application areas including aircraft, vehicles, control rooms and manufacturing processes. He graduated in Ergonomics from Loughborough University of Technology.

E. Jane Fulton is currently Director of Human Factors at IDTWO, an Industrial Design Company in San Francisco, California. She came to IDTWO from the Institute for Consumer Ergonomics, Loughborough, where she worked as a Senior Research Officer for nine years. She has degrees in Psychology and Architecture.

Amanda Griffiths, BA, PhD, MSc, CertEd, CPsychol, AFBPsS, FRSH, is a Research Fellow in Occupational Health Psychology, and Deputy Director of the Centre for Organizational Health & Development in the Department of Psychology at the University of Nottingham. She had postgraduate qualifications in the fields of educational and occupational psychology, and her current research interests include organizational interventions for stress, work-related upper limb disorders, and systems for the management of occupational health.

Helen Haines is a senior ergonomics consultant with the Institute for Occupational Ergonomics, University of Nottingham. She has an MSc in Ergonomics from Loughborough University of Technology and is undertaking PhD research into participatory ergonomics. She has worked with more than 30 industrial clients and is responsible for the IOE's training programme with the UK Health and Safety Executive.

Peter A. Hancock received his BEd and MSc degrees from Loughborough University where his work concerned modelling of physiological systems. He received his PhD from the University of Illinois where his research focused upon time perception. He has followed broad interests in human factors with particular concern for the energetic aspects of operator performance in association with complex systems. He is the editor of *Human Factors Psychology*, and co-editor of *Human Mental Workload* with his colleague Najmedin Meshkati, editor of *Intelligent Interfaces* with Mark Chignell, and editor of a recent series of two volumes with John Flach, Jeff Caird, and Kim Vicente on *Ecological Approaches to Human Factors*. He is currently an Associate Professor at the University of Minnesota where he directs the Human Factors Research Laboratory.

James Hartley is Professor of Applied Psychology at the University of Keele, Staffordshire, UK. He obtained his first degree and PhD in Psychology from the University of Sheffield. His main research interests are in written communication, with especial reference to typography and layout, but he is also well known for his research into teaching and learning in the context of higher education. Professor Hartley is a prolific writer and has so far published twelve books and over 200 papers. He is a Fellow of both the British Psychological Society and the American Psychological Association.

Christine M. Haslegrave is a senior lecturer in Occupational Ergonomics at the University of Nottingham, currently engaged on research into the biomechanical demands of manual handling tasks. Her work with the Institute for Occupational Ergonomics involves occupational risk assessment and investigation of health and safety problems in industry. She was Head of the Ergonomics Section at the Motor Industry Research Association for several years, with interests in vehicle safety and ergonomic legislative testing.

Peter A. Howarth received a BSc in Ophthalmic Optics (Optometry) from The City University, London, an MSc in Ergonomics from Loughborough University and a PhD in Physiological Optics from the University of California at Berkeley. Formerly a Research Scientist at the Lawrence Berkeley

Lighting Laboratory, School of Optometry, UC Berkeley, where he conducted human factors research in lighting, he is now a lecturer in vision and lighting at the University of Technology, Loughborough. As well as visual ergonomics, his vision research interests include the pupillary system, and the chromatic aberration of the eye.

Åsa Kilbom is professor of Work Physiology at the Swedish Institute of Occupational Health in Stockholm, and head of its Applied Work Physiology Division. After a degree in medicine at the Karolinska Institute, she trained as a specialist in clinical physiology. Her main areas of research are work-related musculoskeletal disorders, especially their multicausal origin and the quantitative exposure–response relationships, and the ergonomics of the ageing workforce.

Barry Kirwan is a lecturer at the University of Birmingham, in the Industrial Ergonomics Group, in the School of Manufacturing and Mechanical Engineering. Besides lecturing in Human Reliability Assessment (HRA), Task Analysis, and Control Room Design, he is involved in various research projects in the field of HRA. He has worked in the nuclear power and reprocessing industries, principally on the THORP project, and has been involved in a number of risk assessments in the nuclear as well as the offshore and chemical industries. He has degrees in psychology, ergonomics and a PhD in HRA, and is the developer of the HRA systems HRMS and JHEDI, co-editor of *A Guide to Task Analysis* and author of *A Guide to Practical Human Reliability Analysis* (both Taylor and Francis). He is also a keen diver.

Masaharu Kumashiro is Professor and Head of the Department of Ergonomics at the University of Occupational and Environmental Health, Japan. He is conducting ergonomics research for occupational health and safety. Much of the work in this area has been compounded with Industrial Engineering, Occupational Medicine and Industrial Psychology. He has a PhD in Environmental Sciences, and is a Fellow of the Ergonomics Society, a director of the Japan Ergonomics Research Society and Chair of the IEA Technical Group for Safety and Health.

Steve Mason obtained a first degree in engineering from the University of Leicester and an MSc in Work design and Ergonomics from the University of Birmingham. On completing his Masters he worked for a number of years as an ergonomist in the Methods Study Department of the General Electric Company. He then spent 18 years in mining where he was Deputy Head of Ergonomics for British Coal. He is currently Principal Consultant in Ergonomics and Safety Management at International Mining Consultants Ltd. He is a Fellow of the Ergonomics Society and a member of the Institution of Occupational Safety & Health.

Lynn McAtamney, BSc(anat), Grad Dip Phty, PhD, MCSP, MErgS, graduated from the University of NSW Australia with a BSc (anatomy) and a year later, in 1980, with a diploma in physiotherapy. After practising clinically she joined the Australian National Institute for Occupational Health and Safety as a research officer. In 1994 she was awarded a PhD in ergonomics from the University of Nottingham, the subject of which was work related upper limb disorders. She is a consultant ergonomist and senior partner of COPE.

Ian McClelland C Dip Tech, MSc, F Ergs Soc, first trained as a mechanical engineer and then moved to Loughborough University of Technology in 1970 where he completed the postgraduate course in Ergonomics. In 1972 he received his Master's degree and joined the Institute for Consumer Ergonomics where he undertook a wide variety of research and consultancy work for government, business and industry. In 1986 he joined Philips Corporate Design as Manager of the Applied Ergonomics Group. Interests include the integration of usability engineering principles into the process of user interface design, the involvement of users as 'co-designers' in interaction design, and evaluation methodology.

E.D. Megaw graduated from Cambridge University in 1964 with a degree in Experimental Psychology. He subsequently obtained an MSc in Work Design and Ergonomics from The University of Birmingham where he then stayed to complete a PhD on manual control. He continued research into eye movement control, visual search, industrial inspection, visual fatigue and ergonomics databases. In 1983 he received The Ergonomics Research Award for his work on industrial inspection. Since 1992 he has been head of the Industrial Ergonomics Group at The University of Birmingham where he is also director of the Ergonomics Information Analysis Centre. The Centre was awarded the President's Medal from The Ergonomics Society in 1995 for its significant contributions to original research and application of knowledge in the field of ergonomics, including publication of the print and now CD ROM journal *Ergonomics Abstracts* (Taylor and Francis).

Dr David Meister is presently a Senior Scientist at the Naval Ocean Systems Center, San Diego, California. He is the author of seven textbooks on various aspects of human factors, including system development, testing and methodology. He is a former President of the Human Factors Society and the 1984 winner of the Franklin V. Taylor award for outstanding contributions to engineering psychology, given by the American Psychological Association.

Najmedin Meshkati received a BSc in Industrial Engineering simultaneously with a BA in Political Science and Economics in 1976 in Iran. He received a MSc Engineering Management and a PhD in Industrial and Systems Engineering from the University of Southern California, respectively in 1978 and 1983. He is a Certified Professional Ergonomist. His research interests include: mental workload, human factors of technology transfer, human error, and human and organizational factors of complex, large scale technological systems such as nuclear power and chemical processing plants. Presently he is an Associate Professor of Human Factors and Ergonomics and the Associate Executive Director at the Institute of Safety and Systems Management, University of Southern California. He is a recipient of the

Presidential Young Investigator Award from the National Science Foundation in 1989.

Kenneth C. Parsons is Reader in Climatic Ergonomics at the University of Technology, Loughborough. He has over 20 years experience in the field of human response to the environment, and is presently Chairman of BSI panel in the ergonomics of the thermal environment. He received the 1992 Ralph Nevins award from ASHRAE which was given for significant accomplishments in the study of bioenvironmental engineering and its effect on human comfort and health.

Dr Stephen T. Pheasant formerly lectured in anatomy and ergonomics at the Royal Free Hospital School of Medicine, London. He is currently an independent consultant, and also an honorary consultant at the Robens Institute Industrial and Environmental Safety Centre at Surrey University. His books include *Bodyspace, Anthropometrics: an Introduction* and *Ergonomics – Standards and Guidelines for Designers* (Taylor and Francis).

Dr J. Mark Porter is a Senior Lecturer in Ergonomics in the Department of Human Sciences at Loughborough University of Technology. He is the Head of the Vehicle Ergonomics Group in this department and a Director of SAMMIE CAD Ltd. He has a BSc in Ergonomics and a PhD in psychophysiology and his research interests cover most areas of ergonomics. He is a Fellow of Ergonomics Society and a recipient of their 1994 Otto Edholm Award 'in recognition of the development and application of SAMMIE as an ergonomics design tool'.

Mansour Rahimi has a BSc and an MSc in Industrial Engineering and a PhD in Industrial Engineering and Operations Research (with specialization in Human Factors Engineering) from Virginia Polytechnic Institute and State University. He is an Associate Professor of Safety Science and has been teaching in different areas of system safety and ergonomics. His current research interests are in human reliability, risks associated with computerized technology, and rehabilitation engineering. He is the Program Director of the Occupational Safety and Health for the Southern California Educational Resource Center, funded by the US National Institute for Occupational Safety and Health (NIOSH).

Dr Jane A. Rajan is the Principal of ergonomiQ, an ergonomics consultancy specializing in high hazard industries. Areas of expertise include control room design, task analysis and human error, in particular in the petrochemical, marine and transportation industries. Jane is currently also an Associate Fellow at the University of Hertfordshire. She is a Fellow of the Ergonomics Society, a Certified Professional Ergonomist and member of the Human Factors Society.

Dr Penelope Sanderson is an Associate Professor of Psychology and of Mechanical and Industrial Engineering at University of Illinois at Urbana-Champaign, where she has been since graduating from University of Toronto with a PhD in cognitive psychology in 1985. Dr Sanderson's research interests include cognitive engineering, human–computer interaction, cognitive skill acquisition, and information technology in healthcare environments. She has been active in synthesizing methodologies for observational data analysis under the heading of Exploratory Sequential Data Analysis (ESDA) and has developed MacSHAPA, a software tool for certain kinds of ESDA.

Robert Scane is joint Course Organiser for the MSc and associated postgraduate courses in User–Interface Design at London Guildhall University. He is also a member of the course team for the MSc in Multimedia systems as well as teaching on undergraduate courses in the fields of telecommunications and decision-support systems. His recent research includes work on methods of assessing usability based on objective parameters of user–interface design solutions, and research on multimedia authoring. He has received a BSc (Hons) in Psychology and a PgDip in Information Systems.

Nigel Shadbolt is Lecturer in Psychology at the University of Nottingham. After receiving his PhD in Artificial Intelligence from Edinburgh University, he moved to Nottingham in 1984 when he founded the AI group. He has published and researched extensively in the areas of knowledge acquisition, expert systems and planning.

Andrew Shepherd is an applied psychologist in the Department of Human Sciences at Loughborough University with research interests in task analysis, information requirements specification, training simulation development and the management of human factors in the wider design process. He worked for a number of years in industry dealing with a wide range of human factors issues and has carried out work for various Government agencies and commercial companies. He is a Fellow of the Ergonomics Society and an Associate Fellow of the British Psychological Society.

Pat Shipley is Honorary Research Fellow and Reader Emeritus in the Department of Organizational Psychology at Birkbeck College, University of London. She had several years industrial management experience before taking up an academic career. She is a chartered occupational psychologist and a Fellow of the British Psychological Society. She is a trustee of the Society for the Furtherance of the Critical Philosophy and Associate Editor of the international journal *Work & Stress*. She has researched extensively and published in the field of occupational/organizational health and safety and runs a small stress research and intervention unit at Birkbeck College, which she founded in 1982.

Geoff Simpson is Manager of Ergonomics and Safety Management at International Mining Consultants Ltd where he is responsible for both contract research and consultancy in ergonomics and safety, primarily (though not exclusively) in the mining industry. He has degrees in Occupational Psychology and Ergonomics and was awarded (on behalf of his Department) both the Ergonomics Society Applications Award and the Bartlett Medal.

He began his ergonomics career with British Steel, followed by a period as lecturer in the Psychology Department at Nottingham University. He then spent 18 years in the UK mining industry, taking up his current post in 1994. He is a Fellow of the Ergonomics Society and a Member of the Institution of Occupational Safety & Health.

Murray A. Sinclair is a Principal Scientist in the HUSAT Research Institute and a Senior Lecturer in the Department of Human Sciences at Loughborough University of Technology. He has an extensive background of research and consultancy in manufacturing industry, particularly with regard to ergonomics issues in the product introduction process. He is a registered practitioner and a Fellow of the Ergonomics Society, a member of the IEEE Engineering Management Society, and the UK Representative on CEN/TC 310/WG 4 Advanced Manufacturing Technology (Ergonomics).

Robert B. Stammers has BSc and PhD degrees in psychology and is a Senior Lecturer in Applied Psychology at Aston University, Birmingham. His research interests are in task analysis, systems ergonomics and training. He has been sponsored to carry out research by government agencies, industry and the Economics and Social Research Council. He is a Fellow of the Ergonomics Society and the British Psychological Society. He was presented with the Ergonomics Society's Otto Edholm Award in 1989. He is the General Editor of the journal, *Ergonomics*.

Moira F. Tracy did her PhD research on biomechanics at the Institute for Occupational Ergonomics, Nottingham University, following a degree in Physics. She developed some manual handling applications for the Post Office and the Health and Safety Executive. She now works as a Health and Safety adviser in the Victoria Infirmary, Glasgow.

Professor John R. Wilson has degrees in management and in ergonomics from Loughborough University and a PhD in work design and ergonomics from Birmingham University; he is a Chartered Psychologist and a Chartered Engineer, and is a Fellow of the Ergonomics Society. Having held teaching posts at Birmingham University and the University of California, Berkeley, he is currently Professor of Occupational Ergonomics and Director of Postgraduate Research in the Department of Manufacturing Engineering and Operations Management, Nottingham University. Professor Wilson is Editor-in-Chief of *Applied Ergonomics*. He is Director of The Institute for Occupational Ergonomics, having co-founded it with Nigel Corlett in 1985, and is Director of the Virtual Reality Applications Research Team. He has produced over 200 publications, and more than 120 are in refereed books, journals or collections. He has been awarded the Ergonomics Society Sir Frederic Bartlett Medal for 1995, for services to international ergonomics teaching and research.

Index

Page numbers in bold denote the first page of the chapter on that particular subject. Where subjects are referred to on a number of consecutive pages, only the first page is indicated.